INTRODUCTORY CHEMISTRY

Kevin Revell
Murray State University

Austin • Boston • New York • Plymouth

Senior Vice President, STEM: Daryl Fox
Senior Program Managers: Beth Cole, Jeff Howard
Senior Marketing Manager: Maureen Rachford
Executive Content Development Manager, STEM: Debbie Hardin
Senior Development Editor: Erica Champion
Development Editor: Heather McCoy
Executive Project Manager, Content, STEM: Katrina Mangold
Editorial Project Manager: Karen Gulliver
Director of Content, Physical Sciences: Heather Southerland
Senior Media Editors: Stacy Benson, Kris Hiebner
Media Editor: Kelsey Hughes
Editorial Assistant: George Hajjar
Marketing Assistant: Morgan Psiuk
Director, Content Management Enhancement: Tracey Kuehn
Senior Managing Editor: Lisa Kinne
Senior Content Project Manager: Won McIntosh
Senior Workflow Project Manager: Lisa McDowell
Production Supervisor: Robin Besofsky
Director of Design, Content Management: Diana Blume
Design Services Manager: Natasha A. S. Wolfe
Cover Design Manager and Designer: John Callahan
Text Designer: Lumina Datamatics, Inc.
Art Manager: Matthew McAdams
Illustrations: Troutt Visual Services
Director of Digital Production: Keri deManigold
Media Project Manager: Dan Comstock
Permissions Manager: Jennifer MacMillan
Executive Permissions Editor: Cecilia Varas
Photo Researcher: Krystyna Borgen, Lumina Datamatics, Inc.
Composition: Lumina Datamatics, Inc.
Printing and Binding: LSC Communications
Cover Images: (Front) Raimund Linke/Getty Images; (Back) Kevin Revell

Library of Congress Control Number: 2020944819

Student Edition Paperback:
ISBN-13: 978-1-319-27967-7
ISBN-10: 1-319-27967-8

Student Edition Loose-leaf:
ISBN-13: 978-1-319-33601-1
ISBN-10: 1-319-33601-9

Printed in the United States of America
1 2 3 4 5 6 25 24 23 22 21 20

Macmillan Learning
One New York Plaza
Suite 4600
New York, NY 10004-1562
www.macmillanlearning.com

BRIEF CONTENTS

CONTENTS

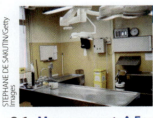

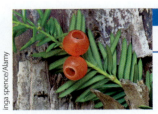

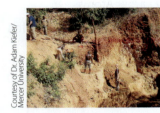

Steve Spohn

CHAPTER 4
Light and Electronic Structure 82
Edging toward Solar Energy

© TopFoto/The Image Works

CHAPTER 5
Chemical Bonds and Compounds 111
Lithium Carbonate and Bipolar Disorder

Index Fototeca Heritage Images/Newscom

CHAPTER 6
Chemical Reactions 139
Lost Cities of the Maya

noomcpkstic/Shutterstock

AP Photo/Philadelphia Evening Bulletin

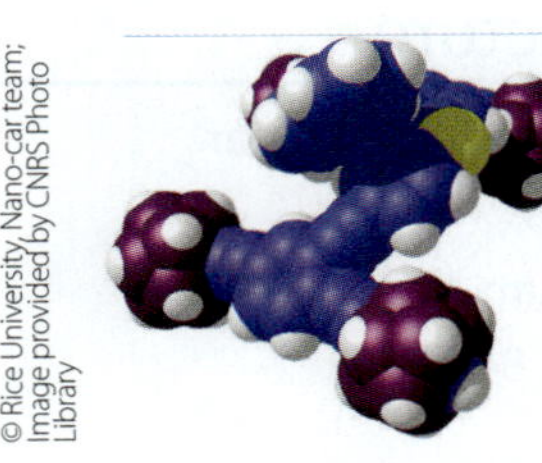

© Rice University, Nano-car team; Image provided by CNRS Photo Library

Vyacheslav Svetlichnyy/Shutterstock

Thomas La Mela/Shutterstock

Gudella/iStock/Getty Images

Caltech

Handout/Reuters/Newscom

CHAPTER 16 Nuclear Chemistry 453

Fukushima

A Word from the Author

Welcome to *Introductory Chemistry!*

For many students, introductory chemistry is a general-education requirement en route to a degree in education, business, or a liberal-arts field. For others, it is a stepping stone toward a challenging general chemistry course and a career in health care, agriculture, or even science and engineering. Some are traditional students, but many others are nontraditional students balancing jobs, family, and the dream of completing a college degree. And for many, chemistry can seem elusive, mystical, and intimidating.

My vision is to make chemistry accessible to these students, not just through a textbook, but through an integrated learning experience that addresses different learning styles and draws on a variety of pedagogical techniques to engage and challenge students.

Let's begin with the text. I've tried to write in a friendly, casual style—using analogies, stories, and images to make the important ideas stick. I've blended this with digital interactives to create an active reading experience. In some of these interactive figures, you can explore chemical changes, choosing different substances to see how they react. In others, you'll be able to practice key knowledge and skills through simple games.

For instructors, *Introductory Chemistry* is more than a textbook—it's a complete curriculum, suited for traditional, flipped, online, or blended active-learning classrooms. I've created video lectures for every section, with corresponding PowerPoint decks that you can modify as you see fit. I've included in-class activities, clicker questions, and speed drills, developed over many semesters. Think of it as a tool belt, equipping you with the curriculum to suit the needs of your classroom.

The pages that follow describe many of these features in more detail. Whether you are teaching a class, taking a class, or just exploring chemistry for the first time, *Introductory Chemistry* is designed to help you achieve your goals. You can do this—let's get started.

Best wishes,

Kevin D. Revell

Carl T. Woods/Murray State University

KEVIN REVELL teaches introductory, general, and organic chemistry at Murray State University. A passionate educator, his teaching experience includes high school, community college, small private, state comprehensive, and state flagship institutions. His work encompasses curriculum, technology-enhanced pedagogy, assessment, and active-learning design. A synthetic chemist by training, his research involves the synthesis and evaluation of functional organic materials. With his wife, Jennifer, Kevin has three children—James, Julianne, and Joshua—and four grandchildren.

Light bulb moments and unrivaled support

Introductory Chemistry is the result of a unique author vision to develop a robust combination of text and digital resources that motivate and build student confidence while providing a foundation for their success. Kevin Revell knows and understands students today. Through his thoughtful and media-rich program, Kevin creates light bulb moments for introductory chemistry students and provides unrivaled support for instructors.

Overview of key features

ACCESSIBLE AND ENGAGING STUDENT RESOURCES

- The text features a **concise, yet compelling narrative** that students understand. Kevin Revell's strong author voice acts as a coach, providing encouragement and strategies for challenging topics. **Engaging analogies** taken directly from Kevin's years in the classroom contextualize chemical concepts, expressing big ideas to students in a framework they will understand. And **personal stories** provide a friendly and welcoming tone to relate content to everyday life and encourage students to draw from their own experiences in scientific thinking.
- Each chapter opens with a story of how real people **apply chemistry concepts** from the chapter to real-world problems. These stories are revisited at the end of the chapter for a deeper understanding using what students have learned.
- The **interactive e-book** includes features for highlighting, note-taking, and accessibility support. In the e-book, chemistry comes alive through a variety of multimedia that enhances visualization, conceptual understanding, and problem-solving skills. **Lecture videos**, created and narrated by Kevin Revell, are an alternative to reading. Foundational concepts and skills are revisited and reinforced with **LearningCurve adaptive quizzing**. **Problem-solving videos** accompany each in-text example problem. **Animations** replace static figures to depict dynamic events or reactions at a molecular level. Embedded **videos** show chemical reactions, simulating a lab experience. A variety of **interactives** enhance the text through lab simulations, problem-solving practice, and exploration of concepts in chemistry.

UNRIVALED INSTRUCTOR SUPPORT

- Built for the traditional lecture or flipped classroom, **Lecture Slides** are easily editable and include the scripts used to create lecture videos. **Lecture Video Outlines** provide students with a pared down set of lecture slides for taking notes. Or build your own lecture using **Image slides** for every in-text figure.
- For the active or flipped classroom, **iClicker Questions** provide instructors with ready-made assessments to evaluate student learning in real time. **Speed Drills** challenge students' memory of foundational knowledge. When used as a timed self-assessment in the classroom, students are motivated to improve.
- In-class or online, **Activities and Worksheets** pair with interactives to keep students engaged by providing a framework and problems for them to try on their own. Worksheets are printable for use in class or available as online assignments in Achieve.
- An **extensive problem library in Achieve** includes select problems from Revell's end-of-chapter questions with hints, targeted feedback, and detailed solutions for all study questions to ensure student mastery.

THE ACHIEVE LEARNING PLATFORM

- The Achieve **design is guided by learning science research** through extensive collaboration and testing by both students and faculty including two levels of Institutional Review Board approval for every study.
- Achieve features a **flexible suite of resources** to support learning core concepts, visualization, problem-solving, and assessment. All student and instructor resources can be assigned or downloaded in Achieve.
- The **iClicker** classroom engagement system syncs with the Achieve **gradebook.**
- **Gradebook integration** can be set up with **your campus LMS** and with **Inclusive Access** programs.
- The Achieve gradebook provides insights for **just-in-time teaching** and reporting on student and full class achievement by learning objective. Powerful **analytics and instructor resources** in Achieve pair with exceptional introductory chemistry content to provide an unrivaled learning experience.

Achieve is the culmination of years of development work put toward creating the most powerful online learning tool for chemistry students. It houses all of our renowned assessments, multimedia assets, e-books, and instructor resources in a powerful new platform.

Achieve supports educators and students throughout the full range of instruction, including assets suitable for pre-class preparation, in-class active learning, and post-class study and assessment. The pairing of a powerful new platform with outstanding introductory chemistry content provides an unrivaled learning experience.

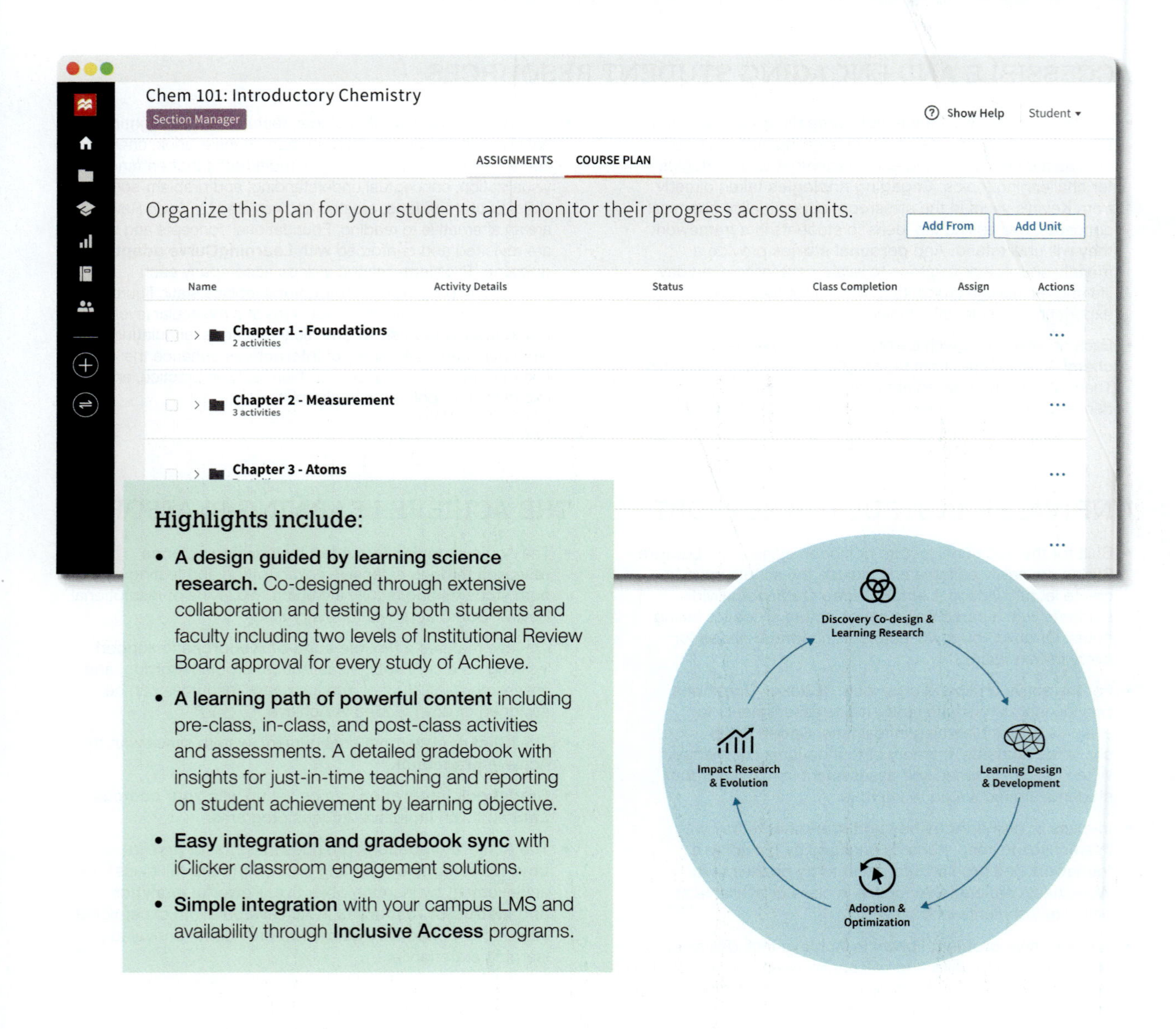

Highlights include:

- **A design guided by learning science research.** Co-designed through extensive collaboration and testing by both students and faculty including two levels of Institutional Review Board approval for every study of Achieve.
- **A learning path of powerful content** including pre-class, in-class, and post-class activities and assessments. A detailed gradebook with insights for just-in-time teaching and reporting on student achievement by learning objective.
- **Easy integration and gradebook sync** with iClicker classroom engagement solutions.
- **Simple integration** with your campus LMS and availability through **Inclusive Access** programs.

For more information or to sign up for a demonstration of Achieve, contact your local Macmillan representative or visit **macmillanlearning.com/achieve**

Engage every student

Achieve supports flexible instruction and engages students. This intuitive platform includes content for pre-class preparation, in-class active learning, and post-class engagement and assessment, providing students with an unparalleled environment and resources for learning.

LearningCurve

LearningCurve's game-like quizzing motivates each student to engage with the course content, and reporting tools help teachers get a handle on what their class needs.

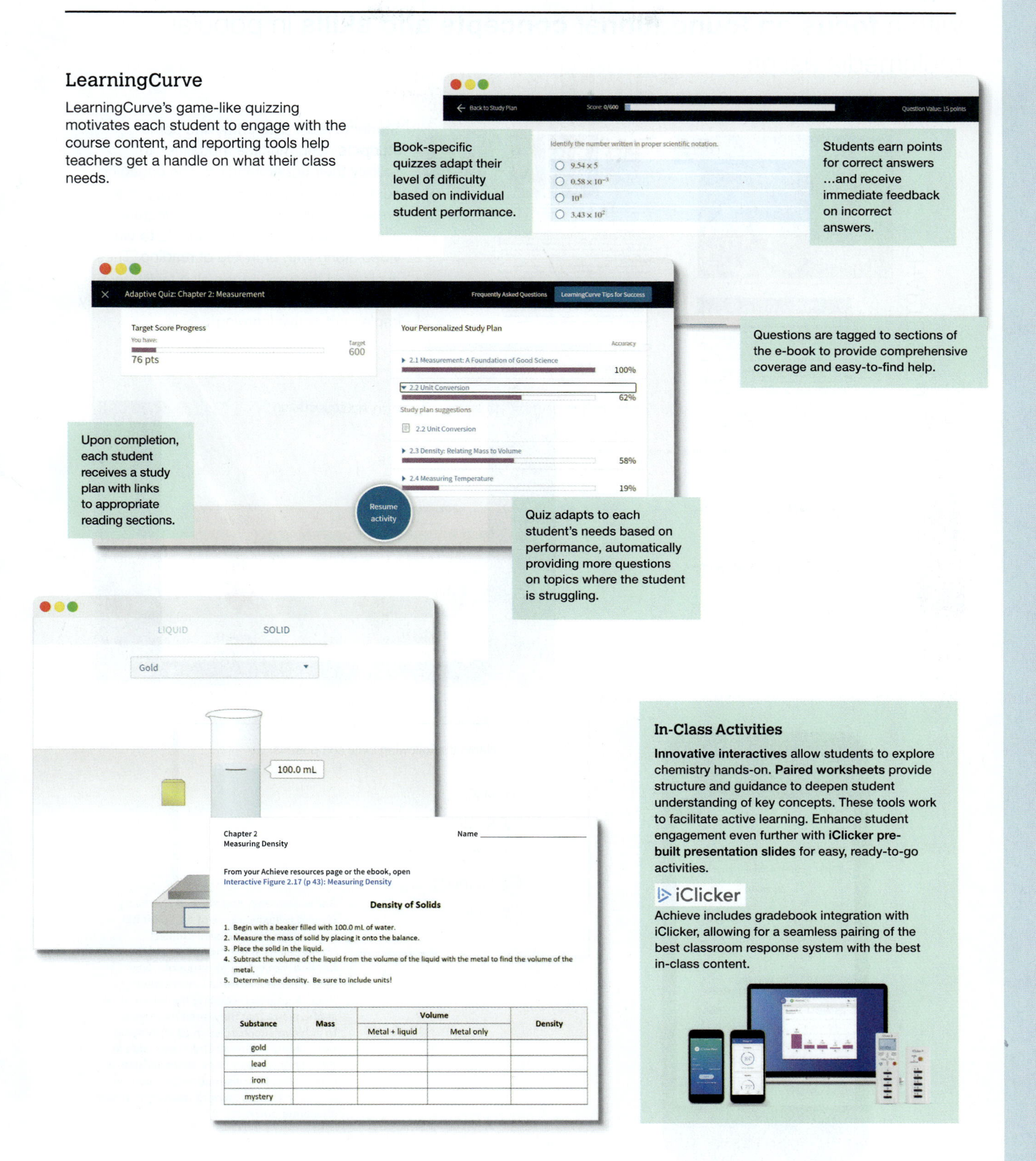

Chapter 2
Measuring Density

Name ______________

From your Achieve resources page or the ebook, open Interactive Figure 2.17 (p 43): Measuring Density

Density of Solids

1. Begin with a beaker filled with 100.0 mL of water.
2. Measure the mass of solid by placing it onto the balance.
3. Place the solid in the liquid.
4. Subtract the volume of the liquid from the volume of the liquid with the metal to find the volume of the metal.
5. Determine the density. Be sure to include units!

Substance	Mass	Volume		Density
		Metal + liquid	Metal only	
gold				
lead				
iron				
mystery				

In-Class Activities

Innovative interactives allow students to explore chemistry hands-on. **Paired worksheets** provide structure and guidance to deepen student understanding of key concepts. These tools work to facilitate active learning. Enhance student engagement even further with **iClicker pre-built presentation slides** for easy, ready-to-go activities.

iClicker

Achieve includes gradebook integration with iClicker, allowing for a seamless pairing of the best classroom response system with the best in-class content.

Focus on Foundations

Digital content is an extension of the author's voice and vision with a **focus on foundational concepts and skills** in popular multimedia assets.

Students learn differently, and some students need multiple modalities to solidify their understanding. The engaging content in ***Introductory Chemistry*** is also presented through an extensive lecture video program, enabling students to watch video along with or in lieu of reading. These acclaimed **lecture videos** also facilitate a flipped classroom and help instructors easily implement active learning.

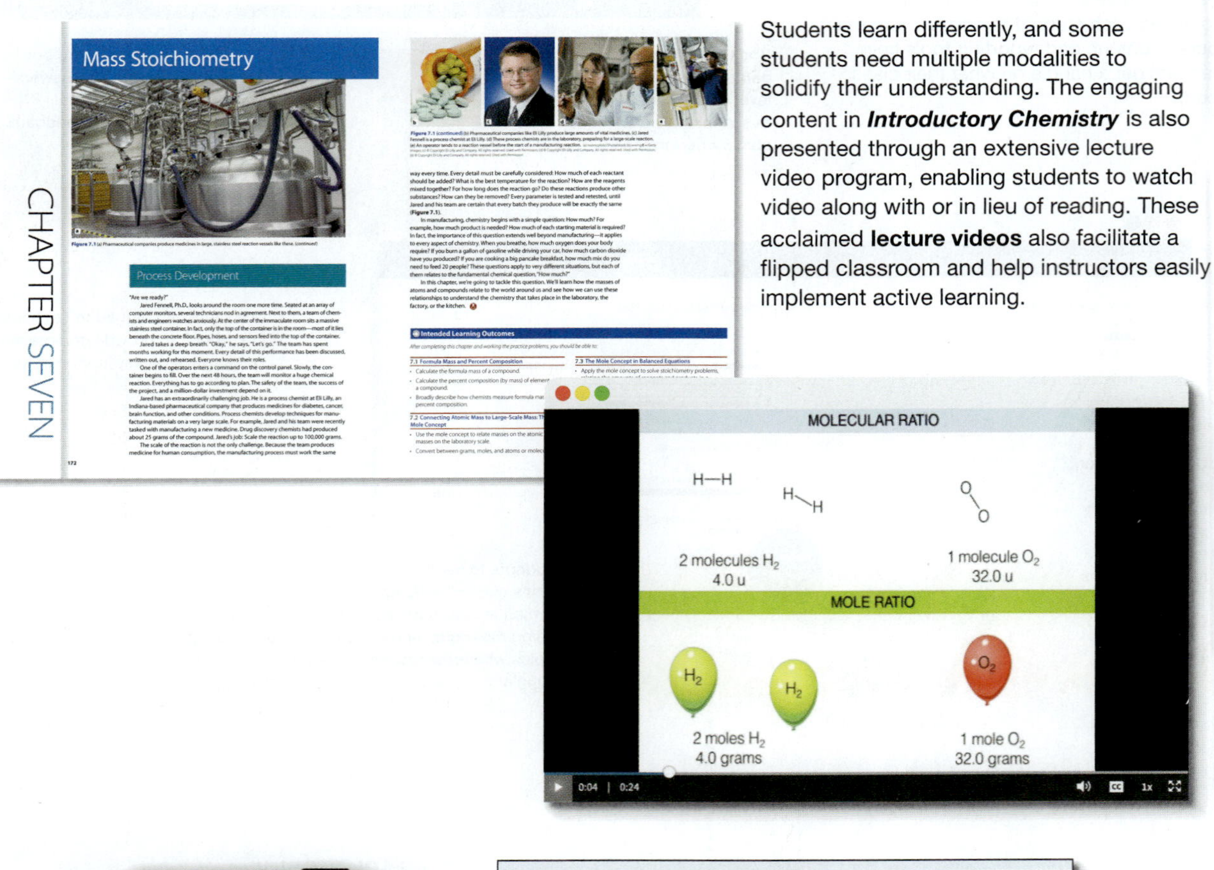

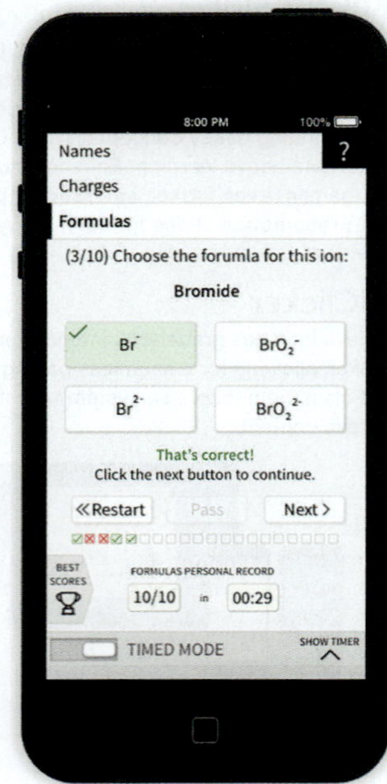

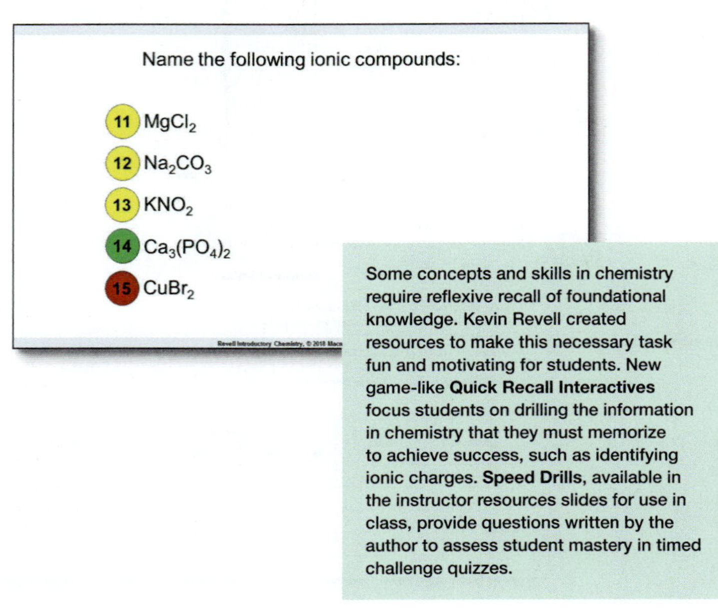

Some concepts and skills in chemistry require reflexive recall of foundational knowledge. Kevin Revell created resources to make this necessary task fun and motivating for students. New game-like **Quick Recall Interactives** focus students on drilling the information in chemistry that they must memorize to achieve success, such as identifying ionic charges. **Speed Drills**, available in the instructor resources slides for use in class, provide questions written by the author to assess student mastery in timed challenge quizzes.

Focus on Visualization

A variety of interactives provide students with resources to help them **focus on understanding through visualization**.

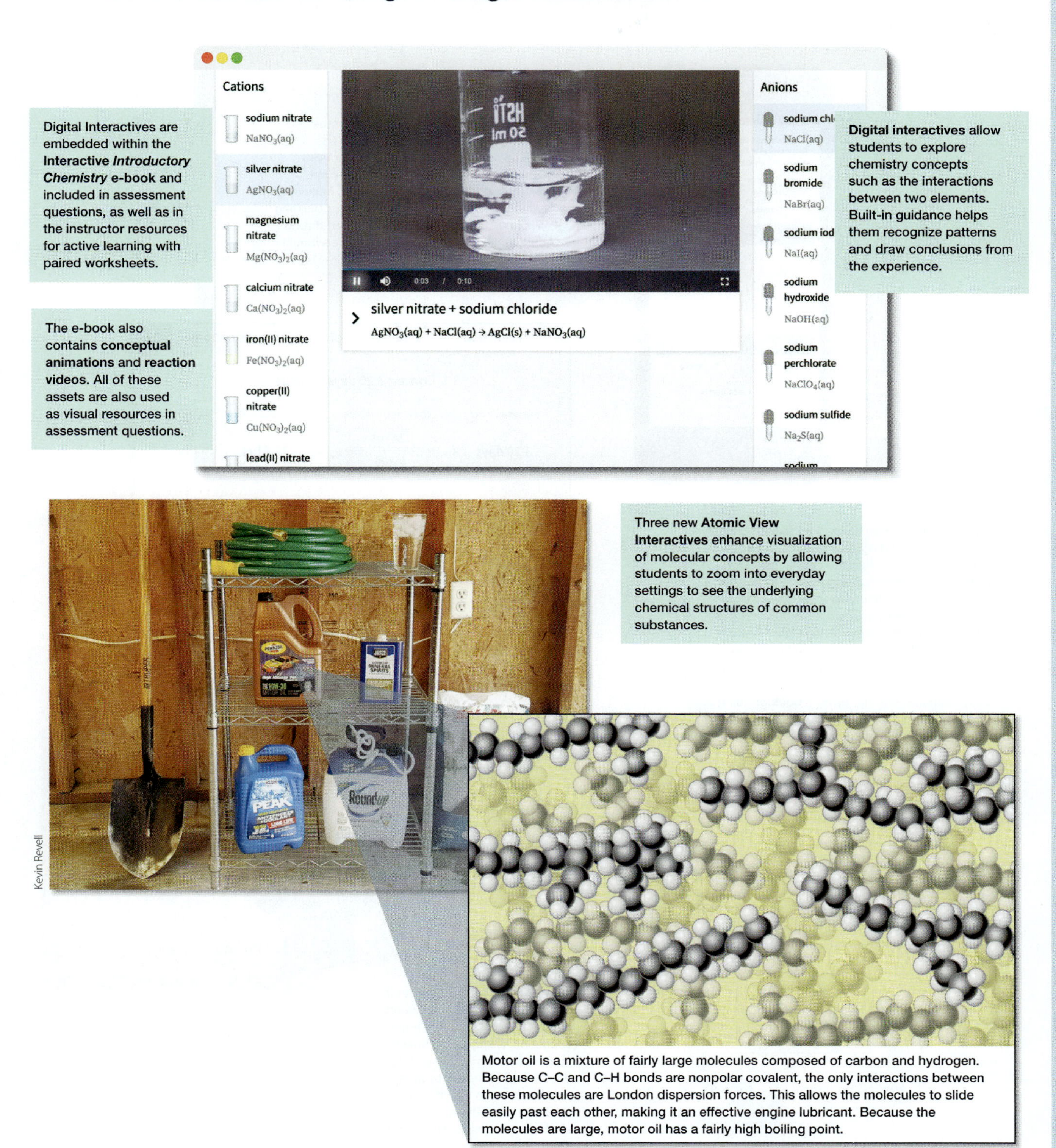

Digital Interactives are embedded within the **Interactive *Introductory Chemistry* e-book** and included in assessment questions, as well as in the instructor resources for active learning with paired worksheets.

The e-book also contains **conceptual animations** and **reaction videos.** All of these assets are also used as visual resources in assessment questions.

Digital interactives allow students to explore chemistry concepts such as the interactions between two elements. Built-in guidance helps them recognize patterns and draw conclusions from the experience.

Three new **Atomic View Interactives** enhance visualization of molecular concepts by allowing students to zoom into everyday settings to see the underlying chemical structures of common substances.

Motor oil is a mixture of fairly large molecules composed of carbon and hydrogen. Because C–C and C–H bonds are nonpolar covalent, the only interactions between these molecules are London dispersion forces. This allows the molecules to slide easily past each other, making it an effective engine lubricant. Because the molecules are large, motor oil has a fairly high boiling point.

Focus on Problem-Solving

The second edition provides extra support for **problem-solving skills and practice** through guided interactives and integrated concept questions.

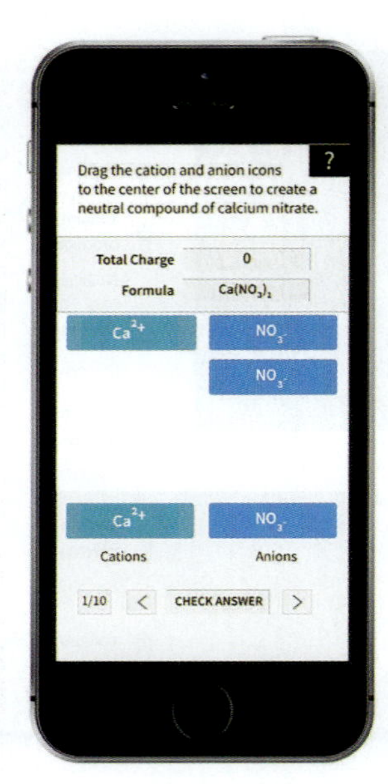

New Guided Practice Interactives offer no/low-stakes practice for the most difficult problem-solving concepts in introductory chemistry. These interactives guide students step by step through unit conversions, rearranging equations, balancing charges, and stoichiometry problems, building skills and confidence. When assigned through **Achieve**, follow-up questions (with feedback) reinforce what they've learned.

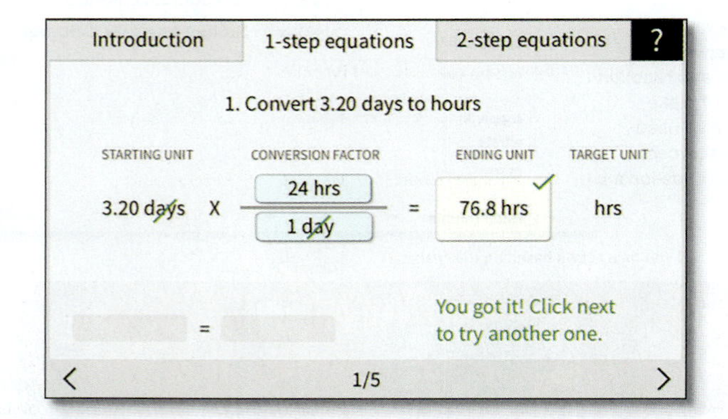

The second edition of ***Introductory Chemistry*** builds on the strengths of the first edition—drawing students into the course through engagement and building their foundational knowledge—while introducing new content and resources to help students build critical thinking and problem-solving skills. Revell's distinct author voice in the text is mirrored in the digital content, allowing students flexibility and ensuring a fully supported learning experience—whether using a book or going completely digital in **Achieve**.

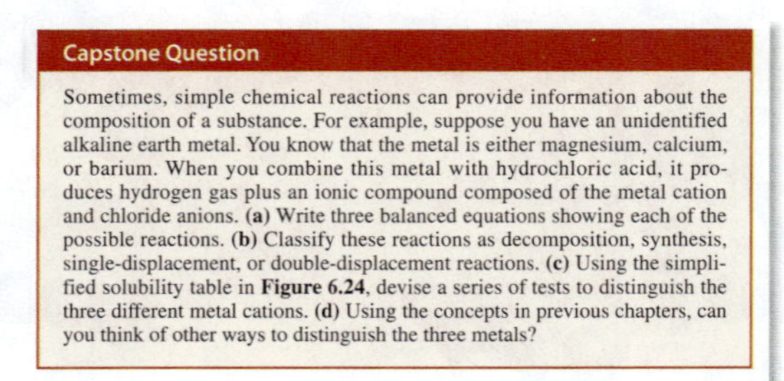

Capstone Question

Sometimes, simple chemical reactions can provide information about the composition of a substance. For example, suppose you have an unidentified alkaline earth metal. You know that the metal is either magnesium, calcium, or barium. When you combine this metal with hydrochloric acid, it produces hydrogen gas plus an ionic compound composed of the metal cation and chloride anions. **(a)** Write three balanced equations showing each of the possible reactions. **(b)** Classify these reactions as decomposition, synthesis, single-displacement, or double-displacement reactions. **(c)** Using the simplified solubility table in **Figure 6.24**, devise a series of tests to distinguish the three different metal cations. **(d)** Using the concepts in previous chapters, can you think of other ways to distinguish the three metals?

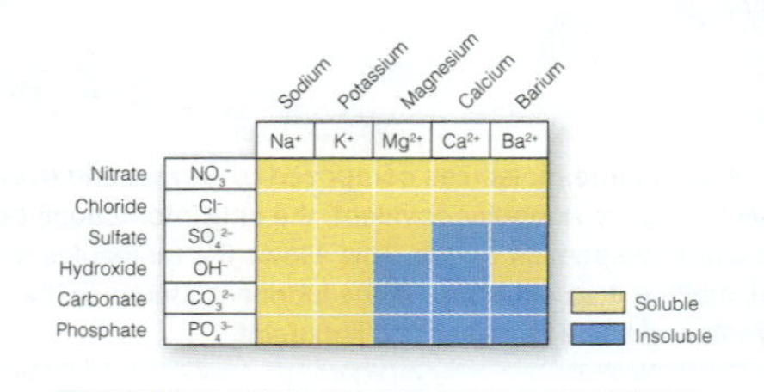

Figure 6.24 Based on this simplified solubility table, how could you distinguish compounds containing magnesium, calcium, and barium ions?

New Capstone Questions at the end of each chapter integrate related concepts in challenging, real-world scenarios. Designed to develop student problem-solving skills, these questions feature accompanying video lessons in which Kevin walks students through problem-solving strategies and processes for these complex questions.

Powerful analytics, viewable in an elegant dashboard, offer instructors a window into student progress. Achieve gives you the insight to address students' weaknesses and misconceptions before they struggle on a test.

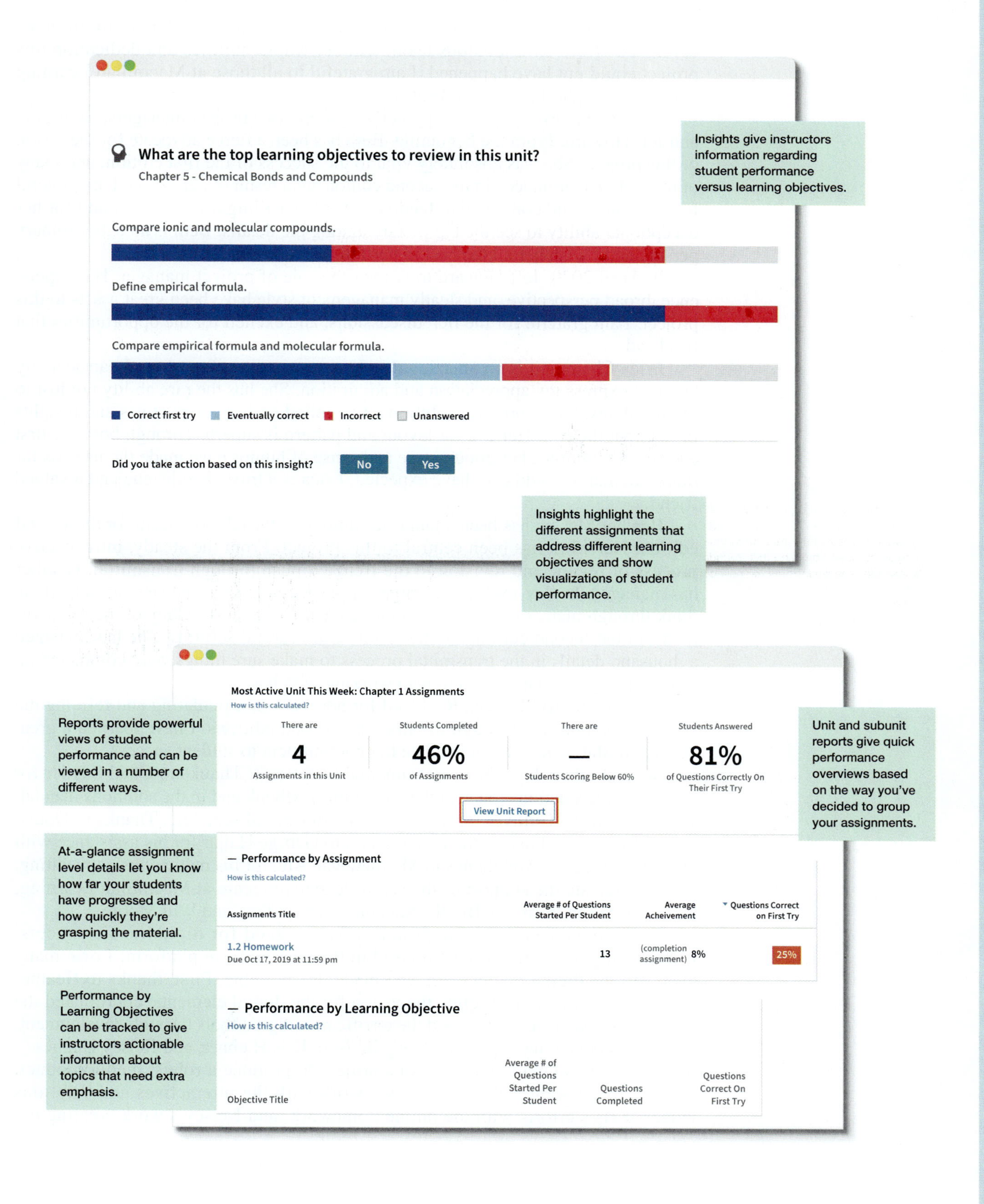

ACKNOWLEDGMENTS

The second edition *Introductory Chemistry* textbook and curriculum result from the combined efforts of a fabulous team, without whose abilities and dedication this project could not have happened. I am grateful to all those at Macmillan Learning that have brought this effort to fruition.

First, I want to express my appreciation to my two project managers, Beth Cole and Jeff Howard. From the beginning, Beth has been a fierce advocate for the vision of this project. She was absolutely vital to the success of the first edition, and many of the features introduced in the second edition are a result of her vision. I am grateful for her honest and constructive feedback, for her willingness to listen, and for her exceptional ability to see the big picture among the details. Beth is an extraordinary person, and I wish her all the best in her future endeavors.

In April 2020, Jeff Howard took over the role of project manager. His experience, broad perspective, and steady management style have been great assets to this project. I am grateful for the rich discussions, and excited for the opportunities that lie ahead.

Erica Champion is an amazing developmental editor, for whom I can scarcely begin to express my appreciation and admiration. She has the rare ability not just to see problems, but to propose solutions. I've relied on her perspectives and insights throughout the text. Her eye for layout and pattern is superb. Through both the first and second editions, her good nature and sense of humor have made the process far more fun than I would ever have expected. Erica is a trusted colleague and a valued friend.

Heather McCoy has been a fantastic addition to the editorial team for the second edition. Her work has been central to this project. From the steady, introspective process of review and revision to the frenzied push through transmittal, Heather has managed the day-to-day challenges of revising the text, and she has helped me think through many of the issues encountered along the way. Many of the improvements in the second edition are the result of her careful scrutiny. She has managed a thousand details in the transmittal process to make sure things were completed on time and in good shape. I've enjoyed working with her.

I'm grateful to Maureen Rachford for her work in articulating and refining the vision of this project through the first and second editions. I have learned a great deal from Maureen, and I appreciate her commitment to student success.

Several others have been instrumental to the text: Thanks to Chad Snyder for producing the solutions found at the end of this textbook and in the solutions manual. Thanks to Karen Gulliver for managing the transmittal schedule. Thanks to Donna Brodman for coordinating the reviews, and to George Hajjar for his assistance with the review process. And thanks to Matthew Van Atta for his meticulous copyediting.

I appreciate the support of the senior leadership team at Macmillan Learning, including Susan Winslow, Brooke Suchomel, Daryl Fox, and Will Moore.

From its conception, this project was designed for digital environments. It is thrilling to see the project housed in the new Achieve platform; I owe many thanks to the digital media team who made this possible. First, thanks to Heather Southerland, who helped refine the vision for the digital elements, coordinated the migration to Achieve, and guided the terrific new digital tools through development. Thanks also to the media team—Stacy Benson, Kris Hiebner, and Kelsey Hughes—who managed so many facets of this project to produce a robust media product. Specifically, I want recognize Stacy's work with the interactives and solutions videos, Kris's efforts with the online homework, and Kelsey's work creating the adaptive quiz assignments. Also, thanks to Kelly Lancaster and the SAVI team for coordinating with vendors. This has been a terrific team.

Many others have played vital roles in the development of the media elements: Thanks to Angela Piotrowski and Jamie Carberry for their voicework in the solutions videos, and to Andrew Waldeck for his work in scripting and animating the solutions. Thanks to Alex Gordon, who produced the excellent artwork in the animations with the help of animators Sara Egner, Clarissa Cochran, Tommy Turner, and Cheryl McCutchan. Thanks to the SAVI (Simulations, Animations, Videos, and Interactives) team at Macmillan: Jenn Ferralli, Brian Banister, Brian Hochhalter, and Jeff Sims. Thanks also to Six Red Marbles for the design and development of the Guided Practice, Quick Recall, and Atomic View interactives. Thanks to Tyler Covington and Steve Lemon for their work in producing the capstone videos.

I appreciate the efforts of those who brought together the artwork for this project. Thanks to Cecilia Varas and Krystyna Borgen for their patient tenacity in obtaining photos for the textbook and ancillaries. Thanks to Troutt Visual Services for the illustrations for this text, to John Callahan who produced the cover, Natasha Wolfe who managed the design, and to Matthew McAdams who served as the art manager.

Many thanks also to the content management team: Won McIntosh, Diana Blume, Tracey Kuehn, Lisa McDowell, Robin Besofsky, and the team at Lumina Datamatics.

Finally, I want to acknowledge Ben Roberts and Leslie Allen. Although they were not involved with the second edition, their vision and efforts on the first edition produced lasting impacts; I am grateful for their history with this project.

Outside of the Macmillan Learning team, there are many others whose contributions I very much appreciate:

I would like to thank my colleagues, friends, and family whose research, stories, insights, and talents have contributed to this project. This includes Lauren Waugh, Adam Kiefer, Isabel Villaseñor, Yareli Jáidar, Jared Fennell, James Tour, Christian Joachim, Leonhard Grill, Francesca Moresco, Alexiane Agullo, Omar Yaghi, Jarrod Eubank, Aaron Blanco, Susan Hurley, Kimberly Revell, Gary and Cindy Deaton, Julie Revell, Daniel Johnson, Robert Grubbs, Richard Hurtt, and Jack Revell.

I would like to extend my thanks to the many colleagues who reviewed the textbook. Your thoughtful feedback has shaped and polished this project. I am also grateful to my colleagues at Murray State: Thanks to Ricky Cox, who continues to inspire my teaching. Thanks to Harry Fannin, Steve Cobb, and Bommanna Loganathan for their unwavering support. And thanks to Sally Sroda and Diane Thiede, who make everything in the office work so smoothly.

I would also like to acknowledge key mentors whose efforts years ago made this possible, including Bob Chandler (Lake Gibson High School), Mark Trudell (University of New Orleans), Walt Trahanovsky (Iowa State University), Bob Herron (Southeastern University), Deborah Hazelbaker (Southeastern University), and especially Ed Turos (University of South Florida).

Finally, thanks to my family, who have supported and encouraged me so much through this process. Thanks especially to Jennifer, for her constant love, encouragement, and sacrificial support to make this dream happen.

We thank the many reviewers who aided in the development of this edition:

Dawood Afzal, Truman State University
John Alexander, Marymount College, Rancho Palos Verdes
Jeffrey Allison, Texas State University
Molly Anderson, Baker University, Baldwin City
Premilla Arasasingham, Moorpark College
Luis Avila, Columbia University, New York
David Baker, Delta College
Rebecca A. L. Baylon, University of Idaho, Moscow
Frank H. Bellevue III, Limestone College
Marguerite H. Benko, Ivy Tech College, Central
Mandy Blackburn, University of Central Missouri
Dina Borysenko, Milwaukee ATC, Downtown
Glenda L. Bossow, Bethany Lutheran College
Ken Capps, College of Central Florida, Ocala
Bernard Castillo, University of the Virgin Islands, Kingshill
Mihaela Chamberlin, Northern Virginia Community College, Annandale
Anthony Chibba, McMaster University
Jeannine M Christensen, Moraine Valley Community College

Robert Doti, Ivy Tech College, Central
Dylan Drake-Wilhelm, Metro Community College, Omaha
Julie Ellefson, William Rainey Harper College
John D. Fields, Fayetteville Technical Community College
Melinda Findlater, Texas Tech University, Lubbock
Jennifer Firestine, Lindenwood University, St. Charles
Mario G. Garcia-Rios, Bristol Community College, New Bedford
Steve Gentemann, Southwestern Illinois College
Thomas Giagou, Merced College, Merced
Matt Gnezda, Ivy Tech College, Central
Jennifer Gotcher, Austin Community College, South Austin
Maru Grant, Ohlone College, Fremont
Carol Green, St. Charles County Community College
Randal Hallford, Midwestern State University
Christopher G. Hamaker, Illinois State University
Bernadette Harkness, Delta College
Xiche Hu, University of Toledo, Toledo
Alan Jircitano, Pennsylvania State University, Erie Behrend
Alandra Kahl, Pennsylvania State University, Greater Allegheny
Sondra Kekec, Mineral Area College
Colleen Kelley, Pima County Community College, Downtown
Robert Killin, Arizona Western College
Josef Kren, Doane University, Crete
Sujatha Krishnaswamy, Chandler Gilbert Community College
Rebecca Laird, University of Iowa, Iowa City
Shawn Llopis, Delgado Community College, New Orleans
Bommanna G. Loganathan, Murray State University
Harpreet Malhotra, Florida State College, Jacksonville/Roosevelt Boulevard
Nicole Mabante, Rio Salado College
Debra Mansperger, Mesa Community College, Red Mountain Campus
Arnulfo Mar, University of Texas, Rio Grande Valley, Brownsville
Marta K. Maron, University of Colorado, Denver
Jason Matthews, Florida State College, Jacksonville/State Street West
Erandi Prashani Mayadunne, Lonestar College, University Park
Beth Michael-Smith, St. Charles County Community College
Mitchel D. Millan, Casper College
Amanda L. Miller, Mohawk Valley Community College
Jeremy Mitchell-Koch, Bethel College, North Newton
Erick Moffett, Pearl River Community College
Rajiv Narula, SUNY College of Technology, Canton
Antonio Nicodemo, John Abbott College
Mya A. Norman, University of Arkansas, Fayetteville
Jung Oh, Kansas State University Polytechnic
Kefa K. Onchoke, Stephen F. Austin State University
Franklin Ow, East Los Angeles College
Jeff Owens, Edmonds Community College
Naresh Pandya, University of Hawaii, Kapiolani Community College
Lynda Peebles, Texas Woman's University, Denton
Jamin Perry, Baker University, Baldwin City
James Ross, East Los Angeles College
Emily Rowland, University of Mississippi, Main
Spencer Russell, University of North Carolina, Greensboro
Sharadha Sambasivan, Suffolk County Community College, Ammerman
Hussein Samha, Southern Utah University
Niladri Sarker, Rio Salado College
Trineshia Sellars-Baxley, Palm Beach State College, Lake Worth
Joe Selzler, Ventura College
Suki Smaglik, Yakima Valley Community College, Yakima
Sally Ann Smesko, Daemen College
Rhodora Snow, John Tyler Community College-Midlothian
Chad Snyder, Liberty University
Rajani Srinivasan, Tarleton State University
Brandon Tenn, Merced College, Merced
Diann Thomas, Ozarks Technical College, Springfield
Jeffrey E. Tinnon, Pearl River Community College
Harold Trimm, SUNY Broome, Broome Community College
Christina Turner, Hillsborough Community College, Dale Mabry
Cyriacus Uzomba, Austin Community College, San Gabriel
Dusty Ventura, Merced College, Merced
Xin Wen, California State University, Los Angeles
V. Wheaton, American River College
Amy Vickers Whiting, North Central Texas, Corinth
William Williams, Hudson Valley Community College
Ashley Wilson, Ivy Tech College, Central
Andrew R. Wolf , Colorado Mesa University
Curtis Zaleski, Shippensburg University of Pennsylvania
Chen Zhou, University of Central Missouri
Alan Zombeck, Delta College

And of the first edition:
Samuel Melaku Abegaz, Columbus State University
Mireille J. Aleman, Palm Beach Atlantic University
John U. Alexander, Marymount California University
Jeffrey Allison, Austin Community College
Sarah Alvanipour, Houston Community College
Vicki Amszi, Blinn College
Ilija Arar, Cayuga Community College
Premilla Arasasingham, Moorpark College, El Camino College
Luis Avila, Columbia University
Yiyan Bai, Houston Community College
C. Eric Ballard, The University of Tampa
Brian F. Bartlett, Central New Mexico Community College
Vladimir A Benin, University of Dayton
Stacey-Ann Benjamin, Broward College
Marguerite Healy Benko, Ivy Tech Community College
Terrence M. Black, Nassau Community College
Simon Bott, University of Houston

Ivanna Campbell, McLennan Community College
Ken Capps, College of Central Florida
Sevada Chamras, Glendale Community College
Kaiguo Chang, Metropolitan Community College
Kristin Clark, Ventura College
J. De Anda, Long Beach City College
Maria Cecilia de Mesa, Baylor University
Milagros Delgado, Florida International University
Dylan D. Drake, Wilhelm, Metropolitan Community College
Daniel J. Dwyer, University of Montana
Sarah Edwards, Western Kentucky University
Jack F. Eichler, University of California, Riverside
Julie Ellefson, Harper College
Melinda Findlater, Texas Tech
Paul Forster, University of Nevada, Las Vegas
C. Karen Fortune, Houston Community College
Steve Gentemann, Southwestern Illinois College
Jennifer N. Gotcher, Austin Community College
Pierre Goueth, MiraCosta College
Carol Green, St. Charles Community College
Lindsay Groce, Big Bend Community College
Yi Guo, Chandler Gilbert Community College
Christopher G. Hamaker, Illinois State University
Zachariah M. Heiden, Washington State University
Prof. Xiche Hu, University of Toledo
Thomas E. Janini, The Ohio State University Agricultural Technical Institute
Eugenio Jaramillo, Penn State Altoona
Crisjoe Joseph, California State University, Channel Islands
Joshua Kellogg, University of North Carolina at Greensboro
Edith Preciosa Kippenhan, University of Toledo
Joy Kobayashi, Ventura College
Sujatha Krishnaswamy, Chandler Gilbert Community College
Allison C. Lamanna, Boston University
Jean-Marie Magnier, Springfield Technical Community College
Arnulfo Mar, University of Texas, Rio Grande Valley
Matt Marlow, Nicholls State University
J. Aaron Matthews, Florida State College at Jacksonville
N. Alpheus Mautjana, Santa Fe College
Troy Milliken, Holmes Community College
Katherine A. Moga, The Ohio State University
Reza Mohseni, East Tennessee State University
Daniel Moriarty, Siena College
Christopher P. Morong, Minnesota State University, Mankato
Tamari Narindoshvili, Blinn College Bryan Campus
Jung Oh, Kansas State University Polytechnic
Bruce Osterby, University of Wisconsin, La Crosse
Franklin Ow, East Los Angeles College
Monica Rabinovich, The University of North Carolina at Charlotte
Bhavna Rawal, Houston Community College
Mario C. Raya, Bristol Community College
Jimmy Reeves, University of North Carolina at Wilmington
Anthony Revis, Saginaw Valley State University
Harrison Rommel, Central New Mexico Community College, Santa Fe Community College
James Ross, East Los Angeles College
Emily Rowland, University of Mississippi
Gerald B. Rowland, University of Mississippi
Niladri Sarker, Rio Salado College
Einhard Schmidt, Santa Monica College
James Selzler, Ventura College
Mary Snow Setzer, University of Alabama in Huntsville
Amanda Charlton-Sevcik, Baylor University
Supriya Sihi, Houston Community College
Chad Snyder, Grace College
Brandon Tenn, Merced College
Susan Tansey Thomas, University of Texas at San Antonio
Harold Trimm, SUNY Broome
Lucas J. Tucker, Siena College
Melanie Veige, University of Florida
Elaine Vickers, Southern Utah University
B. Blairanne Williams, Western Kentucky University
William Williams, Hudson Valley Community College
Shaun M. Williams, Lenoir-Rhyne University
Neil M. Wolfman, Boston College
Kevin A. Wood, North Central Texas College
Marilyn Wooten, Trinity University
Tammy Wooten-Boyd, Community College of Philadelphia
Curtis Zaleski, Shippensburg University
Rong Zhang, University of South Florida
Chen Zhou, University of Central Missouri
Tom Zownir, Manchester Community College

To the cloud of witnesses who urge me on:
To my grandparents, for their courage, dedication, perseverance, and love.
To Mom and Dad, for their constant love and encouragement,
and for the tremendous example they set.
To Gary and Cindy, and to Mom and Dad Kruger,
for their support and prayers through the years.

To my wife and best friend Jennifer, the love of my life,
looking forward to great things to come.
To my children, Jamie and Megan, Julie, and Joshua;
and my grandchildren, Grace, Shawn, Emma, and Elijah:
Run with patience the race set before you.

And to God our Father, who gives life and peace:
I dedicate this book in love and thankfulness.

Foundations

Figure 1.1 (a) The Pacific yew tree contains a substance with powerful anticancer properties. (*continued*)

CHAPTER ONE

Taxol

Along the cool, coastal areas of the Pacific Northwest, a small evergreen tree grows. Called the *Pacific yew*, it is a plain-looking tree with soft needles and red berries (**Figure 1.1**). In 1962, a botanist named Arthur Barclay came across a stand of these trees in the forests near Mount St. Helens, in Washington State. Barclay worked for the U.S. Department of Agriculture, and he was searching for plants that might provide new medicines.

Barclay collected a sample of twigs, needles, and bark from the trees. He sent this sample and others like it to Mansukh Wani and Monroe Wall, two chemists in North Carolina. Using a process much like brewing tea, Wani and Wall ground up the plant matter and then mixed it with hot water and other liquids to pull material out of the crushed wood and leaves. They next purified and tested these substances to see how they would affect the growth of tumor cells. One substance, found only in the bark of the Pacific yew, showed a remarkable ability to destroy the tumor cells.

In 1971, after nearly a decade of work, Wani and Wall published a paper describing the isolation, structure, and anticancer properties of this substance. They called the substance *taxol*, after the scientific name for the Pacific yew, *Taxus brevifolia*. A few years later, Susan Horwitz, a biochemist at Yeshiva University in New York City, discovered that this substance prevented cancer cells from reproducing. It did this in a way that had never been seen before.

Horwitz's findings led other scientists to explore this substance further. By 1984, it had entered clinical trials (that is, testing on human patients). The substance was given the trade name Taxol® and the common name *paclitaxel*.

However, the promise of Taxol created a new problem. The Pacific yew was the only known source of the drug, and harvesting the bark killed the trees. The head of

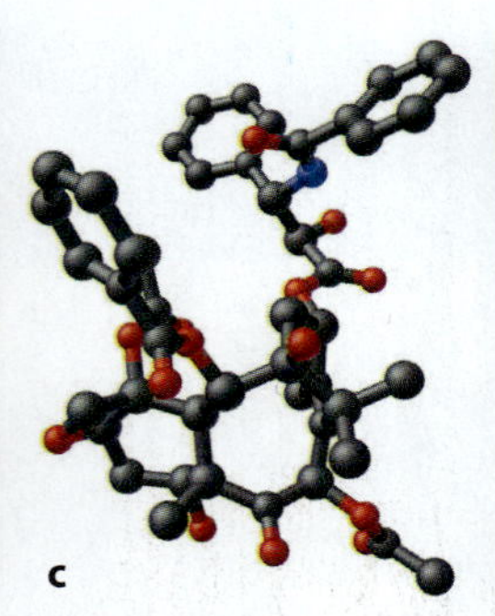

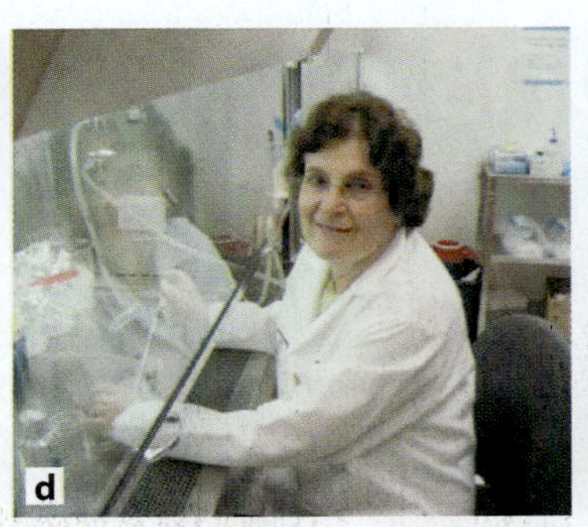

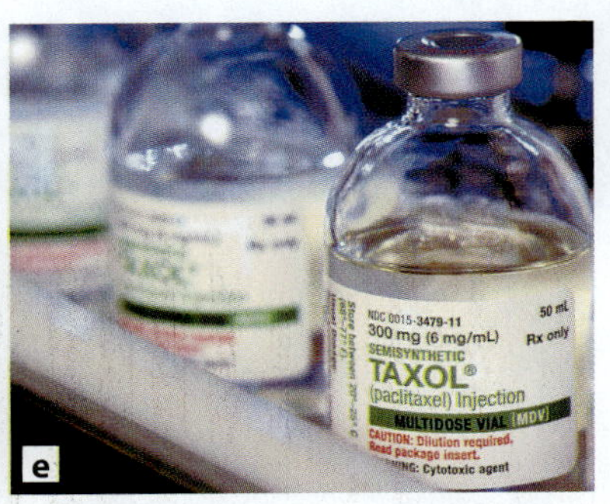

Figure 1.1 (continued) (b) Mansukh Wani (left) and Monroe Wall were the chemists who first identified Taxol®. (c) In this model of Taxol®, black spheres represent carbon. Red spheres represent oxygen, white are hydrogen, and blue is nitrogen. Lines between the spheres illustrate the way the particles are joined together. (d) Susan Horwitz discovered how Taxol® works. (e) Today, Taxol® is the leading treatment for many forms of cancer. (a) inga spence/Alamy; (b) Jimmy W. Crawford/RTI International; (c) [no credit]; (d) Courtesy of Dr. Susan Band Horwitz; (e) Courtesy Bristol-Myers Squibb

the National Cancer Institute estimated that keeping up with the demand for yew bark would require 360,000 trees per year—well beyond the capacity of the slow-growing Pacific yew to replenish itself. What could be done?

This situation led to dozens of questions: How does the Pacific yew produce Taxol? Do other trees also produce this substance? Can Taxol be produced synthetically in a laboratory or factory? Is it possible to produce simpler substances that mimic Taxol's anticancer properties? By the early 1990s research teams across the world were searching for answers to these urgent questions.

A major breakthrough came when scientists discovered that the English yew, a more common variety of the yew tree, produces a similar substance in its needles that can be converted into Taxol in a laboratory. Because this process involves the renewable needles rather than the bark, farmers can harvest the substance from English yews without killing the trees.

More recently, researchers have learned to grow yew cells in the laboratory and to collect the anticancer compound directly from the cells. Today, pharmaceutical companies use a combination of these techniques to produce Taxol, which is among the most widely used medicines for the treatment of cancer.

Think for a moment about the many facets of developing Taxol: From harvesting the first yew bark to using it in hospitals across the world, the Taxol story involved people whose expertise ranged from biology and medicine to business and agriculture. It involved government agencies and private companies. It involved conservation and innovation. And at the center of the story is a single substance, buried in the bark of an obscure little tree. Through careful study, this simple, naturally occurring substance changed the way we treat cancer and improved the lives of people all over the world.

Intended Learning Outcomes

After completing this chapter and working the practice problems, you should be able to:

1.1 Chemistry: Part of Everything You Do

- Describe the impact of chemistry on a variety of other fields.

1.2 Describing Matter

- Describe the difference between composition and structure.
- Differentiate between elements, compounds, homogeneous mixtures, and heterogeneous mixtures.
- Describe the three phases of matter.
- Compare and contrast physical and chemical properties and physical and chemical changes.

1.3 Energy and Change

- Define heat energy in terms of the motion of particles.
- Describe the relationship between the potential energy of a system and its potential for change.

1.4 The Scientific Method

- Describe the key components of the scientific method.
- Explain the differences between a hypothesis, a theory, and a scientific law.

1.1 Chemistry: Part of Everything You Do

The substances around us fill our lives with textures, flavors, sights, and smells. From the aroma of a delicious meal to the color of the sunset, from the hardness of a diamond to the softness of a favorite shirt, our world is filled with materials that have widely varying properties. To understand these properties, we must look closely at the smaller building blocks that make up the materials. We must explore how these building blocks are arranged and how they interact with the world around them.

Understanding what substances are made of and how they behave are the essence of the field of **chemistry**. Regardless of your background, your interests, or your career, your day-to-day activities involve chemistry. Are you interested in working in a healthcare field? Chemistry is essential to understanding how our bodies function and how they interact with their surroundings. Are you interested in physics, engineering, or construction? What about computers and technology? The properties of trade materials—from concrete and steel to semiconductors and solar cells—depend on the composition and properties of the substances they are made of. Are you interested in agriculture or the environment? Everything from exploring rock formations to monitoring air, soil, and water quality involves chemistry. Chemistry is sometimes referred to as "the central science," because it connects with every other field in the sciences (**Figure 1.2**). Whether your

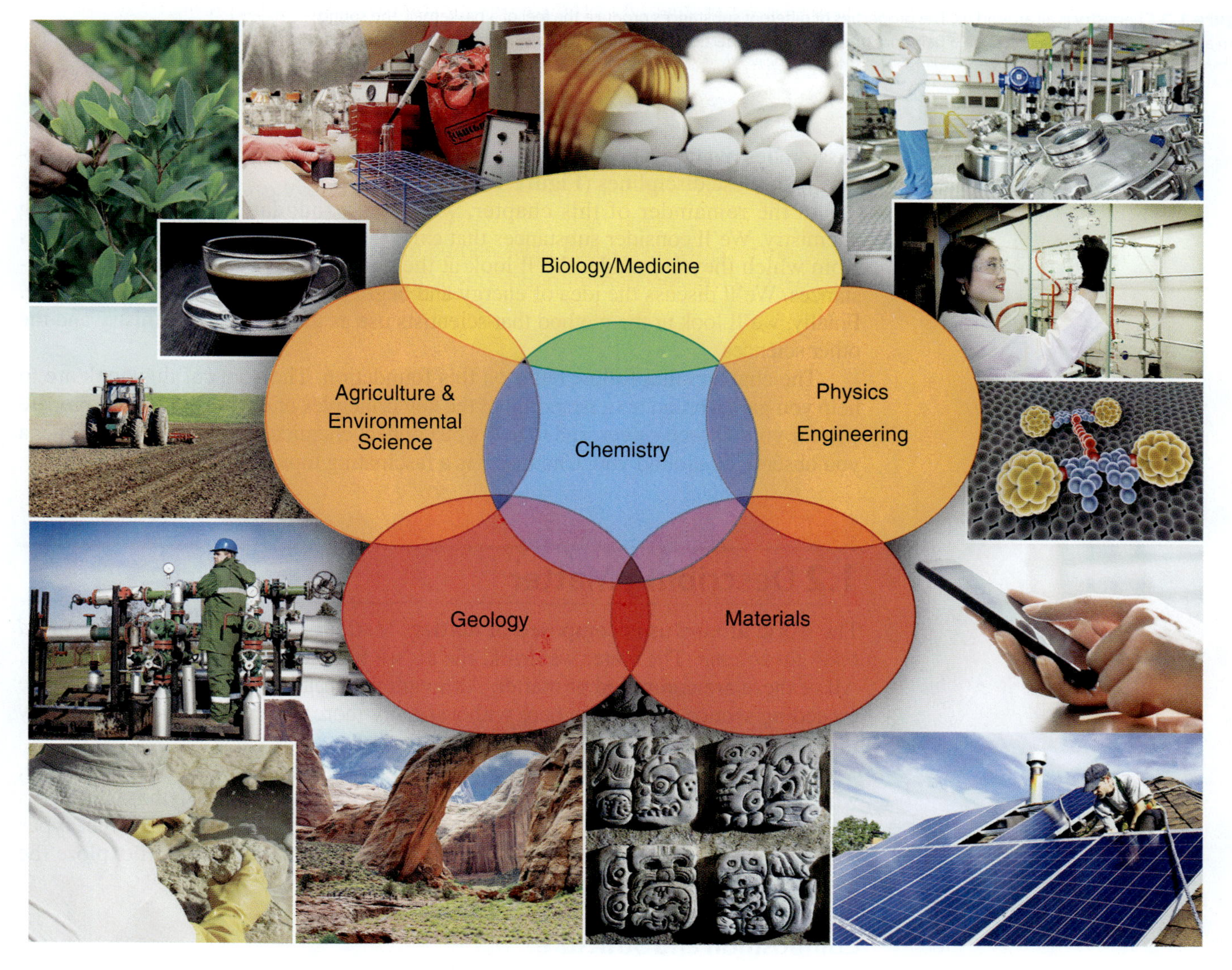

Figure 1.2 Chemistry connects to every other science and to our day-to-day lives. (clockwise from top left) Aizar Raldes/AFP/Getty Images; Chursina Viktoria/Shutterstock; Brad Davis/AP Images; STILLFX/Shutterstock; Dmitry Kalinovsky/Shutterstock; Copyright 2016 Murray State University. All rights reserved; ALFRED PASIEKA/Science Source; LDprod/Shutterstock; Elena Elisseeva/Shutterstock; De Agostini/G. Dagli Orti/Getty Images; Lissandra Melo/Shutterstock; Courtesy of Dr. Adam Kiefer/Mercer University; ZoranOrcik/Shutterstock; oticki/Shutterstock

Figure 1.3 Chemistry is all around us. The properties of different substances produce the feel of a basketball, the sound from a fine violin, the subtle colors on a painting, the lifesaving properties of a new medicine, and the flavor of great coffee. (clockwise from left) Mingo Nesmith/Icon Sportswire/AP Images; skynesher/E+ Collection/Getty Images; Margie Politzer/Getty Images; Bloomberg/Getty Images; Aaron Blanco/The Brown Coffee Company

passion lies in art, music, business, or sports, you can find ways that chemistry overlaps with these disciplines (**Figure 1.3**).

In the remainder of this chapter, we'll lay a foundation for understanding chemistry. We'll consider substances that exist all around us and the building blocks from which they are made. We'll look at the changes that occur within these substances. We'll discuss the idea of energy and begin to see how energy drives change. Finally, we'll look at the method that scientists use as they explore chemistry and the other sciences.

The chapters that follow build on this foundation. The goals of this book are to help you gain a broad understanding of the key principles of chemistry, to help prepare you for your chosen career, and to help you connect chemical principles to phenomena you observe all around you. Chemistry is a fascinating topic; let's explore it!

1.2 Describing Matter

The world around us is composed of matter. Matter is *anything that has mass and takes up volume*. The water we drink, the air we breathe, the ground under our feet: All of these are composed of matter. Although we broadly defined chemistry in the preceding section, let's begin here with a slightly more formal definition: Chemistry is *the study of matter and its changes*.

Our definition of chemistry covers some very broad questions. For example, what is a substance made of? How are its components arranged—that is, what is its structure? How does structure affect the properties of a substance? What types of changes can a substance undergo? In the coming chapters, we will explore the answers to these questions.

Composition and Structure

At the beginning of this chapter, we saw how scientists discovered a substance in the Pacific yew tree that destroyed cancer cells. One of their first challenges was to

Figure 1.4 A deck, canoe, and baseball bat share a similar composition, but each has a different structure.

answer the question, "What is this substance made of?" In other words, they wanted to determine the *composition* and *structure* of the substance.

Composition refers to the simple components that make up the material. **Structure** refers to both the composition and arrangement of those simpler substances. For example, in **Figure 1.4** we see a deck, a canoe, and a baseball bat. Each of these has basically the same composition: They are all made of wood. However, the wood is arranged in different structures that help each item fulfill its own function.

In chemistry, we ask similar questions on a smaller scale: "What is the composition of wood?" That is, what is it made of? Or, "What is the structure of wood?" That is, how are its components arranged? Answering these questions helps us understand why substances behave the way they do.

Pure Substances and Mixtures

To understand more about composition and structure, we begin with *atoms*. **Atoms** are the fundamental units of matter. Scientists represent atoms as tiny spheres, or by using one- or two-letter symbols, or by using both together (**Figure 1.5**). We will discuss atoms in more detail in the following chapters; for now, you can think of an atom as a tiny unit of matter.

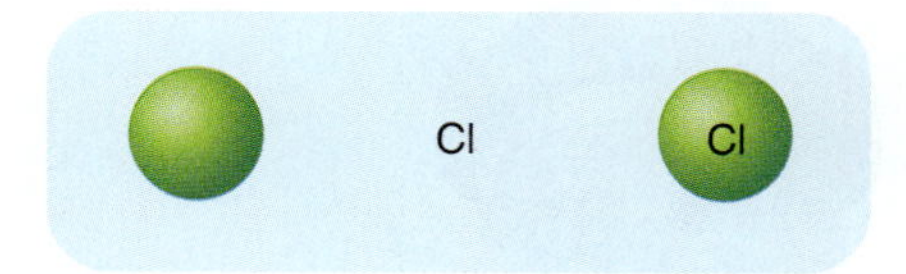

Figure 1.5 We represent atoms using a colored sphere, a one- or two-letter symbol, or a combination of these two.

The simplest form of matter is an **element**. An element is a substance that is made of only one type of atom. Common elements include gold, silver, iron, oxygen, and nitrogen. Each element is composed of a unique type of atom. For example, gold contains only gold atoms, and silver contains only silver atoms.

Compounds are substances composed of more than one element, bound together in fixed ratios. For example, water is made of two parts hydrogen and one part oxygen. If the ratio of hydrogen to oxygen changes at all from 2:1, it is no longer water, but some other substance.

Many compounds form groups of atoms called **molecules**. In molecules, the atoms bind tightly together and behave as a single unit. For example, a water molecule contains two hydrogen atoms and one oxygen atom (**Figure 1.6**). Some elements also exist as molecules. In hydrogen, nitrogen, and oxygen, the atoms of the element pair together to form *diatomic* ("two-atom") molecules (**Figure 1.7**).

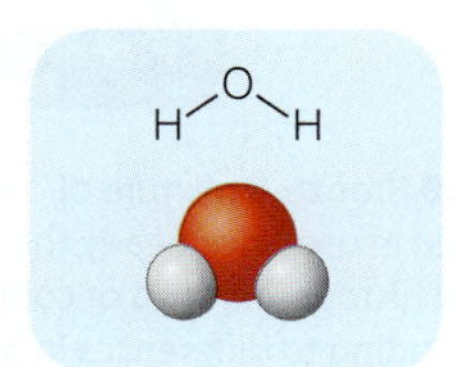

Figure 1.6 Water is a compound composed of hydrogen and oxygen. In a water molecule, two hydrogen atoms bind to one oxygen atom. To represent this idea, we can use elemental symbols connected by lines, or we can use spheres (balls) to signify each atom. Chemists normally use red spheres to represent oxygen atoms and white spheres to represent hydrogen atoms. See Appendix A2 for a list of colors used to represent different atoms.

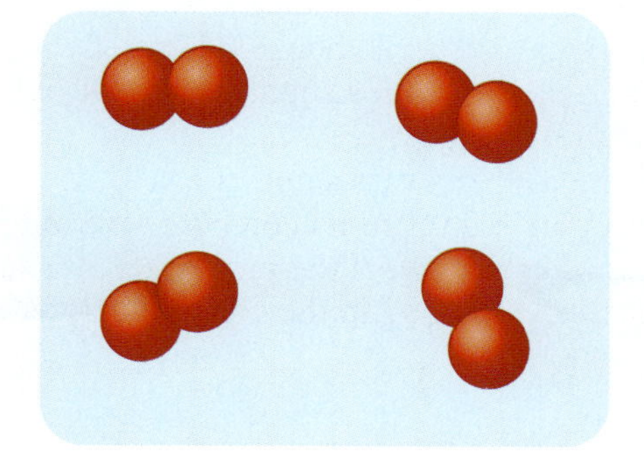

Figure 1.7 Elemental oxygen exists as diatomic molecules. A molecule of oxygen contains two oxygen atoms joined together. This image depicts four oxygen molecules.

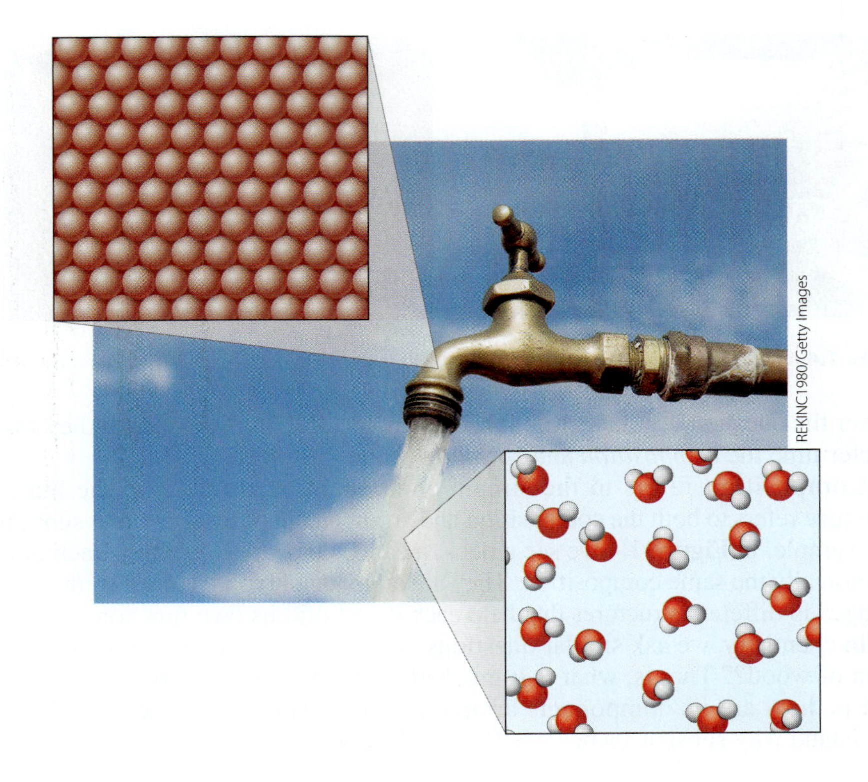

Figure 1.8 Copper is an element (composed of only one type of atom), but water is a compound made of hydrogen and oxygen atoms.

Figure 1.8 shows water flowing from a copper pipe. Copper is an element, so the only type of atom present in copper is the copper atom. In contrast, water is a compound. Each molecule of water is composed of two hydrogen atoms and one oxygen atom. Both the element copper and the compound water are examples of **pure substances**: materials that are composed of only one element or compound.

In contrast to pure substances, **mixtures** contain more than one substance—and the substances are not bound in a fixed ratio. For example, brass and bronze are both mixtures of metals, known as *alloys*. Brass is a mixture of the pure substances copper and zinc, while bronze is a mixture of the pure substances copper and tin (**Figure 1.9**). In these mixtures, the ratio of copper to the other metals is not fixed; it may be altered to affect the properties of the alloy, such as hardness and color.

Figure 1.9 Bronze, a mixture of copper and tin, has been in use since nearly the beginning of recorded history. The ratio of copper to tin may vary, leading to differences in properties such as color, melting point, and hardness. This image is of a mask of an Akkadian ruler, possibly Sargon (Syria, approximately twenty-third century B.C.E.). Inset: An atomic view of bronze, showing a mixture of copper and tin atoms.

Figure 1.10 (a) A mixture of salt and water is homogeneous. (b) A mixture of sand and water is heterogeneous.

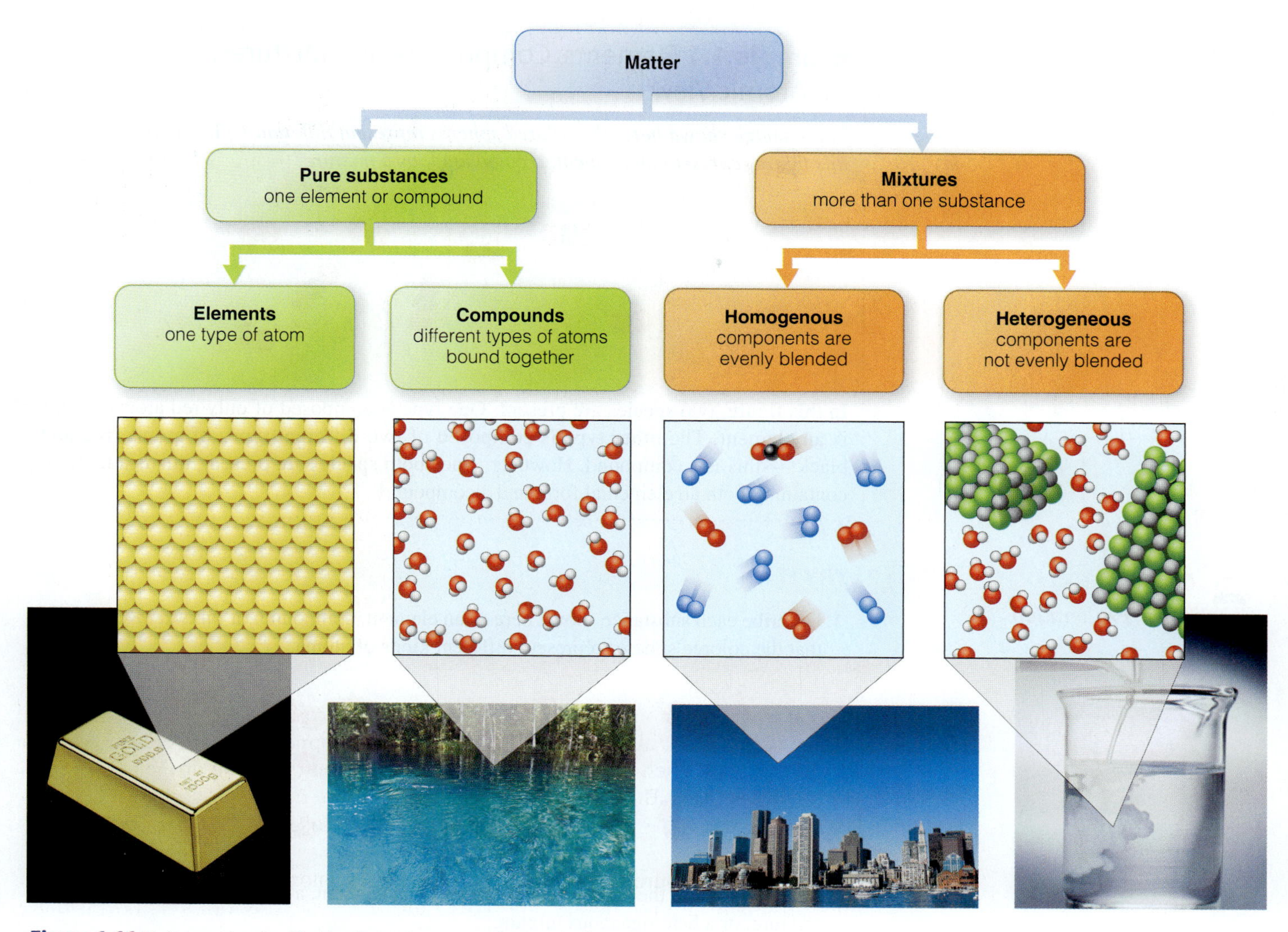

Figure 1.11 Matter can be classified by the arrangement of atoms. In these images, gold and gray spheres represent metal atoms, red represents oxygen, blue represents nitrogen, black represents carbon, white represents hydrogen, and green represents chlorine. sumire8/Shutterstock; Sherri_j's_pics/Shutterstock; Greg Kushmerek/Shutterstock; © 1995 Michael Dalton/Fundamental Photographs, NYC

Mixtures whose components are evenly blended throughout are called **homogeneous mixtures**. Metal alloys, a cup of coffee, and the air around us are all homogeneous mixtures. In contrast, **heterogeneous mixtures** contain regions with significantly different composition (**Figure 1.10**). Sand and water, gravel, and chocolate-chip cookie dough are all heterogeneous mixtures. They each have regions where one component is present in far greater proportion. **Figure 1.11** summarizes the categories of pure substances and mixtures.

A key difference between compounds and mixtures is that mixtures can be separated into their individual components without changing the identity of the substances. For example, chemists commonly use a technique called *filtration* to separate heterogeneous mixtures. If you pour a mixture of sand and water through a filter, the sand remains trapped in the filter while the water passes through (**Figure 1.12**). The mixture has separated, but the substances have not changed. By contrast, compounds cannot be separated without changing the substances into their elemental forms.

Figure 1.12 Mixtures can be separated. Chemists use filtration to separate heterogeneous mixtures like sand and water.

Example 1.1 Elements, Compounds, and Mixtures: An Atomic View

In the image shown here, the colored spheres represent different kinds of atoms. Does this figure represent an element, a compound, or a mixture?

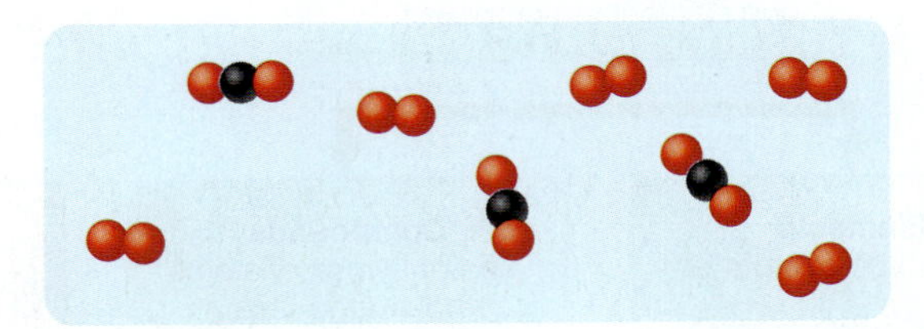

In this figure, two species are present: One type is composed of only red atoms — this is an element. The other type is composed of two different types of atoms (red and black) — this is a compound. However, since both species are present, this is a mixture containing both an elemental form and a compound.

Check It

Watch Explanation

TRY IT

1. Describe each substance shown here as an element, compound, or mixture. Remember that the colored spheres represent different kinds of atoms.

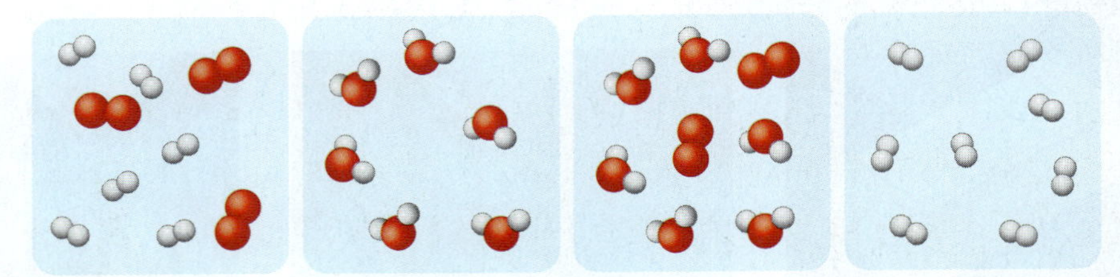

2. Describe each of the following substances as an element, a compound, a homogeneous mixture, or a heterogeneous mixture:
   - **a.** table salt, a substance made up of sodium atoms and chlorine atoms bound in a 1:1 ratio
   - **b.** a fruit drink in which frozen concentrate is combined with water, then stirred until it is evenly blended
   - **c.** a substance made up only of zinc atoms
   - **d.** blue cheese dressing, a creamy white sauce containing small chunks of cheese

States of Matter

Matter typically exists in one of three forms, which we call the three **states of matter**, or sometimes the three *phases of matter*. The three basic states of matter are *solid*, *liquid*, and *gas*. We distinguish between these states by properties that are visible on a *macroscopic* level (visible to the naked eye):

- **Solids** have both a definite shape and a definite volume.
- **Liquids** have a definite volume, but no definite shape. Liquids adopt the shape of their container.
- **Gases** do not have a definite shape or a definite volume. Gases fill any container they occupy.

If we heat a substance, it can transition from solid to liquid (**melting**) or from liquid to gas (**vaporization**). If we cool it, the reverse occurs: The gas changes to a liquid (**condensation**), and the liquid changes to a solid (**freezing**). We can see its visible properties change in an obvious way as the substance transforms between these states, but it also changes on the atomic level.

Let's use the element aluminum as an example. At room temperature, aluminum is a solid. Its atoms are close to each other, packed into an ordered framework (**Figure 1.13**). If we were able to "zoom in" to see the atoms, we would notice something else: Even though this framework holds each atom in place, the atoms are not perfectly still. Rather, they are vibrating, moving within their spaces in the framework.

Next, let's begin to heat the solid aluminum. When we do this, the atoms vibrate faster. The more heat we add, the faster the atoms move. Eventually, they begin to break out of the rigid framework and travel freely past each other. The substance is now in the *liquid* state. The particles in a liquid move randomly, but they remain close to each other.

If we continue heating the liquid aluminum, the atoms move faster and faster until they begin to break out of the liquid phase and enter the *gas* phase. Particles in the gas phase move about freely and have very little interaction with each other.

If we cool the aluminum, we observe the opposite effect: As the temperature drops, the fast-moving gas particles slow and begin clustering together. Particles drop to the bottom of the container, transitioning from the gas phase to the liquid phase. Similarly, as we cool a liquid, the particles move more and more slowly until eventually they settle back into a solid framework.

The states of matter offer an important lesson: *The behavior of any substance is determined by the arrangement of the particles that compose the substance*. As the aluminum in the example transitioned from solid to liquid to gas, the atoms did not change, but their arrangement did—with a huge impact on how the substance behaves. Because of this, it is important to think not just in terms of the macroscopic world, but also to ask, "What is going on at the atomic level?" The better you can visualize motion and interaction at the atomic level, the easier it is to understand and predict the way a substance behaves.

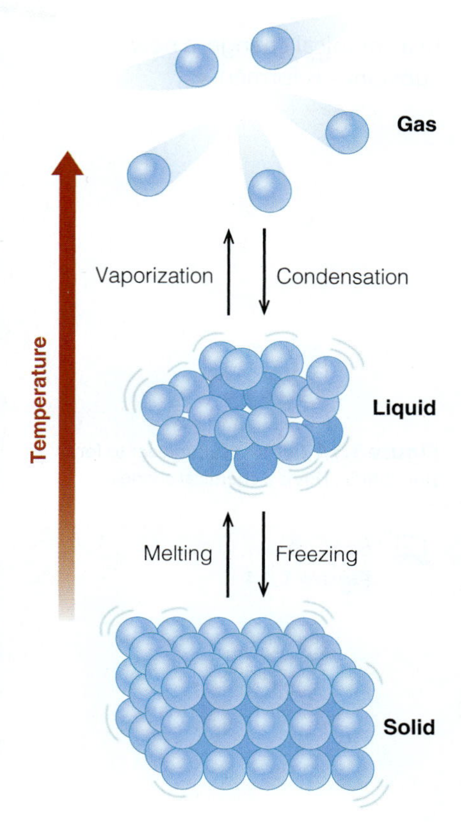

Figure 1.13 As a sample is heated, it transitions from solid to liquid to gas.

Explore Figure 1.13

As a substance is heated, its particles move faster.

The behavior of any substance is determined by the arrangement of the particles in the substance.

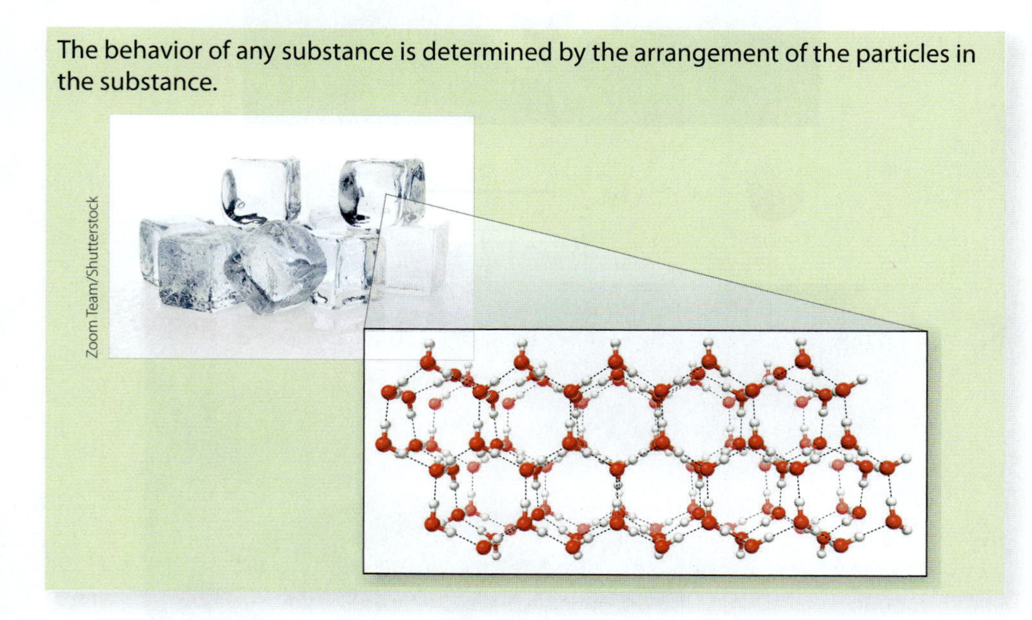

Properties and Changes

Chemists study the properties of matter and also the ways that matter can change. We commonly describe both properties and changes as being *physical* or *chemical*.

Physical properties are the properties of a substance that we can measure without changing the identity of the substance. For example, we can measure color, temperature, mass, volume, shape, hardness, flexibility, and a host of other factors without changing the composition of that substance. Similarly, **physical changes** are changes that occur without altering the identity of the substance. For example, we can take an iron bar, melt it, pour it into a mold, and let it cool, solidifying into a new shape. We have altered the shape and the phase of the iron, but it is still composed of iron, and so each of these changes is a physical change.

In a physical change, the properties change but the composition stays the same.

Phase changes are physical changes.

In a chemical change, a new substance is formed.

Chemical properties are properties that we cannot measure without changing the identity of a substance. For example, flammability is a chemical property. If we want to determine if something is flammable, we must try to burn it. If it burns, it is no longer the original substance. When a substance changes into something different, it has undergone a **chemical change**. Chemical changes are also called **chemical reactions**. In some chemical changes, elements combine to form compounds. For example, when heated, zinc and sulfur combine to form a new compound, zinc sulfide (**Figure 1.14**). In other chemical changes, compounds break apart to produce elemental substances or rearrange to form new compounds. For example, methane gas can react with oxygen gas to produce two new compounds, carbon dioxide and water (**Figure 1.15**). The key idea is that *a chemical change always involves the formation of a different substance*.

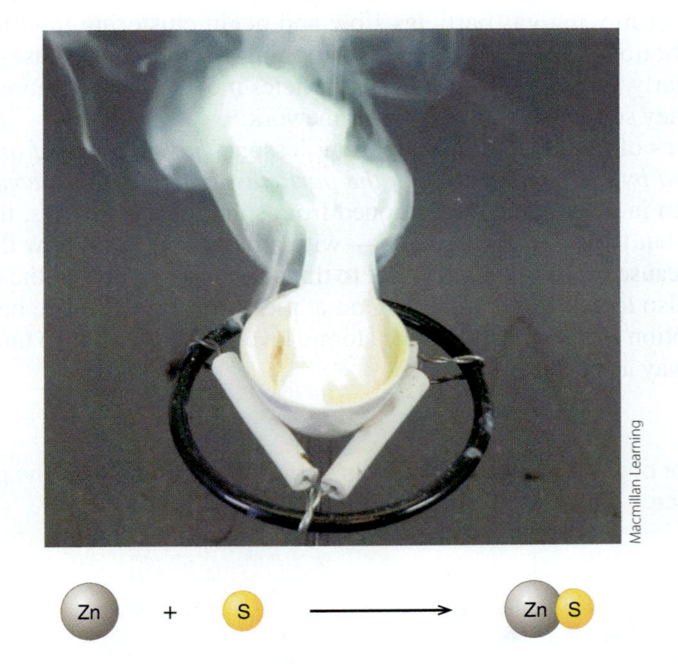

Figure 1.14 Zinc and sulfur react to form zinc sulfide. This is a chemical change.

 *Explore* Figure 1.14

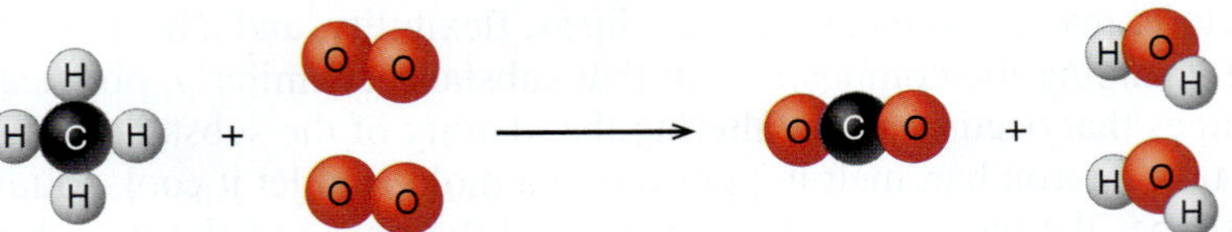

Figure 1.15 The flame of a gas stove results from a chemical change. Methane combines with oxygen gas to produce carbon dioxide gas and water.

Example 1.2 Physical and Chemical Properties

Describe each property as physical or chemical:

a. A paint has a deep blue color.

b. An iron bar heats up quickly if left out in the sun.

c. Zinc metal reacts with hydrogen chloride gas to form zinc chloride and elemental hydrogen.

To determine whether a property is physical or chemical, we must ask whether a new substance is formed. If the property involves the formation of a new substance, it is a chemical property. If not, it is a physical property.

The color of paint is a physical property. Observing its color does not change its identity. Similarly, the fact that an iron bar heats up quickly is a physical property. Although the temperature of the bar may change (a physical change), the bar is still composed of iron.

However, the reaction of zinc metal with hydrogen chloride gas involves the formation of two new substances (zinc chloride and elemental hydrogen). The fact that zinc reacts in this way is a chemical property. The change that results is a chemical change.

Example 1.3 Physical and Chemical Changes

Gold bars are produced by pouring molten gold into a mold, as shown here. As the gold cools, its atoms transition from moving freely past each other into a well-ordered framework. Is this a physical change or a chemical change?

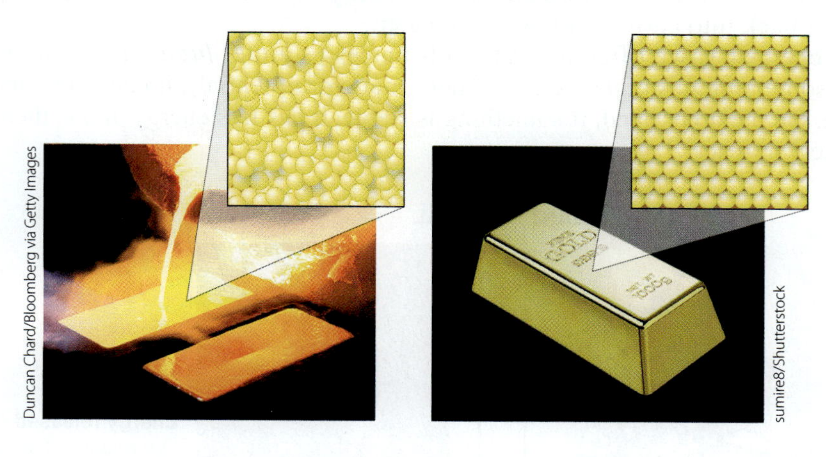

Duncan Chard/Bloomberg via Getty Images

sumire8/Shutterstock

When gold transitions from liquid to solid, its arrangement of atoms changes, but it does not form a new element or compound. Because no new substance has been formed, this is not a chemical change. It is a physical change. Phase changes are physical changes.

TRY IT

3. Each of these statements describes a property or change. Identify each property or change as physical or chemical.

a. The element bromine has a deep red color.

b. Magnesium burns brightly in oxygen to form magnesium oxide.

c. A saltwater solution conducts electricity.

d. A copper slug has a mass of 15.2 grams.

e. Over time, iron hinges rust.

f. Your body converts glucose and oxygen into carbon dioxide and water.

Check It

Watch Explanation

1.3 Energy and Change

Ljupco Smokovski/Shutterstock

Figure 1.16 A bike spontaneously moves downhill to a lower-energy state.

At higher temperatures, particles have more kinetic energy.

Substances with a large amount of potential energy are more likely to react.

We often describe chemical and physical changes in terms of their *energy*. **Energy** is the ability to do work. We discuss energy in more detail in Chapter 8, but a foundational concept of energy is essential for understanding chemistry. Broadly, energy takes two forms: *kinetic* and *potential*. **Potential energy** refers to energy that is stored. **Kinetic energy** is the energy of motion. The faster an object is moving, the greater the kinetic energy it has.

Heat energy is a type of kinetic energy. Heat energy involves *the kinetic energy of the particles within a substance*. When a substance is heated, the particles within the substance vibrate or move more and more quickly. For example, if a stove is hot, the particles on the hot surface vibrate more rapidly than if the stove is cold. On the macroscopic level, the stove doesn't have kinetic energy (it is sitting still). But on the atomic level, the particles on the surface of the stove are moving rapidly.

Physical and chemical changes involve changes in energy. Often, an object or substance changes in a way that releases energy. For example, a bicycle releases energy as it quickly rolls downhill (**Figure 1.16**). Changes that release energy often happen spontaneously. On the other hand, when we push the bicycle up the hill, we move it to a higher-energy state through an input of energy.

Energy changes take many forms. For example, a tree grows by absorbing energy from the Sun to convert two simple chemicals, carbon dioxide and water, into plant material (**Figure 1.17**). The energy that the plant harvests from the Sun is stored in the plant material. If the tree falls, we can use it for firewood—releasing the stored potential energy as heat and converting the plant material back into carbon dioxide and water.

In chemistry, we often describe substances as either *high-energy* or *stable*. When something has high energy (either kinetic or potential), it can bring about a change. On the other hand, if something is stable, it has less energy and is therefore less likely to react.

PLANT MATERIAL

Energy stored
A tree grows by absorbing energy from the sun to convert carbon dioxide and water into plant material.

Morey Milbradt/Getty Images

Evgeny Dubinchuk/Shutterstock

Energy released
Fire releases the stored potential energy as heat, converting the plant material back into carbon dioxide and water.

CARBON DIOXIDE + WATER

Figure 1.17 Energy is stored and released in many forms.

For example, in Zimbabwe there is an extraordinary rock formation (**Figure 1.18**). Time and erosion have carved away much of the soil beneath the rocks, leaving them precariously balanced. Because these rocks sit high off the ground, they have a good deal of potential energy. Eventually, these rocks will fall. When they do, they will release this stored energy. Right now, you probably wouldn't want to camp underneath the rocks. But once the rocks fall, you have nothing more to worry about—the energy has been released, and the rock formation will be stable.

Figure 1.18 The Balancing Rocks in Zimbabwe contain potential energy.

Consider a second example: a mousetrap (**Figure 1.19**). If we put energy in, we can bend back the spring and set the trap. If we touch the trigger—whap! The bar snaps with a large force. Energy is stored in the set trap, but once this energy is released, the trap is stable. We don't have to worry about getting our fingers snapped if the spring on the trap is not coiled. If there's no potential, there's no problem.

Ethylene oxide **(Figure 1.19c)** is a simple compound that behaves a lot like a mousetrap. It contains two carbon atoms and an oxygen atom, but the structure of the compound is strained. If the compound encounters a water molecule, it snaps open like the mousetrap to form a product that contains less potential energy. The new product is lower in energy, so it is less likely to react with other compounds.

In the coming chapters, we talk frequently about the energy or the stability of matter. As a general rule, *systems move toward the lowest energy state possible*. As we just noted, substances that have a low amount of energy are said to be *stable*—they do not react as readily as high-energy substances do.

The potential energy in any substance depends on its structure.

Many chemical changes absorb or release heat energy. An **exothermic** change is one that releases heat energy. Wood burning on a campfire is an example of an exothermic change. Changes that require energy to occur are **endothermic**. For example, we must heat water to convert it to steam; this is an endothermic change.

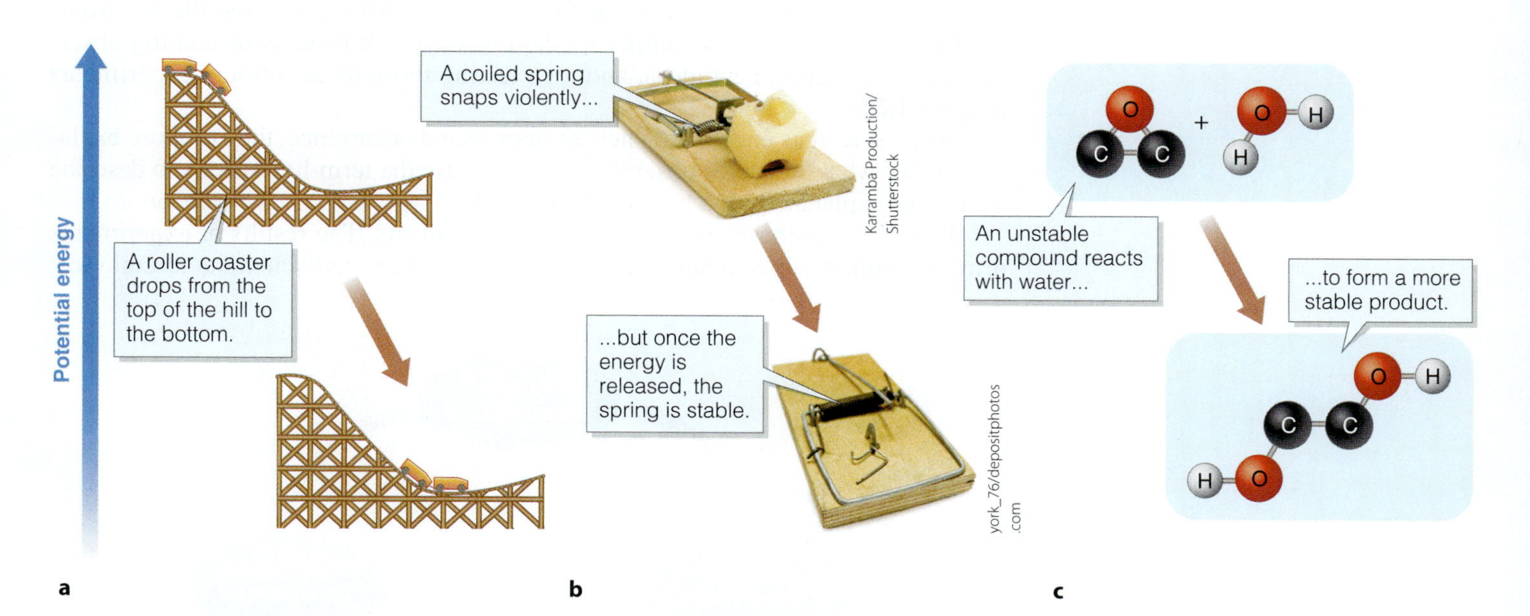

Figure 1.19 Objects tend to move spontaneously from higher to lower potential energy. Objects with lower potential energy are more stable. This happens with everyday objects like roller coasters and mousetraps, but it also happens with atoms and molecules.

Example 1.4 Energy and Chemical Changes

When charcoal burns, the charcoal (which is composed of carbon) reacts with oxygen to produce carbon dioxide and heat, as represented in the equation shown here. Which has the higher potential energy: carbon plus oxygen, or carbon dioxide? Which is more stable? Is this reaction endothermic or exothermic?

$$\text{Carbon} + \text{Oxygen} \rightarrow \text{Carbon Dioxide} + \text{Heat}$$

This chemical change involves the release of energy. This means the substance formed by the change (carbon dioxide) has less potential energy than the substances that were present before the change (carbon and oxygen). Since carbon dioxide has less potential energy, it is more stable. And because this change released energy, we refer to it as an exothermic change.

TRY IT

4. How do the particles in hot coffee differ from those in cold coffee?

5. When we heat water, it becomes steam. Is this an endothermic or exothermic process? Which has the higher energy, liquid water or steam?

6. Which has more stored potential energy, a new battery or a dead battery? Which is more stable?

1.4 The Scientific Method

At the beginning of this chapter, we described the discovery of Taxol, a cancer-fighting compound found in the bark of the Pacific yew tree. From the discovery of the compound to its widespread use today, scientists followed a systematic method to formulate and test their ideas.

The approach that scientists take to solving problems is called the **scientific method**. At its core, the scientific method is a cyclical process of making observations, formulating new ideas, and then testing those ideas through experiments (**Figure 1.20**).

When scientists encounter a new or unexpected occurrence, they propose explanations for what they have observed. Scientists use the term **hypothesis** to describe a tentative explanation that has not been tested.

To test a hypothesis, scientists devise experiments. The results of experiments provide support for or against a hypothesis, and they lead scientists to embrace,

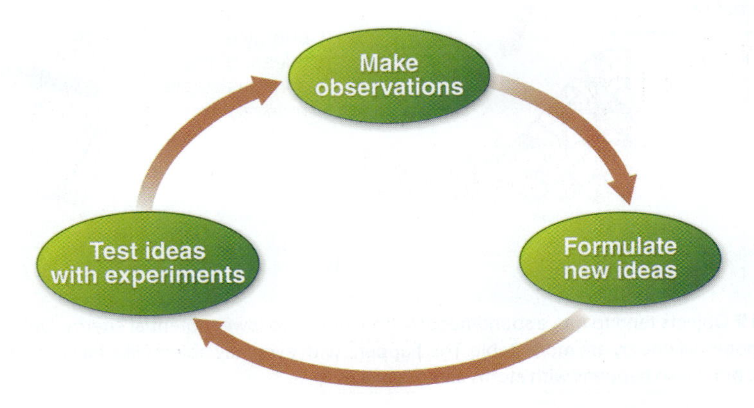

Figure 1.20 The scientific method is a cyclic process of making observations, formulating new ideas, and then testing those ideas.

refine, or discard their ideas. The observations that scientists make from experiments often lead to new hypotheses, which in turn lead to new experiments.

Through this cyclic process, scientists refine their ideas over time. A **theory** is an idea that is supported by experimental evidence. This term can also have a broader meaning: Sometimes it is used to mean a *paradigm*—a way of thinking about or analyzing a particular topic.

To many people, the term *theory* means an idea is unproven, but scientists use this term differently. For example, in the coming chapters we discuss the atomic theory. The evidence for atoms is indisputable, and atoms can be indirectly observed through a number of techniques. Rather than conveying uncertainty, the term *atomic theory* describes our modern understanding of atoms. When we explain how different substances behave, we do so using the atomic theory. Atomic theory is not just a guess supported by evidence—it's a way of thinking about the world around us.

When scientists talk about a theory, they mean an idea that is supported by evidence, or a paradigm for describing the world around us.

Theories often develop from the work of many scientists. Because of this, effective scientists must communicate their findings and ideas to others. This practice enables other scientists to reproduce experiments, verify or refute results, and analyze or challenge ideas. For example, when Susan Horwitz observed how Taxol inhibited the growth of cancer cells, she wrote a paper that shared these findings with other cancer researchers (**Figure 1.21**). Other scientists tested her findings and refined her ideas. Our understanding of how Taxol works is a theory, built on many different experiments, and involving many different scientists.

Scientists use the term **scientific law** to describe observations that are true in widely varying circumstances. A scientific law does not explain *why* something occurs; it simply observes that it is true. Often, a scientific law describes a mathematical relationship. For example, consider the law of gravity: This law describes observations and phenomenon that consistently occur. Scientists use the law of gravity mathematically, to calculate the weight of a building or the trajectory of a rocket. This law predicts *what* will happen, but it does not explain *why* it happens.

A common misconception is that scientific research progresses from hypothesis to theory to law. In fact, theories and laws are two very different ends. Scientific laws provide a concise description of a behavior, while theories explain *how* or *why* things happen.

Scientific laws describe what happens, but not why. Theories describe how or why something happens.

In the chapters ahead, we'll explore the work of scientists past and present. We'll see how experiments and observations led them to formulate ideas. We'll see how new experiments and new findings caused old ideas to be discarded or refined.

Science is never finished. There are always new questions to ask, and new problems to solve. Every day, people across the world add to the story of science. Someday, you may be part of the story, too.

Proc. Natl. Acad. Sci. USA
Vol. 77, No. 3, pp. 1561–1565, March 1980
Cell Biology

Taxol stabilizes microtubules in mouse fibroblast cells

(cell cycle/cytoskeleton/cell migration/antimitotic agents)

PETER B. SCHIFF AND SUSAN BAND HORWITZ

Departments of Cell Biology and Molecular Pharmacology, Albert Einstein College of Medicine, Bronx, New York 10461

Communicated by Harry Eagle, December 18, 1979

ABSTRACT Taxol, a potent inhibitor of human HeLa and mouse fibroblast cell replication, blocked cells in the G_2 and M phase of the cell cycle and stabilized cytoplasmic microtubules. The cytoplasmic microtubules of taxol-treated cells were visualized by transmission electron microscopy and indirect im-

0.5% or less, a concentration that had no effect on control reactions.

Cells. HeLa (human) cells, strain S_3, were grown in suspension culture in Joklik's modified Eagle's minimal essential

Figure 1.21 In 1980, Susan Horwitz published a paper describing how Taxol inhibited the growth of cancer cells. Scientists write papers like this one to share their findings with others.

Check It
Watch Explanation

TRY IT

7. Classify each of the following as a hypothesis, theory, or law:

a. When leaving your house, you notice that your trash has been strewn across the driveway. You guess that the neighbor's dog is responsible.

b. In most coastal areas, high tides occur one to two times each day.

c. After repeated experiments, scientists believe that a class of antibiotics functions by weakening the cell walls of certain bacteria.

8. Since ancient times, humans have used plants for medicinal purposes. In 1962, Arthur Barclay suspected that the Pacific yew might contain substances that could fight disease. At the time, was this a hypothesis, a theory, or a scientific law? What types of experiments were conducted to test this idea?

Capstone Video

Capstone Question

In Washington State, near the site where Arthur Barclay collected the Pacific Yew samples that led to the discovery of Taxol, Mount St. Helens towers over the surrounding hills. In 1980, a volcanic eruption ripped 1,300 ft of rock off the top of the mountain, sending gas and ash miles into the air (**Figure 1.22**). Falling rock mixed with ice from the mountain to produce massive mudslides. The force of the blast and the mudslides destroyed the surrounding forests, buildings, and roads. Fifty-seven people died in the eruption. Over a million tons of a toxic gas called sulfur dioxide spewed into the air. This gas combined with water in the atmosphere to produce a new compound, sulfurous acid. As the eruption subsided, the molten rock that had flowed down the mountain cooled and solidified, reshaping the landscape for miles around.

Think about the changes that took place in this cataclysm. Which of these changes were physical changes? Which were chemical changes? What changes in energy accompanied this event? Is Mount St. Helens more stable or less stable than it was before the eruption?

USGS/Alamy

Hum Images/Alamy

Figure 1.22 Mount St. Helens, (a) the day before and (b) shortly after the eruption.

SUMMARY

This chapter introduced several key ideas that are foundational for the study of chemistry. Because chemistry involves the world all around you, knowledge of chemistry will serve you well throughout your career, whether you work in health care, manufacturing, agriculture, or some other field.

The study of chemistry begins with matter. We define matter as anything that has mass and takes up volume. All matter is composed of tiny units called atoms. The arrangement of atoms within a substance determines its properties.

Elements are substances that are made of only one type of atom. Compounds are substances made of more than one type of atom, bound together in fixed ratios. In many compounds, the atoms form discrete groups called molecules. Mixtures contain more than one element or compound. The ratios of components in mixtures may vary.

We commonly describe matter by its physical or chemical properties. We can measure physical properties without changing the identity of a substance. In contrast, chemical properties are properties of a substance that we cannot measure without changing the identity of a substance. Chemical properties describe the ways in which substances change to form new substances.

Physical and chemical changes involve changes in energy. Energy is the ability to do work. Energy can exist as kinetic energy (including heat energy) or as potential energy. As a general rule, spontaneous changes involve moving from high-energy to low-energy states. Objects or substances that do not have a large amount of stored energy are said to be stable.

When scientists encounter questions, they use a systematic approach called the scientific method to formulate and test their ideas. The scientific method is a cyclical process of making observations, formulating new ideas, and testing ideas through experiments. Scientists often propose untested ideas called hypotheses. An idea that is supported by experimental evidence is called a theory. Theories often explain how or why something works. A scientific law is something that is observed to be true in a variety of circumstances. A scientific law predicts what will happen but does not explain why it happens.

You Can Do This

"It is possible to move a mountain by carrying away small stones."

In my early 30s, with a wife, a full-time job, and three small children, I decided to go back to school. I was more than a little reluctant — my previous university experience had left a bad taste in my mouth, and I dreaded going back. Nonetheless, I knew I had to complete the degree to accomplish my own career goals and to provide for my family's future.

Gandee Vasan/Getty Images

Over the next couple of years, I determined to make a little progress each day. I taped a picture of a bulldog over my desk to remind me of my goal. Bulldogs aren't the prettiest of animals, or the smartest. But they grab on tight, and they don't let go. They win by gaining an inch at a time.

As you begin your own chemistry course, I want to encourage you: You can do this. It's impossible to learn chemistry in one evening, but you can learn a little bit each day. If you will commit to read, to study, and to do practice problems, you can do this. If you will find others who can hold you accountable, and if you will seek help when you need it, you can do this. Start carrying away stones; soon you'll see your mountain begin to move.

Gary Deaton

Yuval Helfman/500px/Getty Images

Key Terms

1.1 Chemistry: Part of Everything You Do

chemistry The study of matter and its changes.

1.2 Describing Matter

matter Anything that has mass and takes up volume.

composition The components that make up a material.

structure The arrangement of simple units within a substance. In chemistry, structure refers to both the composition and arrangement of simple units within a substance.

atoms The fundamental units of matter.

element A substance made of only one type of atom.

compounds Pure substances composed of more than one element in a fixed ratio.

molecules Groups of atoms that are held tightly together.

pure substances Substances composed of only one element or only one compound.

mixtures Substances containing more than one substance.

homogeneous mixtures Mixtures in which the components are evenly blended throughout.

heterogeneous mixtures Mixtures in which the components are not evenly blended throughout.

states of matter The classification of matter as a solid, liquid, or gas; also called the phases of matter.

solid A state of matter having a definite shape and a definite volume. The particles in a solid are held in fixed positions.

liquid A state of matter having definite volume but no definite shape. The particles in a liquid are close together but move freely past each other.

gas A state of matter that does not have a definite shape or a definite volume. The particles in a gas move freely with very little interaction.

melting A transition from the solid phase to the liquid phase.

vaporization A transition from the liquid phase to the gas phase.

condensation A transition from the gas phase to the liquid phase.

freezing A transition from the liquid phase to the solid phase.

physical properties The properties of a substance that can be measured without changing the identity of the substance.

physical changes Changes that do not alter the identity of the substance.

chemical properties Properties of a substance that cannot be measured without changing the identity of a substance.

chemical changes Changes that produce new substances; also called chemical reactions.

chemical reactions Changes that produce new substances; also called chemical changes.

1.3 Energy and Change

energy The ability to do work.

potential energy Energy that is stored.

kinetic energy The energy of motion.

heat energy A type of kinetic energy, involving the movement of particles within a substance.

exothermic change A change that releases heat energy.

endothermic change A change that absorbs energy.

1.4 The Scientific Method

scientific method A cyclical process of making observations, formulating new ideas, and then testing those ideas through experiments; also called the scientific process.

hypothesis A tentative explanation that has not been tested.

theory An idea that has been tested and refined, or a paradigm for describing a particular topic.

scientific law A statement that describes observations that are true in widely varying circumstances. Scientific laws often describe mathematical relationships. However, they do not explain why something occurs; they only observe that it occurs.

Additional Problems

1.1 Chemistry: Part of Everything You Do

9. What are your career goals? What substances are unique or integral to your field of study? In which aspects of your chosen field might knowledge of chemistry help you?

10. The chart shown here includes a number of career pathways and suggests some of the chemistry topics associated with each one. Fill in the chart with some examples where an overlap between each field and chemistry might occur. Feel free to add more lines for other careers that interest you.

Field	Overlap with Chemistry
Human or Veterinary Medicine	Human and animal physiology, drug design, blood and tissue analysis, etc.
Ecology and Environmental Science	Monitoring of air and water quality, plant and animal biology
Engineering	
Forensics	
Computer Science	
Farming	
Geology	
Business and Management	
Psychology	
Manufacturing	

1.2 Describing Matter

11. Determine whether or not the following are made of matter:
a. air b. a football c. sunlight d. water

12. Determine whether or not the following are made of matter:
a. sweet tea b. happiness c. sound d. sand

13. What is the difference between composition and structure?

14. What is the difference between an element and a compound?

15. Describe each of the following as an element, a compound, a homogeneous mixture, or a heterogeneous mixture:
a. Earth's atmosphere, which contains about 78% nitrogen, 21% oxygen, and small amounts of other gases
b. fluorine gas, a substance that contains only fluorine atoms
c. carbon monoxide, a substance containing carbon and oxygen atoms in a fixed 1:1 ratio

16. Describe each of these as an element, a compound, a homogeneous mixture, or a heterogeneous mixture:
a. an alloy of gold and tin
b. phosphorus trichloride, a substance containing phosphorus and chlorine atoms in a 1:3 ratio
c. titanium metal, a substance containing only titanium atoms

17. In the graphics shown here, the colored spheres represent different atoms. Indicate whether each image represents an element, compound, homogeneous mixture, or heterogeneous mixture.

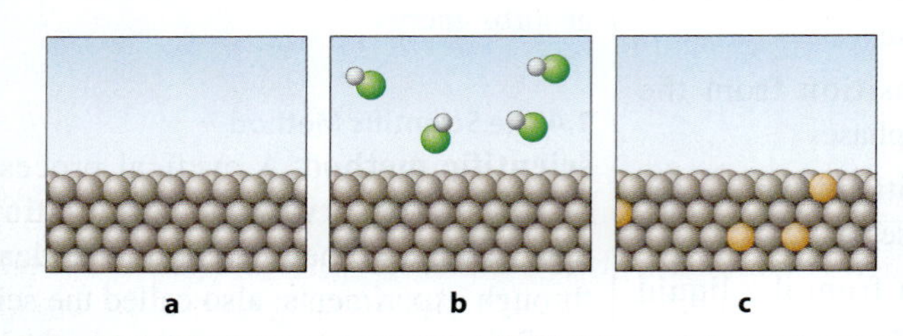

18. An atomic-level representation of saltwater is shown here. Is saltwater best described as an element, compound, homogeneous mixture, or heterogeneous mixture?

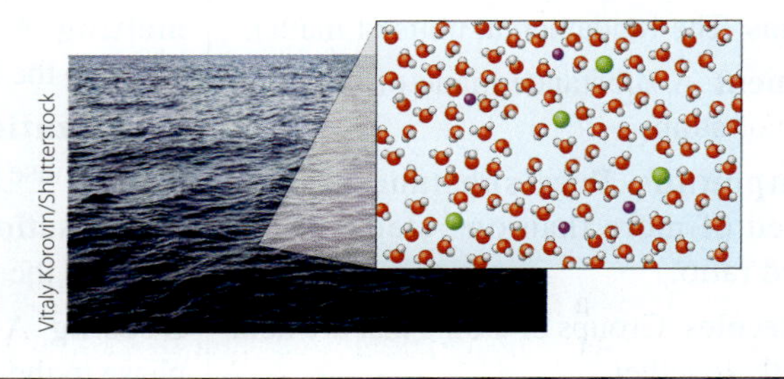

19. How does particle movement differ in a solid, a liquid, and a gas?

20. In the image shown here, colored spheres represent different atoms that are moving freely around a container. Does this image represent an element, a compound, or a mixture? Does it represent a solid, liquid, or gas?

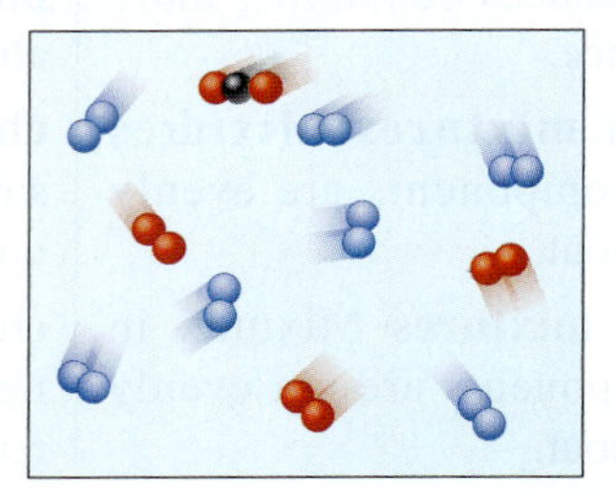

21. Identify each of the following as a solid, liquid, or gas:
a. a substance that has a definite shape and a definite volume
b. a substance that fills its container and takes on the shape of the container
c. a substance in which the particles remain in fixed positions

22. Identify each of the following as a solid, liquid, or gas:
a. a substance made up of particles that are close together but able to move freely past each other
b. a substance made up of particles that are far apart and have almost no interaction with each other
c. a substance with a definite volume but no definite shape

23. Describe each property or change as physical or chemical:
a. The paint on a new truck has a shiny red color.
b. A metal wire conducts electricity.
c. Magnesium metal reacts with hydrogen chloride gas to form magnesium chloride and hydrogen gas.
d. When cooked on a stove, eggs form a fluffy yellow solid.
e. Water is heated and it becomes steam.

24. Describe each change as physical or chemical:
a. Sugar dissolves in water.
b. Charcoal burns, producing carbon dioxide and ash.
c. After you eat, the acids in your stomach break complex molecules into simpler pieces.
d. A chocolate bar is melted down and poured over pretzels.

25. In the introduction to this chapter you read about Taxol, a compound that is commonly used in the treatment of cancer. Describe each of the following properties of Taxol as chemical or physical:

a. Taxol is a white solid.
b. When Taxol enters a patient's liver, it is broken down into simpler compounds.
c. Taxol melts at about 216 °C.

26. Aspirin is a substance that is commonly used as a pain reliever. Describe each of the following properties of aspirin as chemical or physical:

a. Aspirin is a solid at room temperature but melts at 136 °C.
b. Aspirin reacts with the compound sodium hydroxide to form two new compounds.
c. When a person takes an aspirin, a substance in the stomach reacts with the aspirin, breaking it into two simpler compounds.

27. Consider the transition from solid to gas shown here. Is this a physical change or a chemical change?

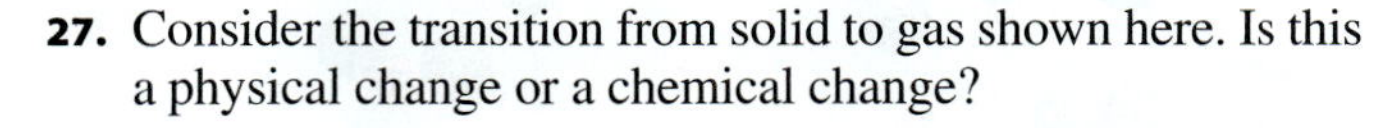

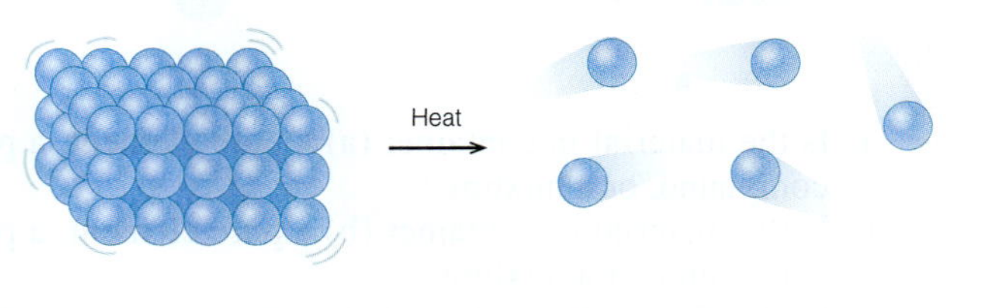

28. A mixture of nitrogen gas and hydrogen gas (a) undergoes a change to form a new compound, ammonia (b). Is this a physical change or a chemical change?

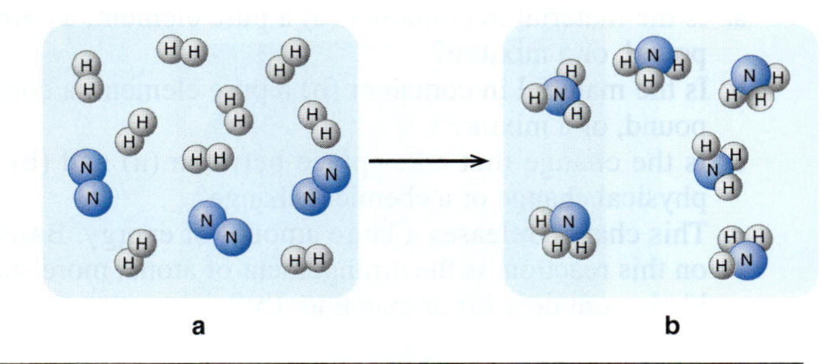

29. Seawater is a mixture of salt and water. Across the world, people isolate salt by adding seawater to shallow ponds, called salt pans (**Figure 1.23**). As the water evaporates, salt crystals form. Does this separation use a physical change or a chemical change?

Figure 1.23 This worker is isolating salt by evaporating the water from a saltwater mixture.

30. When the two leads of a battery are connected to a saltwater solution, it causes the water to be broken into the elements hydrogen and oxygen (**Figure 1.24**). Is this a physical change or a chemical change?

Figure 1.24 An electric current produces bubbles of hydrogen and oxygen gas as it splits water into its elemental components.

1.3 Energy and Change

31. When heat energy is transferred to an object, what happens to the atoms in the object?

32. How do the water molecules in ice differ from the water molecules in liquid water? Which molecules move more quickly?

33. Which item in each pair has more potential energy? Which is more energetically stable?

a. a rock at the top of a hill or a rock at the bottom of a hill
b. wood or ashes
c. a tightly wound spring or a spring that is relaxed

34. Indicate whether each event results in a higher-energy (less stable) or a lower-energy (more stable) product:

a. Heat is added to water, converting it to steam.
b. A firecracker explodes, leaving scattered fragments of ash and paper.
c. A candle burns, producing carbon dioxide gas and water.
d. Sunlight provides energy to convert carbon dioxide and water into woody material, causing a tree to grow.

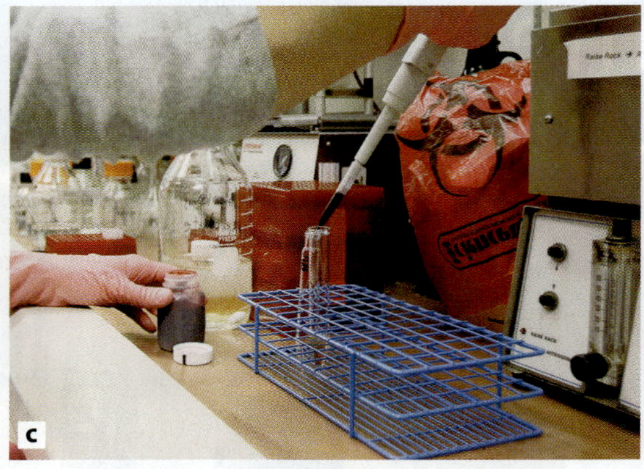

Figure 2.1 (continued) (b) A mysterious container was found in the vehicle of the deceased. (c) A scientist prepares a blood sample for analysis. (d) Dr. Waugh is photographed here with one of the instruments used in the analysis. (a) STEPHANE DE SAKUTIN/Getty Images; (b) Courtesy of Dr. Lauren Richards Waugh; (c) Brad Davis/AP Images; (d) Courtesy of Dr. Lauren Richards Waugh

Using an array of instruments and measurement techniques, the toxicologists identified the substance as MDPV (methylenedioxypyrovalerone), a synthetic drug similar to methamphetamine.

Next they tested the victim's blood and urine, and they found that both contained high levels of MDPV. From these findings, the toxicologists connected the victim's erratic behavior and sudden death to an overdose of MDPV.

At the time, MDPV was legal. Smoke shops and online stores sold it as "bath salts"—a clever ruse to conceal a devastating drug. The toxicologists' discovery marked the first time in West Virginia that MDPV had been identified as the cause of a death. In the following months, a string of MDPV-related deaths occurred across the country. States quickly moved to ban the sale of MDPV and related drugs.

Kraner and Waugh's success in solving this problem hinged on two factors: the quality of their methods and the quality of their measurements. Measurement is critical to the scientific method. It forms the bedrock of chemistry, forensics, and all other sciences. In the pages that follow, we'll explore how scientists make measurements and communicate their findings to others. At the end of the chapter, we'll circle back to this case and examine some of the measurements and techniques that helped to provide the answer.

Whether your career takes you into criminal justice, health sciences, or some other area, the concepts in this chapter will serve you well beyond this course. Knowing how to measure carefully, assess the limits of your information, and think about numerical data and possible outcomes will strengthen your understanding of the world and expand your ability to succeed in any field.

Intended Learning Outcomes

After completing this chapter and working the practice problems, you should be able to:

2.1 Measurement: A Foundation of Good Science

- Convert between standard and scientific notation, and solve multiplication and division problems involving scientific notation.
- Describe the quality of measurements using the terms *accuracy* and *precision*.
- Identify significant digits in a measured number, report measurements to an appropriate number of significant digits, and apply the rules for significant digits to simple calculations.

2.2 Unit Conversion

- Perform unit conversions using the factor-label method.

2.3 Density: Relating Mass to Volume

- Relate the density, mass, and volume of a substance.

2.4 Measuring Temperature

- Convert between Celsius, Fahrenheit, and Kelvin temperature scales.

2.1 Measurement: A Foundation of Good Science

Measurement is a foundation of good science, but the value of good measurement extends beyond science. Almost anything you choose to do—from graphic design to fantasy football, from cooking to construction—involves measurement. In studying chemistry (or any science), it is essential to be able to make good measurements, to understand the limits of measurements, and to communicate clearly what you have found.

Builders sometimes say, "Measure twice, cut once." It is important to make sure your measurement is accurate; otherwise, you may waste valuable time and materials.

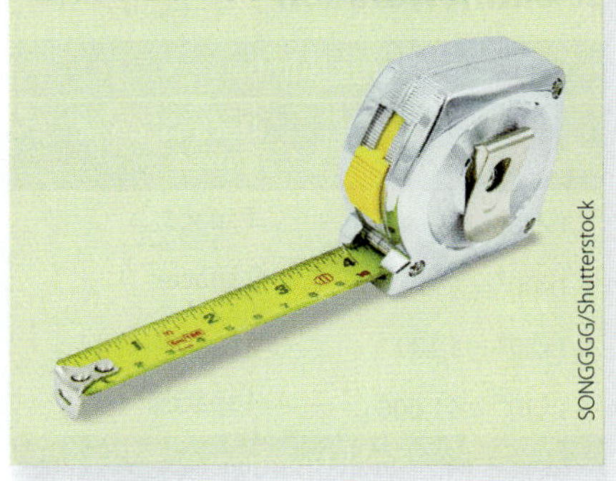

SONGGGG/Shutterstock

Scientific Notation: Working with Very Large and Very Small Numbers

One of the challenges of science is dealing with measurements that are very large or very small. For example, the distance from the Earth to the Sun is approximately 149,600,000,000 meters. On the other hand, in the forensic analysis described at the beginning of this chapter, the toxicologists found that each cubic centimeter of blood contained 0.00000109 grams of MDPV. That's small, but not nearly as small as the mass of a hydrogen atom, which is 0.000000000000000000000001672 grams. How can we work with such unruly numbers?

Scientific notation is a way to show very large and very small numbers in a concise format. Scientific notation expresses numbers as the product of two values, called the *coefficient* and the *multiplier*. In the coefficient, we write a measured value with only one digit before the decimal point. The multiplier is 10 raised to an exponent (10^x). The multiplier shifts the decimal point to the left or right to give the value in standard form. For example, we use scientific notation to express the number 5,100 like this:

$$5.1 \times 10^3$$

Look at the values of the powers of ten in **Table 2.1**. Notice that each time the exponent is increased by one, the value becomes ten times greater. Stated differently, *each unit of the exponent moves the decimal point one place.*

TABLE 2.1 Powers of Ten

Exponent	Value
10^3	1,000.
10^2	100.
10^1	10.
10^0	1.
10^{-1}	0.1
10^{-2}	0.01
10^{-3}	0.001

Converting from Scientific Notation to Standard Notation

When converting from scientific notation to standard notation, we move the decimal the number of units indicated by the exponent in the multiplier. With positive exponents, the decimal point moves to the right:

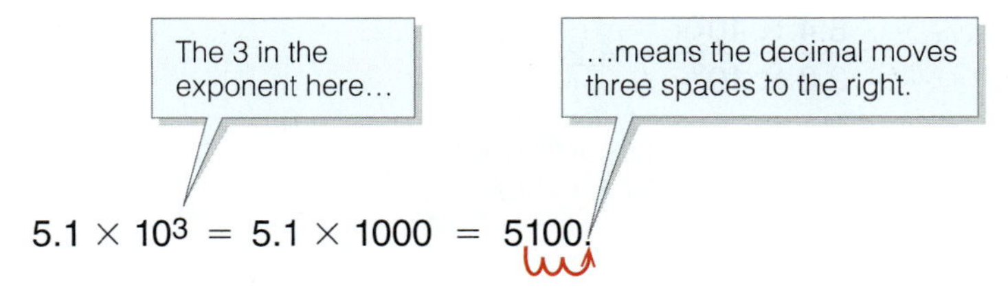

The 3 in the exponent here...
...means the decimal moves three spaces to the right.
5.1 × 10³ = 5.1 × 1000 = 5100.

With negative exponents, we move the decimal the opposite direction:

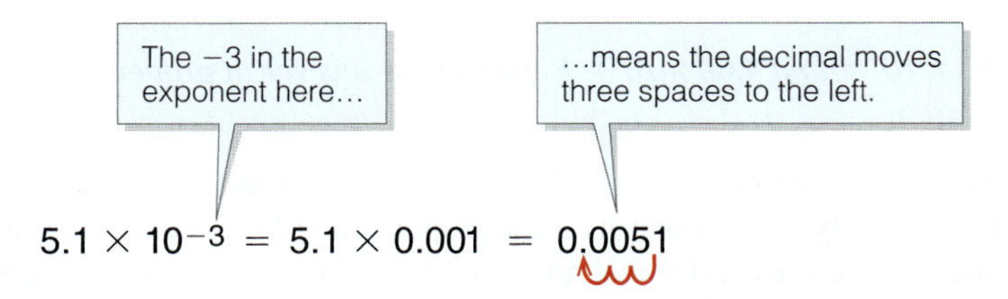

In scientific notation, the exponent tells how many spaces the decimal point must be moved.

Large numbers have a positive exponent in scientific notation. Small numbers have a negative exponent.

TABLE 2.2 Multipliers in Scientific Notation

Large Numbers	
Example	**Decimal point moves**
$5.1 \times 10^1 = 51.$	1 space
$5.1 \times 10^2 = 510.$	2 spaces
$5.1 \times 10^3 = 5{,}100.$	3 spaces
$5.1 \times 10^4 = 51{,}000.$	4 spaces
Small Numbers	
Example	**Decimal point moves**
$5.1 \times 10^{-1} = 0.51$	−1 space
$5.1 \times 10^{-2} = 0.051$	−2 spaces
$5.1 \times 10^{-3} = 0.0051$	−3 spaces
$5.1 \times 10^{-4} = 0.00051$	−4 spaces

Remember that *positive exponents represent large numbers*, and *negative exponents represent small numbers*. **Table 2.2** details the trends for positive and negative exponents.

Converting from Standard Notation to Scientific Notation

To write a number in scientific notation, we move the decimal point in the coefficient to the right or the left until only one nonzero digit remains in front of the decimal point. The number of digits we move the decimal point in the coefficient is the exponent in the 10^x multiplier. Again, we use negative exponents for small values and positive exponents for large values.

For example, we write 0.0000234 in scientific notation as follows:

$$0.0000234 = 2.34 \times 10^{-5}$$

...so the exponent is −5.
We must move the decimal point 5 spaces to be in front of just one nonzero digit...

Calculations Involving Scientific Notation

When multiplying numbers using scientific notation, we multiply the coefficients and add the exponents in the multipliers:

$$3.1 \times 10^4 \times 2.0 \times 10^2 = 6.2 \times 10^6$$

exponents
coefficients
Add exponents: $10^4 + 10^2 = 10^6$
Multiply coefficients: $3.1 \times 2.0 = 6.2$

When dividing numbers using scientific notation, we divide the coefficients and subtract the exponent in the denominator from the exponent in the numerator:

$$\frac{8.4 \times 10^7}{2.0 \times 10^3} = 4.2 \times 10^4$$

Subtract exponents: $10^{(7-3)} = 10^4$
Divide coefficients: $8.4/2.0 = 4.2$

Example 2.1 Multiplication with Scientific Notation

Multiply 3.95×10^{-22} and 4.00×10^{24}. Express your answer in proper scientific notation.

When we multiply the coefficients and add the exponents, we get an answer of 15.8×10^2. This is mathematically correct, but it is not proper scientific notation because there is more than one digit before the decimal point in the coefficient. To convert to proper notation, we need to move the decimal over one space and then add one unit to the exponent. This gives us a final answer of 1.58×10^3.

...and therefore we increase the exponent by one unit.

$$(3.95 \times 10^{-22})(4.00 \times 10^{24}) = 15.8 \times 10^{2} = 1.58 \times 10^{3}$$

After solving, we move the decimal one unit to the right, so there is only one digit in front of the decimal point...

If you use a calculator to solve problems, here's a word of caution: Different calculators input and represent scientific notation differently. I encourage you to practice these techniques on your calculator until you are confident you can do them correctly.

TRY IT

1. Write each number in scientific notation.

a. 130,000,000,000 **b.** 4,700,000 **c.** 0.005 **d.** 4.1

2. Write each number in standard notation.

a. 5.31×10^{5} **b.** 1.213×10^{3} **c.** 4.091×10^{-4}

3. Solve each calculation. See if you can solve them by hand, and then solve them with a calculator. Make sure your answers for both methods agree!

a. $(2.1 \times 10^{9}) \times (3.0 \times 10^{5})$ **b.** $(3.0 \times 10^{5})/(1.5 \times 10^{2})$

c. $(2.0 \times 10^{-3})(6.0 \times 10^{2})$ **d.** $(4.2 \times 10^{-3})/(2.1 \times 10^{-4})$

Check It
Watch Explanation

Units of Measurement

I love to make enchiladas (**Figure 2.2**). When preparing this meal, I follow a recipe that requires two pounds of roasted chili peppers, ¼ cup of lime juice, and 1 teaspoon of salt (as well as other ingredients). After assembling the enchiladas, I bake them at a temperature of 350 °F for 25 minutes.

Courtesy of Kevin Revell

Figure 2.2 Enchilada night in the Revell household is a great tradition. For the meal to come out right, the ingredients must be measured correctly.

Every component of the meal—ingredients, cooking temperature, and cooking time—has a *unit* associated with the measurement, such as pounds, cups, teaspoons, degrees, and minutes. Units of measurement are quantities with accepted values that can be communicated between people. When I follow this recipe, I trust that my ¼ cup measure is the same size as the ¼ cup measure of the person who gave me the recipe. Without units, and without an accepted standard for what units mean, the measurements are worthless.

There are many different units. For example, if you wanted to measure the length of your kitchen table, you could use inches, meters, yards, leagues, miles, or cubits. Which is most appropriate?

Practice
Metric System
It is important to know the common metric prefixes. Test your knowledge with this interactive quiz.

In the United States, people commonly use *English units*, such as inches and feet (length), gallons (volume), and pounds (weight). Most of the world uses a different system, called the *metric system*. **Table 2.3** lists the relationships between some of these measurements.

TABLE 2.3 Relationships between Some Common English and Metric Units

Measurement	Metric Unit	English Unit	Relationship
Length	meter (m)	foot (ft)	1 m = 3.281 ft
	kilometer (km)	mile (mi)	1 km = 0.621 mi
Mass or weight	kilogram (kg)	pound (lb)	1 kg = 2.205 lb
Volume	liter (L)	gallon (gal)	1 L = 0.264 gal

TABLE 2.4 Fundamental (SI) Units of Measurement

Measurement	Unit
Mass	kilogram (kg)
Length	meter (m)
Time	second (s)
Temperature	kelvin (K)
Light intensity	candela (cd)
Electric current	ampere (A)
Amount	mole (mol)

For easier communication, the international scientific community developed an accepted set of *fundamental units*, sometimes called *SI units* (the abbreviation SI comes from the French for "International System"). **Table 2.4** presents these units, which are based on the metric system. From these fundamental units, we can derive a host of other units. For example, a **liter (L)** is a derived unit of volume corresponding to 0.001 cubic meters (m^3). We use meters per second (m/s) to describe velocity, or kilograms per liter (kg/L) to describe density. (We discuss these units in more detail in the sections that follow.)

The metric system uses a series of prefixes to describe larger or smaller amounts. **Table 2.5** summarizes the key metric prefixes. For example, many wireless networks transmit data at 160 million bits (b) every second. Rather than writing this quantity of information as 160,000,000 bits, it is easier to use the metric prefix and express this as 160 megabits (abbreviated as 160 Mb). As with scientific notation, the metric prefixes allow us to express very large or small numbers in a concise way.

Working with Small Units
From Table 2.5, we see that
$1 \text{ mg} = \frac{1}{1{,}000} \text{ g}.$
This means that 1,000 mg = 1 g.

TABLE 2.5 Common Metric Prefixes

Prefix	Symbol	Meaning	
tera-	T	10^{12}	1,000,000,000,000
giga-	G	10^{9}	1,000,000,000
mega-	M	10^{6}	1,000,000
kilo-	k	10^{3}	1,000
deci-	d	10^{-1}	$\frac{1}{10}$
centi-	c	10^{-2}	$\frac{1}{100}$
milli-	m	10^{-3}	$\frac{1}{1{,}000}$
micro-	μ	10^{-6}	$\frac{1}{1{,}000{,}000}$
nano-	n	10^{-9}	$\frac{1}{1{,}000{,}000{,}000}$
pico-	p	10^{-12}	$\frac{1}{1{,}000{,}000{,}000{,}000}$

Even for common measurements such as mass and temperature, scientists don't always use the fundamental units in Table 2.4. For example, the official SI unit of mass is the kilogram. But chemists often work with far smaller amounts than this in the laboratory, so they routinely deal with grams or with milligrams.

TRY IT

4. Refer to Tables 2.3, 2.4, and 2.5 to write the appropriate abbreviations for each unit.
   a. meters b. milliliters c. kilometers d. microamperes
   e. picoseconds f. megacandelas g. millimoles h. nanograms
5. Refer to Table 2.5 to answer the following questions:
   a. How many bytes are in a kilobyte?
   b. How many microseconds are in a second?
   c. How many mg are in a g?
   d. How many MA are in an A?

Check It
Watch Explanation

Describing the Quality of Measurements

Let's go ahead and measure something. **Figure 2.3** shows a ball bearing, a small steel ball used in many machine parts. What is its diameter? (I encourage you to stop, look carefully, and commit to a value before reading the next paragraph.)

When I pose this question to a classroom full of people, I typically get a range of measurements. Often, the range looks something like this:

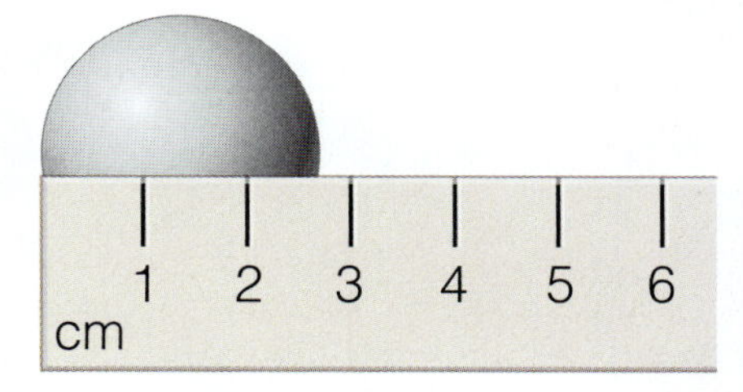

Figure 2.3 What is the diameter of this ball? How sure are you about this?

Measurement	Percentage of Students
2.6 cm	25%
2.7 cm	50%
2.8 cm	25%

Each of these responses is reasonable and is a correct reading. If you looked at the ball in Figure 2.3 carefully, you probably considered more than one of these responses. You know that the ball is between 2.5 and 3 cm. Most likely, it is between 2.6 and 2.8 cm. But how can you know for sure?

In fact, you can't. Uncertainty is a part of science, and it is a part of measurements. You are certain that the ball is two-point-something centimeters, but you can't be sure of the last decimal point. This is perfectly normal. In science, it is acceptable to *estimate* the last number in a measurement. When other scientists read this measurement, they understand that the last measured digit is an estimated value.

The last decimal place in a measurement is an estimated value.

But what if we need to measure something more exactly? Estimating the diameter of a ball to plus or minus a couple tenths of a centimeter is fine in general; but if we are doing more meticulous work, like designing a part for a high-performance engine, that sort of measurement is not sufficient. In this case, we need a tool like a caliper (**Figure 2.4**), which measures much more precisely than a tape measure or a ruler.

Scientists use two terms to describe the quality of measurements: *accuracy* and *precision*. Accuracy refers to how reliable measurements are. That is, does a measurement (or average of measurements) reflect the true value? If it does, it is accurate. On the other hand, precision refers to how finely a measurement is made, or how close groups of measurements are to each other.

It is possible to be accurate without being precise. Accurate but not precise describes our measurement of the ball in Figure 2.3. It's also possible to be precise without being accurate. For example, a laboratory balance can measure the mass of a sample to ±0.001 gram. However, if that balance is not calibrated correctly, numbers read from the balance may be inaccurate even if they are reported to several decimal places. A classic way of visualizing precision and accuracy is with a series of targets, as shown in **Figure 2.5**.

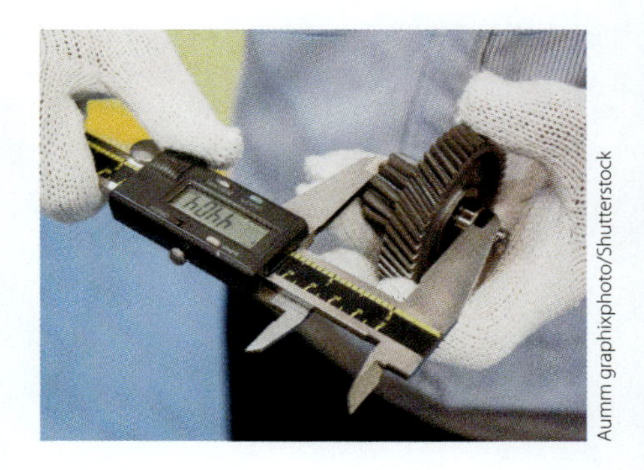

Figure 2.4 This caliper is much more precise than a ruler. It is used to measure to the nearest 0.01 mm.

Both accurate and precise

Precise but not accurate

Accurate but not precise

Neither precise nor accurate

Figure 2.5 Precision means the results are very close together. Accuracy means the average of the results is correct.

TRY IT

6. Your laboratory is conducting tests on the nutritional energy value in different types of candies. At the beginning of this process, you ask a new colleague to record the mass of a set of eight individually wrapped peppermints. Your colleague handles the candies very carefully and records the mass of each candy as shown in the table.

Candy	Mass
1	4.3231 g
2	4.3577 g
3	4.2902 g
4	4.3341 g
5	4.3209 g
6	4.3076 g
7	4.3725 g
8	4.3326 g

When you compare the measured masses with those from previous experiments, the average mass is about 0.1 g greater. When you consult with your colleague, you find that she did not remove the wrappers before measuring the masses. Describe the value of these measurements using the terms *precise* and *accurate.*

Chemists measure mass using balances like the one shown here. They routinely test the balances by measuring objects of known mass. If the balance is not reading correctly, they adjust it so the reading is accurate. This is known as *calibrating* the balance.

GIPhotoStock/Science Source

Describing Precision: Significant Digits

Figure 2.6 shows two different pieces of glassware we use to measure the volume of liquids: a beaker and a graduated cylinder. In this figure, we can see that the volume in the beaker is between 40 and 50 mL, and closer to 50. Based on this observation, we might assert that the volume is 48 mL. We are not certain about the last digit, but it is a reasonable estimate. The graduated cylinder contains the same amount of water as the beaker, but it is marked more finely so we are able to make a more precise measurement. We can estimate the volume of the graduated cylinder to be between 48 and 49 mL, but closer to 48. For example, we might estimate it to be 48.1 mL.

From these examples, we can infer a general rule. When we use a tool with clearly marked units to measure a quantity such as length, volume, or temperature, we can estimate *one digit* between the marked values. The volume markings on the beaker in Figure 2.6 show to the tens place, and so we are able to estimate to the ones place. In the graduated cylinder in Figure 2.6, the volume is marked to the ones place, so we are able to estimate to the tenth place.

Significant digits, which are also sometimes called *significant figures*, are the digits contained in a measured value. The number of significant digits indicates how precisely a measurement is made. In our previous example, we measured the volume of liquid in the beaker (48 mL) to *two* significant digits. We measured the volume of liquid in the graduated cylinder (48.1 mL) to *three* significant digits. The measurement with more significant digits is more precise.

When measuring, estimate one digit between the marked values.

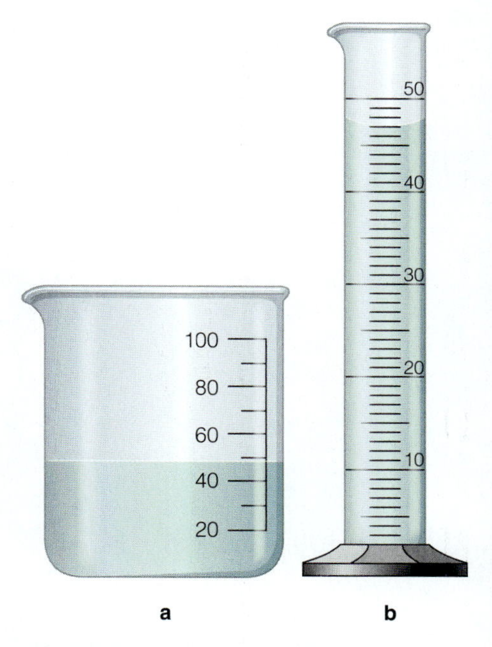

Figure 2.6 We can use a beaker (a) or a graduated cylinder (b) to measure volume. The graduated cylinder gives us a more precise measurement than the beaker.

Example 2.2 Measuring Volume in a Graduated Cylinder

What is the volume of water in the graduated cylinder? Report your answer to the correct number of significant digits.

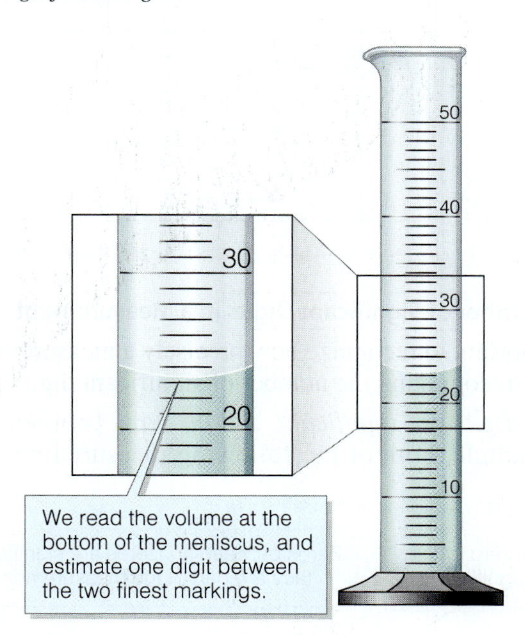

In narrow glass cylinders, water adheres to the glass surface, resulting in a slightly curved surface called a *meniscus*. The correct way to read a graduated cylinder is to report the volume at the bottom of the meniscus. Notice that the bottom of the meniscus lies between the 23-mL and the 24-mL marks. We estimate the volume to the nearest tenth of a milliliter, one decimal place beyond what is marked on the cylinder. By careful inspection, we can estimate the meniscus to be 23.3 mL—although 23.2 mL or 23.4 mL are also acceptable measurements.

Check It

Watch Explanation

TRY IT

7. What is the temperature reading on the thermometer? Make sure your answer includes the correct number of significant digits.

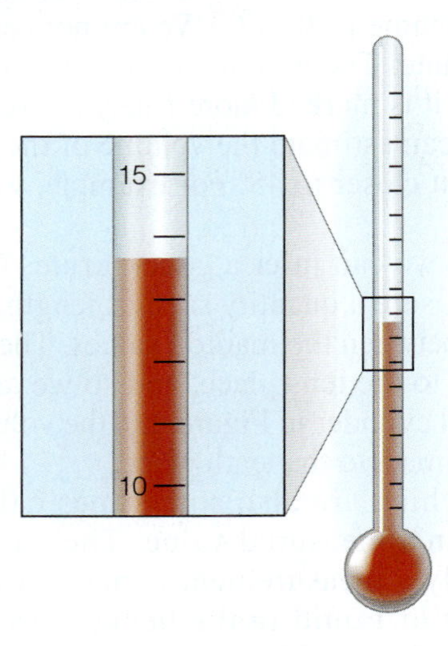
15
10

8. The image shown here is of a liquid sample in a graduated cylinder. What is the volume of the sample, reported to the appropriate number of significant digits?

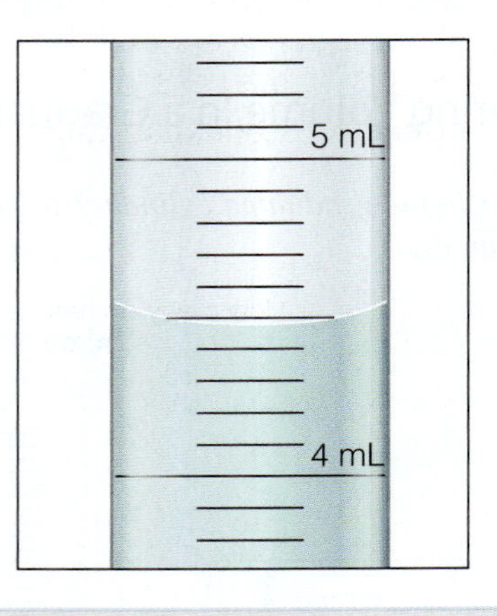
5 mL
4 mL

Determining the Number of Significant Digits in a Measurement

Significant digits indicate the precision of a measurement.

Sometimes it is important to recognize how precisely a measurement is made. There are several rules for recognizing the number of significant digits in a measurement:

1. *All nonzero digits are significant*, and *all zeros between nonzero digits are significant*. For example, both of the following measured numbers contain five significant digits:

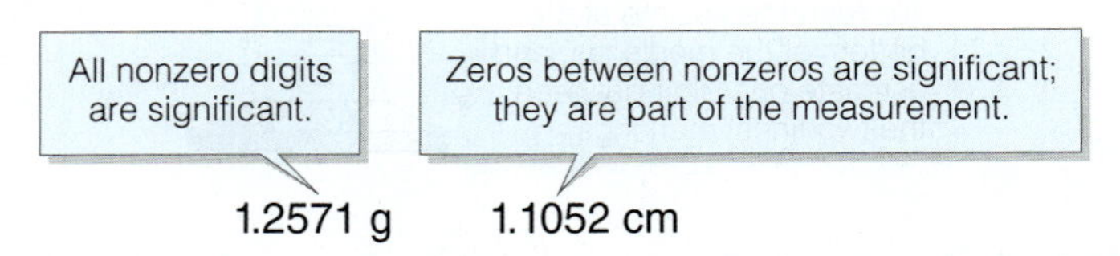
All nonzero digits are significant.
Zeros between nonzero are significant; they are part of the measurement.
1.2571 g
1.1052 cm

2. *If a decimal point is present, zeros to the right of the last nonzero digit are significant.* A number containing zeros after the decimal point signifies that the number has been measured that precisely.

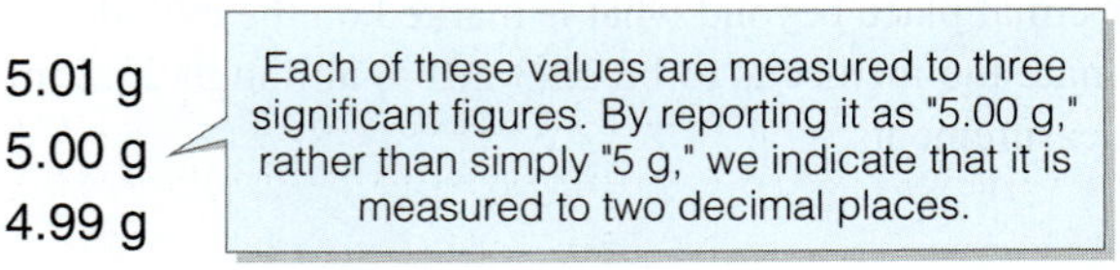

5.01 g
5.00 g
4.99 g

Each of these values are measured to three significant figures. By reporting it as "5.00 g," rather than simply "5 g," we indicate that it is measured to two decimal places.

3. *Zeros to the left of the nonzero numbers are never significant.* Zeros to the left of a number and to the left of the decimal point are meaningless. Zeros to the left of the measured values are useful for values less than one, because they show where the decimal point lies; however, they are not part of the measurement and are not significant. For example, if we measure the width of an object to be 4.5 millimeters, we say we measured to two significant digits. What if we then convert this measurement to meters? A width of 4.5 millimeters is equal to 0.0045 meters. Did our measurement suddenly become more precise? No—we changed the units, but the number of significant digits did not change.

These zeros serve no function, and are not significant digits.

These zeros are placeholders to show where the decimal point goes. They do not count as significant digits.

0000000010 kilograms — significant

0.0045 meters — significant

4. *If there is no decimal point present, zeros to the right of the last nonzero digit may or may not be significant.* If a number contains zeros to the right of the last nonzero digit but does not contain a decimal point, it is hard to know how precisely the number is measured. For example, the cost of a program may be reported as \$10,000,000 per year. Does that mean it costs exactly 10 million dollars? Probably not. But how closely has it been rounded? Is the cost between 9 million and 11 million? Or is the cost plus or minus \$1,000, meaning it is between \$9,999,000 and \$10,001,000? It's not clear.

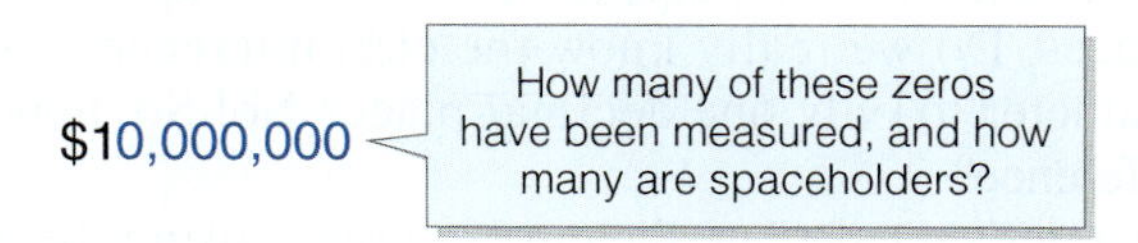

$10,000,000
How many of these zeros have been measured, and how many are spaceholders?

For numbers that end in zeros but do not contain a decimal point, there are two ways we can specify how precisely we've measured. The first approach is to specify the uncertainty right after the number. For example, we might know that a large object has a mass of 10,000 kg, measured to the nearest 100 kg. We write this as $10{,}000 \pm 100$ kg. Alternatively, we can write the value in scientific notation as 1.00×10^4 kg. By doing so, we specify the number of zeros that we are fairly certain about (in this case, two zeros).

Science Source

Figure 2.7 Counted numbers are exact numbers; significant digits do not apply. There are seven pennies in the photo.

Counted values and defined relationships (such as metric prefixes) are exact numbers.

Working with Exact Numbers

Exact numbers are values for which there is no uncertainty. When dealing with exact numbers, we do not need to be concerned about significant digits. For example, **Figure 2.7** shows a group of seven pennies. We don't debate over whether there are 7.0 pennies or 7.1 pennies; clearly, there are exactly seven pennies. *Counted values are exact numbers.*

A second type of exact number is a number that comes from a *definition*. For example, there are 12 items in a dozen. There are 1,000 milligrams in a gram. These are exact relationships, and so we don't worry about significant digits for this type of number.

Let's look at a second example to help clarify this point. Consider the plate of tacos shown in **Figure 2.8**. If we want to count the number of tacos, we can do so exactly: There are three tacos. On the other hand, to determine the mass of the guacamole that was added to the tacos, we would make a measurement, then report this value to as many significant digits as we could measure.

Courtesy of Kevin Revell

Figure 2.8 The number of tacos is exact, but the amount of guacamole on each taco is measured.

Check It
Watch Explanation

TRY IT

9. How many significant digits are in each of the following measured units?

a. a distance of 14.3 kilometers

b. a toxin concentration of 0.0079 milligrams per liter

c. a mass of 4.300 kilograms

d. a time of 000024.3 seconds

10. Identify each of the following values as measured or exact:

a. There are 96.1 nutritional Calories in a jelly donut.

b. The burn marks from a fire cover an area of 24.1 square meters.

c. There are 13 floats at a homecoming parade.

d. There are 1×10^9 bytes of data in a gigabyte.

e. A bag contains 1 pound of coffee.

So your calculator reports the circumference to eight decimal places. Calculators say lots of things. If you type in 0.7734 and turn it upside down, a calculator will even say hello. It is your responsibility to decide how precisely you really know the circumference and not to simply accept what your calculator says.

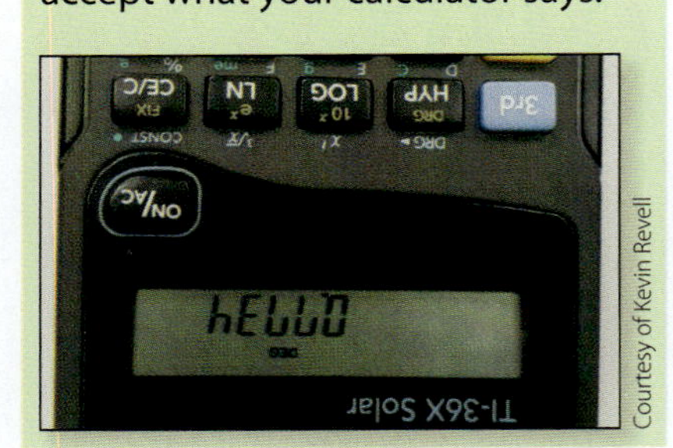
hELLO
TI-36X Solar
Courtesy of Kevin Revell

Using Significant Digits in Calculations

In the previous section, we discussed the measurement of a ball. Let's say that we measured the diameter of the ball to be 2.7 cm, and we were relatively certain to ±0.1 cm. That is, we think the diameter is 2.7 cm, but it could be slightly less (2.6 cm) or slightly more (2.8 cm).

Now suppose we need to calculate the circumference of the ball. The circumference of a sphere is pi (π) times the diameter. If we enter $2.7 * \pi$ into a calculator, we'll get something like the following answer:

$$2.7 \text{ cm} * \pi = 8.48230016 \text{ cm}$$

In this example, the calculator reports the circumference of the ball measured to eight decimal places. Do we really know the circumference this precisely, if we measured the diameter to only one decimal place? No! So, how precisely do we know the circumference?

To answer that, let's recalculate the circumference using the low and high values of 2.6 and 2.8 cm, and see how it affects the value of the circumference:

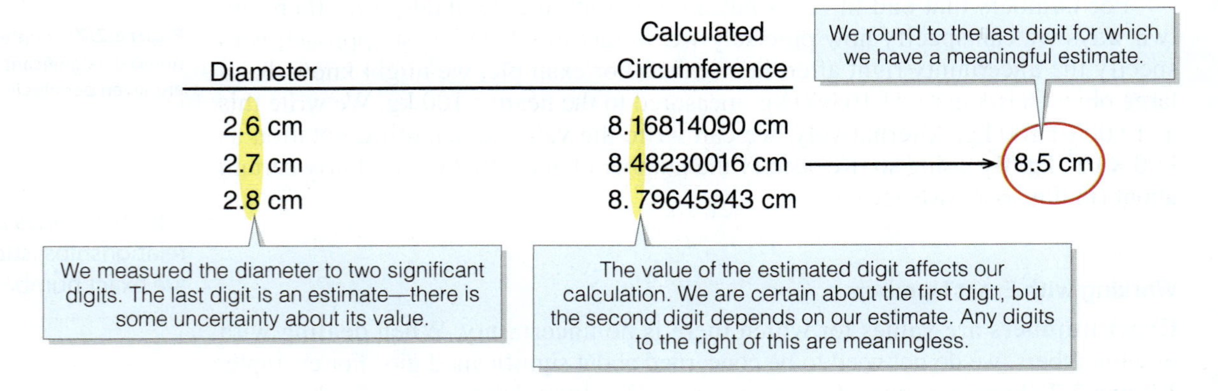
Diameter
Calculated Circumference
2.6 cm
2.7 cm
2.8 cm
8.16814090 cm
8.48230016 cm
8.79645943 cm
8.5 cm
We round to the last digit for which we have a meaningful estimate.
We measured the diameter to two significant digits. The last digit is an estimate—there is some uncertainty about its value.
The value of the estimated digit affects our calculation. We are certain about the first digit, but the second digit depends on our estimate. Any digits to the right of this are meaningless.

If the last dropped digit is 5 or greater, we round up. If the last dropped digit is 4 or less, we round down.

18.3502 g ⟶ 18.4 g

18.3481 g ⟶ 18.3 g

Notice that when we started the calculation, we were not sure about the second significant digit. After the calculation, we're still unsure about the second significant digit. Based on this, we round the answer to the second significant figure, or 8.5 cm.

Based on problems like this, scientists have developed two common rules for reporting calculations with significant digits:

Rule 1: When multiplying or dividing, report the same number of digits as are in the least precise starting measurement.

Rule 2: When adding or subtracting, round to the last decimal place of the least precise starting measurement.

The following examples demonstrate how these rules are applied in different calculations.

Example 2.3 Significant Digits: Multiplying and Dividing

In nutrition, the energy available from a food is often measured in Calories. In one study, researchers find that a 55.023-gram candy bar contains 283.1 Calories. How many Calories are present in each gram of this candy bar? Report your answer with the correct amount of significant digits.

To solve this problem, we divide 283.1 Calories by 55.023 grams. The calculator reports a value of 5.145121131. We round this value to have the same number of significant digits as the starting quantity with the *fewest* significant digits. Although we know the mass to five significant digits (55.023), we know the energy to only four digits (283.1), so we report only four significant digits. Based on this observation, we round the answer to 5.145 Calories/gram.

Example 2.4 Significant Digits: Multiplying by an Exact Number

According to the United States Mint, a quarter is made from a mixture of copper and nickel, and it has a mass of 5.670 grams. What is the mass of eight quarters?

In this case, we are multiplying a measured quantity (5.670 grams) by an integer (8). Because the integer value is an exact number, *it is not included in the significant digit consideration*. Therefore, the fewest measured significant digits is four. We multiply 5.670×8, then round to four significant digits to arrive at the correct answer of 45.36 grams.

Example 2.5 Significant Digits When Adding and Subtracting

A chemist collects three samples of a liquid. The first sample has a volume of 3.62 L, the second has a volume of 0.255 L, and the third has a volume of 21.2 L. What is the total volume, reported to the correct number of significant digits?

When we add these three volumes together, we obtain a value of 25.075 L. However, because our least precise value goes only to the tenth place, we round the number to that value, giving a final answer of 25.1 L.

Volume 1	3.62 L
Volume 2	0.255 L
+ Volume 3	21.2 L
Total	25.075 L
	=25.1 L

Since our least precise measurement only went to the tenth place, we also round our result to the tenth place.

Example 2.6 Significant Digits with Multiple Calculations

Nitrate is a common component of fertilizers that can contaminate water supplies. A chemist tested three samples of water and found they contained nitrate in the amounts shown. Together, the three samples had a volume of 1.514 L. What was the total mass of nitrate in the three samples? What was the average mass of nitrate per liter of water? Answer each question to significant digits.

Sample	Mass of Nitrate
A	12.83 mg
B	11.2 mg
C	14.391 mg

Adding the three masses gives us a value of 38.421 mg. However, since we don't know the mass of sample B more precisely than a tenth of a milligram, it would be incorrect to report the total mass to three decimal places. Therefore, we round our answer to the tenth place and report the mass as 38.4 mg. We are confident of this answer to three significant digits.

	Sample A	12.83 mg
	Sample B	11.2 mg
+	Sample C	14.391 mg
	Total	38.421 mg
		= 38.4 mg

To find the average mass of nitrate per liter of water, we divide the total mass by the volume. However, to minimize rounding errors, we wait until the end of the calculation to round to significant digits. That is, we use the full, unrounded value from the first measurement (38.421 mg), rounding only after our final calculation. We are confident of the mass to three significant digits, and the volume to four significant digits. Because this is a division problem, we keep the least of these values, and therefore round our answer to three significant digits.

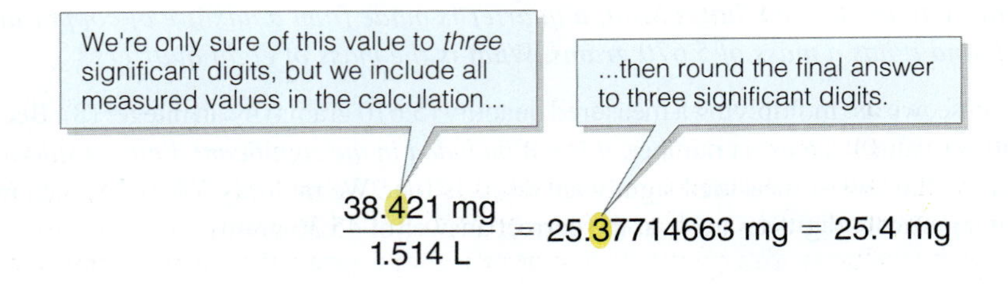

We're only sure of this value to *three* significant digits, but we include all measured values in the calculation...
...then round the final answer to three significant digits.
38.421 mg / 1.514 L = 25.37714663 mg = 25.4 mg

TRY IT

11. In 2002, Paula Radcliffe broke a world record when she ran the Chicago Marathon in 137.30 minutes. A marathon is 26.2 miles. What was her average speed in minutes per mile? Make sure your answer has the correct number of significant digits.

12. Four runners run an 800-m race. Their respective times are 98.12 s, 98.64 s, 101.33 s, and 104.04 s.

a. Find the sum of their times, reported to significant digits.

b. Find the average of their times, reported to significant digits.

13. A toxicologist measures the concentration of an illegal drug in a urine sample. She measures five times and obtains the following values, in milligrams per liter: 12.1, 12.9, 11.5, 13.1, and 12.5. Calculate the average concentration. Make sure your answer has the correct number of significant digits.

Bankoo/Shutterstock

Figure 2.9 This board displays exchange rates between currencies.

2.2 Unit Conversion

When working with measured quantities, it is often necessary to convert between units. If you are traveling internationally, you may have to convert from one currency to another (**Figure 2.9**). If you are measuring a distance between cities, you might need to relate kilometers to miles. If you are working in a lab, it is often necessary to convert between units of mass or volume.

Dimensional Analysis

Sometimes, you can do simple unit conversions in your head. However, as we delve deeper into chemistry and deal with increasingly complex units, it is important to have a systematic approach to unit conversions that allows us to keep careful track of units and to check our work. One common way to approach unit conversion problems is a technique called *dimensional analysis*, also known as the *factor-label method*.

The cornerstone of the factor-label method is this: When performing a mathematical operation involving measured quantities, *whatever we do to the number, we also do to the units*. For example, to calculate the area of a square, we multiply the length and the width (**Figure 2.10**). If the length and width measurements are in feet, then the area is in units of feet times feet, or feet squared (ft^2).

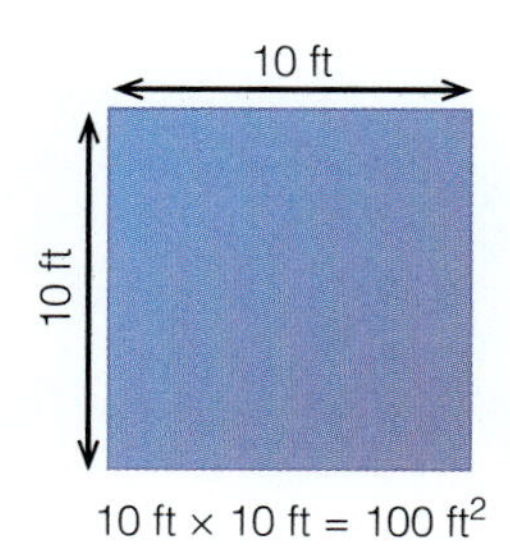

Figure 2.10 When calculating the area of a square, we multiply both the numbers and the units.

How can we use this concept to convert between units? To understand how the factor-label method works, consider an example. Suppose we have a copper pellet with a mass of 0.281 kilograms. We need to express this mass in grams. How do we do this?

First, we need to know the relationship between kilograms and grams. From Table 2.5, we see that 1 kg = 1,000 g. Because these two quantities are equal, we can write them as two fractions, both of which have a value of one:

$$\frac{1\text{ kg}}{1{,}000\text{ g}} = 1 \quad or \quad \frac{1{,}000\text{ g}}{1\text{ kg}} = 1$$

These fractions are examples of **conversion factors**. Conversion factors are fractions that contain equivalent amounts of different units in the numerator and in the denominator. To convert from one unit to another, we multiply by one of these two conversion factors. Because a conversion factor is equal to one, we are not changing the value of the quantity; we are only changing the units. When we multiply quantities having units, we cancel any units that appear in both the numerator and the denominator.

Use conversion factors to change from one unit to another.

To convert between units, we *choose the conversion factor that cancels the old units* and retains the new units. In this example, we want to get rid of kilograms and end up with an answer in grams. We therefore use the conversion factor with grams in the numerator:

$$0.281\ \cancel{\text{kg}} \times \frac{1{,}000\text{ g}}{1\ \cancel{\text{kg}}} = 281\text{ g}$$

Starting unit (kg) — Conversion factor — New unit (g)

This gives us a final answer of 281 grams.

What if we select the wrong conversion factor? If we multiply by $\frac{1\text{ kg}}{1{,}000\text{ g}}$ we get units of $\frac{\text{kg}^2}{\text{g}}$, as the following incorrect calculation shows:

$$0.281\text{ kg} \times \frac{1\text{ kg}}{1{,}000\text{ g}} = 2.81 \times 10^{-4}\text{ kg}^2/\text{g}$$

Wrong conversion factor

If the units come out wrong, we've chosen the wrong conversion factor.

Verifying the units is a good way to check our work. If we work the problem correctly, the units come out right. If the units do not make sense, we probably set up the problem incorrectly.

Example 2.7 Converting between Units

Heat energy is commonly measured in calories or in joules (1 calorie = 4.184 joules). A particular chemical reaction releases 150.0 calories of heat energy. What is this energy in joules?

As in the previous example, we begin with the amount that we want to convert (in calories). To convert this value from calories to joules, we must multiply it by a conversion factor. Because we want to cancel out the calories, we write this unit in the denominator and write joules in the numerator:

$$\text{Starting unit} \times \text{Conversion factor} = \text{New unit}$$

$$150.0 \text{ calories} \times \frac{4.184 \text{ joules}}{1 \text{ calorie}} = 627.6 \text{ joules}$$

We are able to cancel out the calories, leaving us with a final answer of 627.6 joules.

Example 2.8 Converting between Units

A brick has a mass of 815.2 grams. What is this mass in kilograms?

We begin with the amount in grams. To convert this value from grams to kilograms, we must multiply it by a conversion factor. We know from Table 2.5 that 1,000 g = 1 kg. Because we want to cancel out the grams, we write this unit in the denominator and write kilograms in the numerator:

$$815.2 \text{ g} \times \frac{1 \text{ kg}}{1{,}000 \text{ g}} = 0.8152 \text{ kg}$$

This gives us a final answer of 0.8152 kg.

Check It
Watch Explanation

TRY IT

14. For each of the following, indicate the correct conversion factor to use, and find the value in the new unit:

a. Convert 125.31 mg to g.

b. Nanoparticles are tiny capsules that are a promising new way to introduce medicine into the body. These particles are much too small to see with the naked eye. One such capsule has a diameter of 6.0×10^{-8} meters. What is this diameter in nanometers?

c. The average home in Louisiana uses about 15,000 kilowatt hours (kWh) per year. What is this energy consumption in kilojoules (kJ)? (1 kWh = 3,600 kJ)

Practice
Unit Conversions
The way to get good at unit conversions is to practice. Try this activity to test your skills.

Problems Involving Multiple Conversions

To solve many problems, we need to make more than one unit conversion. When this occurs, we can place the unit conversions adjacent to each other. Consider the following examples:

Example 2.9 Problems Requiring Multiple Conversions

How many micrometers are in 0.0129 inches? (2.54 centimeters [cm] = 1 inch and 10,000 micrometers [μm] = 1 cm)

In this example, we need to make two conversions. We first convert from inches to centimeters, then from centimeters to micrometers. For the first conversion, we write 2.54 cm over 1 inch, as shown below. This allows us to cancel out the inches, leaving units of centimeters. In the second step, we write 10,000 micrometers over 1 centimeter, so the centimeters cancel out, leaving micrometers.

$$0.0129\ \text{in} = \frac{2.54\ \text{cm}}{1\ \text{in}} \times \frac{10{,}000\ \mu\text{m}}{1\ \text{cm}} = 328\ \mu\text{m}$$

Convert from inches to cm…

…then from cm to μm.

In this problem, the initial measured value (0.0129 inches) contained only three significant digits. As a result, we round the answer to three digits as well.

A conversion factor contains two equivalent amounts. The old unit cancels out, leaving the new unit.

Example 2.10 Converting Units in the Denominator

In air, sound waves travel at 768 miles per hour. How fast is this in meters per second (m/s)? (1 mile = 1,609.3 meters)

Let's begin by converting miles to meters. Because we need meters in the numerator, we write a conversion factor that has meters in the numerator and miles in the denominator. This allows us to cancel out the miles.

In the second conversion step, we use the relationship between seconds and hours (3,600 s = 1 h). Because we need to have seconds in the denominator of the final expression, we place hours in the numerator of the conversion factor. Rounded to significant digits, this gives us an answer of 343 m/s.

$$\frac{768\ \text{miles}}{1\ \text{hr}} \times \frac{1{,}609.3\ \text{m}}{1\ \text{mile}} \times \frac{1\ \text{hr}}{3{,}600\ \text{s}} = \frac{343\ \text{m}}{1\ \text{s}}$$

This type of unit conversion problem pops up in many of the following chapters, and I encourage you to make sure you can do them efficiently. Before moving on, try to master the problems in the following question box.

TRY IT

15. Amoxicillin is a powerful antibiotic. For mild infections, a doctor may prescribe 0.750 grams/day of amoxicillin. What is this dosage in milligrams per hour?

16. Upon graduating with a good GPA and work experience, you are pleased to receive two job offers. Company A offers a salary of $42,000/year. Company B offers an hourly pay of $25.00/hour. Assuming that you will work 40 hours/week for 50 weeks per year, use the factor-label method to calculate your annual income at Company B. Which offer is more lucrative?

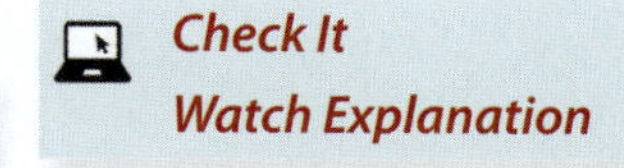

Converting between Volume Units

Chemists commonly make measurements involving volume (that is, the space something occupies). To measure volume, we use units of length raised to the third power, such as cm^3 (cubic centimeters) or m^3 (cubic meters); see **Figure 2.11**. The units we choose depend on the size of the quantity we are measuring. For example, if we are measuring the volume of a shipping container, we probably want to express this as

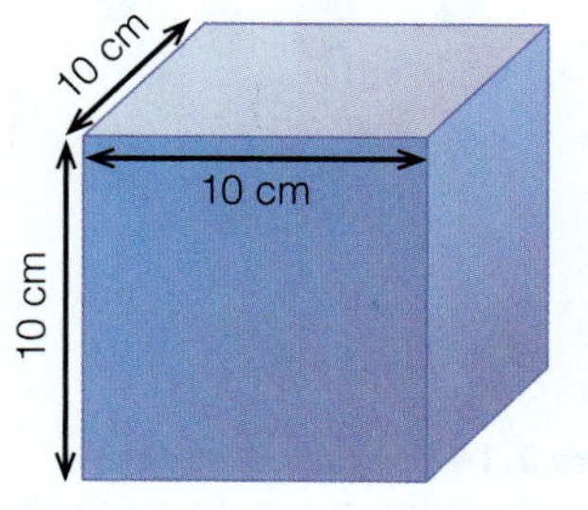

10 cm × 10 cm × 10 cm = 1,000 cm^3

Figure 2.11 The volume of a cube is equal to its length times width times height.

1 liter (L) = 1 dm^3

Yuri_Arcurs/Getty Images

Mega Pixel/Shutterstock

Figure 2.13 One cubic decimeter, also called a liter, is approximately the volume of a tissue box.

naito29/Shutterstock

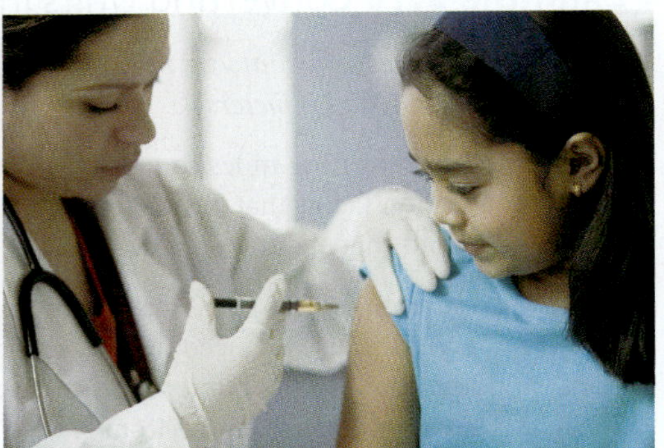

Blend Images/Brand X Pictures/Getty Images

Figure 2.12 We commonly encounter very large and very small volumes. The best units to use depend on how much we are measuring.

cubic meters. To describe the volume of a soft drink or the amount of liquid in a vaccine injection, we would choose a smaller unit (**Figure 2.12**).

The *liter* (L) is a common derived unit of volume. A liter is defined as 1 cubic decimeter (1 dm^3), which is slightly smaller than a box of tissues (**Figure 2.13**).

1 milliliter (mL) = 1 cm^3

Another common unit of measurement is the *cubic centimeter* (cm^3). A cubic centimeter is the same unit of volume as a *milliliter*. A cubic centimeter is about the same size as a standard game die (**Figure 2.14**).

iofoto/Shutterstock

Figure 2.14 One cubic centimeter is approximately the volume of a standard game die. This volume is also referred to as a milliliter.

Converting Units Raised to a Power

We must be careful when converting between units that are raised to a power, like those used to measure area or volume. To illustrate this point, let's ask a question: How many cubic decimeters are in one cubic meter? We know that 1 meter is equal to 10 decimeters:

$$1 \text{ m} = 10 \text{ dm}$$

To find the relationship between cubic meters and cubic decimeters, we must cube both sides. When we do this, we must cube both the numbers and the units

$$(1 \text{ m})^3 = (10 \text{ dm})^3$$
$$1 \text{ m}^3 = 1{,}000 \text{ dm}^3$$

Although 1 meter contains 10 decimeters, 1 *cubic meter* contains 1,000 *cubic decimeters*. The following examples illustrate this idea further.

Example 2.11 Conversions with Cubic Units

How many cubic centimeters (cm^3) are in 1 cubic meter (m^3)?

From the common metric unit conversions in Table 2.5, we know that

$$100 \text{ cm} = 1 \text{ m}$$

To convert to cm^3 and m^3, we cube both sides of the equation:

$$(100 \text{ cm})^3 = (1 \text{ m})^3$$

We cube the numbers, and we also cube the units. This means that

$$1{,}000{,}000 \text{ cm}^3 = 1 \text{ m}^3$$

So although there are just 100 centimeters in 1 meter, there are 1 million cubic centimeters in a cubic meter!

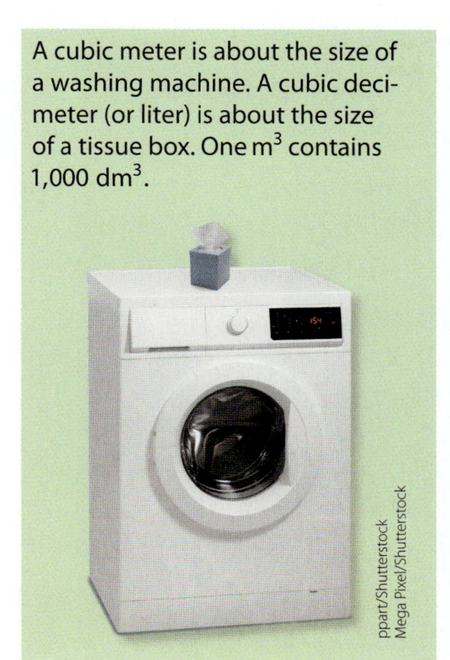

A cubic meter is about the size of a washing machine. A cubic decimeter (or liter) is about the size of a tissue box. One m^3 contains 1,000 dm^3.

ppart/Shutterstock
Mega Pixel/Shutterstock

Example 2.12 Conversions with Volume Units

A liquid has a volume of 15.3 cubic centimeters (cm^3). What is this volume in liters?

To make this conversion, we first identify our unit relationships. We know that 1 cubic centimeter is equal to 1 milliliter (mL), and there are 1,000 mL in 1 L. To solve this problem, we begin with 15.3 cm^3, then multiply by two conversion factors. The first cancels out cm^3 (leaving us with mL). The second cancels out mL, leaving us with liters.

$$15.3 \cancel{\text{cm}^3} \times \frac{1 \cancel{\text{mL}}}{1 \cancel{\text{cm}^3}} \times \frac{1 \text{ L}}{1{,}000 \cancel{\text{mL}}} = 0.0153 \text{ L}$$

Example 2.13 Conversions with Volume Units

*A hospital IV (**Figure 2.15**) is set to drip at a rate of 125 mL/h. How many liters of fluid is this patient receiving per day?*

In this problem, we need to convert both our volume and time units. We begin with the drip rate in mL/h, then use two conversion factors. The first converts volume from mL to L, and the second converts the time unit from hours to days. This gives us a final answer of 3.00 L/d:

$$\frac{125 \cancel{\text{mL}}}{\cancel{\text{hr}}} \times \frac{1 \text{ L}}{1{,}000 \cancel{\text{mL}}} \times \frac{24 \cancel{\text{hr}}}{1 \text{ day}} = \frac{3.00 \text{ L}}{\text{day}}$$

Notice that the answer has three significant digits. The two unit conversions are both exact numbers, so they do not limit the number of significant digits in the calculations. Because the only measured quantity (the drip rate of 125 mL/h) has three significant digits, we can keep this many digits in our final answer.

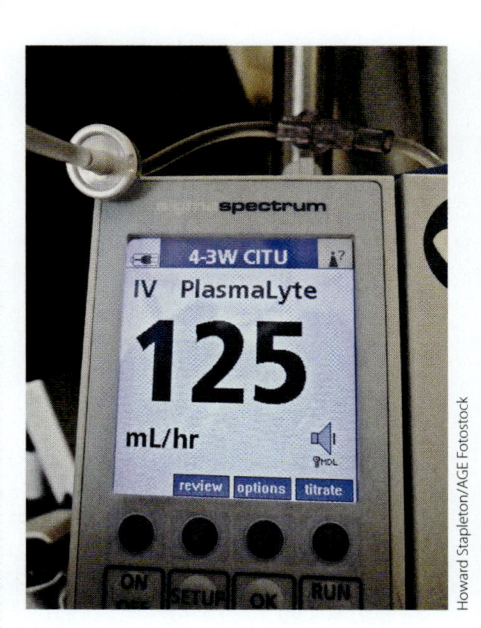

Howard Stapleton/AGE Fotostock

Figure 2.15 This intravenous (IV) drip is set to deliver fluid at 125 mL/hour.

Check It
Watch Explanation

TRY IT

17. A raindrop has a volume of about 0.05 cm^3. Express this volume in the following units:

a. milliliters **b.** liters **c.** cubic meters

18. On average, a volume of 84,760 ft^3 of water flows over Niagara Falls each second. At this rate, how many m^3 of water flow over the falls in an hour? (1 meter = 3.281 feet)

615 collection/Alamy

Figure 2.16 Prosthetic limbs, such as the one worn by U.S. Army sergeant and 100-m gold medalist Jerrod Fields, must be very strong but also very lightweight.

2.3 Density: Relating Mass to Volume

Materials are usually measured by mass or by volume. It is often necessary to convert between the mass and volume of a substance. To do this, we use a property of the material called the *density*. **Density** is defined as the mass of a substance per unit volume. That is:

$$\text{density} = \frac{\text{mass}}{\text{volume}}$$

Or, written in abbreviated form:

$$d = \frac{m}{V}$$

Density is an important physical property. For example, aluminum and titanium metal are highly valued for uses from automobiles to laptop computers to prosthetic limbs (**Figure 2.16**). They are strong, and because they have low densities, an object made from one of these metals weighs less than the same object made from other metals, like iron or lead. **Table 2.6** summarizes the densities of several common materials.

To measure the density of a sample, we must first determine the mass and volume. For example, if we want to measure the density of a solution of sodium chloride, we could measure the mass of the solution on a balance, and measure the volume of the solution using a graduated cylinder (**Figure 2.17**).

If the mass is 11.29 grams, and the volume is 10.4 milliliters, we calculate the density this way:

$$d = \frac{m}{V} = \frac{11.29\text{ g}}{10.4\text{ mL}} = 1.09\text{ g/mL}$$

Notice that as before, we've rounded the answer to the correct number of significant digits.

TABLE 2.6 Densities of Common Materials

Material	Density (g/cm^3)
Aluminum	2.70
Titanium	4.51
Iron	7.87
Copper	8.96
Lead	11.34
Gold	19.31
Water*	1.00
Seawater*	1.02
Air*	0.001

*At 25 °C and standard atmospheric pressure

Figure 2.17 A balance measures mass. A graduated cylinder measures the volume of a liquid. By measuring both quantities, we can determine the density of a liquid.

Converting between Mass and Volume

One of the most important uses of density is in unit conversions. If we know the density of a sample, we can relate its mass and volume. This process is illustrated in the following examples.

We use density to convert between mass and volume.

Example 2.14 Conversions Using Density I

A sample of iron metal has a volume of 8.83 cm³. If the density of iron is 7.87 g/cm³, what is the mass of the iron sample?

By rearranging the density equation, we see that the mass is equal to density times volume. Substituting in the values of the density and the volume, we make the following calculation:

$$m = dV = \left(7.87 \frac{\text{g}}{\cancel{\text{cm}^3}}\right)(8.83\ \cancel{\text{cm}^3}) = 69.5\ \text{g}$$

The mass of the iron sample is 69.5 grams. In this calculation, the units of volume (cm^3) canceled out, leaving grams as the unit of mass. It is important to make sure that the units of volume are the same in the density and volume terms. If the units do not cancel out correctly, the answer will not be correct.

Example 2.15 Conversions Using Density II

Mercury is a liquid with a density of 13.534 g/mL. What is the volume of a 1.213-kg sample of mercury?

To solve this problem, we first need to make sure that our mass units are the same. Because the density of mercury is given in units of g/mL, we need to convert the mass of the sample from 1.213 kg to 1,213 g. This will allow us to cancel out the units in the next step.

We next rearrange the density equation to show that volume is equal to mass over density. We then substitute in the values of the mass and the density, and solve. The grams cancel out, leaving us with units of milliliters, as follows:

$$V = \frac{m}{d} = \frac{1{,}213\ \text{g}}{13{,}534\ \text{g/mL}} = 89.63\ \text{mL}$$

Rounding to four significant digits, we get an answer of 89.63 milliliters.

Will It Float?

The density of water is 1 g/mL.

The density of an object determines whether it floats or sinks in water. Pure water has a density of 1 g/cm^3. Objects that are less dense than water ($d < 1\ g/cm^3$) float, while objects more dense than water ($d > 1\ g/cm^3$) sink (**Figure 2.18**).

Example 2.16 Will It Float?

A toy submarine has a volume of 14.3 mL and a mass of 15.2 grams. Will the toy float or sink?

To solve this problem, we first find the density:

$$d = \frac{m}{V} = \frac{15.2\ \text{g}}{14.3\ \text{mL}} = 1.06\ \text{g/mL}$$

The density of this toy is greater than the density of water (1 g/mL), so the toy will sink.

Check It

Watch Explanation

TRY IT

19. An unknown liquid has a volume of 15.0 mL and a mass of 11.3 g. What is the density of this liquid? Based on this information, what mass would you expect for a 2.50-L sample of this liquid?

20. You recently began working for a rock quarry that produces gravel. The company has purchased a truck with a capacity of 10.0 m^3, and you have been asked to determine how many kilograms of gravel this truck will carry. To answer this question, you measure 1.0 L of gravel and find that it has a mass of 1.8 kg. Based on this result, what is the density of gravel in kg/L? In kg/m^3? What is the mass of a 10.0-m^3 load of this gravel? (1,000 L = 1 m^3)

21. A cylindrical steel tank with a diameter of 20.0 cm, a length of 150.0 cm, and an overall mass of 150.0 kg is accidentally dropped into a lake. What is the cylinder's density in g/cm^3? Will it sink or float?

Explore

Figure 2.18

Courtesy of Gary Deaton

Figure 2.18 In order to float, an object's density must be less than that of water.

2.4 Measuring Temperature

In chemistry, we often need to measure and report temperature. There are three common temperature scales. In the United States, the **Fahrenheit scale** is most commonly used. Most of the world uses the **Celsius scale**. On the Celsius scale, the freezing point of water is 0 °C, and the boiling point is 100 °C. Notice in **Figure 2.19** that the difference between the freezing point and boiling point of water is 180 °F, but only 100 °C. This means that a degree Fahrenheit is a smaller unit than a Celsius degree. To convert between Celsius and Fahrenheit, we use the following relationships:

$$°\text{F} = \frac{9}{5}\,°\text{C} + 32$$

$$°\text{C} = \frac{5}{9}\,(°\text{F} - 32)$$

A third temperature scale is the **Kelvin scale**. Scientists use this scale to describe events at very low temperatures and also to predict the way gases behave. The Celsius and Kelvin scales have the following relationship:

$$\text{K} = °\text{C} + 273.15$$

A unit on the Kelvin scale is the same as a degree Celsius. The difference between the two scales is the zero point. A temperature of 0 °C corresponds to the freezing point of water, while a temperature of 0 K corresponds to absolute zero (the lowest possible temperature, at which particles have zero kinetic energy).

When reporting the temperature in Fahrenheit or Celsius, the word *degrees* follows the number. For example, "32 degrees Fahrenheit equals 0 degrees Celsius." In contrast, with the Kelvin scale, the word *degrees* is not included, so it is correct to say, "Zero degrees Celsius is equal to 273.15 Kelvin."

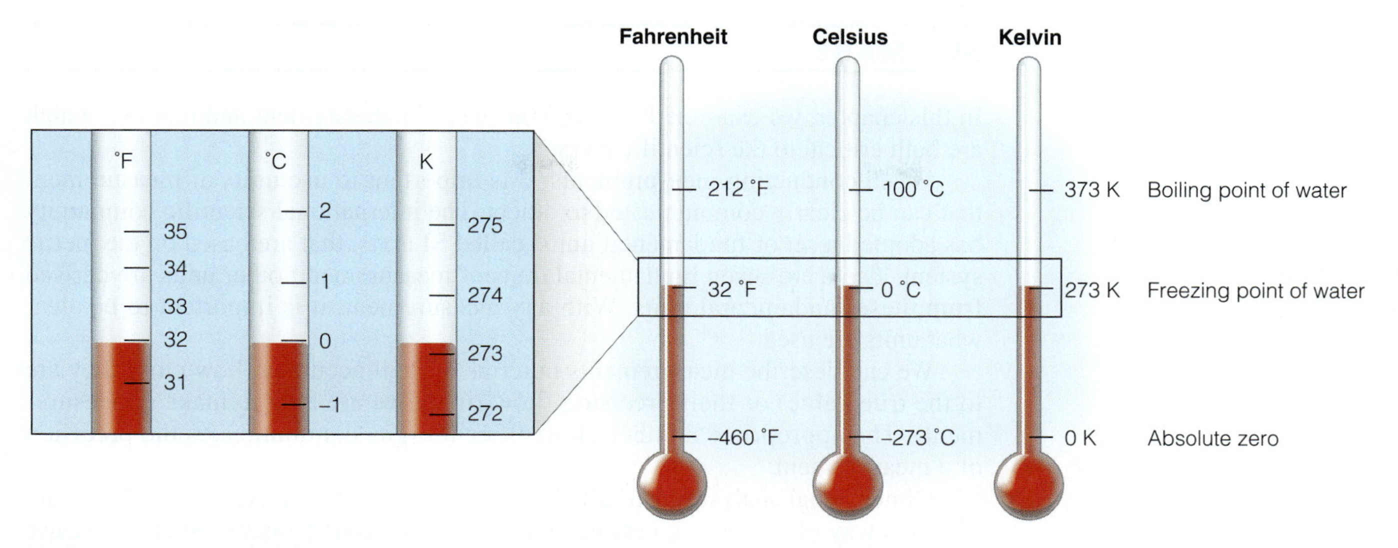

Figure 2.19 Degrees in the Celsius and Kelvin scales are the same size, but a degree Fahrenheit is a smaller unit.

TRY IT

22. While on a business trip to Brussels, Belgium, you hear a weather forecast predicting a high temperature of 15 °C. What is this temperature in degrees Fahrenheit? Should you take a jacket when you leave your hotel?

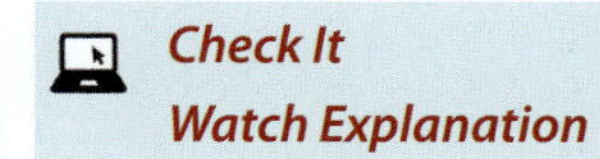

Capstone Question

You are on the management team of a company that is considering purchasing a tanker truck. The truck you want has a volume of 30,000 liters and an empty mass/weight of 15,800 kg. However, the route you wish to travel has a bridge with a weight limit of 25 tons (**Figure 2.20**). If you purchase the truck and fill it to capacity with a liquid whose density is 0.80 kg/L, what will the mass of the truck be? Will it be too heavy to cross the bridge? What is the maximum volume you will be able to legally transport? (1 kg = 2.20 lb)

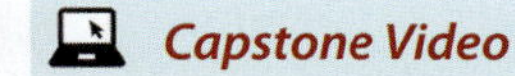

Figure 2.20 Many older bridges have weight limits. Shipping companies must keep this in mind when planning travel routes.

SUMMARY

In this chapter, we examined the key concepts of measurement and method, which are both critical to the scientific process.

When conducting measurements, it is important to use units of measurement that can be clearly communicated to others. The international scientific community has adopted a set of fundamental units, called SI units, that are based on the metric system. There are seven fundamental units of measurement; other units are derived from these fundamental units. With any measurement, it is important to be clear what units are used.

We can describe measurements in terms of their accuracy (how close they are to the true value) or their precision (how finely we are able to make a measurement). The appropriate number of significant digits communicates the precision of a measurement.

Dimensional analysis, also called the factor-label method, is a standard, broadly effective way of converting between units. When converting between units, we use a fractional relationship containing equivalent amounts (but different units) in the numerator and the denominator. In these problems, we cancel units in the same way we cancel numerical values.

Often, we need to convert between measurements of mass and volume of a substance. To do this, we use the density of the substance. Density is defined as the mass of a substance divided by its volume. In addition to its usefulness as a conversion factor, density is an important physical property that we use for many applications.

When measuring temperature, we use the Celsius, Kelvin, or Fahrenheit scales. The Celsius scale is the most commonly used scale for scientific measurements, although the Kelvin scale is important for calculations involving gases or very low temperatures.

A Strange Death

At the beginning of this chapter, we described the case of a man who had died as a result of consuming MDPV, a synthetic drug similar to methamphetamine. This was the first death in West Virginia to be linked to MDPV. Following the findings in this case, however, investigators attributed the abuse of this substance to a series of other deaths. The case provides an interesting example of the role of measurement and scientific method in solving problems. Let's briefly review how the scientists approached this problem, and examine their findings in more detail.

The toxicologists assigned to the case, James Kraner and Lauren Waugh, first determined whether other common drugs were involved. They used routine tests for detecting the presence of popular street drugs, such as cocaine and methamphetamine, in the victim's blood and urine. One of the keys to effective science is to understand and apply what is already known. Sometimes a day in the library can save a year in the lab.

These routine tests indicated that none of these more common drugs were present at the time of death. Having ruled out the other drugs, the toxicologists hypothesized that the substance in the container might have caused the erratic behavior. To identify the substance, they used an instrument called a mass spectrometer, which enables scientists to measure the masses of individual molecules (we will describe this instrument in more detail in Chapter 7). By combining this measurement with other tests, Kraner and Waugh were able to positively identify the substance as MDPV.

After identifying the unknown compound, they needed to determine how much (if any) MDPV was in the body when the person died. To do this, Dr. Waugh prepared a set of samples of blood serum (a liquid component of blood) that contained different known amounts of the drug (**Figure 2.21**). She also tested samples that had no drug present, to make sure the test would give a negative response in the absence of the drug.

After all of this preparatory work, Dr. Waugh was finally ready to test the victim's blood and urine for MDPV levels. By comparing the results of these tests with known solutions, she determined that the blood of the deceased contained 1.09 milligrams of MDPV per liter of blood (that is, 1.09 mg/L). The MDPV concentration in the urine was 60.3 mg/L.

Analysis of blood and urine samples is a common way to test for illegal drugs or other toxic substances. In combination with preparation, carefully planned methods, and meticulous measurement, tests like this can yield a wide range of forensic and health information.

Brad Davis/AP Images

Figure 2.21 Forensic chemists measure the amount of drugs such as MDPV in blood samples such as those shown here.

Key Terms

2.1 Measurement: A Foundation of Good Science

scientific notation A way to show very large and very small numbers in a concise format. Scientific notation expresses numbers as the product of two values, called the *coefficient* and the *multiplier*.

units of measurement Quantities with accepted values.

liter (L) A common unit of volume, defined as 1 cubic decimeter (1 dm^3).

accuracy A measure of how reliable measurements are—that is, how closely they reflect the true value.

precision A measure of how finely a measurement is made, or how close a group of measurements are to each other. Precision is often denoted by significant digits.

significant digits The digits contained in a measured value. The number of significant digits indicates how precisely a measurement is made. Also called *significant figures*.

exact numbers Numbers for which there is no uncertainty. Counted integers and defined relationships (such as metric prefixes) are exact numbers.

2.2 Unit Conversion

conversion factors Fractions that are used to convert from one unit to another. A conversion factor contains equivalent amounts of different units in the numerator and the denominator.

2.3 Density: Relating Mass to Volume

density A physical property of a substance, defined as the mass per unit volume.

2.4 Measuring Temperature

Fahrenheit scale (°F) A temperature scale commonly used in the United States. On the Fahrenheit scale, water freezes at 32 °F and boils at 212 °F.

Celsius scale (°C) A temperature scale commonly used throughout the world. On the Celsius scale, water freezes at 0 °C and boils at 100 °C. Sometimes called the *centigrade scale*.

Kelvin scale (K) The official SI scale for measuring temperature. The Kelvin scale sets absolute zero at 0 K, the freezing point of water at 273.15 K, and the boiling point of water at 373.15 K.

Additional Problems

2.1 Measurement: A Foundation of Good Science

23. Write the following values in scientific notation:

a. 1,500,000 km
b. 0.000398 g
c. 1,200,000,000 J
d. 0.02019 s

24. Write the following values in scientific notation:

a. 13,201 kg
b. 0.00000593 g
c. 1,400,000,000 km
d. 0.00322 s

25. Convert the following values to standard notation:

a. 5.192×10^{-4} kg
b. 1.23×10^{3} J
c. 4.2×10^{8} °C
d. 6.02×10^{23} atoms

26. Convert the following values to standard notation:

a. 8.13×10^{-3} ng
b. 3.4×10^{7} lb
c. 4.2×10^{5} molecules
d. 1.301×10^{2} m

27. Each of these numbers is written in exponential notation, but not in proper scientific notation. Write each number correctly. (*Hint*: The coefficients do not have one nonzero digit in front of the decimal point. Shift the decimal point to correct this, and adjust the exponents to correspond to the change.)

a. 52.1×10^{9} min
b. 0.83×10^{9} g
c. 435×10^{-9} m

28. Each of these numbers is written in exponential notation, but not in proper scientific notation. Write each number correctly. (*Hint*: The coefficients do not have one nonzero digit in front of the decimal point. Shift the decimal point to correct this, and adjust the exponents to correspond to the change.)

a. 533.1×10^{-9} m
b. 0.083×10^{3} J
c. 435×10^{-16} kg

29. Complete each calculation:

a. $(4.4 \times 10^{-3})(1.2 \times 10^{-5})$
b. $(1.4 \times 10^{-3}) \div (1.2 \times 10^{-5})$
c. $(5.4 \times 10^{5})(2.2 \times 10^{3})$
d. $(8.132 \times 10^{-2})/(4.19 \times 10^{3})$

30. Complete each calculation:

a. $(1.4 \times 10^{6})(9.1 \times 10^{-5})$
b. $(1.4 \times 10^{-3})^2$
c. $(5.4 \times 10^{5})/(3.7 \times 10^{-5})$
d. $(2.1 \times 10^{4}) \div 3$

31. Using the units and prefixes in Tables 2.3 and 2.4, determine the name of the unit represented by each of these abbreviations:

a. A
b. K
c. ms
d. mg
e. kA

32. Using the units and prefixes in Tables 2.3 and 2.4, determine the name of the unit represented by each abbreviation:

a. µA
b. nm
c. kg
d. mm
e. µs

33. Refer to Table 2.5 to answer the following:

a. How many grams are in a megagram?
b. How many calories are in a kilocalorie?
c. How many ms are in 1 s?
d. How many A are in 1 GA?

34. Refer to Table 2.5 to answer the following:

a. How many nanoseconds are in a second?
b. How many watts are in a gigawatt?
c. How many µg are in a g?
d. How many J are in a kJ?

35. Which is a more precise measurement of distance: a number that is rounded to the nearest gigameter, or a number that is rounded to the nearest nanometer?

36. Which is a more precise measurement of mass: a number that is rounded to the nearest milligram, or a number that is rounded to the nearest microgram?

37. In the laboratory, you measure out a compound to be used in a chemical reaction using two different balances. The first balance reports the mass as 10.23 grams. The second balance reports the mass as 11.1925 grams. Can you tell which balance is more accurate? Can you tell which measurement is more precise?

38. In a forensic laboratory, a toxicologist uses an analytical balance (an instrument for measuring small masses) to determine that a skin sample has a mass of 0.30 milligrams. Nearby, at the local salvage yard, a truck pulls onto a scale, and the operator determines that the scrap metal in the truck has a mass of 4,813 kilograms. Which measurement has more significant digits? Which measurement is more precise?

39. Consider this image of a test tube and ruler. Should the length of the test tube be reported as 5 cm, as 5.1 cm, or as 5.12 cm? Why?

40. This graduated cylinder shows a volume that is more than 5 mL but less than 6 mL. How precisely can you estimate this volume? Report the volume to the correct number of significant digits.

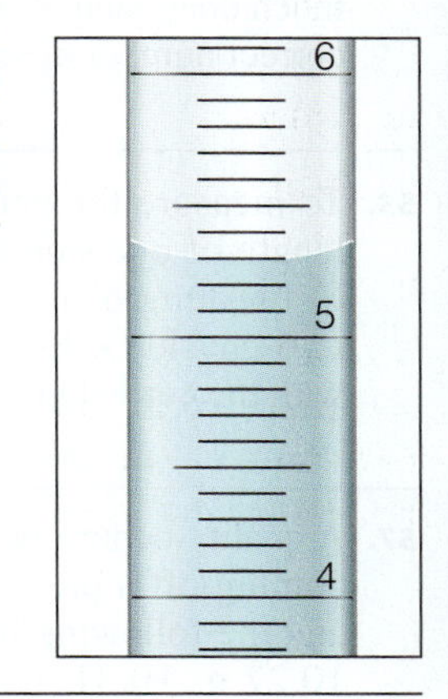

41. Identify the number of significant digits in the following measured numbers:

a. 0.0005123 g
b. 32.01 lb
c. 43.300 m
d. 4.0 GPA

42. Identify the number of significant digits in the following measured values:

a. 2.304 km
b. 0.002000 g
c. 12.0 mA
d. 3.023×10^4 m

43. A builder measures the length of a wall for a new home and finds it is 20.31 meters. How many significant digits are in this measurement?

44. The mass of a wood shard recovered from a fire is 0.0397 g. How many significant digits are in this measurement?

45. The following values are written in standard notation. Convert each value to scientific notation, using the correct number of significant digits.

a. 5,200,000 ng, measured to four significant digits
b. 0.0003920 L, measured to four significant digits
c. $1,800,000, measured to six significant digits

46. The following values are written in standard notation. Convert each value to scientific notation, using the correct number of significant digits.

a. 4,260,000,000 bytes of data, measured to four significant digits
b. 0.000320 L, measured to three significant digits
c. 52,000,000,000 km, measured to six significant digits

47. What is the difference between a measured number and an exact number?

48. A dairy farmer reads a study that says lactating cows (cows that are producing milk) consume an average of 27 pounds of dry hay each day. The farmer has 44 cows and projects hay costs based on these numbers. Which of these values is measured, and which is exact?

49. Identify each value as measured or exact:

a. There are 40 quarters in a $10.00 quarter roll.
b. The distance from your house to your workplace is 6.2 miles.
c. There are 105 nutritional Calories in a medium banana.
d. An octagon has 8 sides.

50. Identify each value as measured or exact:

a. A cyclist rides 82 kilometers in one day.
b. The cyclist's bicycle has 2 wheels.
c. The month of July contains 31 days.
d. There are 1,000 mg in 1 gram.

51. Carry out each calculation, rounding to the appropriate number of significant digits.

a. the area of a property: 103.2 ft $\times$ 59.3 ft
b. the perimeter of a triangle: 11.502 cm + 18.00 cm + 1.92312 cm
c. the density of a block of iron: 25.3261 grams $\div$ 3.6 cm^3

52. Carry out each calculation, rounding to the appropriate number of significant digits:

a. the total time spent running during a week: 25.24 minutes + 28.1 minutes + 56 minutes
b. the circumference of a circle: $C = 2\pi r$, where r = 1.32 inches and the number 2 is an exact number
c. the wavelength of blue light (in meters), found by dividing the speed of light by the frequency: $(3.00 \times 10^8$ m/s$) \div (6.911 \times 10^{14}$ $s^{-1})$

53. You and a friend win $500.00 and divide it evenly. How much does each of you get? Make sure your answer has the correct number of significant digits.

54. Your sales team wins a bonus of $900.00, to be divided evenly. If there are seven people on your sales team, how much will each of you get? Make sure your answer has the correct number of significant digits.

55. To measure the perimeter of a square, you measure the length of one side to be 12.6 cm. The perimeter is equal to the length of one side times the number of sides: 12.6 cm/side $\times$ 4 sides. Report the perimeter of the square with the correct number of significant digits.

56. You are considering a new job with a long commute. To measure the distance, you drive from your home to the worksite and back. Your car's odometer reports the round-trip distance as 53.8 km. If you made exactly 20 trips to this new job in a month, how many miles would you commute? Report your answer to the correct number of significant digits.

57. A quality-control process measures the mass of new screws coming off a press. A sample of ten screws is collected, and the following masses are recorded: 10.31 g, 10.33 g, 10.27 g, 10.31 g, 10.50 g, 10.31 g, 10.28 g, 10.39 g, 10.41 g, and 10.36 g. Calculate the total mass and the average mass of these screws. Make sure your answer has the correct number of significant digits.

58. The presence of drugs (both legal and illegal) in public wastewater systems can provide valuable information for scientists who study public health issues. A chemist analyzes four samples of wastewater and finds they contain the following amounts of heroin: 98.5 ng/L, 109.2 ng/L, 114.3 ng/L, and 102.1 ng/L. Calculate the average concentration of heroin in these samples. Make sure your answer has the correct number of significant digits.

2.2 Unit Conversion

59. Perform the following metric conversions using the factor-label method:

a. How many ng are in 0.0213 g?
b. Convert 17,397.4 m to km.
c. Express 0.000310 L in μL.

60. Perform the following metric conversions using the factor-label method:

a. Convert 1.5×10^{14} A to GA.
b. Convert 0.0870 m to cm.
c. Convert 25.4 ng to g.

61. Complete the following metric conversions:

a. Convert 23.21 μL to L.
b. Convert 50,000 g to kg.
c. Convert 5.40×10^{-7} m to nm.

62. Complete the following metric conversions:

a. Convert 4.3 L to mL.
b. Convert 4,320 mg to g.
c. Convert 1.53×10^{-7} g to ng.

63. Complete the following metric conversions:

a. Convert 0.081 mL to μL.
b. Convert 112,507 mg to kg.
c. Convert 5.40×10^{-2} μm to nm.

64. Complete the following metric conversions:

a. Convert 0.00310 mg to μg.
b. Convert 153,719,218 mW to kW.
c. Convert 3.214×10^8 nA to mA.

65. The unit of electrical resistance is the ohm (Ω). A circuit has a resistance of 5.4×10^5 Ω. What is this amount in megaohms (MΩ)?

66. The explosive power of a nuclear weapon is usually expressed as the number of tons of TNT that would be required to produce the same explosion. The atomic bomb used in Hiroshima, Japan, in 1945 had a power of approximately 1.5×10^4 tons of TNT. In 1961, the Soviet Union detonated a test bomb that measured a power of 5.0×10^{10} tons of TNT. Express the strength of these two weapons in kilotons and in megatons.

67. Refer to **Table 2.7** to perform the following conversions:

a. 15.2 inches to centimeters
b. 47.23 kilocalories to joules
c. 1.2 tons to kilograms
d. 3.55 miles to kilometers

68. Refer to Table 2.7 to answer the following conversion questions:

a. How many milliliters are in 2.0 gallons?
b. How many calories are in 8.14×10^6 joules?
c. How many meters are in 4.23×10^3 miles?
d. How many kilocalories are in 3.2×10^4 joules?

TABLE 2.7 Common Conversions

Length
1 inch = 2.54 cm
1 mile = 1.609 km
Volume
1 gallon = 3.785 liters (L)
Mass/Weight (at Earth's gravity)
1 kg = 2.20 pounds (lb)
1 ton = 2,000 lb
Energy
1 calorie (cal) = 4.184 joules (J)

69. While trying to remove a bolt with a socket wrench, you find that the bolt is sized in units of millimeters and is slightly smaller than a ½-inch socket. What size metric socket should you use? (This will be an integer value.)

70. Allen wrenches are small, L-shaped wrenches used to assemble bicycles, furniture, and other common items. Allen wrenches come in both metric and English sizes. A 6-mm wrench is almost, but not exactly, the same size as a ¼-inch wrench. Which is smaller?

71. A plane flies at a speed of 540 miles per hour. What is this speed in meters per second?

72. A leaky faucet drips water at a rate of 35 mL per hour. Express this rate in terms of liters per day.

73. A small car has a fuel efficiency of 34.3 miles per gallon. What is this fuel efficiency in kilometers per liter?

74. A large pickup truck has a fuel efficiency of 16.2 miles per gallon. Express this fuel efficiency in kilometers per liter.

75. Rocephin® is an antibiotic sometimes used to treat skin infections. In children, the maximum recommended daily dose of this medicine is 75 mg/kg of body mass. Express this dosage in g/lb, using the conversion (2.20 kg = 1 lb).

76. Adderall® is used to treat attention-deficit hyperactivity disorder (ADHD). In children, a dosage of 10 mg per day is fairly common. At this rate, how many grams of Adderall does a child take over the course of a year?

77. Large amounts of the herbicide glyphosate have been linked to kidney problems and reproductive difficulties. The U.S. Environmental Protection Agency limits the levels of glyphosate in drinking water to a maximum of 0.7 mg/L. What is this value in g/m^3?

78. Dioxin is a highly toxic compound that has been shown to increase the risk of cancer in humans. Dioxin is sometimes produced from waste incineration. The U.S. Environmental Protection Agency limits the levels of dioxin in drinking water to a maximum of 30 ng per liter. If a 1,500-mL sample of water contains 1.2×10^{-8} g of dioxin, is it above the maximum level?

79. How many square meters are in a square kilometer? (*Hint*: It's not 1,000. Watch your units.)

80. How many square feet are in a square yard?

81. The length, width, and height of a box are each 1 yard. What is the volume of this box in cubic feet (ft^3)?

82. How many dm^3 are in 1 m^3? How many cm^3 are in 1 m^3?

83. How many liters are in 1 m^3? How many milliliters are in 1 m^3?

84. How many cm^3 are in 1 milliliter? How many cm^3 are in 1 liter?

85. Electrical power is commonly measured in watts (W) or in kilowatts (kW). A commercial solar panel generates about 0.10 watts per square inch of surface area. Based on this, how many kilowatts of power could be generated by a residential solar panel array with a surface area of 685 ft^2?

Figure 2.22 Silver carp commonly jump as boats approach.

86. Silver carp are large fish known for jumping out of the water as boats pass by (**Figure 2.22**). They are an invasive species, and over the past several decades they have spread through much of the middle United States. Many people worry about the impact they have on other plant and animal life. Recently, a team of scientists sought to determine whether the arrival of silver carp has impacted plant plankton levels. To do this, they measured the amount of chlorophyll (a pigment produced by the plankton) in a large lake, and compared levels of chlorophyll before the carp arrived to current chlorophyll levels. The scientists found that average chlorophyll levels in the shallow regions of the lake had dropped from 23.62 μg/L to 18.50 μg/L during this time period. What is the overall change in chlorophyll concentrations in μg/L? What is the overall change in g/m^3?

2.3 Density: Relating Mass to Volume

87. Cobalt has a density of 8.90 g/cm^3 at room temperature. What is this density in kg/m^3?

88. Aluminum has a density of 2.70 g/cm^3 at room temperature. What is this density in kg/L?

89. A sample of an unknown solid has a mass of 15.23 g and a volume of 23.7 mL. What is the density of this material? Report your answer to the correct number of significant digits.

90. A sample of an unknown liquid has a mass of 20.365 g and a volume of 28.1 mL. What is the density of this material? Report your answer to the correct number of significant digits.

91. To determine the density of a liquid, you measure out a volume of 9.3 mL. The mass of the liquid is 9.649 g. Report your answer to the correct number of significant digits.

92. A small metal cube is 1.34 cm on each side and has a mass of 18.936 grams. What is the density of this cube?

93. Indicate whether each of the following would float or sink in water:

a. a solid having a mass of 42.3 grams and a volume of 40.1 cm^3

b. a solid having a mass of 42.3 grams and a volume of 44.1 cm^3

94. Indicate whether each of the following would float or sink in water:

a. a rectangular block of length 5.0 cm, width 4.0 cm, and height 2.0 cm, with a mass of 0.583 kg

b. a diving tank with a volume of 80 cubic feet filled with air; the filled tank has a weight of 21 pounds

*Questions 95–98 refer to **Table 2.8**.*

95. Calculate each of the following:

a. the volume of a bar of pure gold with a mass of 5.00 grams

b. the volume of an iron statue with a mass of 1.8 kg

96. Calculate each of the following:

a. the mass of a copper pipe in which the copper has a volume of 70.5 cm^3

b. the mass of a titanium rod having a volume of 0.81 cm^3

TABLE 2.8 Densities of Common Metals

Material	Density (g/cm^3)
Aluminum	2.70
Titanium	4.51
Iron	7.87
Copper	8.96
Lead	11.34
Gold	19.31

97. You have three blocks—one made of iron, one made of aluminum, and one made of lead. Each block is 5.0 cm × 4.0 cm × 1.0 cm. What is the mass of each block? Make sure your answer has the correct number of significant digits.

98. To measure the density of a small statue, you determine the mass of the statue to be 2.219 kg. Next, you pour 400 milliliters of water into a graduated cylinder (measured to the nearest mL). You then gently lower the statue into the water and observe that the volume rises to 682 milliliters. Calculate the density of the statue. Based on Table 2.8, what element might the statue be made of?

2.4 Measuring Temperature

99. What is absolute zero? What are the values of absolute zero on the Fahrenheit, Celsius, and Kelvin temperature scales?

100. The average temperature of the human body is 98.6 °F. What is this temperature in °C and K?

101. Which is a larger temperature change: a change of 1.3 °C or a change of 2.5 °F?

102. Which is larger, 1 degree Celsius or 2 degrees Fahrenheit?

103. Room temperature is typically around 72 °F. Convert this temperature to degrees Celsius and to kelvins.

104. Convert 400 K to degrees Celsius and degrees Fahrenheit.

Challenge Questions

105. A family needs to replace the flooring in their house. The local hardware store has two options that they like: carpet, which sells for $5.80 per square yard, or hardwood, which sells for $1.40 per square foot. Which option is less expensive? If they need to replace 600 ft^2 of flooring, how much will each option cost?

106. For your new candle-making business, you need to purchase a large amount of a unique scented wax. You plan to charge $9.95 per large candle. The wax is available from a U.S. supplier for $24.00/lb and also from a German supplier for €9.20/kg. If the current exchange rate is $1 = €0.76, and 1 kg = 2.20 lb, which supplier is giving the better price?

107. Zebra mussels are small, freshwater animals with a striped shell (**Figure 2.23**). In 2017, scientists observed a dramatic increase in the population of these creatures in Kentucky Lake. These scientists sampled different regions of the lake bottom, and found an average of $1,100 \pm 100$ mussels/m^2.

a. Express this value in scientific notation, using the correct number of significant digits.

b. Kentucky Lake covers an area of 250 square miles. Based on the measurements and uncertainty the scientists reported, calculate low and high estimates for the total number of mussels in this lake. (1 mile = 1,609 meters)

RLSPHOTO/iStock/Getty Images

Figure 2.23 Zebra mussels like the one shown here have sharply increased in numbers in many lakes and rivers.

CHAPTER THREE

Atoms

Figure 3.1 (a) Community mines, like this one in Mozambique, are found in gold-rich regions of South America and East Africa. *(continued)*

Mercury and Small-Scale Gold Mining

Every summer, Adam Kiefer selects a team of students, packs supplies and equipment, and boards a plane bound for the gold-rich regions of South America and East Africa. But for Kiefer—a chemistry professor at Mercer University—gold is not the objective. It's the problem. Mining practices in these regions produce toxic mercury vapors that contaminate the soil, air, and water, causing severe health problems for the people who live nearby. Kiefer and his team work with the miners, residents, and governmental officials, measuring mercury levels, teaching safer techniques, and working to clean up some of the most highly polluted regions in the world (**Figure 3.1**).

Why does gold mining release toxic mercury? To answer that question, we need to know a bit more about the mining process. Despite the abundance of gold, millions of people in these regions are desperately poor. Many make their living as artisanal miners in small-scale gold mines. Each day, miners descend into community mines to collect coarse gold-containing ore. They carry the ore to a grinding facility, where they mill the hard rock down to fine particles. Miners then place the particles in large bowls, add water, and stir. This causes the lighter soil to rise to the top, where it can be drained off. The heavier ore remains in the bottom of the bowl. After removing the soil, the miners stir in elemental mercury. The gold particles dissolve in the mercury, forming small, shiny pellets containing a mixture of the two elements, called an *amalgam*. To separate the gold from the mercury, the miners place the solid amalgam onto a smoldering log. The heat evaporates the mercury, leaving behind tiny pellets of purified gold. The miners sell the gold to local dealers, earning just enough to survive.

Mercury is highly toxic, affecting the central nervous system and producing tremors, anxiety, memory loss, and eventually insanity. Small-scale gold-mining operations release more mercury into the environment than any other human activity.

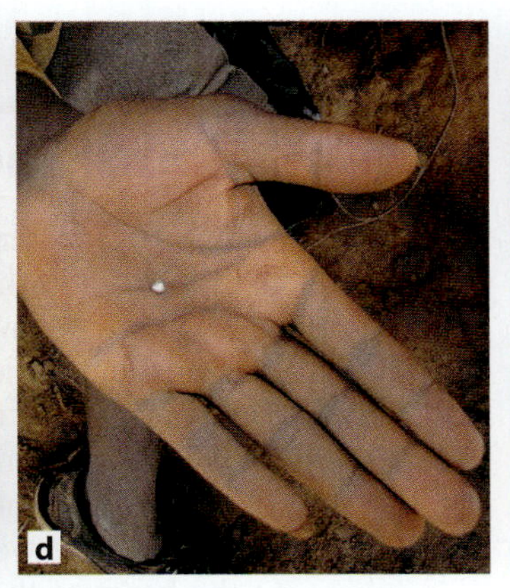

Figure 3.1 (continued) (b) A miner grinds ore into fine pieces. (c) This worker is sifting the ore from the lighter soil. Mercury is added during this phase. (d) The mixture of mercury and gold that forms this tiny pellet is called an *amalgam*. (e) A miner heats the amalgam on hot coals to evaporate the mercury. (a–e): Courtesy of Dr. Adam Kiefer/Mercer University

Kiefer and his team work to make measurements and raise awareness in the mining communities. As he puts it, "These miners don't pollute the environment because they don't care about their community; they do it because they need to feed their families. By using science and technology, we teach miners about the hazards of elemental mercury and how to protect themselves, their families, and the environment. We also train governmental officials in the techniques that we use, so that in the future they can work with the communities themselves."

Gold and mercury have many similarities. Both elements are metals. Both are shiny, melt at fairly low temperatures, and conduct electricity. However, gold is a solid at room temperature while mercury is a liquid. And although many people have died fighting over gold, gold itself does not pose the health hazards that mercury does.

Why are these elements similar? What causes their differences? As we saw in Chapter 1, the properties of any substance are connected to its structure. In this chapter, we'll explore the discovery and structures of the elements. We'll lay the groundwork that will allow us to answer the question, "Why does this element behave the way it does?" These are important concepts; let's begin to dig.

Intended Learning Outcomes

After completing this chapter and working the practice problems, you should be able to:

3.1 Atoms: The Essential Building Blocks

- Describe the development of atomic theory and its key observations about atoms.
- Apply the law of conservation of mass to solve mass problems related to chemical reactions.
- Describe chemical changes using atomic theory.

3.2 The Periodic Table of the Elements

- Describe the organization of the periodic table.

3.3 Uncovering Atomic Structure

- Describe the behavior of charged particles.
- Describe how the discovery of the battery and Rutherford's gold foil experiment shaped our understanding of atomic structure.
- Describe the relative mass and charge of protons, neutrons, and electrons and their arrangement within an atom.

3.4 Describing Atoms: Identity and Mass

- Relate the number of protons to atomic number and the sum of nuclear particles to mass number.
- Describe the nuclear structure of isotopes and calculate average atomic mass from a distribution of isotopes and relative abundances.
- Differentiate between mass number and average atomic mass.

3.5 Electrons—A Preview

- Contrast the description of electrons in the Bohr model and the quantum mechanical model.
- Identify the overall charge of an atom or ion based on the number of protons and electrons.

3.1 Atoms: The Essential Building Blocks

Right now, somewhere, a breeze is blowing across turquoise-blue water. A boat creaks as it rocks in the gentle waves. Somewhere else, snow is falling. Somewhere a person is sitting down with a cup of coffee and reading the news on her tablet. Somewhere a masterpiece is being painted.

The world around us—everything we can see, touch, taste, hear, or smell—arises from matter. And all matter is composed of a basic set of elements. At present, there are 118 known elements, but only about 98 of those are naturally occurring. The stunning diversity of the world around us arises from a very small set of building blocks.

At the heart of these elements is the *atom*. The atom is the fundamental unit of matter. Each of the elements is composed of a different type of atom, which gives each element its unique properties. As atoms combine into *compounds* and *mixtures*, materials with new and different behaviors are produced (**Figure 3.2**).

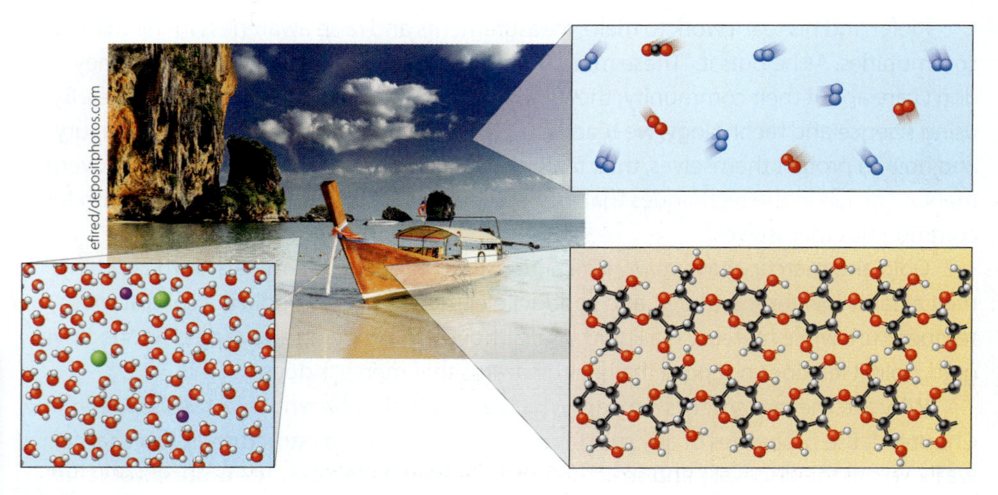

Figure 3.2 The world around us is made of atoms. The type and arrangement of atoms give each material its unique properties.

Explore
Figure 3.2
Seeing Atoms
Use this interactive for an up-close view of the structures of different substances.

Uncovering the Atom: From Democritus to Dalton

The word *atom* derives from the Greek word *atomos*, meaning "indivisible." The Greek philosopher Democritus (**Figure 3.3**), who lived around 400 B.C.E., argued that if you cut a substance in half, what results is two pieces of the same substance. He said that if you cut something in half over and over, you would eventually reach a point where you could no longer divide it and still have the same substance. At that point it was *atomos*, "indivisible."

Figure 3.3 This bust of Democritus is housed in the Naples National Museum in Italy.

The idea of the atom remained dormant until about the 1700s. Then, building on the technological innovations of the Renaissance, scientists began exploring chemical behavior more systematically and unlocking clues about atomic structure.

In the late 1700s, Antoine Lavoisier cataloged a large number of chemical reactions. He observed that when a chemical change takes place, the total mass before and after the reaction is the same. Based on this research, he formulated the **law of conservation of mass**, which states that *in chemical reactions, matter is neither created nor destroyed.*

For example, hydrogen and oxygen react explosively to form water (**Figure 3.4**). Suppose we combined 4 grams of hydrogen with 32 grams of oxygen in a sealed container and then caused the two elements to react. The water produced would have a mass of 36 grams. That is, we have the same amount of mass at the end of the reaction as we had at the beginning.

We can use the law of conservation of mass to predict the amount of substances that will be consumed or produced in chemical reactions. The following example illustrates this principle.

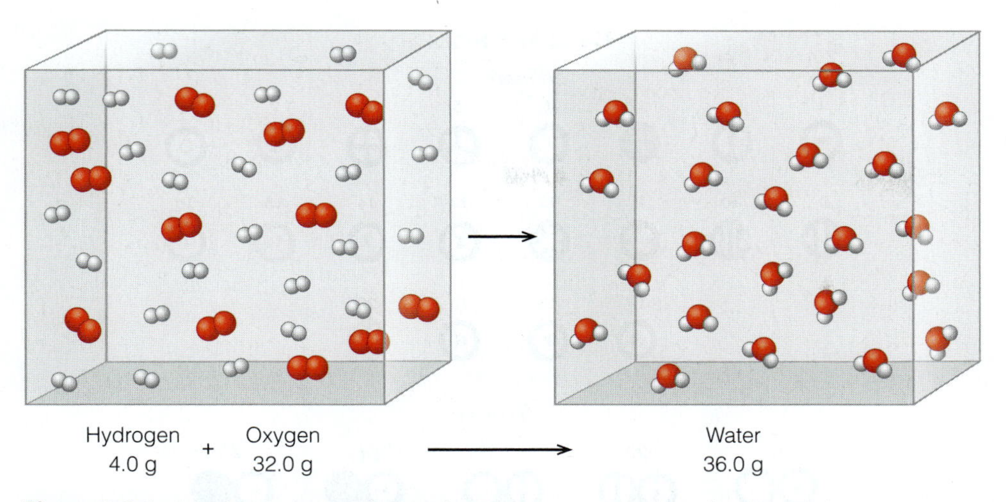

Figure 3.4 The law of conversation of mass tells us that matter is not created or destroyed in a reaction. The mass before the reaction is the same as the mass afterward.

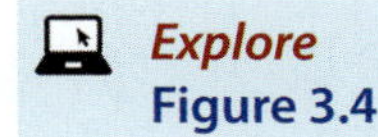

Example 3.1 Using the Law of Conservation of Mass

The bright light of a signal flare is caused by a chemical change in which magnesium combines with oxygen to produce a compound known as magnesium oxide. If 48.6 grams of magnesium combines with 32.0 grams of oxygen, how many grams of magnesium oxide will form?

We can represent this change as follows:

$$\underset{48.6\text{ g}}{\text{Magnesium}} + \underset{32.0\text{ g}}{\text{oxygen}} \xrightarrow{\text{change}} \text{magnesium oxide}$$

The law of conservation of mass states that the total amount of matter before and after a chemical change is the same. Together, the masses of magnesium and oxygen are 80.6 grams. Based on this law, 80.6 grams of magnesium oxide will form.

TRY IT

1. In an acetylene torch, acetylene gas reacts with oxygen to produce carbon dioxide and water (and a lot of heat). If 26 kg of acetylene is burned, 88 kg of carbon dioxide and 18 kg of water are produced. How many kilograms of oxygen gas are consumed in this reaction?

In 1808, John Dalton, an English schoolteacher and weather enthusiast, published a paper that articulated a framework for the modern **atomic theory** (**Figure 3.5**). Dalton's theory described several concepts that laid the foundation for our understanding of chemistry:

- Elements are made of tiny, indivisible particles called atoms.
- The atoms of each element are unique.
- Atoms can join together in whole-number ratios to form compounds.
- Atoms are unchanged in chemical reactions.

Atoms are not created or destroyed in chemical reactions.

To appreciate the impact of Dalton's ideas, let's consider an example. When charcoal burns (**Figure 3.6**), the charcoal—which is essentially the element carbon—reacts with oxygen in the air to form a new compound, carbon dioxide. From Dalton we know that carbon atoms are distinct from oxygen atoms. When they combine, they do so in a whole-number ratio. One carbon atom combines with

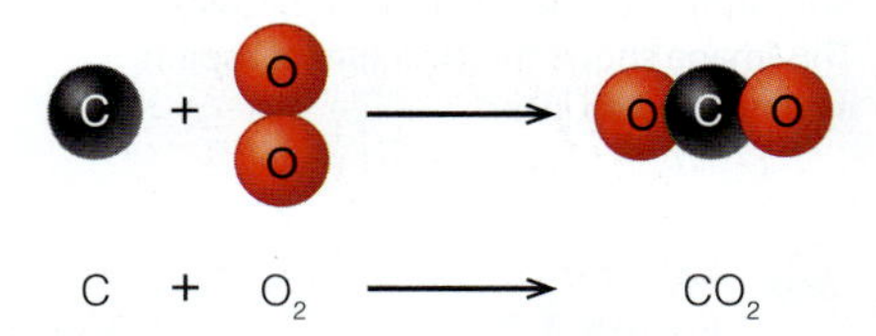

Figure 3.6 Carbon (charcoal) reacts with oxygen to form carbon dioxide.

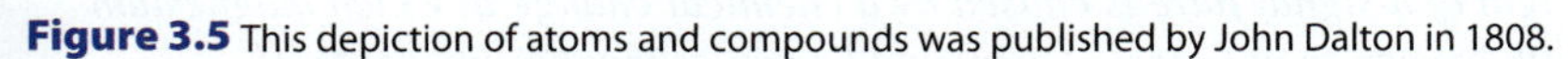

The Granger Collection

Figure 3.5 This depiction of atoms and compounds was published by John Dalton in 1808.

two oxygen atoms to produce one molecule of carbon dioxide. The atoms themselves are not changed, but the properties of carbon dioxide are different from the properties of carbon and oxygen.

From this brief history, here are three fundamental ideas that you should be sure to understand:

1. All matter is composed of atoms.
2. The atoms of each element have unique characteristics and properties.
3. In chemical reactions, atoms are not changed but combine in whole-number ratios to form compounds.

The study of chemistry springs from these important concepts.

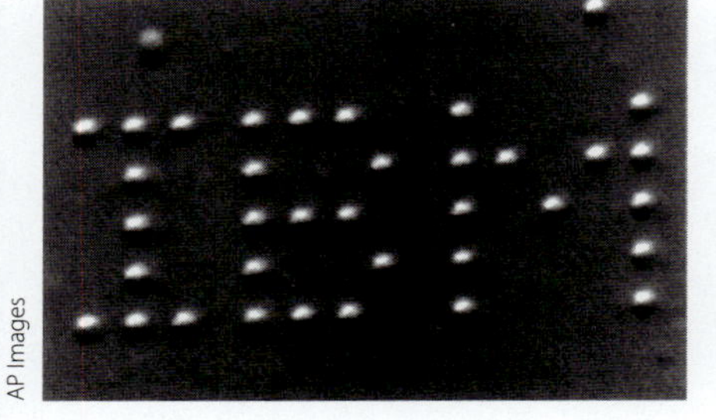

AP Images

Figure 3.7 IBM visualized and manipulated atoms using a technique called scanning tunneling microscopy.

Explore Figure 3.7

Can We See Atoms?

Today the ideas that Dalton developed are essential to our understanding of chemistry. But given the importance of atoms, is it possible to actually see them? Historically, the answer has been no. Atoms are too small to see with the naked eye or even with a microscope, and so we've always had to gather information about atoms indirectly. However, in recent decades, scientists have developed several techniques that make it possible to visualize atomic structure. A dramatic example took place in the early 1990s, when scientists at IBM labs used a technique called *scanning tunneling microscopy* not only to visualize atoms on a metal surface, but even to move them to form the letters *IBM* (**Figure 3.7**). Recently, IBM scientists took this technique one step further, producing a simple stop-motion movie made from individual atoms.

A more common technique for visualizing atoms is *X-ray crystallography*. In this technique, a solid is bombarded with a form of energy called X-rays. Based on the patterns the X-rays form as they pass through the material, scientists are able to visualize the arrangement of atoms in the solid. For example, **Figure 3.8** shows the structure of aspirin, a common painkiller. Because the properties of any substance depend on how its atoms are arranged, the ability to "see" atomic arrangements through X-ray crystallography helps scientists study diseases, develop new medicines, and design materials with unique new properties.

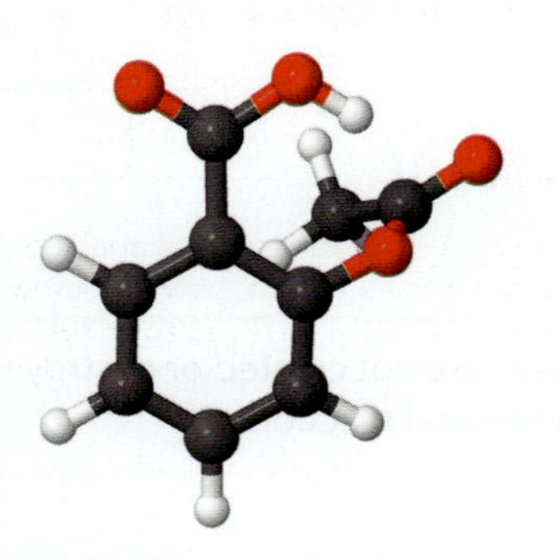

Figure 3.8 Using X-ray crystallography, scientists can construct images showing how atoms combine to form molecules. This image shows the structure of aspirin, a common pain killer.

TRY IT

2. One of the three chemical reactions shown here is drawn incorrectly, because it does not follow the law of conservation of mass. Which one is incorrect? Explain your answer.

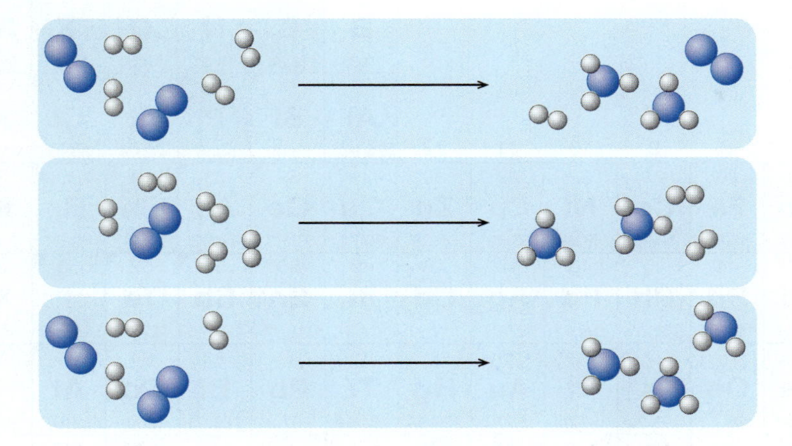

3. Consider the following statement: "Phosphorus oxide is formed when 1 phosphorus atom combines with 2.5 oxygen atoms." How does this statement conflict with the concepts of atomic theory? What is a better way to express the idea?

Check It

Watch Explanation

3.2 The Periodic Table of the Elements

By the late 1860s approximately 60 elements were known, and more were being discovered each year. A Russian scientist named Dmitri Mendeleev organized the known elements into a table based on their atomic masses. He also arranged elements with similar properties, such as chlorine, bromine, and iodine, into columns. By doing so he was able to predict the existence of several elements, such as germanium and gallium, that had not yet been discovered. Mendeleev's table became the basis for the modern **periodic table of the elements** (**Figure 3.9**), which organizes all known elements based on their properties.

The term *periodic* means "cyclical," and it indicates how the table is organized. You've probably used a different periodic table—a calendar (**Figure 3.10**). On a calendar, each horizontal row represents an entire week and covers the whole spectrum of activities that are part of your normal routine. For example, on Sundays I usually attend church and enjoy a quiet afternoon with my family. On Monday my work week begins. I have certain duties that must be done each day: a lab on Tuesday, a meeting on Thursday, and so on. Friday night is family night, and on Saturdays I do yard work and hang out with my kids. From Sunday to Saturday, I've covered an entire spectrum of activity. The next Sunday, the cycle starts again. Although no week is exactly the same, certain days tend to be similar.

Sun	Mon	Tue	Wed	Thu	Fri	Sat
	1	2	3	4	5	6
7	8	9	10	11	12	13
14	15	16	17	18	19	20
21	22	23	24	25	26	27
28	29	30	31			

Figure 3.10 A calendar is organized like the periodic table. A horizontal row (red arrow) represents a full cycle of days, and a column (blue arrow) groups days that are similar.

The periodic table is organized in much the same way. As we move from left to right across a horizontal *row* of the periodic table, we cover the entire spectrum of chemical properties. The rows in the periodic table are called **periods**.

On a calendar, particular days (such as Monday) are arranged into columns. These days have similar characteristics. Likewise, atoms that are in the same column have similar properties. The columns of the periodic table are called **groups**, or *families*.

Elements in the same column of the periodic table have similar properties.

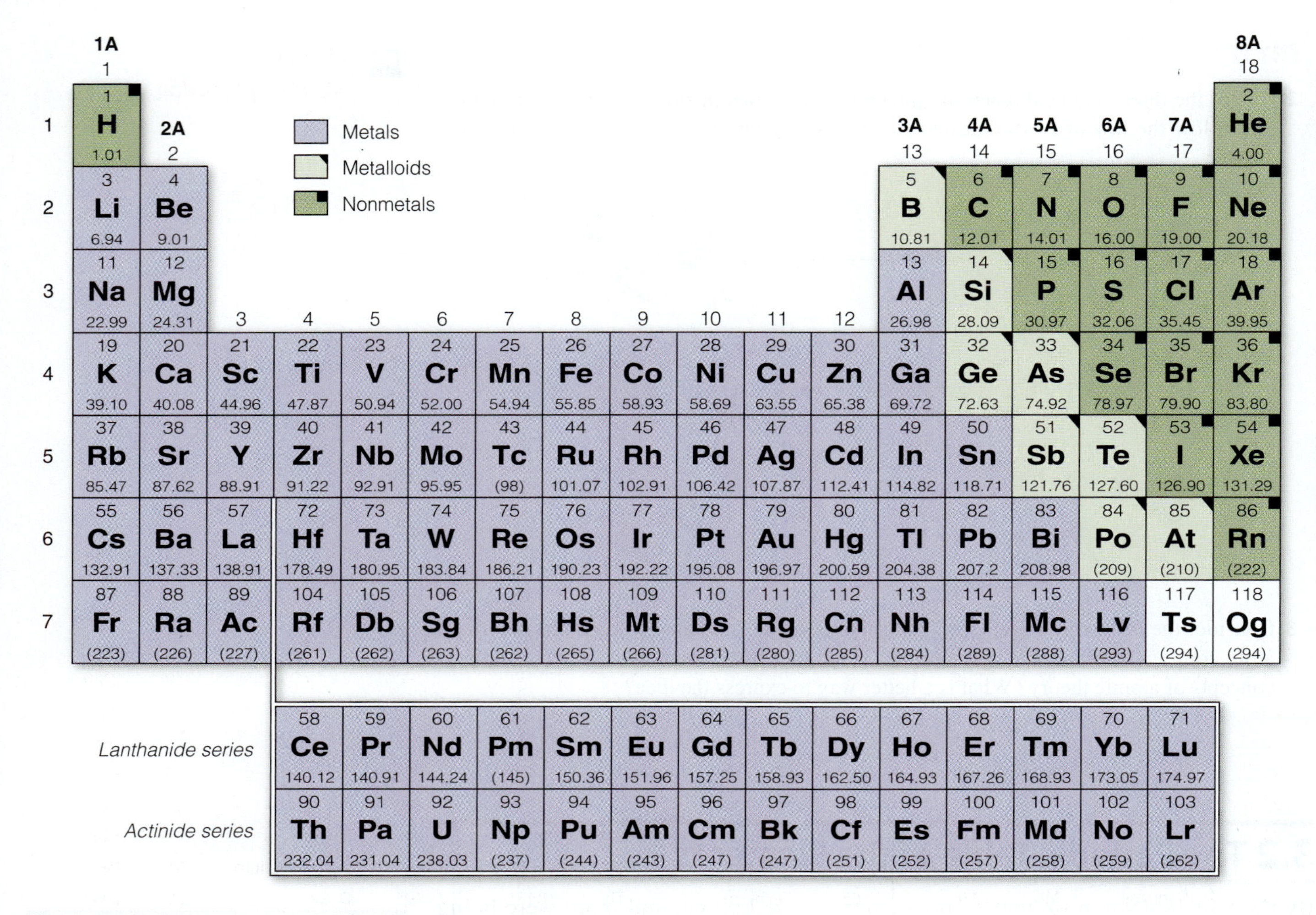

Figure 3.9 The modern periodic table contains 118 known elements.

> **Co *versus* CO**
> Each of the atomic symbols has either a capitalized one-letter symbol or a two-letter symbol whose second letter is lowercase. Be careful with capitalization! For example, Co refers to the elemental metal cobalt, but CO is a compound composed of carbon and oxygen—the toxic gas carbon monoxide.

The periodic table uses an accepted set of one- and two-letter abbreviations for different elements. Chemists routinely use these chemical symbols to describe molecules, compounds, and chemical reactions. To understand these chemical descriptions, it is essential for you to know the symbols and names of the most common or important elements. The elements and their symbols are listed in **Table 3.1**. Most of the symbols are based on the English words; but some—such as iron (Fe), gold (Au), silver (Ag), sodium (Na), and potassium (K)—are derived from the Latin words for these elements.

Regions of the Periodic Table

The blocks of elements on the left and right sides of the periodic table are the **main-group elements** (**Figure 3.11**). We can predict many properties of these elements based on their location on the periodic table.

> The left- and right-hand blocks of the periodic table contain the main-group elements. ■

The elements in the middle of the table are called the *transition elements*. At the bottom of the periodic table are two additional rows, called the *inner transition elements*. Although these elements actually belong in the last two rows of the table, they are typically shown below the table to make it easier to read.

The periodic table contains seven periods (horizontal rows), which we number from the top of the table. There are two ways to identify the columns in the periodic table. In the modern system, the numbers 1–18 designate the columns that contain the main-group and transition elements. An older but very useful system numbers the main-group columns as 1A–8A. We will use both numbering systems in the chapters to come.

TABLE 3.1 The Elements

Atomic Number	Atom	Symbol	Discovery Year	Atomic Number	Atom	Symbol	Discovery Year	Atomic Number	Atom	Symbol	Discovery Year
1	Hydrogen	H	1776	41	Niobium	Nb	1801	81	Thallium	Tl	1861
2	Helium	He	1895	42	Molybdenum	Mo	1781	82	Lead	Pb	ancient
3	Lithium	Li	1817	43	Technetium	Tc	1937	83	Bismuth	Bi	ancient
4	Beryllium	Be	1797	44	Ruthenium	Ru	1844	84	Polonium	Po	1898
5	Boron	B	1808	45	Rhodium	Rh	1803	85	Astatine	At	1940
6	Carbon	C	ancient	46	Palladium	Pd	1803	86	Radon	Rn	1900
7	Nitrogen	N	1772	47	Silver	Ag	ancient	87	Francium	Fr	1939
8	Oxygen	O	1774	48	Cadmium	Cd	1817	88	Radium	Ra	1898
9	Fluorine	F	1886	49	Indium	In	1863	89	Actinium	Ac	1899
10	Neon	Ne	1898	50	Tin	Sn	ancient	90	Thorium	Th	1829
11	Sodium	Na	1807	51	Antimony	Sb	ancient	91	Protactinium	Pa	1913
12	Magnesium	Mg	1755	52	Tellurium	Te	1783	92	Uranium	U	1789
13	Aluminum	Al	1825	53	Iodine	I	1811	93	Neptunium	Np	1940
14	Silicon	Si	1824	54	Xenon	Xe	1898	94	Plutonium	Pu	1940
15	Phosphorus	P	1669	55	Cesium	Cs	1860	95	Americium	Am	1944
16	Sulfur	S	ancient	56	Barium	Ba	1808	96	Curium	Cm	1944
17	Chlorine	Cl	1774	57	Lanthanum	La	1839	97	Berkelium	Bk	1949
18	Argon	Ar	1894	58	Cerium	Ce	1803	98	Californium	Cf	1950
19	Potassium	K	1807	59	Praseodymium	Pr	1885	99	Einsteinium	Es	1952
20	Calcium	Ca	1808	60	Neodymium	Nd	1885	100	Fermium	Fm	1952
21	Scandium	Sc	1879	61	Promethium	Pm	1945	101	Mendelevium	Md	1955
22	Titanium	Ti	1791	62	Samarium	Sm	1879	102	Nobelium	No	1958
23	Vanadium	V	1830	63	Europium	Eu	1901	103	Lawrencium	Lr	1961
24	Chromium	Cr	1797	64	Gadolinium	Gd	1880	104	Rutherfordium	Rf	1964
25	Manganese	Mn	1774	65	Terbium	Tb	1843	105	Dubnium	Db	1967
26	Iron	Fe	ancient	66	Dysprosium	Dy	1886	106	Seaborgium	Sg	1974
27	Cobalt	Co	1735	67	Holmium	Ho	1867	107	Bohrium	Bh	1981
28	Nickel	Ni	1751	68	Erbium	Er	1842	108	Hassium	Hs	1984
29	Copper	Cu	ancient	69	Thulium	Tm	1879	109	Meitnerium	Mt	1982
30	Zinc	Zn	ancient	70	Ytterbium	Yb	1878	110	Darmstadtium	Ds	1987
31	Gallium	Ga	1875	71	Lutetium	Lu	1907	111	Roentgenium	Rg	1994
32	Germanium	Ge	1886	72	Hafnium	Hf	1923	112	Copernicum	Cn	1996
33	Arsenic	As	ancient	73	Tantalum	Ta	1802	113	Nihonium	Nh	2003
34	Selenium	Se	1817	74	Tungsten	W	1783	114	Flerovium	Fl	1998
35	Bromine	Br	1826	75	Rhenium	Re	1925	115	Moscovium	Mc	2003
36	Krypton	Kr	1898	76	Osmium	Os	1803	116	Livermorium	Lv	1999
37	Rubidium	Rb	1861	77	Iridium	Ir	1803	117	Tennessine	Ts	2010
38	Strontium	Sr	1790	78	Platinum	Pt	1735	118	Oganesson	Og	2005
39	Yttrium	Y	1794	79	Gold	Au	ancient				
40	Zirconium	Zr	1789	80	Mercury	Hg	ancient				

Metals, Nonmetals, and Metalloids

The elements on the left-hand side of the periodic table are **metals**. Metals are usually solid at room temperature. They can be molded into different shapes, and they conduct heat and electricity. The transition elements in columns 3–12 of the periodic table are often called the **transition metals**. These tend to be harder and less reactive than the metals of columns 1–2. The transition metals include common building materials like iron and copper as well as precious metals such as gold and silver (**Figure 3.12**).

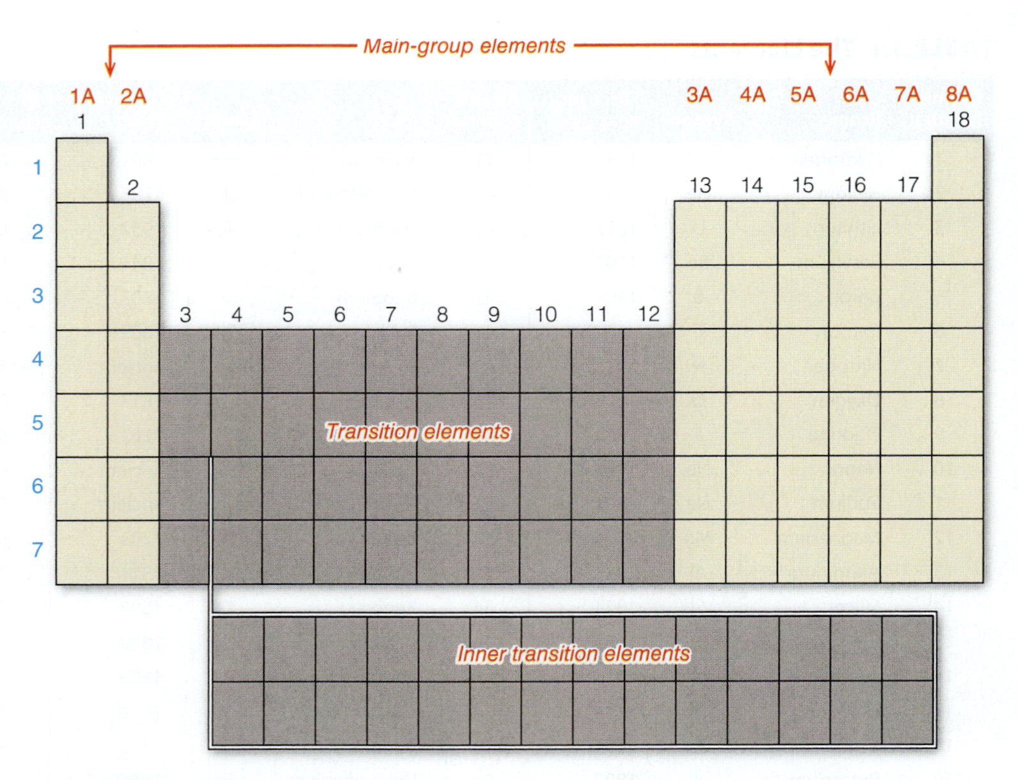

Figure 3.11 The rows of the periodic table are numbered 1 through 7. Two numbering systems identify the columns in the periodic table. An older system (shown in red) numbers the main-group columns as 1A–8A. The modern system numbers all 18 columns that include the main-group and transition elements.

The inner transition elements are also metals. The first row of these elements is the *lanthanide series.* This row contains heavier, naturally occurring metals called the *rare earth metals.* The second row is called the *actinide series.* The first few elements in this series (up to uranium) are naturally occurring. The elements that occur after uranium are mostly human-made (we'll discuss this in Chapter 16), although scientists have identified trace amounts of a few of these elements in nature.

The left side of the periodic table contains the metals. The upper right side contains the nonmetals.

The **nonmetals** are found on the upper right side of the periodic table. The physical properties of the nonmetals vary widely. For example, carbon and sulfur are both solids; bromine is a liquid; and nitrogen, oxygen, and fluorine are all gases. Nonmetals form a rich variety of compounds ranging from simple gases to plastics to biomolecules. These elements form the atomic basis for life (**Figure 3.13**).

a
© amnachphoto/depositphotos.com

b
© scanrail/depositphotos.com

c
sumire8/Shutterstock

d
Daniel D Malone/Shutterstock

Figure 3.12 The transition metals include (a) iron; (b) copper; (c) gold; and (d) silver.

Figure 3.13 Nonmetals such as carbon, nitrogen, oxygen, and hydrogen are the atomic building blocks for plant and animal life.

Figure 3.14 The metalloid element silicon is a critical component in computer processors.

Between the metals and nonmetals, there is a stairstep pattern of elements called the **metalloids**. Metalloids are *semiconductors*—they conduct electricity, but not as efficiently as metals do. The metalloids, especially silicon, are essential components of modern electronics (**Figure 3.14**).

Groups (Families) of Elements

Elements in the same column on the periodic table tend to have similar properties. Let's briefly examine the properties of several important families of main-group elements.

Alkali metals lie on the far left of the periodic table, in column 1A. These elements, such as lithium, sodium, and potassium, are very soft metals that react violently with air or even with moisture (**Figure 3.15**).

Just to the right of the alkali metals, in column 2A, are the **alkaline earth metals** (**Figure 3.16**). Magnesium and calcium are the most common elements in this group. These elements are also reactive, but less violently so than the alkali metals. They react slowly with water but burn brightly when combined with oxygen.

The elements in column 7A are the **halogens** (**Figure 3.17**). These elements all exist as diatomic ("two-atom") molecules in their elemental form. The halogens react quickly with metals and other nonmetals to form many different compounds. Substances as diverse as bleach, Teflon®, fire retardants, antiseptics, and table salt include halogens in their composition (**Figure 3.18**).

Explore Figure 3.15

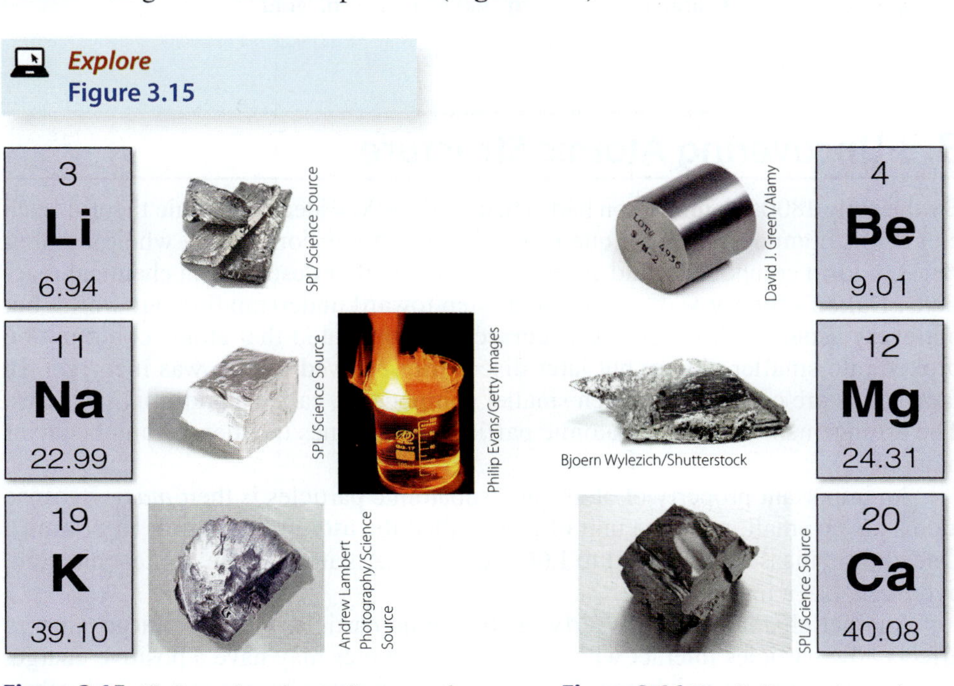

Figure 3.15 Alkali metals such as sodium are soft metals that react violently with water.

Figure 3.16 The alkaline earth metals include beryllium, magnesium, and calcium.

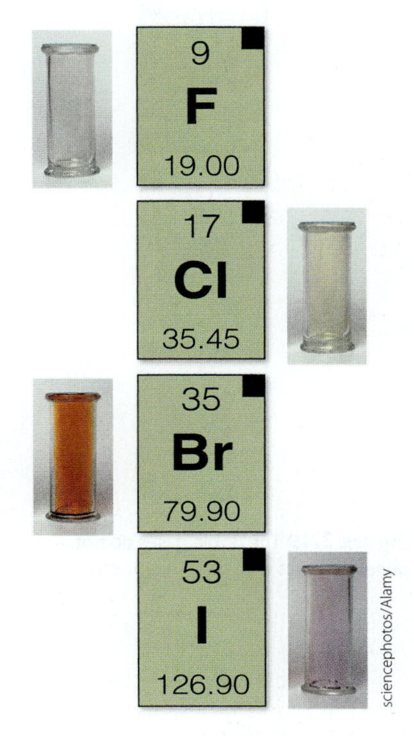

Figure 3.17 The halogens are reactive nonmetals.

Kevin Revell

Figure 3.18 Halogens are used in a wide variety of compounds and applications.

Finally, the elements in column 8A are the noble gases (**Figure 3.19**). This family is much different from the other elements. The noble gases are very stable and generally do not react with other elements to form compounds. There are no known compounds involving helium or neon and only a handful involving the heavier atoms in this family. These elements are all gases at room temperature.

The noble gases are in the rightmost column of the periodic table.

Example 3.2 Navigating the Periodic Table

Identify each of the following elements as a metal, a nonmetal, or a metalloid:

a. *a transition element*
b. *the lightest element in the halogen family*
c. *the element in row 4, column 14*
d. *an element in group 8A*

To solve these problems, we can refer to the periodic table in Figure 3.9.

a. The transition elements all have metallic properties. They are commonly called the transition metals.

b. The lightest element in the halogen family is fluorine (F). This element is a nonmetal.

c. The element in row 4, column 14 is germanium (Ge). This element is a metalloid.

d. The elements in group 8A are the noble gases. These are nonmetals.

Check It
Watch Explanation

TRY IT

4. Identify the elements located in each position on the periodic table:
   a. row 3, column 14 **b.** row 2, column 4A **c.** row 6, column 9
5. Based on its location in the periodic table, which of these elements would behave most similarly to calcium?
   a. potassium **b.** magnesium **c.** scandium **d.** plutonium
6. Using the periodic table in Figure 3.9, identify each element as a main-group metal, a transition metal, an inner-transition metal, a metalloid, or a nonmetal.
   a. sodium **b.** sulfur **c.** copper **d.** samarium
   e. silicon **f.** argon **g.** oxygen **h.** gold

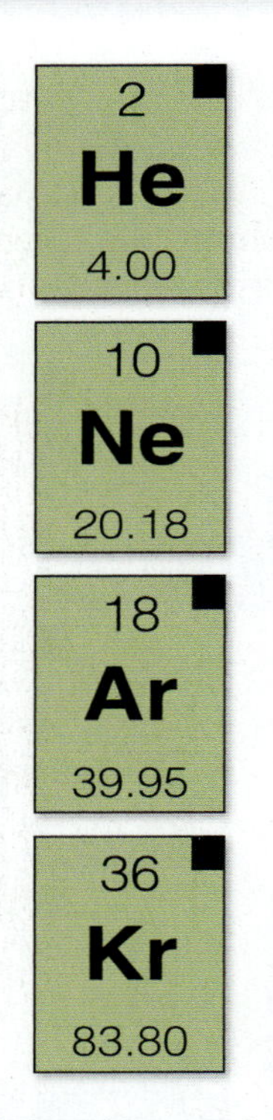

Figure 3.19 The noble gases do not generally react with other elements.

3.3 Uncovering Atomic Structure

By the early 1800s, John Dalton had articulated the key ideas of atomic theory: Each element is composed of a unique type of atom; atoms combine in whole-number ratios to form compounds; and atoms are not created or destroyed in chemical reactions. Dalton's theory was an enormous step toward understanding our world, but not every aspect of his model was correct. Dalton stated that atoms could not be broken into smaller pieces, but later discoveries showed Dalton was incorrect. In fact, atoms are composed of even smaller components, called subatomic particles. The arrangement of these subatomic particles determines the identity and behavior of atoms.

An important property of atoms and subatomic particles is their *mass*. Because atoms are so small, we use a unit of mass called the atomic mass unit (u or amu). One atomic mass unit is equal to 1.66×10^{-27} kg. An atom of the lightest element, hydrogen, has a mass of 1.0 u.

$1\ \text{u} = 1.66 \times 10^{-27}$ kg

Another characteristic property of subatomic particles is their charge. Charge affects how particles interact with each other. Particles may have a positive charge, a negative charge, or no charge. Particles with opposite charges attract each other; those with like charges repel each other. We see the effects of charged particles through the familiar phenomenon called *electricity* or *electrical energy*, a form of energy that involves the motion of charged particles.

The Discovery of Charged Particles

In 1800 the Italian scientist Alessandro Volta built the first battery (more formally called an *electrochemical cell*)—a device capable of moving electrically charged particles (**Figure 3.20**). Volta built his device by placing alternating plates of zinc and copper on either side of pieces of cardboard soaked in sulfuric acid. When he connected the two sides with a metal wire, the result was an *electrical current*, the flow of charged particles from one side of the battery to another.

Electricity involves the flow of electrons.

Explore **Figure 3.20**

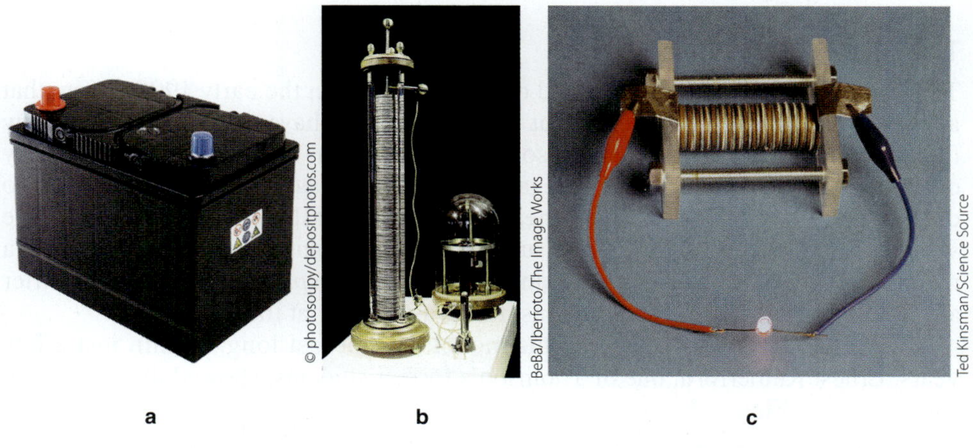

Figure 3.20 (a) A battery, like the car battery shown here, produces a flow of charged particles called an electric current. (b) Alessandro Volta built the first battery using plates of zinc and copper, separated by cardboard soaked in acid. (c) A battery like Volta's can be made at home using aluminum foil, zinc washers, copper pennies, and paper or cardboard that has been soaked in saltwater or vinegar.

The invention of the battery helped scientists discover many of the elements in a few short years. Much of the information on Mendeleev's periodic table was a direct result of the battery.

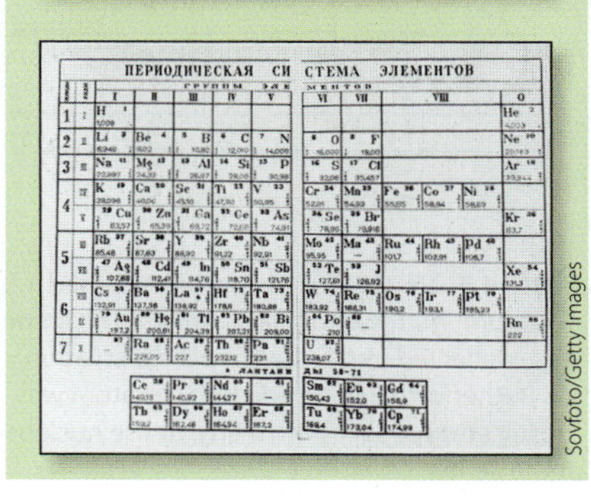

Volta's battery was a critical milestone in the development of science. Using the controlled electrical energy of the battery, scientists soon discovered they could use electrical energy to separate some compounds, such as water, into their elements (**Figure 3.21**). In addition, scientists began to identify charged particles and to measure how charged particles interact with each other (**Figure 3.22**)

By the end of the nineteenth century, scientists had discovered that elements could produce both positive and negative charges. In 1897 J. J. Thomson, an English scientist, discovered the **electron**—a tiny, negatively charged particle that was 2,000 times smaller than the lightest atom. Thomson showed that electricity involves the flow of electrons.

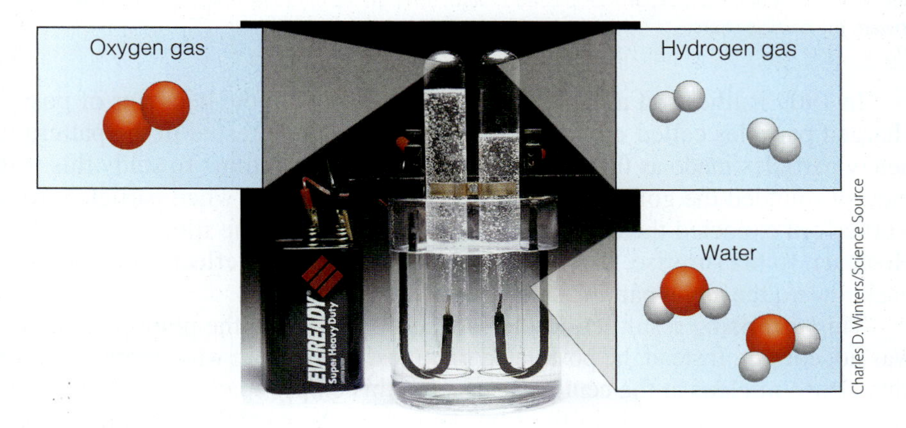

Figure 3.21 The electrical energy from a battery can split water into the elements hydrogen and oxygen.

Explore
Figure 3.22

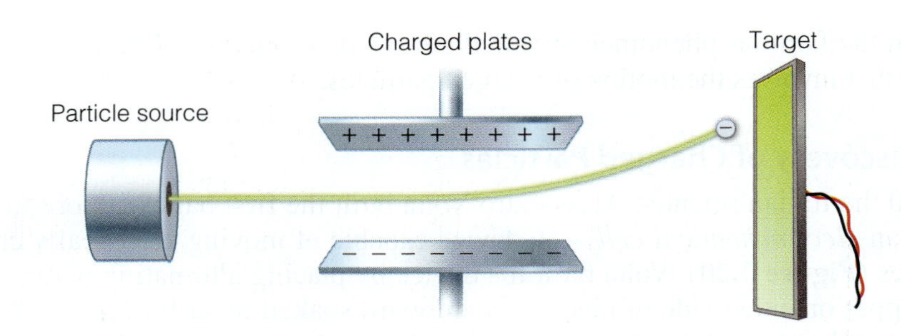

Figure 3.22 Scientists like J. J. Thomson studied charged particles by passing them between two charged plates and then measuring where they hit a target on the other side. This technique became the basis for early television screens.

The Discovery of the Nucleus

Thanks to the work of Thomson and others, scientists in the early 1900s knew that atoms contain charged particles. But they still did not have a clear picture of how these particles are arranged. Thomson knew that atoms contain small, negatively charged electrons. To balance the negative charge, he assumed that atoms must also contain positive charges. He hypothesized that the small, negative electrons were spread throughout a positive atomic substance the way blueberries are scattered in a muffin (**Figure 3.23**). This model came to be called the *plum pudding model*, after a dessert (similar to blueberry muffins) that was popular at the time.

However, like good muffins, this model did not last long. Within just a few years, Ernest Rutherford, one of Thomson's former students, showed that the model was incorrect (**Figure 3.24**).

Figure 3.23 The "plum pudding model" envisioned atoms as negative electrons spread throughout a positively charged material, like blueberries in a muffin.

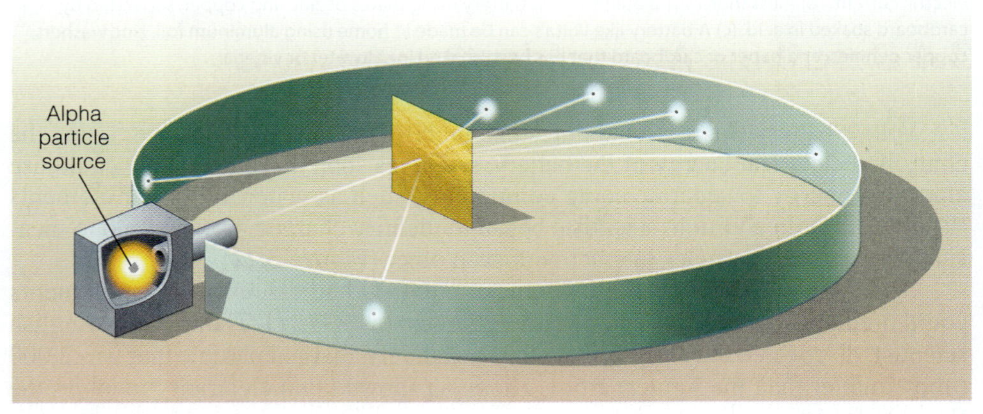

Figure 3.24 In the gold foil experiment, Rutherford and his students fired alpha particles into a gold film. While most of the alpha particles passed straight through the film, a few were deflected back toward the particle source.

In 1909 Rutherford and his students were studying the behavior of positively charged particles called *alpha particles*. They were interested in the pattern these heavy particles made as they passed through a thin gold film. To study this pattern, they surrounded the gold film with a material that glowed when particles struck it. Rutherford expected the alpha particles to pass through the film and hit behind it. However, to his surprise, a small number of the particles reflected off the film and back toward the alpha particle source.

This surprising result led Rutherford to conclude that the plum pudding model was not correct. Instead, he postulated that most of the atom was empty space with a very dense **nucleus** at the center. Because of this empty space, most alpha particles passed through the atoms in the film. However, the particles that hit the nucleus deflected back toward the source (**Figure 3.25**).

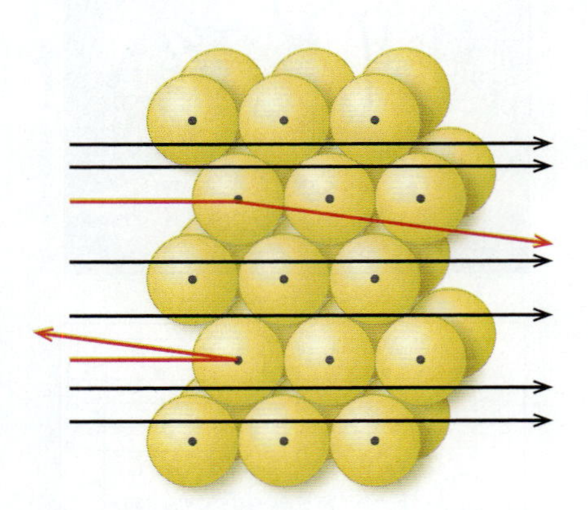

Figure 3.25 Most of the particles passed right through the film (black arrows), but some were deflected (red arrows). This observation led Rutherford to conclude that the atom was mostly empty space with a tiny, dense nucleus at the center.

We now know that Rutherford was correct. Atoms are mostly empty space. We often draw atoms as a nucleus surrounded by orbiting electrons; however, these pictures are never drawn to scale. Although the nucleus contains almost all of the mass of the atom, it is incredibly dense, packing the mass into a very tiny volume. The volume of the nucleus compared to the volume of the atom has been compared to the size of an insect inside a football stadium.

The nucleus contains two particles, **protons** and **neutrons**. Protons have a mass of about 1 atomic mass unit (u) and a charge of +1. Neutrons also have a mass of about 1 u, but they have no charge (**Table 3.2**).

TABLE 3.2 Subatomic Particles

Particle	Mass (u)	Charge	Location
Proton	1.0073	+1	Nucleus
Neutron	1.0087	–	Nucleus
Electron	0.0005	−1	Electron cloud

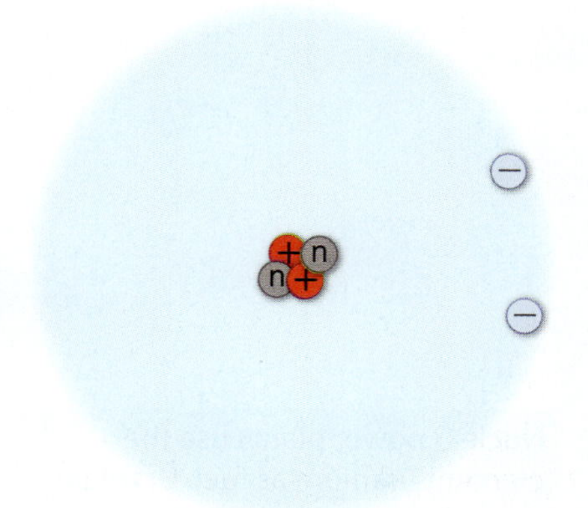

Figure 3.26 Protons and neutrons make up the nucleus of an atom, such as the helium atom shown here. The electrons occupy the space around the nucleus.

The rest of the atom is occupied by the tiny, negatively charged electrons. Electrons have a charge of −1. The space around the nucleus is called the **electron cloud** (**Figure 3.26**). The electrons are much lighter than the nuclear particles. Their mass is only about 1/2000 of the mass of the proton or neutron.

As an example, consider the helium atom in Figure 3.26. A helium nucleus contains four particles—two protons and two neutrons. Around the nucleus, two electrons occupy the electron cloud. To be electrically neutral, the number of protons and electrons in an atom must be the same.

The nucleus contains nearly all the mass of the atom. ■

3.4 Describing Atoms: Identity and Mass

Atomic Number and Mass Number

So what makes a hydrogen atom a hydrogen atom? In the previous section, we saw that atoms contain three subatomic particles—protons, neutrons, and electrons. But which of these particles determines the identity of the atom?

The answer to this question is that *the number of protons determines the identity of the atom*. As we will see, hydrogen atoms can have different numbers of neutrons. Similarly, hydrogen can gain or lose electrons. What defines hydrogen is the number of protons in its nucleus.

The number of protons defines the atom. ■

Because the number of protons determines the identity of an atom, we call this number the **atomic number**. The atomic number is the integer value normally seen above the atomic symbol on the periodic table (**Figure 3.27**). Because neutral atoms contain the same number of protons and electrons, the atomic number also tells us the number of electrons in a neutral atom.

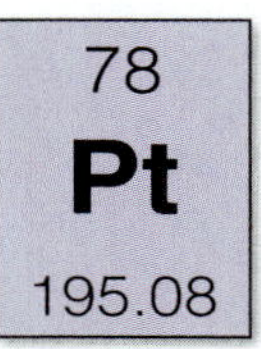

Figure 3.27 The atomic number is found above the symbol on the periodic table. For this element, platinum, the atomic number is 78.

In addition to the atomic number, we also commonly describe atoms by their mass number. The **mass number** is *the sum of the number of protons and the number of neutrons in an atom*. Note that both the atomic number and the mass number are integers. Typically, the periodic table does not show the mass number.

Sometimes, atoms have the same atomic numbers but different mass numbers. For example, the hydrogen atom has three forms (**Figure 3.28**). By definition, every hydrogen atom has one proton. Most hydrogen atoms contain no neutrons, so they have a mass number of one. However, some hydrogen atoms have one neutron and therefore a mass number of two. A tiny percentage of hydrogen atoms even have two neutrons, and therefore a mass number of three. Atoms that have the same atomic number but different mass numbers are called **isotopes**. The three isotopes of hydrogen are referred to as protium, deuterium, and tritium.

The atomic number is the number of protons in an atom. The mass number is the number of protons plus the number of neutrons. ■

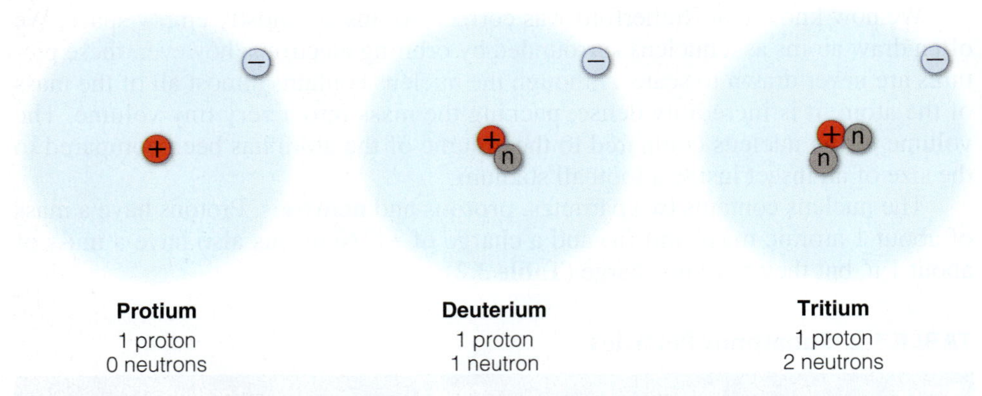

Figure 3.28 Hydrogen exists as three isotopes. The lightest of these, protium, is the most common.

Nuclear power plants use the element uranium as fuel. Uranium exists as two common isotopes: ^{235}U and ^{238}U.

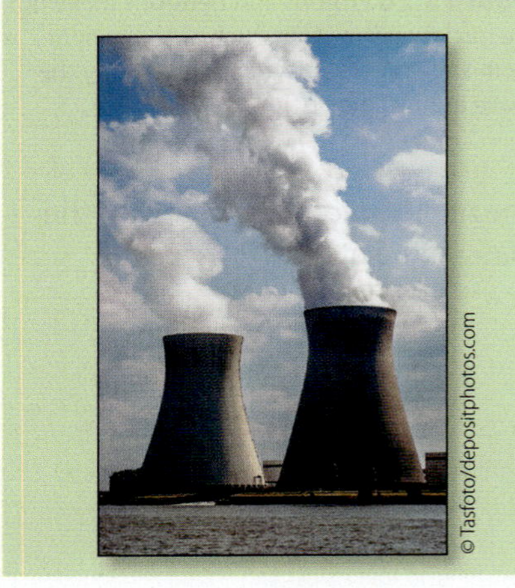

If we want to show the atomic number and mass number of an atom, we typically write these numbers alongside the chemical symbols. We place atomic numbers at the lower left-hand side of the atomic symbol and mass numbers at the upper left-hand side of the atomic symbol. For example, chlorine has two common isotopes. The first isotope has 17 protons and 18 neutrons; the second isotope has 17 protons and 20 neutrons. The two symbols may therefore be written like this:

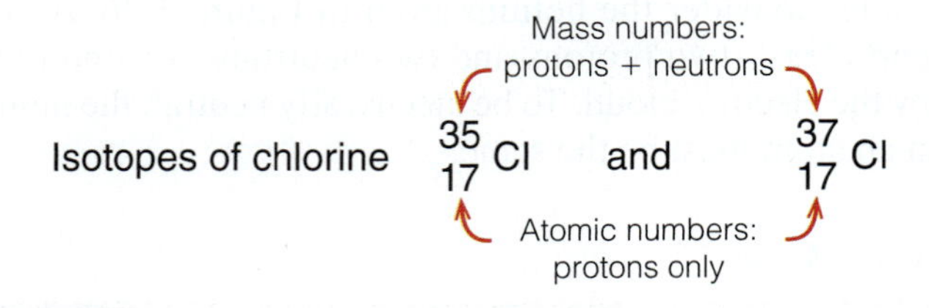

Example 3.3 Writing Atomic Symbols with Atomic Number and Mass Number

Uranium has two common isotopes: The lighter isotope has a mass number of 235; the heavier isotope has a mass number of 238. Write atomic symbols for both of these isotopes, including the atomic number and mass number.

To write these symbols, we must identify the symbol and atomic number. The symbol for uranium is U, and the atomic number is 92. We write these symbols with the mass number in the upper left and the atomic number in the lower left:

$$^{235}_{92}U \quad \text{and} \quad ^{238}_{92}U$$

When writing chemical symbols, the atomic number and the chemical symbol are redundant: Atomic number 92 is always uranium, and uranium is always atomic number 92. While it is sometimes helpful to write both the atomic and mass numbers, scientists often write just the mass number and symbol. For example, using just the mass number and symbol, we can write the two isotopes of uranium as ^{235}U (typically read as "U-235") and as ^{238}U (read as "U-238").

Example 3.4 Determining the Nuclear Structure from the Atomic Number and Mass Number

One isotope of cadmium has a mass number of 114. How many protons and neutrons are in a nucleus of ^{114}Cd*?*

To solve this problem, we need to identify the atomic number. Using a periodic table, we determine that cadmium is atomic number 48. This means that cadmium has 48 protons. To find the number of neutrons, we subtract the atomic number from the mass number:

$$\text{Mass number} - \text{Atomic number} = \text{Number of neutrons}$$
$$114 - 48 = 66$$

A ^{114}Cd atom contains 48 protons and 66 neutrons. Because the number of protons and electrons are equal in a neutral atom, we also know that this atom contains 48 electrons.

TRY IT

7. Write the chemical symbol for each of these atoms, including the mass number and the atomic number.

a. an atom of gold, containing 79 protons and 118 neutrons

b. an atom of mercury, containing 80 protons and 122 neutrons

8. The most common isotope of silicon has a mass number of 28. How many neutrons are in this isotope?

9. Find the number of protons and neutrons in each of the following:

a. an atom with atomic number 15 and mass number 31

b. an isotope of copper that has 34 neutrons

c. an isotope of carbon with an equal number of protons and neutrons

Check It
Watch Explanation

Average Atomic Mass

In the previous section, we noted that the mass number is not usually included on the periodic table. Because the periodic table is a quick-reference guide, it contains the information that we most often need. Although the mass number is important, chemists more often use another value, called the *average atomic mass*.

Most periodic tables show each element's average atomic mass, not its mass number.

The average atomic mass is the *weighted average* of the different isotopes. To understand what a weighted average means, consider a simple example involving poker chips (**Figure 3.29**). Suppose we have a mixture of chips whose values are either \$1 or \$2. If we have equal numbers of each chip, then our average is \$1.5—right in

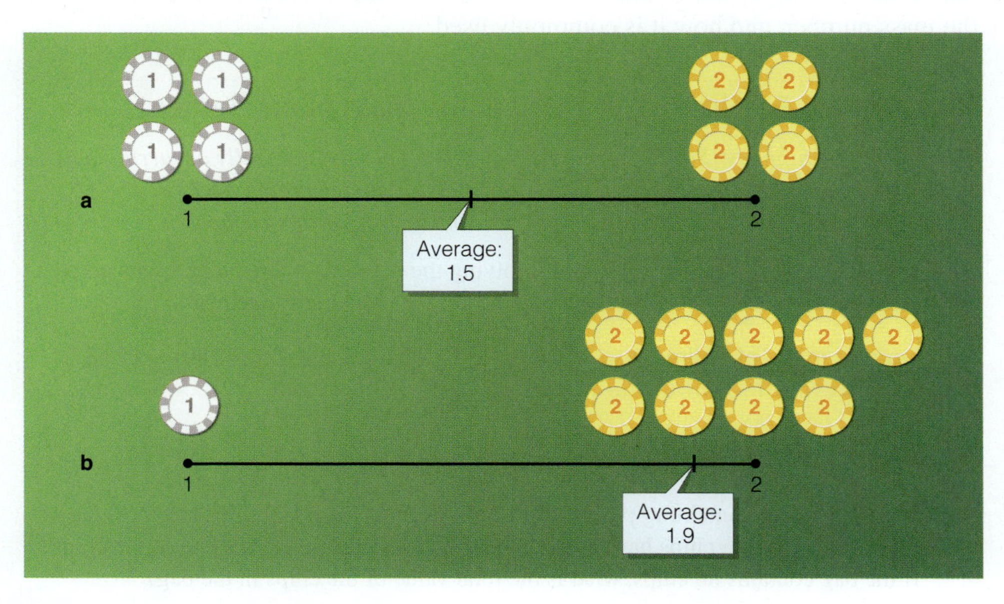

Figure 3.29 A weighted average considers how many units of each value are present.

the middle of the two values. However, if we have more of the \$2 chips, then our average is closer to \$2 than to \$1. Notice that the values of the chips are integer values, but the average value is a decimal value somewhere between the two.

For a small number of chips, we can calculate the average simply by adding up the values of each chip and dividing by the total number of chips. However, if the number of chips is very large, this approach is not practical. An easier way of doing this calculation is to use the *percentage* of each type of chip present.

For example, imagine we have a dump truck full of poker chips. There are too many chips to count, but we know that 10% of the chips are \$1 chips, and 90% of the chips are \$2 chips. To find the average value of the chips, we do the following calculation:

$$\text{Weighted average value} = (\text{fraction A} \times \text{value A}) + (\text{fraction B} \times \text{value B})$$

where fraction A and fraction B are the percentages of chips that are type A and B, expressed as a decimal:

$$\text{Average value of chips} = (0.10 \times \$1) + (0.90 \times \$2) = \$1.9$$

We use this same approach to calculate the average mass of isotopes. For example, carbon exists primarily as two isotopes: ^{12}C has a mass of 12.000 u, and 98.93% of all naturally occurring carbon is carbon-12. ^{13}C has a mass of 13.0034 u, and 1.07% of all naturally occurring carbon is carbon-13. What is the average atomic mass for carbon?

Before working this question mathematically, stop and think about it: Out of 100 carbons, 99 have a mass of 12 u, and only 1 has a mass of 13 u. Thus the average mass is much closer to 12 than to 13. Now let's calculate the answer. We find the average atomic mass by multiplying the mass of each isotope by its percent abundance (expressed as a decimal) and adding these values together:

$$\text{Average mass of carbon} = \underbrace{(12.0000\ \text{u})}_{\text{Mass of 1st isotope}}\underbrace{(0.9893)}_{\text{abundance}} + \underbrace{(13.0034\ \text{u})}_{\text{Mass of 2nd isotope}}\underbrace{(0.0107)}_{\text{abundance}} = 12.01\ \text{u}$$

Remember, *u* stands for "atomic mass units."

The average atomic masses are normally included on the periodic table, just beneath the atomic symbol. We don't often have to calculate the average atomic mass, but it is important to understand what this value means, how it differs from the mass number, and how it is commonly used.

Example 3.5 Calculating the Average Mass of an Element

The element antimony (Sb) exists mainly as two isotopes. Of naturally occurring antimony, 57.21% is ^{121}Sb, which has a mass of 120.9 u. The remaining 42.79% is ^{123}Sb, whose mass is 122.9 u. What is the average atomic mass for this element?

We find the average atomic mass by multiplying the mass of each isotope by its percent abundance, expressed as a decimal, and then adding these values together.

$$\text{Average mass of antimony} = (120.9\ \text{u})(0.5721) + (122.9\ \text{u})(0.4279) = 121.8\ \text{u}$$

Check It
Watch Explanation

TRY IT

10. You have a bag containing poker chips. In the bag, 25% of the chips have a value of \$1, and 75% of the chips have a value of \$3. What is the average value of the chips? If the bag contains 80 chips, what is the total value of the chips in the bag?

11. Indium has two isotopes, ^{113}In and ^{115}In. The average atomic mass of indium is 114.8 u. Which isotope of indium is more common?

12. Silicon exists as a mixture of three isotopes, shown in the table below. What is the average atomic mass for silicon? Check your answer by comparing it to the average atomic mass on the periodic table (you might find slight differences in rounding).

Isotope	Mass (u)	Abundance
^{28}Si	27.9769	92.2297
^{29}Si	28.9765	4.6832
^{30}Si	29.9738	3.0872

3.5 Electrons—A Preview

Rutherford's gold foil experiment demonstrated that the nucleus, composed of protons and neutrons, occupies a very tiny, very dense space at the center of the atom. The remaining volume of the atom is occupied by electrons. As we'll see in the chapters ahead, the arrangement of electrons around the nucleus determines how atoms combine to form compounds. In Chapter 4, we will explore electronic structure within a single atom. In Chapter 5, we will see how atoms combine to form compounds by gaining, losing, or sharing electrons. For now, let's complete our survey of atomic structure with a broad overview of electron behavior and then preview some of the concepts that lie ahead.

The Bohr Model and the Quantum Model

In the early twentieth century, Ernest Rutherford and Niels Bohr introduced a new model for atomic structure, commonly called the **Bohr model**. This model treated the atom like a tiny solar system, with the nucleus at the center and the electrons orbiting the nucleus, much like the different planets orbit the Sun (**Figure 3.30**).

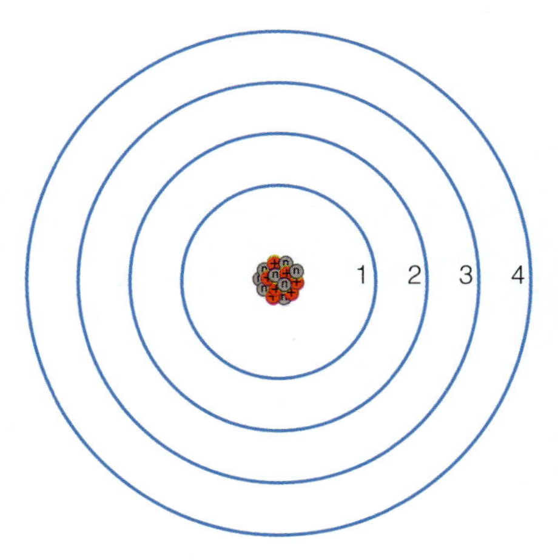

Figure 3.30 In the Bohr model, electrons orbit the nucleus much like planets orbit the Sun in fixed orbits.

The Bohr model was a significant advance over the plum pudding model of atomic structure, and it enabled chemists to explain the properties of some elements. However, the Bohr model does not give a complete picture of electron behavior. In the early 1900s, scientists began to realize that electrons cannot be described simply as negatively charged particles. Sometimes their behavior can be understood only by describing the electrons as waves moving at different energies. This description of electrons is called the **quantum model**. We will explore this model in more detail in the next chapter.

Table 3.3 summarizes the different models of atomic structure. Notice that each new model built on the concepts that came before it. Dalton's original atomic theory described the behavior of atoms and compounds, but it treated atoms as indivisible particles. A century later, J. J. Thomson's plum pudding model was a first attempt at describing the structure of the atom. After Rutherford's gold foil experiment showed this model was incorrect, the Bohr model emerged. The Bohr model correctly described the dense, positively charged nucleus with orbiting electrons, but it treated electron motion too simply. Although the Bohr model could explain some common trends, it left many questions unanswered. Finally, the modern quantum model emerged. This model explains a wide variety of physical and chemical properties and has led to many technological advances. Technologies ranging from solar panels to supercomputers rely on our understanding of the arrangement and behavior of electrons.

TABLE 3.3 Development of Models of Atomic Structure

Model	Year	Key Ideas
Dalton's atomic theory	1808	Atoms are indivisible particles.
Plum pudding model	1904	Atoms are solid, with negative electrons spread throughout a positively charged matrix.
Bohr model	1913	Electrons orbit the nucleus like planets orbit the Sun.
Quantum model	1920s	Electrons behave both as particles and as waves. They occupy an electron cloud around the nucleus.

The Formation of Ions

Atoms frequently gain or lose electrons. To understand why this is so, consider the structure of the atom: Unlike the protons and neutrons, which pack tightly into the nucleus, the electrons occupy the outer volume of the atom. As a result, atoms sometimes lose electrons, or they pick up electrons from neighboring atoms.

Atoms can gain or lose electrons to form *ions*.

When this happens, the number of protons and electrons is no longer the same. As a result, the atom gains a net positive or negative charge. Atoms or groups of atoms that contain an overall charge are called **ions**. For example, a lithium atom often loses one electron. When this happens, it becomes a positively charged lithium ion (**Figure 3.31**):

	Protons	Electrons	Overall Charge
Lithium atom	3	3	0
Lithium ion	3	2	+1

On the other hand, fluorine atoms tend to gain one electron. When they do, they become negatively charged fluoride ions:

	Protons	Electrons	Overall Charge
Fluorine atom	9	9	0
Fluoride ion	9	10	−1

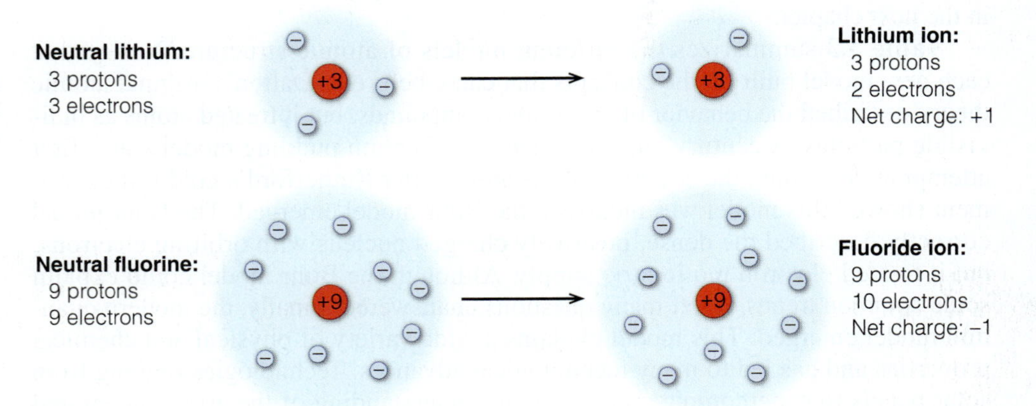

Figure 3.31 Some atoms lose or gain electrons to form ions.

Example 3.6 Finding the Charge on an Ion

A calcium ion has 20 protons but only 18 electrons. What is the charge on the ion?

Each proton has a charge of +1, and each electron has a charge of −1. This calcium ion has a total charge of (+20 − 18), or +2.

Example 3.7 Relating Ion Charge to Subatomic Particles

Sulfur is atomic number 16. Sulfur atoms commonly form sulfide ions, which have a charge of −2. How many electrons are in the electron cloud of a sulfide ion?

The atomic number gives us the number of protons in the nucleus and also the number of electrons in the cloud of a neutral atom. Because the ion has a charge of −2, we know that it has two extra electrons present. Based on this, the sulfide ion has 18 electrons present.

16 electrons (neutral atom) + 2 additional electrons = 18 electrons total

Two atoms are sitting in a bar. Suddenly, one atom says, "Ouch! I just lost an electron!"

The other atom looks over dubiously. "Are you sure?"

"Yes," the first replies, "I'm positive!"

TRY IT

13. Determine the net charge on the following atoms and ions:

a. an atom with 5 protons, 6 neutrons, and 5 electrons

b. an ion with 16 protons, 16 neutrons, and 18 electrons

c. a potassium ion with 18 electrons

d. an ion with atomic number 13, but only 10 electrons

14. Magnesium has atomic number 12. The magnesium ion has a charge of +2. How many electrons are in this ion?

Check It
Watch Explanation

Explore
Atom Builder
Try this interactive to see how protons, neutrons, and electrons combine in atoms and ions.

Capstone Question

In nature, mercury commonly exists as *cinnabar*, a beautiful red compound containing mercury and sulfur (**Figure 3.32**). In this compound, mercury forms an ion with a charge of +2, and sulfur forms an ion with a charge of −2. (a) Complete the table below, showing the number of protons, neutrons, and electrons for the two most common isotopes of mercury (^{200}Hg and ^{202}Hg) in both the elemental and ion forms. (b) Heating cinnabar in the presence of oxygen gas produces elemental mercury and new compounds called sulfur oxides. Is this a physical or a chemical change? (c) A manufacturer begins with 10.00 kg of cinnabar and produces 8.62 kg of elemental mercury. What was the mass of sulfur in the original cinnabar? Express the mass of sulfur in cinnabar as a percentage.

Isotope	Form	Protons	Neutrons	Electrons
^{220}Hg	Atom (charge = 0)			
	Ion (charge = +2)			
^{222}Hg	Atom (charge = 0)			
	Ion (charge = +2)			

Capstone Video

a

b

Figure 3.32 (a) This rock contains large amounts of cinnabar, a deep-red compound composed of mercury and sulfur. (b) For centuries, cinnabar has been used as a pigment, such as this antique Chinese vase.

SUMMARY

Although the Greek philosopher Democritus postulated the existence of the atom in ancient times, it was not until the eighteenth century that scientists began to explore chemical reactivity more systematically. In the late 1700s, Antoine Lavoisier expressed the law of conservation of mass, which states that matter is not created or destroyed in chemical changes. In the early 1800s, John Dalton developed the foundation for the modern atomic theory. Key facets of this theory are that all matter is made of atoms; that atoms of different elements are unique; that atoms combine in whole-number ratios to form compounds; and that atoms are not created or destroyed in chemical reactions.

The invention of the battery in 1800 spawned the discovery of many new elements. This wealth of new knowledge enabled Dmitri Mendeleev to develop the periodic table. The periodic table is organized based on atomic number and also on chemical reactivity. A row on the periodic table represents a full cycle of chemical behavior, while elements in a single column share physical and chemical traits.

In the nineteenth and early twentieth centuries, technical innovations enabled scientists to probe the structure of the atom. J. J. Thomson first described the electron. Ernest Rutherford, building on the work of Thomson and others, discovered that atoms are almost entirely empty space and have a very dense nucleus at the center.

The nucleus is composed of two particles, positively charged protons and uncharged neutrons. The *atomic number* is the number of protons in the nucleus, as well as the number of electrons in an electrically neutral atom. The number of protons determines the identity of the atom. Atoms with the same number of protons but different numbers of neutrons are isotopes. The mass number of any isotope is the sum of the protons and neutrons in the nucleus. The mass that is usually displayed on periodic tables is the average atomic mass. This is a weighted average of the masses of the different isotopes of that element.

The discovery of the nucleus led to the development of the Bohr model of the atom, which describes the atom as a very dense nucleus with electrons orbiting around it. The Bohr model gave way to the quantum mechanical model, which treats electrons as waves rather than simply as orbiting particles.

In an atom the negatively charged electrons occupy the region around the nucleus, called the electron cloud. Although the electron cloud accounts for most of the atom's volume, electrons have a much smaller mass than the particles in the nucleus, so that nearly all the mass of the atom arises from the protons and neutrons. Atoms can gain or lose electrons, giving them an overall charge. Atoms or groups of atoms with an overall charge are called ions.

Mercury and Small-Scale Gold Mining, Continued

At the beginning of this chapter, we saw how the use of mercury in small-scale gold mining has led to disastrous effects for many miners and the surrounding communities. As we wrap up our discussion of atoms and elements, let's look more closely at mercury, and the problems associated with this intriguing element.

Mercury's physical properties are mesmerizing. It is the only naturally occurring metal that is liquid at room temperature (**Figure 3.33**). It shines like silver, yet flows effortlessly. It is dense enough that lead floats on it.

Mercury is a key part of small-scale gold mining. Miners isolate gold by first mixing gold ore with mercury to form a solid mixture called an amalgam. Heating the amalgam causes the mercury to evaporate, leaving pure gold.

But mercury is a powerful toxin, and it is especially dangerous in its gaseous form. When people inhale gaseous mercury, the atoms enter the bloodstream, where they lose electrons to form mercury ions (Hg^{2+}). These ions are especially toxic to the kidneys and the central nervous system. Extensive exposure to mercury can lead to tremors, mood changes, dementia, and ultimately death.

Dr. Adam Kiefer and his students work with communities affected by small-scale gold mining practices. They track the amount of mercury in the air, educate local leaders about the risks of mercury, and even help to reduce airborne mercury levels.

One of the challenges Dr. Kiefer faces is that gaseous mercury is odorless and invisible. To help people better understand how mercury evaporates, he uses a demonstration involving ultraviolet (UV) light and a fluorescent background (**Figure 3.34**). (*Ultraviolet light* is high-energy light that we cannot see. If something is *fluorescent*, it glows when ultraviolet light shines on it.) As mercury evaporates, the vapors absorb the ultraviolet light, producing shadows on the fluorescent background. This demonstration helps people understand that mercury mixtures constantly release gaseous atoms into the air.

Using a device called an *atomic absorbance spectrometer*, scientists can actually measure the amount of mercury in the air by measuring the amount of UV light the air absorbs. By linking the measuring device to a GPS system, Dr. Kiefer and his team can even map the concentrations of mercury in different neighborhoods (**Figure 3.35**).

Recently, Dr. Kiefer designed a device that miners can use to separate the mercury from the gold without releasing mercury into the environment. While this invention has the potential to drastically reduce mercury pollution, there is still much to be done.

videophoto/Getty Images

Figure 3.33 Mercury is the only metal that is a liquid at room temperature.

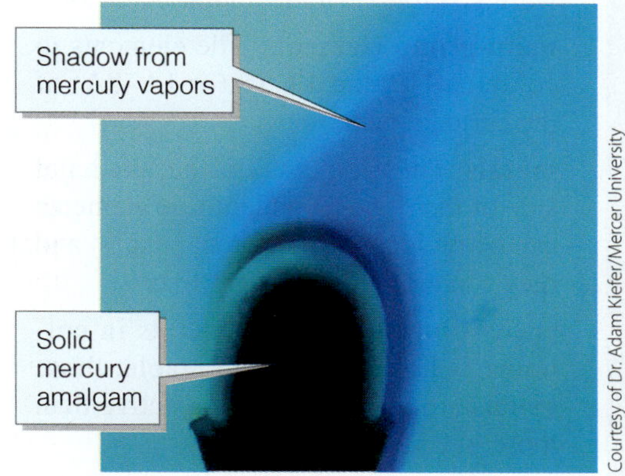

Courtesy of Dr. Adam Kiefer/Mercer University

Figure 3.34 By placing a mercury amalgam in front of a fluorescent screen with a UV light source, we can see the shadows from the gaseous mercury atoms as they evaporate off the metal surface.

Explore Figure 3.34

a

b

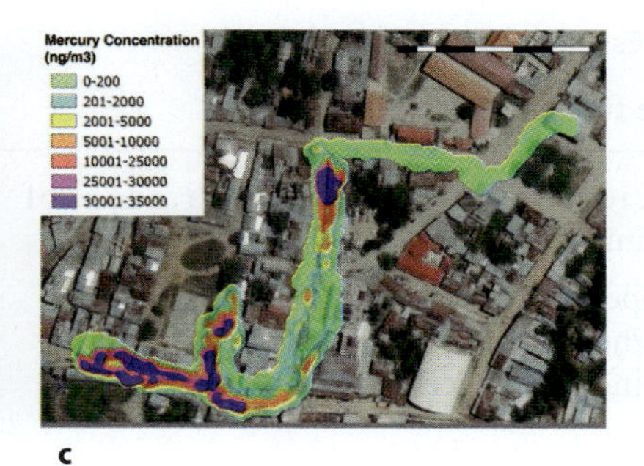

c

Courtesy of Dr. Adam Kiefer/Mercer University

Figure 3.35 (a) This image shows Dr. Adam Kiefer (left) with his students in Peru. (b) The team works with local officials to monitor mercury levels in the air. (c) By linking the measuring device to a GPS system, the Kiefer team can produce maps like this one that shows the concentration of mercury vapor along each street. The dark purple region indicates an area wherein miners separate the mercury and gold. The mercury concentration in this area exceeds 30,000 nanograms per cubic meter of air.

Key Terms

3.1 Atoms: The Essential Building Blocks

law of conservation of mass In chemical reactions, matter is neither created nor destroyed.

atomic theory A theory describing matter in terms of fundamental units called atoms.

3.2 The Periodic Table of the Elements

periodic table of the elements A chart that organizes all the known elements based on their masses and properties.

period A horizontal row on the periodic table; a period encompasses a range of behavior from metallic to nonmetallic.

group A vertical column on the periodic table; also called a family. Elements within a group exhibit similar behaviors.

main-group elements The elements in columns 1–2 and 13–18 (or 1A–8A) of the periodic table.

metals The elements on the left-hand side of the periodic table; these elements can be molded into different shapes, and they conduct heat and electricity.

transition metals The metals in columns 3–12 of the periodic table; these metals are harder and less reactive than those in columns 1–2.

nonmetals Elements on the upper right-hand side of the periodic table; these elements have widely varying properties and form many different compounds.

metalloids Elements whose properties lie between those of metals and nonmetals.

alkali metals Metal elements in column 1 (or 1A) of the periodic table; these metals are very reactive.

alkaline earth metals Metal elements in column 2 (or 2A) of the periodic table; these metals are very reactive.

halogens Nonmetal elements in column 17 (or 7A) of the periodic table; these elements form many different types of compounds.

noble gases The nonmetal elements in column 18 (or 8A) of the periodic table; these elements usually do not form compounds.

3.3 Uncovering Atomic Structure

subatomic particles The particles from which atoms are composed. The three major subatomic particles are protons, neutrons, and electrons.

atomic mass unit (u or amu) A unit of mass equal to 1.66×10^{-27} kg.

charge A characteristic property of subatomic particles that affects how particles interact with each other.

electron A negatively charged subatomic particle; the electrons occupy the space around the nucleus.

nucleus The tiny, dense center of an atom; the nucleus contains protons and neutrons.

proton A positively charged subatomic particle that resides in the nucleus of the atom.

neutron A subatomic particle having no charge that resides in the nucleus of the atom.

electron cloud The space around the nucleus; the electron cloud accounts for nearly the entire volume of the atom.

3.4 Describing Atoms: Identity and Mass

atomic number The number of protons in an atom; also the number of electrons in a neutral atom.

mass number The sum of protons and neutrons in an atom.

isotopes Atoms that have the same atomic number, but different mass numbers.

3.5 Electrons—A Preview

Bohr model An early model of atomic structure that treated the atom like a tiny solar system, with the nucleus at the center, and the electrons orbiting the nucleus.

quantum model The modern description of electronic behavior that treats electrons both as particles and as waves.

ion An atom or group of atoms with an overall charge.

Additional Problems

3.1 Atoms: The Essential Building Blocks

15. What is the fundamental unit of matter?

16. How many naturally occurring elements are there?

17. Which scientist is credited with each of these accomplishments?

a. first proposing the idea of atoms
b. developing the law of conservation of mass
c. developing the modern atomic theory

18. List four key features of Dalton's atomic theory.

19. Alka-Seltzer® is a common treatment for upset stomach. When the tablets are mixed with water, a chemical change produces gas bubbles (**Figure 3.36**). To test Lavoisier's law, a chemist freezes a small amount of water in a plastic bottle. She then adds an Alka-Seltzer tablet and seals the bottle tightly. She measures the mass of the bottle. She then sets the bottle on the counter. As the ice melts, it begins to react with the tablet, producing bubbles inside the bottle. After all of the ice has melted and the tablet has dissolved, she measures the mass of the bottle again. Assuming the bottle does not burst, would you expect the mass afterward to be more than, less than, or the same as the mass before? How do you explain this?

Figure 3.36 Alka-Seltzer® is a common treatment for upset stomach. When the tablets are mixed with water, a chemical change produces gas bubbles.

20. On a crisp October night, you build a campfire using four heavy logs. The next morning, you notice that the charred remains of the logs are much lighter than the logs were the night before. Does this contradict the law of conservation of mass? What happened to the atoms in the logs?

21. Sodium hydroxide reacts with hydrogen chloride to form two new compounds, sodium chloride and water. When 200 grams of sodium hydroxide are combined in a sealed flask with 100 grams of hydrogen chloride, a reaction takes place. How many grams of material will be present in the flask after the reaction takes place?

22. When natural gas (methane) burns, it reacts with oxygen in the air to form two new compounds, carbon dioxide and water. If 100 grams of methane and 100 grams of air are combined in a sealed container, and the mixture is allowed to react, what mass will be present in the container after the reaction takes place?

23. When it burns, 24.3 grams of magnesium can react completely with 16.0 grams of oxygen gas to form a new compound, magnesium oxide. No other compounds are formed. What mass of magnesium oxide is formed in this reaction?

24. Zinc metal reacts with hydrochloric acid to form zinc chloride and hydrogen gas, as shown in this equation:

$$\text{Zinc} + \text{hydrochloric acid} \rightarrow \text{zinc chloride} + \text{hydrogen gas}$$

If 65.4 grams of zinc reacts in this way, it will consume 72.9 grams of hydrochloric acid and produce 2.0 grams of hydrogen gas. Based on this information, how many grams of zinc chloride will be produced?

25. Strontium and chlorine combine in only one ratio: one strontium atom for every two chlorine atoms. Based on this information, indicate whether the following combinations are possible:

a. 10 strontium atoms and 20 chlorine atoms
b. 10 strontium atoms and 30 chlorine atoms
c. 140 billion strontium atoms and 220 billion chlorine atoms
d. 3.8×10^{21} strontium atoms and 7.6×10^{21} chlorine atoms

26. Aluminum and bromine atoms combine in only one ratio: one aluminum atom for every three bromine atoms. Based on this information, indicate whether the following combinations are possible:

a. 3 aluminum atoms and 9 bromine atoms
b. 4.2×10^4 aluminum atoms and 8.4×10^4 bromine atoms
c. 50,000 aluminum atoms and 1.5×10^5 bromine atoms
d. 9 dozen aluminum atoms and 27 dozen bromine atoms

27. Each of these panels depicts a chemical reaction. Which of them does *not* follow the law of conservation of mass? Explain your answer.

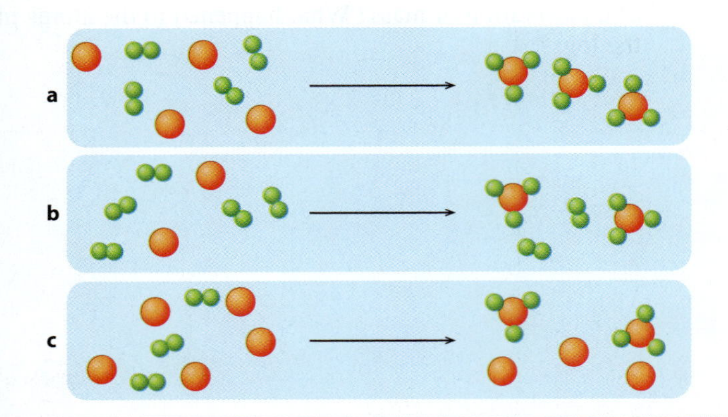

28. Each of these panels depicts a chemical reaction. Which of them does *not* follow the law of conservation of mass? Explain your answer.

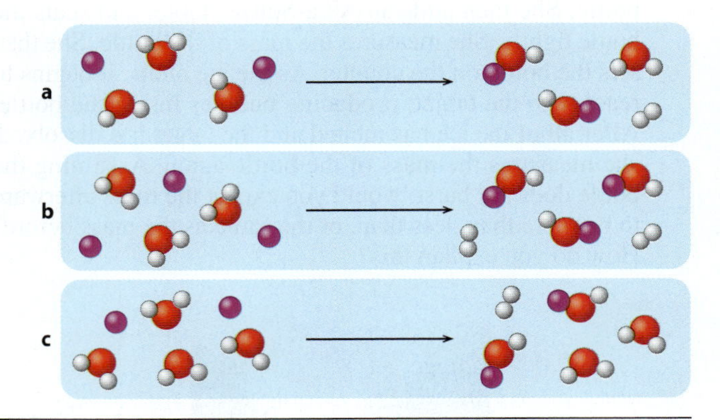

3.2 The Periodic Table of the Elements

29. Write the name of the element that corresponds with each of the following symbols:

a. Cs b. Ti c. K d. B e. Xe

30. Write the name of the element that corresponds with each of the following symbols:

a. Np b. O c. Fe d. S e. Th

31. Write the name of the element that corresponds with each of the following symbols:

a. W b. B c. U d. S e. C

32. Write the name of the element that corresponds with each of the following symbols:

a. O b. Ba c. Mg d. W e. Pb

33. Write the atomic symbol for each of the following elements:

a. fluorine b. germanium c. tin d. magnesium

34. Write the atomic symbol for each of the following elements:

a. potassium b. argon c. ruthenium d. lithium

35. Write the atomic symbol for each of the following elements:

a. iodine b. copper c. silver d. hydrogen

36. Write the atomic symbol for each of the following elements:

a. iron b. nickel c. gold d. titanium

37. On the periodic table, how are atoms with similar chemical properties grouped?

38. As you move across a horizontal row of the periodic table, where do you find the metals, metalloids, and nonmetals?

39. Use the periodic table to identify each of the following elements as a main-group metal, a transition metal, an inner-transition metal, a metalloid, or a nonmetal:

a. boron b. gallium c. chromium d. barium
e. uranium f. oxygen g. tin h. iron

40. Use the periodic table to identify each of the following elements as a main-group metal, a transition metal, an inner-transition metal, a metalloid, or a nonmetal:

a. carbon b. lithium c. vanadium d. calcium
e. plutonium f. platinum g. einsteinium h. argon

41. What are the main-group elements?

42. What series of elements contains most of the rare earth elements?

43. Identify each of the following elements as a metal, metalloid, or nonmetal:

a. platinum b. oxygen c. lithium d. uranium

44. Identify each of the following elements as a metal, metalloid, or nonmetal:

a. carbon b. mercury c. californium d. tellurium

45. Potassium, rubidium, and cesium all belong to which family of elements?

46. Calcium, strontium, and barium all belong to which family of elements?

47. Identify the family of elements to which each of these elements belongs:

a. chlorine b. krypton c. rubidium d. strontium

48. Identify the family of elements to which each of these elements belongs:

a. francium b. fluorine c. helium d. calcium

49. Based on its location on the periodic table, which of these elements would behave most like magnesium?

a. sodium b. beryllium c. aluminum d. argon

50. Based on its location on the periodic table, which of these elements would behave most like oxygen?

a. nitrogen b. phosphorus c. sulfur d. fluorine

51. Based on its location on the periodic table, which two of these elements would *not* behave like chlorine?

a. fluorine b. bromine c. argon d. sulfur

52. Based on its location on the periodic table, which two of these elements would *not* behave like strontium?

a. rubidium b. calcium c. yttrium d. barium

53. Identify each element based on its location on the periodic table:

a. row 4, column 5 b. row 3, group 7A
c. row 2, column 18

54. Identify each element based on its location on the periodic table:

a. row 1, column 1 b. row 4, group 12
c. row 2, group 3A

3.3 Uncovering Atomic Structure

55. Charge is a fundamental property of subatomic particles. What types of charges are possible in a subatomic particle?

56. What types of charges attract each other? What types of charges repel each other?

57. Match each of the scientists in the first column with the discovery in the second column:

(1) Democritus	a. the periodic table
(2) Antoine Lavoisier	b. first theorized about the existence of atoms
(3) John Dalton	c. the law of conservation of mass
(4) Dmitri Mendeleev	d. the modern atomic theory

58. Match each of the scientists in the first column with the discovery in the second column:

(1) Alessandro Volta	a. discovered the electron
(2) J. J. Thomson	b. model in which electrons orbit the nucleus like planets orbit the sun
(3) Ernest Rutherford	c. idea that the atom has a dense nucleus
(4) Niels Bohr	d. invented the battery

59. When Rutherford and his students conducted the gold foil experiment, what did they expect to see? How did their actual observations differ from their hypothesis?

60. What conclusions was Rutherford able to draw from his gold foil experiment?

61. Which has more volume, the nucleus or the electron cloud? Which has more mass?

62. How does the density of the nucleus compare to the density of the rest of the atom?

63. John Dalton's original atomic theory contained the following key ideas. Which part(s) of these ideas was/were incorrect?

a. Elements are made of tiny, indivisible particles called atoms.
b. The atoms of each element are unique.
c. Atoms can join together in whole-number ratios to form compounds.
d. Atoms are unchanged in chemical reactions.

64. What findings led J. J. Thomson to articulate the plum pudding model rather than considering atoms to be indivisible particles as John Dalton did?

65. Which two subatomic particles each have a mass of about one atomic mass unit? Which two particles are located in the nucleus? Which two particles have an overall charge?

66. Of the three major subatomic particles (protons, neutrons, and electrons), which is the smallest? Which has no charge? Which particle occupies the space outside the nucleus?

3.4 Describing Atoms: Identity and Mass

67. Using the periodic table, identify the atomic number for each of these elements:

a. Ti b. S c. Te d. Li

68. Using the periodic table, identify the atomic number for each of these elements:

a. Rf b. Ar c. He d. Es

69. Using the periodic table, find the symbols for the elements that correspond to each atomic number:

a. atomic number 13
b. atomic number 86
c. atomic number 99

70. Using the periodic table, find the symbols for the elements that correspond to each atomic number:

a. atomic number 29
b. atomic number 5
c. atomic number 16

71. Give the atomic number and the mass number for each of the following:

a. an atom with 14 protons and 16 neutrons
b. an atom with 27 protons and 32 neutrons
c. an atom with 20 protons and 26 neutrons

72. Give the atomic number and the mass number for each of the following:

a. an atom with 35 protons and 44 neutrons
b. an atom with 35 protons and 46 neutrons
c. an atom with 82 protons and 124 neutrons

73. Write the atomic symbol, including the atomic number and the mass number, for each of the following:

a. a potassium atom with 20 neutrons
b. an argon atom with 22 neutrons
c. a fermium atom with 157 neutrons

74. Write the atomic symbol, including the atomic number and the mass number, for each of the following:

a. a silicon atom with 14 neutrons
b. a gallium atom with 39 neutrons
c. a hafnium atom with 106 neutrons

75. The most common isotope of chromium has a mass number of 52. How many neutrons are in an atom of ^{52}Cr?

76. The most common isotope of nickel has a mass number of 58. How many neutrons are in an atom of ^{58}Ni?

77. Using the periodic table, complete the following table:

Atom	Symbol	Protons	Neutrons	Atomic Number	Mass Number
Hydrogen	H	1	0		
Sulfur			16		
		52			128
	He		2		
			51	40	

78. Using the periodic table, complete the following table:

Atom	Symbol	Protons	Neutrons	Atomic Number	Mass Number
	Fr	87			223
Tungsten			110		
	U				238
		86	136		
			57	44	

79. You are trying to find enough change in your room to buy a candy bar. After a thorough search, you come up with 8 pennies, 5 nickels, 4 dimes, and 3 quarters. What is the total value of the coins you found? What is the average value of the coins you found?

80. You are given a bag of coins that contains 70% dimes and 30% nickels. What is the average value of the coins in the bag?

81. Gallium exists as two isotopes. The first has 31 protons and 38 neutrons. The second has 31 protons and 40 neutrons. Write an atomic symbol for each of these isotopes, including the atomic number and mass number.

82. Indium exists as two isotopes. The first has 49 protons and 64 neutrons. The second has 49 protons and 66 neutrons. Write an atomic symbol for each of these isotopes, including the atomic number and mass number.

83. Silver exists as two isotopes, ^{107}Ag and ^{109}Ag. How many protons and neutrons are in the nucleus of each isotope?

84. Iridium exists as two isotopes, ^{191}Ir and ^{193}Ir. How many protons and neutrons are in the nucleus of each isotope?

85. Bromine exists as a mixture of two isotopes. Of bromine atoms, 50.7% are ^{79}Br, and 49.3% are ^{81}Br. Use the mass numbers and the relative abundance to estimate the average atomic mass of bromine.

86. Boron exists as a mixture of two isotopes. Of boron atoms, 19.80% are ^{10}B, and 80.20% are ^{11}B. Use the mass numbers and the relative abundance to estimate the average atomic mass of boron.

87. Lead exists as a mixture of four major isotopes, shown in the table. Calculate the average atomic mass for lead.

Isotope	Mass (u)	% Abundance
^{204}Pb	203.9730	1.40
^{206}Pb	205.9745	24.10
^{207}Pb	206.9759	22.10
^{208}Pb	207.9766	52.40

88. Zinc exists as a mixture of five major isotopes, shown in the table. Calculate the average atomic mass for zinc.

Isotope	Mass (u)	% Abundance
^{64}Zn	63.9291	48.63
^{66}Zn	65.9260	27.90
^{67}Zn	66.9271	4.10
^{68}Zn	67.9248	18.75
^{70}Zn	69.9253	0.62

89. What is the difference between the *mass number* and the *average atomic mass*?

90. Why does the periodic table commonly show the average atomic mass rather than the mass number?

3.5 Electrons — A Preview

91. How did the Bohr model differ from the plum pudding model of atomic structure?

92. How does the quantum mechanical model differ from the Bohr model of atomic structure?

93. How many electrons are in each of the following neutral atoms?

a. hydrogen **b.** helium **c.** carbon **d.** selenium

94. How many electrons are in each of the following neutral atoms?

a. copper **b.** gold **c.** mercury **d.** xenon

95. Find the number of protons and electrons in each of the following neutral atoms:

a. Li **b.** Ge **c.** Bi **d.** Ba

96. Find the number of protons and electrons in each of the following neutral atoms:

a. Y **b.** N **c.** O **d.** Cf

97. A beryllium ion has 4 protons but only 2 electrons. What is the charge on this ion?

98. An oxide ion has 8 protons and 10 electrons. What is the charge on this ion?

99. Find the net charge on each of the following:

a. an ion with 7 protons, 7 neutrons, and 10 electrons
b. an ion with 38 protons, 50 neutrons, and 36 electrons
c. an atom with 35 protons, 44 neutrons, and 35 electrons

100. Find the net charge on each of the following:

a. a phosphorus ion with 18 electrons
b. an argon atom with 18 electrons
c. a potassium ion with 18 electrons
d. a calcium ion with 18 electrons

101. How many electrons are in a neutral silver atom? How many electrons are in a silver ion with a charge of +1?

102. How many electrons are in a neutral zinc atom? How many electrons are in a zinc ion with a charge of +2?

103. Titanium commonly forms two ions. The first ion has a charge of +2. The second ion has a charge of +4. In total, how many electrons are present in each of these ions?

104. Iron commonly forms two ions. The first ion has a charge of +2. The second ion has a charge of +3. In total, how many electrons are present in each of these ions?

105. Under extreme conditions, gold will lose one electron to form a +1 ion. Mercury often loses two electrons to form a +2 ion. Find the number of protons and electrons in each of these ions.

106. Sodium atoms easily lose one electron to form a +1 ion. Magnesium atoms easily lose two electrons to form a +2 ion. How many electrons are in each of these ions?

107. Fluorine, chlorine, and bromine all gain one electron to form ions with a −1 charge. Find the number of protons and electrons in each of these ions.

108. Lithium, potassium, and rubidium all lose one electron to form ions with a +1 charge. Find the number of protons and electrons in each of these ions.

CHAPTER FOUR

Light and Electronic Structure

Figure 4.1 (a) Solar cells now provide part of the electrical energy for many homes. (*continued*)

Edging toward Solar Energy

What if the power company paid *you* for electricity?

This seemingly far-fetched dream is edging closer to reality, thanks to recent advances in solar energy technology. We'll get to your power bill in just a moment. First, a little history: In 1839 Edmond Becquerel, a French scientist, discovered that shining light on certain materials produced an electric current. Over a century later, in 1954, Bell Labs produced the first *solar cell*, a device that could convert sunlight into electricity.

In the years since, many have hailed solar energy as the power source of the future: clean, renewable energy that would finally replace fossil fuels. Scientists and engineers dreamed of creating homes that were energetically self-sufficient, disconnected from the power grid.

Despite these dreams, large-scale solar power has been agonizingly slow to develop. One problem is location: Solar power is much more promising for sun-drenched areas like Southern California or Central Africa than for northern areas like Minnesota or Finland. Then there's the issue of cost: Even though the price of solar cells has dropped dramatically in the past decades (**Figure 4.1**), solar power is still more expensive than coal or natural gas.

But perhaps the most challenging issue has been energy storage. Because our homes are not exposed to sunlight all the time, solar power requires either a way of storing the energy for later use or a backup power supply.

Enter Elon Musk. He is the founder of SpaceX, a leader in private space exploration, and the CEO of Tesla Motors, which produces high-performance electric cars. He also helped create Solar City, a company that produces and installs home solar energy systems.

For years, batteries were too big, expensive, and inefficient to handle the energy needs of an entire house. But over the last few years, companies like Tesla have driven improvements in battery technology and performance. In 2015, Musk introduced a

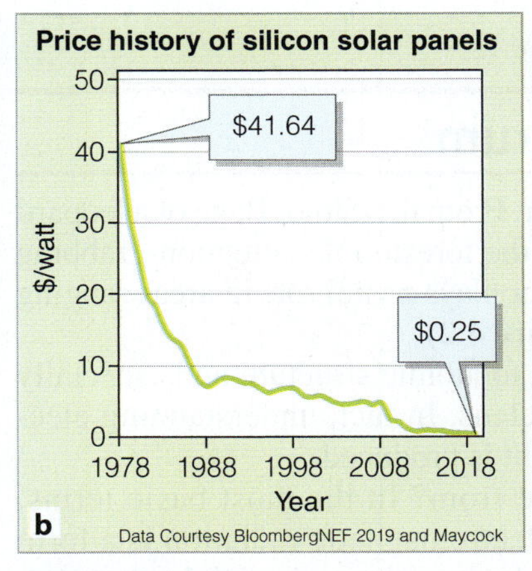

Figure 4.1 (continued) (b) The price of solar cells has dropped dramatically in recent years, but solar energy still costs more than other energy sources. (c) In 2015, Tesla CEO Elon Musk unveiled his new home battery, based on the battery used in the Tesla electric cars. (d) A worker installs a Tesla battery system in the garage of a California home. Steve Spohn; Data from BloombergNEF 2019 and Maycock; Tribune Content Agency LLC/Alamy; Ian Thomas Jansen-Lonnquist/Bloomberg via Getty Images

rechargeable battery system capable of powering a small home. Based on the Tesla automotive technology, these long-term battery systems can connect to solar panels. During the day, solar panels charge the battery; at night the battery helps power the home, reducing monthly power bills. In some places, homeowners can even sell excess power back to the power company.

The Tesla battery system is an impressive achievement and an important step in the development of solar energy. And though it's not likely to completely replace fossil fuels in our lifetimes, solar power is an important part of our energy future.

So, how does solar power work? To begin answering that question, we must first understand the relationship between light and the structure of the atom. In this chapter, we'll see that both the absorption and production of light are intimately connected to the configuration of electrons around the nucleus. And light is just the beginning: As we delve deeper, we'll see how the properties of the elements relate to their electronic structure and learn how to determine a great deal of information with a quick glance at the periodic table. Let's take a look.

Intended Learning Outcomes

After completing this chapter and working the practice problems, you should be able to:

4.1 The Electromagnetic Spectrum

- Qualitatively and quantitatively describe the relationships between the wavelength, frequency, and energy of electromagnetic radiation.

4.2 Color, Line Spectra, and the Bohr Model

- Describe line spectra, the Bohr model, and how they are related.
- Describe the absorption or emission of light as a function of electron transitions.

4.3 The Quantum Model and Electron Orbitals

- Describe Heisenberg's uncertainty principle and the wave nature of electrons.
- Identify the number of orbitals and the maximum electron capacity of the *s*, *p*, *d*, and *f* sublevels.
- Correlate each primary energy level with the available sublevels.

4.4 Describing Electron Configurations

- Write electron configurations for atoms and ions, using either full notation or noble gas shorthand.
- Identify the inner, outer, and valence electrons in an atom or ion.
- Apply the octet rule to explain the exceptional stability of the noble gases.

4.5 Electron Configuration and the Periodic Table

- Use the periodic table to quickly identify the highest-occupied energy level and sublevel of an element.
- Use the periodic table to identify the number of valence electrons for main-group elements.

4.1 The Electromagnetic Spectrum

Ewais/Shutterstock

Figure 4.2 Light and color are closely connected to atomic structure.

Our world is immersed in light and soaked in color. From the camouflage of a leopard to the glow of a sunset, from the placid green of the forest to the attention-grabbing text of a billboard, the perception of light and color plays a vital role in areas ranging from survival to fashion to communication (**Figure 4.2**).

The production of light is closely connected to atomic structure and especially to the configuration of electrons around the nucleus. In fact, understanding electronic structure is a key to understanding how light is produced.

So, what is light? And where does it come from? In the most basic terms, light is a type of **electromagnetic radiation**. Electromagnetic radiation is a form of energy that travels in waves that are produced when charged particles move or vibrate relative to each other. These energy waves exist in small increments called **photons**. We can think of a photon as a *packet* of light (**Figure 4.3**).

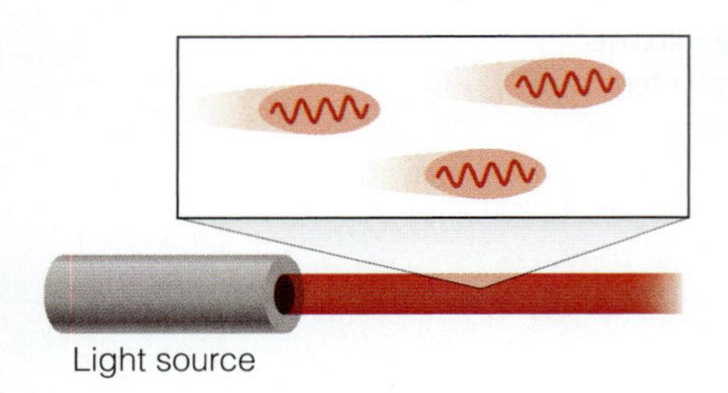

Figure 4.3 Light travels in packets of energy called photons.

Electromagnetic radiation ranges from very low-energy waves (TV and radio waves) to very high-energy waves (X-rays and gamma rays). This broad continuum of electromagnetic energy is referred to as the **electromagnetic spectrum** (**Figure 4.4**).

In the middle of this spectrum is a small range of radiation that our eyes can detect—we perceive it as visible light. This narrow range of radiation is the **visible spectrum**. If we look more closely at the visible spectrum (**Figure 4.5**), we can see that it is made up of the colors of the rainbow: red, orange, yellow, green, blue, and violet.

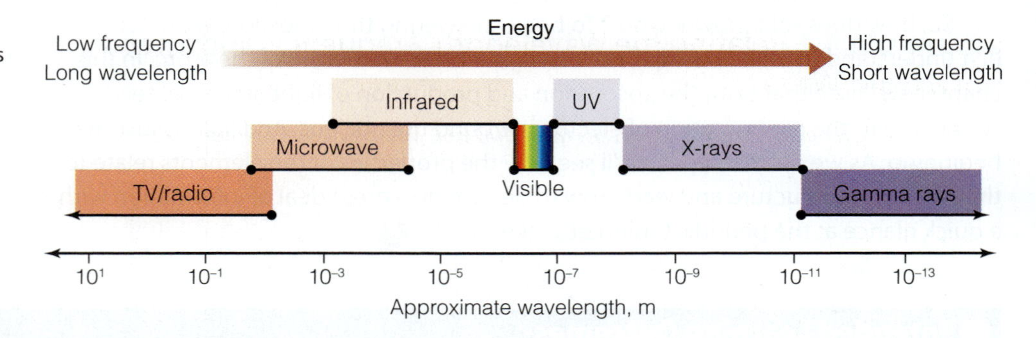

Figure 4.4 The electromagnetic spectrum covers a wide range of energies. Visible light is a narrow sliver of the entire spectrum.

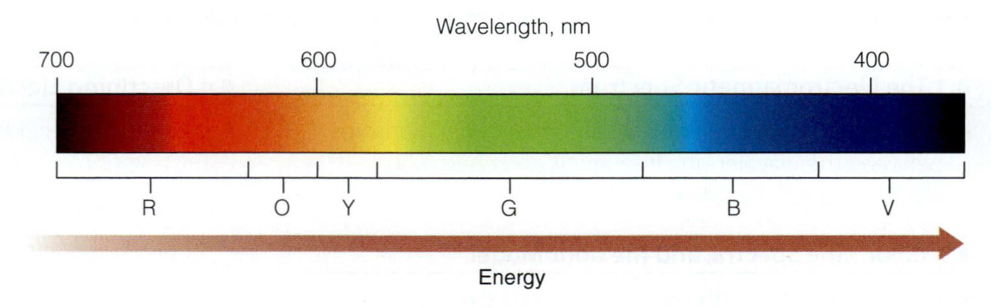

Figure 4.5 The visible spectrum ranges from about 700 nm to about 370 nm. The colors of the spectrum are red (R), orange (O), yellow (Y), green (G), blue (B), and violet (V).

Wavelength, λ

Figure 4.6 Wavelength is the distance from a point on one wave cycle to the same point on the next cycle.

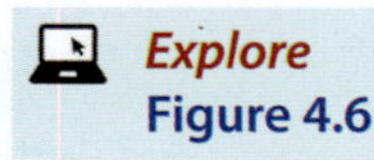
Explore Figure 4.6

Wavelength and Frequency

We use several parameters to describe electromagnetic waves. **Wavelength**, symbolized by the Greek letter *lambda* (λ), is the distance from a point on one wave to the same point on the next wave (**Figure 4.6**). Typically, we measure wavelength in meters or nanometers (Recall that 1 nm = 10^{-9} m).

The color of light is related to its wavelength. On the low-energy end of the visible spectrum, red light has a wavelength of about 700 nm. Wavelengths longer than this fall into the *infrared* (IR) region. On the high-energy end of the visible spectrum, violet light has a wavelength of about 400 to 370 nm. Wavelengths shorter than this fall into the *ultraviolet* (UV) region.

Frequency, symbolized by the Greek letter *nu* (ν), is the number of waves that pass through a point in one second. A frequency of one wave cycle per second is a **hertz**. The units corresponding to hertz are written as **Hz**, as 1/s, or sometimes as s^{-1}. For example, if a wave has a frequency of 10,000 hertz, we can write this in three different ways: as 10,000 Hz, as 10,000/s, or as 10,000 s^{-1}. Each of these notations indicates 10,000 cycles per second.

$1/s = 1\ s^{-1} = 1$ Hz

Wavelength and the frequency are inversely related to each other. This means that as the wavelength decreases, the frequency increases, and vice versa. Mathematically, we describe this principle using the following relationship, where λ is the wavelength and ν is the frequency:

$$c = \lambda\nu$$

Before we define c, let's look at the units involved in this expression: The wavelength, λ, is often measured in meters. The frequency, ν, is the number of cycles per second. If we multiply m × 1/s, we get m/s, which is a unit of speed (**Figure 4.7**). Because we are describing light waves, *c is the speed of light*. In a vacuum, the speed of light has a constant value of 3.00×10^8 m/s. Example 4.1 illustrates how we use this relationship to convert between wavelength and frequency.

$$c = \lambda\nu$$

Speed of Light = Wavelength × Frequency

$$\frac{m}{s} = m \times \frac{1}{s}$$

Figure 4.7 This equation gives the relationship between speed, wavelength, and frequency.

Example 4.1 Relating the Wavelength, Frequency, and Speed of a Light Wave

A beam of green light has a wavelength of 500 nm. What is the frequency of this light?

To solve this problem, we rearrange the equation shown earlier to solve for frequency:

$$\nu = \frac{c}{\lambda}$$

Because *c* is given in units of meters per second, we also need to express the wavelength in meters so the units will cancel. One nanometer is equal to 10^{-9} m, so we can say that

$$500 \text{ nm} = 500 \times 10^{-9} \text{ m}$$

$1 \text{ nm} = 10^{-9}$ m

We then insert the values into the equation:

$$\nu = \frac{c}{\lambda} = \frac{3.00 \times 10^8 \frac{\cancel{m}}{s}}{500 \times 10^{-9} \cancel{m}} = 6.00 \times 10^{14}\ s^{-1} = 6.00 \times 10^{14} \text{ Hz}$$

Notice that the meters cancel out, leaving us with units of 1/s. We can also write this as s^{-1}, or as Hz. This is the standard unit for frequency.

TRY IT

1. A beam of red light has a wavelength of 720 nm. What is the frequency of this light?
2. A beam of blue light has a frequency of 6.41×10^{14} Hz. What is the wavelength of this light in meters? In nanometers?

Check It
Watch Explanation

Figure 4.8 Blue waves have a shorter wavelength but higher frequency than red waves. The blue wave has higher energy.

Explore Figure 4.8

Planck's constant is named for Max Planck (1858–1947), a German physicist who was instrumental in developing quantum theory (described in Section 4.3). For his work, Planck was awarded the 1918 Nobel Prize in Physics.

akg-images/bilwissedition

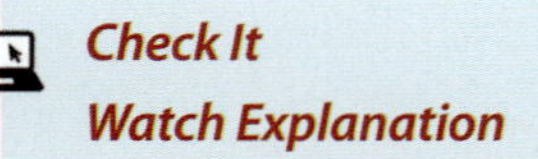

Check It
Watch Explanation

The Energy of a Photon

We often describe light in terms of its energy. The energy of a photon of light depends on its frequency and wavelength. For example, compare the two light waves shown in **Figure 4.8**. Notice that the red wave has a longer wavelength than the blue one does. This means that the red wave oscillates up and down more slowly, resulting in fewer cycles per second (that is, a lower frequency). On the other hand, the blue wave oscillates with a higher frequency. The energy of a wave depends on how quickly the wave oscillates. The red wave (with its longer wavelength) has lower energy, while the blue wave (with its shorter wavelength) has higher energy.

We can relate the energy of a single photon to the frequency using this relationship:

$$E = h\nu$$

where E is the energy (measured in *joules*), ν is the frequency, and h, which is referred to as *Planck's constant*, has a value of 6.63×10^{-34} J · s. To relate energy to wavelength, we can also write this expression as

$$E = hc/\lambda$$

Example 4.2 illustrates how we use these relationships to convert between frequency, wavelength, and energy.

Example 4.2 Relating the Wavelength, Frequency, and Energy of Light

A photon has a frequency of 7.50×10^{14} Hz. What is the wavelength of this light? What color is this light? What is the energy of the photon?

To find the wavelength, we rearrange the equation $c = \lambda\nu$ and substitute the given values for c and ν:

$$\lambda = \frac{c}{\nu} = \frac{3.00 \times 10^{8}\ \frac{m}{\cancel{s}}}{7.50 \times 10^{14}\ \frac{1}{\cancel{s}}} = 4.00 \times 10^{-7}\ m = 400\ nm$$

By comparing this answer to the visible spectrum in Figure 4.5, we can see that light with a wavelength of 400 nm is violet (or purple). Finally, we can solve for the energy of the wave.

$$E = h\nu = (6.63 \times 10^{-34}\ J \cdot \cancel{s}) \times (7.50 \times 10^{14}\ \cancel{s^{-1}}) = 4.97 \times 10^{-19}\ J$$

TRY IT

3. In the original *Star Wars* movie, Luke Skywalker was given a blue light saber. After losing his hand (and saber) in *The Empire Strikes Back*, he returned in the third movie with a green light saber. Which light saber emitted light waves with higher energy, the blue or the green?
4. What is the energy in joules of a beam of light whose frequency is 6.98×10^{14} Hz?
5. What is the wavelength of a photon with an energy of 3.75×10^{-19} J? Based on the visible spectrum shown in Figure 4.5, approximately what color of light is this?

4.2 Color, Line Spectra, and the Bohr Model

Color and Line Spectra

Our study of light and electronic structure begins with a technology that is over a thousand years old: fireworks (**Figure 4.9**). The explosive powder in fireworks often contains metals that produce distinctive colors as the mixture burns.

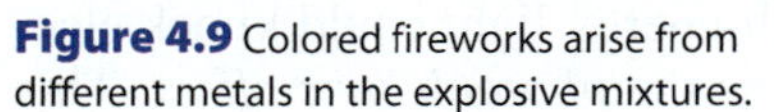

Figure 4.9 Colored fireworks arise from different metals in the explosive mixtures.

Figure 4.10 Different elements give off characteristic colors when heated in a flame. GIPhotoStock/Science Source

Explore
Figure 4.10

We can easily observe this behavior in the laboratory by conducting an experiment called a *flame test*. In this experiment, a wire is dipped in a solution containing metal ions (recall that ions are charged particles). The ions of each element give off a characteristic color when heated (**Figure 4.10**). For example, when a wire is dipped in a solution containing calcium, the flame burns bright orange. When dipped in a solution of copper, the flame burns bright green.

We see a similar effect in gas lamps, such as those used in neon signs. These lamps produce light by passing an electric current through a tube filled with a gas such as neon, helium, argon, or krypton (**Figure 4.11**). Like the metals in a flame test, each gas in a lamp produces a characteristic color.

Figure 4.11 Gas lamps produce characteristic colors depending on which element is present. Each of these lamps contains a different noble gas.

We can use a glass *prism* to analyze the light from these lamps. When light passes through a prism, it separates into its constituent colors. Many sources of white light produce all of the visible colors. If we pass this light through a prism, we see the complete rainbow of colors (**Figure 4.12**). But if we do the same thing with light from a gas lamp, we see a fascinating result (**Figure 4.13**): Rather than producing a continuous spectrum of colors, gas lamps produce colors of only certain energies. These bands of color are called *spectral lines*. For example, the light from a hydrogen lamp contains only four lines of color: red, light blue, deep blue, and violet. These distinctive patterns are unique to each element and are called **line spectra**.

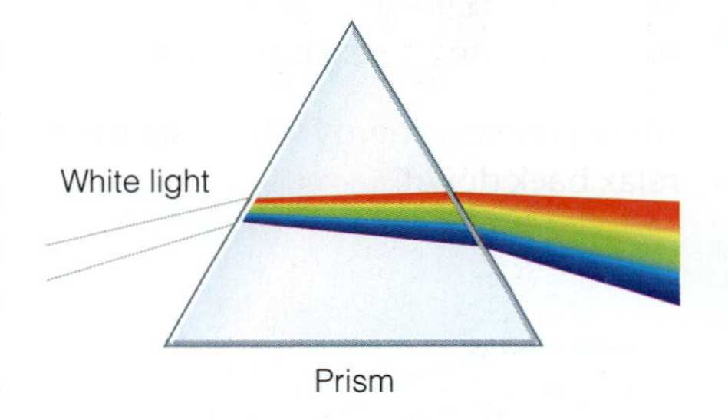

Figure 4.12 When white light passes through a prism, it produces all the colors of the rainbow.

Scientists often use line spectra as "fingerprints" to identify elements. For example, when scientists analyze light from the Sun, they find lines that correspond to the spectral lines of hydrogen and helium. From this, they know that the Sun and other stars are composed largely of these two elements.

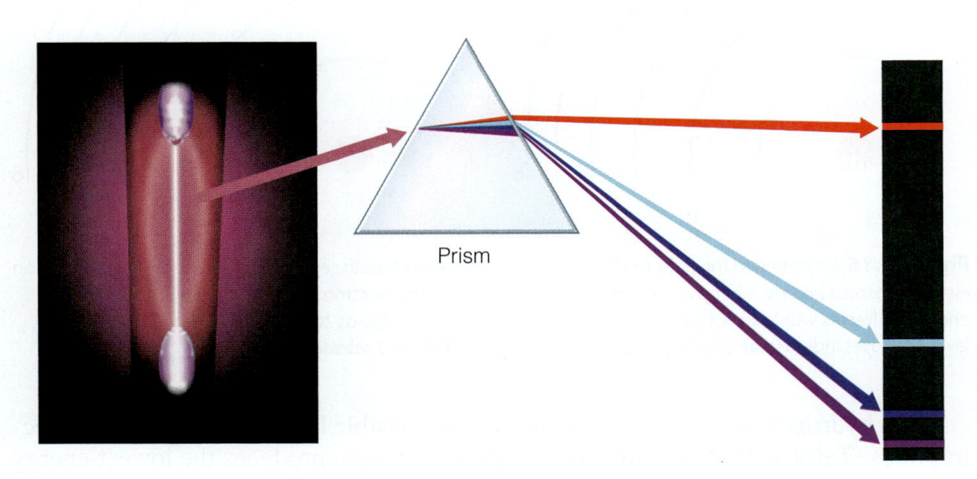

Figure 4.13 When light from a gas lamp passes through a prism, it separates into distinct lines of color that are called *spectral lines*. In this figure the lamp is filled with hydrogen, and the spectrum is unique to that element.

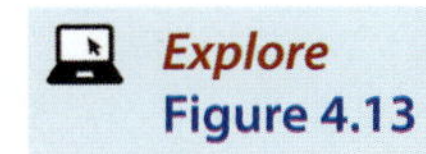

Explore
Figure 4.13

Light can be separated into its constituent colors by using either a prism or a *diffraction grating*—a surface containing very close parallel lines. You can see this effect on the surface of a CD or DVD. Modern instruments use diffraction gratings rather than prisms to separate light.

The Bohr Model

By the early twentieth century, scientists knew that atoms contain a dense nucleus surrounded by electrons. They also knew that high-energy light could knock electrons off of some atoms—a phenomenon known as the *photoelectric effect*. This effect showed that light energy was somehow connected to electron structure. But how did it relate to the line spectra?

In 1913, the Danish scientist Niels Bohr unveiled a theory (now called the **Bohr model**) that connected line spectra with electron structure. Bohr proposed that electrons orbit the nucleus in much the same way that planets orbit the Sun (**Figure 4.14**). He suggested that orbits closer to the nucleus are lower in energy than those that are farther away, and he posited that only certain orbits, or *energy levels*, are "allowed."

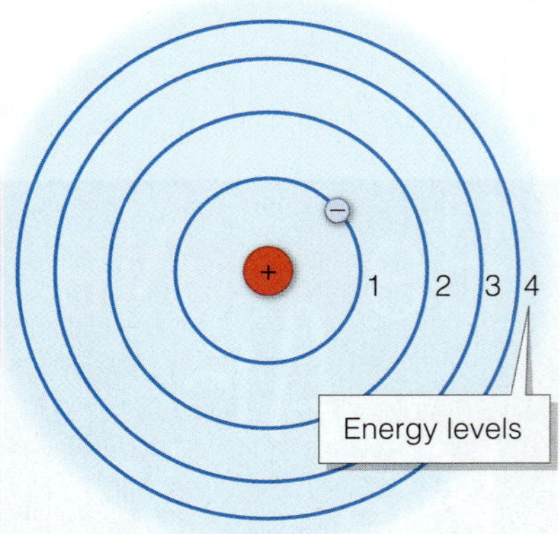
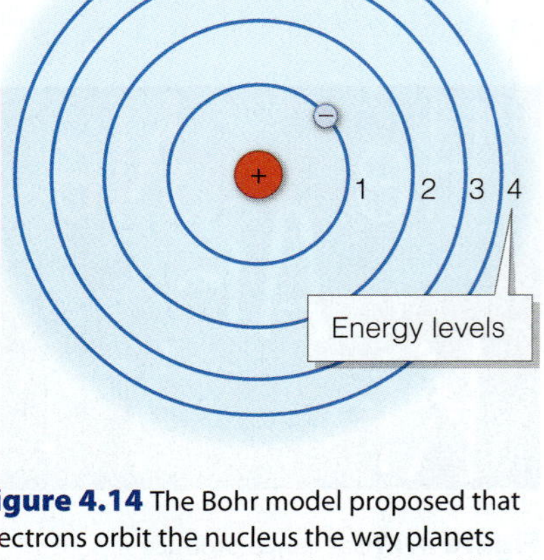

Figure 4.14 The Bohr model proposed that electrons orbit the nucleus the way planets orbit the Sun and that the electrons could jump from one energy level to another.

Bohr also proposed that electrons can jump from one energy level to another. He theorized that when an electron absorbs light, it jumps to a higher energy level. When it drops to a lower energy level, it releases that energy as light. When an atom's electrons are in the lowest possible levels, the atom is said to be in the *ground state*. If electrons jump to higher levels, the atom is in the *excited state*.

When atoms absorb energy, electrons jump to excited states.

Atoms release energy when electrons relax back down.

The Bohr model neatly explained the line spectra of hydrogen (**Figure 4.15**). In its ground state, hydrogen's one electron occupies the lowest possible level—energy level 1. However, this electron can absorb energy (in the form of heat, light, or electrical energy). When this happens, the electron jumps from level 1 (the ground state) to a higher energy level (an excited state). Eventually, the electron relaxes from the excited state back down to lower energy states, and ultimately to the ground state. As it does so, it releases energy as electromagnetic radiation.

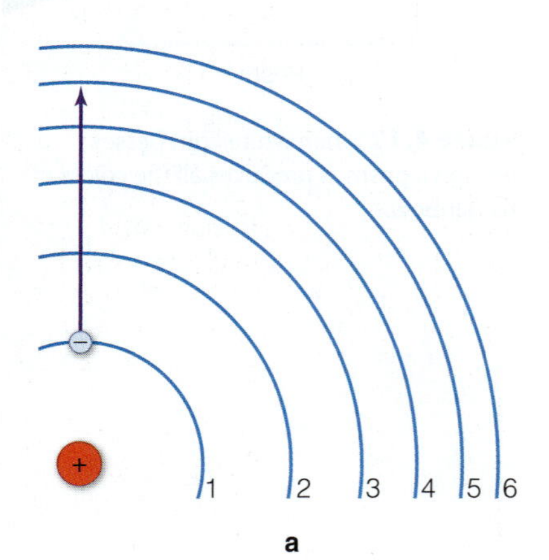

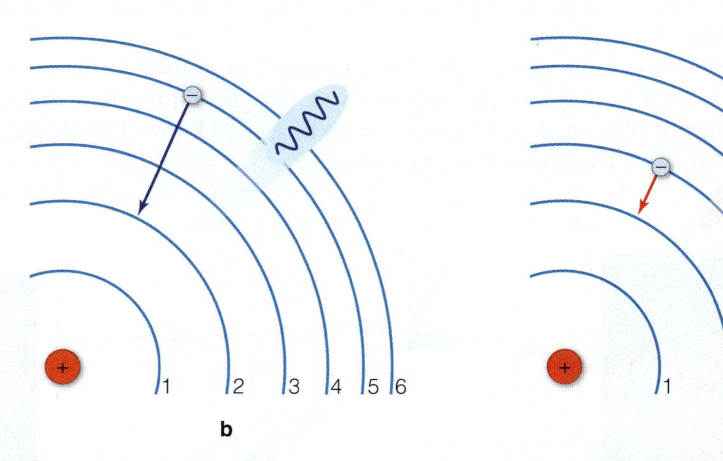

Figure 4.15 According to the Bohr model, electrons orbit the nucleus in certain allowed energy levels. When an electron absorbs energy, it jumps to a higher energy level. When the electron comes back down, it releases that energy—often as visible light. (a) An electron absorbs energy and jumps up to level 5. (b) A drop from level 5 to level 2 releases indigo (dark blue) light. (c) A drop from level 3 to level 2 releases red light.

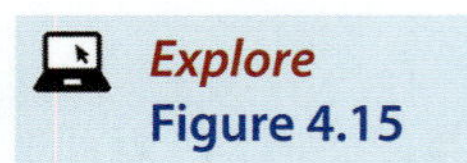

Explore Figure 4.15

In a hydrogen atom, four transitions produce visible light, resulting in four spectral lines (**Table 4.1**). Notice that the smallest transition produces the lowest-energy light (red), while the biggest transition produces the highest-energy light (purple). Other transitions also take place, but they produce radiation that is either too high in energy (ultraviolet) or too low in energy (infrared) for our eyes to detect.

Imagine standing on a staircase. You can jump from one step to another, but you can't hover between steps—you will always land on one of the steps. Similarly, electrons occupy only specific energy "steps." Electrons can absorb energy to move to a higher level (step) or release energy to move to a lower level.

From bonfires to smartphones, light is produced when electrons are excited to higher levels and then relax back down.

Johan Mard/Folio Images/Media Bakery

TABLE 4.1 Transitions in the Hydrogen Line Spectrum

Transition	Color Produced
$3 \rightarrow 2$	Red
$4 \rightarrow 2$	Light blue
$5 \rightarrow 2$	Indigo (deep blue)
$6 \rightarrow 2$	Purple (violet)

Bohr's ideas about light and electron energy levels were a critical advance. They provided a foundation for understanding the interplay of light and electrons that occurs all around us. Is your shirt colored? It is absorbing some colors, but reflecting others. The absorption of light energy causes electrons to jump to higher levels. Do you have a light on as you're reading this? The electrical current is continually exciting the electrons to higher energy levels. As they relax back down, this energy is released as visible light. Here are a few other examples related to this concept:

1. *Why does the Sun give off light?* The Sun is powered by a type of reaction called *nuclear fusion*, which releases an incredible amount of energy (we'll discuss nuclear fusion in Chapter 16). This energy continually excites electrons (and other charged particles), which release this energy as electromagnetic radiation. The Sun radiates electromagnetic energy across the spectrum, not just visible light.
2. *Why does a fire give off light?* When a substance like wood burns, the carbon and hydrogen in the substance react with oxygen in the air. This reaction releases heat energy as well as gaseous products such as carbon dioxide and water vapor. The electrons within these compounds are excited to higher levels. As these gaseous products exit the reaction, their electrons release this energy as visible light that we see as flames.
3. *Do you give off light?* People don't give off visible light; however, as your body produces heat, you release this energy in the form of infrared radiation. Specialized cameras can convert infrared energy into visible images, making it possible to "see" someone in the dark (**Figure 4.16**). This technique is called *infrared imaging* or *thermal imaging*.

a

Monty Rakusen/Getty Images

b

Collection Christophel/Alamy

Figure 4.16 (a) This image shows the infrared energy given off by two factory workers. (b) In the classic action movie *Predator*, Arnold Schwarzenegger covered himself in mud to mask the infrared radiation produced by his body heat.

Fluorescence

Not every electronic transition involves visible light. Sometimes, a substance absorbs invisible UV energy and then releases it in lower-energy steps as visible light. This phenomenon is known as *fluorescence*. Ultraviolet lamps (also called black lights) cause many substances to fluoresce. For example, bark scorpions are nearly impossible to spot with the naked eye, but they contain substances that fluoresce and show up easily under a black light.

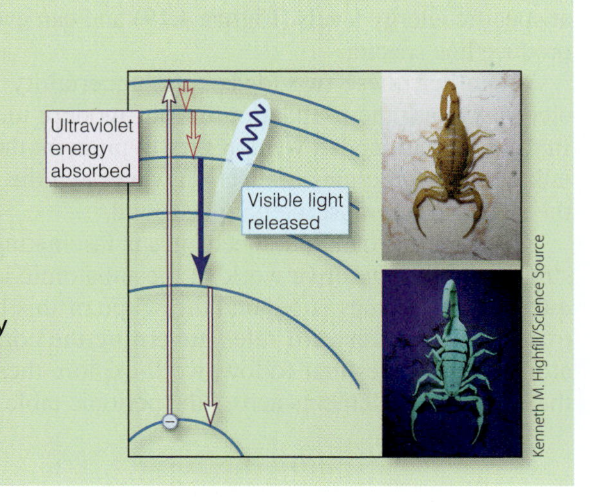

Kenneth M. Highfill/Science Source

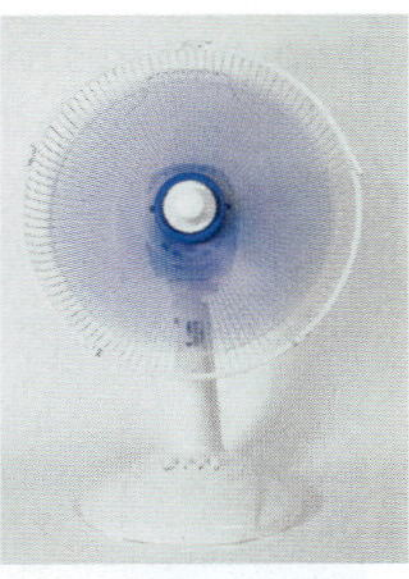

Ted Kinsman/Science Source

Figure 4.17 As the blades on a fan move faster and faster, we can no longer pinpoint their location. Instead we describe them by the shape of the region they occupy.

4.3 The Quantum Model and Electron Orbitals

The Bohr model proved useful for describing the chemical properties of the main-group elements, and the concept of electron energy levels was a key advance that elegantly explained the line spectrum of hydrogen. However, Bohr's model was unable to explain the properties of the transition elements or the complex line spectra of elements larger than hydrogen.

In the 1920s and 1930s, a series of discoveries led scientists to abandon much of the Bohr model in favor of a more nuanced description of electron behavior, called the **quantum model**, which describes electrons as both particles and waves. These early discoveries marked the dawn of *quantum mechanics*, a field of study that deals with the unique and surprising behavior of subatomic particles. In this section we will examine some key ideas from quantum mechanics, and we'll use the quantum model to describe electron behavior more completely.

We never know the exact location of electrons.

The Uncertainty Principle and the Wave Nature of Electrons

In 1927 the German scientist Werner Heisenberg introduced a startling idea called the **uncertainty principle**. This principle deals with the mass, velocity, and location of subatomic particles. A central idea of this principle is that *it is impossible to know the exact velocity and location of a particle*. Now, this doesn't make much sense to us in our everyday world—but as we deal with tiny, fast-moving particles, uncertainty becomes more significant.

As a crude analogy to help think about this concept, consider an electric fan (**Figure 4.17**). When the fan is off, we know exactly where the blades are located. However, if the fan is turned on, the blades move so quickly that we no longer know exactly where they are—we just know they are moving in an area that occupies a circle. We don't stick a finger in that circle, because we know the blades occupy that area.

Similarly, in quantum mechanics, we never talk about the exact position of an electron. According to Heisenberg, this position is impossible for us to know. Instead, we talk about the *most probable locations* of the electrons or the *energies* that the electrons possess.

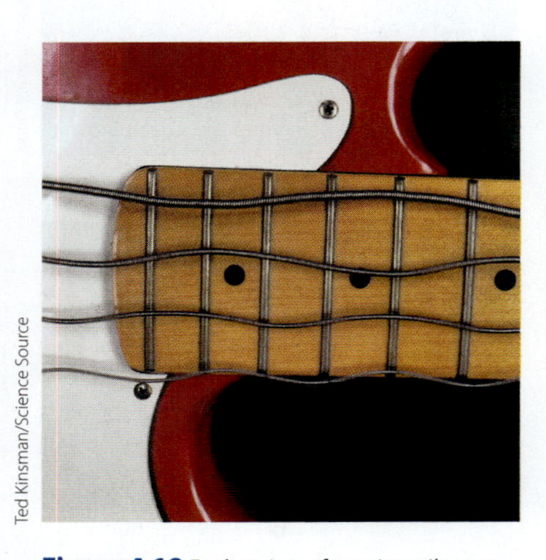

Ted Kinsman/Science Source

Figure 4.18 Each string of a guitar vibrates with a specific wavelength and frequency, producing a unique note.

A second principle of quantum mechanics is equally surprising: When dealing with tiny particles (such as electrons), we can't describe them simply as particles, because many of their behaviors more closely resemble energy waves. For example, the line spectra of the elements clearly show that electrons exist only at certain energies. If electrons are simply negatively charged particles, we can't explain why this is so. But what if electrons behave like waves? Many types of waves, such as those produced by a guitar string (**Figure 4.18**), oscillate at specific energies. By treating electrons as energy waves, quantum mechanics can explain why electrons exist only at specific energy levels (**Figure 4.19**) and can even predict the energy changes that produce line spectra.

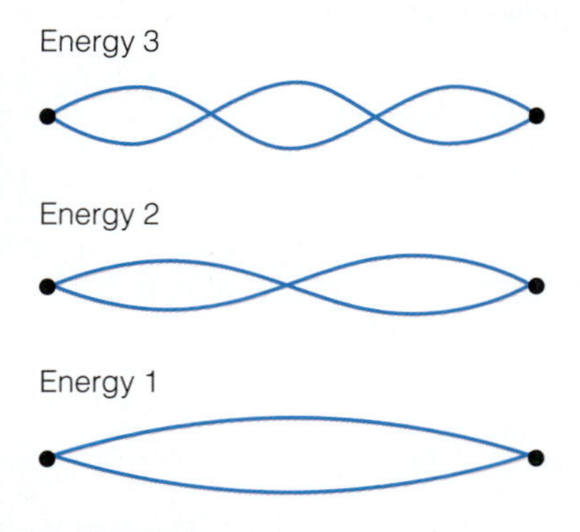

Figure 4.19 Describing electrons as waves helps explain why they exist only at specific energy levels.

Together, these two ideas—the uncertainty principle and the wave nature of electrons—are the foundation for our modern understanding of electron structure. In the quantum model, we don't try to pinpoint the location of electrons. Rather, we talk about the energies of electron waves and the regions around the nucleus where the electrons are most likely to be found.

The story of quantum mechanics is fascinating and full of surprises. It is a wild, strange, counterintuitive world at the subatomic level. And though the math behind quantum mechanics is beyond the scope of this book, the complex equations lead to some straightforward rules governing the configuration of electrons within an atom. In the pages that follow, we'll explore these rules and see how they explain the behavior of elements across the periodic table.

TABLE 4.2 Electrons in Each Energy Level

Level	Possible Electrons
1	2
2	8
3	18
4	32

TABLE 4.3 Energy Sublevels

Sublevel	Number of Orbitals	Electron Capacity
s	1	2
p	3	6
d	5	10
f	7	14

Energy Levels and Sublevels

The quantum model describes how electrons are arranged within an atom. To describe these arrangements, let's begin with four basic rules:

1. *Electrons occupy different energy levels.* In quantum mechanics, the energy level is identified by a whole number called the **principal quantum number** (**Figure 4.20**). The lowest energy level (level 1) lies closest to the nucleus. Higher energy levels (2, 3, 4, etc.) lie farther from the nucleus. Each energy level can hold a maximum number of electrons (**Table 4.2**). Higher energy levels can hold more electrons than lower energy levels.
2. *Each energy level contains one or more sublevels.* There are four common energy **sublevels**, which are designated by the letters *s*, *p*, *d*, and *f*. Each sublevel can hold a set number of electrons, as summarized in **Table 4.3**.
3. *Each sublevel contains one or more orbitals.* An **orbital** is the region where the electrons are most likely to be found. Each sublevel contains a different number of these orbitals. An *s* sublevel contains 1 orbital; a *p* contains 3, a *d* contains 5, and an *f* contains 7.
4. *Each orbital can hold up to two electrons.* Electrons have a tiny magnetic field, called *spin*. When electrons pair together in orbitals, their spins orient in opposite directions. An orbital is filled if it contains two paired electrons.

Now that we have the fundamental rules of electron arrangements, let's look at each of the energy levels and sublevels.

Energy level 1 contains only the *s* sublevel. The **s sublevel** is the simplest of the sublevels. The *s* sublevel has only one orbital, shaped like a sphere (**Figure 4.21**).

Energy level 2 contains both *s* and *p* sublevels. The 2*s* sublevel is spherical in shape, but it lies farther from the nucleus than the 1*s* sublevel (**Figure 4.22**). The **p sublevel** contains three different orbitals. Each orbital has a shape similar to

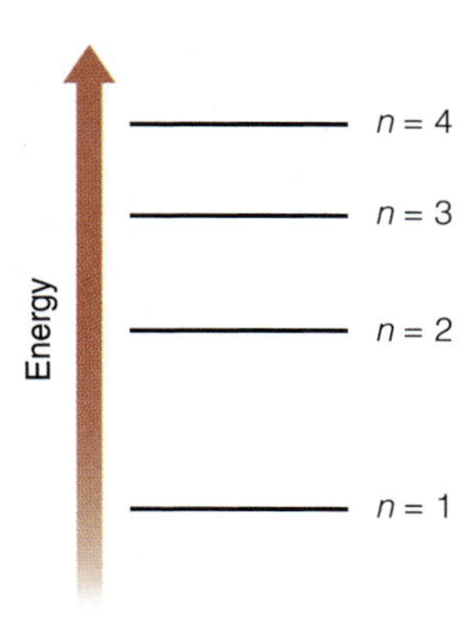

Figure 4.20 Electrons occupy different energy levels, which are represented by the principal quantum number, *n*. Energy level 1 is the lowest in energy, and it is closest to the nucleus.

Energy levels and sublevels are sometimes referred to as *energy shells* and *subshells*, respectively.

The difference between *orbits* and *orbitals* can be confusing. The Bohr model described electrons as particles orbiting the nucleus, like planets orbit the Sun. The quantum model states that we can never actually know the location of the electron. The term *orbital* describes the region around the atom where the electron is most likely to be.

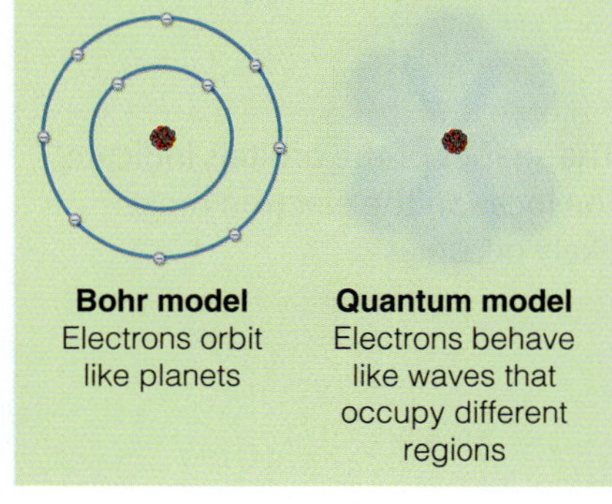

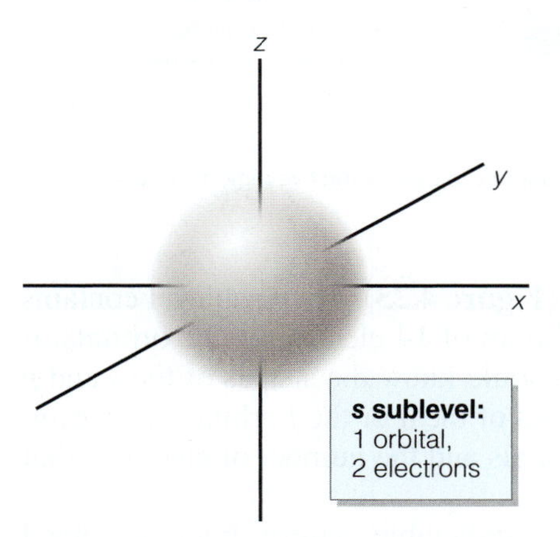

Figure 4.21 The *s* sublevel can hold two electrons. The shading represents the region the electron will most likely occupy.

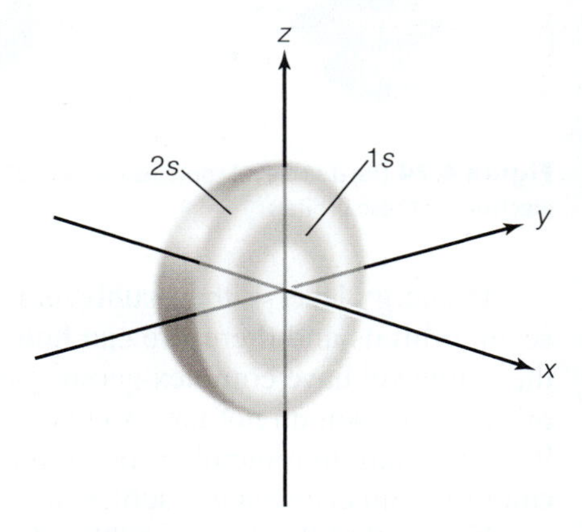

Figure 4.22 This image shows a cutaway view of the 1*s* and 2*s* sublevels. Both sublevels have a spherical shape, but the 2*s* sublevel is farther from the nucleus.

an infinity symbol (**Figure 4.23**). The three orbitals can be thought of as orienting along the *x*, *y*, and *z* axes. Because each orbital can hold up to two electrons, a *p* sublevel can contain a total of six electrons.

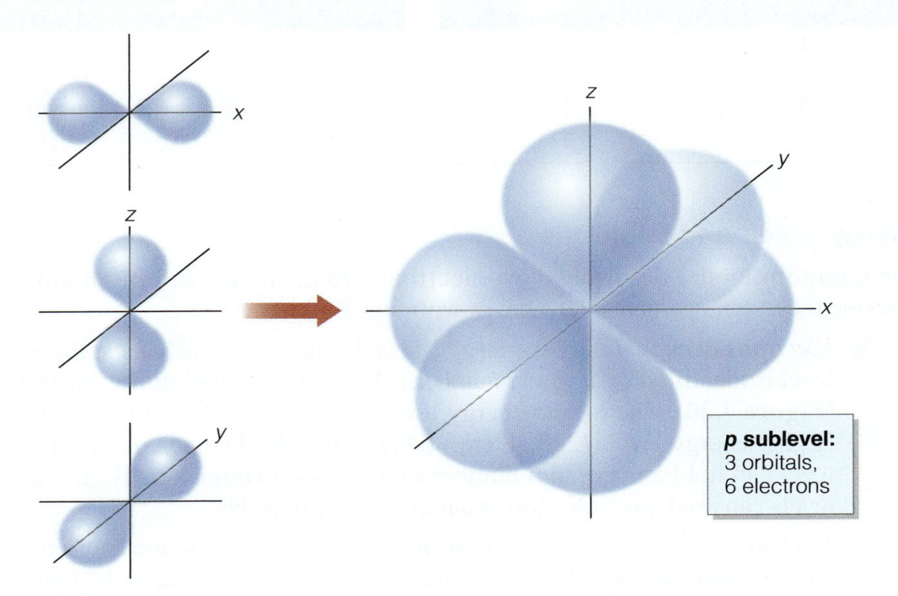

Figure 4.23 The *p* sublevel is composed of three orbitals occupying the regions around the *x*, *y*, and *z* axes.

If you place two bar magnets side by side, they will pair up with their poles facing in opposite directions. Similarly, when two electrons occupy the same orbital, their magnetic fields (that is, their spins) orient in opposite directions. We often show this as two adjacent, half-headed arrows—one pointing up and one pointing down.

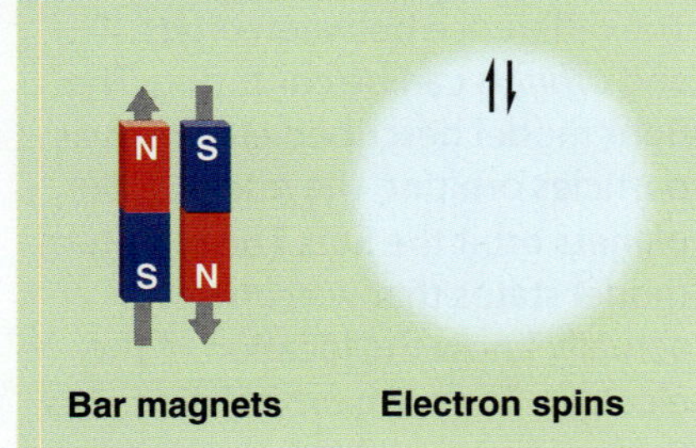

Energy level 3 contains not only *s* and *p* sublevels but also the *d* sublevel. The *d* sublevel contains five orbitals, and therefore it can hold a maximum of 10 electrons. The shapes of the five *d* orbitals are shown in **Figure 4.24**.

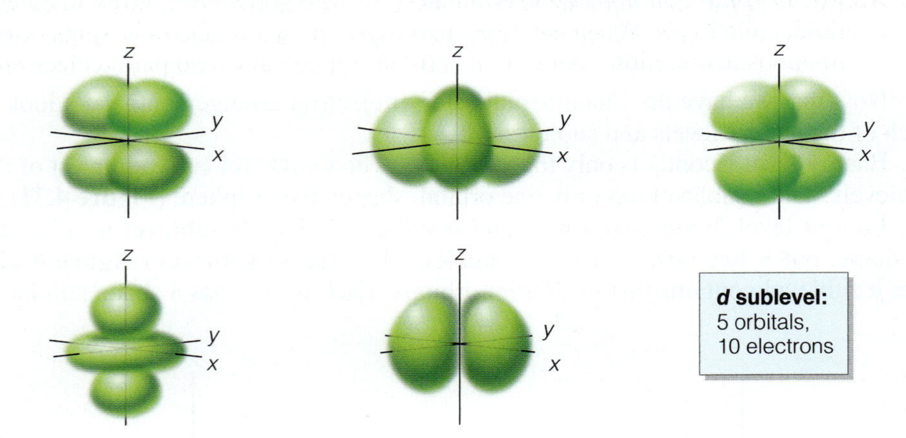

Figure 4.24 The *d* sublevel contains five orbitals. These diagrams represent the five regions occupied by electrons in these orbitals.

The shapes of the orbitals indicate the location the electron most likely occupies.

In energy level 4, the *f* sublevel is added (**Figure 4.25**). The *f* sublevel contains seven orbitals and therefore can hold a maximum of 14 electrons. The orbitals of the *f* sublevel have complex geometries. You should know the shapes of the *s* and *p* orbitals, but you do not have to know the shapes of the *d* or the *f* orbitals. However, it is important to remember the number of orbitals and the number of electrons that can fit within each energy sublevel.

Notice that the energy sublevels follow a predictable pattern: Each new level adds a new sublevel, and each new sublevel has two more orbitals (four more electrons) than the one before. **Table 4.4** is a summary of the sublevels for energy levels 1 through 4.

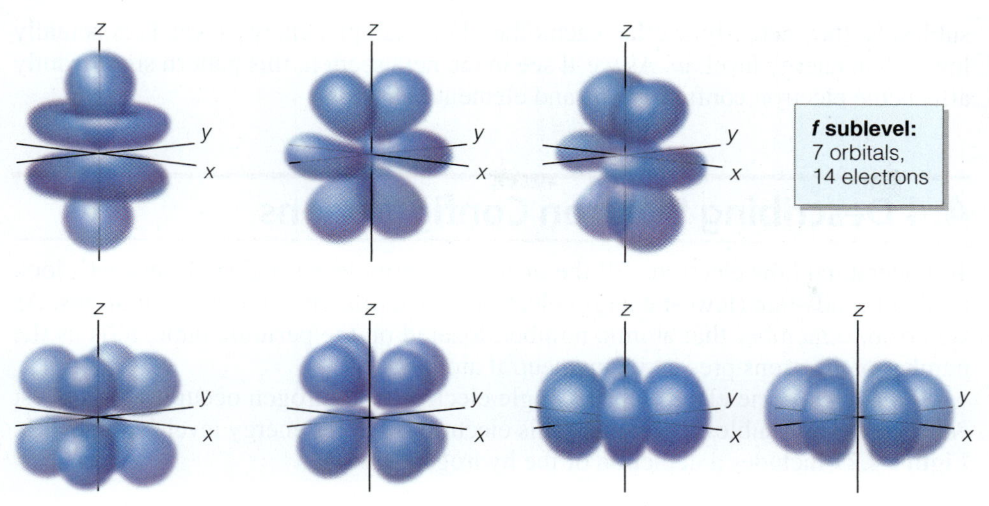

Figure 4.25 The *f* sublevel contains seven orbitals. These diagrams represent the seven regions occupied by electrons in these orbitals.

TABLE 4.4 Energy Levels, Sublevels, and Electron Capacity

Energy Level	1	2	3	4
Sublevels				*f* (14 e^-)
			d (10 e^-)	*d* (10 e^-)
		p (6 e^-)	*p* (6 e^-)	*p* (6 e^-)
	s (2 e^-)	*s* (2 e^-)	*s* (2 e^-)	*s* (2 e^-)
Electron Capacity	2	8	18	32

Note: the symbol e^- means electrons.

Above energy level 4, the trend continues: Level 5 has five sublevels, level 6 has six sublevels, and so on. However, even the largest elements on the periodic table fit all of their electrons within the *s*, *p*, *d*, and *f* sublevels, so we don't need to worry about any sublevels beyond these four.

Figure 4.26 shows the relative energy differences between the different energy levels and sublevels. Notice that within an energy level, the *s* orbital is the lowest in energy, followed by the *p*, *d*, and *f* orbitals. Also notice that as the energy levels get higher, they group together more closely. As the energy levels branch out into more

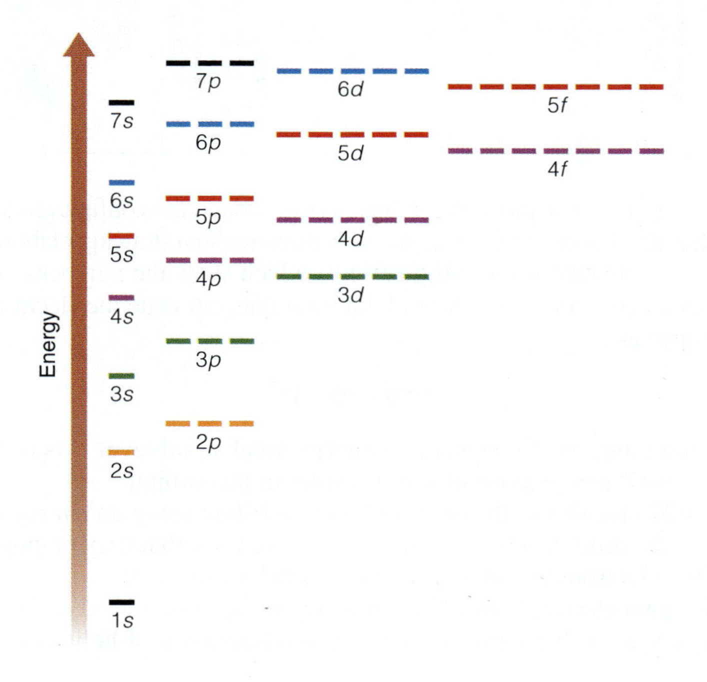

Figure 4.26 This diagram represents the energy differences among the energy sublevels. Each horizontal line represents an orbital that can hold two electrons. Notice that the higher energy levels overlap each other.

sublevels, they actually overlap each other. For example, energy level 4*s* is actually lower than energy level 3*d*. As we'll see in the next section, this pattern significantly affects the electron configurations and element behaviors.

4.4 Describing Electron Configurations

To understand how electrons fill the different energy levels and sublevels, let's look at the ground-state (lowest energy) electron configurations of a series of atoms. As we do so, remember that atomic number, located on the periodic table, tells us the number of electrons present in any neutral atom.

Hydrogen—one electron. The single electron in hydrogen occupies the lowest energy level and sublevel possible. This electron occupies energy level 1, sublevel *s*. **Figure 4.27** includes a depiction of the hydrogen atom.

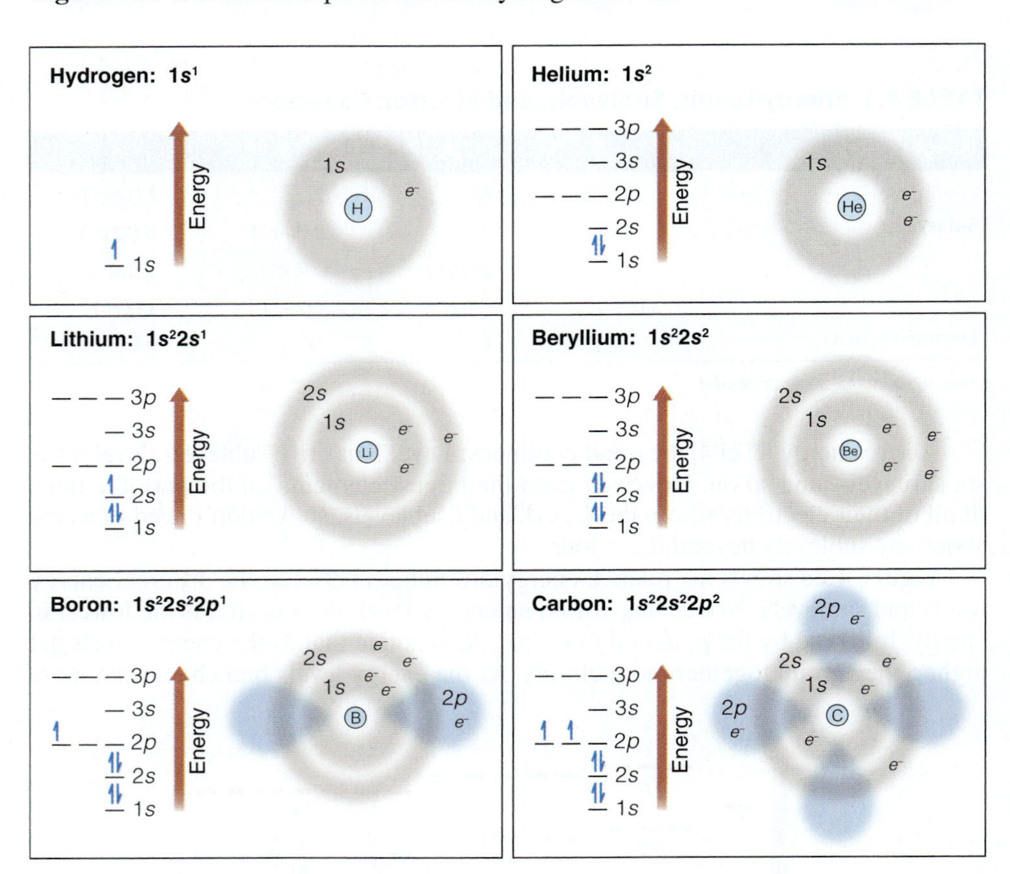

Figure 4.27 Electron configurations may be represented by a written electron configuration (at the top of each box), by an energy diagram (lower left in each box), or as a sketch showing the shape of the occupied orbitals (lower right).

Although graphical depictions can help us understand the configuration of an atom, we sometimes need to express electronic structures without drawing elaborate pictures. To do this, we write **electron configurations**, which show the number of electrons in each occupied energy level and sublevel. For example, we write the electron configuration of hydrogen as

$$\text{Hydrogen: } 1s^1$$

In this notation, the 1*s* means that energy level 1, sublevel *s*, is occupied. The superscript 1 indicates that one electron resides in this orbital.

Figure 4.27 also shows the order of electron filling using an *energy diagram*. In this diagram, the orbitals are represented by blank lines that may be populated with electrons. The electrons are shown as half-headed arrows.

Helium—two electrons. Both electrons go in the lowest level and sublevel (1*s*), filling energy level 1. We write the electron configuration of helium as $1s^2$. Notice in the energy diagram of helium (see Figure 4.27) that when we represent two electrons in one orbital, we draw one with the half-arrowhead up (↿) and the other with the half-arrowhead down (⇂). This notation signifies that the spins (that is, the magnetic fields of the electrons) are oriented in opposite directions.

Lithium—three electrons. The first two electrons fill the first energy level. The third electron occupies the next-lowest sublevel, which is 2*s*. We write the electron configuration for lithium as $1s^2 2s^1$, to indicate that there are two electrons in level 1*s* and one electron in level 2*s*.

Beryllium—four electrons. The electron configuration for beryllium is $1s^2 2s^2$.

Boron—five electrons. After filling sublevels 1*s* and 2*s*, the next-lowest energy sublevel is 2*p*. The electron configuration for boron is $1s^2 2s^2 2p^1$.

Carbon—six electrons. The first four electrons fill sublevels 1*s* and 2*s*. The two remaining electrons occupy level 2*p*. However, this configuration introduces a new question: Because three identical *p* orbitals are available, do the two electrons pair up in a single orbital, or does each electron singly occupy its own orbital? In this case, the electrons occupy their own orbitals rather than pairing up (**Figure 4.28**). This idea is referred to as *Hund's rule*: If empty orbitals of the same energy are available, electrons will singly occupy orbitals rather than pair together.

The electron configurations of the remaining elements in row 2 of the periodic table are shown in **Figure 4.29**. Notice that as we move across the periodic table in row 2, the second energy level fills with electrons. Neon, the last element in row 2, has a completely filled energy level.

Imagine that you are a manager for a dormitory. Each room can hold two people, but the dorm is not full. If you gave the residents a choice of having their own room or sharing a room, what would they choose? If there is no difference in rent, most people would choose their own room. Similarly, if empty orbitals are available, each electron singly occupies its own orbital before pairing up to share orbital space.

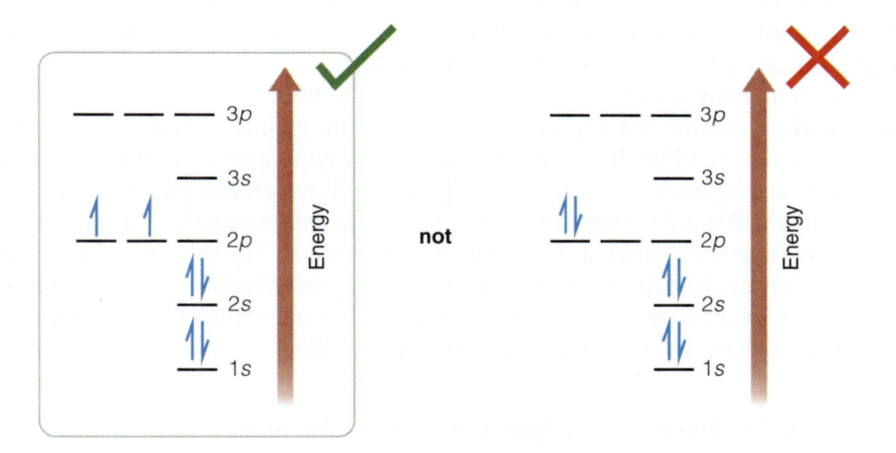

Figure 4.28 If empty orbitals of the same energy are available, electrons will singly occupy orbitals rather than pairing together. This is referred to as *Hund's rule*.

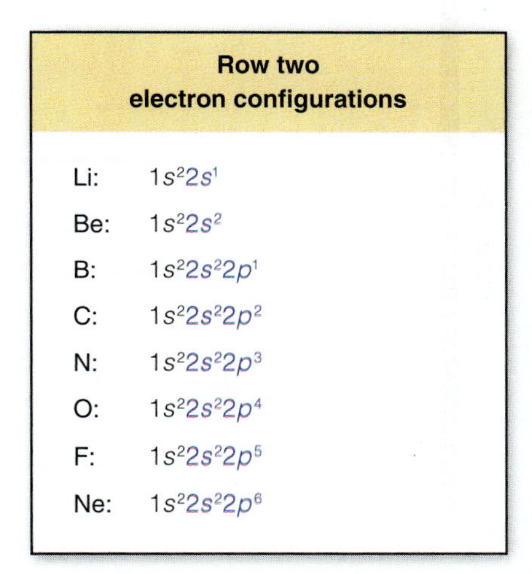

Row two electron configurations	
Li:	$1s^2 2s^1$
Be:	$1s^2 2s^2$
B:	$1s^2 2s^2 2p^1$
C:	$1s^2 2s^2 2p^2$
N:	$1s^2 2s^2 2p^3$
O:	$1s^2 2s^2 2p^4$
F:	$1s^2 2s^2 2p^5$
Ne:	$1s^2 2s^2 2p^6$

Figure 4.29 Electron configurations for the elements in row 2 of the periodic table are shown here.

Row 3 of the periodic table contains the elements whose highest-energy electrons occupy level 3. Example 4.3 illustrates the electron configuration for a row 3 element.

Example 4.3 Writing Electron Configurations

Silicon is a vital component of electronic devices. What is the electron configuration of silicon? Refer to the periodic table and Figure 4.26 as needed.

To answer this question, we first need to identify the number of electrons in a silicon atom. From the periodic table, we can see that silicon is atomic number 14: This means there are 14 electrons in a neutral silicon atom.

The atomic number tells us both the number of protons and the number of electrons in an atom. ■

We begin by placing two electrons in the lowest energy level (1*s*). Sublevel 2*s* can hold two more electrons. Sublevel 2*p* can hold six. The next-lowest level is 3*s*, which can hold two electrons. The remaining two electrons go in the next-lowest level, which is 3*p*. Based on this information, we can write the electron configuration for silicon as $1s^2 2s^2 2p^6 3s^2 3p^2$. The superscript numbers indicate the number of electrons in each level.

The superscripts show the number of electrons in each sublevel. These numbers should add up to equal the number of electrons in the atom. ■

Check It
Watch Explanation

TRY IT

6. Without referring to Figure 4.29, write the electron configurations for each atom.

He N O F Ne Na

Valence Electrons and the Octet Rule

Electron configurations are important because they help determine how an atom behaves. As we'll see in the chapters to come, chemical changes involve the gain, loss, or sharing of electrons. These changes take place at the outermost (highest energy) levels and sublevels. Understanding the electron configuration enables us to explain and predict an element's chemical properties.

Valence electrons occupy *s* and *p* sublevels.

The highest-occupied electron energy level is the **valence level** (sometimes called the *valence shell*). As a general rule, *atoms hold up to eight electrons in their valence (highest-occupied) level*. To see why this is so, consider the simplified representation of the electron energy levels in **Figure 4.30**. Notice that the level above 2*p* is 3*s*. The level above 3*p* is 4*s*, and the level above 4*p* is 5*s*. If an atom has its *s* and *p* sublevels filled, the next electron goes into a higher energy level.

For example, argon has 18 electrons. Therefore, we write its electron configuration as $1s^2 2s^2 2p^6 3s^2 3p^6$. An argon atom has eight electrons in its valence level (energy level 3). The next atom on the periodic table, potassium, has an electron configuration of $1s^2 2s^2 2p^6 3s^2 3p^6 4s^1$. Potassium has one electron in its valence level (energy level 4). Once the *s* and *p* orbitals of the valence level are filled, additional electrons go in the next-lowest energy level.

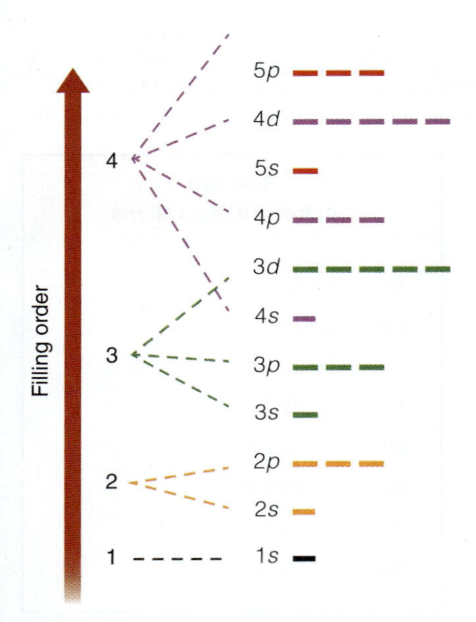

Figure 4.30 Because of the way sublevels overlap, we always move to the next energy level once the *p* sublevel is filled.

The elements with filled valence levels are the noble gases, which are located in the far-right column of the periodic table. Helium, with two electrons, has energy level 1 completely filled. The other noble gases (neon, argon, krypton, xenon, and radon) all have eight electrons in their valence level, which completely fill the *s* and *p* sublevels (**Table 4.5**). These elements are stable and unreactive—they generally do not combine with other atoms to form compounds. We describe this stability using the **octet rule**, which states that *an atom is stabilized by having its highest-occupied (valence) energy level filled*. If an atom has eight electrons (an octet) in its valence shell, the *s* and *p* sublevels are completely filled.

TABLE 4.5 Electron Configurations for Noble Gases

Element	Electron Configuration	Valence Electrons
He	$1s^2$	2
Ne	$1s^2 2s^2 2p^6$	8
Ar	$1s^2 2s^2 2p^6 3s^2 3p^6$	8
Kr	$1s^2 2s^2 2p^6 3s^2 3p^6 4s^2 3d^{10} 4p^6$	8

As you practice writing electron configurations for larger atoms, you will probably want to use the chart in Figure 4.30. However, you don't need to memorize it: As you'll see later in this chapter, the periodic table contains all the information required—you just have to know how to look for it.

Electron Configurations for Larger Atoms

Let's look at the electron configurations for some larger atoms. For example, the following three atoms lie in row 3 of the periodic table:

Sodium (11 electrons): $1s^2 2s^2 2p^6 3s^1$

Phosphorus (15 electrons): $1s^2 2s^2 2p^6 3s^2 3p^3$

Chlorine (17 electrons): $1s^2 2s^2 2p^6 3s^2 3p^5$

Notice that in each of these configurations, the first two levels (sometimes called the *inner electrons*, shown in red) do not change. If we add electrons, the change affects only the valence level. Because the inner electrons do not change, we often represent them in a simpler form, called *noble gas notation*. For example, $1s^2 2s^2 2p^6$ is the configuration of neon. Rather than writing this configuration out, we represent it by enclosing the symbol for neon in square brackets: [Ne]. Using this shorthand notation makes the electron configurations simpler to write:

Sodium (11 electrons): $[\text{Ne}]3s^1$

Phosphorus (15 electrons): $[\text{Ne}]3s^2 3p^3$

Chlorine (17 electrons): $[\text{Ne}]3s^2 3p^5$

In this notation, we always use the noble gases, which have complete octets in the inner levels.

Besides convenience, using noble gas notation has another advantage. It allows us to focus on *outer electrons* (those beyond the largest filled noble gas configuration). The outer electrons include the valence level and partially filled *d* and *f* sublevels (**Figure 4.31**). Chemical bonds involve changes in the outer electrons. We will explore these changes in Chapter 5.

Chemical bonds involve outer electrons.

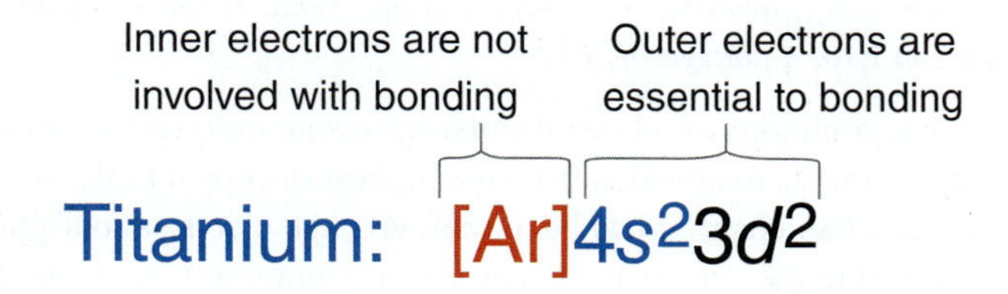
Inner electrons are not involved with bonding
Outer electrons are essential to bonding
Titanium: [Ar]4s23d2

Figure 4.31 Chemical bonds involve the outer electrons (those beyond the largest filled noble gas configuration). These include both valence electrons and *d* and *f* orbital electrons.

Example 4.4 Identifying Valence Electrons

How many valence electrons are present in oxygen? How many are present in sulfur?

Oxygen has the electron configuration $1s^2 2s^2 2p^4$. Its valence level (level 2) contains a total of six electrons.

Sulfur has the electron configuration $1s^2 2s^2 2p^6 3s^2 3p^4$. Its valence level (level 3) also contains a total of six electrons.

Notice that both elements contain the same number of valence electrons, and they are located in the same column of the periodic table. As we'll see in the coming section, the columns of the periodic table contain elements with similar valence configurations.

Example 4.5 Identifying Inner, Outer, and Valence Electrons

Write the electron configuration for selenium using noble gas shorthand. Identify the inner electrons, the outer electrons, and the valence electrons.

The full electron configuration of selenium is $1s^2 2s^2 2p^6 3s^2 3p^6 4s^2 3d^{10} 4p^4$. Because the first 18 electrons ($1s^2 2s^2 2p^6 3s^2 3p^6$) correspond to the electron configuration of argon, we can rewrite this using noble gas shorthand:

$$\text{Selenium}: \quad [\text{Ar}]4s^2 3d^{10} 4p^4$$

The electrons within the [Ar] configuration are the inner electrons. The remaining electrons in the 4*s*, 3*d*, and 4*p* sublevels are the outer electrons. Selenium has a total of 16 outer electrons.

The valence electrons are only those in the highest energy level—that is, energy level 4. Selenium has six valence electrons.

Check It
Watch Explanation

TRY IT

7. Write electron configurations for each atom, using noble gas notation.

P I Fe Sr

8. Identify the inner, outer, and valence electrons in the electron configurations of each of the atoms in Question 7.

Electron Configurations for Ions

In Chapter 3, we said that atoms are able to gain or lose electrons to form *ions*—that is, particles with an overall positive or negative charge. Ions are a vital part of chemistry, so it is important to understand and be able to describe the electron configuration of ions as well as neutral atoms. Let's begin with two examples.

The octet rule: Eight electrons in the valence shell is a very stable configuration.

Example 4.6 Writing Electron Configurations for Ions

What is the electron configuration of a sodium atom? What is the electron configuration of a sodium ion with a charge of +1?

The sodium ion has a charge of +1, so it must have one electron fewer than neutral sodium. The electron that is removed is from the highest-occupied (valence) energy shell (3*s*). Notice that once this electron has been removed, the electron configuration of the sodium ion is identical to the very stable electron configuration of neon, as shown here.

Species	Symbol	Full Configuration	Noble Gas Shorthand
Sodium atom	Na	$1s^2 2s^2 2p^6 3s^1$	$[Ne]3s^1$
Sodium ion (+1 charge)	Na^+	$1s^2 2s^2 2p^6$	[Ne]

Example 4.7 Writing Electron Configurations for Ions

What is the electron configuration of an oxide ion, which is an oxygen ion with a charge of −2?

Because the oxide ion (that is, a negatively charged oxygen atom) has a charge of −2, it must have two more electrons than neutral oxygen. The electrons that are gained by the atom fill the lowest-available energy shell. Notice that when this happens, the electron configuration of the oxygen ion is identical to the very stable electron configuration of neon.

Species	Symbol	Full Configuration	Noble Gas Shorthand
Oxygen atom	O	$1s^2 2s^2 2p^4$	$[He]2s^2 2p^4$
Oxide ion (−2 charge)	O^{2-}	$1s^2 2s^2 2p^6$	[Ne]

In Example 4.6, the sodium atom lost an electron and ended up with the same electron configuration as neon. In Example 4.7, the oxygen atom gained two electrons and also ended up with the same electron configuration as neon. Na^+, O^{2-},and Ne are **isoelectronic**, meaning they have the same electron configurations.

The gain, loss, and sharing of electrons is central to the understanding of chemistry. The idea that atoms will gain or lose electrons and either fill or empty their highest-occupied electron energy levels is a recurring theme. You will see it again and again as we study patterns of bonding and reactivity in future chapters.

TRY IT

9. Write electron configurations for these ions.

Li^+ Cl^- S^{2-}

10. In Section 4.4, Na^+, O^{2-}, and Ne are described as *isoelectronic*. Which of the following atoms or ions are also isoelectronic with those ions?

He F F^- Mg^{2+} P

Check It
Watch Explanation

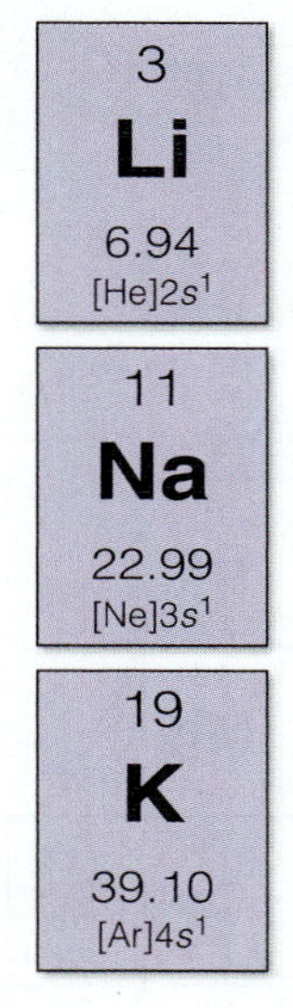

Figure 4.32 Lithium, sodium, and potassium have similar properties because they have similar electronic structure. They are grouped in a column on the periodic table.

4.5 Electron Configuration and the Periodic Table

When Mendeleev organized the periodic table of the elements in the late 1800s, he based it on the chemical and physical properties he observed. He organized elements with similar behaviors into columns. Earlier in this chapter, we said that electron configuration is critically important to the properties of atoms. So how do electron configurations relate to the periodic table?

To answer that question, let's take a look at three atoms that behave very similarly: lithium, sodium, and potassium. All react quickly—even violently—with water, and they all tend to form ions with a +1 charge. Here are the electron configurations of these three atoms:

Lithium (3 electrons): [He]$2s^1$

Sodium (11 electrons): [Ne]$3s^1$

Potassium (19 electrons): [Ar]$4s^1$

Do you see a pattern? In their highest-occupied shell, these atoms all have an s^1 configuration. *These elements exhibit similar behaviors because their electron configurations are similar*. Now think about where we find these elements on the periodic table. They are located in a single column, on the far left-hand side of the table (**Figure 4.32**).

Let's look at another example: fluorine, chlorine, and bromine all form ions with a charge of −1. Here are their electron configurations:

Fluorine (9 electrons): [He]$2s^2 2p^5$

Chlorine (17 electrons): [Ne]$3s^2 3p^5$

Bromine (35 electrons): [Ar]$4s^2 3d^{10} 4p^5$

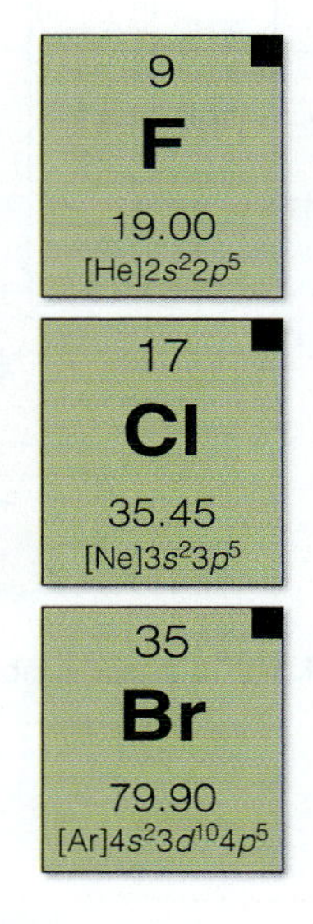

Figure 4.33 Fluorine, chlorine, and bromine have similar properties. They all have a p^5 electron configuration.

What are the similarities this time? In their valence shell, they all have a configuration of s^2p^5. (Bromine has the 3*d* level filled between the *s* and *p* sublevels, but this does not affect the overall trend.) How are these elements organized on the periodic table? As in the previous example, they are located in a single column; but these elements are on the right-hand side of the periodic table (**Figure 4.33**).

The periodic table is organized like a calendar. A row on a calendar shows the whole range of days—Sunday through Saturday. Days that are similar—for example, Mondays—are grouped into columns.

Family Similarities
Families (groups) of elements share similar electron configurations:
Alkali metals: s^1
Alkaline earth metals: s^2
Halogens: s^2p^5
Noble gases: s^2p^6

Now let's take a close look at the periodic table in **Figure 4.34**. The organization of the periodic table is based on element masses and properties, but it is also based on electronic structure. For example, elements in the first row (H and He) have electrons only in energy level 1, elements in the second row (Li through Ne) have electrons in energy level 2, and so forth. For any atom, we can determine its highest-energy electron shell from the row it occupies on the periodic table.

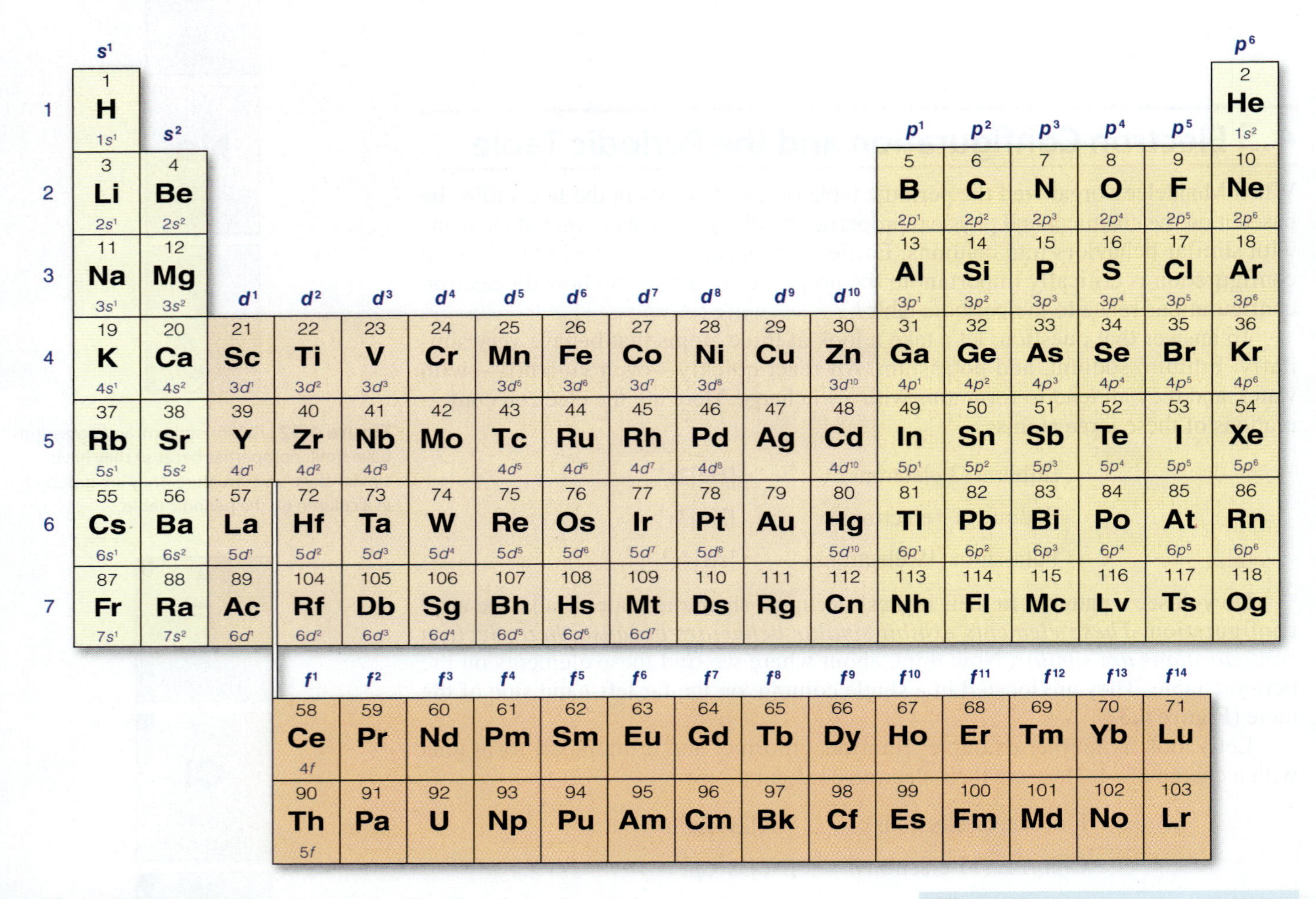

Figure 4.34 The periodic table is organized based on electronic structure.

Explore Figure 4.34

Elements within a group have similar valence configurations. ■

Columns of the periodic table correspond to the sublevel electron configuration. Every atom in the first column has an electron configuration of s^1, every atom in the second column has an electron configuration of s^2, and so on. Once you understand the periodic table, you can quickly determine the electron configuration of the highest-occupied energy levels.

Understanding periodic table organization also makes it unnecessary to memorize the filling sequence of the sublevels. With a glance at the periodic table, you should be able to tell the sequence in which the sublevels are filled. For example, look at the diagram of energy levels in **Figure 4.35**. What energy sublevels come after 5*s*? Moving left to right across the table, you can see that the next sublevels are 4*d*, then 5*p*, then 6*s*.

Helium has its valence energy level filled, and it behaves like a noble gas. On the periodic table, helium is placed above the p^6 noble gases even though its electron configuration is s^2.

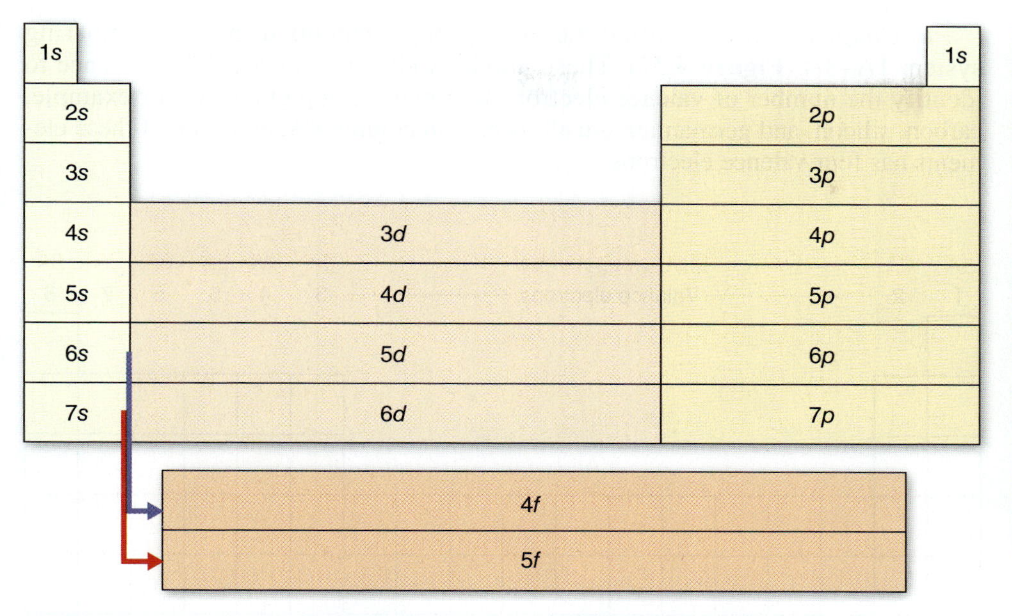

Figure 4.35 We can use the periodic table to determine the energy sublevel filling sequence.

Exceptions to the Rule

A few exceptions to the filling sequence rules are not shown on the periodic table in Figure 4.34. For example, copper (Cu) is expected to have an electron configuration of [Ar]$4s^2 3d^9$, but its actual configuration is [Ar]$4s^1 3d^{10}$. And because the 4*f* and 5*f* orbitals are so close in energy to the 5*d* and 6*d* orbitals, respectively, there are multiple exceptions to the filling sequences in the *f* blocks. Do not worry about remembering these exceptions; instead, focus on understanding the general trends.

As you become more comfortable with how the periodic table is organized, you should be able to quickly identify the valence configuration of an atom. For example, what is the valence configuration of sulfur? The strategy for solving this question is shown in **Figure 4.36**: The row indicates the energy level (3), and the column shows the sublevel and filling (s^2p^4).

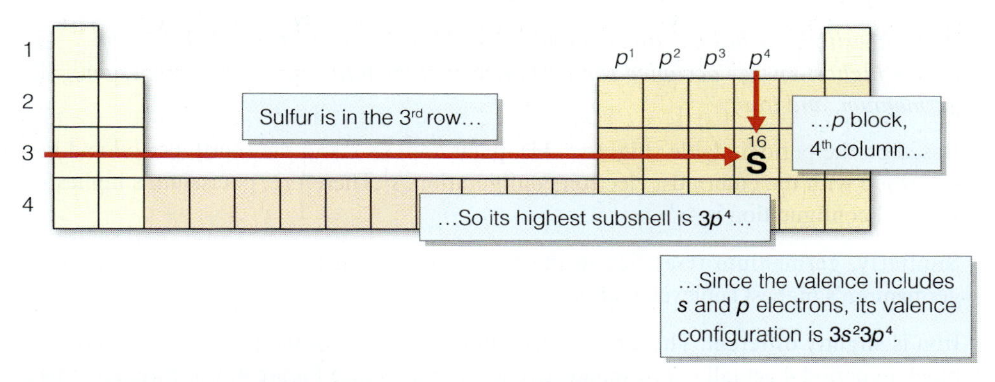

Figure 4.36 The periodic table allows us to quickly identify the valence electron configuration of an atom based on its location.

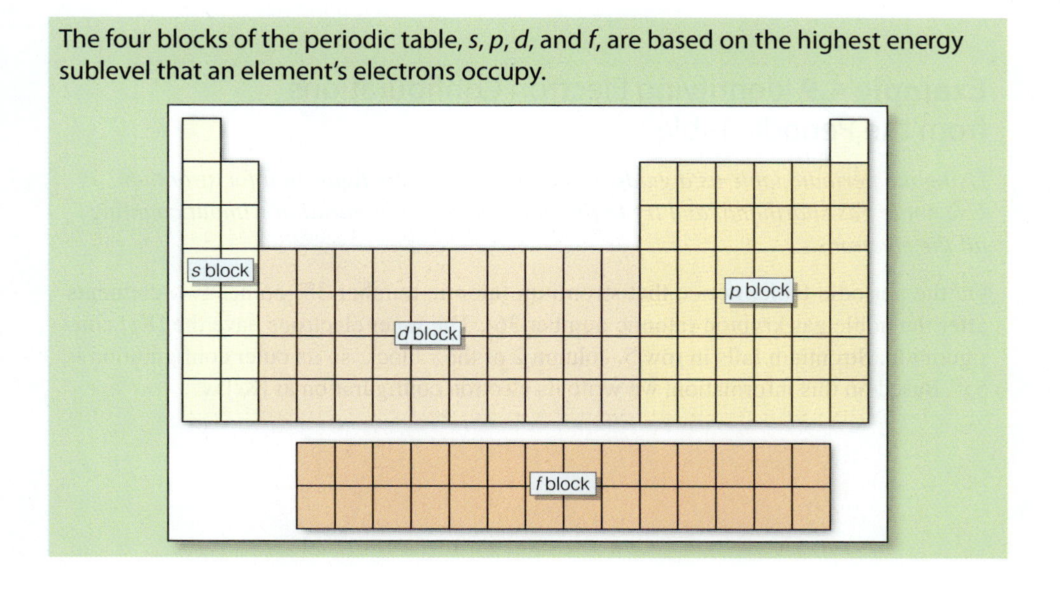

In Chapter 3, we identified the main-group elements using the numbering system 1A–8A (**Figure 4.37**). These numbers also give us a quick reference to identify the number of valence electrons in a main-group element. For example, carbon, silicon, and germanium are all located in column 4A, and each of these elements has four valence electrons.

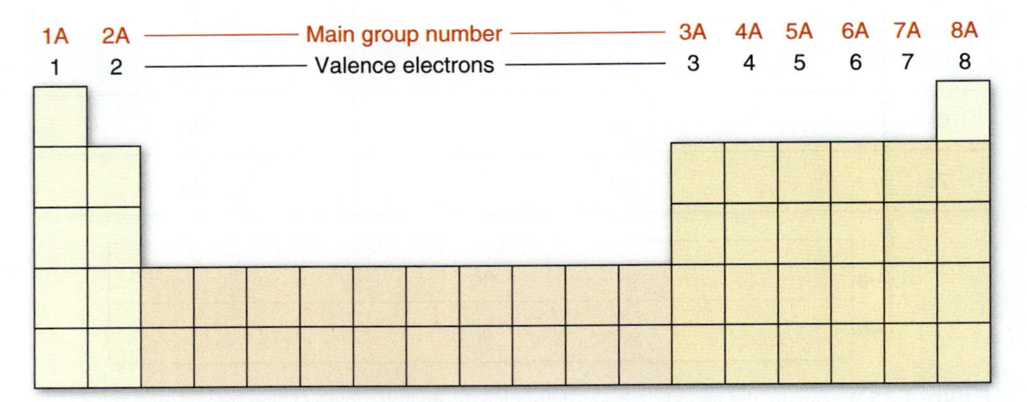

Figure 4.37 The main-group numbers 1A–8A tell us the number of valence electrons in each main-group element.

Example 4.8 Identifying the Highest-Shell Electron Configuration from the Periodic Table

Without writing the full electron configuration, identify the electron configuration for the highest-energy occupied sublevel of each of the following atoms: potassium, germanium, and iron.

Based on the periodic table (Figure 4.34), potassium (K) is in the fourth period and in the group with the outermost electron configuration, s^1.Therefore, potassium's highest-sublevel configuration is $4s^1$.

Similarly, germanium (Ge) lies in the fourth period, in the p^2column. Therefore, germanium's highest sublevel configuration is $4p^2$.

Iron is slightly different. The *d* block fills after the *s* block of the level higher, so the *d* block in period 4 actually corresponds to energy level 3 (see Figure 4.35). Based on this information, we can say that iron has a highest-energy electron configuration of $3d^6$.

Example 4.9 Identifying Electron Configurations from the Periodic Table

Using the periodic table as a guide, write the electron configuration for strontium. Use noble gas shorthand, and try to find the electron configuration without counting all the electrons.

On the periodic table, notice that strontium (atomic number 38) comes two elements after the noble gas krypton (atomic number 36). The inner electrons have the [Kr] configuration. Strontium falls in row 5, column 2 of the *s* block, so its outer configuration is $5s^2$. Based on this information, we write its electron configuration as [Kr]$5s^2$.

TRY IT

11. Oxygen has a valence-level electron configuration of s^2p^4. What are the other atoms that have an s^2p^4 valence-level configuration?

12. Write correct electron configurations for the following atoms. Try to do this using the periodic table as a reference, without adding up the electrons.

    Mg Si Zr Sb

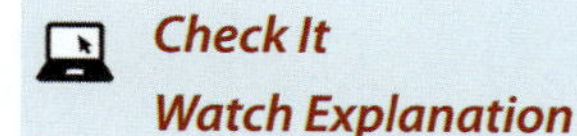

Capstone Question

The soft yellow glow of many streetlights is produced by small amounts of sodium metal present in the bulbs (**Figure 4.38**). In the line spectrum of sodium, we see that this light has a wavelength of 589 nm. When sodium absorbs energy, its highest-energy electron jumps from the ground state to energy level 3*p*. When the electron relaxes back down, it releases this energy as 589-nm yellow light. Write the electron configuration for sodium in its ground state, and then write the electron configuration for the excited sodium atom. What is the energy difference in joules between energy sublevels 3*s* and 3*p* in a sodium atom? How many of these electronic transitions must occur to produce 1.00 kJ of light energy?

ken biggs/Alamy Stock Photo

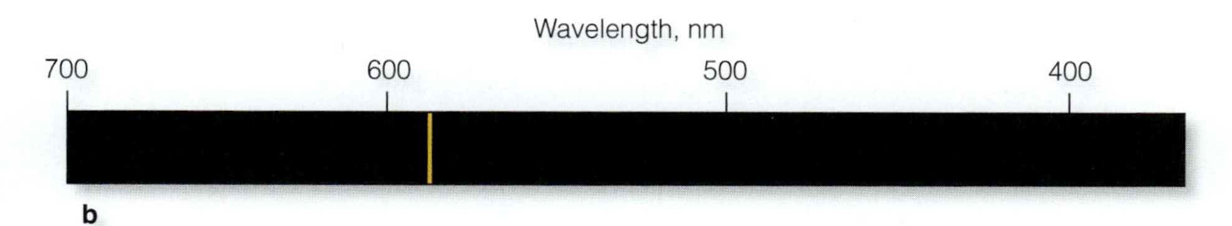

Figure 4.38 (a) The yellow color of these streetlights is produced by small amounts of sodium metal. (b) The line spectrum of sodium contains two closely spaced yellow lines at a wavelength of 589 nm.

SUMMARY

In this chapter we explored electronic structure in detail. Beginning with the electromagnetic spectrum, we saw that the absorption and emission of light (or other electromagnetic radiation) provides critical clues about electronic structure. In the early 1900s, Niels Bohr theorized that (1) electrons orbit the nucleus in much the same way that planets orbit the Sun; (2) electrons transition from one orbit to another by absorbing or releasing energy; and (3) line spectra result from these transitions. Although the planetary model has been discarded, the idea that line spectra result from electron transitions from one energy level to another is critical to understanding electron structure.

The Bohr model was replaced by the quantum model, which describes electronic structure in terms of the *energies* and most probable locations of electrons. Electrons do not occupy planet-like orbits (as the Bohr model suggested) but occur in more complex, wave-like patterns called *orbitals*. Each orbital can hold up to two electrons. These orbitals belong to energy *levels* and *sublevels*.

Energy level 1 has only one sublevel, but each additional energy level includes a new sublevel. The sublevels are *s*, *p*, *d*, and *f*, and they hold 2, 6, 10, and 14 electrons, respectively. Because chemical reactions involve the gain, loss, or sharing of electrons, it is important to understand and be able to express the electron configurations of atoms. Electron configurations in the highest-occupied energy level are especially important. These electrons are referred to as the *valence electrons*.

The periodic table is organized by electronic structure as well as by atomic number and chemical reactivity. We can quickly determine the electronic structure of elements on the periodic table based on the row (primary energy level) and the column (sublevel and electron count) where they lie. Columns on the periodic table contain elements with similar electronic structures and therefore similar patterns of behavior.

Solar Cells: Converting Light into Electric Current

At the beginning of this chapter, we introduced the concept of solar energy. We saw that light and electron structure are closely connected. When a material absorbs light, electrons are excited to higher energy levels. But how does this produce electricity?

Most solar cells use a design called a *p–n junction* to convert sunlight into electrical current. In this design, two slightly different semiconductors (called *p-type* and *n-type*) are placed next to each other. The metalloid element silicon is the main component of both types of semiconductors; the difference is in which other elements are present. An *n*-type semiconductor contains tiny amounts of an element with an s^2p^3 valence configuration, such as arsenic. These elements have high-energy electrons that are easily removed from the atom. (The term *n-type* refers to the negative charges [electrons] that are available in the solid.) On the other hand, *p*-type semiconductors contain tiny amounts of an element with an s^2p^1valence configuration, such as gallium. These elements have empty orbitals in their highest energy level, which can hold additional electrons.

When a *p*-type material is placed next to an *n*-type material, some of the electrons from the *n*-type move to the *p*-type, creating a barrier of positive and negative charges (**Figure 4.39**).

When a photon of light hits the semiconductor, it excites an electron to a higher energy level. In a semiconductor, excited electrons can move from atom to atom. When this happens, the charges along the *p–n* junction push electrons away from the negative charge and toward the positive charge. This push produces a movement of electrons—an electric current.

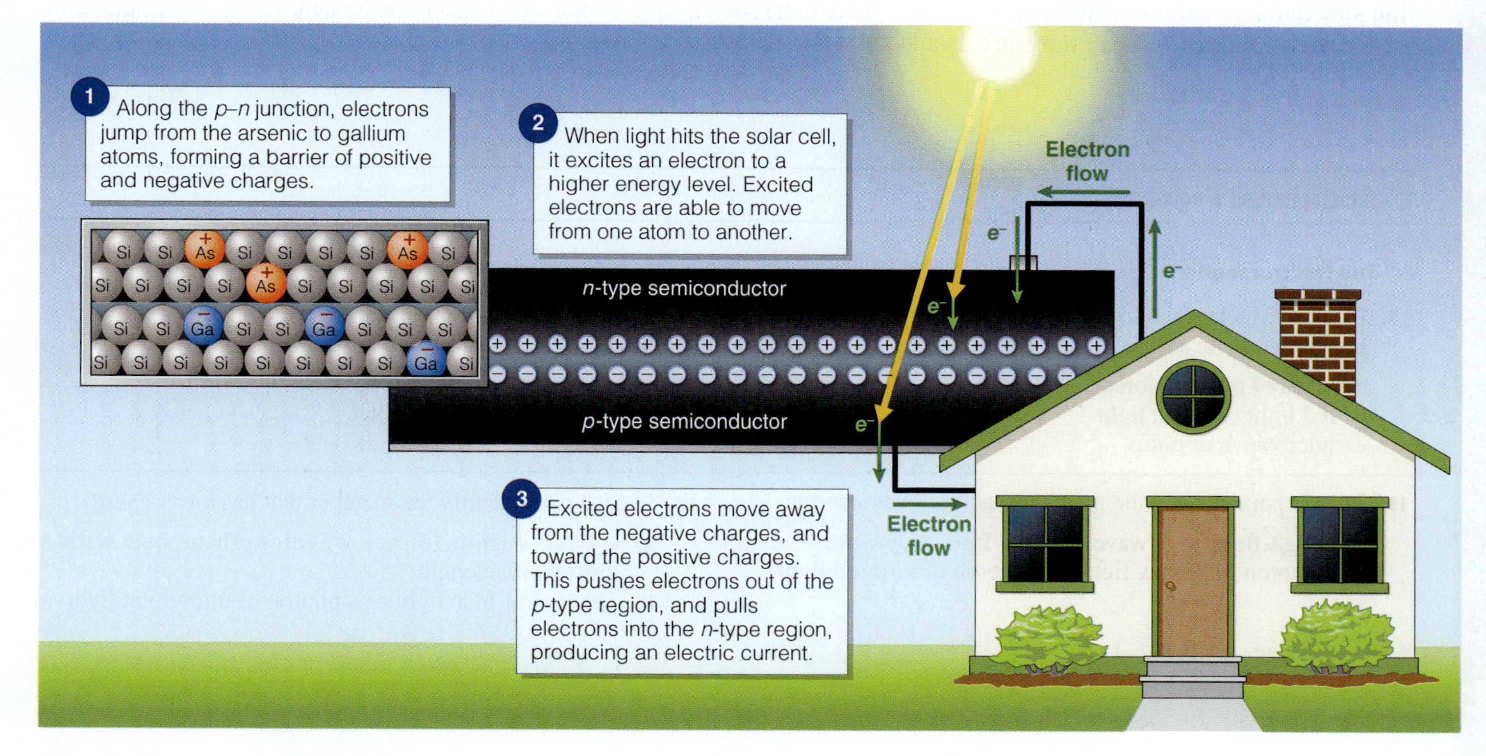

1
Along the p–n junction, electrons jump from the arsenic to gallium atoms, forming a barrier of positive and negative charges.
Si
As
Ga
2
When light hits the solar cell, it excites an electron to a higher energy level. Excited electrons are able to move from one atom to another.
n-type semiconductor
p-type semiconductor
e⁻
Electron flow
Electron flow
3
Excited electrons move away from the negative charges, and toward the positive charges. This pushes electrons out of the p-type region, and pulls electrons into the n-type region, producing an electric current.

Figure 4.39 Most solar panel semiconductors use silicon doped with elements like gallium and arsenic. When the sunlight hits the cells, electrons are excited to a higher energy level, which allows them to move from one atom to another. A line of charge along the *p–n* junction pushes electrons, creating an electric current.

Key Terms

4.1 The Electromagnetic Spectrum

electromagnetic radiation A form of energy produced when charged particles move or vibrate relative to each other; electromagnetic radiation exists as waves.

photon A small increment or packet of electromagnetic energy (often visible light).

electromagnetic spectrum All forms of electromagnetic energy, ranging from low-energy waves (TV and radio) to visible light to high-energy waves such as gamma rays.

visible spectrum The narrow range of electromagnetic energy that we perceive as light.

wavelength (λ) The distance from a point on one wave to the same point on the next wave.

frequency (ν) The number of waves that pass through a point in one second; typically measured in hertz.

hertz (Hz) A unit of frequency equal to one wave cycle per second.

4.2 Color, Line Spectra, and the Bohr Model

line spectrum A pattern of light energies called *spectral lines*; they are formed when gas-phase elements release energy. Each element has a characteristic line spectrum.

Bohr model An early model of atomic structure that treated the atom like a tiny solar system, with the nucleus at the center, and the electrons orbiting the nucleus.

4.3 The Quantum Model and Electron Orbitals

quantum model The modern description of electronic behavior that treats electrons as both particles and waves.

uncertainty principle The idea that it is impossible to know the exact velocity and location of a particle; this principle becomes important when studying electrons.

principal quantum number An integer number that identifies the energy level an electron occupies.

sublevel A set of electron orbitals that occurs in an electron energy level; the four main sublevels are *s*, *p*, *d*, and *f*.

orbital A region where electrons are most likely to be found; each orbital can hold up to two electrons.

***s* sublevel** A sublevel that contains one orbital and can hold up to two electrons; the *s* sublevel is present in every energy level.

***p* sublevel** A sublevel that contains three orbitals and can hold up to six electrons; the *p* sublevel is present in energy levels 2 and higher.

***d* sublevel** A sublevel that contains five orbitals and can hold up to 10 electrons; the *d* sublevel is present in energy levels 3 and higher.

***f* sublevel** A sublevel that contains seven orbitals and can hold up to 14 electrons; the *f* sublevel is present in energy levels 4 and higher.

4.4 Describing Electron Configurations

electron configuration The number of electrons in each energy level and sublevel.

valence level The highest-occupied electron energy level in an atom.

octet rule The principle that an atom is stabilized by having its highest-occupied (valence) energy level filled.

isoelectronic Describes atoms or ions that have the same electron configuration.

Additional Problems

4.1 The Electromagnetic Spectrum

13. Identify which region of the electromagnetic spectrum has higher energy in each of the following pairs:

a. infrared or ultraviolet
b. red light or green light
c. microwave or radio

14. Identify which region of the electromagnetic spectrum has higher energy in each of the following pairs:

a. green light or microwave
b. gamma rays or ultraviolet
c. X-rays or blue light

15. In each pair, identify the member that has lower energy:

a. a high-frequency wave or a low-frequency wave
b. a photon of yellow light or a photon of infrared light

16. In each pair, identify the member that has lower energy:

a. a wave with a longer wavelength or one with a shorter wavelength
b. a photon of blue light or a photon of ultraviolet light

17. Which of the following has a lower frequency than green light?

a. red light
b. light with a wavelength of 400 nm
c. light with a wavelength of 1.0×10^{-6} m

18. Which of the following has higher energy than blue light?

a. ultraviolet light
b. light with a frequency higher than blue light
c. light with a wavelength longer than blue light

19. In each pair, which one has the longest wavelength?

a. a radio wave or a microwave
b. a microwave or visible light
c. infrared light or ultraviolet light

20. In each pair, which one has the longest wavelength?

a. visible light or gamma rays
b. microwaves or X-rays
c. visible light or infrared light

21. Determine the wavelength of each wave. Assume each wave is traveling at 3.00×10^{8} m/s.

a. a wave with a frequency of $750,000\ s^{-1}$
b. a wave with a frequency of 20 Hz
c. a wave with a frequency of 2.5×10^{5} Hz

22. Find the frequency for each wave. Assume each wave is traveling at 3.00×10^{8} m/s.

a. a photon of green light with a wavelength of 5.5×10^{-7} m
b. a photon of red light with a wavelength of 700 nm
c. a photon of light with a wavelength of 5.2 cm

23. The highest intensity of photons from the Sun have a wavelength of about 5.00×10^{-7} m. What color is this light? Refer to Figure 4.5.

24. While traveling through an intersection, you encounter a traffic light with a frequency of 4.30×10^{14} Hz. What color is this light? Should you stop, keep going through the intersection, or proceed with caution? Refer to Figure 4.5.

25. A beam of light has a frequency of 4.17×10^{14} Hz. What is the energy per photon of this light?

26. A beam of light has a frequency of 6.29×10^{14} Hz. What is the energy per photon of this light?

27. A medical X-ray machine has a wavelength of 2.10×10^{-11} m. What is the frequency (in Hz) and energy (in J) of this wave?

28. A microwave has a wavelength of 1.20 cm. What is the frequency (in Hz) and energy (in J) of this wave?

29. Carbon dioxide readily absorbs radiation with an energy of 4.67×10^{-20} J. What is the wavelength and frequency of this radiation? Does this radiation fall in the ultraviolet, visible, or infrared range?

30. Determine the wavelength and frequency for each of the following:

a. a photon of green light with an energy of 3.32×10^{-19} J
b. a photon of ultraviolet radiation with an energy of 3.49×10^{-18} J

4.2 Color, Line Spectra, and the Bohr Model

31. What are line spectra? How are line spectra observed?

32. How did the Bohr model describe electron motion? How did this model explain line spectra?

33. When an atom absorbs energy, what happens to its electrons?

34. When an atom releases light energy, what happens to its electrons?

35. When an electric current passes through a tube filled with neon, the tube begins to glow. How are electrons involved in producing this light?

36. Molten iron glows with a reddish-orange color. How is heat energy converted into light energy in this process?

37. For each of these processes, indicate whether light is absorbed or emitted:

a. An electron moves from level 2 to level 4.
b. An electron moves from level 6 to level 3.
c. An electron jumps from level 3 to level 6.
d. An electron relaxes from level 7 to level 2.

38. For each of these processes, indicate whether light is absorbed or emitted:

a. An electron moves from level 1 to level 5.
b. An electron moves from level 4 to level 2.
c. An electron is excited to a higher energy level.
d. An electron relaxes to a lower energy level.

39. A hydrogen atom absorbs a photon of ultraviolet light, exciting an electron from energy level 1 to energy level 5. The excited electron drops back down in three steps, as shown. Using the following table, identify the energy released in each transition as ultraviolet, visible, or infrared. The first transition has been done as an example.

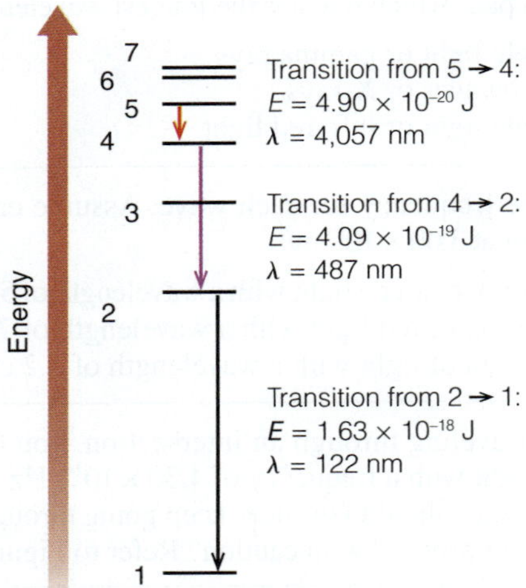

Transition	Wavelength	Region
5 → 4	4,057 nm	infrared
4 → 2	487 nm	
2 → 1	122 nm	

40. A hydrogen atom absorbs a photon of ultraviolet light, exciting an electron from energy level 1 to energy level 7. The excited electron drops back down in three steps, as shown. Using the following table, identify the energy released in each transition as ultraviolet, visible, or infrared.

Energy
7
6
5
4
3
2
1
Transition from 7 → 5:
$E = 4.23 \times 10^{-20}$ J
$\lambda = 4{,}668$ nm
Transition from 5 → 2:
$E = 4.58 \times 10^{-19}$ J
$\lambda = 434$ nm
Transition from 2 → 1:
$E = 1.63 \times 10^{-18}$ J
$\lambda = 122$ nm

Transition	Wavelength	Region
7 → 5	4,668 nm	infrared
5 → 2	434 nm	
2 → 1	122 nm	

4.3 The Quantum Model and Electron Orbitals

41. Why do we describe electrons in terms of orbitals or energy levels instead of describing the exact location of the electron?

42. How is an *orbital* in the quantum model different from the orbits described in the Bohr model?

43. How many electrons can occupy a single orbital?

44. When an orbital contains two electrons, how do the electron spins align?

45. For each of these energy levels, list all sublevels that are present:

a. level 1 **b.** level 2 **c.** level 3 **d.** level 4

46. In total, how many electrons can fit in energy level 4?

47. How many electrons can fit in a *p* sublevel?

48. How many electrons can fit in a *d* sublevel?

49. What is the difference between an orbital and a sublevel?

50. Sketch the shape of an *s* orbital and a *p* orbital.

51. How many electrons can occupy an atom's lowest energy level? How many can occupy its second-lowest energy level?

52. Complete this table to show the number of orbitals and maximum number of electrons in each energy sublevel.

Sublevel	Number of Orbitals	Number of Electrons
s		2
	3	
	5	10
f		

53. Indicate whether each of these level/sublevel combinations are possible:

a. level 1, sublevel *s*
b. level 4, sublevel *s*
c. level 1, sublevel *p*

54. Indicate whether each of these level/sublevel combinations are possible:

a. level 2, sublevel *d*
b. level 3, sublevel *f*
c. level 3, sublevel *d*

4.4 Describing Electron Configurations

55. Complete the orbital energy diagrams, using arrows to represent the electrons. The first one (nitrogen) is filled out as an example.

56. Complete the orbital energy diagrams, using arrows to represent the electrons:

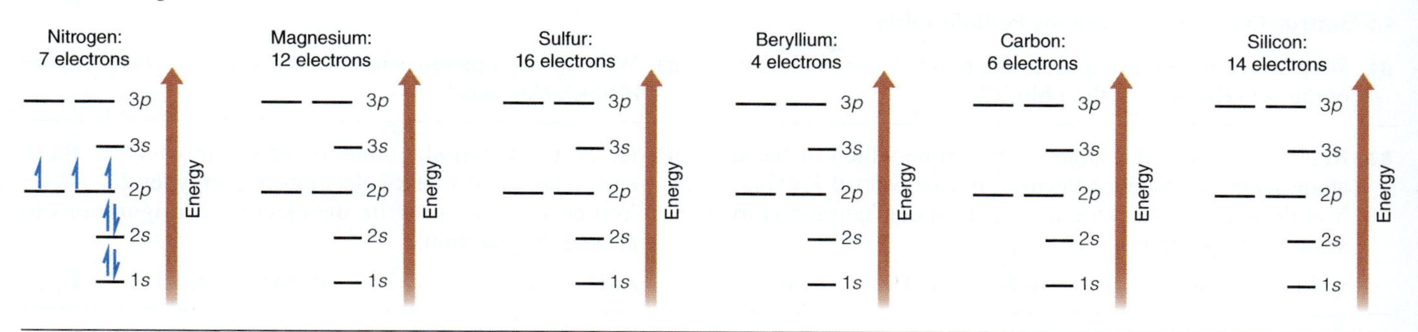

57. Write electron configurations for each of these atoms:

a. H b. Be c. C d. Ne

58. Write electron configurations for each of these atoms:

a. N b. Co c. Ru d. I

59. What is the valence level? Which two sublevels are involved with an atom's valence?

60. Why is energy level 3 considered "filled" with only eight electrons?

61. What is the octet rule? What family of elements illustrates the octet rule?

62. If an atom fulfills the octet rule, which sublevels are filled?

63. Rewrite the following electron configurations using noble gas shorthand:

a. $1s^2 2s^1$
b. $1s^2 2s^2 2p^6 3s^2$
c. $1s^2 2s^2 2p^6 3s^2 3p^5$

64. Rewrite the following electron configurations using noble gas shorthand:

a. $1s^2 2s^2 2p^3$
b. $1s^2 2s^2 2p^6 3s^2 3p^6 4s^2 3d^{10} 4p^6 5s^2 4d^{10} 5p^6 6s^1$
c. $1s^2 2s^2 2p^6 3s^2 3p^6 4s^2 3d^{10} 4p^3$

65. Use noble gas shorthand to write electron configurations for the following halogens: F, Cl, Br, and I.

66. Use noble gas shorthand to write electron configurations for the following alkaline earth metals: Be, Mg, Ca, and Sr.

67. Write the electron configuration for magnesium. How many inner-shell and valence electrons does magnesium have?

68. Write the electron configuration for phosphorus. How many inner-shell and valence electrons does phosphorus have?

69. Identify the inner electrons, outer electrons, and valence electrons in each electron configuration:

a. nitrogen: $1s^2 2s^2 2p^3$
b. potassium: $1s^2 2s^2 2p^6 3s^2 3p^6 4s^1$
c. gallium: $[Ar]4s^2 3d^{10} 4p^1$
d. tungsten: $[Xe]6s^2 4f^{14} 3d^4$

70. Identify the inner electrons, outer electrons, and valence electrons in each electron configuration:

a. lithium: $1s^2 2s^1$
b. chlorine: $1s^2 2s^2 2p^6 3s^2 3p^5$
c. promethium: $[Xe]\ 6s^2 4f^5$
d. iodine: $[Kr]5s^2 4d^{10} 5p^5$

71. Write electron configurations for each of these ions:

a. O^{2-} b. K^+ c. Br^- d. N^{3-}

72. Write electron configurations for each of these ions:

a. Na^+ b. Mg^{2+} c. F^- d. S^{2-}

73. Each of these atoms can gain or lose electrons to become isoelectronic with neon. Write the number of electrons each atom would need to gain or lose, and indicate the charge that would result:

a. O b. F c. Na d. Mg

74. Each of these atoms can gain or lose electrons to become isoelectronic with argon. Write the number of electrons each atom would need to gain or lose, and indicate the charge that would result:

a. S b. Cl c. K d. Ca

75. List three ions that are isoelectronic with helium.

76. List three ions that are isoelectronic with krypton.

77. Fluorine tends to gain one electron while sodium metal tends to lose one electron. Why is this so? What rule can be used to describe this behavior?

78. Potassium metal tends to lose one electron while calcium metal tends to lose two electrons. Why is this so? What rule can be used to describe this behavior?

79. Based on the octet rule, which of these atoms or ions have the stability of a filled valence level?

a. K b. K^+ c. S^- d. S^{2-}

80. Based on the octet rule, which of these atoms or ions have the stability of a filled valence level?

a. Ca^{2+} b. Al c. Xe d. I^-

4.5 Electron Configuration and the Periodic Table

81. Why do we sometimes refer to the two left-hand columns of the periodic table as the *s* block?

82. Why do we sometimes refer to the transition metals as the *d*-block elements?

83. Refer to the periodic table to determine which of these atoms have their valence electrons in energy level 3. (*Note*: You do not need to write the electron configurations to answer this question.)

a. Li b. Si c. P d. Se e. He f. Ar

84. Refer to the periodic table to determine which of these atoms have their valence electrons in energy level 4. (*Note*: You do not need to write the electron configurations to answer this question.)

a. K b. Na c. Y d. Mn e. Cl f. Br

85. What is the inner electron configuration of elements in row 2?

86. What is the inner electron configuration of elements in row 4?

87. The elements in group 2A have how many valence electrons?

88. The elements in group 6A have how many valence electrons?

89. Without writing electron configurations, identify the number of valence electrons in each of these main-group elements:

a. calcium b. aluminum c. bromine d. neon

90. Without writing electron configurations, identify the number of valence electrons in each of these main-group elements:

a. carbon b. selenium c. xenon d. rubidium

91. Refer to the periodic table to identify the number of valence electrons (outer-shell *s* and *p* electrons) present in each of these atoms:

a. Mg b. N c. P d. As e. Cl f. Ca

92. Refer to the periodic table to identify the number of valence electrons (outer-shell *s* and *p* electrons) present in each of these atoms:

a. Be b. Sr c. Ne d. Br e. As f. Si

93. Using the periodic table as a reference, answer these questions:

a. What atom has an electron configuration of $[Ar]4s^1$?
b. What atom has an electron configuration of $[He]2s^2 2p^3$?
c. What atom has an electron configuration of $[Kr]5s^2 4d^8$?

94. Using the periodic table as a reference, answer these questions:

a. What atom has an electron configuration of $[He]2s^1$?
b. What atom has an electron configuration of $[Ne]2s^2 2p^5$?
c. What atom has an electron configuration of $[Ar]4s^2 3d^6$?

95. Using the periodic table as a reference, answer these questions:

a. What subshell is filled after 4*s*?
b. What subshell is filled after 4*d*?
c. What subshell is filled after 3*p*?

96. Using the periodic table as a reference, answer these questions:

a. What subshell is filled after 3*s*?
b. What subshell is filled after 3*d*?
c. What subshell is filled after 6*p*?

97. Using the periodic table, find the electron configuration of the highest-filled sublevel for each of these elements. Try to do this without writing the full electron configuration. The first answer is given as an example.

a. germanium (answer: $4p^2$) b. tellurium
c. technetium d. boron

98. Using the periodic table, find the electron configuration of the highest-filled sublevel for each of these elements. Try to do this without writing the full electron configuration. The first answer is given as an example.

a. iron (answer: $3d^6$) b. astatine
c. gold d. barium

99. Identify the group of elements with these characteristics:

a. This group has a valence electron configuration of s^2.
b. This group of elements does not react to form compounds.
c. This group of elements has two valence electrons.

100. Identify the group of elements with these characteristics:

a. This group of elements has a valence electron configuration of s^2p^5.
b. This group of elements has a completely filled valence.
c. This group of elements has one valence electron.

101. Oxygen, sulfur, and selenium belong to a family of elements that is sometimes called the *chalcogens*. What is the valence electron configuration of this group?

102. What is the valence electron configuration of the group of elements that includes boron, aluminum, and gallium?

103. What is the outermost electron configuration for the elements in group 12 of the periodic table?

104. What is the outermost electron configuration for the elements in group 6A of the periodic table?

Chemical Bonds and Compounds

Figure 5.1 (a) Allied soldiers captured by the Japanese, February 1942. (*continued*)

CHAPTER FIVE

Lithium Carbonate and Bipolar Disorder

Sometimes, combinations produce surprising results. Consider the story of lithium carbonate — a simple compound used worldwide to treat bipolar disorder — and the combination of bad luck, good luck, and keen observation that led to its discovery.

In February 1942, during the heart of World War II, Japanese forces attacked the Allied stronghold at Singapore. After a week of fighting, the Allies surrendered (**Figure 5.1**). Among those captured was John Cade, a young psychiatrist serving in the Australian Army Medical Corps. He spent the next three years in a prisoner-of-war camp.

While imprisoned, Cade observed a number of prisoners who suffered from bipolar disorder: They fluctuated between wildly aggressive behavior (called the manic phase) and deep depression. Cade began to suspect that a toxic chemical caused the prisoners' erratic behavior and that their moods stabilized after the toxin was expelled through their urine.

After his release at the end of the war, Cade returned to Australia and resumed his career in psychiatry. On the side, he began exploring the ideas he had developed in captivity. He collected urine samples from bipolar patients and injected the urine into guinea pigs. Interestingly, the guinea pigs treated with urine from bipolar patients died faster than those treated with urine from healthy people. Cade delved deeper. He suspected that a compound called uric acid might be the mysterious toxin. He began to study the effects of pure uric acid and related compounds on the guinea pigs. He found that one such compound, lithium urate, reduced the toxic effects of the other compounds present. Intrigued by this result, he decided to test a simpler lithium-containing compound: lithium carbonate. When he injected guinea pigs with pure lithium carbonate, the animals became sedate.

Ultimately Cade's ideas about toxins in the urine were discarded, but the effects he observed from lithium carbonate opened a new door. Cade wondered if lithium carbonate would also sedate patients suffering from the manic phase of bipolar disorder. To see if it was safe, he first tested it on himself. Finding no long-term effects, he treated the manic

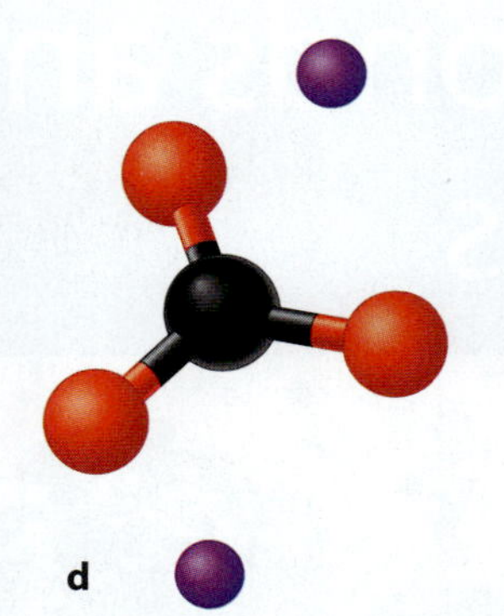

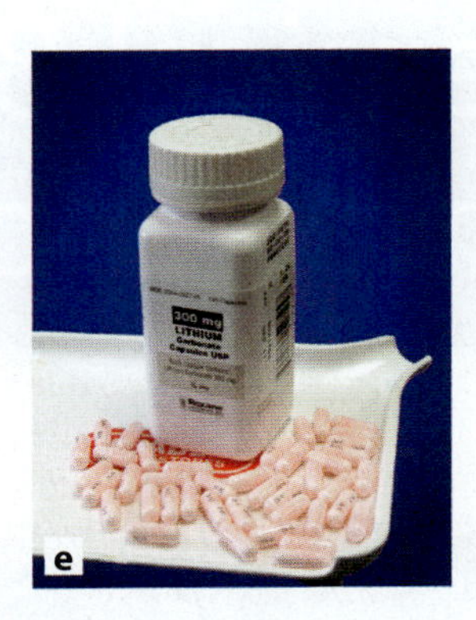

Figure 5.1 (continued) (b) John Cade discovered that lithium carbonate could treat bipolar disorder. (c) Mogens Schou carried on Cade's work, extensively studying and promoting the effects of lithium carbonate. (d) Lithium carbonate is made from lithium, carbon, and oxygen atoms. (e) Today millions of people take lithium carbonate for the treatment of bipolar disorder. © TopFoto/The Image Works; Newspix/Getty Images; AP Photo/Marty Lederhandler; Charles D. Winters/Science Source

patients in his ward with lithium carbonate. This human testing was remarkably successful, and in 1949 he published his results. In the decades that followed, another psychiatrist, Mogens Schou, extensively studied the effects of lithium carbonate. Today, lithium carbonate remains one of the most common and least expensive treatments for bipolar disorder.

Like most science—and most other human activities—this story is messy. Cade's ideas were conceived in the harshest of circumstances. His initial ideas were incorrect. And by today's standards, Cade's experiments seem reckless. But his careful observations, both in the prison camp and the laboratory, led him to insights that changed the way we treat mental illness. Out of all the messy pieces, a beautiful discovery emerged.

In this chapter, we'll begin to study how chemical bonds bring atoms together to form compounds. Just as a complete story can be much different from the pieces that comprise it, compounds behave much differently from the elements they are composed of. As atoms combine to form compounds, new and intriguing properties emerge—properties that create new materials, new medicines, and new opportunities.

Intended Learning Outcomes

After completing this chapter and working the practice problems, you should be able to:

5.1 Lewis Symbols and the Octet Rule

- Use the periodic table to identify the number of valence electrons in an atom.
- Represent valence electrons using Lewis dot symbols.

5.2 Ions

- Describe and predict the formation of main-group ions using the octet rule.
- Identify common monatomic and polyatomic ions by name, symbol or formula, and charge.

5.3 Ionic Bonds and Compounds

- Predict ionic formulas based on cation and anion charges.
- Broadly describe the arrangement of ions in an ionic solid.
- Convert between the name and formula for an ionic compound.

5.4 Covalent Bonding

- Describe how nonmetals fulfill the octet rule through covalent bonds.
- Differentiate between empirical and molecular formulas.
- Name binary covalent compounds.

5.5 Distinguishing Ionic and Covalent Compounds

- Distinguish ionic and covalent compounds based on their chemical formulas.

5.6 Aqueous Solutions: How Ionic and Covalent Compounds Differ

- Contrast the behavior of ionic compounds and covalent compounds in aqueous solutions.

5.7 Acids—An Introduction

- Describe the ionization of acids in aqueous solution.
- Name binary acids and oxyacids.

5.1 Lewis Symbols and the Octet Rule

In Chapter 4, we saw that families of atoms such as the alkali metals, the halogens, and the noble gases exhibit similar behaviors because they have similar electronic configurations. In this chapter, we will explore how these electronic configurations lead to the formation of chemical bonds.

Chemical bonding involves changes in an atom's outer or *valence electrons*. Recall that valence electrons are the electrons in the highest-occupied energy level of an atom. Because of the sublevel filling sequence, the valence level involves only the *s* and *p* sublevels. Since two electrons can fit in an *s* sublevel, and six electrons can fit in the *p* sublevel, up to eight electrons can occupy the valence level. For main-group elements, we can quickly determine the number of valence electrons from the periodic table: The column (group) number for the main groups is also the number of valence electrons (**Figure 5.2**). For example, nitrogen (N) is in group 5A, so it has five valence electrons. Neon (Ne) is in group 8A, so it has eight valence electrons.

The valence level holds up to eight electrons. ■

Group	1A	2A	3A	4A	5A	6A	7A	8A
Valence electrons	1	2	3	4	5	6	7	8
Configuration	s^1	s^2	s^2p^1	s^2p^2	s^2p^3	s^2p^4	s^2p^5	s^2p^6
	H							He
	Li	Be	B	C	N	O	F	Ne
	Na	Mg	Al	Si	P	S	Cl	Ar

Figure 5.2 The main-group numbers (1A–8A) also indicate the number of electrons in each atom's valence level.

To visualize chemical bonding, it is often helpful to draw **Lewis dot symbols**. These symbols represent the number of valence electrons in an atom as dots drawn around the atomic symbol. Here are the Lewis symbols for each of the row 2 elements:

Li• Be: •B: •C: :N: :O: :F: :Ne:

In Chapter 4 we introduced the *octet rule*, which states that *an atom is stabilized by having its valence energy level filled.* For elements in row 2 and below, eight electrons are required to fill the valence level. The octet rule explains why the noble gases are so stable, and it also allows us to predict how main-group elements form chemical bonds. These elements fulfill the octet rule by gaining or losing electrons to form *ions*, or by sharing electrons between two atoms. We will explore these behaviors in the sections that follow.

Main-group elements can fulfill the octet rule by gaining, losing, or sharing electrons. ■

5.2 Ions

Cations: Ions with a Positive Charge

Main-group metals fulfill the octet rule by losing electrons to form positively charged ions, called **cations** (pronounced *cat-***eye***-uns*). For example, consider sodium metal: Sodium has an electron configuration of $1s^2 2s^2 2p^6 3s^1$. Because $1s^2 2s^2 2p^6$ is the same configuration as neon, we often write it as $[Ne]3s^1$. To fill its valence level (level 3), sodium would have to gain seven electrons—an unlikely occurrence. However, by losing just one electron, sodium becomes electronically identical to neon—a very stable arrangement that fulfills the octet rule. As a result,

Na^{+}, Mg^{2+}, and Ne are all *isoelectronic*—meaning they have the same electron configuration.

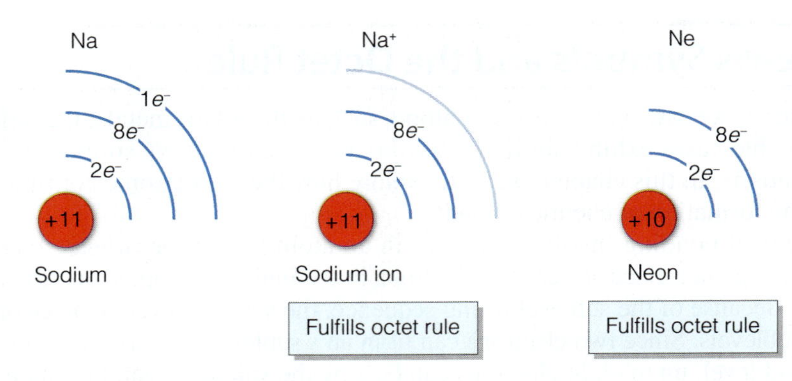

Figure 5.3 Sodium has one valence electron. By losing its outermost electron, sodium becomes electronically identical to the noble gas neon and fulfills the octet rule.

Alkali metals form +1 ions.

Alkaline earth metals form +2 ions.

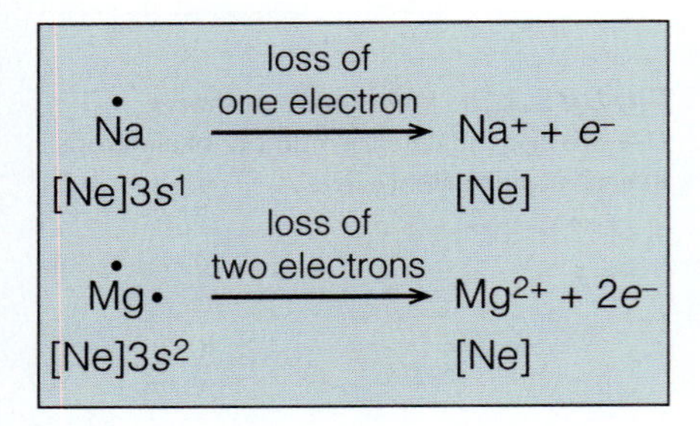

Figure 5.4 Sodium and magnesium both lose their valence electrons to become isoelectronic with neon.

sodium easily loses one electron to form Na^{+}, a common ion (**Figure 5.3**). All of the alkali metals (metals in group 1A of the periodic table) lose one electron to form +1 ions.

Magnesium has an electron configuration of [Ne]$3s^2$. Just as sodium lost one electron to reach the [Ne] electron configuration, magnesium can reach this electron configuration by losing two electrons (**Figure 5.4**). Because it loses two electrons, it has a charge of +2, which is written as Mg^{2+}. Each of the alkaline earth metals (group 2A) loses two electrons to form +2 ions (**Figure 5.5**).

The *transition metals* (elements in the *d* block of the periodic table) also tend to lose electrons to form positively charged ions. But unlike main-group metals, transition metal ions do not follow a simple pattern. Transition metals typically form ions having a charge between +1 and +4, and some transition metals form multiple charged ions. Metals in the lower part of the *p* block also behave this way (**Figure 5.6**).

Transition metal ions may have multiple charges.

Naming Cations

In general, metal cations are given the same name as the neutral metal. For example, the cation produced from sodium metal is simply called the *sodium ion*.

As mentioned earlier, some metals can have more than one charge. For example, iron commonly forms both +2 and +3 ions. Historically, these two ions were named as *ferrous* and *ferric* ions, respectively. Similarly, copper commonly forms

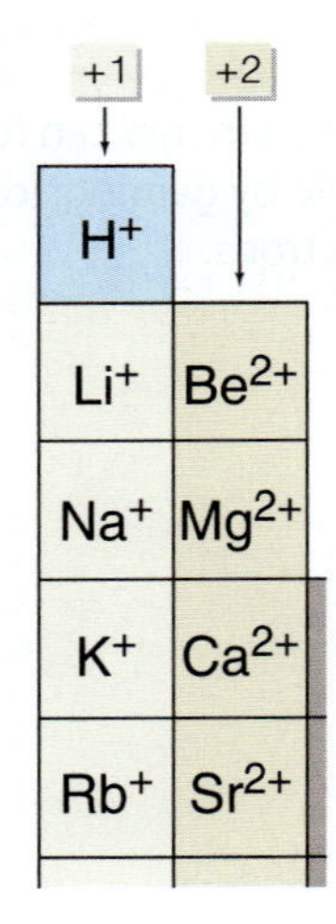

Figure 5.5 The group 1A elements (hydrogen and the alkali metals) form +1 ions. Group 2A elements (the alkaline earth metals) form +2 ions.

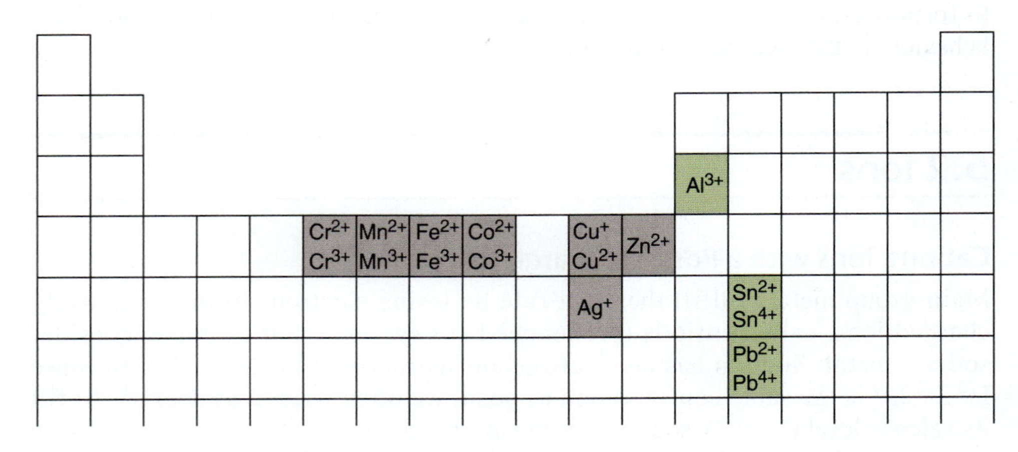

Figure 5.6 Many common transition and *p*-block metals form ions with multiple charges.

TABLE 5.1 Naming Ions with More Than One Charge

Atom	Ion	Older Name	Modern Name
Iron	Fe^{2+}	Ferrous	Iron(II)
	Fe^{3+}	Ferric	Iron(III)
Copper	Cu^{+}	Cuprous	Copper(I)
	Cu^{2+}	Cupric	Copper(II)

both +1 (*cuprous*) and +2 (*cupric*) ions. Although you will encounter these names occasionally, the modern style of naming these ions puts the charge in Roman numerals within parentheses immediately after the atom name. For example, the ferrous ion (Fe^{2+}) is named as iron(II), which is read as "iron-two"; the ferric ion (Fe^{3+}) is named as iron(III), read as "iron-three" (**Table 5.1**).

If an atom can form more than one cation, use Roman numerals after the atom name to specify the charge. ■

Example 5.1 Naming Cations

Name each of the following ions: Ag^{+}, Pb^{2+}, and Pb^{4+}.

From Figure 5.6, we see that silver (Ag) forms only one ion. Therefore, we refer to Ag^{+} as a *silver* ion. However, lead (Pb) forms two different ions. To distinguish them, we refer to Pb^{2+} as a lead(II) ion and Pb^{4+} as a lead(IV) ion.

TRY IT

1. Provide names for each of these cations:

Ca^{2+} Cr^{2+} Cr^{3+} Al^{3+}

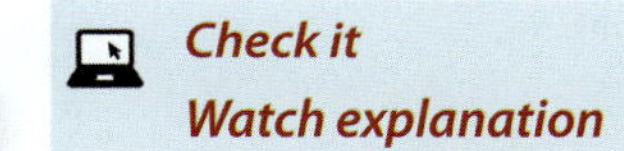

Anions: Ions with a Negative Charge

The nonmetals lie on the right-hand side of the periodic table. Unlike metals, the valence shells of most nonmetals are nearly full. To fulfill the octet rule, most nonmetals gain electrons to form negatively charged ions, called **anions** (pronounced *an-**eye**-uns*).

Nonmetals gain electrons to form anions. ■

For example, fluorine has an electron configuration of $1s^2 2s^2 2p^5$. By gaining one electron, fluorine can achieve an electron configuration of $1s^2 2s^2 2p^6$, the same electron configuration as neon. This configuration fulfills the octet rule and provides tremendous stability. As a result, fluorine tends to aggressively "grab" an electron, forming a very stable ion with a charge of –1 (**Figure 5.7**). The other halogens (chlorine, bromine, iodine) also form –1 ions.

Halogens form –1 ions. ■

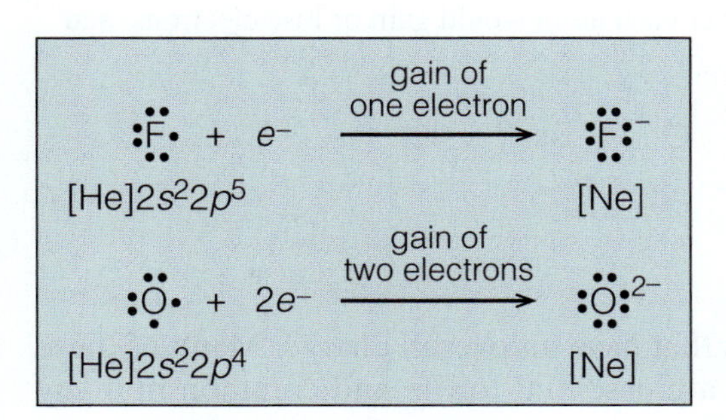

Figure 5.7 Fluorine and oxygen both gain electrons to become isoelectronic with neon.

Chalcogens form –2 ions.

Oxygen, sulfur, and the atoms below them on the periodic table comprise a family called the *chalcogens* (group 6A). Each of these elements is two electrons short of a noble gas configuration. For example, oxygen has an electron configuration of $1s^2 2s^2 2p^4$. It needs two electrons to fill its outer valence level, and so it tends to gain two electrons, resulting in a charge of –2. Sulfur also forms a stable ion with a charge of –2.

What about group 5A elements, such as nitrogen and phosphorus? Consistent with the pattern just described, these atoms gain three electrons to fill their valence level and so form ions with a charge of –3 (**Figure 5.8**).

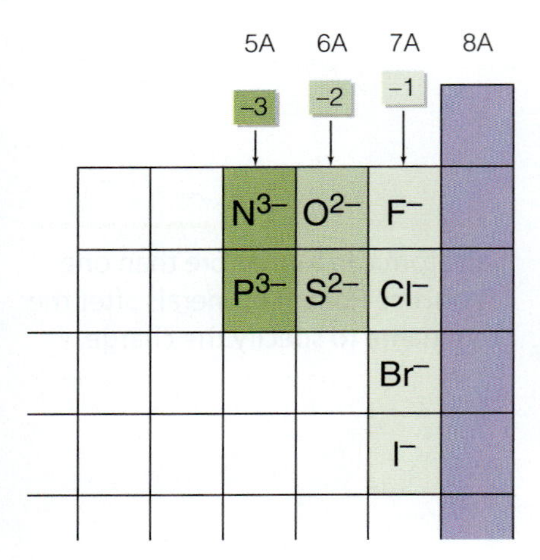

Figure 5.8 The halogens form –1 ions. Oxygen and sulfur form –2 ions. Nitrogen and phosphorus form –3 ions. The noble gases (shaded violet) have complete valence shells and do not form ions.

Naming Anions

When an atom gains electrons, we name the resulting anion by changing the end of the atom name to *–ide*. For example, chlorine atoms form chlor*ide* ions, oxygen atoms form ox*ide* ions, and sulfur atoms form sulf*ide* ions. A list of common anions is given in **Table 5.2**.

TABLE 5.2 Common Anions

Atom	Anion Symbol	Anion Name
Nitrogen	N^{3-}	Nitride
Phosphorus	P^{3-}	Phosphide
Oxygen	O^{2-}	Oxide
Sulfur	S^{2-}	Sulfide
Fluorine	F^{-}	Fluoride
Chlorine	Cl^{-}	Chloride
Bromine	Br^{-}	Bromide

Sports drinks contain ions that are commonly lost during exercise. They include sodium, potassium, chloride, and phosphate.

Example 5.2 Naming Ions and Predicting Charges

Predict the ions that would be formed from an atom of calcium and from an atom of sulfur. Name each ion.

Calcium belongs to the alkaline earth metal family. It has an electron configuration of $[Ar]4s^2$. Calcium loses its two valence electrons, resulting in a charge of +2. Cations are given the same name as the parent atom, so we refer to Ca^{2+} as the *calcium ion*.

Sulfur is a nonmetal with an electron configuration of $[Ne]3s^2 3p^4$. To fill its valence shell, sulfur gains two electrons, giving the ion a charge of –2. We refer to S^{2-} as the *sulfide ion*.

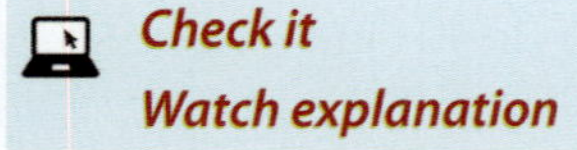

Check it

Watch explanation

TRY IT

2. Use the periodic table to predict whether each atom would gain or lose electrons, and write the charge on the ion formed:

Cl Br O Be K

Polyatomic Ions

Polyatomic ions are groups of atoms that have an overall charge. Many of these ions, such as acetate and phosphate, are essential to life and common in many different materials and applications. Formulas and names for the most common

TABLE 5.3 Common Polyatomic Ions

Formula	Name	Formula	Name
NH_4^+	Ammonium		
NO_3^-	Nitrate	SO_4^{2-}	Sulfate
CO_3^{2-}	Carbonate	SO_3^{2-}	Sulfite
HCO_3^-	Bicarbonate (also called hydrogen carbonate)	HSO_4^-	Bisulfate (also called hydrogen sulfate)
NO_2^-	Nitrite	ClO_4^-	Perchlorate
PO_4^{3-}	Phosphate	ClO_3^-	Chlorate
HPO_4^{2-}	Hydrogen phosphate	ClO_2^-	Chlorite
$C_2H_3O_2^-$	Acetate	ClO^-	Hypochlorite
OH^-	Hydroxide	CrO_4^{2-}	Chromate
CN^-	Cyanide	$Cr_2O_7^{2-}$	Dichromate
O_2^{2-}	Peroxide	MnO_4^-	Permanganate

polyatomic ions are given in **Table 5.3**. Notice that this table contains only one common polyatomic cation (ammonium). All others are anions.

Naming Polyatomic Ions

Although Table 5.3 contains many ion names, there are patterns that will help you keep these names organized. Notice that most of the polyatomic ions contain oxygen—these are called **oxyanions**. We name oxyanions by adding the suffix *–ate* to the root of the element. For example, the oxyanion from carbon (CO_3^{2-}) is carbon*ate*, and the oxyanion formed from phosphorus (PO_4^{3-}) is phosph*ate*.

Some elements form more than one oxyanion. In these cases we use the suffix *–ate* to indicate the ion with more oxygen atoms present, and the suffix *–ite* to indicate the ion with fewer oxygen atoms present. For example, there are two common nitrogen oxyanions:

"*–ate* is great, and *–ite* is lite"
More oxygen atoms: *–ate*
Fewer oxygen atoms: *–ite*

NO_3^- nitrate
NO_2^- nitrite

Chlorine forms four oxyanions. In this case, we use the prefix *per–* (meaning "more than") to indicate the largest number of oxygen atoms, and the prefix *hypo–* (meaning "below") to indicate the least number of oxygen atoms:

ClO_4^- perchlorate
ClO_3^- chlorate
ClO_2^- chlorite
ClO^- hypochlorite

A Summary of the Common Ions

As you continue studying chemistry, you will find it essential to know the structure, formula, and charge of common monatomic and polyatomic ions. **Figure 5.9** summarizes the most common ions. You should be very familiar with these ions, because you will use them regularly throughout this course.

Monatomic atoms

H^+																	
Li^+	Be^{2+}													N^{3-}	O^{2-}	F^-	
Na^+	Mg^{2+}											Al^{3+}		P^{3-}	S^{2-}	Cl^-	
K^+	Ca^{2+}				Cr^{2+} Cr^{3+}	Mn^{2+} Mn^{3+}	Fe^{2+} Fe^{3+}	Co^{2+} Co^{3+}		Cu^+ Cu^{2+}	Zn^{2+}					Br^-	
Rb^+	Sr^{2+}									Ag^+			Sn^{2+} Sn^{4+}			I^-	
													Pb^{2+} Pb^{4+}				

Polyatomic atoms

NH_4^+ Ammonium			
NO_3^-	Nitrate	SO_4^{2-}	Sulfate
CO_3^{2-}	Carbonate	SO_3^{2-}	Sulfite
HCO_3^-	Bicarbonate (Hydrogen carbonate)	HSO_4^-	Bisulfate (Hydrogen sulfate)
NO_2^-	Nitrite	ClO_4^-	Perchlorate
PO_4^{3-}	Phosphate	ClO_3^-	Chlorate
HPO_4^{2-}	Hydrogen phosphate	ClO_2^-	Chlorite
$C_2H_3O_2^-$	Acetate	ClO^-	Hypochlorite
OH^-	Hydroxide	CrO_4^{2-}	Chromate
CN^-	Cyanide	$Cr_2O_7^{2-}$	Dichromate
O_2^{2-}	Peroxide	MnO_4^-	Permanganate

Figure 5.9 It is important to know the names, formulas, and charges for these common ions.

Practice

Common Ions

How well do you know the common ions? Try this interactive game to practice and test your knowledge.

Example 5.3 Gathering Information from Ion Names

The four ions named below are less common and are not listed in Table 5.3. Which of these are polyatomic? Identify each one as a cation or an anion.

a. bromate **b.** bromite **c.** palladium(II) **d.** selenide

From the suffixes *–ate* and *–ite*, we know that both bromate and bromite are oxyanions of the element bromine. Further, we know that bromate contains more oxygen atoms than bromite. The actual formula for bromate is BrO_3^-, and the formula for bromite is BrO_2^-.

Palladium is a transition metal, so it forms a cation. The (II) indicates that this ion is Pd^{2+}.

Finally, the ending *–ide* indicates that selenide is a monatomic anion formed from the element selenium. Selenium lies just below sulfur on the periodic table, so we predict the charge of this ion to be –2.

Check it

Watch explanation

TRY IT

3. Write the symbol and the charge for each ion listed. Refer to the periodic table as needed.

calcium nitrate scandium(III) telluride

4. Name each of these ions:

Cs^+ Fe^{2+} SO_4^{2-} As^{3-}

5.3 Ionic Bonds and Compounds

Ionic Bonds and Ionic Lattices

Opposite charges attract each other. When positive and negative ions come near each other, they stick tightly together. The force of attraction between oppositely charged ions is called an **ionic bond**. A compound composed of oppositely charged ions is an **ionic compound**. Because metals form cations and nonmetals form anions, the compounds formed between metals and nonmetals are ionic compounds.

Metal cations and nonmetal anions form ionic bonds.

Ionic compounds contain many cations and anions, joined through ionic bonds. To understand how these ions fit together, let's consider the structure of a compound composed of sodium cations (Na^+) and chloride anions (Cl^-). A single Na^+ ion and a single Cl^- ion adhere to each other in an ionic bond. But what happens if additional ions are present? They pack together in a structure of alternating positive and negative charges that stretch out in three dimensions (**Figure 5.10**). This array of positive and negative ions is called an **ionic lattice**.

An empirical formula gives the smallest whole-number ratio of atoms in a compound.

To represent the composition of compounds like this one, we use a **chemical formula** that indicates the type and amount of each element present. We represent ionic compounds using a specific type of chemical formula, called an **empirical formula**. An empirical formula gives the smallest whole-number ratio of atoms in a compound. Subscripts written after each atom indicate the number of that atom present. If no subscript is written, we understand the number of atoms to be one. In this instance, the formula is written simply as NaCl.

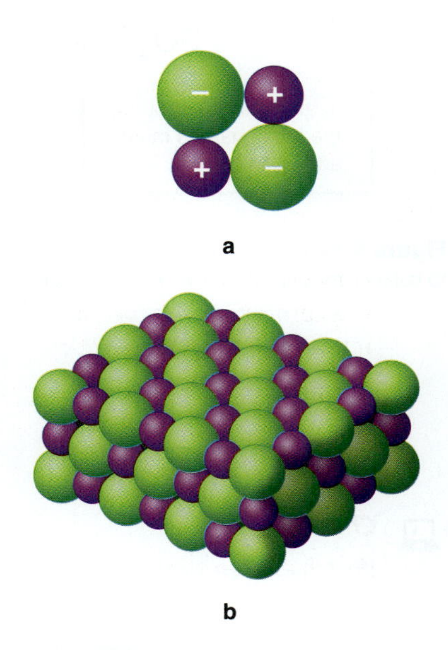

Figure 5.10 (a) Ions pack together in a framework of alternating positive and negative charges. (b) This packing results in a three-dimensional framework called an ionic lattice.

The empirical formula gives the smallest number of ions necessary to form a compound. This number of ions is called the **formula unit**. For example, the empirical formula for sodium chloride is NaCl; a formula unit of sodium chloride contains one sodium ion and one chloride ion.

When writing empirical formulas for ionic compounds, we write the symbol or formula for the cation, followed by the anion. In the next section, we'll look at several more examples of empirical formulas and formula units.

Predicting Formulas for Ionic Compounds

Some ionic compounds contain cations and anions with different charges. For example, consider the solid composed of potassium and sulfide ions. The potassium cation has a charge of +1 while the sulfide anion has a charge of –2. To form a neutral solid, *the positive charges must equal the negative charges*. To balance the charges, the solid must contain two potassium ions for every one sulfide ion:

In an ionic compound, the total charge must equal zero.

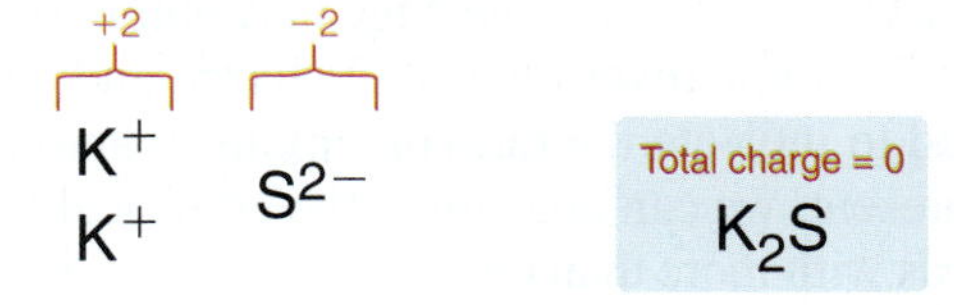

We therefore write the empirical formula for this compound as K_2S. Put another way, a formula unit of potassium sulfide contains two potassium ions and one sulfide ion.

We can also predict the formulas for ionic solids containing polyatomic ions. For example, consider the ionic compound produced from calcium (Ca^{2+}) and nitrate (NO_3^-) ions: For the positive and negative charges to balance, there must be two nitrate ions for every one calcium ion (**Figure 5.11**). We could write this formula as CaN_2O_6. But it is better to write the formula as $Ca(NO_3)_2$ because this shows that two nitrates are attached, rather than some other arrangement of nitrogen and oxygen. If we have more than one polyatomic ion in the formula, we write that ion inside parentheses to show that the entire unit is repeating.

Practice
Balancing Charges
To write an ionic compound formula correctly, you must balance the charges on the ions. Try this interactive to practice this skill.

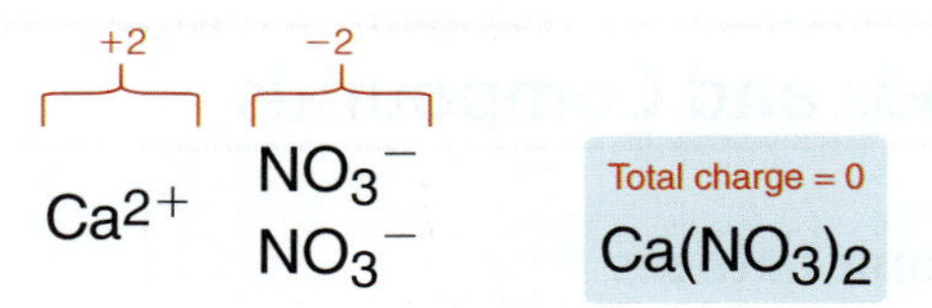

Figure 5.11 To balance the charges, this compound requires two nitrate ions for every one calcium ion.

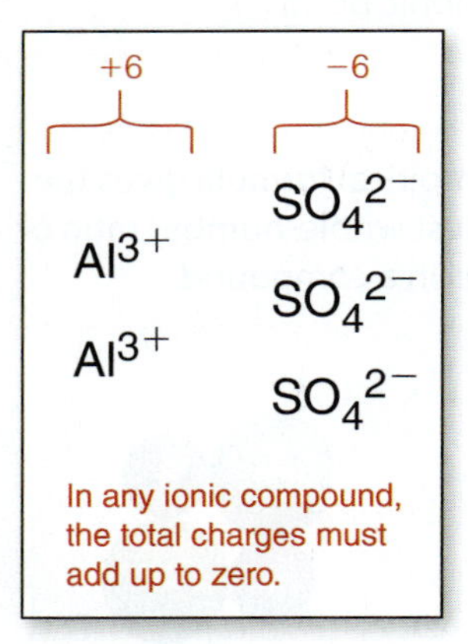

Figure 5.12 Three sulfate ions are required to balance the charge on two aluminum ions.

Example 5.4 Writing Formulas for Ionic Compounds

A compound is composed of two ions, aluminum and sulfate. What is the formula for this compound?

We know the aluminum ion has a charge of +3, and the sulfate ion has a charge of −2 (see Figure 5.9). For the charges of these ions to balance, we must have two aluminum ions for every three sulfate ions, as shown in **Figure 5.12**. Therefore, we write the formula for this compound as $Al_2(SO_4)_3$. As before, we put the repeating polyatomic ion in parentheses.

Example 5.5 Writing Formulas for Ionic Compounds

What is the formula for a compound composed of iron(III) and bromide ions?

Recall that iron is a transition metal, and it can have more than one possible charge. The name *iron(III)* indicates that this ion is Fe^{3+}. The bromide ion is Br^-. For the charges to balance, there must be three bromide ions for every one iron(III) ion. Therefore, we write this formula as $FeBr_3$.

Check it
Watch explanation

TRY IT

5. Predict the empirical formulas for compounds formed from these ions:

a. magnesium and chloride
b. potassium and phosphate
c. lead(II) and oxide
d. ammonium and carbonate

Naming Ionic Compounds

To name an ionic compound, we give the cation name followed by the anion name. For example, NaCl is *sodium chloride*, and $MgCl_2$ is *magnesium chloride*. Because we know that a magnesium ion always has a +2 charge and a chloride ion has a −1 charge, there is no need to indicate the ratio of cations to anions. Given the name of the cation and anion present, we can determine the empirical formula.

For transition metals with more than one possible charge, it is important to include the charge of the ion in parentheses with the name. For example, copper and chloride ions form two different compounds, CuCl and $CuCl_2$. In CuCl, the copper ion must have a charge of +1 to balance the charge from the chloride ion. In $CuCl_2$, the copper ion must have a charge of +2. Therefore, we name these compounds as follows:

CuCl	copper(I) chloride
$CuCl_2$	copper(II) chloride

Compounds containing polyatomic ions are named in the same way as those containing monatomic ions. For example, the ionic compound $MgSO_4$ consists of a monatomic cation (magnesium) and a polyatomic anion (sulfate). Therefore, the name of this compound is magnesium sulfate.

Table 5.4 summarizes the names, formulas, and uses of several common ionic compounds. It is important to be able to convert between the name and empirical formulas for ionic compounds. This process is further illustrated in the examples that follow.

TABLE 5.4 Common Ionic Compounds

Compound	Formula	Application
Sodium chloride	$NaCl$	Table salt
Sodium fluoride	NaF	Fluoride treatment
Sodium bicarbonate	$NaHCO_3$	Baking soda
Calcium oxide	CaO	Cement mix
Lithium carbonate	Li_2CO_3	Treatment of bipolar disorder
Ammonium nitrate	NH_4NO_3	Fertilizer

Example 5.6 Naming Ionic Compounds

Name the following compound: $Fe(NO_2)_2$.

The keys to solving this problem are to identify the ions present and to know their charges. The anion in this formula is nitrite, which has a charge of −1. Because two NO_2^- ions are present, the charge on the iron (Fe) cation must be +2. Fe^{2+} is named as iron(II), and so the total compound is iron(II) nitrite.

Example 5.7 Writing the Formula for an Ionic Compound

Write the empirical formula for ammonium sulfide.

We know that ammonium is NH_4^+ and that sulfide is S^{2-}. For the charges to balance, there must be two ammonium ions for each sulfide ion. To show this, we put the ammonium formula in parentheses with a two on the outside. Listing the cation first, we write the formula for this compound as $(NH_4)_2S$.

TRY IT

6. Name each of these compounds:

a. $RbCl$ **b.** $CuBr_2$ **c.** $ZnCO_3$ **d.** K_2SO_4

7. Write the empirical formula for each compound named:

a. zinc sulfide **b.** iron(III) oxide **c.** ammonium phosphate

8. Titanium is a transition metal that can have multiple ionic charges. The titanium compound TiO_2 is commonly used as an additive in paints. In this compound, what is the charge on the cation? What is the name of this compound?

Check it
Watch explanation

5.4 Covalent Bonding

Nonmetal–Nonmetal Bonds

In the previous section, we saw how ionic bonds form between metal cations and nonmetal anions. When two nonmetal atoms come together, a different type of bond occurs, called a **covalent bond**. In a covalent bond, two electrons are shared between two atoms.

Covalent bonds form between nonmetal atoms.

For example, consider the bond that forms between two hydrogen atoms. Each hydrogen atom has one proton and one electron. To form a covalent bond, the two electrons "pair up" in the space between the two nuclei (**Figure 5.13**).

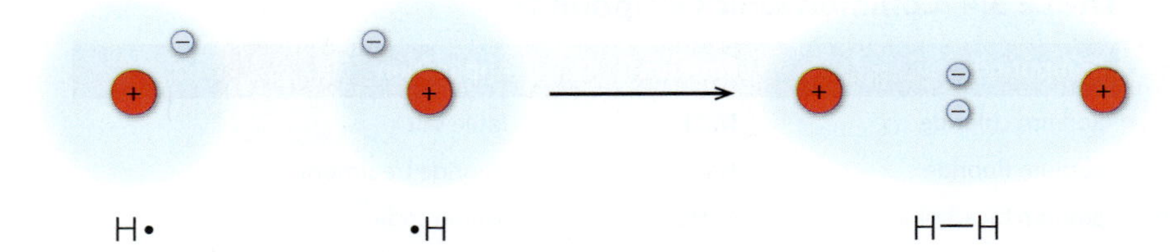

Figure 5.13 Two atoms form a covalent bond by sharing a pair of electrons.

A dash between two chemical symbols indicates a covalent bond.

The force of attraction between the nuclei and the two electrons holds the atoms together. We represent these shared electrons by drawing a dash between the symbols of the two atoms:

H• + •H —(Two electrons shared)→ H:H = H—H

Molecules are held together by covalent bonds.

Remember that the first energy level holds only two electrons. By forming a covalent bond, each hydrogen atom completes its valence level.

When two hydrogen atoms combine, they form a *molecule*. In earlier chapters, we defined molecules as groups of atoms that bind together and behave as a unit. The bonds that hold molecules together are covalent bonds. In its elemental form, hydrogen is a gas composed entirely of these two-atom molecules (**Figure 5.14**).

Figure 5.14 This balloon contains elemental hydrogen gas. The gas is composed of molecules containing two atoms each.

:Ö=Ö:

:N≡N:

Atoms sometimes share two or even three pairs of electrons in covalent bonds. We represent double covalent bonds using two dashes between the atoms, and triple covalent bonds using three dashes. We will discuss covalent bonding in more detail in Chapter 9.

As a second example, let's look at the bonding that occurs between two fluorine atoms. Each fluorine atom contains seven valence electrons and therefore needs only one more electron to complete its valence level. Two fluorine atoms can form a single covalent bond. By forming this bond, the atoms fill their valence shell with eight electrons and satisfy the octet rule.

:F̤̈• + •F̤̈: —(Two electrons shared)→ :F̤̈:F̤̈: = :F̤̈—F̤̈:

As in the hydrogen example, we use a dash to represent two shared electrons. We call this type of drawing a **Lewis structure**. Lewis structures depict the arrangement of valence electrons within a molecule or polyatomic ion. We will explore Lewis structures further in Chapter 9, when we take a more detailed look at the bonding and properties of molecules.

Hydrogen and fluorine are two of seven elements that exist as *diatomic* ("two-atom") molecules in their elemental forms. The others are nitrogen, oxygen, and the rest of the halogens (**Figure 5.15**).

The Magnificent Seven
Elements that form Diatomic Molecules

Hydrogen: H_2
Nitrogen: N_2
Oxygen: O_2
Fluorine: F_2
Chlorine: Cl_2
Bromine: Br_2
Iodine: I_2

Figure 5.15 Seven elements exist as diatomic molecules.

Covalent Compounds

Covalent compounds form when different elements combine through covalent bonds, forming discrete molecules. Water is an example of a covalent compound. In a water molecule, an oxygen atom covalently bonds to two hydrogen atoms:

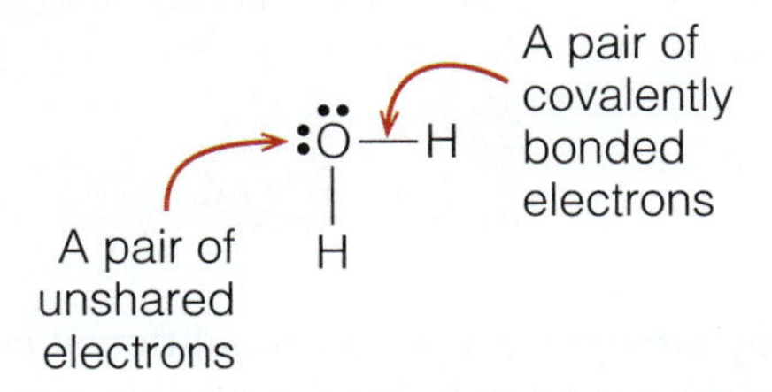

The valence level of each hydrogen atom is filled with two electrons. What about the oxygen atom? It has four unshared electrons and two covalent bonds. Between the unshared and the shared electrons, the oxygen atom has eight electrons in its valence level and fulfills the octet rule.

Covalent compounds fulfill the octet rule by sharing electrons.

To describe covalent compounds, we often use **molecular formulas**. This type of chemical formula gives the actual number of atoms in the molecule rather than the simplest whole-number ratio.

For example, consider hydrazine, a fuel used for rocket thrusters (**Figure 5.16**). A hydrazine molecule contains two nitrogen atoms and four hydrogen atoms. The empirical formula for this compound is the smallest whole-number ratio, or NH_2. However, chemists usually prefer to write this compound using the molecular formula: N_2H_4.

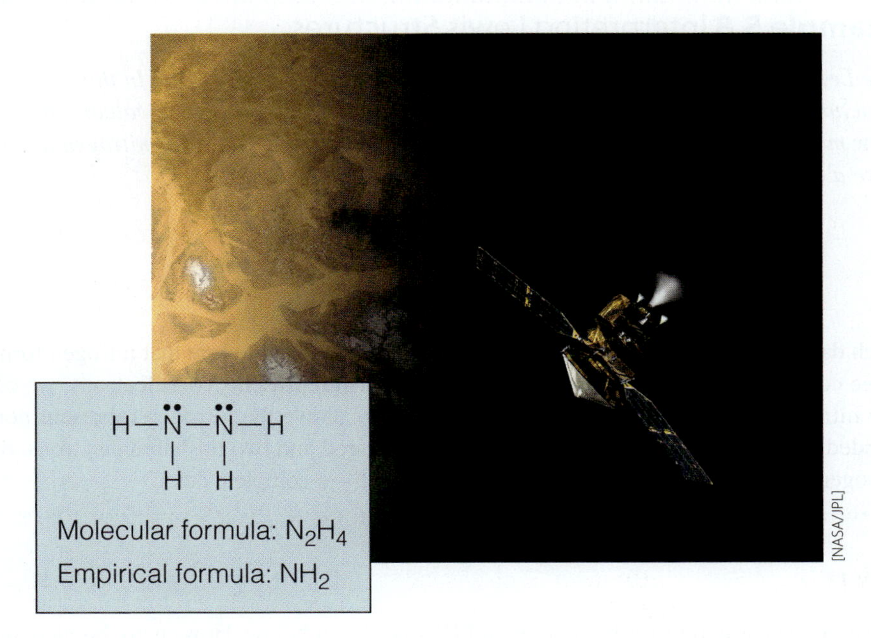

Figure 5.16 We usually describe covalent molecules, such as the rocket fuel hydrazine, by their molecular formula rather than their empirical formula.

Covalent bonds often lead to complex structures. Consider the molecule octane, a component of gasoline (**Figure 5.17**): One molecule of octane contains 25 different covalent bonds. Larger compounds may contain hundreds or thousands of covalent bonds. Because of this complex bonding, elements can often combine in many different ratios.

Figure 5.17 A molecule of octane is composed of hydrogen and carbon atoms, held together by covalent bonds.

Lithium carbonate is an ionic compound used to treat bipolar disorder. In polyatomic ions like carbonate (CO_3^{2-}), the atoms are held together with covalent bonds.

Each unique bonding arrangement produces a different compound. For example, **Table 5.5** lists several of the compounds that form between phosphorus and oxygen.

TABLE 5.5 Covalent Compounds Containing Phosphorus and Oxygen

Compound Name	Formula
Phosphorus monoxide	PO
Diphosphorus trioxide	P_2O_3
Diphosphorus tetroxide	P_2O_4
Tetraphosphorus decoxide	P_4O_{10}

Example 5.8 Interpreting Lewis Structures

The Lewis structure for a molecule of ammonia (NH_3) is shown below. In this structure, how many electrons does the nitrogen atom share through covalent bonds? How many of the valence nitrogen electrons are not shared? Does this nitrogen atom have a complete octet?

Each dash represents two shared electrons. We see from the structure that nitrogen forms three covalent bonds to hydrogen. Because each dash represents two electrons, we can say nitrogen has six shared electrons. The two dots above the nitrogen represent non-bonded (unshared) electrons. Combining the six shared and two unshared electrons, the nitrogen atom has eight electrons in its valence shell—a complete octet.

Check it

Watch explanation

TRY IT

9. The Lewis structure for the compound HCl is shown below. How many bonded and nonbonded electrons are in the valence of the chlorine atom? Does this atom fulfill the octet rule?

10. Consider the Lewis structure shown below. What is the molecular formula for this compound? What is the empirical formula?

```
   H   H
   |   |
H—C—C—H
   |   |
   H   H
```

Naming Covalent Compounds

Covalent compounds containing only two elements are called *binary covalent compounds*. These compounds are named in a manner that is similar to ionic compounds. The element that is lower and farther to the left on the periodic table is named first, and the full element name is used. The element that is nearer to the upper right on the periodic table is named as though it were an anion, by changing the end of the atom name to *–ide*.

However, there is one complicating factor: Because covalent compounds can form in many different ratios, covalent compounds use a series of prefixes (**Table 5.6**) to indicate the number of atoms present. A prefix is assigned to both the first and second part of the name. If the molecule contains only one atom of the first element, the prefix *mono–* is not used.

For example, phosphorus and chlorine commonly form two compounds that have the formulas PCl_3 and PCl_5. How do we name these compounds? Phosphorus is to the left of chlorine on the periodic table (**Figure 5.18**), so we name phosphorus first and then name chlorine as the anion (*chloride*). Using the prefixes in Table 5.6, we refer to PCl_3 as *phosphorus trichloride*, and PCl_5 as *phosphorus pentachloride*.

When naming covalent compounds, use prefixes to indicate how many atoms are present.

TABLE 5.6 Prefixes for Naming Covalent Compounds

Atoms	Prefix
1	mono–
2	di–
3	tri–
4	tetra–
5	penta–
6	hexa–
7	hepta–
8	octa–
9	nona–
10	deca–

Pent– or Penta–
If the root name of the atom begins with a vowel, we remove the ***–a*** from the end of the prefix to make it easier to pronounce. For example, PCl_5 is phosphorus **penta**chloride, but P_2O_5 is diphosphorus **pent**oxide.

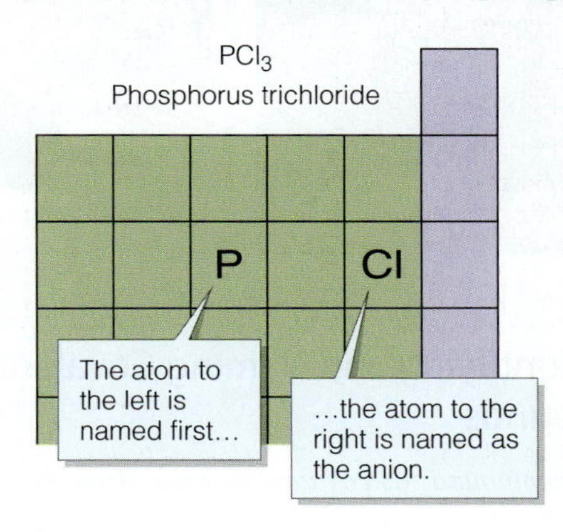
PCl_3
Phosphorus trichloride
P
Cl
The atom to the left is named first...
...the atom to the right is named as the anion.

Figure 5.18 When naming a covalent compound, the atom that lies farthest left on the periodic table comes first.

Example 5.9 Naming Covalent Compounds

Nitrogen and oxygen form two covalent compounds, NO_2 and N_2O_4. Name each of these compounds.

Because nitrogen is to the left of oxygen on the periodic table, we name nitrogen first (*nitrogen*) and then oxygen as the anion (*oxide*). The first compound, NO_2, is called *nitrogen dioxide*. The second compound, N_2O_4, is called *dinitrogen tetroxide*. Notice that we use the prefix on the first name only if more than one atom is present.

TRY IT

11. Write the names of these covalent compounds:

N_2O_3 SO_2 CF_4 P_4O_9

Check it
Watch explanation

5.5 Distinguishing Ionic and Covalent Compounds

Let's briefly review what we've covered so far: To fulfill the octet rule, atoms can either gain or lose electrons to form ions, or they can share electrons through covalent bonds.

Covalent compounds form between nonmetal atoms. These compounds form distinct units called *molecules*. We generally describe covalent compounds using molecular formulas that indicate the exact number of each atom contained in one molecule.

Ionic compounds form between oppositely charged ions. We describe an ionic compound by its *formula unit* or its *empirical formula*—that is, the simplest whole-number ratio of cations to anions in the compound. We avoid using the term *molecule* to describe these compounds, because an ionic solid has no molecular unit.

As we'll see in the sections and chapters ahead, the differences in ionic and covalent bonding lead to many unique physical properties (**Figure 5.19**). Because of this, it is important to be able to distinguish between ionic and covalent compounds. The key to doing this is to identify the elements present. Is the compound composed entirely of nonmetals? If so, it is most likely a covalent compound. Is it composed of a metal and a nonmetal? This indicates that the compound is ionic. Does it contain any of the common polyatomic ions described in Table 5.3? Again, this suggests it is ionic. Example 5.10 illustrates how we can differentiate between covalent and ionic compounds.

Metal + Nonmetal: Ionic bond ■

Nonmetal + Nonmetal: Covalent bond ■

© drewrawcliffe/depositphotos.com

Luis Carlos Jimenez del rio/Shutterstock

Figure 5.19 Limestone is composed of calcium carbonate, an ionic compound. Olive oil is composed of covalent molecules containing carbon, hydrogen, and oxygen. The properties of any compound are determined by the types of elements and bonds that are present.

Example 5.10 Identifying and Naming Covalent and Ionic Compounds

Identify each of these compounds as covalent or ionic. Provide an appropriate name for each compound.

a. $MgBr_2$ **b.** $FeCl_3$ **c.** SF_6

Ionic compounds form between a metal and a nonmetal. Covalent compounds form between two nonmetals. To solve this problem, the first step is to identify each element as a metal or a nonmetal; the next step is to decide whether the compound is covalent or ionic. Note where these elements fall on the periodic table:

F
Mg S Cl
Fe Br

In the first example, $MgBr_2$, magnesium is a metal and bromine is a nonmetal—so this is an ionic compound. We therefore name the compound simply by naming the cation first and then the anion. This compound is magnesium bromide.

In the second example, $FeCl_3$, iron is a metal and chlorine is a nonmetal. Again, this is an ionic compound. Remember that iron forms cations with more than one charge, so we must specify the charge in parentheses. Because the cation is bound to three chloride ions, this ion is Fe^{3+}, or iron(III). This compound is iron(III) chloride.

In the third example, SF_6, both sulfur and fluoride are nonmetals. Therefore this is a covalent compound, and we must use prefixes to indicate the number of each atom present. Because sulfur is to the left of chlorine on the periodic table, it is named first. This compound is sulfur hexafluoride.

TRY IT

12. Identify each of these compounds as ionic or covalent, and write its name:

LiCl ICl BCl_3 Al_2O_3

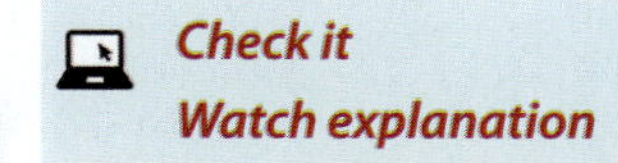
Check it
Watch explanation

5.6 Aqueous Solutions: How Ionic and Covalent Compounds Differ

One of the most important differences between ionic and covalent compounds is how they behave when combined with water. To understand this critical difference, let's begin with some fundamental ideas: When a substance such as salt or sugar mixes with water, it disperses through the liquid, forming a homogeneous mixture called a solution (**Figure 5.20**). (If the liquid is water, we call it an *aqueous solution*.) When this happens, we say that the solid has *dissolved*. Compounds that dissolve in water are said to be soluble in water; those that do not are *insoluble*.

Figure 5.20 Ocean water contains many dissolved compounds. It is an aqueous solution.

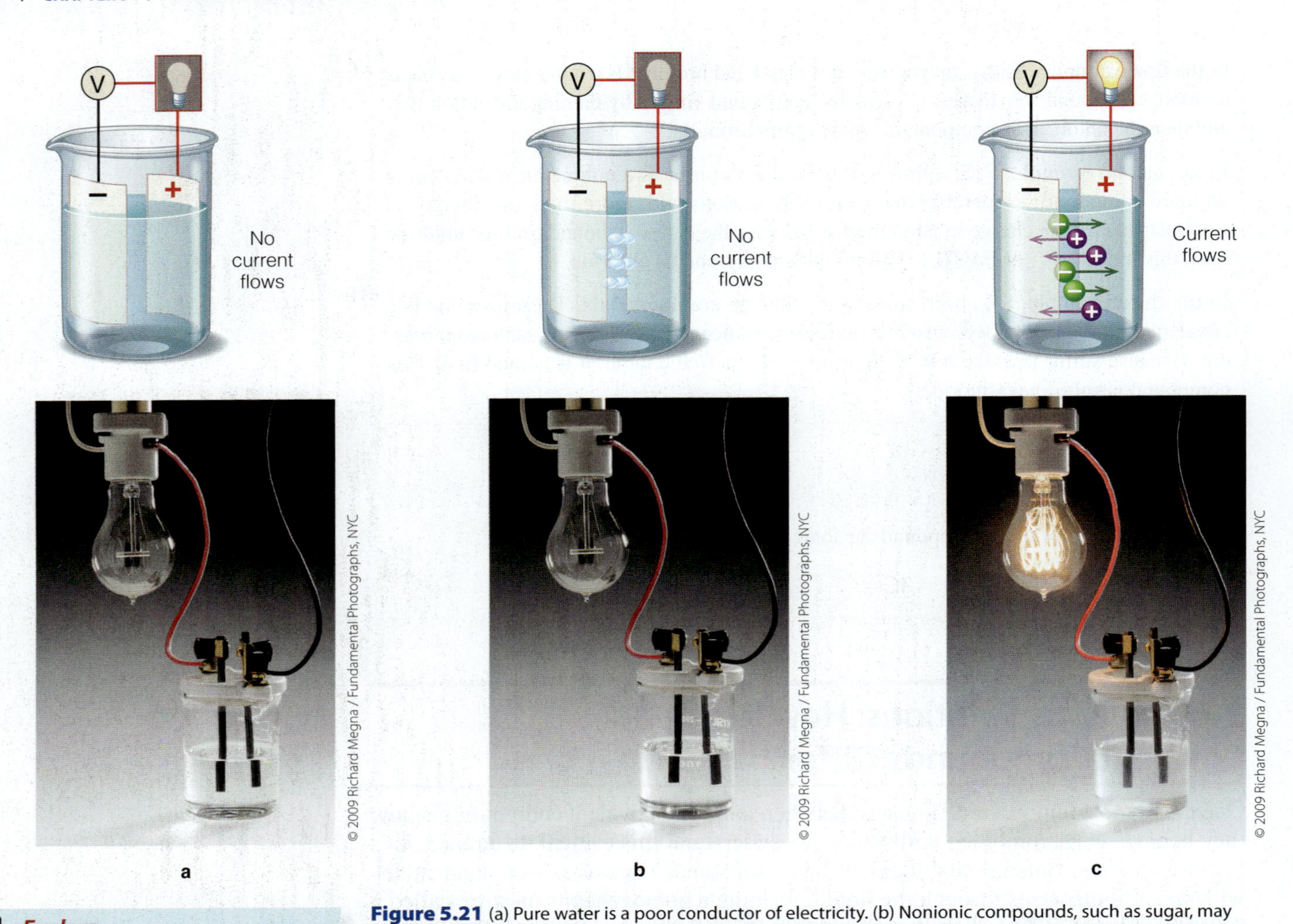

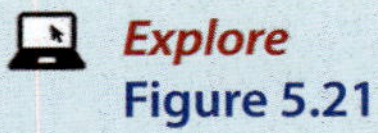

Explore Figure 5.21

Figure 5.21 (a) Pure water is a poor conductor of electricity. (b) Nonionic compounds, such as sugar, may dissolve in water, but they do not increase the solution's ability to conduct electricity. (c) Ionic compounds, like salt, dissociate into ions; the resulting solution conducts electricity.

Devices that test for water purity often test how well the water conducts electricity. If a water sample conducts electricity well, we know that ionic compounds are present.

Pure water is a poor conductor of electricity (**Figure 5.21**). However, if ionic compounds are dissolved in water, the resulting solutions conduct electricity much more efficiently. Because of this property, we refer to aqueous ionic solutions as **electrolyte solutions**, and we call the ionic compounds *electrolytes*.

When ionic compounds dissolve in water, the positive and negative ions are pulled away from each other and surrounded by water ions (**Figure 5.22**). This

Figure 5.22 When an ionic solid like salt dissolves in water, the water molecules pull the ions away from the solid and into solution.

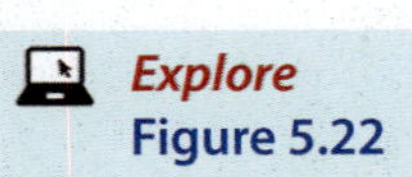

Explore Figure 5.22

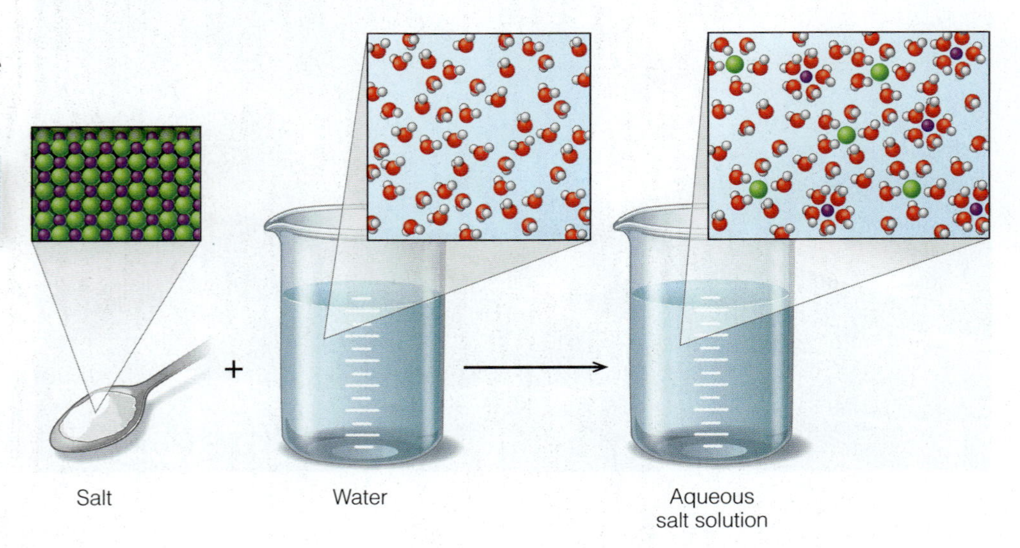

process of pulling apart the ions in an ionic solid is called **dissociation**. The dissolved ions help carry electric current through the aqueous solutions.

As a general rule, covalent compounds do not form ions in water. Because of this, aqueous solutions containing only covalent compounds are not electrolytic (Figure 5.21b).

5.7 Acids—An Introduction

Most covalent compounds do not form ions when dissolved in water, but this rule has one important exception: **Acids** are covalent compounds that produce H^+ ions in aqueous solution. Most acids contain a covalent bond between hydrogen and a species that can form a stable anion. When dissolved in water, this bond breaks to produce a hydrogen cation and a corresponding anion.

Acids produce H^+ ions in water.

For example, HCl and HNO_3 are both acidic molecules. When dissolved in water, these compounds *ionize* (form two ions):

- HCl ionizes to form H^+ and Cl^- in aqueous solution.
- HNO_3 ionizes to form H^+ and NO_3^- in aqueous solution.

We will explore the behavior of acids in Chapters 6 and 12. For now, it is important that you be able to identify and name common acids. The most common acids are listed in **Table 5.7**. When writing the formulas for acids, we typically write the formula with H first, as though it were the cation, followed by the anion.

Acids are *corrosive*, meaning they destroy many substances, including metal surfaces. They can also cause severe burns to the skin and should be handled with care.

TABLE 5.7 Common Acids

Formula	Name
HF	Hydrofluoric acid
HCl	Hydrochloric acid
HBr	Hydrobromic acid
HI	Hydroiodic acid
H_2CO_3	Carbonic acid
HNO_3	Nitric acid
HNO_2	Nitrous acid
H_2SO_4	Sulfuric acid
H_3PO_4	Phosphoric acid
$HC_2H_3O_2$	Acetic acid

Naming Acids

Binary Acids

Binary acids consist of H^+ and a single nonmetal element. The most common of these acids are those formed from the halogens: HF, HCl, HBr, and HI. These acids are named by combining the prefix *hydro–*, the root name of the halogen, and the suffix *–ic acid*:

HF	hydrofluoric acid
HCl	hydrochloric acid
HBr	hydrobromic acid
HI	hydroiodic acid

Oxyacids

Oxyacids are compounds that dissociate to form H^+ and an oxyanion. There are two rules for naming acids that dissociate to form oxyanions:

1. If the anion ends in *–ate*, name the acid by changing the suffix to *–ic acid.* For example:

NO_3^-	nitrate ion	HNO_3	nitric acid
CO_3^{2-}	carbonate ion	H_2CO_3	carbonic acid

Derivatives of the sulfur and phosphorus oxyanions deviate slightly from this rule:

SO_4^{2-}	sulfate ion	H_2SO_4	sulfuric acid
PO_4^{3-}	phosphate ion	H_3PO_4	phosphoric acid

2. If the anion ends in *–ite*, name the acid by changing the suffix to *–ous acid.*

NO_2^-	nitrite ion	HNO_2	nitrous acid

Example 5.11 Naming Acids of Oxyanions

Name each of these acids, using the guidelines described earlier:

a. $HClO_4$ **b.** H_2CrO_4

In water, $HClO_4$ ionizes to form H^+ and ClO_4^- ions. Because ClO_4^- is the perchlorate ion (see Table 5.3), $HClO_4$ is named *perchloric acid*. Similarly, CrO_4^{2-} is the chromate ion, so H_2CrO_4 is *chromic acid*.

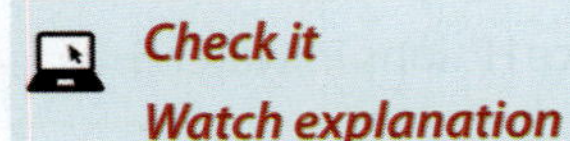

Check it
Watch explanation

TRY IT

13. Name these acids:

a. HF **b.** HClO **c.** $HC_2H_3O_2$

14. Write a formula for the acidic, ionic, and covalent compounds shown here.

a. chlorous acid **b.** zinc chlorate **c.** boron trichloride

Capstone Video

Capstone Question

Ascorbic acid, more commonly known as vitamin C, is an essential part of your diet (**Figure 5.23**). A related compound, calcium ascorbate, is a common food additive and vitamin supplement with the chemical formula $Ca(C_6H_7O_6)_2$. Based on this, what is the formula and charge of the ascorbate ion? Using this information, predict (a) the empirical formula for sodium ascorbate, (b) the empirical formula for aluminum ascorbate, and (c) both the molecular and empirical formulas for ascorbic acid.

Kevin Revell

Figure 5.23 Vitamin C is a common component of citrus. A related compound, calcium citrate, is a common vitamin supplement.

SUMMARY

Chemical bonding involves the gain, loss, or sharing of valence electrons. A key factor in chemical bonding is the octet rule, which states that atoms are stabilized by the presence of eight electrons in their valence shells. Atoms that fulfill the octet rule have completely filled *s* and *p* sublevels in their valence shell.

To fulfill the octet rule, many atoms gain or lose electrons, forming ions. Metals tend to lose electrons to form positive ions (cations) while nonmetals tend to gain electrons to form negative ions (anions). Polyatomic ions are groups of atoms that contain an overall charge.

Ionic compounds are a combination of positive ions (cations) and negative ions (anions). In any ionic compound, the total charge must be equal to zero. When naming an ionic compound, we give the name of the cation first, followed by the name of the anion. Ionic compounds bind together in lattices of alternating charges. We describe an ionic compound by its empirical formula, which is the lowest whole-number ratio of atoms in that compound.

In covalent bonds, electrons are shared between two nonmetal atoms. Covalent compounds form discrete units called molecules. We typically describe a covalent solid by its molecular formula, which gives the number of each type of atom present in the molecule. When naming covalent compounds, we use prefixes to indicate the number of each type of atom present.

Ionic and covalent compounds behave differently in water. When ionic compounds dissolve in water, they dissociate into their component cations and anions. Dissolved ions enhance water's ability to conduct electricity. Because of this trait, ionic compounds are sometimes referred to as electrolytes. In contrast, most covalent compounds remain intact when dissolved in water.

Acids are covalent compounds that ionize in water to produce H^+ ions and a corresponding anion. The names of acids derive from the names of the anions they produce in solution.

Continuing Cade's Work

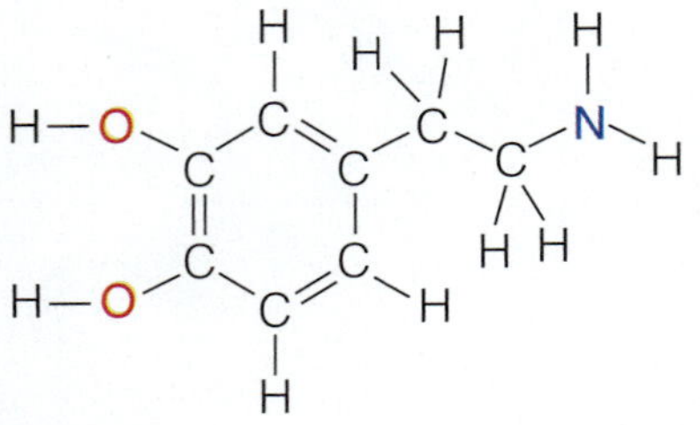

Dopamine

Figure 5.24 Dopamine is connected to mood, memory, and motor control.

A lot has changed since John Cade began using lithium carbonate to treat bipolar disorder. Today we have a much better (though far from complete) understanding of how ions and compounds affect our brain's function. For example, scientists now know that the covalent compound dopamine (**Figure 5.24**) plays a critical role in the working of the brain. Dopamine conveys signals between nerve cells, and it affects brain functions such as mood, memory, and motor control. Parkinson's disease (a degenerative disorder affecting muscle control) arises from a drop in dopamine levels. Other medical and cognitive issues, including drug addiction, perception of pain, appetite, and sexual gratification, all involve dopamine levels.

To perform its function, dopamine binds to cells in the central nervous system at special locations on the cell surface called *receptor sites*. When dopamine docks to a receptor site, it activates the site in much the same way that a key activates a lock. Like a key, the molecule's *size* and *shape* (along with other features) are critically important to its function. The shape of a molecule depends on the electronic structure of its atoms and on the covalent bonds that hold the atoms together. We'll explore the shape of molecules in much more detail in Chapter 9.

Medicinal chemists often search for molecules that can mimic the function of biological molecules like dopamine. They explore how slight changes in molecular structure (and therefore in molecule size and shape) affect the molecule's ability to bind to a receptor site. For molecules in the brain, these small differences in structure create profound differences in function.

For example, look at the three molecules in **Figure 5.25**. Do you notice their similarity to dopamine? The first molecule is adrenaline, a hormone that stimulates the nervous system. The second is ephedrine, a commercial decongestant and appetite suppressant. The third is methamphetamine, a devastatingly addictive, mood-altering drug. Like dopamine, each of these molecules affects brain function. But their small differences in size and shape affect how they bind to receptors, causing different responses in mood and behavior.

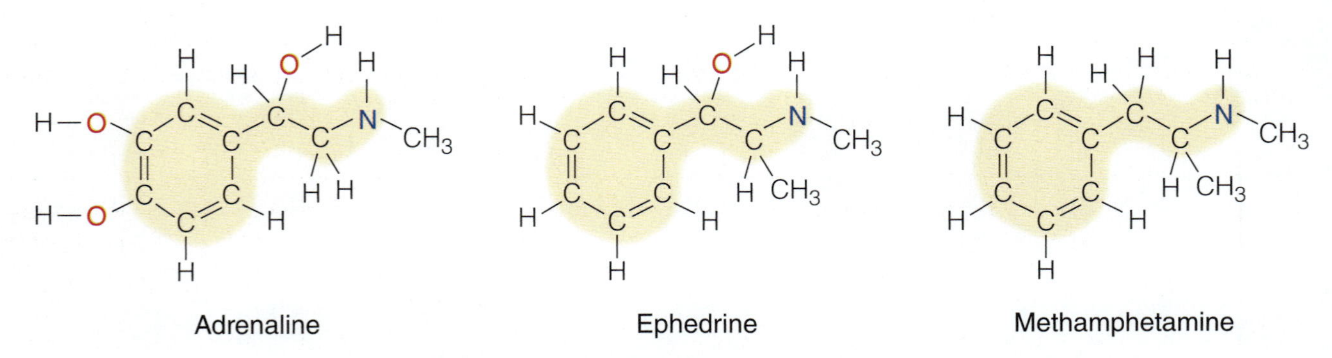

Figure 5.25 These compounds are similar to dopamine, and they also affect brain function. The yellow shading highlights their structural similarities.

Key Terms

5.1 Lewis Symbols and the Octet Rule

Lewis dot symbol A method of representing the valence structure of an atom or ion that involves using dots around the atomic symbol to indicate valence electrons.

5.2 Ions

cation A positively charged ion.

anion A negatively charged ion.

polyatomic ion A group of covalently bonded atoms with an overall charge.

oxyanion A negatively charged polyatomic ion that contains oxygen.

5.3 Ionic Bonds and Compounds

ionic bond A force of attraction between oppositely charged ions.

ionic compound A compound composed of oppositely charged ions.

ionic lattice A tightly packed array of alternating positive and negative charges; the characteristic arrangement of ions in an ionic solid.

chemical formula A representation of the type and amount of each element present in a compound.

empirical formula A chemical formula that gives the smallest whole-number ratio of atoms in a compound.

formula unit In ionic compounds, the smallest number of ions necessary to form a compound; the combination of atoms described by an empirical formula.

5.4 Covalent Bonding

covalent bond A bond in which two electrons are shared between atoms; covalent bonds typically form between nonmetals.

Lewis structure A depiction of the arrangement of valence electrons in a molecule or polyatomic ion, in which the Lewis symbols for atoms are shown connected by dashes representing covalent bonds.

covalent compounds Compounds formed by covalent bonds; these compounds form discrete groups of atoms called molecules.

molecular formula A formula that gives the actual number of atoms in the molecule.

5.6 Aqueous Solutions: How Ionic and Covalent Compounds Differ

solution A homogeneous mixture; for example, a solid mixed in a liquid.

soluble Having the ability to be dissolved in a liquid.

electrolyte solution An aqueous solution containing dissociated ions; this type of solution conducts electricity more effectively than pure water.

dissociation The process by which ions are pulled apart from a solid lattice when an ionic compound dissolves in water.

5.7 Acids—An Introduction

acid A covalent compound that produces H^+ ions in aqueous solution.

oxyacid A covalent compound that dissociates in aqueous solution to form H^+ and an oxyanion.

Additional Problems

5.1 Lewis Symbols and the Octet Rule

15. Using the periodic table, predict the number of valence electrons in each of these atoms:

Li C Si Kr Se

16. Using the periodic table, predict the number of valence electrons in each of these atoms:

Be Mg Ca Ge I

17. Write Lewis dot symbols to show the valence structures of each of these atoms:

Na N H As Sb

18. Write Lewis dot symbols to show the valence structures of each of these atoms:

Be F Ar Cs S

19. Write the electron configuration for the following atoms. Indicate which electrons are the valence electrons.

Mg N P I

20. Write the electron configuration for the following atoms. Indicate which electrons are the valence electrons.

Be S Ge Br

21. Indicate whether each of these species fulfills the octet rule:

a. a sodium atom
b. a Na^+ ion
c. a fluorine nucleus with 9 electrons
d. a fluorine nucleus with 10 electrons

22. Indicate whether each of these species fulfills the octet rule:

a. a magnesium nucleus surrounded by 10 electrons
b. a phosphorus atom
c. an argon atom

5.2 Ions

23. What family of elements forms only +1 ions?

24. What family of elements forms only +2 ions?

25. Potassium has an electronic structure of [Ar]$4s^1$. What is the electronic structure of the potassium ion (K^+)?

26. Calcium has an electronic structure of [Ar]$4s^2$. What is the electronic structure of the calcium ion (Ca^{2+})?

27. Write the electronic structure for each of these atoms and ions:

a. a lithium atom
b. a lithium ion, Li^+
c. a sodium atom
d. a sodium ion, Na^+

28. Write the electronic structure for each of these atoms and ions:

a. a magnesium atom
b. a magnesium ion, Mg^{2+}
c. a beryllium atom
d. a beryllium ion, Be^{2+}

29. Using the periodic table as a reference, predict the charge for each of these ions:

a. a beryllium ion
b. a strontium ion
c. a sodium ion
d. a cesium ion

30. Using the periodic table as a reference, predict the charge for each of these ions:

a. a potassium ion
b. a barium ion
c. a calcium ion
d. a lithium ion

31. What two charges are most common for a copper ion?

32. What two charges are most common for an iron ion?

33. Name each of the following cations:

a. Na^+ b. Mg^{2+} c. Cr^{2+} d. Cr^{3+}

34. Name each of the following cations:

a. K^+ b. Ca^{2+} c. Co^{2+} d. Co^{3+}

35. Name each of the following cations:

a. Fe^{2+} b. Fe^{3+} c. Rb^+ d. Ba^{2+}

36. Name each of the following cations:

a. Sn^{2+} b. Sn^{4+} c. Ag^+ d. Be^{2+}

37. Using the periodic table as a reference, write the symbol and charge for each cation:

a. strontium
b. zinc
c. copper(II)
d. manganese(III)

38. Using the periodic table as a reference, write the symbol and charge for each cation:

a. aluminum
b. lead(II)
c. lead(IV)
d. magnesium

39. What family of elements forms only −1 ions? What family of elements typically forms −2 ions?

40. Unlike the other nonmetals, the noble gases do not form stable ions. Why is this so?

41. The electronic structure of fluorine is [He]$2s^2 2p^5$. What is the electronic structure of the fluoride ion (F^-)?

42. The electronic structure of oxygen is [He]$2s^2 2p^4$. What is the electronic structure of the oxide ion (O^{2-})?

43. Write the electronic structure for each of these atoms and ions:

a. a chlorine atom
b. a chloride ion, Cl^-
c. a bromine atom
d. a bromide ion, Br^-

44. Write the electronic structure for each of these atoms and ions:

a. a nitrogen atom
b. a nitride ion, N^{3-}
c. a sulfur atom
d. a sulfide ion, S^{2-}

45. Indicate whether each atom would gain or lose electrons to fulfill the octet rule:

a. Na b. S c. Mg d. Br

46. Indicate whether each atom would gain or lose electrons to fulfill the octet rule:

a. Ba b. O c. K d. F

47. Determine whether the following would gain or lose electrons to fulfill the octet rule:

a. a calcium atom
b. an atom in the halogen family
c. an atom with an electron configuration of [Ar]$4s^2 3d^{10} 4p^4$
d. an atom with an electron configuration of [Xe]$6s^2$

48. Determine whether the following would gain or lose electrons to fulfill the octet rule:

a. an alkaline earth metal
b. an oxygen atom
c. an atom with an electron configuration of [Ar]$4s^2 3d^{10} 4p^5$
d. an atom with an electron configuration of [Kr]$5s^1$

49. Using the periodic table as a reference, write the symbol and charge for each of these ions:

a. fluoride
b. iodide
c. oxide
d. selenide

50. Using the periodic table as a reference, write the symbol and charge for each of these ions:

a. chloride
b. bromide
c. sulfide
d. phosphide

51. Name each of the following anions:

a. F^- b. S^{2-} c. O^{2-} d. I^-

52. Name each of the following anions:

a. Cl^- b. Br^- c. P^{3-} d. Te^{2-}

53. Using the periodic table as a reference, predict the charge of each of these ions:

a. beryllium ion b. oxide ion c. chloride ion

54. Using the periodic table as a reference, predict the charge of each of these ions:

a. bromide ion b. sodium ion c. barium ion

55. What charges would you expect on each of these ions?

a. a halogen ion
b. an alkali metal ion
c. an ion formed from a neutral atom with electron configuration $[Ne]3s^2 3p^5$

56. What charges would you expect on each of these ions?

a. an ion formed from a calcium atom
b. an alkaline earth metal ion
c. an ion formed from a neutral atom with electron configuration $[Ne]3s^1$

57. Write the name and the charge of the ion formed from each of these atoms:

a. K b. Rb c. Cl d. Br

58. Write the name and the charge of the ion formed from each of these atoms:

a. Mg b. Ca c. O d. S

59. Write the symbol and charge for each of these ions:

a. a fluoride ion
b. a strontium ion
c. a beryllium ion
d. a phosphide ion

60. Write the symbol and charge for each of these ions:

a. a ferrous ion
b. a copper(II) ion
c. a nitride ion
d. a rubidium ion

61. Identify each of these anions as monatomic or polyatomic:

a. nitride
b. nitrate
c. sulfite
d. sulfide

62. Identify each of these anions as monatomic or polyatomic:

a. bromate
b. bromite
c. bromide
d. perbromate

63. What two suffixes commonly indicate oxyanions?

64. Four common oxyanions are formed from bromine: BrO_4^-, BrO_3^-, BrO_2^-, and BrO^-. Name each of these ions.

65. Write the formula and charge for each of these polyatomic ions:

a. ammonium
b. carbonate
c. hydroxide
d. acetate

66. Write the formula and charge for each of these polyatomic ions:

a. nitrate
b. nitrite
c. sulfate
d. bicarbonate

67. Write the formula and charge for each of these polyatomic ions:

a. chlorate
b. sulfite
c. hypochlorite
d. permanganate

68. Write the formula and charge for each of these polyatomic ions:

a. cyanide
b. peroxide
c. dichromate
d. bisulfate

69. Write the symbol or formula and charge for each of these ions:

a. tin(IV)
b. cupric ion
c. fluoride
d. sulfate

70. Write the symbol or formula and charge for each of these ions:

a. lead(II)
b. aluminum
c. bromide
d. chlorate

71. Write the symbol or formula and charge for each of these ions:

a. zinc
b. chromate
c. sulfite
d. phosphide

72. Write the symbol or formula and charge for each of these ions:

a. iodide
b. hydrogen phosphate
c. iodate
d. chromium(II)

5.3 Ionic Bonds and Compounds

73. Predict the empirical formulas for compounds formed from these ions:

a. lithium and chloride
b. calcium and bromide
c. oxide and calcium
d. iron(II) and phosphide

74. Predict the empirical formulas for compounds formed from these ions:

a. sodium and fluoride
b. chromium(III) and chloride
c. silver and sulfide
d. lithium and nitrite

75. Write the empirical formula for each of these compounds:

a. aluminum chloride
b. iron(II) sulfide
c. calcium sulfate
d. aluminum oxide

76. Write the empirical formula for each of these compounds:

a. iron(III) nitrate
b. copper(II) nitrate
c. ammonium phosphate
d. ammonium phosphide

77. Write the empirical formula for each of these compounds:

a. chromium(III) acetate
b. zinc chlorate
c. silver nitrate
d. lead(II) carbonate

78. Write the empirical formula for each of these compounds:

a. tin(IV) chloride
b. ammonium chlorite
c. lithium bicarbonate
d. cobalt(III) hydroxide

79. Write the empirical formula for each of these compounds:

a. chromium(III) hypochlorite
b. potassium permanganate
c. sodium cyanide
d. lead(II) perchlorate

80. Write the empirical formula for each of these compounds:

a. tin(IV) chloride
b. ammonium chlorate
c. lithium bisulfate
d. sodium hydrogen phosphate

81. In these compounds, determine the charge on the transition metal cation:

a. $SnCl_2$ b. $SnCl_4$
c. $Pb(NO_3)_2$ d. $FeCO_3$

82. In these compounds, determine the charge on the transition metal cation:

a. Ag_3PO_4 b. $Cu(C_2H_3O_2)_2$
c. $InBr_3$ d. $Cr_2(SO_4)_3$

83. Name each of these ionic compounds:

a. NaBr b. K_2O c. $FeBr_3$ d. CuS

84. Name each of these ionic compounds:

a. KOH b. $Cu(C_2H_3O_2)_2$ c. K_2CrO_4 d. NH_4Cl

85. Name each of these ionic compounds:

a. $FeCO_3$ b. $Al(NO_2)_3$
c. $Ba(NO_3)_2$ d. $(NH_4)_2SO_4$

86. Name each of these ionic compounds:

a. $Na_2Cr_2O_7$ b. AgOH
c. $ZnCO_3$ d. $Cr_2(SO_4)_3$

5.4 Covalent Bonding

87. When two nonmetals bond together, why do they form a covalent bond rather than an ionic bond?

88. What seven elements exist as diatomic molecules in their elemental forms?

89. How many electrons are shared in a covalent bond? When drawing structures, how do we typically represent covalent bonds?

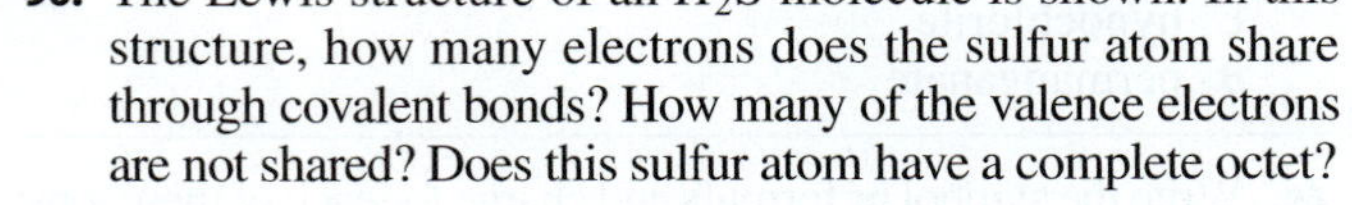

90. The Lewis structure of an H_2S molecule is shown. In this structure, how many electrons does the sulfur atom share through covalent bonds? How many of the valence electrons are not shared? Does this sulfur atom have a complete octet?

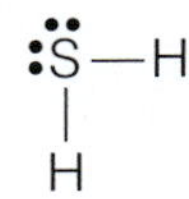

91. This figure shows a group of water molecules. How many water molecules are in this image? How many covalent bonds are present in each molecule?

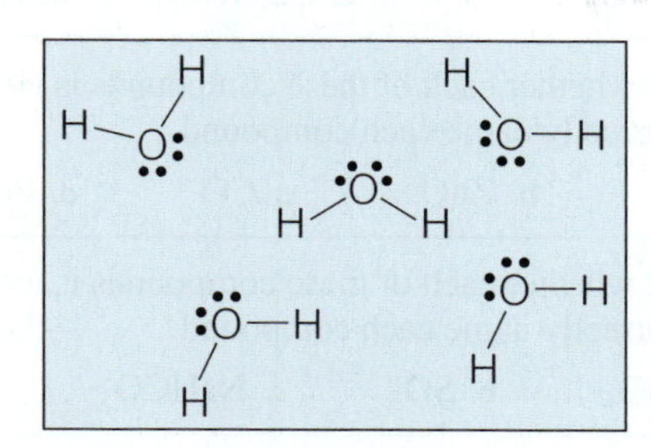

92. This figure shows a group of CH_4 molecules. How many CH_4 molecules are in this image? How many covalent bonds are present in each molecule?

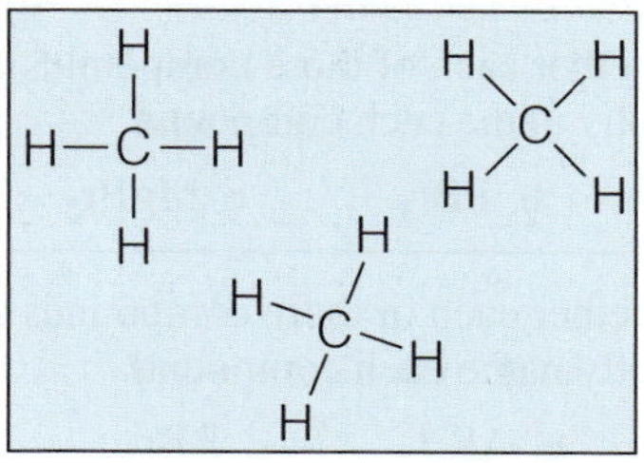

93. Acetic acid, shown here, is the main component of vinegar. Give the molecular formula and the empirical formula for this compound.

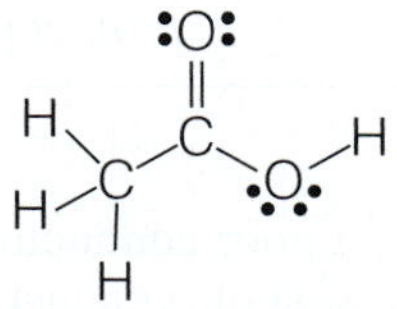

Acetic Acid

94. The structure of oxalic acid is shown here. Give the molecular formula and the empirical formula for this compound.

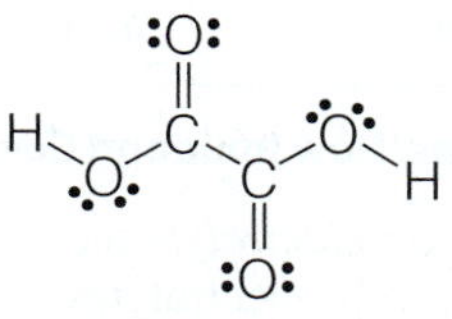

Oxalic Acid

95. Compounds such as Freon® 112 (referred to as chlorofluorocarbons, or CFCs) were used as refrigerants for many years, but they were phased out because of their harmful effects on Earth's atmosphere. Write the molecular and empirical formulas for Freon 112.

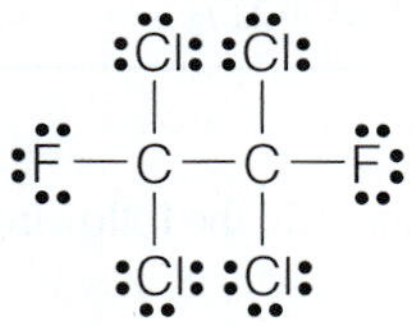

Freon 112

96. Propane is a natural gas that is widely used as a heating fuel. Give the molecular and empirical formulas for propane.

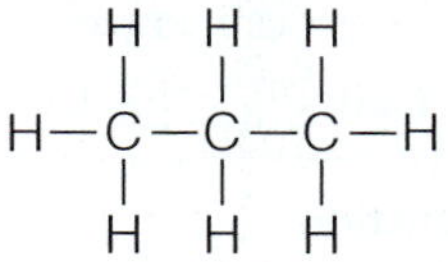

Propane

97. Why is it necessary to use prefixes when naming binary covalent compounds?

98. When naming binary covalent compounds, when is a prefix not used?

99. Name each of these covalent compounds:

a. SCl_2 b. NF_3 c. N_2O_4 d. P_4O_{10}

100. Name each of these covalent compounds:

a. SO_3 b. CCl_4 c. N_2F_4 d. S_2Cl_2

101. Write molecular formulas for each of these covalent compounds:

a. arsenic tribromide
b. dinitrogen pentoxide
c. disulfur dioxide

102. Write molecular formulas for each of these covalent compounds:

a. disulfur dioxide
b. selenium tetrafluoride
c. tetraphosphorus trisulfide

5.5 Distinguishing Ionic and Covalent Compounds

103. When looking at a binary compound (one that has just two elements), how can we tell if it is ionic or covalent?

104. Why is it acceptable to write the formula of acetylene as C_2H_2 but not acceptable to write the formula of magnesium oxide as Mg_2O_2?

105. Determine whether these compounds contain ionic bonds or covalent bonds:

a. NaBr b. PCl_3 c. MnF_2

106. Determine whether these compounds contain ionic bonds or covalent bonds:

a. CO_2 b. N_2 c. KCl

107. Indicate whether these compounds would form an ionic lattice or discrete molecules:

a. KCl b. CCl_4 c. P_4O_{10} d. Na_2S

108. Indicate whether these compounds would form an ionic lattice or discrete molecules:

a. CO_2 b. MgF_2 c. $Ca(NO_3)_2$ d. Na_3PO_4

109. Indicate whether each of these compounds is ionic or covalent. Correctly name each compound.

a. NaBr b. PBr_3 c. $MgBr_2$ d. SBr_2

110. Indicate whether each of these compounds is ionic or covalent. Correctly name each compound.

a. SO_3 b. ZnO c. CO d. Fe_2O_3

111. Indicate whether each of these compounds is ionic or covalent. Correctly name each compound.

a. $SiCl_4$ b. $AlCl_3$ c. BBr_3 d. Na_2SO_3

112. Indicate whether each of these compounds is ionic or covalent. Correctly name each compound.

a. $MgSO_4$ b. SO_3 c. $NaHCO_3$ d. CO_2

113. Write the correct chemical formula for each compound, using empirical formulas for ionic compounds and molecular formulas for covalent compounds.

a. manganese(III) chloride b. phosphorus trichloride
c. sulfur dioxide d. titanium(IV) oxide

114. Write the correct chemical formula for each compound, using empirical formulas for ionic compounds and molecular formulas for covalent compounds.

a. silver bromide b. selenium dibromide
c. sulfur trioxide d. copper(II) sulfite

5.6 Aqueous Solutions: How Ionic and Covalent Compounds Differ

115. What does the term *electrolyte* mean? What types of compounds are likely to be electrolytes?

116. By itself, water is a poor conductor of electricity. What must be present for water to conduct electricity efficiently?

117. When sodium sulfate is dissolved in water, it dissociates. What ions are present in an aqueous solution of sodium sulfate?

118. Ethylene glycol ($C_2H_6O_2$) is a covalent compound. Describe what happens to the $C_2H_6O_2$ molecules as this compound dissolves in water.

119. Which of these compounds are likely to dissociate in an aqueous solution? How can you tell?

a. KCl b. $CaBr_2$ c. CO_2 d. C_2H_6O

120. Which of these compounds are likely to dissociate in an aqueous solution? How can you tell?

a. SO_3 b. $Zn(ClO_4)_2$ c. CS_2 d. MgF_2

5.7 Acids—An Introduction

121. What are acids? How do acids differ from most covalent compounds?

122. Place these compounds in the following Venn diagram: NaCl, CCl_4, and HCl.

Covalent compounds
Electrolytes

123. Name the following acids:

a. HCl b. HBr c. HI

124. H_2S is an acidic, foul-smelling gas. It is often called *hydrogen sulfide*. Name this compound using the rules for naming binary acids.

125. Name the following acids:

a. HNO_3 b. HNO_2 c. $HClO_4$ d. $HClO_2$

126. Name the following acids:

a. H_2SO_4 b. H_2SO_3 c. $HClO_3$ d. HClO

127. The formate ion is a biologically important ion with the formula CHO_2^-. Based on this information, what is the name of the acid having the formula $HCHO_2$?

128. The selenate ion has the formula SeO_4^{2-}. Based on this information, what is the name of the acid having the formula H_2SeO_4?

129. Classify each of these compounds as an ionic compound, a covalent compound, or an acid. Name each compound.

a. $NaNO_2$ b. N_2O_4 c. HNO_2 d. KNO_2

130. Classify each of these compounds as an ionic compound, a covalent compound, or an acid. Name each compound.

a. K_2SO_4 b. SO_2 c. H_2SO_4 d. $NaHSO_4$

Chemical Reactions

Figure 6.1 (a) This image shows the palace ruins at Palenque. *(continued)*

CHAPTER SIX

Lost Cities of the Maya

For thousands of years, the Mayan civilization dominated the Yucatán region in southeast Mexico, Guatemala, and Belize. Massive cities like Tikal and Palenque glistened with temples, palaces, and stadiums. Their culture flourished, with remarkable achievements in architecture and mathematics. Mayan sculpture and writing, which used intricate figures called *hieroglyphs*, left a stunning record of their history, culture, and beliefs (**Figure 6.1**).

But something happened. Around 900 C.E., the cities of the inner Yucatán began to decline. Their populations quickly diminished. By the time the Spanish arrived in the early 1500s, most of the Maya had migrated north and east to the coastal areas. The great cities were abandoned, lost to the jungle.

What caused the decline?

For years, archaeological conservationist Isabel Villaseñor studied the ruins of Mayan cities like Palenque. Her work focused on a specific feature — the white plaster covering the great temples and palaces. In many places, ancient artisans even molded this plaster to create intricate artwork and hieroglyphs.

Over the past several decades, Dr. Villaseñor and other archaeologists have learned how the Mayans produced plaster for their buildings. The story involves chemical reactions that are still used today to produce plaster and cement.

The Maya began with limestone, a type of rock composed of calcium carbonate ($CaCO_3$), which is abundant in the Yucatán Peninsula. They crushed the limestone into small pieces and then heated it over huge fires. Above about 900 °C, calcium

Figure 6.1 (continued) (b) These plaster sculptures are from the Mayan city Ek Balam. (c) Mayan writing used hieroglyphs like these, which were often made from molded plaster. (d) Isabel Villaseñor overlooks the Temple of the Foliated Cross, Palenque. (e) Villaseñor and coworkers worked inside the crypt of the Mayan King Tikal, located inside the temple. The sarcophagus lid in front of her is carved limestone. (a) Index Fototeca Heritage Images/Newscom; (b) John Mitchell/Alamy; (c) De Agostini/G. Dagli Orti/ Getty Images; (d) Courtesy of Dr. Isabel Villasenor; (e) Courtesy of Dr. Isabel Villaseñor

carbonate decomposes to form two new compounds—carbon dioxide (CO_2) and calcium oxide (CaO). The powdery calcium oxide was then mixed with water to produce a new compound, today called *slaked lime*. Combined with volcanic ash, this material formed a soft putty that was easily molded. As the lime mixture slowly dried, a third chemical reaction took place, producing a hard and beautiful white surface.

But the plaster came at a price: Producing lime created huge demands on the local resources. The Maya used freshly cut trees to generate the hot, slow-burning fires needed to convert limestone to calcium oxide. Quarrying the limestone, cutting down the trees, and maintaining the fires required immense human effort—but the environmental cost may have been greater still.

The story of Mayan plaster is a story of chemical changes. In this chapter, we will explore common chemical changes. We'll see how these changes are expressed as chemical equations and how atomic structure determines the way different elements and compounds react. We'll explore patterns of chemical change and use them to predict the outcomes of common reactions. And, at chapter's end, we will return to Palenque, to revisit the reactions that produced the great plastered surfaces, and examine clues that suggest how cities like Palenque were transformed from thriving cultural centers into abandoned ruins, hidden for centuries in the Yucatán jungles.

Intended Learning Outcomes

After completing this chapter and working the practice problems, you should be able to:

6.1 Chemical Equations

- Write chemical equations to express the identity and ratio of species in a chemical change.
- Use a balanced equation to describe the ratio in which atoms or compounds react.
- Correctly balance an equation.

6.2 Classifying Reactions

- Classify synthesis, decomposition, single-displacement, and double-displacement reactions.

6.3 Reactions between Metals and Nonmetals

- Predict the products formed from the reaction of metals and nonmetals.
- Identify the species that are oxidized and reduced in a metal-nonmetal combination reaction.

6.4 Combustion Reactions

- Predict the products formed from the combustion of metals and from the combustion of hydrocarbons.

6.5 Reactions in Aqueous Solution

- Apply the solubility rules to determine whether common ionic compounds are water soluble and predict the products of precipitation reactions.
- Predict the products of acid-base neutralization reactions.
- Describe precipitation and neutralization reactions using molecular, complete ionic, and net ionic equations.

6.1 Chemical Equations

Scientists use **chemical equations** to describe chemical changes. For example, we just saw how the Mayans heated limestone (calcium carbonate) to convert it to lime (calcium oxide) and carbon dioxide gas. We could represent this change with a chemical equation:

$$CaCO_3 \rightarrow CaO + CO_2$$

In this reaction, $CaCO_3$ is the **reactant** (sometimes called the *reagent* or the *starting material*). The reactants are shown on the left side of the arrow. The **products** that form in the reaction (CaO and CO_2) are shown on the right side of the arrow.

Chemical equations also convey the *ratios* of reactants and products. For example, look carefully at the molecular representation of the reaction in **Figure 6.2**. In this reaction, two molecules of hydrogen gas react with one molecule of oxygen gas to produce two molecules of water. Notice that this reaction follows the law of conservation of mass: Atoms are not created or destroyed in this chemical change, and the total mass of the starting materials is equal to the mass of the products (**Figure 6.3**).

We describe the reaction of hydrogen and oxygen using this chemical equation:

$$2\,H_2 + O_2 \rightarrow 2\,H_2O$$

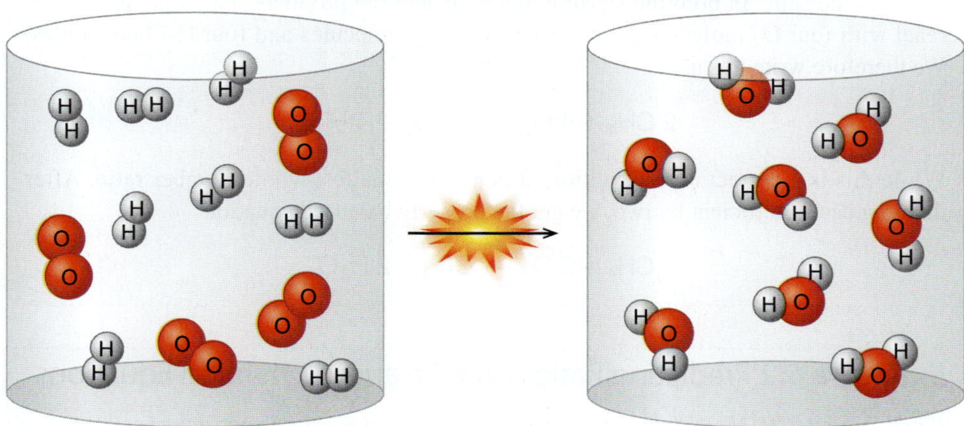

Figure 6.2 (Top) When ignited with a flame, a balloon filled with hydrogen reacts in a dramatic fireball. (Bottom) The reaction involves the combination of hydrogen with oxygen, as shown. Each oxygen molecule combines with two hydrogen molecules to produce two new water molecules.

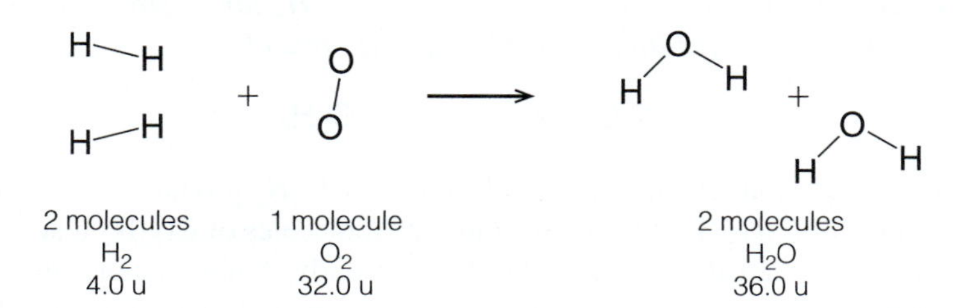

Figure 6.3 The reaction of hydrogen with oxygen to form water follows the conservation of mass. The total mass of the reactants is equal to the total mass of the products.

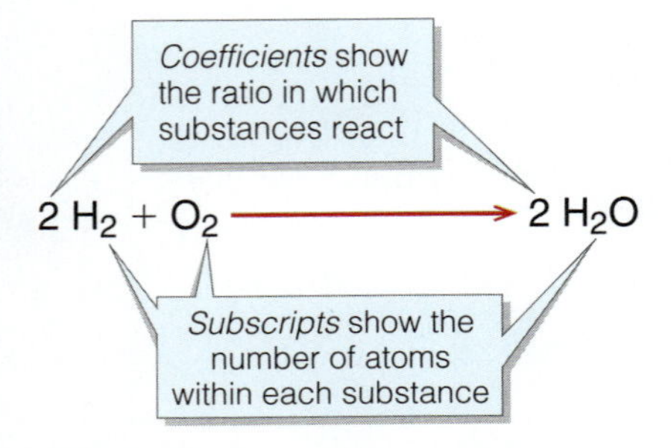

Figure 6.4 It is important to understand the difference between coefficients and subscripts.

Recall from Chapter 5 that the **subscripts** (for example, H_2, O_2, H_2O) show the number of atoms in each molecule or formula unit. The numbers that precede the chemical formulas are called **coefficients**. These numbers show the ratio in which compounds react or are formed (**Figure 6.4**). The coefficients are written to show the smallest whole-number ratio in which molecules react. If a reactant or product does not have a coefficient, we assume that coefficient to be one.

Chemical equations describe what happens at the atomic level, but they also provide a ratio for larger reactions. For example, what if 2 million molecules of hydrogen reacted with 1 million molecules of oxygen? Based on the ratio in the equation, 2 million molecules of water would form.

If we have correctly described the ratios of reactants and products in a chemical reaction, the equation is *balanced*. In a **balanced equation**, the number and type of each atom are the same in the reactants as in the products. A properly balanced equation shows the smallest whole-number ratio of reactants to products.

A balanced equation contains the same number and type of atom on each side of the arrow.

Example 6.1 Writing Chemical Equations

When methane gas (CH_4) burns, it reacts with molecular oxygen to form carbon dioxide and water, as represented in the image below. Write a balanced equation to describe this reaction.

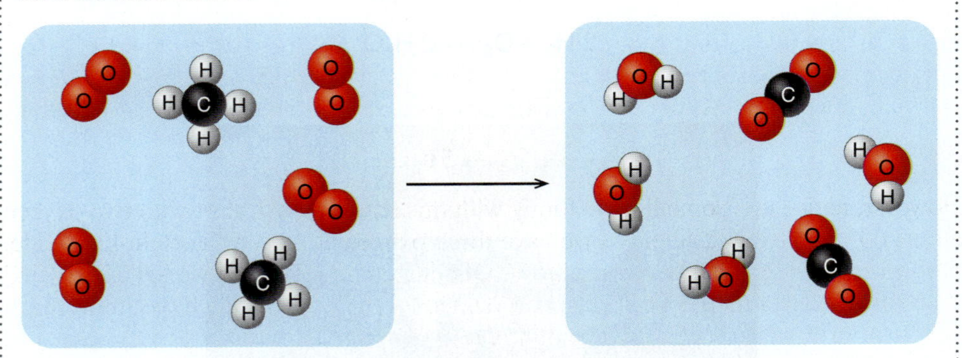

In this chemical reaction, the reactants are CH_4 and O_2. The products for this reaction are H_2O and CO_2. Notice that in this picture, the number of carbon, hydrogen, and oxygen atoms is the same in both the starting materials and the products. Two CH_4 molecules react with four O_2 molecules to produce two CO_2 molecules and four H_2O molecules. We therefore write this as

$$2\,CH_4 + 4\,O_2 \rightarrow 2\,CO_2 + 4\,H_2O$$

While this is a correct representation, it is not the smallest whole-number ratio. After dividing each coefficient by two, we get the properly balanced equation.

$$CH_4 + 2\,O_2 \rightarrow CO_2 + 2\,H_2O$$

Example 6.2 Predicting Ratios from Balanced Chemical Equations

When heated, CH_2N_2 undergoes an explosive chemical reaction to produce two new substances, N_2 and C_2H_4, as shown in the balanced equation below. If 20 molecules of CH_2N_2 react in this way, how many molecules of C_2H_4 form? How many molecules of N_2 and C_2H_4 form if 2 million molecules of C_2H_4 react?

$$2\,CH_2N_2 \rightarrow 2\,N_2 + C_2H_4$$

The balanced equation tells us that 2 molecules of CH_2N_2 produce 2 molecules of N_2 and 1 molecule of C_2H_4. Following this ratio, 20 molecules of CH_2N_2 react to produce 20 molecules of N_2 and 10 molecules of C_2H_4. Similarly, 2 million molecules of CH_2N_2 react to produce 2 million molecules of N_2 and 1 million molecules of C_2H_4.

TRY IT

1. Write a balanced equation to describe the reaction of nitrogen and hydrogen shown here:

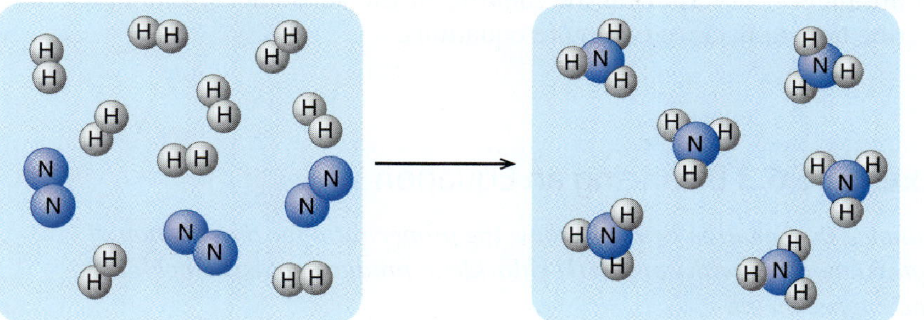

2. Based on the balanced equation in Question 1, how many molecules of NH_3 could be made from the reaction of 3,000 molecules of hydrogen gas with 1,000 molecules of nitrogen gas?

Balancing Equations

Sometimes, when describing a reaction, we begin with the identities of the starting materials and products without knowing the ratio in which the reaction takes place. For example, iron can react with oxygen to form iron(III) oxide—the reddish-brown compound we know as rust. To describe this process, we could write the equation as

$$Fe + O_2 \rightarrow Fe_2O_3$$

However, notice that something is funny with this equation: We start with two oxygen atoms on the left-hand side, but we have three oxygen atoms on the right-hand side. Did we conjure up another oxygen atom? Of course not! Although we have identified the compounds correctly, we have not accounted for the *ratio* in which the atoms react.

To more accurately describe this reaction, we must balance the equation to account for each atom in the starting materials and products. To do this, we add coefficients to the elements and compounds in the equation until the number and type of atom on each side are the same. This process is shown in **Figure 6.5**.

The Maya used iron oxide to produce the red-brown colors in their artwork.

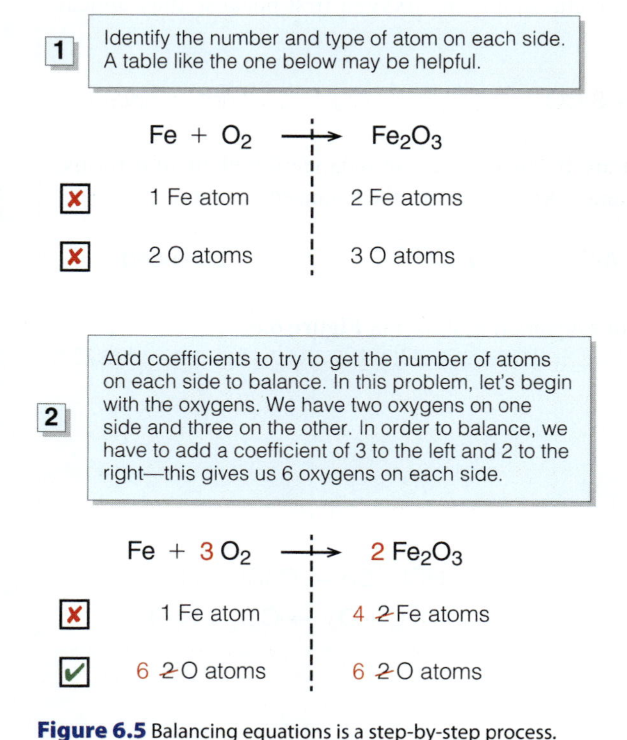

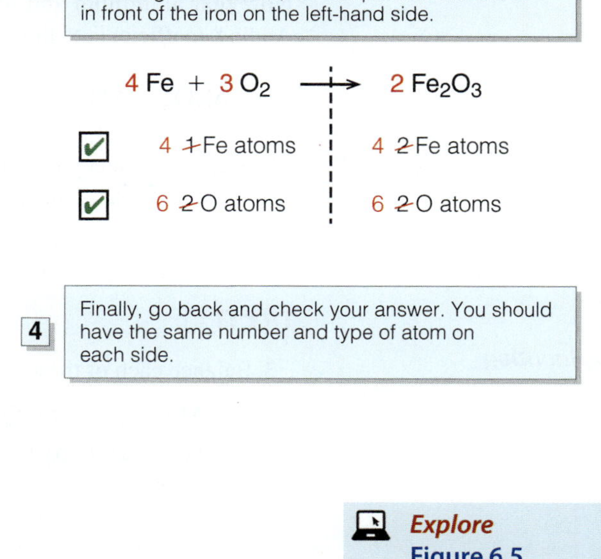

Figure 6.5 Balancing equations is a step-by-step process.

Explore
Figure 6.5

Notice that when balancing this equation, *we never touched the subscripts*. The subscripts show the chemical identity of each molecule or formula unit present. When we balance equations, we are simply adjusting the ratio of starting materials and products—not altering the identity of the substances. Examples 6.3 and 6.4 describe how to balance two more equations.

Balance equations by adjusting the coefficients, not the subscripts.

Example 6.3 Balancing an Equation

Balance the equation below to show the proper ratio for the reaction of potassium metal with copper(II) chloride to produce potassium chloride and copper metal:

$$K + CuCl_2 \rightarrow Cu + KCl \qquad \text{not balanced}$$

Let's begin by balancing the chlorine atoms. To do this, we put the coefficient 2 in front of KCl. This balances the chlorine atoms but not the potassium atoms. However, we can balance the potassium atoms by putting a 2 in front of the K.

$$2\ K + CuCl_2 \rightarrow Cu + 2\ KCl \qquad \text{balanced}$$

It is helpful to balance elemental forms last.

Example 6.4 Balancing an Equation

Balance the following equation:

$$Al_2O_3 + C + Cl_2 \rightarrow AlCl_3 + CO \qquad \text{not balanced}$$

Notice that in this equation, aluminum and oxygen always appear as part of a compound while carbon and chlorine appear as elemental forms. This means that we can add coefficients to the carbon or chlorine without disrupting the balance of any of the other compounds. *In general, it is easier to balance elemental forms last.* So in this example, we would balance aluminum and oxygen first because they appear only in compounds:

$$Al_2O_3 + C + Cl_2 \rightarrow 2\ AlCl_3 + 3\ CO \qquad \text{C, Cl not balanced}$$

Now that aluminum and oxygen are balanced, we can balance the elemental forms. Adding coefficients to the carbon and chlorine gives us the balanced equation:

$$Al_2O_3 + 3\ C + 3\ Cl_2 \rightarrow 2\ AlCl_3 + 3\ CO \qquad \text{balanced}$$

The stepwise strategy for solving this equation is shown in **Figure 6.6**.

Check It
Watch Explanation

TRY IT

3. Balance each of these equations:

$Mg + O_2 \rightarrow MgO$ $\qquad$ $HCl + Co \rightarrow CoCl_2 + H_2$

$Br_2 + Al \rightarrow AlBr_3$ $\qquad$ $C_4H_8 + O_2 \rightarrow CO_2 + H_2O$

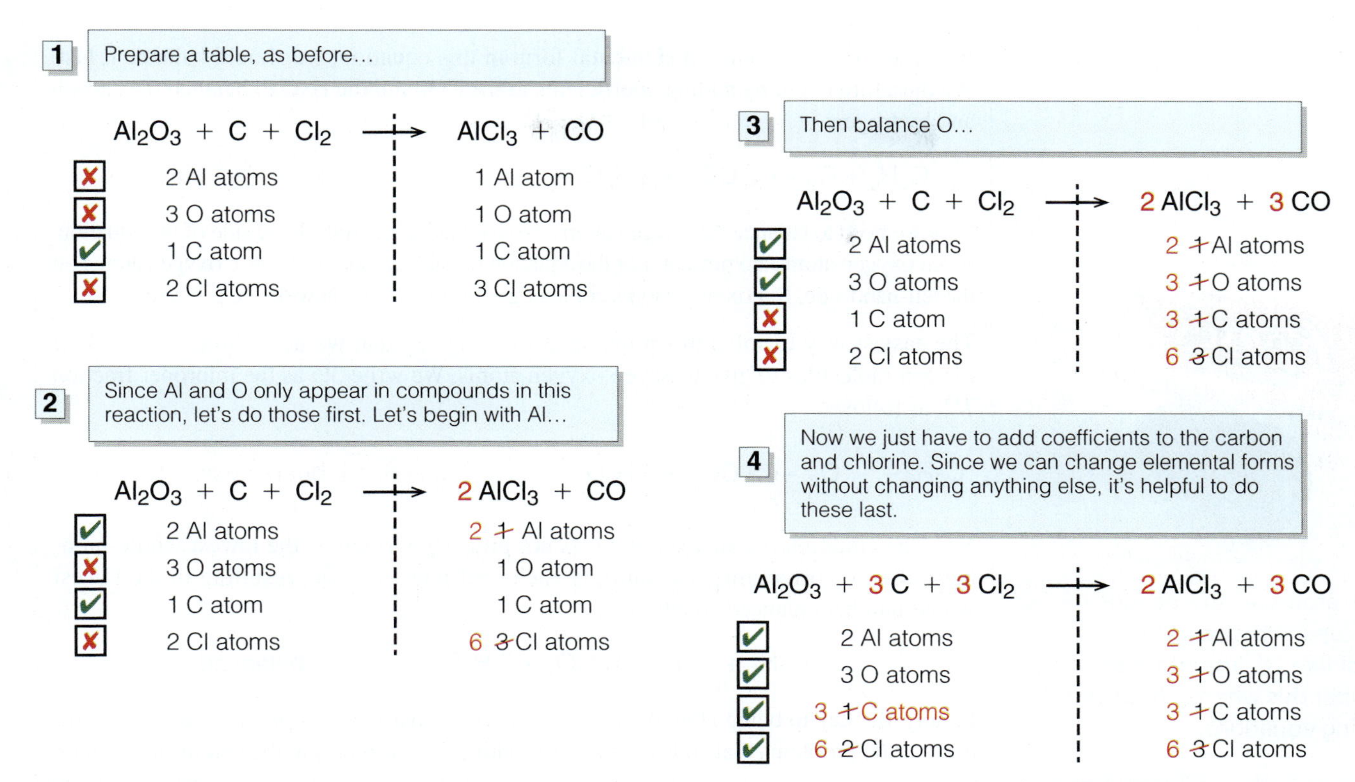

Figure 6.6 When balancing an equation, it is helpful to balance the elemental forms last.

***Explore* Figure 6.6**

Strategies for Balancing Equations

We can use a few other techniques to balance complicated equations more easily. In some equations, it is possible to balance polyatomic ions rather than atoms. In other equations, using a fractional coefficient may be helpful as an intermediate step. Examples 6.5 and 6.6 illustrate these techniques.

Example 6.5 Balancing Equations with Polyatomic Ions

Balance the following equation:

$$Ni(NO_3)_2 + NaOH \rightarrow Ni(OH)_2 + NaNO_3 \qquad \text{not balanced}$$

We could try to balance this equation by making a table of all the elements present, as we did before. However, notice that this equation contains two common polyatomic ions—nitrate (NO_3^-) and hydroxide (OH^-). Although it is possible to balance the atoms individually, it is much easier to simply balance the nitrate and hydroxide ions. We have two hydroxide ions on the right, so we add the coefficient 2 in front of NaOH to balance them on the left. Similarly, we place the coefficient 2 in front of $NaNO_3$, to balance the nitrate ions on the left and right:

$$Ni(NO_3)_2 + 2\ NaOH \rightarrow Ni(OH)_2 + 2\ NaNO_3 \qquad \text{balanced}$$

This approach works as long as the ions in the reaction do not change. We will see several examples of this type of reaction in the sections that follow.

Example 6.6 Balancing with Fractional Coefficients

Balance the following equation:

$$C_2H_6 + O_2 \rightarrow CO_2 + H_2O \qquad \text{not balanced}$$

Because oxygen appears in elemental form in this equation, we should balance it last. We therefore begin by adding coefficients to the CO_2 and the H_2O to balance the carbon and hydrogen:

$$C_2H_6 + O_2 \rightarrow 2\ CO_2 + 3\ H_2O \qquad \text{oxygen not balanced}$$

Now we need to balance the oxygen atoms. Notice that on the right-hand side of the equation, seven oxygen atoms are present. For the equation to balance, we need seven oxygen atoms on the left-hand side. But oxygen atoms come in groups of two. So how do we get seven?

The easiest way to solve this problem is to recognize that we need three and one-half oxygen molecules to give us seven oxygen atoms. We write 3½ as the improper fraction 7/2, as follows:

$$C_2H_6 + \frac{7}{2}\ O_2 \rightarrow 2\ CO_2 + 3\ H_2O \qquad \text{balanced, but not whole numbers}$$

Now the equation is balanced, but it is not properly written as the lowest whole-number ratio. To correct this, we multiply each coefficient by two, resulting in the lowest whole-number balanced equation:

$$2\ C_2H_6 + 7\ O_2 \rightarrow 4\ CO_2 + 6\ H_2O \qquad \text{balanced}$$

Finally, the key to being able to balance equations consistently is *practice*. Be sure to try each of the problems that follow and do the additional problems at the end of the chapter.

Photo Researchers, Inc./Science Source

Oxygen atoms are like peanut butter cups—they come in packs of two. It's important to remember this when writing and balancing equations.

Check It

Watch Explanation

TRY IT

4. Balance each of these equations, using the techniques described in this section:

$$Pb(ClO_3)_2 + NaI \rightarrow NaClO_3 + PbI_2$$
$$Al(NO_2)_3 + Ca \rightarrow Al + Ca(NO_2)_2$$
$$C_4H_{10} + O_2 \rightarrow CO_2 + H_2O$$

Equations with Phase Notations

For many chemical processes, it is important to know not only the identity of the components but also the *phase* or *state* of the components. Because of this, *phase notations* are sometimes used in chemical equations (**Table 6.1**). If we wish to indicate that a compound is a solid, we write (*s*) after the chemical formula. Similarly, we represent liquids by the symbol (*l*) and gases by the symbol (*g*).

TABLE 6.1 Phase Symbols

Symbol	Meaning
(*s*)	Solid
(*l*)	Liquid
(*g*)	Gas
(*aq*)	Aqueous solution (dissolved in water)

Many reactions involve substances that are dissolved in water. A substance that is dissolved in water is said to be in an *aqueous solution*. We represent this by using the symbol (*aq*) in the balanced equation. For example, consider the change that occurs when you pour a soft drink (**Figure 6.7**): The dissolved carbonic acid (H_2CO_3) reacts to produce two new compounds, water and carbon dioxide gas. We can write this equation using phase notation as follows:

$$H_2CO_3\ (aq) \rightarrow H_2O\ (l) + CO_2\ (g)$$

We sometimes use equations with phase symbols to represent physical changes, such as changes of state. For example, we could use the following equation to show water freezing:

$$H_2O\ (l) \rightarrow H_2O\ (s)$$

In the sections that follow, we will use the phase symbols extensively to describe reactions.

6.2 Classifying Reactions

As you begin to study chemical reactions, you'll encounter a large number of examples. You may find it overwhelming—I certainly did when I started learning chemistry, and sometimes I still do. So how can you begin to make sense of it all? One of the keys is to organize reactions into similar types. This makes it easier to recognize patterns in chemical behavior and to predict similar reactions that may take place.

Let's illustrate this idea with an example from the grocery. To make shopping easier, the store organizes food by groups: meats, fresh vegetables, breads. It also arranges food by origin: Italian, Mexican, Thai. These simple organizational systems help us remember, understand, and communicate with one another.

In this section and those that follow, we'll explore broad patterns of chemical behavior. We'll classify reactions by the types of products, or by the way the reaction takes place. No single classification system is universal, but they all serve one purpose—to organize and simplify the world around us. As we begin to explore chemical reactions, let's consider four broad types of reactions: decomposition, synthesis, single displacement, and double displacement (**Figure 6.8**).

studiovin/Shutterstock

Figure 6.7 The bubbles that form when a soft drink is poured arise from the reaction of carbonic acid to produce carbon dioxide and water.

In **decomposition** reactions, a single reactant forms two or more products. As an example of this type of reaction, let's return to the Mayan production of lime plaster. The first step in this process was the decomposition of calcium carbonate into two simpler compounds, calcium oxide and carbon dioxide:

$$CaCO_3\ (s) \rightarrow CaO\ (s) + CO_2\ (g)$$

Decomposition:
One forms two or more

$$2\ H_2O \longrightarrow 2\ H_2 + O_2$$
$$CaCO_3 \longrightarrow CaO + CO_2$$

Single Displacement:
One element replaces another

$$Zn + CuCl_2 \longrightarrow ZnCl_2 + Cu$$
$$Ca + 2\ HBr \longrightarrow CaBr_2 + H_2$$

Synthesis (Combination):
Two form one

$$H_2 + Cl_2 \longrightarrow 2\ HCl$$
$$CaO + H_2O \longrightarrow Ca(OH)_2$$

Double Displacement:
Two ions replace each other

$$NaI + AgNO_3 \longrightarrow AgI + NaNO_3$$
$$MgBr_2 + Pb(ClO_4)_2 \longrightarrow PbBr_2 + Mg(ClO_4)_2$$

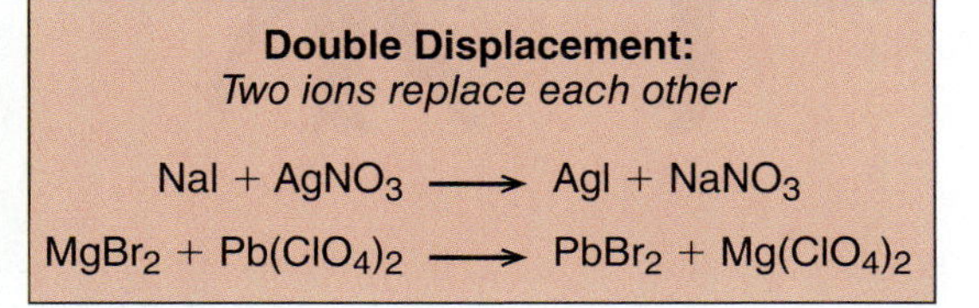

Figure 6.8 Many reactions fall under the four categories shown here.

Although it is an ancient technology, this decomposition reaction is still vitally important today. Calcium oxide is the key ingredient in cement and mortar, and manufacturers use modern kilns to convert limestone (calcium carbonate) into calcium oxide (**Figure 6.9**).

Synthesis (or *combination*) reactions occur when two reactants join together to form a single product. The second step in the production of lime plaster is an example of a synthesis reaction: In this reaction, calcium oxide combines with water to produce a single new compound, calcium hydroxide:

$$CaO\ (s) + H_2O\ (l) \rightarrow Ca(OH)_2\ (s)$$

Another example of a synthesis reaction is the combination of hydrogen gas and bromine gas to form hydrogen bromide:

$$H_2\ (g) + Br_2\ (g) \rightarrow 2\ HBr\ (g)$$

We will look extensively at synthesis reactions in Section 6.3.

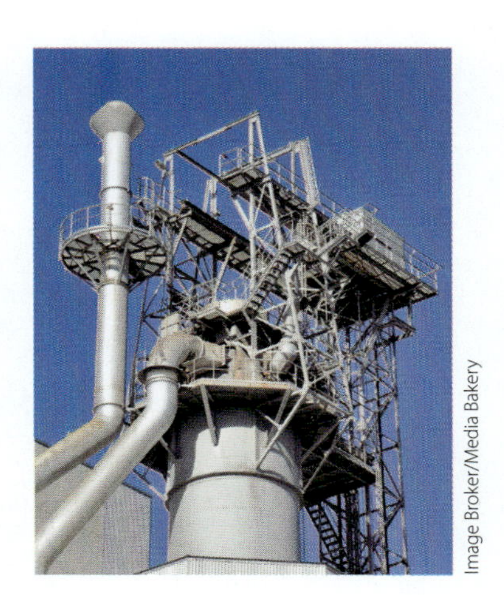

Image Broker/Media Bakery

Figure 6.9 Lime kilns like this one convert limestone into calcium oxide.

In a **single-displacement** reaction, one element replaces another element in a compound. For example, when zinc is combined with aqueous copper(II) sulfate, elemental zinc displaces copper to form a new compound. This process converts copper into its elemental form:

$$Zn\ (s) + CuSO_4\ (aq) \rightarrow ZnSO_4 (aq) + Cu\ (s)$$

Similarly, the compound HCl reacts with elemental tin to produce a new compound, tin(II) chloride, and elemental hydrogen:

$$Sn\ (s) + 2\ HCl\ (aq) \rightarrow SnCl_2\ (aq) + H_2\ (g)$$

In a **double-displacement** reaction, two compounds rearrange to form two new compounds. These reactions involve a "swap" of cation-anion pairs. For example, the reaction of KCl with $AgNO_3$ in aqueous solution produces two new compounds, KNO_3 and AgCl. This reaction can be thought of as simply a swap of anions:

$$KCl + AgNO_3 \rightarrow KNO_3 + AgCl$$

The anions "swap" positions.

A simple way to think about single- and double-displacement reactions is to consider the cations and anions as dance partners (**Figure 6.10**). In a single displacement, an element "cuts in" on the dance—producing a new element and a new compound (new dance partners). In a double displacement, the two dancing couples switch partners, producing two new compounds.

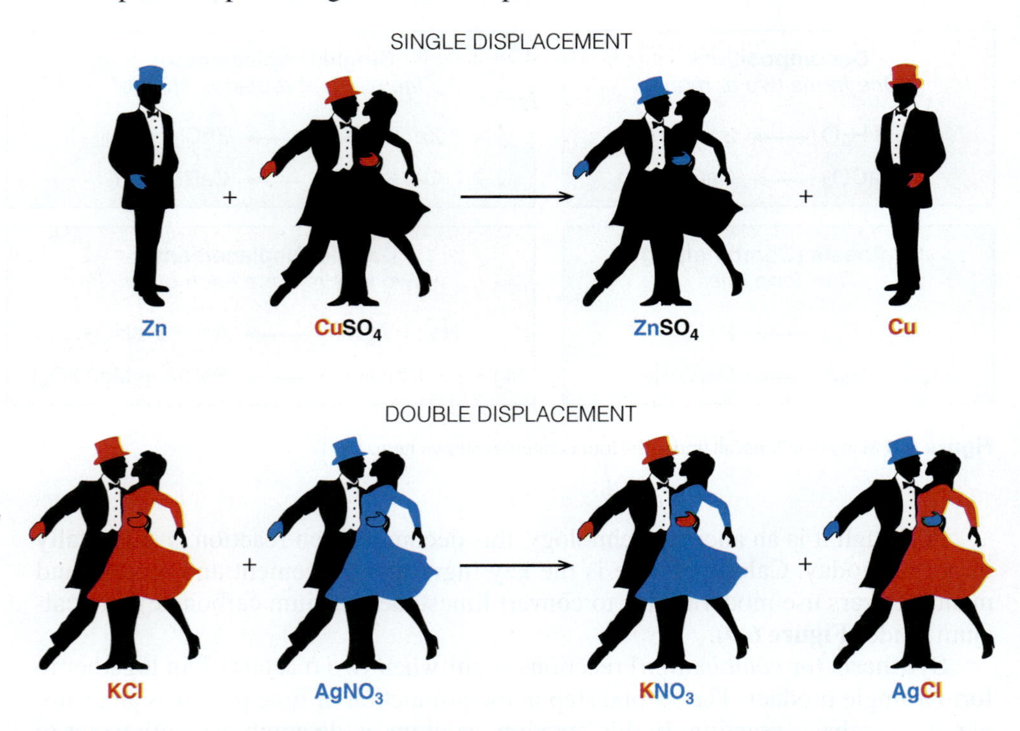

Figure 6.10 Single- and double-displacement reactions are similar to changes in dance partners.

Example 6.7 Classifying a Chemical Reaction

An aqueous solution of hydrochloric acid reacts with solid zinc metal to produce hydrogen gas and a new compound, zinc chloride, that dissolves in water. Write a balanced equation to show this reaction, then classify it as a synthesis, decomposition, single-displacement, or double-displacement reaction. Finally, add phase symbols to describe the phase of each reactant and product.

Our two reactants are hydrochloric acid and elemental zinc. The two products are zinc chloride and hydrogen gas. From Chapter 5, we know that the formula for hydrochloric acid is HCl. The formula for zinc chloride is $ZnCl_2$, and hydrogen gas is H_2. We begin with this unbalanced equation:

$$HCl + Zn \rightarrow ZnCl_2 + H_2 \qquad \text{not balanced}$$

Next, we balance this equation. In this case, addition of the coefficient 2 in front of HCl is all that is needed:

$$2\ HCl + Zn \rightarrow ZnCl_2 + H_2 \qquad \text{balanced}$$

Notice that this reaction contains one element and one compound in the reactants, and a different element and compound in the products. This is an example of a single-displacement reaction. From the question, we know that HCl and $ZnCl_2$ are both dissolved in water, zinc is a solid, and H_2 is a gas. In the final step, we represent this using phase symbols.

$$2\ HCl\ (aq) + Zn\ (s) \rightarrow ZnCl_2\ (aq) + H_2\ (s) \qquad \text{balanced, with phase symbols}$$

TRY IT

5. Classify the following changes as decomposition, synthesis, single-displacement, or double-displacement reactions:

$$Pb(NO_3)_2 + 2\ KI \rightarrow 2\ KNO_3 + PbI_2$$
$$4\ Al + 3\ O_2 \rightarrow 2\ Al_2O_3$$
$$2\ H_2O_2 \rightarrow 2\ H_2O + O_2$$
$$Fe + SnCl_2 \rightarrow Sn + FeCl_2$$

6. When solid lead(II) sulfite is heated, it forms two new compounds, solid lead(II) oxide and sulfur dioxide gas. What type of reaction is this? Write a balanced equation for this reaction, including phase symbols.

Check It
Watch Explanation

6.3 Reactions between Metals and Nonmetals

Now that we have a method for classifying reactions, let's look at some ways that specific substances react. We'll begin with one of the most common types of synthesis reactions—the combination of a metal and a nonmetal.

Metals and nonmetals react to form ionic compounds. Many of these reactions are intense, producing bright flames and tremendous heat (**Figure 6.11**).

In these reactions, the metal loses electrons to form a positively charged ion (cation) while the nonmetal gains electrons to form a negatively charged ion (anion). The result is an ionic compound:

$$\text{Metal} + \text{Nonmetal} \rightarrow \text{Ionic compound}$$

The reactions of metals and nonmetals are examples of an **oxidation-reduction reaction**. When an atom is **oxidized**, it loses electrons. When an atom is **reduced**, it gains electrons. We will explore oxidation-reduction reactions in more detail in Chapter 14.

Let's look at some examples of this type of reaction. We'll begin with the reaction of elemental calcium with sulfur to form calcium sulfide. The calcium loses two electrons to form Ca^{2+} while the sulfur gains two electrons to form the

Reacting with O_2
Most metals react with oxygen to form new compounds. In these reactions, the metal loses electrons. The term *oxidized* arises from these reactions, but its meaning is broader. In any reaction where an atom loses electrons, we say the atom is oxidized.

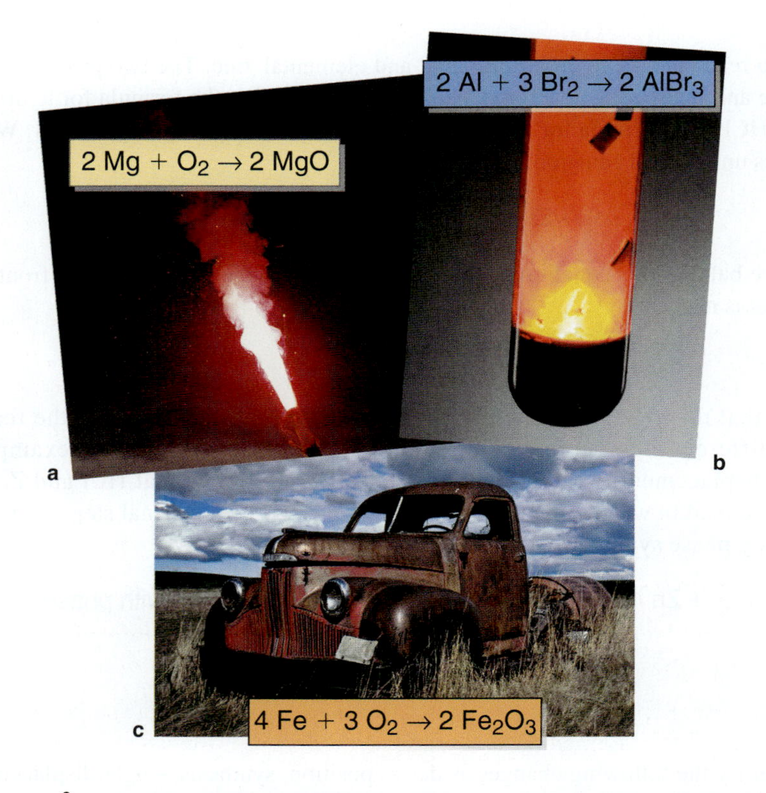

Figure 6.11 Metals and nonmetals react to form ionic compounds. (a) The bright light of a signal flare results from the reaction of metallic magnesium with oxygen gas. (b) Sodium and chlorine gas react violently to produce sodium chloride. (c) Rust forms by the slow reaction of iron with oxygen gas. Jason Edwards/Getty Images; © 2008 Richard Megna/Fundamental Photographs; Jeff Zenner Photography/Shutterstock

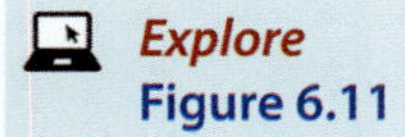
Explore **Figure 6.11**

sulfide ion, S^{2-}. We can illustrate this reaction using Lewis symbols, or we can represent it using a chemical equation, as shown here:

$$\dot{Ca}\cdot + :\ddot{\underset{\cdot}{S}}\cdot \longrightarrow Ca^{2+} + :\ddot{\underset{\cdot\cdot}{S}}:^{2-}$$

Ca loses two electrons   S gains two electrons

$$Ca + S \rightarrow CaS$$

Consider another example: Magnesium metal reacts with chlorine gas to produce magnesium chloride (**Figure 6.12**).

Charles D. Winters/Science Source

Figure 6.12 Magnesium reacts violently with chlorine gas.

The metal atom (magnesium) loses two electrons while each chlorine atom gains one electron. In this reaction, a covalent bond breaks as the ions form:

$$Mg\cdot + :\ddot{C}l-\ddot{C}l: \longrightarrow Mg^{2+} + 2\,:\ddot{C}l:^{-}$$

Mg loses two electrons

Each Cl gains one electron

$$Mg + Cl_2 \rightarrow MgCl_2$$

So, how can we predict the compounds that will form in metal-nonmetal reactions? Recall from Chapter 5 that metals and nonmetals form specific, stable ions. By knowing the charges of the common ions, we can often predict the ionic compounds that result from these reactions. This process is illustrated in the following two examples.

Many metal-nonmetal reactions produce common ions.

Example 6.8 Predicting the Products when Metals and Nonmetals React

What compound is formed when aluminum metal reacts with chlorine gas? Write a balanced equation for this reaction.

We know that aluminum forms a +3 ion, while a chloride ion has a charge of −1. Therefore, the compound formed must have the formula $AlCl_3$. We can write an unbalanced equation as shown:

$$Al + Cl_2 \rightarrow AlCl_3 \qquad \text{not balanced}$$

To balance the equation, we add coefficients to make sure the number and type of atom on each side of the equation are the same, as shown.

$$2\,Al + 3\,Cl_2 \rightarrow 2\,AlCl_3 \qquad \text{balanced}$$

Example 6.9 Predicting the Products When Metals and Nonmetals React

When tin metal reacts with bromine, it is oxidized to the tin(IV) ion while bromine is reduced to form bromide ions. Write a balanced equation for this reaction.

The question indicates that the tin(IV) ion is formed. Therefore, the neutral compound produced will have the formula $SnBr_4$. We can write an unbalanced equation as shown:

$$Sn + Br_2 \rightarrow SnBr_4 \qquad \text{not balanced}$$

To balance the equation, we only need to add a coefficient in front of the bromine. This gives us the final balanced equation.

$$Sn + 2\,Br_2 \rightarrow SnBr_4 \qquad \text{balanced}$$

TRY IT

7. If zinc metal reacts with bromine gas, which species is oxidized? Which species is reduced?
8. Predict the products from these reactions:
   a. the reaction of potassium and sulfur
   b. the reaction of silver and oxygen

Check It
Watch Explanation

6.4 Combustion Reactions

Many common reactions involve elemental oxygen. Oxygen is a major component of Earth's atmosphere, and it reacts with nearly every other element and with many compounds. Both metals and nonmetals react with oxygen. For example, consider the way that oxygen reacts with tin (a metal) and also with carbon and with sulfur (both nonmetals):

$$Sn + O_2 \rightarrow SnO_2$$
$$C + O_2 \rightarrow CO_2$$
$$S + O_2 \rightarrow SO_2$$

The first product, tin(IV) oxide, is an ionic compound. The other two products, carbon dioxide and sulfur dioxide, are molecular compounds with covalent bonds. Yet all three elements react with oxygen in similar ways. Each of these reactions releases a large amount of heat and forms an oxide compound. These rapid and exothermic reactions with oxygen are called **combustion** reactions.

Some of the most important combustion reactions involve compounds composed of hydrogen and carbon, called *hydrocarbons*. These reactions are important because fossil fuels, such as coal, oil, and natural gas, are composed primarily of hydrocarbons (**Table 6.2**). The combustion reactions of fossil fuels produce most of the energy used worldwide (**Figure 6.13**).

franckreporter/Getty Images

Figure 6.13 The combustion of coal and natural gas produces most of our electricity, and the combustion of gasoline powers most of our transportation.

When hydrocarbons react with oxygen, they form two main products, carbon dioxide and water. For example, methane (CH_4) is the main component of natural gas (**Figure 6.14**).

Chepko Danil Vitalevich/Shutterstock

Figure 6.14 Gas stoves use the combustion of methane gas to produce heat.

TABLE 6.2 Common Hydrocarbons

Formula	Name	Use
CH_4	Methane	Natural gas
C_2H_2	Acetylene	Torches for cutting and welding
C_2H_4	Ethylene	Manufacture of plastic
C_3H_8	Propane	Natural gas component; used for heating and power
C_4H_{10}	Butane	Lighter fluid
C_6H_6	Benzene	Solvent; precursor for many pharmaceutical compounds
C_8H_{18}	Octane	Component of gasoline

The combustion reaction of methane follows this balanced equation:

$$CH_4 + 2\,O_2 \rightarrow CO_2 + 2\,H_2O$$

Similarly, octane, C_8H_{18}, is a key component of gasoline. This compound also reacts with oxygen to produce carbon dioxide and water:

$$2\,C_8H_{18} + 25\,O_2 \rightarrow 16\,CO_2 + 18\,H_2O$$

The combustion of hydrocarbons produces carbon dioxide and water. ■

Nonmetal elements often react with oxygen to form many different compounds. For example, the combustion of elemental sulfur produces several different oxides, including sulfur dioxide (SO_2) and sulfur trioxide (SO_3). These compounds, commonly referred to as *sulfur oxides* (SO_x), are toxic gases that contribute to environmental issues such as acid rain. Because coal contains a small amount of sulfur (in addition to hydrocarbon compounds), the combustion of coal leads to the release of sulfur oxides into the atmosphere (**Figure 6.15**).

Sulfur Oxides
(SO_x)
SO
SO_2
SO_3
SO_4
S_2O_7
Rudmer Zwerver/Shutterstock

Figure 6.15 When oxygen reacts with sulfur, it produces a set of compounds known as sulfur oxides. Because coal contains small amounts of sulfur, sulfur oxides are produced when coal is burned for fuel.

Example 6.10 Predicting the Products of Combustion Reactions

Write a balanced equation for the combustion of propane gas, a common fuel used for home heating, cooking, and so on. The formula for propane is C_3H_8.

Hydrocarbons react with oxygen to produce carbon dioxide and water. We can write an unbalanced equation for this reaction first:

$$C_3H_8 + O_2 \rightarrow CO_2 + H_2O$$

We then balance the equation to give us the final answer:

$$C_3H_8 + 5\ O_2 \rightarrow 3\ CO_2 + 4\ H_2O$$

For another example of balancing a combustion reaction, see Example 6.6 earlier in this chapter.

TRY IT

9. Write a balanced equation that describes the combustion of magnesium.

10. Write a balanced equation that describes the combustion of C_5H_{12}.

6.5 Reactions in Aqueous Solution

Representing Dissociation: Molecular and Ionic Equations

In Chapter 5 we described the behavior of different types of compounds when dissolved in water (Section 5.6). Recall that when an ionic compound dissolves in water, it dissociates into cations and anions. That is, the ions that make up the compound are pulled apart and surrounded by water molecules (**Figure 6.16**). This behavior enables different ions to react in different ways.

Ionic compounds dissociate when they dissolve in water.

When describing the chemical processes that take place in solution, we use two different equation styles. In a **molecular equation**, we write ions together as compounds. In an **ionic equation**, we show dissociated ions as separate species. For

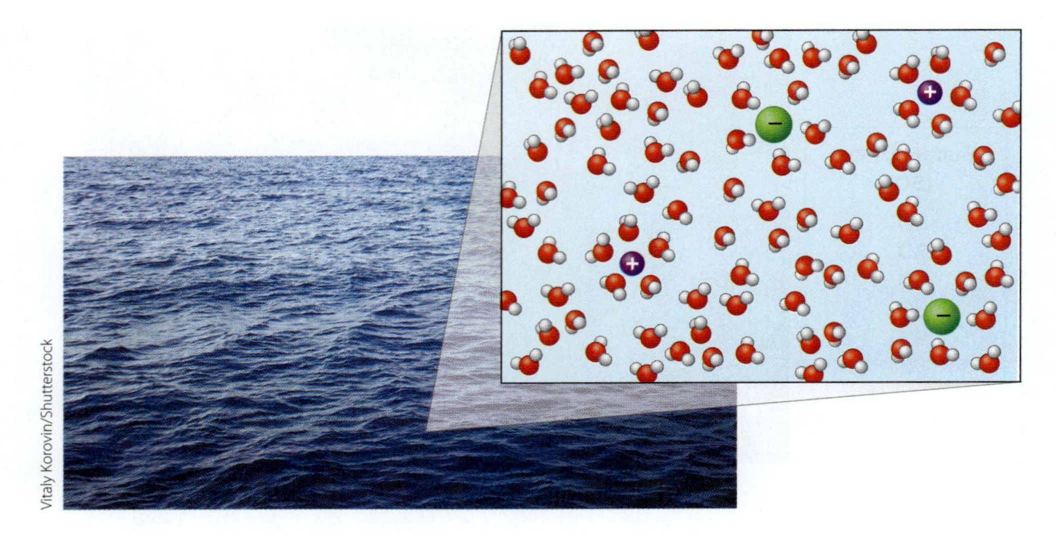

Figure 6.16 When an ionic compound dissolves in water, the cations and anions dissociate.

example, potassium bromide dissociates in water. We can write this as a molecular equation (shown in blue) or as an ionic equation (shown in red):

Molecular equation: $KBr\ (s) \rightarrow KBr\ (aq)$

Ionic equation: $KBr\ (s) \rightarrow K^+\ (aq) + Br^-\ (aq)$

Similarly, we could show the dissociation of magnesium nitrate in solution. In this example, the formula unit $Mg(NO_3)_2$ dissociates to form one magnesium ion and two nitrate ions:

Molecular equation: $Mg(NO_3)_2\ (s) \rightarrow Mg(NO_3)_2\ (aq)$

Ionic equation: $Mg(NO_3)_2\ (s) \rightarrow Mg^{2+}\ (aq) + 2\ NO_3^-\ (aq)$

Notice that the ionic equation shows only the dissociated (aqueous) ions separately; we show solid ionic compounds as a complete unit.

Practice
Ion Charges

Do you remember the names, symbols, and charges for the common ions covered in Chapter 5? If not, you should go back and relearn the list in Figure 5.9 . You will use them extensively in the sections ahead. Try the Common Ions interactive game to test your knowledge!

Example 6.11 Writing an Ionic Equation

Write an ionic equation showing how solid potassium sulfate dissociates when dissolved in water. Include phase symbols.

Potassium sulfate has the formula K_2SO_4. When dissolved in water, this compound dissociates into the component potassium and sulfate ions. We write this as follows:

$$K_2SO_4\ (s) \rightarrow 2\ K^+\ (aq) + SO_4^{2-}\ (aq)$$

In the reactant, we show the two potassium ions with a subscript because they are part of a compound. However, when the compound dissociates, the potassium ions also separate. Because the two aqueous ions are no longer together, they are shown with a coefficient in the products.

Check It
Watch Explanation

TRY IT

11. Write ionic equations to show the following two changes:

a. Solid ammonium bicarbonate dissolves in water.

b. Solid aluminum nitrate dissolves in water.

Solubility Rules and Precipitation Reactions

Not all ionic solids dissolve in water. Many ionic compounds are insoluble. What determines whether a compound is water soluble?

To answer this question, let's think about the energy changes that occur when an ionic solid dissociates in water (**Figure 6.17**). In an ionic lattice, ions are stabilized by the oppositely-charged ions that surround them. When they dissociate, the ions are stabilized by water molecules. Which situation is more energetically favorable? The answer to this question determines whether a compound will dissolve.

Ions with a charge of +1 or −1 are usually soluble. ■

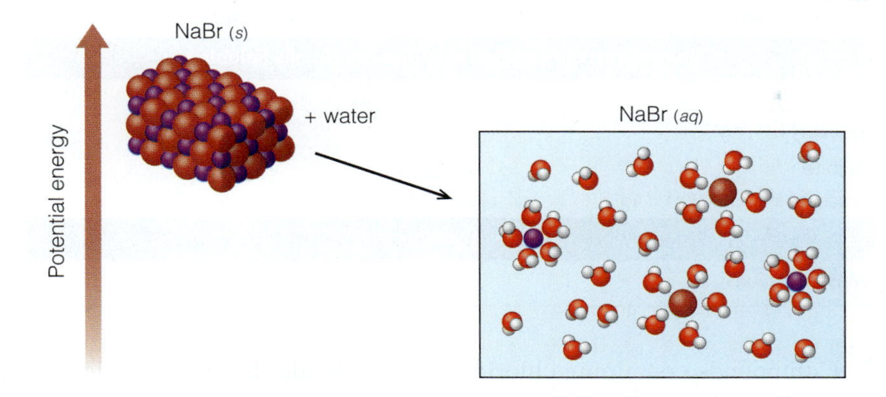

Figure 6.17 Sodium bromide dissolves in water because the energy gain from the interaction between the ions and water is greater than the energy gain between the ions in the lattice.

Have you ever thought about leaving one job to take another? There are many factors to consider. How does the pay compare? Are the hours better? Is the work more enjoyable or rewarding? In the same way, solubility is determined by competing factors. Which is more stable—the ionic lattice, or the dissociated ions surrounded by water molecules? The answer to this question varies for different compounds.

The solubility of ionic compounds depends on many factors, including the charge of the ions, the size of the ions, and the way they pack together. The more strongly the ions bind to each other, the less likely they are to dissolve in water. For example, the greater the charge on an ion, the more tightly it sticks to oppositely charged ions. As a result, compounds composed of ions with a charge of two or three tend to be insoluble in water. Iron(III) oxide (Fe_2O_3), lead(II) sulfide (PbS), and barium carbonate ($BaCO_3$) are all insoluble in water.

The way ions fit together can be hard to predict, and sometimes ionic compounds are surprisingly stable. For example, ionic compounds containing the halogen ions (chloride, bromide, and iodide) are nearly always water soluble—unless they are bonded to silver (Ag^+) or lead(II) (Pb^{2+}).

The solubilities of common ionic compounds are summarized in **Figure 6.18** and in **Table 6.3**. Note some of the key trends:

1. Compounds containing alkali metals (Li^+, Na^+, K^+, Rb^+), ammonium (NH_4^+), or the large oxyanions nitrate (NO_3^-), chlorate (ClO_3^-), perchlorate (ClO_4^-), or acetate ($C_2H_3O_2^-$), are nearly always soluble.

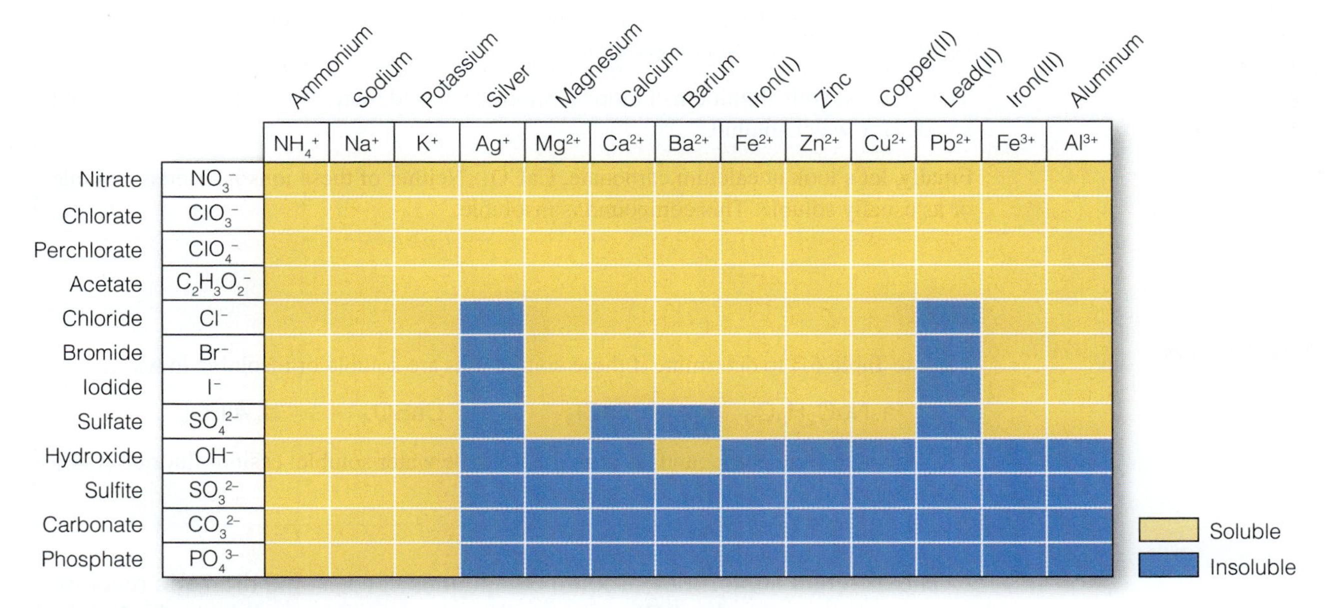

Ammonium
Sodium
Potassium
Silver
Magnesium
Calcium
Barium
Iron(II)
Zinc
Copper(II)
Lead(II)
Iron(III)
Aluminum
NH_4^+
Na^+
K^+
Ag^+
Mg^{2+}
Ca^{2+}
Ba^{2+}
Fe^{2+}
Zn^{2+}
Cu^{2+}
Pb^{2+}
Fe^{3+}
Al^{3+}
Nitrate NO_3^-
Chlorate ClO_3^-
Perchlorate ClO_4^-
Acetate $C_2H_3O_2^-$
Chloride Cl^-
Bromide Br^-
Iodide I^-
Sulfate SO_4^{2-}
Hydroxide OH^-
Sulfite SO_3^{2-}
Carbonate CO_3^{2-}
Phosphate PO_4^{3-}
Soluble
Insoluble

Figure 6.18 A graphical approach to the solubility rules is presented here. Notice that compounds containing the alkali metals, NH_4^+, NO_3^-, ClO_3^-, ClO_4^-, and $C_2H_3O_2^-$ are always soluble.

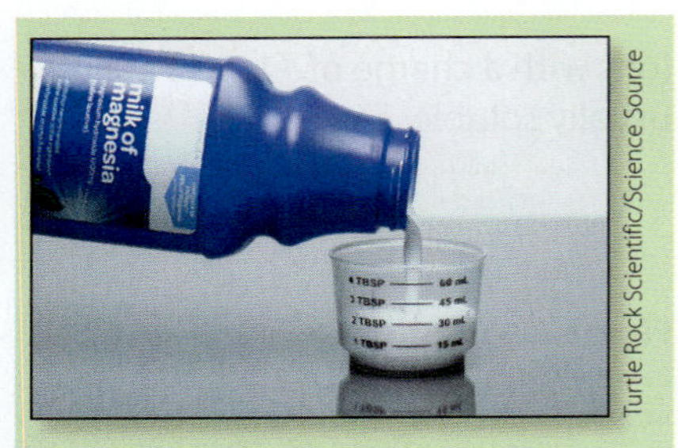

Turtle Rock Scientific/Science Source

Magnesium hydroxide is a common treatment for acid indigestion. Because magnesium hydroxide is insoluble in water, the mixture has a white, milky appearance, leading to the name *milk of magnesia*.

TABLE 6.3 Solubility Rules

Compounds Containing These Ions Are Nearly Always Soluble	
Alkali metals	Li^+, Na^+, K^+, Rb^+
Ammonium	NH_4^+
Large −1 oxyanions	NO_3^-, ClO_3^-, ClO_4^-, $C_2H_3O_2^-$
Compounds Containing These Ions Are Usually Soluble	
Halides (except Pb^{2+}, Ag^+)	F^-, Cl^-, Br^-, I^-
Sulfate (except Ba^{2+}, Ca^{2+}, Pb^{2+}, Ag^+)	SO_4^{2-}
Not Soluble	
Most other ions	

2. Compounds containing chloride (Cl^-), bromide (Br^-), and iodide (I^-) are soluble, except when they are bonded to silver (Ag^+) or lead(II) (Pb^{2+}).
3. Compounds containing sulfate (SO_4^{2-}) are usually soluble, except when they are bonded to Ba^{2+}, Ca^{2+}, Pb^{2+}, or Ag^+.
4. Most other ionic compounds are insoluble.

These guidelines can help to determine whether compounds are water soluble. However, many compounds are slightly soluble—meaning that a small number of ions will dissolve, but most of the ionic solid will not. In some applications, you may need to know the solubilities more precisely than is given here. Solubility also depends on temperature: Many ionic compounds that are insoluble at room temperature become quite soluble at higher temperatures.

Example 6.12 Predicting the Solubility of an Ionic Compound

Using the information in Figure 6.18 and Table 6.3, determine whether these compounds are soluble or insoluble in water:

a. Na_3PO_4 b. $AlCl_3$ c. $CaCO_3$

Let's begin with sodium phosphate, Na_3PO_4. Notice that Na^+ is on the list of ions that are always soluble. Even though phosphate is not typically a soluble ion, this compound is soluble because it contains sodium.

The situation with aluminum chloride is similar: Chlorides are nearly always soluble, and so $AlCl_3$ is water soluble.

Finally, let's look at calcium carbonate, $CaCO_3$. Neither of these ions is listed as soluble or as usually soluble. This compound is insoluble.

Check It
Watch Explanation

TRY IT

12. Use Table 6.3 to determine if these compounds are soluble or insoluble in water:

a. $NaC_2H_3O_2$ b. $FeCl_3$ c. $CaSO_4$ d. $AlPO_4$

13. Which of these compounds is more likely to be water soluble: cesium fluoride (CsF) or barium chromate ($BaCrO_4$)? Why?

The different solubilities of ionic compounds cause some dramatic reactions. For example, aqueous lead(II) nitrate and aqueous sodium iodide are both colorless solutions. However, if we combine these solutions, a yellow solid immediately forms (**Figure 6.19**). The yellow solid is lead(II) iodide.

The other product from this reaction, sodium nitrate, remains in solution. We describe this reaction using the following equation:

$$Pb(NO_3)_2\,(aq) + 2\,NaI(aq) \rightarrow 2\,NaNO_3\,(aq) + PbI_2\,(s)$$

This type of reaction, in which two aqueous solutions combine to produce a solid (insoluble) product, is called a **precipitation reaction**. The solid product formed in the reaction is the **precipitate**. In the equation for this reaction, the precipitate is indicated by the (*s*) notation after the chemical formula. Compounds that remain in solution are shown using the (*aq*) notation. Precipitation reactions are examples of double-displacement reactions.

It is helpful to think of this reaction in terms of ionic equations. Before the solutions are mixed, each contains dissolved ions:

$$\text{Lead(II) nitrate solution: } Pb^{2+}\,(aq) + 2\,NO_3^-\,(aq)$$

$$\text{Sodium iodide solution: } Na^+\,(aq) + I^-\,(aq)$$

Figure 6.19 Lead(II) iodide is insoluble in water. When a solution containing dissolved Pb^{2+} is combined with a solution containing dissolved I^-, PbI_2 immediately forms as a yellow solid.

Alexandre Dotta/Science Source

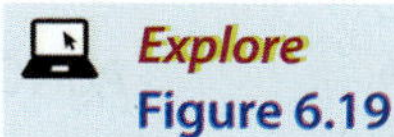

Explore Figure 6.19

When these solutions are combined, all four ions are present in solution. However, as the lead and iodide ions encounter each other (**Figure 6.20**), they bind together and drop out of solution as a solid (precipitate).

We can write this as an ionic equation. Because all of the ions are shown, this is sometimes called a **complete ionic equation**:

$$Pb^{2+}(aq) + 2\,NO_3^-(aq) + 2\,Na^+(aq) + 2\,I^-(aq) \rightarrow 2\,Na^+(aq) + 2\,NO_3^-(aq) + PbI_2(s)$$

The driving force for a precipitation reaction is the formation of the solid. Notice in the ionic equation that the nitrate and the sodium ions don't actually react: They are present in the starting materials and are unchanged in the products. The sodium and nitrate ions are paired together only because the other ions (Pb^{2+} and I^-, shown in red in the equation) have dropped out of the solution. The sodium and nitrate ions are **spectator ions** because they are not changed in the reaction.

Precipitation reactions are double-displacement reactions.

Sometimes it is helpful to show precipitation reactions as a **net ionic equation**. In this type of equation, we omit the spectator ions and include only the ions directly involved in the precipitation (**Figure 6.21**). For our reaction example, the net ionic equation is

When ions combine to form an insoluble compound, a precipitation reaction occurs.

$$Pb^{2+}\,(aq) + 2\,I^-\,(aq) \rightarrow PbI_2\,(s)$$

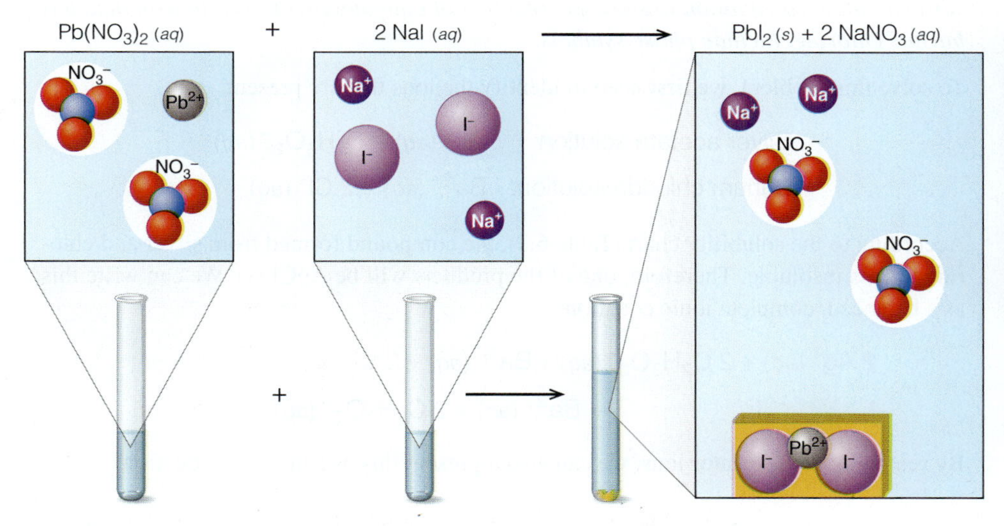

Figure 6.20 The driving force for a precipitation reaction is the formation of the solid. If a cation and anion precipitate out of solution, the other two ions are left in solution together.

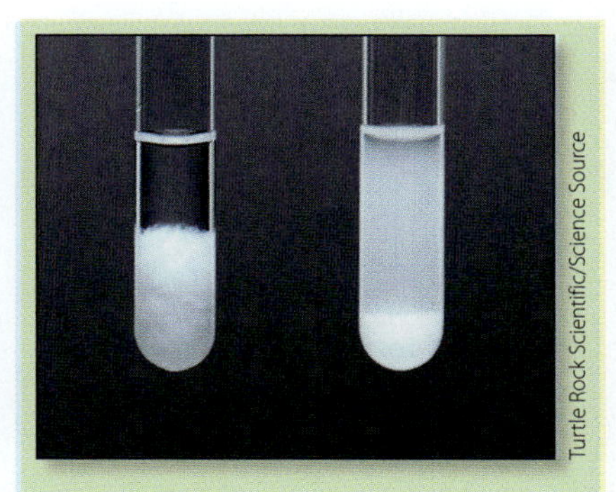

Turtle Rock Scientific/Science Source

Sometimes precipitates form distinctive solids, but at other times they appear as "cloudy" mixtures.

Molecular Equation
Shows neutral compounds

$$Pb(NO_3)_2\ (aq) + 2\ KCl\ (aq) \longrightarrow PbCl_2\ (s) + 2\ KNO_3\ (aq)$$

Complete Ionic Equation
Shows all ions present

$$Pb^{2+}\ (aq) + 2\ NO_3^-\ (aq) + 2\ K^+\ (aq) + 2\ Cl^-\ (aq) \longrightarrow PbCl_2\ (s) + 2\ K^+\ (aq) + 2\ NO_3^-\ (aq)$$

Net Ionic Equation
Omits spectator ions; only shows ions that react

$$Pb^{2+}\ (aq) + 2\ Cl^-\ (aq) \longrightarrow PbCl_2\ (s)$$

Figure 6.21 We use three types of equations to describe aqueous reactions.

We use the solubility rules to predict precipitation reactions.

The solubility rules are the key to predicting precipitation reactions. If a cation and an anion can combine to form an insoluble product, a precipitation reaction will occur. **Figure 6.22** contains several additional examples of precipitation reactions. Can you identify the products of these reactions?

Figure 6.22 Each of the reactions shown produces a precipitate. Can you identify the solid formed in each example?

Explore Figure 5.22

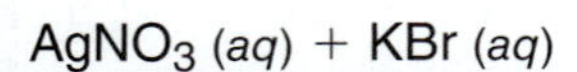

$AgNO_3\ (aq) + KBr\ (aq)$

$Cu(NO_3)_2\ (aq) + 2\ NaOH\ (aq)$

$Pb(NO_3)_2\ (aq) + Na_2SO_4\ (aq)$

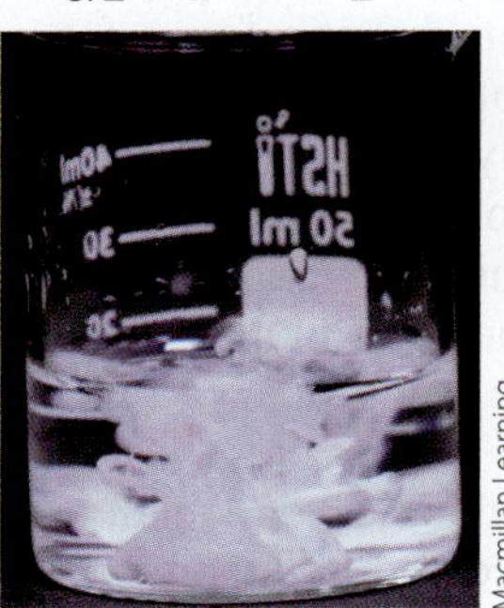

Example 6.13 Predicting Precipitation Reactions

Write a balanced equation to show the reaction of aqueous silver acetate with aqueous barium chloride. Include phase symbols.

To solve this problem, we first need to identify the ions that are present:

Silver acetate solution: $Ag^+\ (aq) + C_2H_3O_2^-\ (aq)$

Barium chloride solution: $Ba^{2+}\ (aq) + 2\ Cl^-\ (aq)$

According to the solubility chart (Table 6.3), the compound formed from silver and chloride ions is insoluble. Therefore, one of the products will be AgCl (*s*). We can write this as a balanced, complete ionic equation:

$$2\ Ag^+\ (aq) + 2\ C_2H_3O_2^-\ (aq) + Ba^{2+}\ (aq) + 2\ Cl^-\ (aq) \rightarrow Ba^{2+}\ (aq) + 2\ C_2H_3O_2^-\ (aq) + 2\ AgCl\ (s)$$

By removing the spectator ions, we can also represent this as a net ionic equation:

$$Ag^+\ (aq) + Cl^-\ (aq) \rightarrow AgCl\ (s)$$

Or as a balanced molecular equation:

$$2\ AgC_2H_3O_2\ (aq) + BaCl_2\ (aq) \rightarrow Ba(C_2H_3O_2)_2\ (aq) + 2\ AgCl\ (s)$$

If we tried this reaction in the lab, silver chloride would form as a white precipitate (**Figure 6.23**).

Figure 6.23 Silver and chloride ions combine to form a white precipitate.

Example 6.14 Predicting and Balancing Precipitation Reactions

Write a balanced molecular equation to show the precipitation reaction of calcium chloride with sodium phosphate. Include phase symbols.

To solve this problem, we first write out the empirical formulas for the two starting materials:

$$CaCl_2 + Na_3PO_4$$

These ionic compounds are both water soluble, so they exist in solution as dissociated ions. Based on Table 6.3, two of these ions, calcium and phosphate, form an insoluble solid. The formula for this compound is $Ca_3(PO_4)_2$:

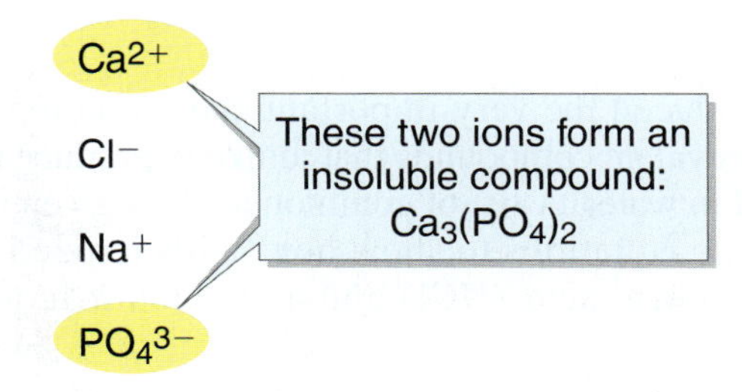

The other two ions, sodium and chloride, remain in solution. We can describe this reaction with an unbalanced equation, showing the precipitate as a solid:

$$CaCl_2\ (aq) + Na_3PO_4\ (aq) \rightarrow Ca_3\ (PO_4)_2\ (s) + NaCl\ (aq) \qquad \text{unbalanced}$$

Finally, we balance this equation by balancing the ions on each side:

$$3\ CaCl_2\ (aq) + 2\ Na_3PO_4\ (aq) \rightarrow Ca_3(PO_4)_2\ (s) + 6\ NaCl\ (aq) \qquad \text{balanced}$$

Another approach to solving this problem is to recognize that precipitation reactions are double-displacement reactions. For a precipitation reaction to occur, the cation-anion pairs must swap. We must keep in mind the charges on the ions and write empirical formulas that result in an overall charge of zero. In this instance, the anions swap positions, producing calcium phosphate and sodium chloride:

$$3\ CaCl_2\ (aq) + 2\ Na_3PO_4\ (aq) \rightarrow 3\ Ca_3(PO_4)_2\ (s) + 6\ NaCl\ (aq)$$

The anions "swap" positions.

Referring again to Table 6.3, we see that one of these two products, calcium phosphate, forms an insoluble solid.

While this approach may make the problem simpler, remember that a precipitation reaction cannot occur unless one of the two product compounds is insoluble. If neither product is insoluble, all of the ions remain dissolved in water, and no reaction takes place.

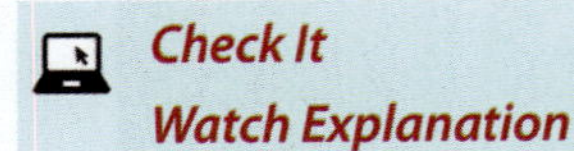

TRY IT

14. Write a complete ionic equation showing the reaction of aqueous barium nitrate with aqueous sodium phosphate to form barium phosphate and sodium nitrate. Include phase symbols.

15. Refer to Table 6.3 to identify the precipitate formed in the following reactions. Balance the reactions.

$$KOH\ (aq) + Pb(NO_3)_2\ (aq) \rightarrow$$
$$FeCl_2\ (aq) + Li_2S\ (aq) \rightarrow$$
$$MgBr_2\ (aq) + K_3PO_4\ (aq) \rightarrow$$

16. In which of these combinations would no precipitation reaction occur?

a. Ammonium iodide is combined with silver acetate.

b. Potassium chloride is combined with sodium hydroxide.

c. Zinc sulfate is combined with calcium chloride.

Acid-Base Neutralization Reactions

Acids and Bases

In Chapter 5 we introduced the very important class of compounds called *acids*. Recall that acids are covalent compounds that ionize to produce H^+ ions and a stable anion when dissolved in water. A list of common acids is given in **Table 6.4**.

TABLE 6.4 Common Acids

Formula	Name
HF	Hydrofluoric acid
HCl	Hydrochloric acid
HBr	Hydrobromic acid
HI	Hydroiodic acid
H_2CO_3	Carbonic acid
HNO_3	Nitric acid
HNO_2	Nitrous acid
H_2SO_4	Sulfuric acid
H_3PO_4	Phosphoric acid
$HC_2H_3O_2$	Acetic acid

We often use ionic equations to show how acids ionize in water. For example, aqueous hydrochloric acid (HCl) and aqueous nitric acid (HNO_3) ionize as follows:

$$HCl\ (aq) \rightarrow H^+\ (aq) + Cl^-\ (aq)$$
$$HNO_3\ (aq) \rightarrow H^+\ (aq) + NO_3^-\ (aq)$$

Bases are compounds that produce hydroxide (OH^-) ions in aqueous solution. Many common bases are soluble metal hydroxides (**Table 6.5**). In water, these compounds dissociate to give metal cations and hydroxide anions. For example, sodium hydroxide dissociates in water to produce sodium and hydroxide ions:

$$NaOH\ (s) \rightarrow Na^+\ (aq) + OH^-\ (aq)$$

TABLE 6.5 Common Hydroxide Bases

Formula	Name
$LiOH$	Lithium hydroxide
$NaOH$	Sodium hydroxide
KOH	Potassium hydroxide
$Ba(OH)_2$	Barium hydroxide

In aqueous solution, acids produce H^+ ions and bases produce OH^- ions.

Neutralization Reactions

Acids and bases undergo **neutralization reactions**. In these reactions, the H^+ from the acid combines with the OH^- from the base to form water. We can show this as a net ionic equation:

$$H^+\ (aq) + OH^-\ (aq) \rightarrow H_2O\ (l)$$

For example, hydrochloric acid reacts with sodium hydroxide. We can write this as a molecular equation or as a complete ionic equation:

$$HCl\ (aq) + NaOH\ (aq) \rightarrow H_2O\ (l) + NaCl\ (aq)$$
$$H^+\ (aq) + Cl^-\ (aq) + Na^+\ (aq) + OH^-\ (aq) \rightarrow H_2O\ (l) + Na^+\ (aq) + Cl^-\ (aq)$$

The spectator ions, Na^+ and Cl^-, combine to form an ionic compound. Similarly, nitric acid reacts with lithium hydroxide:

$$HNO_3\ (aq) + LiOH\ (aq) \rightarrow H_2O\ (l) + LiNO_3\ (aq)$$

Again, the H^+ from the acid reacts with the OH^- from the base to form water. The remaining ions end up together as $LiNO_3$. The ionic compound that is formed from the spectator ions is often called a *salt*.

In general, the following expression describes the neutralization of an acid with a hydroxide base:

$$\text{Acid} + \text{Base} \longrightarrow \text{Salt} + \text{Water}$$

Like precipitation reactions, acid-base neutralizations are double-displacement reactions: The H^+ and OH^- combine to form water, while the other two ions combine to form the salt. The driving force of the reaction is the formation of water; the salt forms from the other two ions.

Chemists sometimes refer to ionic compounds as *salts*.

The formation of water is the driving force for a neutralization reaction.

Acid-base neutralizations are double-displacement reactions. The H^+ and OH^- combine to form water (HOH), and the two spectator ions combine to form a salt:

$$HCl + NaOH \rightarrow HOH + NaCl$$

Example 6.15 Predicting the Products from Acid-Base Neutralization Reactions

Write a balanced equation to show the reaction of sulfuric acid with sodium hydroxide. Include phase symbols.

We begin by writing an unbalanced equation to describe this reaction. One product is water, and the other product is the ionic compound produced by the spectator ions:

$$H_2SO_4 + NaOH \rightarrow H_2O + Na_2SO_4 \qquad \text{unbalanced}$$

A simple way to balance neutralization reactions is to balance the number of H^+ and OH^- ions that combine. Because H_2SO_4 produces two H^+ ions, there must be two OH^- ions to neutralize them, and two water molecules will be produced. Adding phase symbols gives us the balanced form shown here.

$$H_2SO_4\ (aq) + 2\ NaOH\ (aq) \rightarrow 2\ H_2O\ (l) + Na_2SO_4\ (aq) \qquad \text{balanced}$$

TRY IT

17. Predict the products for the following neutralization reactions. Make sure the equations are balanced.

$$HBr + NaOH \rightarrow$$
$$HClO_4\ (aq) + Ba(OH)_2\ (aq) \rightarrow$$
$$H_3PO_4\ (aq) + KOH\ (aq) \rightarrow$$

18. Write a complete ionic equation and a net ionic equation for the following neutralization reaction:

$$LiOH\ (aq) + HNO_3\ (aq) \rightarrow LiNO_3\ (aq) + H_2O\ (l)$$

Check It
Watch Explanation

Capstone Video

Capstone Question

Sometimes, simple chemical reactions can provide information about the composition of a substance. For example, suppose you have an unidentified alkaline earth metal. You know that the metal is either magnesium, calcium, or barium. When you combine this metal with hydrochloric acid, it produces hydrogen gas plus an ionic compound composed of the metal cation and chloride anions. **(a)** Write three balanced equations showing each of the possible reactions. **(b)** Classify these reactions as decomposition, synthesis, single-displacement, or double-displacement reactions. **(c)** Using the simplified solubility table in **Figure 6.24**, devise a series of tests to distinguish the three different metal cations. **(d)** Using the concepts in previous chapters, can you think of other ways to distinguish the three metals?

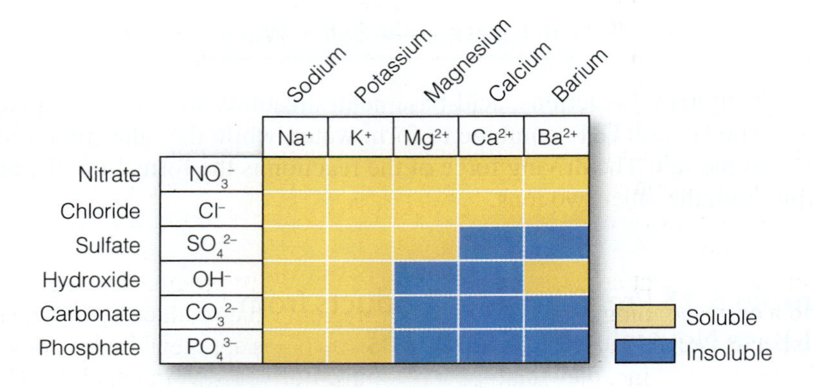

Figure 6.24 Based on this simplified solubility table, how could you distinguish compounds containing magnesium, calcium, and barium ions?

SUMMARY

Scientists describe chemical changes, or reactions, using chemical equations. Chemical equations show the substances present before the reaction (called the reactants) on the left side of the arrow, and the substances present after the reaction (called the products) on the right side of the arrow.

Atoms are not created or destroyed in chemical reactions. To reflect this, we write balanced equations. An equation is balanced if it contains the same number and type of atom in the reactants and in the products. Balancing equations often involves trial and error and requires practice to do it efficiently. Sometimes chemical equations also show the phase of the reactants and products in parentheses after each element or compound.

There are many types of chemical reactions and different ways of organizing reactions. Many reactions can be classified as decomposition, synthesis, single-displacement, and double-displacement reactions. These classifications are based on the number and type of reactant and product formed.

We also classify reactions by patterns of chemical behavior. In this chapter, we examined four broad classes of reactions: metal-nonmetal reactions, combustion reactions, precipitation reactions, and acid-base neutralization reactions (**Table 6.6**).

Oxidation-reduction reactions involve the gain and loss of electrons. An atom or ion that loses electrons is oxidized; a species that gains electrons is reduced. Metals and nonmetals combine in these reactions to produce ionic compounds.

TABLE 6.6 Patterns of Chemical Behavior

Reaction Type	Classification	Starting Materials	Products
Metal-Nonmetal	Combination	Metal + Nonmetal	Ionic compound
Combustion		Element + Oxygen	Oxide compound
		Hydrocarbon + Oxygen	$CO_2 + H_2O$
Precipitation reaction	Double displacement	Two soluble compounds	Insoluble compound (precipitate)
Acid-Base neutralization	Double displacement	Acid + Base	Water + Salt

Many elements react with oxygen to produce oxide compounds. Similarly, hydrocarbons (compounds composed of hydrogen and carbon) react with oxygen to form carbon dioxide and water. Collectively, reactions with oxygen are called *combustion reactions*.

When an ionic compound dissolves in water, the ions separate in a process called *dissociation*. Not all ionic compounds dissolve in water. Solubility rules provide broad guidelines for determining which compounds do and do not dissolve in water.

Precipitation reactions occur when two aqueous solutions combine to produce an insoluble product. The insoluble product drops out of solution (precipitates), leaving the other ions—called spectator ions—remaining in solution. We commonly describe precipitation reactions using ionic equations. Complete ionic equations show all ions present, while net ionic equations show only those directly involved in the central reaction.

Acids are covalent compounds that ionize in water to produce hydrogen cations (H^+) and a corresponding anion. Bases are compounds that produce hydroxide anions (OH^-) when dissolved in water. The most common bases are metal hydroxides. When acids and bases combine, they undergo a neutralization reaction in which the H^+ from the acid reacts with the OH^- from the base to form water. Like precipitation reactions, acid-base reactions are examples of double-displacement reactions.

The Mayan Lime Cycle

We began this chapter with the great Mayan ruins of Palenque. Now that we have a broader understanding of chemical reactions, let's look again at the specific process the Mayans used to make plaster for their temples and palaces (**Figure 6.25**):

The first step in lime production is heating. Above about 900 °C, calcium carbonate decomposes to produce calcium oxide and carbon dioxide:

$$CaCO_3\,(s) \rightarrow CaO\,(s) + CO_2\,(g)$$

Calcium oxide is a fine white powder, commonly called *quicklime*. The calcium oxide is then mixed with water to produce calcium hydroxide, also called *slaked lime*, in a synthesis reaction:

$$CaO\,(s) + H_2O\,(l) \rightarrow Ca(OH)_2\,(s)$$

The slaked lime is often mixed with some other material, such as volcanic ash, that strengthens the overall structure. As this lime plaster dries, it reacts with carbon dioxide in the air to return to calcium carbonate:

$$Ca(OH)_2\,(s) + CO_2\,(g) \rightarrow CaCO_3\,(s) + H_2O\,(l)$$

This three-step process is called the *lime cycle*. But what does this have to do with the decline of Mayan cities?

Evidence from the hieroglyphs at Palenque suggests that a major building project was completed just before the population began to decline. To produce enough lime for this project, the Maya needed massive amounts of firewood. To obtain the firewood, they cleared huge swaths of the Yucatán jungles. Many archaeologists believe that this deforestation harmed the water supplies and promoted widespread drought. This drought may have led to the decline in population and ultimately to the abandonment of cities. Years later, the jungles slowly grew back and overtook the ancient structures.

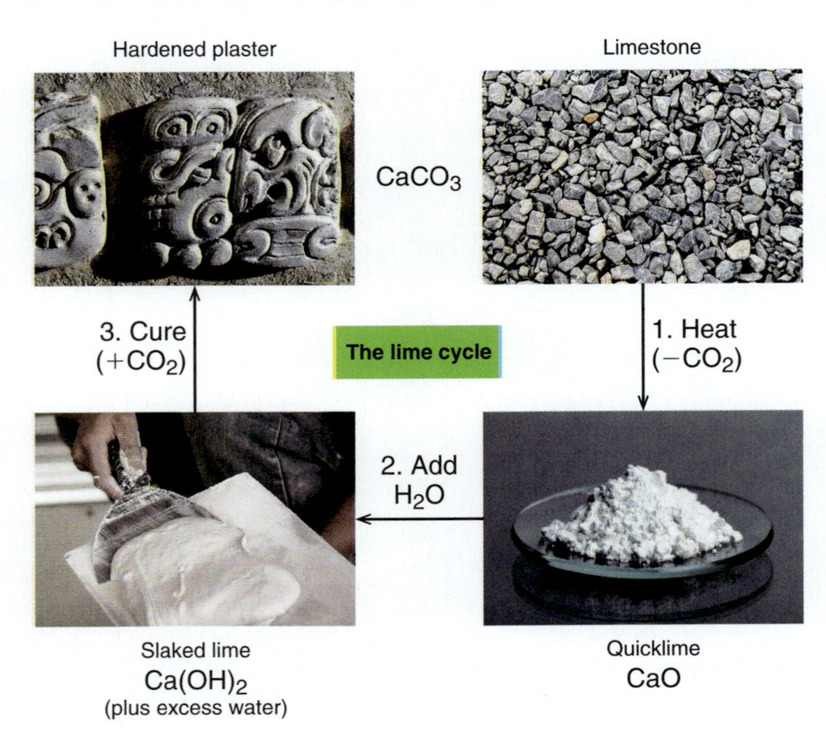

Figure 6.25 Through the lime cycle, natural limestone ($CaCO_3$) is converted into CaO, then into $Ca(OH)_2$. $Ca(OH)_2$ can be molded or shaped before it cures, reforming the hard $CaCO_3$. De Agostini/G. Dagli Orti/Getty Images; © tendo23/depositphotos.com; Turtle Rock Scientific/Science Source; Chaiyaporn Baokaew/Shutterstock

Figure 6.26 (a) Dr. Yareli Jáidar applies a coat of nanolime to a Mayan structure. (b) This image shows a jaguar painting that has degraded over time. (c) This is the same painting after restoration.

Today, archaeological conservators like Yareli Jáidar work to preserve archaeological and artistic treasures (**Figure 6.26**). She often uses a modern material called *nanolime.* Nanolime is composed of very tiny particles of calcium hydroxide, often just 100 nanometers across (from which the name originates). When conservationists coat a plaster structure with nanolime, the tiny lime particles penetrate the painting surface to fill pores and cracks. As the lime dries, it fuses with the ancient lime, sealing the structures against further damage.

Key Terms

6.1 Chemical Equations

chemical equation A symbolic representation of a chemical change. Such equations consist of reactants and products separated by an arrow.

reactant The starting material in a chemical change, shown on the left-hand side of a chemical equation.

product The compounds produced in a chemical change, shown on the right-hand side of a chemical equation.

subscript In a chemical formula, subscript numbers show the number of each atom or ion present.

coefficient In a chemical formula, the numbers written before each reactant or product to indicate the ratios in which components of the reaction are consumed or produced.

balanced equation A chemical equation in which the number and type of atoms are the same for the reactants and the products.

6.2 Classifying Reactions

decomposition A reaction in which a single reactant forms two or more products.

synthesis A reaction in which two reactants join together to form a single product; also called a *combination reaction.*

single displacement A reaction in which one element replaces another element in a compound.

double displacement A reaction in which two compounds swap cation-anion pairs to form two new compounds.

6.3 Reactions between Metals and Nonmetals

oxidation-reduction reaction A chemical change in which one species loses electrons (oxidation) while another gains electrons (reduction).

oxidation The loss of electrons.

reduction The gain of electrons.

6.4 Combustion Reactions

combustion A reaction in which oxygen gas combines with elements or compounds to produce oxide compounds.

6.5 Reactions in Aqueous Solution

molecular equation A chemical equation in which all species are written as neutral compounds.

ionic equation A chemical equation that shows dissociated ions as separate species.

precipitation reaction A type of chemical change in which two aqueous solutions combine to produce an insoluble product.

precipitate A solid product formed from the combination of two solutions.

complete ionic equation An equation that shows all ions present in a solution.

spectator ion An ion that is present in a solution but not directly involved in a chemical change.

net ionic equation An equation in which the only ions shown are those directly involved in the chemical change, and spectator ions are omitted.

base A compound that produces hydroxide ions in aqueous solutions. In later chapters, bases will be further defined as proton (H^+) acceptors.

neutralization reaction A reaction in which an acid and a base combine to produce water and an ionic compound (called a *salt*).

Additional Problems

6.1 Chemical Equations

19. Write a balanced equation to represent the reaction shown. Write the coefficients as the lowest whole-number ratio.

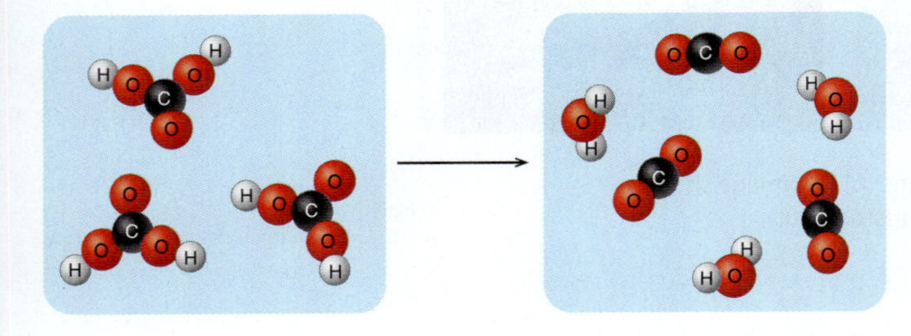

20. Write a balanced equation to represent the reaction shown. Write the coefficients as the lowest whole-number ratio.

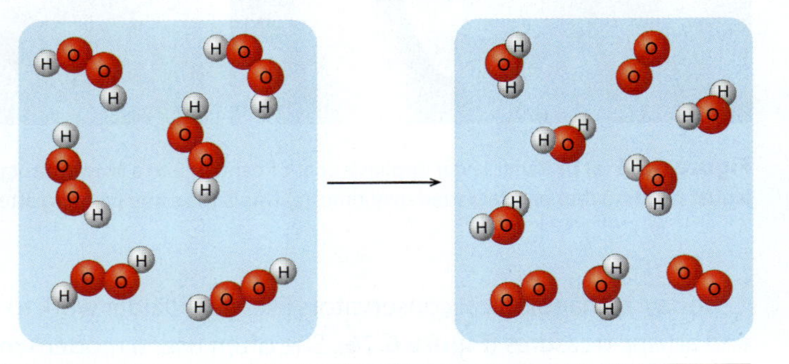

21. Write a balanced equation to represent the reaction shown. Write the coefficients as the lowest whole-number ratio.

22. Write a balanced equation to represent the reaction shown. Write the coefficients as the lowest whole-number ratio.

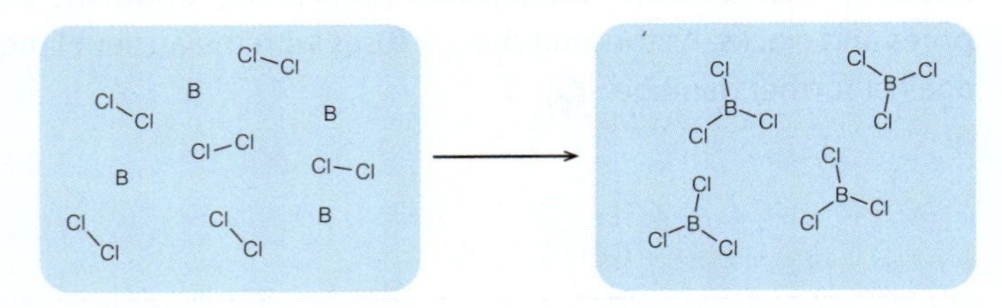

23. Describe each of the following changes using chemical equations:

a. One molecule of hydrogen reacts with one molecule of bromine to produce two molecules of hydrogen bromide.

b. Two atoms of sodium and one molecule of fluorine react to produce two molecules of sodium fluoride.

24. Describe each of the following changes using chemical equations:

a. One unit of calcium hydroxide reacts to form one unit of calcium oxide and one molecule of water.

b. Four iron atoms react with three oxygen molecules to form two units of iron(III) oxide.

25. In this reaction, how many units of C form when 30 units of A react?

$$2\,A + B \rightarrow C$$

26. In this reaction, how many units of B are required to form 40 units of D?

$$A + 2\,B \rightarrow C + D$$

27. The balanced equation here shows the reaction of hydrogen gas with chlorine gas. Based on this equation,

$$H_2 + Cl_2 \rightarrow 2\,HCl$$

a. How many molecules of HCl form from 30 molecules of H_2?

b. How many molecules of Cl_2 are required to produce 12 molecules of HCl?

28. The balanced equation here shows the reaction of calcium with oxygen to produce calcium oxide. Based on this equation,

$$2\,Ca + O_2 \rightarrow 2\,CaO$$

a. How many molecules of oxygen react with 18 calcium atoms?

b. If 30 calcium atoms react with 15 oxygen molecules, how many units of calcium oxide form?

29. The reaction of aluminum with chlorine gas is shown here. Based on this equation,

$$2\,Al + 3\,Cl_2 \rightarrow 2\,AlCl_3$$

a. How many molecules of chlorine gas are needed to react with 10 aluminum atoms?

b. How many units of $AlCl_3$ can be produced from 10 aluminum atoms?

30. Hydrobromic acid (HBr) can react with barium hydroxide as shown here. Based on this equation,

$$2\,HBr + Ba(OH)_2 \rightarrow BaBr_2 + 2\,H_2O$$

a. If 5 units of barium hydroxide react in this way, how many water molecules form?

b. How many HBr molecules are needed to produce 40 units of $BaBr_2$?

31. Balance each of the following equations:

a. $PCl_3 + F_2 \rightarrow PF_3 + Cl_2$
b. $SO_2 + O_2 \rightarrow SO_3$
c. $B + F_2 \rightarrow BF_3$

32. Balance each of the following equations:

a. $Zn + HCl \rightarrow ZnCl_2 + H_2$
b. $Li + CuCl_2 \rightarrow LiCl + Cu$
c. $C_3H_4O + H_2 \rightarrow C_3H_8O$

33. Balance the following equations:

a. $PCl_3 + H_2O \rightarrow H_3PO_3 + HCl$
b. $O_2 + H_2S \rightarrow SO_2 + H_2O$
c. $HCl + MnO_2 \rightarrow MnCl_2 + 2H_2O + Cl_2$

34. Balance the following equations:

a. $TiCl_4 + H_2O \rightarrow TiO_2 + HCl$
b. $VCl_3 \rightarrow VCl_2 + VCl_4$
c. $Cr_2O_3 + Al \rightarrow Al_2O_3 + Cr$

35. Balance the following equations by balancing the ions:

a. $Hg(NO_3)_2 + KCl \rightarrow KNO_3 + HgCl_2$
b. $MgBr_2 + NaOH \rightarrow NaBr + Mg(OH)_2$
c. $AgC_2H_3O_2 + BaCl_2 \rightarrow AgCl + Ba(C_2H_3O_2)_2$

36. Balance the following equations by balancing the ions:

a. $Cr(NO_3)_2 + KOH \rightarrow KNO_3 + Cr(OH)_2$
b. $(NH_4)_2CO_3 + Ba(C_2H_3O_2)_2 \rightarrow NH_4C_2H_3O_2 + BaCO_3$
c. $AgC_2H_3O_2 + BaCl_2 \rightarrow AgCl + Ba(C_2H_3O_2)_2$

37. Balance the following equations. For these equations, it may be helpful to write a fractional coefficient for the diatomic element before converting to whole numbers.

a. $N_2 + H_2 \rightarrow NH_3$
b. $C_2H_6 + O_2 \rightarrow CO_2 + H_2O$
c. $HCl + Al \rightarrow AlCl_3 + H_2$

38. Balance the following equations. For these equations, it may be helpful to write a fractional coefficient for the diatomic element before converting to whole numbers.

a. $P + Cl_2 \rightarrow PCl_5$
b. $IrF_5 + H_2 \rightarrow IrF_4 + HF$
c. $C_5H_{10} + O_2 \rightarrow CO_2 + H_2O$

39. Balance the following equations:

a. $Na + O_2 \rightarrow Na_2O$
b. $Pb(NO_3)_2 + KCl \rightarrow PbCl_2 + KNO_3$
c. $C_2H_2 + O_2 \rightarrow CO_2 + H_2O$

40. Balance the following equations:

a. $NaOCl + HCl \rightarrow Cl_2 + NaCl + H_2O$
b. $AgC_2H_3O_2 + MgCl_2 \rightarrow AgCl + Mg(C_2H_3O_2)_2$
c. $N_2H_6 + O_2 \rightarrow NO_2 + H_2O$

41. Write a balanced equation to describe each of these chemical changes. Include phase symbols.

a. Zinc metal reacts with aqueous copper(II) chloride to produce copper metal and aqueous zinc chloride.
b. Propane gas, C_3H_8, reacts with molecular oxygen to produce carbon dioxide gas and water vapor.
c. Solid iron reacts with chlorine gas to produce solid iron(II) chloride.

42. Write a balanced equation to describe each of these chemical changes. Include phase symbols.

a. When calcium metal is heated in water, aqueous calcium hydroxide and hydrogen gas form.
b. When solid magnesium burns (reacts with oxygen gas), it forms solid magnesium oxide.
c. Aqueous silver nitrate reacts with aqueous sodium chloride to produce solid silver chloride and aqueous sodium nitrate.

43. Provide the name and the phase of the reactants for each reaction:

a. $2\ HCl\,(aq) + Na_2CO_3\,(s) \rightarrow 2\ NaCl\,(aq) + CO_2\,(g) + H_2O\,(l)$
b. $Fe\,(s) + 2\ HNO_3\,(aq) \rightarrow Fe(NO_3)_2\,(aq) + H_2\,(g)$
c. $ZnCl_2\,(aq) + Pb(ClO_4)_2\,(aq) \rightarrow PbCl_2\,(s) + Zn(ClO_4)_2\,(aq)$

44. Provide the name and phase of the products for each reaction:

a. $P\,(s) + Cl_2\,(g) \rightarrow PCl_3\,(g)$
b. $Na_2SO_4\,(aq) + BaBr_2\,(aq) \rightarrow BaSO_4\,(s) + 2\ NaBr\,(aq)$
c. $Cr\,(s) + 2\ HCl\,(aq) \rightarrow CrCl_2\,(aq) + H_2\,(g)$

6.2 Classifying Reactions

45. Classify each of these chemical reactions as a synthesis, decomposition, single-displacement, or double-displacement reaction:

a. $3\ NH_3 \rightarrow 3\ N_2 + 3\ H_2$
b. $CaO + H_2O \rightarrow Ca(OH)_2$
c. $AgNO_3\,(aq) + KBr\,(aq) \rightarrow KNO_3\,(aq) + AgBr\,(s)$

46. Classify each of these chemical reactions as a synthesis, decomposition, single-displacement, or double-displacement reaction:

a. $Sn + 2\ Br_2 \rightarrow SnBr_4$
b. $2\ HBr + Sn \rightarrow SnBr_2 + H_2$
c. $HBr + NaOH \rightarrow NaBr + H_2O$

47. Classify each of these chemical reactions as a synthesis, decomposition, single-displacement, or double-displacement reaction:

a. $Ca + O_2 \rightarrow 2\ CaO$
b. $Co\ (s) + 2\ HCl\ (aq) \rightarrow CoCl_2\ (aq) + H_2\ (g)$
c. $PbNO_3\ (aq) + 2\ KBr\ (aq) \rightarrow 2\ KNO_3\ (aq) + PbBr_2\ (s)$

48. Classify each of these chemical reactions as a synthesis, decomposition, single-displacement, or double-displacement reaction:

a. $K_2SO_4\ (aq) + Pb(C_2H_3O_2)_2\ (aq) \rightarrow 2\ KC_2H_3O_2\ (aq) + PbSO_4\ (s)$
b. $2\ Zn + O_2 \rightarrow 2\ ZnO$
c. $PCl_5 \rightarrow PCl_3 + Cl_2$

6.3 Reactions between Metals and Nonmetals

49. What types of compounds form when metals and nonmetals react?

50. What is oxidation? Do metals or nonmetals oxidize more easily?

51. In these reactions, identify the element that is oxidized and the element that is reduced:

a. $2\ Mg + O_2 \rightarrow 2\ MgO$
b. $Fe + S \rightarrow FeS$
c. $Br_2 + Ca \rightarrow CaBr_2$

52. In these reactions, identify the element that is oxidized and the element that is reduced:

a. $2\ Cu + O_2 \rightarrow 2\ CuO$
b. $S + Hg \rightarrow HgS$
c. $Sn + 2\ Cl_2 \rightarrow SnCl_4$

53. Predict the products from these reactions, and balance the equations:

a. $Be + Cl_2 \rightarrow$
b. $K + Cl_2 \rightarrow$
c. $Co\ (s) + Cl_2\ (g) \rightarrow$
d. $Cu\ (s) + O_2\ (g) \rightarrow$

54. Predict the products from these reactions, and balance the equations:

a. $Ca + O_2 \rightarrow$
b. $Sr + Br_2 \rightarrow$
c. $Pb\ (s) + O_2\ (g) \rightarrow$
d. $Cr\ (s) + O_2\ (g) \rightarrow$

6.4 Combustion Reactions

55. What element is always involved in combustion reactions?

56. Each of the following elements reacts with oxygen in combustion reactions. Identify whether the products of these reactions would be ionic compounds or covalent compounds.

a. magnesium b. calcium c. carbon d. nitrogen

57. What are hydrocarbons? What products form from the combustion of hydrocarbons?

58. Fossil fuels like coal contain primarily hydrocarbons, but they also contain a small amount of other elements, like sulfur. What products result from the combustion of hydrocarbons? What products result from the combustion of sulfur?

59. Predict the products from these combustion reactions, and balance the equations:

a. $Mg + O_2 \rightarrow$
b. $C_2H_4 + O_2 \rightarrow$
c. $C_4H_{10}\ (g) + O_2\ (g) \rightarrow$

60. Predict the products from these combustion reactions, and balance the equations:

a. $Zn + O_2 \rightarrow$
b. $C_5H_{12} + O_2 \rightarrow$
c. $C_8H_{18}\ (l) + O_2\ (g) \rightarrow$

61. Write a balanced equation for each of these reactions:

a. combustion of C_2H_2
b. combustion of C_4H_8
c. combustion of C_9H_{18}

62. Write a balanced equation for each of these reactions:

a. combustion of CH_4
b. combustion of C_3H_8
c. combustion of C_5H_{10}

6.5 Reactions in Aqueous Solution

63. The following molecular equations each show an ionic compound dissolving in water. Show the ionization process by rewriting each one as an ionic equation:

a. $KCl\ (s) \rightarrow KCl\ (aq)$
b. $Li_2SO_4\ (s) \rightarrow Li_2SO_4\ (aq)$
c. $(NH_4)_2CO_3\ (s) \rightarrow (NH_4)_2CO_3\ (aq)$

64. The following molecular equations each show an ionic compound dissolving in water. Show the ionization process by rewriting each one as an ionic equation:

a. $CaBr_2\ (aq) \rightarrow CaBr_2\ (aq)$
b. $LiNO_3\ (s) \rightarrow LiNO_3\ (aq)$
c. $Cu(ClO_4)_2\ (s) \rightarrow Cu(ClO_4)_2\ (aq)$

65. Write ionic equations to show the dissociation of each solid as it is dissolved in water:

a. sodium bicarbonate
b. aluminum chloride
c. chromium(III) chlorite

66. Write ionic equations to show the dissociation of each solid as it is dissolved in water:

a. ammonium cyanide
b. tin(IV) bromide
c. cobalt(II) chlorate

67. Using the general solubility rules in Section 6.5, state whether each of these compounds is likely to be soluble or insoluble:

a. an ionic compound containing the ammonium ion
b. an ionic compound containing a +2 cation and a −3 anion
c. an ionic compound containing a nitrate ion

68. Using the general solubility rules in Section 6.5, state whether each of these compounds is likely to be soluble or insoluble:

a. an ionic compound composed of an alkaline earth metal and the oxide ion
b. an ionic compound containing a transition metal ion and an acetate ion
c. an ionic compound containing a silver ion and a halogen ion

69. Using the general solubility rules in Section 6.5, state whether each of these compounds is likely to be soluble or insoluble:

a. an ionic compound composed of ammonium and a chlorate ion
b. an ionic compound containing a transition metal ion and a phosphate ion
c. an ionic compound composed of a transition metal cation and a perchlorate anion

70. Using the general solubility rules in Section 6.5, state whether each of these compounds is likely to be soluble or insoluble:

a. an ionic compound composed of an alkaline earth metal and the sulfide ion
b. an ionic compound containing an aluminum cation and a halogen anion
c. an ionic compound containing a lead(II) cation and a halogen anion

71. Refer to the general solubility rules or Table 6.3 to determine whether each of these compounds is soluble or insoluble in water:

a. LiOH
b. K_2S
c. $CaSO_4$
d. zinc sulfate
e. copper(II) nitrate
f. $Fe(C_2H_3O_2)_3$

72. Refer to the general solubility rules or Table 6.3 to determine whether each of these compounds is soluble or insoluble in water:

a. NH_4OH
b. $BaCl_2$
c. $FePO_4$
d. iron(II) chloride
e. lead(II) chloride
f. aluminum sulfide

73. In general, which would you expect to be more water soluble: a salt with a +1 cation and a −2 anion, or a salt with a +2 cation and a −2 anion?

74. Which would you expect to be more soluble, potassium nitrite (KNO_2) or aluminum oxide (Al_2O_3)? Why?

75. These equations represent precipitation reactions. Rewrite them as complete ionic equations.

a. $AgNO_3(aq) + KCl(aq) \rightarrow KNO_3(aq) + AgCl(s)$
b. $Ba(ClO_4)_2(aq) + K_2SO_4(aq) \rightarrow BaSO_4(s) + 2\ KClO_4(aq)$

76. These equations represent precipitation reactions. Rewrite them as complete ionic equations.

a. $MgBr_2(aq) + Pb(ClO_3)_2(aq) \rightarrow PbBr_2(s) + Mg(ClO_3)_2(aq)$
b. $K_3PO_4(aq) + AlCl_3(aq) \rightarrow 3\ KCl(aq) + AlPO_4(s)$

77. Identify the spectator ions in this ionic equation:

$$Ca^{2+}(aq) + 2\ NO_3^-(aq) + 2\ Cs^+(aq) + CO_3^{2-}(aq) \rightarrow CaCO_3(s) + 2\ Cs^+(aq) + 2\ NO_3^-(aq)$$

78. Identify the spectator ions in this ionic equation:

$$2\ NH_4^+(aq) + S^{2-}(aq) + Zn^{2+}(aq) + 2\ ClO_3^-(aq) \rightarrow 2\ NH_4^+(aq) + 2\ ClO_3^-(aq) + ZnS(s)$$

79. Identify the spectator ions in this equation, and rewrite it as a net ionic equation:

$$Na^+(aq) + Br^-(aq) + Ag^+(aq) + NO_3^-(aq) \rightarrow AgBr(s) + Na^+(aq) + NO_3^-(aq)$$

80. Identify the spectator ions in this equation, and rewrite it as a net ionic equation:

$$Fe^{2+}(aq) + 2\ NO_3^-(aq) + 2\ K^+(aq) + 2\ OH^-(aq) \rightarrow Fe(OH)_2(s) + 2\ K^+(aq) + 2\ NO_3^-(aq)$$

81. Aqueous solutions of potassium hydroxide and iron chloride react in a precipitation reaction to form insoluble iron(II) hydroxide, as shown here:

$$2\ KOH\ (aq) + FeCl_2\ (aq) \rightarrow 2\ KCl\ (aq) + Fe(OH)_2\ (s)$$

a. Rewrite this equation as an ionic equation.
b. What are the spectator ions in this equation?
c. Remove the spectator ions, and rewrite this as a net ionic equation.
d. What is the driving force for this reaction?

82. Aqueous solutions of silver nitrate and potassium bromide react in a precipitation reaction to form insoluble silver bromide, as shown here:

$$AgNO_3\ (aq) + KBr\ (aq) \rightarrow KNO_3\ (aq) + AgBr\ (s)$$

a. Rewrite this equation as an ionic equation.
b. What are the spectator ions in this equation?
c. Remove the spectator ions, and rewrite this as a net ionic equation.
d. What is the driving force for this reaction?

83. Each of the following reactions results in one water-soluble product and one precipitate. Complete and balance each reaction, and show phases to indicate whether the products are aqueous or solid.

a. $KCl\ (aq) + Pb(NO_3)_2\ (aq) \rightarrow$
b. $KOH\ (aq) + FeCl_3\ (aq) \rightarrow$
c. $BaCl_2\ (aq) + K_3PO_4\ (aq) \rightarrow$

84. Each of the following reactions results in one water-soluble product and one precipitate. Complete and balance each reaction, and show phases to indicate whether the products are aqueous or solid.

a. $K_2SO_4\ (aq) + FeBr_3\ (aq) \rightarrow$
b. $ZnCl_2\ (aq) + AgC_2H_3O_2\ (aq) \rightarrow$
c. $Pb(C_2H_3O_2)_2\ (aq) + MgBr_2\ (aq) \rightarrow$

85. Write balanced equations for the following precipitation reactions, including phase symbols to indicate the insoluble product. If no precipitate forms, indicate "No reaction."

a. Iron(III) acetate reacts with barium sulfide.
b. Potassium phosphate reacts with copper(II) sulfate.
c. Calcium iodide reacts with ammonium acetate.
d. Iron(II) sulfate reacts with barium hydroxide.

86. Write balanced equations for the following precipitation reactions, including phase symbols to indicate the insoluble product. If no precipitate forms, indicate "No reaction."

a. Sodium chloride reacts with silver nitrate.
b. Lead(II) nitrate reacts with sodium iodide.
c. Ammonium phosphate reacts with aluminum chloride.
d. Barium hydroxide reacts with iron(III) chloride.

87. Write molecular equations, complete ionic equations, and net ionic equations to describe the following precipitation reactions. Include phase symbols.

a. reaction of aqueous lithium iodide with aqueous silver nitrate
b. reaction of aqueous lead(II) acetate with aqueous zinc sulfate
c. reaction of aqueous calcium iodide with aqueous lead(II) nitrate

88. Write molecular equations, complete ionic equations, and net ionic equations to describe the following precipitation reactions. Include phase symbols.

a. reaction of aqueous lead(II) nitrate with aqueous sodium bromide
b. reaction of aqueous iron(II) perchlorate with aqueous sodium carbonate
c. reaction of aqueous aluminum sulfate with aqueous lead(IV) nitrate

89. What is the difference between an acid and a base?

90. What are the two common products of acid-base neutralization reactions?

91. Write ionic equations to show the dissociation of each of these acids:

a. HCl b. HNO_3

92. Write ionic equations to show the dissociation of each of these acids:

a. HBr b. HNO_2

93. Balance these neutralization reactions:

a. $H_2SO_4 + NaOH \rightarrow Na_2SO_4 + H_2O$
b. $HNO_3 + Al(OH)_3 \rightarrow Al(NO_3)_3 + H_2O$
c. $H_2SO_4 + Cr(OH)_3 \rightarrow Cr_2(SO_4)_3 + H_2O$

94. Balance these neutralization reactions:

a. $HNO_3 + Ba(OH)_2 \rightarrow Ba(NO_3)_2 + H_2O$
b. $HF + Ba(OH)_2 \rightarrow BaF_2 + H_2O$
c. $NaOH + H_3PO_4 \rightarrow Na_3PO_4 + H_2O$

95. Show the products from these neutralization reactions. Balance the equations if necessary.

a. $HCl + KOH \rightarrow$
b. $H_2SO_4 + 2\ LiOH \rightarrow$
c. $Ba(OH)_2\ (aq) + 2\ HNO_2\ (aq) \rightarrow$

96. Show the products from these neutralization reactions. Balance the equations if necessary.

a. $HBr + NaOH \rightarrow$
b. $H_3PO_4\ (aq) + 3\ KOH\ (aq) \rightarrow$
c. $Ba(OH)_2 + H_2SO_4 \rightarrow$

97. Write a balanced, complete ionic equation showing the reaction of sodium hydroxide with hydroiodic acid.

98. Write a balanced, complete ionic equation showing the neutralization of sulfuric acid with potassium hydroxide.

99. Write a net ionic equation for a double-displacement reaction in which lead(II) iodide is a precipitate.

100. Write a net ionic equation for the neutralization of any acid with a hydroxide base.

Cumulative Questions

101. Classify the reactions below as synthesis, decomposition, single displacement, double displacement, precipitation, acid-base neutralization, oxidation-reduction, or combustion. Some reactions can be classified with more than one of these terms.

a. $Hg\,(l) + S\,(s) \rightarrow HgS\,(s)$
b. $Fe\,(s) + Cu(C_2H_3O_2)_2\,(aq) \rightarrow Cu\,(s) + Fe(C_2H_3O_2)_2\,(aq)$
c. $Pb(NO_3)_2\,(aq) + 2\,KI\,(aq) \rightarrow PbI_2\,(s) + 2\,KNO_3\,(aq)$
d. $2\,K\,(s) + 2\,H_2O\,(l) \rightarrow 2\,KOH\,(aq) + H_2\,(g)$
e. $H_2SO_4\,(aq) + Ba(OH)_2\,(aq) \rightarrow 2\,H_2O\,(l) + BaSO_4\,(s)$
f. $C_8H_{16}\,(l) + 12\,O_2\,(g) \rightarrow 8\,CO_2\,(g) + 16\,H_2O\,(g)$

102. Classify the reactions below as synthesis, decomposition, single displacement, double displacement, precipitation, acid-base neutralization, oxidation-reduction, or combustion. Some reactions can be classified with more than one of these terms.

a. $Fe\,(s) + 2\,AgC_2H_3O_2\,(aq) \rightarrow Fe(C_2H_3O_2)_2\,(aq) + 2\,Ag\,(s)$
b. $AgNO_3(aq) + KCl\,(aq) \rightarrow AgCl\,(s) + KNO_3(aq)$
c. $HBr(aq) + KOH(aq) \rightarrow KBr(aq) + H_2O(l)$
d. $2\,C_2H_2\,(g) + 5\,O_2\,(g) \rightarrow 4\,CO_2\,(g) + 2\,H_2O\,(g)$
e. $Fe\,(s) + 2\,HCl\,(aq) \rightarrow FeCl_2(aq) + H_2\,(g)$
f. $2\,K\,(s) + Br_2\,(g) \rightarrow 2\,KBr\,(s)$

---

103. The reactions here draw from all the reaction types introduced in this chapter. Predict the products, and balance each equation.

a. $AgC_2H_3O_2\,(aq) + NaCl\,(aq) \rightarrow$
b. $Na\,(s) + Cl_2\,(g) \rightarrow$
c. $HCl\,(aq) + Ba(OH)_2\,(aq) \rightarrow$
d. $C_{10}H_{20}\,(l) + O_2\,(g) \rightarrow$

104. The reactions here draw from all the reaction types introduced in this chapter. Predict the products, and balance each equation.

a. $Co\,(s) + Br_2\,(l) \rightarrow$
b. $MgBr_2\,(aq) + K_3PO_4\,(aq) \rightarrow$
c. $Mg\,(s) + O_2\,(g) \rightarrow$
d. $LiOH\,(aq) + H_3PO_4\,(aq) \rightarrow$

CHAPTER SEVEN

Mass Stoichiometry

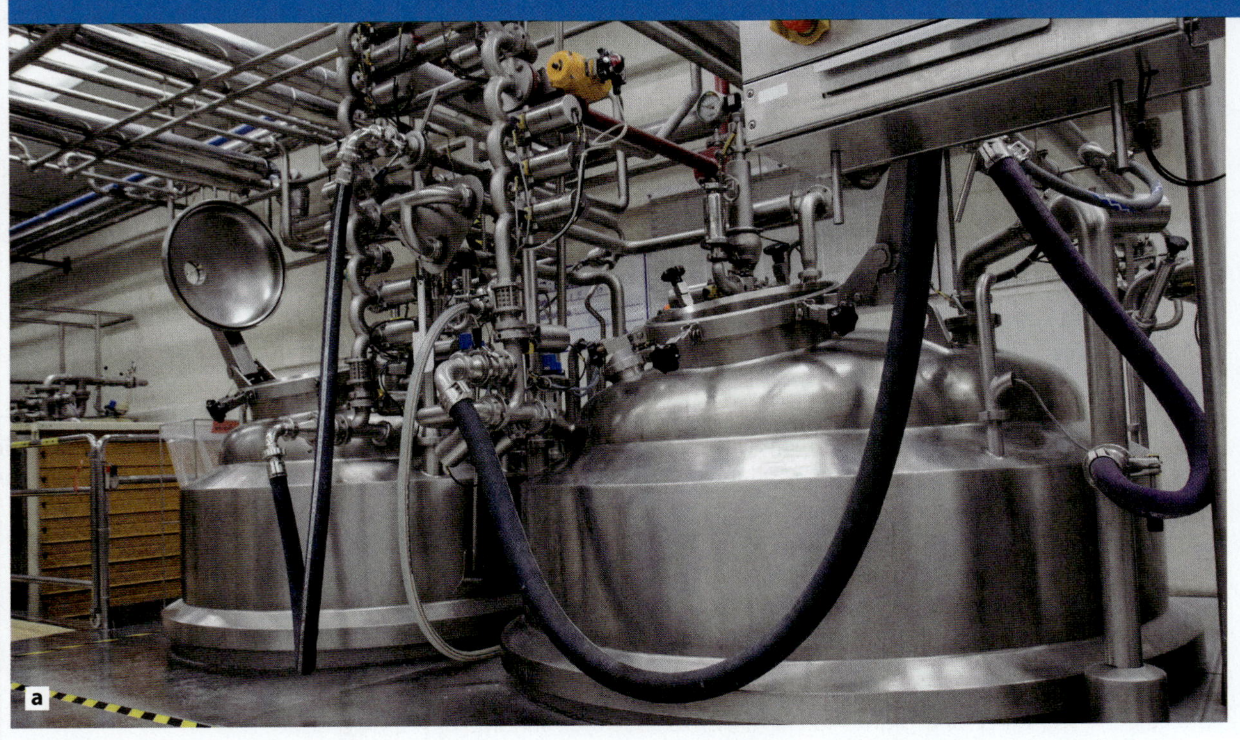

Figure 7.1 (a) Pharmaceutical companies produce medicines in large, stainless steel reaction vessels like these. (*continued*)

Process Development

"Are we ready?"

Jared Fennell, Ph.D., looks around the room one more time. Seated at an array of computer monitors, several technicians nod in agreement. Next to them, a team of chemists and engineers watches anxiously. At the center of the immaculate room sits a massive stainless steel container. In fact, only the top of the container is in the room—most of it lies beneath the concrete floor. Pipes, hoses, and sensors feed into the top of the container.

Jared takes a deep breath. "Okay," he says, "Let's go." The team has spent months working for this moment. Every detail of this performance has been discussed, written out, and rehearsed. Everyone knows their roles.

One of the operators enters a command on the control panel. Slowly, the container begins to fill. Over the next 48 hours, the team will monitor a huge chemical reaction. Everything has to go according to plan. The safety of the team, the success of the project, and a million-dollar investment depend on it.

Jared has an extraordinarily challenging job. He is a process chemist at Eli Lilly, an Indiana-based pharmaceutical company that produces medicines for diabetes, cancer, brain function, and other conditions. Process chemists develop techniques for manufacturing materials on a very large scale. For example, Jared and his team were recently tasked with manufacturing a new medicine. Drug discovery chemists had produced about 25 grams of the compound. Jared's job: Scale the reaction up to 100,000 grams.

The scale of the reaction is not the only challenge. Because the team produces medicine for human consumption, the manufacturing process must work the same

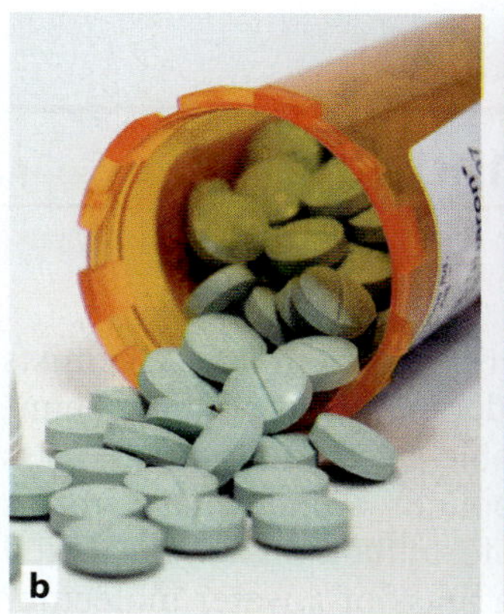

Figure 7.1 (continued) (b) Pharmaceutical companies like Eli Lilly produce large amounts of vital medicines. (c) Jared Fennell is a process chemist at Eli Lilly. (d) These process chemists are in the laboratory, preparing for a large-scale reaction. (e) An operator tends to a reaction vessel before the start of a manufacturing reaction. (a) noomcpkstic/Shutterstock; (b) wwing/E+/Getty Images; (c) © Copyright Eli Lilly and Company. All rights reserved. Used with Permission; (d) © Copyright Eli Lilly and Company. All rights reserved. Used with Permission; (e) © Copyright Eli Lilly and Company. All rights reserved. Used with Permission.

way every time. Every detail must be carefully considered: How much of each reactant should be added? What is the best temperature for the reaction? How are the reagents mixed together? For how long does the reaction go? Do these reactions produce other substances? How can they be removed? Every parameter is tested and retested, until Jared and his team are certain that every batch they produce will be exactly the same (**Figure 7.1**).

In manufacturing, chemistry begins with a simple question: How much? For example, how much product is needed? How much of each starting material is required? In fact, the importance of this question extends well beyond manufacturing—it applies to every aspect of chemistry. When you breathe, how much oxygen does your body require? If you burn a gallon of gasoline while driving your car, how much carbon dioxide have you produced? If you are cooking a big pancake breakfast, how much mix do you need to feed 20 people? These questions apply to very different situations, but each of them relates to the fundamental chemical question, "How much?"

In this chapter, we're going to tackle this question. We'll learn how the masses of atoms and compounds relate to the world around us and see how we can use these relationships to understand the chemistry that takes place in the laboratory, the factory, or the kitchen.

Intended Learning Outcomes

After completing this chapter and working the practice problems, you should be able to:

7.1 Formula Mass and Percent Composition

- Calculate the formula mass of a compound.
- Calculate the percent composition (by mass) of elements in a compound.
- Broadly describe how chemists measure formula mass and percent composition.

7.2 Connecting Atomic Mass to Large-Scale Mass: The Mole Concept

- Use the mole concept to relate masses on the atomic scale to masses on the laboratory scale.
- Convert between grams, moles, and atoms or molecules.

7.3 The Mole Concept in Balanced Equations

- Apply the mole concept to solve stoichiometry problems, relating the amounts of reagents and products in a chemical change.
- Identify the limiting and excess reagents in chemical reactions.

7.4 Theoretical and Percent Yield

- Differentiate between theoretical, actual, and percent yields.
- Describe chemical and physical occurrences that can lead to an actual yield that is less than the theoretical yield.
- Using the theoretical and actual yields, correctly calculate th percent yield for a chemical reaction.

7.1 Formula Mass and Percent Composition

Formula Mass

The masses of elements and compounds play a vital role in chemistry. Chemists use mass to identify unknown compounds and to understand and predict chemical reactions. As we saw in Chapter 3, the periodic table shows the average mass for atoms of each element. When working with compounds, we use these atomic masses to calculate the mass of a single molecule or formula unit. This is called the **formula mass**. For example, what is the formula mass of a water molecule? We know that a molecule of water, H_2O, contains two hydrogen atoms and one oxygen atom. Using the masses from the periodic table (**Figure 7.2**), we calculate the mass of water as follows:

1
H
1.01

6
C
12.01

8
O
16.00

Figure 7.2 We use the average atomic masses from the periodic table to calculate formula mass.

$$\text{Formula mass of } H_2O = 2(1.01\text{ u}) + 1(16.00\text{ u}) = 18.02\text{ u}$$

We found the formula mass by adding up the number and type of each atom, using the masses given in the periodic table. Chemists sometimes refer to the formula mass as the *formula weight*, or as the *molecular mass* or *molecular weight*.

Recall that we measure the mass of atoms in atomic mass units, abbreviated as *u* or sometimes as *amu*. ■

Example 7.1 Calculating the Formula Mass of a Compound

Potassium carbonate, K_2CO_3, is a common water-softening agent. What is the formula mass of this compound?

To solve this problem, we add the atomic masses of each atom in the formula unit.

$$\text{Formula mass of } K_2CO_3 = 2(39.10\text{ u}) + 1(12.01\text{ u}) + 3(16.00\text{ u}) = 138.21\text{ u}$$

TRY IT

1. Find the formula mass for each of these compounds:
   a. H_2CO **b.** Na_2CO_3 **c.** lithium nitrate **d.** iron(II) bromide

Percent Composition

Chemists sometimes describe compounds by the percentage (by mass) of each element in a compound. This is called the **percent composition** of a compound. We calculate percent composition by dividing the mass of one element in a compound by the total mass of a compound and then converting to percent:

$$\text{Percent composition of one element} = \frac{\text{mass of one element}}{\text{mass of entire compound}} \times 100\%$$

Examples 7.2 and 7.3 illustrate this idea.

Example 7.2 Finding the Percent Composition of a Compound

Octane, a component of gasoline, has the molecular formula C_8H_{18}. What is the percent composition of carbon and hydrogen in octane?

To solve this problem, we find the mass of the carbon and the hydrogen atoms in the formula:

$$\text{Mass carbon} = \text{mass}(C_8) = 8(12.01\text{ u}) = 96.08\text{ u}$$
$$\text{Mass hydrogen} = \text{mass}(H_{18}) = 18(1.01\text{ u}) = 18.18\text{ u}$$

Next, we find the total formula mass of the compound:

$$\text{Formula mass of } C_8H_{18} = 8(12.01\text{ u}) + 18(1.01\text{ u}) = 114.26\text{ u}$$

The percent by mass of each element is the mass of that element divided by the total formula mass, expressed as a percentage.

$$\text{Percent carbon} = \frac{\text{mass carbon}}{\text{total formula mass}} \times 100\% = \frac{96.08\text{ u}}{114.26\text{ u}} \times 100\% = 84.09\%$$

$$\text{Percent hydrogen} = \frac{\text{mass hydrogen}}{\text{total formula mass}} \times 100\% = \frac{18.18\text{ u}}{114.26\text{ u}} \times 100\% = 15.91\%$$

Example 7.3 Using Percent Composition

Silicon dioxide (SiO_2) is the main component of glass. This compound contains 46.75% silicon and 53.25% oxygen by mass. What mass of silicon is in a 2.32-kilogram sample of SiO_2?

To find the mass of silicon in the 2.32-kilogram sample, we multiply the mass of the total sample by the percentage of silicon in the sample, expressed as a decimal.

$$\begin{aligned}\text{Mass Si} &= \text{total mass} \times \text{percent silicon} \\ &= 2.32\text{ kg} \times 0.4675 = 1.08\text{ kg}\end{aligned}$$

TRY IT

2. Acetaminophen is a common painkiller. This compound has a molecular formula of $C_8H_9NO_2$. What is the percentage by mass of carbon, hydrogen, nitrogen, and oxygen in this compound?

3. Table sugar is a covalent compound with the formula $C_{12}H_{22}O_{11}$. What is the percentage by mass of carbon in table sugar? What mass of carbon is present in 2.00 kg of table sugar?

Check It

Watch Explanation

How Chemists Measure Formula Mass and Percent Composition

In Example 7.2, we began with the formula for a compound and then calculated its formula mass and percent composition. However, chemists often have the opposite problem: They encounter a new compound—perhaps a natural product with exciting medicinal properties—and must determine its structure and composition. Both formula mass and percent composition are important in identifying unknown compounds. In this section, we'll briefly look at how chemists measure these important properties.

Each year a team of scientists from the University of Alabama at Birmingham collects and studies sponges from the waters around Antarctica. Some compounds discovered in these sponges block the growth of cancer cells. Chemists use the formula mass and percent composition to help determine the structure of these compounds.

Courtesy of Julie B. Schram and UAB News/The University of Alabama at Birmingham

Mass Spectrometry

Chemists measure the masses of compounds using a technique called **mass spectrometry**. A mass spectrometer contains three vital components: an ionizing chamber, an electric field, and a detector (**Figure 7.3**). The ionizing chamber

Explore Figure 7.3

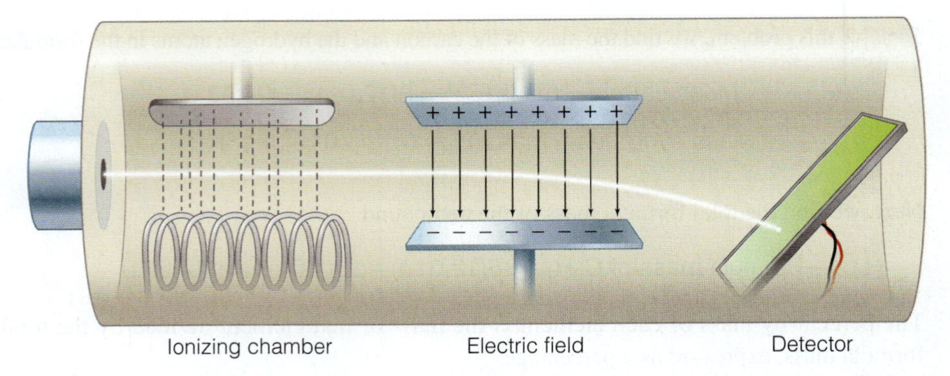

Figure 7.3 In a mass spectrometer, a molecule passes through the ionizing chamber and then into an electric field. The path of the charged molecule curves as it moves through the electric field. The spectrometer calculates the mass of the molecule based on how sharply the molecule's path changes as it passes through the electric field.

Figure 7.4 A forensic chemist uses a mass spectrometer to analyze a mixture.

converts neutral molecules into ions (often by removing an electron). The ions are then accelerated into an electric field, which causes the pathways of the charged particles to curve. Lighter particles curve more sharply than heavier particles. After exiting the electric field, the particles strike a detector. The spectrometer calculates the mass of the compound based on the size of the electric field, the charge on the particle, and the exact point where the particle hits the detector.

Mass spectrometry is routinely used in chemical analysis (**Figure 7.4**). For example, forensic chemists use mass spectrometry to monitor for hazardous or illegal substances. **Figure 7.5** shows the mass spectrum of a sample taken from a dollar bill. Cocaine ($C_{17}H_{22}NO_4$) has a formula mass of 304.2 u—the signal at this mass indicates the presence of cocaine on the surface of the bill.

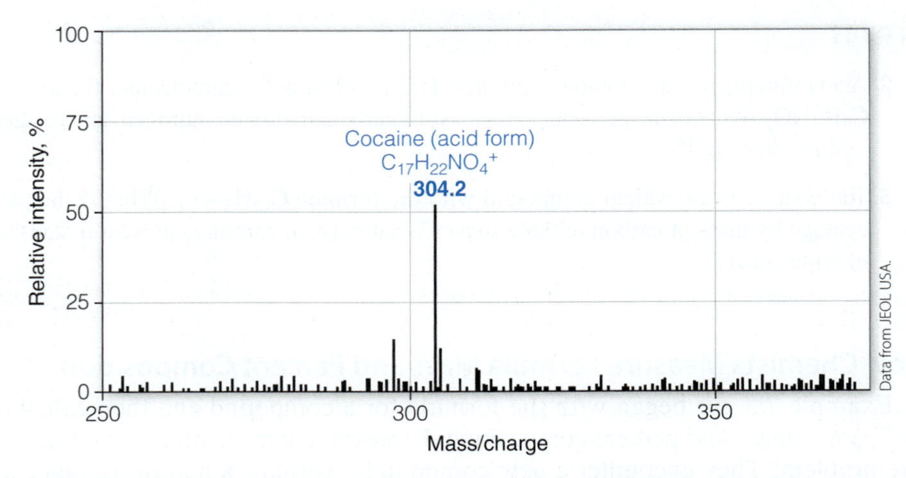

Figure 7.5 This mass spectrum was taken from the surface of a dollar bill. The peak at 304.2 indicates the presence of cocaine.

Elemental Analysis

Chemists measure percent composition through a technique called **elemental analysis**. This technique uses combustion reactions to convert compounds into simpler products (carbon dioxide, water, etc.). The mass of each product indicates the percentage of each element in a sample (**Figure 7.6**). Using a combination of elemental analysis and mass spectrometry, chemists can often determine both the percent composition and molecular formula of an unknown compound.

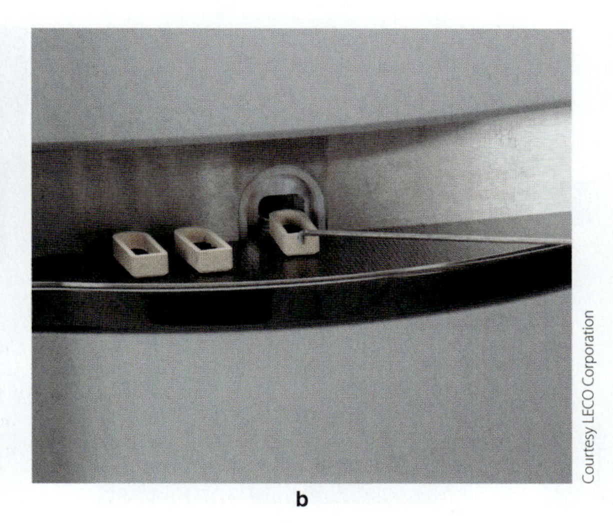

a b

Courtesy LECO Corporation

Figure 7.6 Elemental analysis uses combustion to separate and measure the percentage of different elements in a compound. This image shows (a) a chemist preparing the sample, and (b) the sample being slid into the combustion chamber.

7.2 Connecting Atomic Mass to Large-Scale Mass: The Mole Concept

Avogadro's Number and the Mole

So far, we've expressed the masses of atoms and molecules using atomic mass units. While this measure is useful for calculating formula mass or percent composition, it presents a problem: Atomic mass units are incredibly small ($1\ u = 1.66 \times 10^{-24}$ grams), far too small to be useful in laboratory measurements. So how can we relate these masses to larger amounts—such as grams, kilograms, or tons—that can be measured in the lab or the factory (**Figure 7.7**)?

Guillaume Plisson/Bloomberg/Getty Images

Figure 7.7 Manufacturers use large-scale batch reactors like the ones shown here. It is important to relate reactions between a few atoms to reactions of this scale.

To answer this question, we're going to use a quantity called a **mole** (often abbreviated as *mol*). A mole is a number of items, similar to the term *dozen*. A *dozen* simply means a quantity of 12, regardless of the item. If you have 12 planets, you have a dozen planets. If you have 12 toothpicks, you have a dozen toothpicks. If you have 12 donuts, you have a dozen donuts. It doesn't matter what you're measuring; the term *dozen* refers to the quantity 12.

Like the term dozen, a mole describes the number of units present. A *mole* is 6.02×10^{23} *units of anything*. If you have 6.02×10^{23} donuts, you have a mole of donuts. If you have 6.02×10^{23} carbon atoms, you have a mole of carbon atoms. If you have 6.02×10^{23} oxygen molecules, you have a mole of oxygen molecules. We refer to the number of particles in a mole, 6.02×10^{23}, as **Avogadro's number**.

One mole = 6.02×10^{23} units. ■

The power of the mole concept is that it allows us to express the masses of atoms and compounds in grams (**Figure 7.8**). Here's how it works: One atom of carbon has a mass of 12.01 atomic mass units. One mole of carbon has a mass of 12.01 grams. We can't measure out one atom of carbon in a lab, but we can measure out 12.01 grams. So, we describe the atomic mass in one of two ways: We can say that the mass of carbon is 12.01 u (atomic scale), or we can say that the mass of carbon is 12.01 grams per mole (laboratory scale). When a mass is expressed in grams per mole, we refer to it as the **molar mass**.

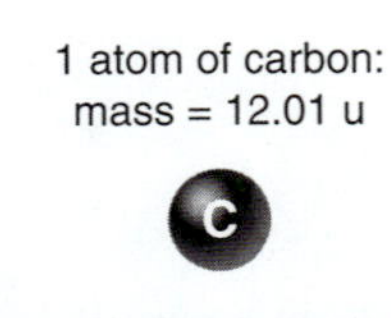

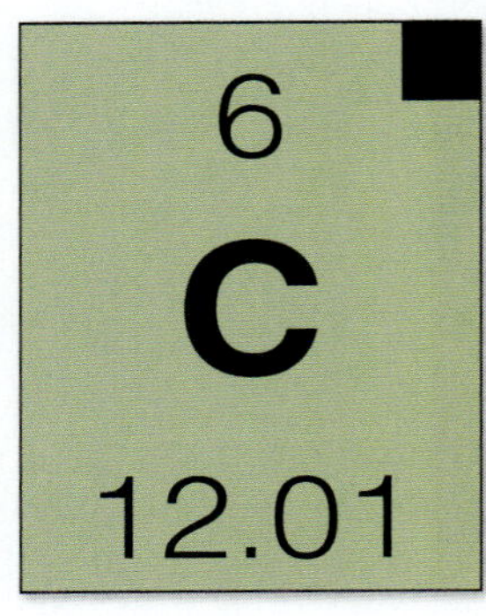

a

Andrew Lambert Photography/Science Source

b

© 2001 Richard Megna/Fundamental Photographs, NYC

c

Explore Figure 7.8

Figure 7.8 The mole concept connects atoms to grams. (a) The average mass of one carbon atom is 12.01 u. (b) The mass of one mole of carbon is 12.01 grams. (c) This image shows one mole of several different elements.

We describe mass in atomic mass units (u) or in grams per mole (g/mol).

Here's another example: The mass of one molecule of CO_2 is 44.01 u, and the mass of one mole of CO_2 is 44.01 grams. So, the *molar mass* of carbon dioxide is 44.01 g/mol. Using moles connects the atomic world to a scale we can measure in the lab.

Converting between Grams, Moles, and Particles

In chemistry we frequently need to convert between grams and moles. To do this we use the molar mass as the conversion factor. For example, how many moles of sodium chloride are present in a 305-gram sample? Based on the formula mass of NaCl, we know that 58.44 grams of NaCl = 1 mole of NaCl. We use this as our conversion factor:

$$305 \text{ g } \cancel{\text{NaCl}} \times \frac{1 \text{ mole NaCl}}{58.44 \text{ g } \cancel{\text{NaCl}}} = 5.22 \text{ moles NaCl}$$

Use the molar mass to convert between grams and moles.

We often need to convert from moles to grams. For example, to prepare a solution that contains 1.20 moles of NaCl, how many grams of NaCl do we need? In this example, we again use the molar mass as the conversion factor:

$$1.20 \cancel{\text{ moles NaCl}} \times \frac{58.44 \text{ g NaCl}}{1 \cancel{\text{ mole NaCl}}} = 70.1 \text{ grams NaCl}$$

Notice that when we write the conversion factor, the unit we wish to cancel (grams or moles) goes on the bottom, and the unit we wish to keep goes on the top. This concept is very important, and you should practice these calculations until you can do them comfortably. Examples 7.4 and 7.5 further illustrate this idea.

Example 7.4 Converting from Grams to Moles

How many moles of aluminum oxide, Al_2O_3, are present in a 57.3-gram sample of this compound?

We first calculate the formula mass of Al_2O_3 and find that 101.96 grams of Al_2O_3 = 1 mole of Al_2O_3. We use this relationship as a conversion factor. We place the unit we wish to keep (moles) on top and the unit we wish to cancel (grams) on the bottom. Rounding to the correct number of significant digits, we get a final answer of 0.562 moles of Al_2O_3.

$$57.3 \text{ g } \cancel{Al_2O_3} \times \frac{1 \text{ mole } Al_2O_3}{101.96 \text{ g } \cancel{Al_2O_3}} = 0.562 \text{ moles } Al_2O_3$$

Example 7.5 Converting from Moles to Grams

A sample contains 0.281 moles of potassium sulfate, K_2SO_4. What is the mass of potassium sulfate in this sample?

As in Example 7.4, we use the molar mass to convert between grams and moles. We first determine that the molar mass of K_2SO_4 is 174.26 grams/mole. As before, we use this mass as a conversion factor. Because we want to move from moles to grams, we write the conversion factor with grams on top and moles on the bottom. Rounding to the correct number of significant figures gives us a final answer of 49.0 g of K_2SO_4.

$$0.281 \text{ }\cancel{\text{moles } K_2SO_4} \times \frac{174.26 \text{ g } K_2SO_4}{1 \text{ }\cancel{\text{mole } K_2SO_4}} = 49.0 \text{ g } K_2SO_4$$

TRY IT

4. Convert the following amounts from grams to moles:
   a. 22.7 g of silver
   b. 43.6 g of sodium nitrate
   c. 974.3 g of boron trifluoride
5. Find the mass of each of the following:
   a. 6.20 moles of silicon
   b. 5.3 moles of molecular bromine (Br_2)
   c. 1.83 moles of ammonium carbonate

Check It
Watch Explanation

Use Avogadro's number to convert between moles and particles.

Because we know the number of particles in a mole, we can also convert between moles and particles, such as atoms or molecules. To do this, we use *Avogadro's number* as the conversion factor. For example, how many atoms are in 4.20 moles of gold? To solve this problem, we multiply the moles of gold by Avogadro's number:

$$4.20 \text{ }\cancel{\text{moles}}\text{ Au} \times \frac{6.02 \times 10^{23} \text{ atoms}}{\cancel{1 \text{ mole}}} = 2.53 \times 10^{24} \text{ atoms Au}$$

The mole concept allows us to directly relate the quantity of atoms and the mass in grams. For example, what is the mass in grams of 2.53×10^{23} iron atoms? To solve this problem, we begin with atoms, then convert to moles, and finally convert to grams:

$$2.53 \times 10^{23} \text{ }\cancel{\text{atoms Fe}} \times \frac{1 \text{ }\cancel{\text{mole Fe}}}{6.02 \times 10^{23} \text{ }\cancel{\text{atoms Fe}}} \times \frac{55.85 \text{ grams Fe}}{1 \text{ }\cancel{\text{mole Fe}}} = 23.5 \text{ grams Fe}$$

Notice in these problems that we *use Avogadro's number to convert between moles and atoms, and we use the molar mass to convert between moles and grams.* To relate grams to atoms, we must first convert to moles (**Figure 7.9**).

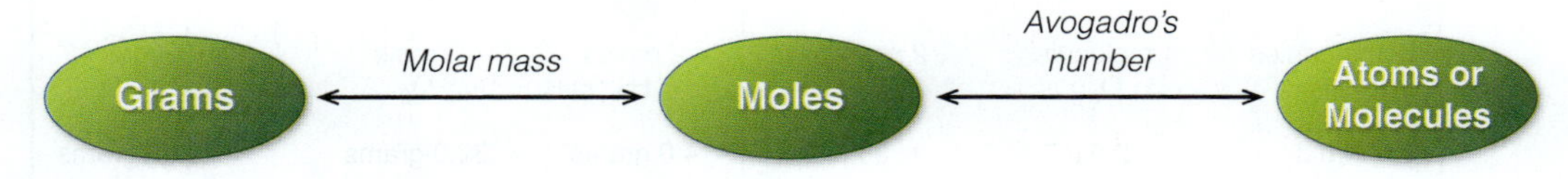

Figure 7.9 To convert between grams and atoms or molecules, you must go through moles.

Example 7.6 Converting from Grams to Molecules

A chemist has a sample containing 42,300 grams of sulfur trioxide, SO_3. How many molecules of SO_3 are present in this sample?

To solve this problem, we must first convert from grams to moles (as in the previous examples) and then convert to molecules. To convert from grams to moles, we use the molar mass of SO_3 (80.06 g/mol) as the conversion factor. We then use Avogadro's number to convert from moles to molecules. Rounding to three significant figures, we find that this mass corresponds to 3.18×10^{26} molecules of SO_3.

$$42{,}300 \text{ g } SO_3 \times \frac{1 \text{ mole } SO_3}{80.06 \text{ g } SO_3} \times \frac{6.02 \times 10^{23} \text{ molecules}}{1 \text{ mole}} = 3.18 \times 10^{26} \; SO_3 \text{ molecules}$$

Check It
Watch Explanation

TRY IT

6. A sample contains 0.0623 moles of titanium. How many titanium atoms are present in this sample? What is the mass in grams of this sample?

7. Find the mass of each of the following:

a. 1.83×10^{24} iron atoms **b.** 3.21×10^{21} CH_4 molecules

8. How many molecules of CS_2 are in a 15.83-gram sample?

In 1937 the airship *Hindenburg* caught fire and exploded over New Jersey, killing 36 people. The airship was filled with hydrogen gas, which reacted explosively with the oxygen gas in the air.

Charles Hoff/NY Daily News Archive/Getty Images

7.3 The Mole Concept in Balanced Equations

In Chapter 6 we discussed the explosive reaction between hydrogen and oxygen:

$$2\,H_2 + O_2 \rightarrow 2\,H_2O$$

We can read this equation in one of two ways. First, we can interpret it in a *molecular* sense: Two molecules of hydrogen react with one molecule of oxygen. But we can also interpret this equation as a ratio of moles: *Two moles* of hydrogen react with *one mole* of oxygen (**Figure 7.10**).

A balanced equation gives both an atomic ratio and a mole ratio.

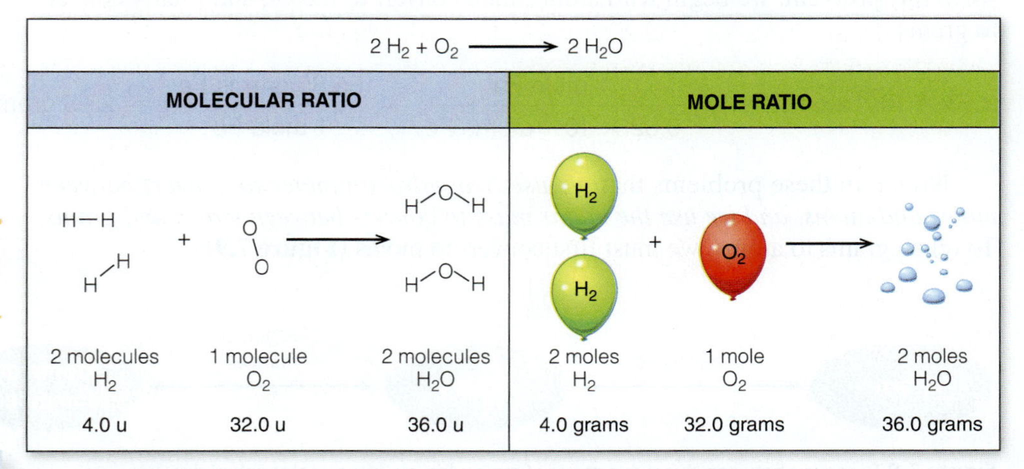

Figure 7.10 A balanced equation represents both the ratio of molecules and the ratio of moles that are involved in a reaction.

Explore
Figure 7.10

Stoichiometry Problems

Because a balanced equation relates the moles of substances in a reaction, we can use it to connect the amount of one reactant or product to the amounts of every other species in the equation. For example, **Figure 7.11** poses a series of questions about a reaction. See if you can answer each question, then look at the explanations that follow.

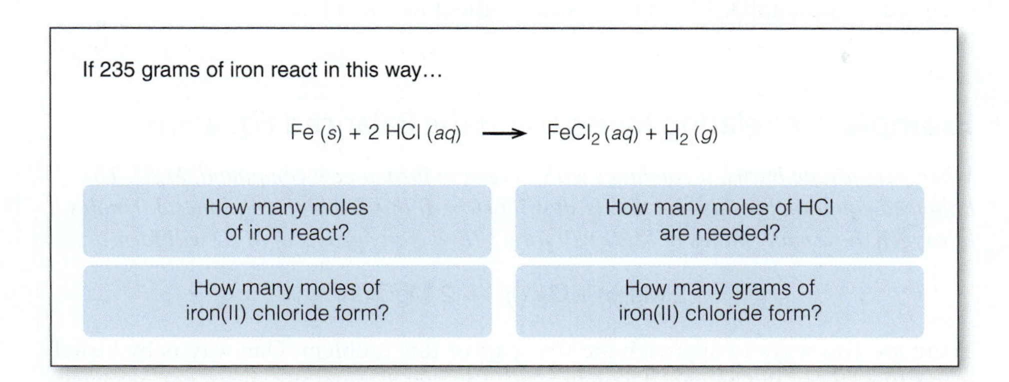

If 235 grams of iron react in this way...
Fe (s) + 2 HCl (aq) ⟶ FeCl2 (aq) + H2 (g)
How many moles of iron react?
How many moles of HCl are needed?
How many moles of iron(II) chloride form?
How many grams of iron(II) chloride form?

Figure 7.11 Using moles, we can relate the amounts of substances in a chemical equation. Can you figure out the answers to these questions from the equation and amount given?

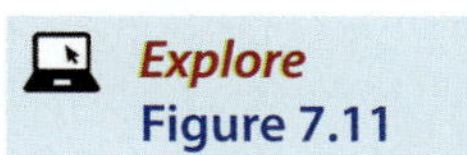
Explore Figure 7.11

Here are the solutions to each of the four questions:

1. *How many moles of iron react?* To answer this question, we need to convert from grams of iron to moles of iron. As before, we use the molar mass of iron (55.85 g/mole) as a conversion factor:

$$235 \text{ g } \cancel{\text{Fe}} \times \frac{1 \text{ mole Fe}}{55.85 \cancel{\text{ g Fe}}} = 4.21 \text{ moles Fe}$$

2. *How many moles of HCl are needed to react with the iron?* Based on the balanced equation, two moles of HCl are required for every one mole of Fe. Because we have 4.21 moles of Fe, this means we need twice that many moles of HCl, or 8.42 moles. More formally, we could solve this problem by unit conversions:

$$4.21 \cancel{\text{ moles Fe}} \times \frac{2 \text{ moles HCl}}{1 \cancel{\text{ mole Fe}}} = 8.42 \text{ moles of HCl}$$

Use the coefficients in the balanced equation to convert between moles of compounds.

Notice how we did this: We created a conversion factor (2 moles HCl/1 mole Fe) from the balanced equation. When relating the moles of one substance to the moles of another, *we use the coefficients in the balanced equation as a conversion factor.*

3. *How many moles of iron(II) chloride form in this reaction?* In the balanced equation, there is one mole of $FeCl_2$ per one mole of Fe. Therefore, the moles of $FeCl_2$ produced are the same as the moles of iron that react (4.21 moles). Again, we can do this more formally by writing a conversion factor from the balanced equation (1 mole $FeCl_2$/1 mole Fe):

$$4.21 \cancel{\text{ moles Fe}} \times \frac{1 \text{ mole FeCl}_2}{1 \cancel{\text{ mole Fe}}} = 4.21 \text{ moles of FeCl}_2$$

4. *How many grams of iron(II) chloride form in this reaction?* In this final question, we have to convert from moles of $FeCl_2$ to grams of $FeCl_2$. Anytime we convert between grams and moles, we use the molar mass of the substance in question. Based on the masses on the periodic table, the molar mass of $FeCl_2$ is 126.75 grams/mole. Because we know there are 4.21 moles of $FeCl_2$ (from question 3), we can set up the final problem like this:

$$4.21 \cancel{\text{ moles FeCl}_2} \times \frac{126.75 \text{ g FeCl}_2}{1 \cancel{\text{ mole FeCl}_2}} = 534 \text{ g of FeCl}_2$$

I live in the corn country of western Kentucky. Planes flying from our small regional airport go only to Chicago O'Hare. I can go anywhere I want, but I must go through Chicago. Like O'Hare, moles are a travel hub in many calculations. If you want to connect one substance to another, you have to travel through moles.

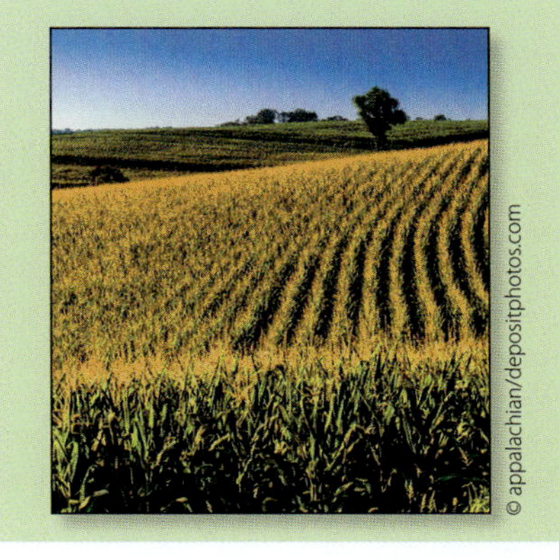

We can relate the amount of one substance to any other substance in a balanced equation. ■

This type of problem is called a **stoichiometry problem**. These problems use the amount of one substance to predict the amount of another substance that is consumed or produced, according to the balanced equation. Once we found the moles of Fe (in question 1), we were able to relate this value to both the moles of HCl and the moles and grams of $FeCl_2$. Given the amount in grams or moles of one substance in the balanced equation, we can find the moles or grams of any other substance in the equation. Examples 7.7 and 7.8 further illustrate this idea.

Example 7.7 Relating Moles Using the Balanced Equation

When magnesium burns, it combines with oxygen to form a new compound, MgO. The balanced equation for this reaction is shown below. If this reaction consumes 3.0 moles of oxygen, how many moles of MgO will form? How many grams of MgO will form?

$$2\,Mg\,(s) + O_2\,(g) \rightarrow 2\,MgO\,(s)$$

There are two ways to approach the first part of this problem. One way is by visual inspection: From the balanced equation, we see that 1 mole of oxygen gas produces 2 moles of magnesium oxide. Based on this ratio, 3.0 moles of oxygen gas will produce 6.0 moles of MgO.

For simple ratios, this sort of visual inspection is very helpful. But we can also solve the problem by using the balanced equation to create a conversion factor. We begin with 3.0 moles of oxygen gas and then multiply by a conversion factor, showing two moles of MgO for every one mole of oxygen gas:

$$3.0\ \cancel{\text{moles } O_2} \times \frac{2\ \text{moles MgO}}{1\ \cancel{\text{mole } O_2}} = 6.0\ \text{moles MgO}$$

Once we have found the moles of MgO, we use the molar mass of MgO (40.31 g/mol) to convert from moles to grams. The calculator gives an answer of 241.9 grams; but we can keep only two significant digits, so we round the answer to 240 grams.

$$6.0\ \cancel{\text{moles MgO}} \times \frac{40.31\ \text{g MgO}}{1\ \cancel{\text{mole MgO}}} = 240\ \text{g MgO}$$

Example 7.8 Relating Grams and Moles Using the Balanced Equation

Sodium metal reacts violently with water, as shown in the balanced equation below. How many moles of H_2 gas are produced by the reaction of 11.0 grams of sodium with water?

$$2\,Na\,(s) + 2\,H_2O\,(l) \rightarrow 2\,NaOH\,(aq) + H_2\,(g)$$

We are given the grams of sodium and asked to relate this amount to the moles of H_2. To solve this problem, we must first convert the grams of sodium to moles of sodium and then relate the moles of sodium to the moles of hydrogen. Graphically, we can express our strategy this way:

$$\text{Grams Na} \Rightarrow \text{Moles Na} \Rightarrow \text{Moles } H_2$$

To convert grams of sodium to moles of sodium, we use the molar mass of sodium: 22.99 g/mol. In order to convert moles of sodium to moles of H_2, we use the balanced equation to relate these two species (1 mole H_2/2 moles Na). We could write this process in two individual steps:

Step 1: Convert g Na to mol Na

$$11.0 \text{ g Na} \times \frac{1 \text{ mol Na}}{22.99 \text{ g Na}} = 0.478 \text{ mol Na}$$

Step 2: Relate mol Na to mol H_2

$$0.478 \text{ mol Na} \times \frac{1 \text{ mol H}_2}{2 \text{ mol Na}} = 0.239 \text{ mol H}_2$$

For simplicity, we can also write the two steps side by side.

$$11.0 \text{ g Na} \times \frac{1 \text{ mol Na}}{22.99 \text{ g Na}} \times \frac{1 \text{ mol H}_2}{2 \text{ mol Na}} = 0.239 \text{ mol H}_2$$

TRY IT

9. Lithium metal reacts with nitrogen gas according to the balanced equation shown here. Based on this equation, how many moles of N_2 are required to react with 12.0 moles of lithium? How many moles of Li_3N are produced in this reaction? How many grams of Li_3N are produced?

$$6 \text{ Li } (s) + \text{N}_2 \ (g) \rightarrow 2 \text{ Li}_3\text{N } (s)$$

10. Sulfur dioxide can react with oxygen gas as shown in this balanced equation:

$$2 \text{ SO}_2 \ (g) + \text{O}_2 \ (g) \rightarrow 2 \text{ SO}_3 \ (g)$$

How many moles of SO_2 are required to produce 40 moles of SO_3? How many moles of O_2 are required?

Check It
Watch Explanation

Gram-to-Gram Questions

One common type of stoichiometry problem is the gram-to-gram problem. In this type of problem, we are given the mass of one reagent or product (usually in grams) and asked to find the mass of another reagent or product (also in grams). Since the balanced equation relates amounts by moles, we must convert to moles to solve this problem. For any question relating the grams of substance A to the grams of substance B, the strategy is to convert from grams of A to moles of A, then to moles of B, then finally to grams of B. This approach is shown graphically in **Figure 7.12**. Example 7.9 describes how we used this strategy to solve a problem of this type.

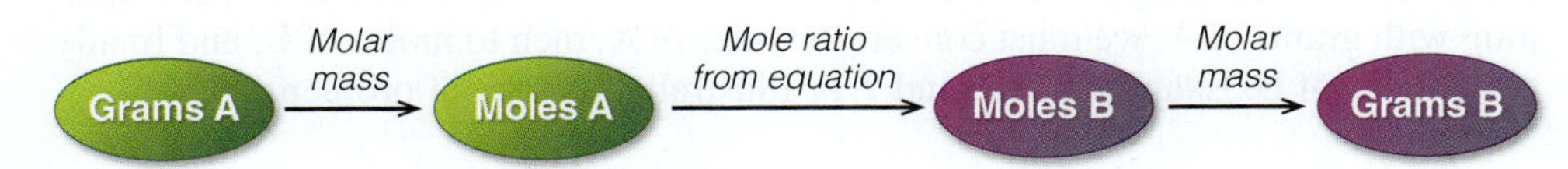

Figure 7.12 We use this strategy for solving gram-to-gram stoichiometry problems.

Example 7.9 Gram-to-Gram Conversion

When heated with a Bunsen burner, $MgCO_3$ decomposes to MgO and CO_2, as shown in the equation. If 5.24 grams of $MgCO_3$ are heated in this manner, how many grams of MgO can be produced?

$$\text{MgCO}_3 \ (s) \rightarrow \text{MgO } (s) + \text{CO}_2 \ (g)$$

To solve this problem, we must first convert from grams of $MgCO_3$ to moles of $MgCO_3$. Next, we use the balanced equation to relate the moles of $MgCO_3$ to the moles of MgO. Finally, we convert moles of MgO into grams. We can do one conversion at a time, or we can string these conversions together, like this.

$$5.24\text{ g }\cancel{MgCO_3} \times \frac{1\text{ }\cancel{\text{mole }MgCO_3}}{84.32\text{ }\cancel{\text{g }MgCO_3}} \times \frac{1\text{ }\cancel{\text{mole }MgO}}{1\text{ }\cancel{\text{mole }MgCO_3}} \times \frac{40.31\text{ g }MgO}{1\text{ }\cancel{\text{mole }MgO}} = 2.51\text{ g }MgO$$

Start with grams of $MgCO_3$	Convert to moles of $MgCO_3$	Relate to moles of MgO	Convert to grams of MgO

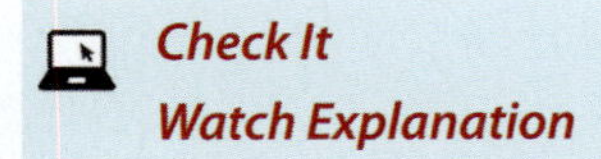

TRY IT

11. Consider the reaction of copper with bromine to produce copper(I) bromide:

$$2\ Cu\ (s) + Br_2\ (l) \rightarrow 2\ CuBr\ (s)$$

How many grams of copper(I) bromide are produced from the reaction of 24.6 grams of copper in this reaction? How many grams of bromine are required to react with this amount of copper?

Strategies for Solving Stoichiometry Problems

For many people, stoichiometry problems can seem overwhelming. Don't worry—you can do this. The key to solving these problems is to recognize the simple patterns; then follow the patterns every time. Notice that these problems rely on just three types of conversions, each with its own conversion factor.

Conversion Type	Conversion Factor
1. Grams and moles of one substance	Molar mass
2. Moles and particles of one substance	Avogadro's number
3. Moles of two different substances	Mole ratio from the balanced equation

Once we understand the conversions, we need to identify where we are starting from and where we are going. Stoichiometry problems provide the amount of one compound (A) and ask us to relate this to the amount of another compound (B). These problems can be solved by using the "mole map" shown in **Figure 7.13**. Like any map, this guide helps us locate where we are, where we are going, and how to get there. For example, how do we get from grams of compound A to particles of compound B? Looking at the mole map, we can see this requires three steps: Beginning with grams of A, we must convert to moles of A, then to moles of B, and finally to particles of B. Examples 7.10 and 7.11 illustrate this type of problem.

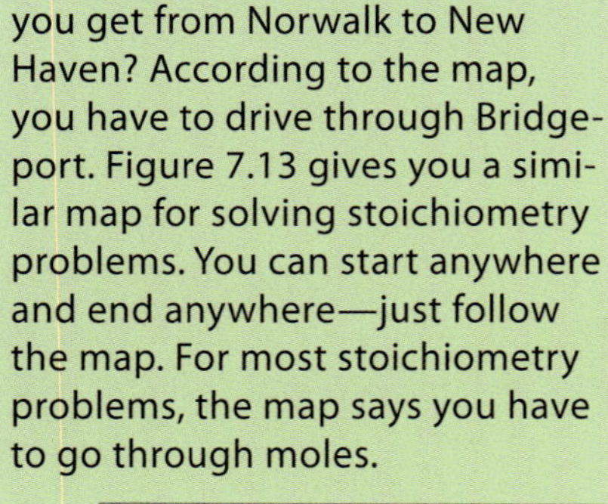
If you're driving along I-95 in southern Connecticut, how do you get from Norwalk to New Haven? According to the map, you have to drive through Bridgeport. Figure 7.13 gives you a similar map for solving stoichiometry problems. You can start anywhere and end anywhere—just follow the map. For most stoichiometry problems, the map says you have to go through moles.

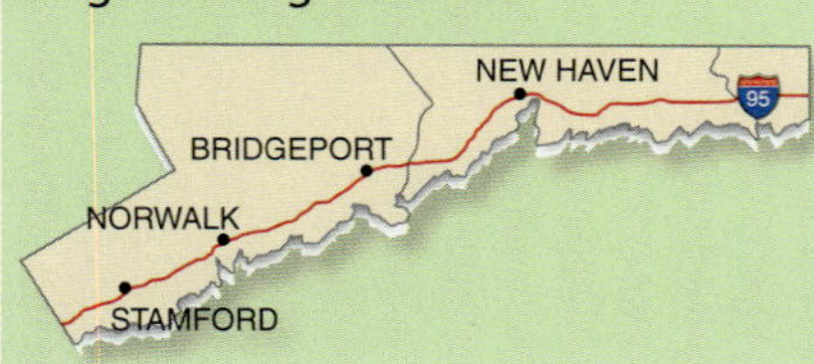

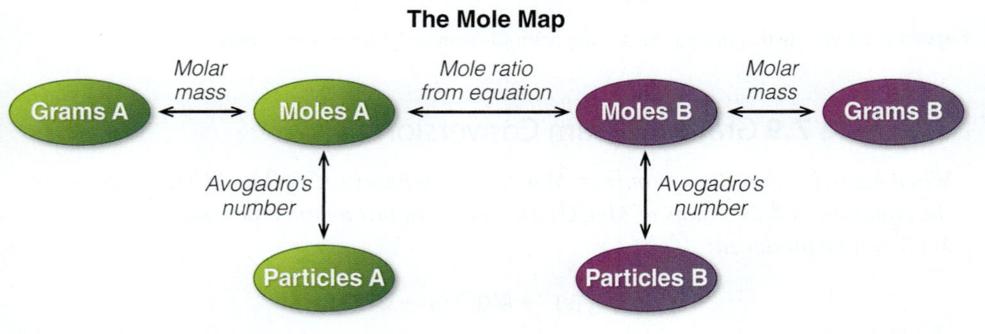

Figure 7.13 This "mole map" shows the strategy for solving any stoichiometry problem involving grams, moles, and atoms or molecules.

Example 7.10 Strategies for Stoichiometry Problems

Zinc metal reacts with aqueous copper(II) chloride, as shown in this equation. If 3.0×10^{21} atoms of zinc react, how many grams of $ZnCl_2$ will form? Show the sequence of conversions necessary, then calculate the numerical answer.

$$Zn\ (s) + CuCl_2\ (aq) \rightarrow ZnCl_2\ (aq) + Cu\ (s)$$

We can look at the map in Figure 7.12 to solve this problem. We know the number of atoms of Zn, and we are asked to find the mass of $ZnCl_2$. By replacing A in the map with Zn and B with $ZnCl_2$, we see that we can do the calculation by taking this route:

$$\text{atoms Zn} \Rightarrow \text{moles Zn} \Rightarrow \text{moles } ZnCl_2 \Rightarrow \text{grams } ZnCl_2$$

Now we set up the conversions to move from atoms of Zn to grams of $ZnCl_2$:

$$3.0 \times 10^{21}\ \cancel{\text{atoms Zn}} \times \frac{1\ \cancel{\text{mole Zn}}}{6.02 \times 10^{23}\ \cancel{\text{atoms Zn}}} \times \frac{1\ \cancel{\text{mole } ZnCl_2}}{1\ \cancel{\text{mole Zn}}} \times \frac{136.28\ \text{g } ZnCl_2}{1\ \cancel{\text{mole } ZnCl_2}} = 0.68\ \text{g } ZnCl_2$$

After rounding to the correct number of significant digits, we find that this reaction can produce 0.68 g of $ZnCl_2$.

Example 7.11 Strategies for Stoichiometry Problems

Aluminum metal reacts with nitric acid (HNO_3) as in the reaction below. If this reaction consumes 5.20 moles of aluminum, how many grams of H_2 will form?

$$2\ Al\ (s) + 6\ HNO_3\ (aq) \rightarrow 2\ Al(NO_3)_3\ (aq) + 3\ H_2\ (g)$$

This example gives us the number of moles of Al and asks for the number of grams of H_2. Following the map in Figure 7.12, we can write a strategy for this problem:

$$\text{moles Al} \Rightarrow \text{moles } H_2 \Rightarrow \text{grams } H_2$$

Now that the route is mapped out, we can do the conversions.

$$5.20\ \cancel{\text{moles Al}} \times \frac{3\ \cancel{\text{moles } H_2}}{2\ \cancel{\text{moles Al}}} \times \frac{2.02\ \text{g } H_2}{1\ \cancel{\text{mole } H_2}} = 15.8\ \text{g } H_2$$

TRY IT

12. Iron reacts with oxygen gas as shown in this balanced equation. How many moles of oxygen are needed to react with 82.1 grams of iron?

$$4\ Fe\ (s) + 3\ O_2\ (g) \rightarrow 2\ Fe_2O_3\ (s)$$

13. When heated, calcium carbonate decomposes to form calcium oxide and carbon dioxide, as shown in this equation. How many moles of $CaCO_3$ are required to produce 71.5 grams of CaO?

$$CaCO_3\ (s) \rightarrow CaO\ (s) + CO_2\ (g)$$

Practice

Stoichiometry Conversions

The way to get good at stoichiometry conversions is to practice. This interactive exercise will help you improve your skills.

Check It

Watch Explanation

Calculations with Limiting Reagents

Before going any further, let's stop and make a sandwich. We can write a "recipe" for a sandwich that looks something like this:

$$2 \text{ slices of bread} + 1 \text{ slice turkey} + 1 \text{ slice cheese} \rightarrow 1 \text{ sandwich}$$

Now imagine that you have 18 slices of turkey, 15 slices of cheese, and 80 slices of bread. Using this recipe, how many turkey-and-cheese sandwiches could you make?

If you answered 15 sandwiches, you are correct. You have plenty of bread—enough to make 40 sandwiches. You have enough turkey to make 18 sandwiches. But you only have enough cheese to make 15 sandwiches. Because the cheese runs out first, it *limits* the number of sandwiches that can be made.

This situation is common in chemistry as well as in cooking. For example, consider the combustion of natural gas, as shown in this reaction:

$$CH_4\ (g) + 2\ O_2\ (g) \rightarrow CO_2\ (g) + 2\ H_2O\ (g)$$

What happens if there is more O_2 than CH_4 available to react? The reaction will take place until all of the CH_4 is consumed. However, there will still be O_2 left over (**Figure 7.14**). Because an excess of oxygen is available, O_2 is called the **excess reagent**. On the other hand, the amount of CH_4 present limits the amount of CO_2 and water that can be produced in this reaction. The compound that is consumed first is called the **limiting reagent**. The limiting reagent determines the amount of product that can form in a reaction.

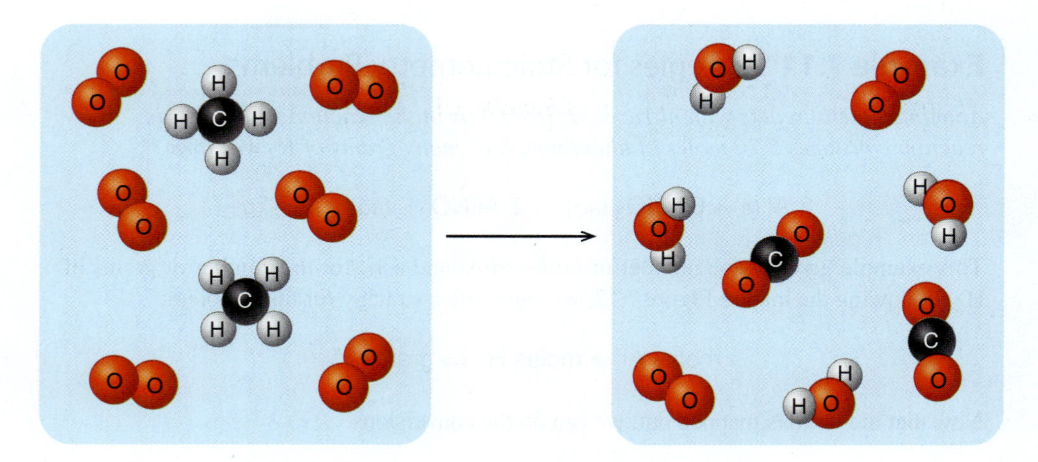

Figure 7.14 The limiting reagent (CH_4) is completely consumed, but the excess reagent (O_2) is not.

The limiting reagent is the one that can form the least amount of product.

Often, just like in the sandwich example, we are given amounts of different starting materials. To find the limiting reagent, we calculate the amount of product that each starting material can produce. The starting material that can produce the least amount of product is the limiting reagent. Examples 7.12 and 7.13 illustrate this important idea.

Example 7.12 Finding the Limiting Reagent

Potassium reacts violently with chlorine gas to produce potassium chloride, as in the equation below. If 1.2 moles of potassium are combined with 15 moles of chlorine gas, how many moles of potassium chloride can form? Which reagent is the limiting reagent?

$$2\ K\ (s) + Cl_2\ (g) \rightarrow 2\ KCl\ (s)$$

To solve this problem, we determine how much potassium chloride can form from the available amount of potassium and the available amount of chlorine:

$$1.2\ \cancel{\text{mol K}} \times \frac{2\ \text{mol KCl}}{2\ \cancel{\text{mol K}}} = 1.2\ \text{mol KCl}$$

$$15\ \cancel{\text{mol Cl}_2} \times \frac{2\ \text{mol KCl}}{1\ \cancel{\text{mol Cl}_2}} = 30\ \text{mol KCl}$$

Since less KCl can be produced from the potassium, it is the limiting reagent.

The potassium limits the amount of KCl that can be produced—it is the limiting reagent. Even though there is enough chlorine to produce a large amount of KCl, there is only enough potassium to produce 1.2 moles.

Example 7.13 Finding the Limiting Reagent

Uranium reacts with fluorine gas according to the equation shown. If 30 moles of uranium combine with 75 moles of F_2, how many moles of UF_6 will form?

$$U + 3\,F_2 \rightarrow UF_6$$

To find the limiting reagent, we calculate the amount of UF_6 that could be produced from the uranium and the amount that could be produced from the fluorine:

$$30\ \cancel{\text{moles U}} \times \frac{1\ \text{mole UF}_6}{1\ \cancel{\text{mole U}}} = 30\ \text{moles UF}_6$$

$$75\ \cancel{\text{moles F}_2} \times \frac{1\ \text{mole UF}_6}{3\ \cancel{\text{moles F}_2}} = 25\ \text{moles UF}_6$$

Since less UF_6 can be produced from the fluorine gas, F_2 is the limiting reagent.

This is the amount of UF_6 that can actually form in this reaction.

Notice that even though fewer moles of uranium are present, fluorine is the limiting reagent. Because the reaction consumes three fluorine molecules for each uranium atom, the fluorine is consumed before the uranium.

TRY IT

14. Phosphorus reacts with chlorine gas according to this equation:

$$2\,P\,(s) + 3\,Cl_2\,(g) \rightarrow 2\,PCl_3\,(l)$$

If 25.0 moles of phosphorus are combined with 50.0 moles of chlorine gas, what is the limiting reagent? How many moles of PCl_3 can be produced?

Check It
Watch Explanation

Finding the Leftovers

In the previous section we used a sandwich recipe as an analogy for a chemical reaction:

2 slices of bread + 1 slice turkey + 1 slice cheese → 1 sandwich

Based on this equation, we said that if we have 18 slices of turkey, 15 slices of cheese, and 80 slices of bread, we can make 15 sandwiches. The cheese is the limiting reagent. But here's another question: If we make all of the sandwiches, how many slices of turkey and of bread will be left over? I encourage you to take a moment and try to figure this out before moving forward.

Practice
Limiting Reagents
This interactive exercise will help you improve your skills.

I love sandwiches.

© balbaz/depositphotos.com

To solve this problem, we first find the amount of each reagent (ingredient) that we used. Following the ratios in the balanced equation, we could summarize our sandwich making as follows:

Used: 30 slices of bread, 15 slices of turkey, 15 slices of cheese
Produced: 15 sandwiches

To find the leftovers, we subtract what we used from our initial amounts:

Bread: 80 slices – 30 slices = 50 slices left over
Turkey: 18 slices – 15 slices = 3 slices left over
Cheese: 15 slices – 15 slices = 0 slices left over (limiting reagent)

Chemical reactions work the same way: If we subtract the amount of a starting material used in the reaction from the amount available, we can determine the amounts of starting material that are left over. Limiting reagents are completely consumed in chemical reactions, but some of the excess reagent will always be left over. Example 7.14 illustrates this idea.

Example 7.14 Finding the Leftover Moles

In an acid-base neutralization reaction, a chemist combines 15 moles of HCl with 20 moles of NaOH. How many moles of H_2O form in this reaction? How many moles of the excess reagent are left over?

To solve this problem, we first write a balanced equation for the reaction:

$$HCl\ (aq) + NaOH\ (aq) \rightarrow NaCl\ (aq) + H_2O\ (l)$$

We have enough NaOH to produce 20 moles of water, but we only have enough HCl to produce 15 moles of water. Therefore HCl is the limiting reagent, and the reaction produces 15 moles of water.

How many moles of NaOH are left over? We started with 20 moles of NaOH and consumed 15 in the reaction. Therefore, 5 moles are left over.

Sometimes it helps to make a table showing the moles of each species present at the beginning of the reaction, the amounts consumed or produced, and then the amounts of each species at the end of the reaction.

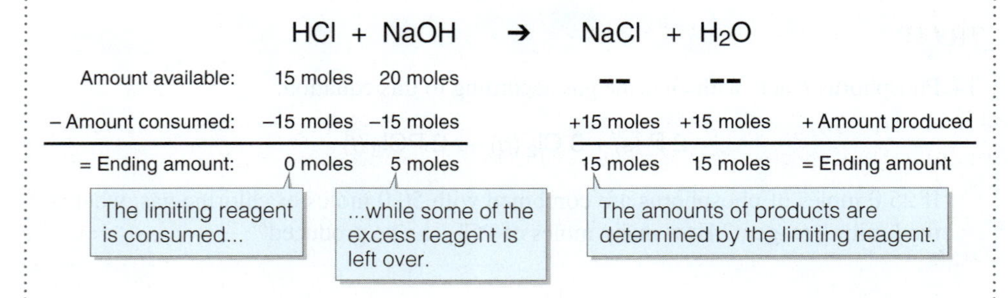

	HCl +	NaOH	→	NaCl +	H_2O	
Amount available:	15 moles	20 moles		--	--	
– Amount consumed:	–15 moles	–15 moles		+15 moles	+15 moles	+ Amount produced
= Ending amount:	0 moles	5 moles		15 moles	15 moles	= Ending amount

The limiting reagent is consumed...
...while some of the excess reagent is left over.
The amounts of products are determined by the limiting reagent.

Example 7.15 Finding the Leftover Mass

Calcium metal reacts with water as shown in this reaction. If 3.04 grams of calcium react with 10.0 grams of water, what is the limiting reagent? What mass of the excess reagent will be left over?

$$Ca\ (s) + 2\ H_2O\ (l) \rightarrow Ca(OH)_2\ (s) + H_2\ (g)$$

To identify the limiting reagent, we first need to identify the reagent that will produce the fewest moles of product.

Calcium is the limiting reagent...

$$3.04\ \text{g Ca} \times \frac{1\ \text{mole Ca}}{40.08\ \text{g Ca}} \times \frac{1\ \text{mole Ca(OH)}_2}{1\ \text{mole Ca}} = 0.0758\ \text{mol Ca(OH)}_2$$

This is the amount of $Ca(OH)_2$ that can form in this reaction.

...Water is the excess reagent.

$$10.0\ \text{g H}_2\text{O} \times \frac{1\ \text{mole H}_2\text{O}}{18.02\ \text{g H}_2\text{O}} \times \frac{1\ \text{mole Ca(OH)}_2}{2\ \text{moles H}_2\text{O}} = 0.277\ \text{mol Ca(OH)}_2$$

Since Ca produces fewer moles of $Ca(OH)_2$, it is the limiting reagent. This means that H_2O is the excess reagent. To find the mass of H_2O that is consumed in this reaction, we can do another stoichiometry calculation:

$$3.04\ \text{g Ca} \times \frac{1\ \text{mol Ca}}{40.08\ \text{g Ca}} \times \frac{2\ \text{mol H}_2\text{O}}{1\ \text{mol Ca}} \times \frac{18.02\ \text{g H}_2\text{O}}{1\ \text{mol H}_2\text{O}} = 2.73\ \text{g H}_2\text{O}$$

Finally, we find the amount of water left over by subtracting the amount of water that was consumed from the starting amount. After rounding to significant figures, we reach a final answer of 7.3 g H_2O left over.

$$10.0\ \text{g H}_2\text{O} - 2.73\ \text{g H}_2\text{O} = 7.3\ \text{g H}_2\text{O}$$

TRY IT

15. A reinforced steel cylinder is filled with 20.0 moles of hydrogen and 12.0 moles of oxygen. The mixture is detonated inside the cylinder, causing this reaction to take place:

$$2\ H_2\ (g) + O_2\ (g) \rightarrow 2\ H_2O\ (g)$$

What is the limiting reagent for this reaction? Calculate the number of moles and grams of hydrogen, oxygen, and water present in the cylinder after the reaction takes place.

Check It

Watch Explanation

7.4 Theoretical and Percent Yield

Many chemists, like the process chemist described at the beginning of this chapter, conduct chemical reactions to create valuable new materials. When carrying out a reaction of this type, we often calculate the **theoretical yield** of the products—that is, the amount of product that can form, based on the amount of starting materials and the balanced equation.

Stoichiometry problems predict the theoretical yield of a reaction.

In practice, however, chemists often obtain less than the theoretical yield. The amount of a product that a chemist actually isolates from an experiment is called the **actual yield**. We report the efficiency of a reaction as the **percent yield**, which we calculate as follows:

$$\text{Percent yield} = \frac{\text{actual yield}}{\text{theoretical yield}} \times 100\%$$

Percent yields are often much lower than 100%. This happens for several reasons: Sometimes part of the material adheres to the container walls. (If you've ever made a cake, you know that it's impossible to get all the batter into the pan—the same holds true in chemical reactions.) Sometimes a competing reaction forms unwanted side

Due to the scale and cost of chemical manufacturing, industrial chemists work hard to optimize the percent yields of reactions. A change in yields of even 1% can drastically reduce costs and increase profits.

products in a reaction. Other times, product is lost during the purification process. The percent yield is a measure of how efficiently a reaction takes place.

For example, suppose a team of chemists run a reaction in which the theoretical yield is 240 grams. However, after the completion of the reaction, they are only able to isolate 180 grams. In this case, the percent yield for the reaction is 75%:

$$\text{Percent yield} = \frac{\text{actual yield}}{\text{theoretical yield}} \times 100\% = \frac{180\ \cancel{g}}{240\ \cancel{g}} \times 100\% = 75\%$$

Example 7.16 illustrates the relationships between theoretical, actual, and percent yields.

Example 7.16 Finding the Theoretical and Percent Yield

Sulfur hexafluoride, SF_6, is widely used in the power industry. It is produced through the following reaction:

$$S\ (s) + 3\ F_2\ (g) \rightarrow SF_6\ (g)$$

A manufacturer reacts 120.0 kilograms of sulfur with excess fluorine gas. What mass of SF_6 is theoretically possible for this conversion? After the reaction is complete, the manufacturer isolates 480.2 kg of SF_6. What was the percent yield for this process?

The theoretical yield is the amount of product possible, based on the balanced equation and the amount of starting material used. This is a stoichiometry problem. As in the earlier examples, we must go through moles to relate the mass of the starting materials and products:

$$120{,}000\ \cancel{g\ S} \times \frac{1\ \cancel{\text{mole S}}}{32.06\ \cancel{g\ S}} \times \frac{1\ \cancel{\text{mole } SF_6}}{1\ \cancel{\text{mole S}}} \times \frac{146.06\ g\ SF_6}{1\ \cancel{\text{mole } SF_6}}$$

$$= 546{,}700\ g \text{ or } 546.7\ kg\ SF_6$$

The theoretical yield for this reaction is 546.7 kg. The actual yield is 480.2 kg. From this, we calculate the percent yield.

$$\text{Percent yield of } SF_6 = \frac{\text{actual yield}}{\text{theoretical yield}} \times 100\%$$

$$= \frac{480.2\ \cancel{kg}}{546.7\ \cancel{kg}} \times 100\% = 87.84\%$$

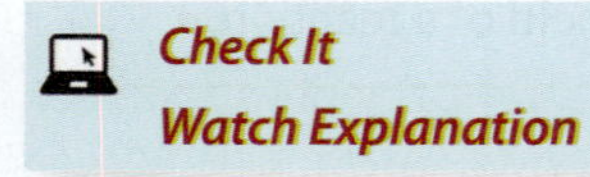

TRY IT

16. A chemist carried out the following reaction, beginning with 5.25 grams of $MgCO_3$. After the reaction was complete, she obtained 2.37 grams of MgO.

$$MgCO_3\ (s) \rightarrow MgO\ (s) + CO_2\ (g)$$

a. What was the theoretical yield of magnesium oxide (MgO) in this reaction?

b. What was the percent yield?

Capstone Question

As part of a team of chemists at a food flavoring company, you need to produce *n*-butyl acetate ($C_6H_{12}O_2$), a sweet-smelling food additive that occurs naturally in apples, pears, and bananas (**Figure 7.15**). Your team plans to prepare this compound from the reaction of acetic acid ($HC_2H_3O_2$) and *n*-butyl alcohol ($C_4H_{10}O$):

$$HC_2H_3O_2\,(l) + C_4H_{10}O\,(l) \rightarrow C_6H_{12}O_2\,(l) + H_2O\,(l)$$

a. Using the data in the table, calculate both the mass and volume of *n*-butyl alcohol required to completely react with 200.0 L of acetic acid.

b. Chemists often use a large excess of *n*-butyl alcohol for this reaction. If you begin with 850.0 L of this reagent, what volume will be left over after the reaction is complete?

c. What is the theoretical yield of *n*-butyl acetate in this reaction? Report your answer in grams, kilograms, and liters.

d. After completing this process and purifying the products, your team isolates 434.9 L of *n*-butyl acetate. Calculate your percent yield for this reaction.

Name	Formula	Boiling Point	Molar Mass	Density
Acetic acid	$HC_2H_3O_2$	119 °C	60.06 g/mol	1.049 g/mL
n-Butyl alcohol	$C_4H_{10}O$	118 °C	74.14 g/mol	0.810 g/mL
n-Butyl acetate	$C_6H_{12}O_2$	126 °C	116.18 g/mol	0.883 g/mL
Water	H_2O	100 °C	18.02 g/mol	1.00 g/mL

Capstone Video

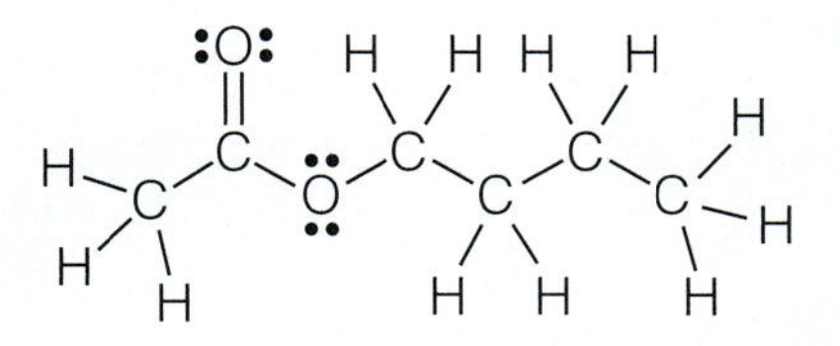

Peter Hatter/Age Fotostock/Media Bakery

Figure 7.15 *n*-Butyl acetate ($C_6H_{12}O_2$) is a fragrant substance that occurs naturally in many types of fruit. It is commonly used in artificial fruit flavorings.

SUMMARY

Chemistry often deals with the question, "How much?" When conducting a chemical reaction, chemists need to know how much starting material to use and how much product to expect. In this chapter we've explored how the atomic masses relate to the masses we can measure in the laboratory or the factory.

We describe compounds by their formula mass. The formula mass is the sum of the masses of the atoms in a molecule or formula unit. Scientists often determine formula mass experimentally by using mass spectrometry.

We can also describe compounds by their percent composition. Scientists determine percent composition using a technique called elemental analysis.

The mole concept allows us to connect the masses of atoms and molecules to masses that we can measure on the laboratory scale. Avogadro's number, 6.02×10^{23}, is the number of particles in a mole. The mass of one mole of a substance in grams has the same numerical value as the mass of one atom or molecule of that substance in atomic mass units. That is, the mass of any chemical substance can be described in atomic mass units or in grams per mole. The measurement of mass in grams per mole is called the molar mass.

The mole concept allows us to relate large amounts using chemical equations. By converting to moles, it is possible to relate the amount of any material involved in a chemical reaction to the amounts of all other materials consumed or produced in the reaction.

Frequently, we are not able to mix chemicals in a perfect ratio; one reagent will be in excess. The limiting reagent determines the amount of product that can form in a reaction.

Chemists commonly describe reactions in terms of their theoretical and percent yields. The theoretical yield is the amount of material that would be produced if the reaction went perfectly. However, in practice, material can be lost because of competing reactions, incomplete transfers between containers, or other issues. The percent yield is an indicator of how closely the actual yield aligns with the theoretical yield. When working in a laboratory or industrial setting, the percent yield is very important.

Process Development, Continued

At the beginning of this chapter, we introduced Jared Fennell, a process chemist at the pharmaceutical company Eli Lilly, who optimizes chemical reactions for large-scale manufacturing. Process chemistry is much different from the earlier stages of pharmaceutical research. In the early phases of a drug development project, scientists try to identify compounds with useful medicinal properties. Chemists prepare and test a large number of compounds to understand how different structures behave in living systems. They don't worry much about percent yield; they need only a small amount of each compound, and good yields are often less important than getting the compounds quickly. However, as compounds move from laboratory testing to animal testing and ultimately to human trials, more and more compound is needed, and the purity of the compound becomes paramount. This is where process chemists come in.

The Eli Lilly team recently published results that offer an insight into the challenges that process chemists face. Their project objective was to prepare a promising new medicine. One of the final steps involved a reaction that combined two molecules by forming a new bond to a nitrogen atom (**Figure 7.16**). Chemists in the early and intermediate stages had produced 25 grams of the compound; the process team had to produce 100,000 grams.

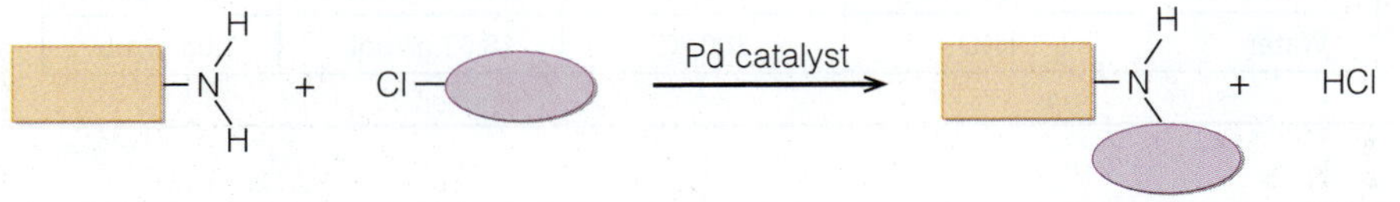

Figure 7.16 A simplified representation of the bond-forming reaction. Shaded areas represent the rest of the molecules.

To make this reaction work, small-scale chemists used a palladium compound as a catalyst (a catalyst helps a reaction take place more quickly but is not used up in the reaction). Palladium is very expensive and must be separated from the products after the reaction is complete. In addition, the catalyst slowly reacted with oxygen. This caused two problems: First, it produced potentially harmful side products. Second, trace amounts of oxygen in the mixture could cause the catalyst to fail, resulting in very low yields and costing the company hundreds of thousands of dollars.

The process team set to work on the problem. They carefully tested the time, temperature, amounts of starting materials, amounts of catalyst, and mixing techniques used in the reaction. They identified the side products. They even developed a unique method to monitor the oxygen levels in the reaction, ensuring that they stayed below the harmful threshold.

Finally, after months of testing and planning, the time came for a large-scale run. This pivotal experiment required 150 kilograms of starting material, valued at nearly $1 million. Everything had to be right. The entire team assembled. Every step of the process had been rehearsed. The reaction started. Over the next 48 hours, the team monitored each step, watching for unexpected changes. At the end of the arduous process, they were rewarded with success. They isolated the pure product at 87% yield—a very good yield for a challenging reaction.

Key Terms

7.1 Formula Mass and Percent Composition

formula mass The mass of a molecule or formula unit.

percent composition The percentage (by mass) of each element in a compound.

mass spectrometry A technique used to measure the mass of a molecule.

elemental analysis A technique used to determine the percent composition of a substance.

7.2 Connecting Atomic Mass to Large-Scale Mass: The Mole Concept

mole A quantity consisting of 6.02×10^{23} units.

Avogadro's number The number of particles in a mole; 6.02×10^{23}.

molar mass The formula mass of an element or compound, expressed in grams per mole.

7.3 The Mole Concept in Balanced Equations

stoichiometry problem A problem that relates the amount of one reagent or product to another in a chemical reaction, using a balanced equation.

excess reagent In a chemical reaction, a reagent that is present in larger stoichiometric quantities than the other reagents; an excess reagent is not completely consumed.

limiting reagent In a chemical reaction, a reagent that is completely consumed and limits the amount of product that can form.

7.4 Theoretical and Percent Yield

theoretical yield The amount of product that can form in a chemical reaction, based on the balanced equation and the amount of starting materials present.

actual yield The amount that a chemist actually recovers from an experiment.

percent yield A measure of the efficiency of a reaction; the actual yield divided by the theoretical yield, expressed as a percentage.

Additional Problems

7.1 Formula Mass and Percent Composition

17. Calculate the formula mass for each of these compounds:

a. $BaSO_4$
b. K_2S
c. $C_6H_{12}O_6$
d. $FeCl_3$

18. Calculate the formula mass for each of these compounds:

a. $NaNO_2$
b. PCl_3
c. UF_6
d. $Mg_3(PO_4)_2$

19. What is the percent by mass of oxygen in each of these compounds?

a. glucose, $C_6H_{12}O_6$
b. ethanol, $C_2H_{12}O_6$
c. amoxicillin, $C_{16}H_{19}N_3O_5S$

20. What is the percent by mass of nitrogen in each of these compounds?

a. urea, CH_4N_2O
b. nitric acid, HNO_3
c. ammonium perchlorate, NH_4ClO_4

21. What is the percent by mass of carbon, hydrogen, and oxygen in aspirin, $C_9H_8O_4$?

22. What is the percent by mass of carbon, hydrogen, nitrogen, and oxygen in acetaminophen, $C_8H_9NO_2$?

23. What is the percent by mass of silver in silver nitrate? What mass of silver is present in a 100.0-gram sample of this compound?

24. What is the percent by mass of iron in iron(III) oxide? What mass of iron is present in a 314.5-gram sample of this compound?

25. What technique is commonly used to measure the formula mass of a compound?

26. What technique is commonly used to measure the percent composition of a compound?

27. The mass spectrum of an unknown explosive (shown below) indicates that the formula mass of the unknown substance is 227 u. Which of these molecular formulas could correspond to this mass?

a. nitroglycerin, $C_3H_5N_3O_9$
b. methyl nitrate, CH_3NO_3
c. ethylene glycol dinitrate, $C_2H_4N_2O_6$
d. trinitrotoluene, $C_7H_5N_3O_6$

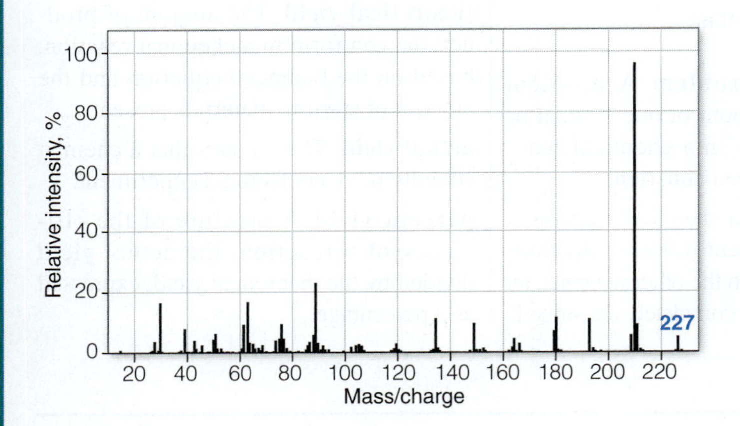

28. The mass spectrum of an illegal drug is shown below. The peak showing a mass of 369 u corresponds to which of the following compounds?

a. cocaine, $C_{17}H_{21}NO_4$
b. methamphetamine, $C_{10}H_{15}N$
c. tetrahydrocannabinol, $C_{21}H_{30}O_2$
d. heroin, $C_{21}H_{23}NO_5$

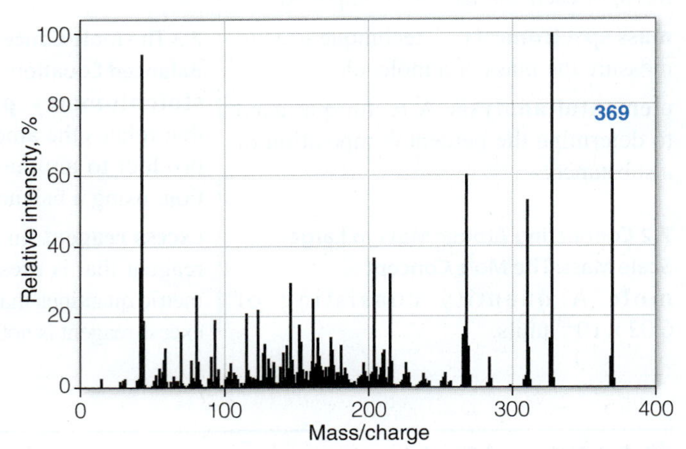

29. A forensics lab analyzes a small amount of a substance containing carbon, hydrogen, and nitrogen. The compound is suspected to be either heroin, a highly addictive narcotic with the formula $C_{21}H_{23}NO_5$, or morphine—a related compound with the formula $C_{17}H_{19}NO_3$. After conducting an elemental analysis, the lab finds the sample to contain 71.6% carbon and 6.7% hydrogen by mass. Which conclusion does this evidence support?

30. A forensics lab analyzes a small amount of a substance containing carbon, hydrogen, and nitrogen. The compound is suspected to be either methamphetamine, a highly addictive narcotic with the formula $C_{10}H_{15}N$, or methylenedioxypyrovalerone (MDPV)—a related compound with the formula $C_{16}H_{21}NO_3$. After conducting an elemental analysis, the lab finds the sample to contain 69.8% carbon and 7.7% hydrogen by mass. Which conclusion does this evidence support?

7.2 Connecting Atomic Mass to Large-Scale Mass: The Mole Concept

31. Calculate the number of grams in one mole of each of the following:

a. propane, C_3H_8
b. calcium chloride, $CaCl_2$
c. ethylene glycol, C_2H_6O
d. sucrose, $C_{12}H_{22}O_{11}$

32. Calculate the number of grams in one mole of each of the following:

a. butane, C_4H_{10}
b. magnesium bromide, $MgBr_2$
c. urea, CH_4N_2O
d. potassium bicarbonate, $KHCO_3$

33. Chemists express the mass of elements and compounds either in atomic mass units (u) or in grams/mole. Which approach is appropriate in each of these situations?

a. You need to measure out 14.2 moles of a compound.
b. You need to report the mass of a single molecule, as measured on a mass spectrometer.

34. Chemists express the mass of elements and compounds either in atomic mass units (u) or in grams/mole. Which approach is appropriate in each of these situations?

a. You need to calculate the amount of reactant to add to a 200-liter industrial reaction vessel.
b. You need to calculate the mass of a single protein molecule.

35. Calculate the number of grams in each of the following:

a. 15.2 moles of KCl
b. 0.319 moles of $MgSO_4$
c. 41.3 moles of neon

36. Calculate the number of grams in each of the following:

a. 1.10 moles of iron(II) nitrate
b. 0.201 moles of sodium phosphate
c. 1.132×10^{-3} moles of $PtCl_4$

37. Calculate the number of grams in each of the following:

a. 1.50 moles of gold
b. 24.3 moles of BF_3
c. 0.131 moles of helium

38. Calculate the number of grams in each of the following:

a. 1.29 moles of cobalt(II) chloride
b. 11.1 moles of potassium carbonate
c. 4.922×10^{-2} moles of $CuCl_2$

39. Calculate the number of moles in each of the following:

a. 182.5 grams of Mg
b. 29.3 grams of Cl_2
c. 304.1 grams of Pb

40. Calculate the number of moles in each of the following:

a. 32.3 grams of NaCl
b. 11.6 grams of CO_2
c. 508.3 grams of C_2H_2

41. Calculate the number of moles in each of the following:

a. 12.6 g of Fe
b. 22.4 g of LiF
c. 1.9 mg of Br_2

42. Calculate the number of moles in each of the following:

a. 123.5 g of KClO
b. 17.3 g of H_2
c. 11.6 mg of $Ca(ClO_4)_2$

43. Convert each of the following to moles and to atoms:

a. 23.4 grams of sodium metal
b. 18.2 grams of phosphorus
c. 192.3 grams of carbon

44. Convert each of the following to moles and to atoms:

a. 18.7 grams of calcium metal
b. 51.9 grams of tellurium, Te
c. 178.4 grams of uranium, U

45. Convert each of the following to moles and to molecules:

a. 11.3 grams of H_2O
b. 77.3 grams of PCl_3
c. 129.4 kilograms of CO_2

46. Convert each of the following to moles and to molecules:

a. 55.2 grams of O_2
b. 68.2 grams of CH_4
c. 31.2 kilograms of SO_2

47. Solve each of these problems:

a. How many molecules of HCl are present in 2.0 moles of HCl?
b. How many gold atoms are present in 0.12 moles of gold?
c. How many atoms are present in 140.2 grams of zinc metal?

48. Solve each of these problems:

a. How many molecules of HBr are present in 3.0 moles of HBr?
b. How many atoms are present in 0.032 moles of silver?
c. How many atoms are present in 109.4 g of helium?

49. How many moles are in a 15.3-gram sample of elemental copper? How many atoms of copper are present in this sample?

50. How many moles are in a 201.5-gram sample of chlorine gas, Cl_2? How many molecules of Cl_2 are present in this sample?

51. A sample contains 205.2 moles of water. How many molecules of water are present in this sample? What is the mass of this sample in grams?

52. A room contains 19.4 moles of oxygen gas. How many O_2 molecules are present in this room? What is the mass of this amount of oxygen in grams?

53. What is the mass in grams of 3.192×10^{22} atoms of zinc?

54. What is the mass in grams of 1.191×10^{24} atoms of vanadium?

55. What is the mass in grams of 1.372×10^{23} molecules of hydrogen gas?

56. What is the mass in grams of 1.314×10^{24} molecules of sulfur dioxide?

57. Sucrose (table sugar) has a molar mass of 342.30 g/mole. How many sucrose molecules are in a 5-pound bag of sugar? (1 pound = 453.6 grams.)

58. Methanol (CH_4O) has a molar mass of 32.04 g/mole. It has a density of 0.791 g/mL. How many molecules are in a 500-mL bottle of methanol?

7.3 The Mole Concept in Balanced Equations

59. Lead and oxygen gas combine to form lead(IV) oxide, as shown here. If 20 moles of Pb react with oxygen in this way, how many moles of O_2 are consumed? How many moles of PbO_2 are produced?

$$Pb\ (s) + O_2\ (g) \rightarrow PbO_2\ (s)$$

60. Lithium and bromine combine to form lithium bromide, as shown here. If 30 moles of Li react with bromine in this way, how many moles of Br_2 are consumed? How many moles of LiBr are produced?

$$2\ Li\ (s) + Br_2\ (l) \rightarrow 2\ LiBr\ (s)$$

61. Zinc metal reacts with hydrochloric acid, as shown in this balanced equation. If 3.0 moles of Zn react in this way, how many moles of HCl are consumed? How many moles of $ZnCl_2$ and H_2 are produced?

$$Zn\ (s) + 2\ HCl\ (aq) \rightarrow ZnCl_2\ (aq) + H_2\ (g)$$

62. Tin metal reacts with silver nitrate solution, as shown in the reaction here. If 5.0 moles of tin react in this way, how many moles of $AgNO_3$ are consumed? How many moles of $Sn(NO_3)_2$ and Ag are formed?

$$Sn\ (s) + 2\ AgNO_3\ (aq) \rightarrow Sn(NO_3)_2\ (aq) + 2\ Ag\ (s)$$

63. Potassium metal reacts with water according to this balanced equation:

$$2\ K\ (s) + 2\ H_2O\ (l) \rightarrow 2\ KOH\ (aq) + H_2\ (g)$$

a. If one mole of potassium reacts in this manner, how many moles of water are consumed?
b. If one mole of potassium reacts in this manner, how many moles of H_2 are produced?
c. How many moles of potassium are required to produce 14.0 moles of H_2?
d. How many moles of KOH are produced if 3,014.2 moles of H_2O are consumed?

64. Elemental boron reacts with fluorine gas according to this balanced equation:

$$2\ B\ (s) + 3\ F_2\ (g) \rightarrow 2\ BF_3\ (g)$$

a. How many moles of BF_3 are produced if 10 moles of boron are consumed?
b. How many moles of F_2 are required to react with 10 moles of boron?
c. How many moles of F_2 are required to produce 60 moles of BF_3?
d. How many moles of B are required to react with 18 moles of F_2?

65. Zinc reacts with hydrochloric acid to form zinc chloride and hydrogen gas, as shown in this equation:

$$Zn\ (s) + 2\ HCl\ (aq) \rightarrow ZnCl_2\ (aq) + H_2\ (g)$$

If 14.2 g of zinc reacts in this way,
a. How many moles of zinc are consumed?
b. How many moles of HCl are consumed?
c. How many moles of $ZnCl_2$ can be produced?
d. How many grams of $ZnCl_2$ can be produced?

66. In aqueous solution, sodium iodide reacts with lead(II) nitrate to give solid lead(II) iodide and aqueous sodium nitrate, as shown in this equation. In one experiment, it is found that 8.34 grams of PbI_2 were produced.

$$2\ NaI\ (aq) + Pb(NO_3)_2\ (aq) \rightarrow PbI_2\ (s) + 2\ NaNO_3\ (aq)$$

Based on this reaction,
a. How many moles of PbI_2 were produced?
b. How many moles of $NaNO_3$ were produced?
c. How many moles of NaI were consumed?
d. How many grams of NaI were consumed?

67. Copper(I) bromide reacts with magnesium metal according to this equation:

$$2\ CuBr\ (aq) + Mg\ (s) \rightarrow 2\ Cu\ (s) + MgBr_2\ (aq)$$

If 0.253 moles of magnesium react in this way,
a. How many grams of CuBr will be consumed?
b. How many grams of Cu will be produced?
c. How many grams of $MgBr_2$ will be produced?

68. Silver nitrate reacts with calcium chloride according to this equation:

$$2\ AgNO_3\ (aq) + CaCl_2\ (aq) \rightarrow Ca(NO_3)_2\ (aq) + 2\ AgCl\ (s)$$

If 0.403 moles of AgCl are produced in this reaction,
a. How many grams of $Ca(NO_3)_2$ are produced?
b. How many grams of $AgNO_3$ are consumed?
c. How many grams of $CaCl_2$ are consumed?

69. The reaction of iron metal with copper(II) bromide is shown in this balanced equation. How many grams of iron(II) bromide result from the reaction of 23.6 grams of iron metal?

$$CuBr_2\ (aq) + Fe\ (s) \rightarrow Cu\ (s) + FeBr_2\ (aq)$$

70. The reaction of sulfuric acid with potassium hydroxide is shown in this balanced equation. If 19.07 grams of KOH react with excess sulfuric acid, how many grams of water are produced?

$$H_2SO_4\ (aq) + 2\ KOH\ (aq) \rightarrow 2\ H_2O\ (aq) + K_2SO_4\ (aq)$$

71. Nitric acid reacts with ammonia to produce ammonium nitrate, as shown here. In this reaction, how many grams of NH_3 are required to produce 1,000 grams of NH_4NO_3?

$$HNO_3\ (aq) + NH_3\ (aq) \rightarrow NH_4NO_3\ (aq)$$

72. Water reacts with sulfur trioxide to produce sulfuric acid, as shown here. In this reaction, how many grams of SO_3 are required to produce 500 grams of H_2SO_4?

$$H_2O\ (l) + SO_3\ (g) \rightarrow H_2SO_4\ (aq)$$

73. Calcium reacts with oxygen gas, as shown in this balanced equation. How many grams of calcium oxide can be produced from 15.0 grams of oxygen gas?

$$2\ Ca\ (s) + O_2\ (g) \rightarrow 2\ CaO\ (s)$$

74. At high temperatures, nitrogen gas and hydrogen gas react to form ammonia, as shown in this balanced equation. How many grams of hydrogen gas are required to react with 200.0 grams of nitrogen gas?

$$N_2\ (g) + 3\ H_2\ (g) \rightarrow 2\ NH_3\ (g)$$

75. When heated, calcium carbonate decomposes to form calcium oxide and carbon dioxide, as shown in this equation. If 150.0 grams of $CaCO_3$ react in this way, how many grams of CaO and CO_2 are produced?

$$CaCO_3\ (s) \rightarrow CaO\ (s) + CO_2\ (g)$$

76. Hydrogen peroxide, H_2O_2, decomposes to form water and oxygen gas, as shown in this reaction. If 10.0 grams of H_2O_2 react in this way, how many grams of H_2O and O_2 are produced?

$$2\ H_2O_2\ (aq) \rightarrow 2\ H_2O\ (l) + O_2\ (g)$$

77. Potassium hydroxide and iron(II) chloride react as shown here. If 3.1 moles of KOH are consumed in this reaction, how many grams of $Fe(OH)_2$ are produced?

$$2\ KOH\ (aq) + FeCl_2\ (aq) \rightarrow 2\ KCl\ (aq) + Fe(OH)_2\ (s)$$

78. Hydrofluoric acid neutralizes magnesium hydroxide as shown here. In this reaction, how many moles of $Mg(OH)_2$ are needed to react with 0.4 grams of HF?

$$2\ HF\ (aq) + Mg(OH)_2\ (aq) \rightarrow MgF_2\ (aq) + 2\ H_2O\ (l)$$

79. Metallic tin reacts with chlorine gas as shown here:

$$Sn\ (s) + 2\ Cl_2\ (g) \rightarrow SnCl_4\ (s)$$

If 0.253 moles of tin react in this way,

a. How many grams of Cl_2 are consumed?
b. How many molecules of Cl_2 are consumed?
c. How many grams of $SnCl_4$ are produced?

80. This balanced equation shows the combustion of propane gas (C_3H_8):

$$C_3H_8\ (g) + 5\ O_2\ (g) \rightarrow 3\ CO_2\ (g) + 4\ H_2O\ (g)$$

If 100.0 g of C_3H_8 reacts in this way,

a. How many moles of O_2 are consumed?
b. How many grams of CO_2 are produced?
c. How many molecules of H_2O are produced?

81. The reaction of sulfur dioxide with oxygen gas is represented in this balanced equation and the molecular depiction here. In this depiction, which reagent is the excess reagent? Which is the limiting reagent?

$$2\ SO_2 + O_2 \rightarrow 2\ SO_3$$

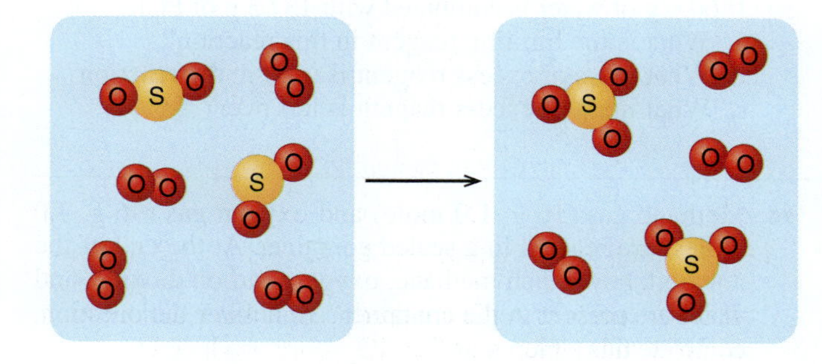

82. The reaction of boron with chlorine gas is represented in this balanced equation and the molecular depiction here. In this depiction, which reagent is the excess reagent? Which is the limiting reagent?

$$2\ B + 3\ Cl_2 \rightarrow 2\ BCl_3$$

83. Consider the following generic reaction:

$$3\ A + B \rightarrow C + D$$

If this reaction is carried out using 8.0 moles of A and 3.0 moles of B,

a. What is the limiting reagent in the reaction?
b. How many moles of C are formed in this reaction?

84. Consider the following generic reaction:

$$A + 2\ B \rightarrow C + D$$

If this reaction is carried out using 3.0 moles of A and 5.0 moles of B,

a. What is the limiting reagent in the reaction?
b. How many moles of C are formed in this reaction?

85. Calcium oxide reacts with water to form calcium hydroxide, as shown in this equation:

$$CaO\ (s) + H_2O\ (l) \rightarrow Ca(OH)_2\ (s)$$

If a bag of calcium oxide is dropped into a lake, which reagent likely will be the limiting reagent?

86. Ethanol is a common component in lighter fluid, which is used to ignite charcoal in a barbecue grill. Here is the reaction for the combustion of ethanol:

$$2\ C_2H_6O\ (l) + 6\ O_2\ (g) \rightarrow 4\ CO_2\ (g) + 6\ H_2O\ (g)$$

If you apply the lighter fluid to charcoal in an open grill and ignite it, which reagent likely will be the limiting reagent? Why is this so?

87. Iron and sulfur react according to this equation:

$$Fe\ (s) + S\ (s) \rightarrow FeS\ (s)$$

If 0.10 moles of iron are combined with 0.13 moles of sulfur, what is the limiting reagent? How many moles of iron(II) sulfide can form in this reaction?

88. Copper and bromine react according to this equation:

$$Cu\ (s) + Br_2\ (l) \rightarrow CuBr_2\ (s)$$

If 0.39 moles of copper are combined with 0.25 moles of bromine, what is the limiting reagent? How many moles of copper(II) bromide can form in this reaction?

89. Potassium reacts aggressively with water, as shown in this equation. If 0.20 grams of potassium are added to 15.0 grams of water, what is the limiting reagent? What mass of KOH can be produced in this reaction?

$$2\ K\ (s) + 2\ H_2O\ (l) \rightarrow 2\ KOH\ (aq) + H_2\ (g)$$

90. Phosgene ($COCl_2$) reacts with ammonia (NH_3), as shown in this reaction. If 1.2 grams of $COCl_2$ are added to 10.5 grams of NH_3, what is the limiting reagent? What mass of CH_4N_2O can be produced in this reaction?

$$COCl_2\ (g) + 2\ NH_3\ (g) \rightarrow CH_4N_2O\ (g) + 2\ HCl\ (g)$$

91. Sodium metal reacts with elemental iodine to form sodium iodide, as shown here:

$$2\ Na\ (s) + I_2\ (s) \rightarrow 2\ NaI\ (s)$$

If 10.0 moles of sodium are combined with 6.0 moles of I_2,

a. What is the limiting reagent?

b. How many moles of the excess reagent are used in the reaction?

c. How many moles of the excess reagent are left over?

92. Aluminum metal reacts with elemental chlorine to form aluminum chloride, as shown here:

$$2\ Al\ (s) + 3\ Cl_2\ (g) \rightarrow 2\ AlCl_3\ (s)$$

If 12.0 moles of aluminum are combined with 15.0 moles of Cl_2,

a. What is the limiting reagent?

b. How many moles of the excess reagent are used in the reaction?

c. How many moles of the excess reagent are left over?

93. Thionyl chloride ($SOCl_2$) reacts violently with water, as shown here:

$$H_2O\ (l) + SOCl_2\ (g) \rightarrow SO_2\ (g) + 2\ HCl\ (aq)$$

If 20.0 g of water is combined with 50.0 g of ($SOCl_2$),

a. What is the limiting reagent in this reaction?

b. What mass of excess reagent is used in the reaction?

c. What mass of excess reagent is left over?

94. Phosphorus trichloride reacts with water, as shown here:

$$PCl_3\ (l) + 3\ H_2O\ (l) \rightarrow H_3PO_3\ (aq) + 3\ HCl\ (aq)$$

If 60.0 g of water is combined with 137.3 g of PCl_3,

a. What is the limiting reagent in this reaction?

b. What mass of excess reagent is used in the reaction?

c. What mass of excess reagent is left over?

95. Silicon (28 g, 1.0 mole) and oxygen gas (128 g, 4.0 moles) are reacted in a sealed container. At the end of the reaction, how much silicon, oxygen, and silicon dioxide are present in the container? To answer the question, complete this table:

	Si +	O_2	→	SiO_2
Starting Moles	1.0 mol	4.0 mol		0 mol
Change	−1.0 mol			
Ending Moles				
Ending Grams				

96. Methane gas (16 g, 1.0 mole) and oxygen gas (96 g, 3.0 moles) are reacted in a sealed container. At the end of the reaction, how much methane, oxygen, carbon dioxide, and water are present in the container? To answer the question, complete this table:

	CH_4 +	2 O_2	→	CO_2 +	2 H_2O
Starting Moles	1.0 mol	3.0 mol		0 mol	0 mol
Change	−1.0 mol				
Ending Moles					
Ending Grams					

97. This reaction shows the neutralization of phosphoric acid (H_3PO_4) with potassium hydroxide (KOH):

$$H_3PO_4\ (aq) + 3\ KOH\ (aq) \rightarrow K_3PO_4\ (aq) + 3\ H_2O\ (l)$$

If 96.3 grams of H_3PO_4 are combined with 150.0 g of KOH,

a. What is the limiting reagent in this reaction?

b. What mass of water can be produced in this reaction?

c. What mass of excess reagent will be left over in this reaction?

98. This equation shows the reaction of ethylene with molecular bromine:

$$C_2H_4\ (g) + Br_2\ (l) \rightarrow C_2H_4Br_2\ (l)$$

If 32.0 grams of C_2H_4 are combined with 172.0 grams of Br_2,

a. What is the limiting reagent in this reaction?

b. What mass of $C_2H_4Br_2$ can be produced in this reaction?

c. What mass of excess reagent will be left over in this reaction?

7.4 Theoretical and Percent Yield

99. What are two of the most common causes for a percent yield that is below 100%?

100. Is it possible for a reaction to occur with a yield that is greater than 100%? Why or why not?

101. Based on the amounts of starting materials used, a chemist calculates a possible yield of 210.3 grams in a reaction. However, after isolating her purified product, she finds that she has only 194.1 g of product. What is her percent yield for this reaction?

102. Based on the amounts of starting materials used, a chemist calculates a possible yield of 51.3 grams in a reaction. However, after isolating his purified product, he finds that he has only 45.1 g of product. What is his percent yield for this reaction?

103. Consider this generic reaction:

$$A + 2\,B \rightarrow C + D$$

If this reaction is carried out using 1.5 moles of A and 4.0 moles of B,

a. What is the limiting reagent in this reaction?
b. What is the theoretical yield (in moles) of compound D?
c. If 1.2 moles of compound D were isolated, what would the percent yield of D be for this reaction?

104. Consider this generic reaction:

$$A + 2\,B \rightarrow C + 2\,D$$

If this reaction is carried out using 2.5 moles of A and 4.0 moles of B,

a. What is the limiting reagent in this reaction?
b. What is the theoretical yield (in moles) of compound C?
c. If 1.6 moles of compound C is isolated, what is the percent yield of C for this reaction?

105. A chemist carries out this reaction, starting with 14.2 g of C_6H_6 and an excess of Br_2:

$$C_6H_6\ (l) + Br_2\ (l) \rightarrow C_6H_5Br\ (g) + HBr\ (g)$$

a. What is the theoretical yield for the reaction?
b. If the chemist actually isolates 16.3 grams of C_6H_6Br, what is the percent yield for this reaction?

106. A chemist carries out this reaction, beginning with 2.10 grams of NaBr and 3.25 grams of $Pb(NO_3)_2$:

$$2\,NaBr\ (aq) + Pb(NO_3)_2\ (aq) \rightarrow PbBr_2\ (s) + 2\,NaNO_3\ (aq)$$

a. Which starting material is the limiting reagent?
b. What is the theoretical yield of $PbBr_2$ in this reaction?
c. If the chemist actually isolates 3.5 grams of $PbBr_2$, what is the percent yield of the reaction?

107. A chemist carries out the following reaction, beginning with 4.20 g of $MgCl_2$ and 3.43 g of Na_3PO_4. After the reaction is complete, he isolates 2.14 grams of $Mg_3(PO_4)_2$ as a white solid.

$$3\,MgCl_2\ (aq) + 2\,Na_3PO_4\ (aq) \rightarrow Mg_3(PO_4)_2\ (s) + 6\,NaCl\ (aq)$$

a. Which starting material is the limiting reagent?
b. What is the theoretical yield of $Mg_3(PO_4)_2$ in this reaction?
c. What is the percent yield of $Mg_3(PO_4)_2$ in this reaction?

108. A chemist dissolves 0.401 g of $AgNO_3$ in a beaker of water and 0.253 g of $MgCl_2$ in a second beaker of water. She then combines the two solutions, which together form AgCl as a white precipitate. She isolates the AgCl and finds its mass to be 0.292 g. Here is the balanced equation for this reaction:

$$2\,AgNO_3\ (aq) + MgCl_2\ (aq) \rightarrow 2\,AgCl\ (s) + Mg(NO_3)_2\ (aq)$$

a. Which starting material is the limiting reagent?
b. What is the theoretical yield of AgCl in this reaction?
c. What is the percent yield of AgCl in this reaction?

Challenge Questions

109. You are conducting a set of experiments to see if it is possible to determine the type of wood used in a campfire by analyzing the ashes. In the laboratory, you weigh out 500 grams of hickory wood chips. You then burn the wood in a simulated campfire. After collecting the ashes, you find that only 37 grams of ashes are present. Does this finding conflict with the law of conservation of mass? How can you explain your findings?

110. You are conducting a series of experiments on mass changes in different chemical reactions. In each reaction you take two compounds dissolved in water in separate tubes (labeled A and B), measure the mass of both tubes before the reaction, and then combine the contents of the tubes. You observe any qualitative changes, and you also measure the mass of the two tubes after the reaction. Your observations are as follows:

Test	Tube A Aqueous Solution	Tube B Aqueous Solution	Observation	Combined Masses before Reaction	Combined Masses after Reaction
1	Sodium chloride	Potassium bromide	Nothing observed	20.07 g	20.06 g
2	Lead(II) nitrate	Sodium iodide	Yellow precipitate	17.84 g	17.81 g
3	Silver nitrate	Sodium chloride	White precipitate	20.44 g	20.42 g
4	Sodium carbonate	Hydrochloric acid	Bubbles formed	18.51 g	16.92 g

In tests 1, 2, and 3, the law of conservation of mass seems to be followed. However, in test 4, you see a drop in mass. Based on your observations, can you explain why the mass in this reaction appears to have decreased?

111. Coal is a leading source of energy for the world. While coal is primarily composed of carbon and hydrogen, a small amount of sulfur is also present. This leads to the production of pollutants such as sulfur dioxide (SO_2) through the following reaction:

$$S\ (s) + O_2\ (g) \rightarrow SO_2\ (g)$$

A sample of coal was found to contain 1.35% sulfur by mass. If 1,250 kg of this coal were burned, what mass of sulfur dioxide could theoretically be formed?

CHAPTER EIGHT

Energy

Figure 8.1 (a) The oil shortage of 1973 sparked renewed interest in using ethanol as fuel. (*continued*)

The Corn Ethanol Debate

Is it wise to use corn to power our cars?

Over the past decade this question has become more and more pressing, and more and more heated. At the heart of the issue is a single compound—ethanol—and the conflicting interests that surround it.

Throughout recorded history, people have produced ethanol (also called *ethyl alcohol* or simply *alcohol*) from grains such as wheat and corn. While most people think of alcohol in terms of beverages, this colorless liquid burns easily and can be used as fuel. Henry Ford's original Model T could run on gasoline or on ethanol.

For years gasoline was far cheaper than ethanol, and it became the fuel of choice for automobiles. But in the 1970s things began to shift. Gasoline is produced from crude oil, and many experts feared the supply of oil would dwindle. More urgently, an alliance of oil-producing countries called OPEC controlled the prices and supply of oil to the rest of the world. In a show of force in October 1973, OPEC blocked shipments of oil to the United States and several of its allies. Crippling gasoline shortages plagued the U.S. until the blockade ended in early 1974 (**Figure 8.1**).

These factors produced a surge of interest in renewable fuels such as solar and wind energy and biofuels made directly from plant matter. In 1978 the U.S. government began to provide subsidies (monetary incentives) to encourage farmers to produce corn for ethanol. Slowly, companies began producing gasoline blends that contained a small percentage of ethanol.

Figure 8.1 (continued) (b) Ethanol is a colorless liquid that burns easily. (c) For millennia, people have produced ethanol from grains, especially for use as alcoholic beverages. The process involves a chemical reaction (fermentation), often followed by purification using a technique called distillation. This image shows a simple still used to make "moonshine"—a mixture containing mainly ethanol and water. (d) Industrial distilleries like this one produce large amounts of ethanol from corn. (e) Today, most vehicles accommodate gasoline/ethanol blends. (a) AP Photo/Philadelphia Evening Bulletin; (b) imagenavi/sozaijiten/AGE Fotostock; (c) Pat Canova/Alamy; (d) BanksPhotos/Getty Images; (e) Courtesy Kevin Revell

In 2005 and 2007, the U.S. Congress passed two laws that created a Renewable Fuel Standard (RFS). These laws forced oil companies to blend renewable fuels (especially ethanol) with gasoline. The new regulations dramatically increased the amount of corn used for ethanol production. Today about 40% of all corn grown in the United States is used to produce ethanol for fuel.

Not everyone is happy about these changes. Due to increased demand, the price of corn doubled between 2005 and 2008. And because farmers rely on corn to feed their livestock, the prices of meat and dairy products have increased sharply.

Advocates of the renewable fuel standard say that it reduces dependence on foreign sources of oil and provides jobs for rural communities. Critics of the policy argue that it creates an undue financial burden on people across the nation while pouring money into a few corn-producing states. Some insist that ethanol is a clean-burning fuel that improves air quality. Others counter that ethanol production actually damages the environment. Who is right?

The answers to these issues are complex, so let's begin with some simpler questions: How well does ethanol actually work? Does a gallon of ethanol produce the same amount of energy as a gallon of gasoline? Do you get better gas mileage with pure gasoline, or with an ethanol blend? To answer these questions, we must examine the energy changes that accompany the combustion of ethanol or gasoline.

In this chapter, we will explore the relationships between energy and chemical reactions. We'll see how scientists measure the energy stored in fuels like gasoline and ethanol. We'll see how to predict energy changes using the rules of stoichiometry developed in Chapter 7. Finally, we'll return to the ethanol/gasoline debate as we answer the question, "Which is better?"

Intended Learning Outcomes

After completing this chapter and working the practice problems, you should be able to:

8.1 Energy, Work, and Heat

- Describe the relationships between heat, work, and total energy change.
- Describe the exchange of energy between the system and surroundings that accompanies a physical or chemical change.

8.2 Heat Energy and Temperature

- Explain the difference between heat and temperature.
- Use calorimetry measurements to determine energy changes.
- Apply the heat capacity or specific heat of a system to solve problems relating to heat energy and temperature.

8.3 Heat Energy and Chemical Reactions

- Use fuel values and reaction enthalpies to calculate the heat absorbed or released in a chemical reaction.

8.1 Energy, Work, and Heat

Most chemical and physical changes are accompanied by changes in energy and temperature. A campfire releases energy as it burns. When an ice cube melts, the water around it becomes colder. As our bodies break down food, we absorb energy stored in the chemical bonds (**Figure 8.2**). To fully describe these changes, we must also describe the changes in energy. The study of energy and temperature changes is called **thermodynamics**. *Thermochemistry* is the part of this field that deals specifically with chemical changes.

Figure 8.2 Winter campers like this one rely on heat energy from the burning wood to warm themselves, melt snow into drinkable water, and cook their food. Most physical and chemical changes also involve changes in heat and temperature.

We first discussed energy in Chapter 1. Recall that **energy** is the ability to do work—that is, the ability to bring about a change. Energy can take several forms, including **potential energy** (stored energy) and **kinetic energy** (the energy of motion).

Heat energy involves the kinetic energy of particles within a substance.

Changes in energy take place in two forms: *heat* and *work*. When a substance absorbs **heat energy** (often simply called *heat*), the particles within the substance vibrate or move faster. Heating a substance increases the kinetic energy of the particles that compose it. **Work** is the transfer of energy from one form to another. Work often involves a force that moves an object.

It's helpful to think of energy, heat, and work using the analogy of *wealth* and *spending* (**Figure 8.3**). What does it mean to have wealth? It means the ability to spend money. What is spending? Spending transfers money from one person to another. For example, if I have some money, I have the ability to spend

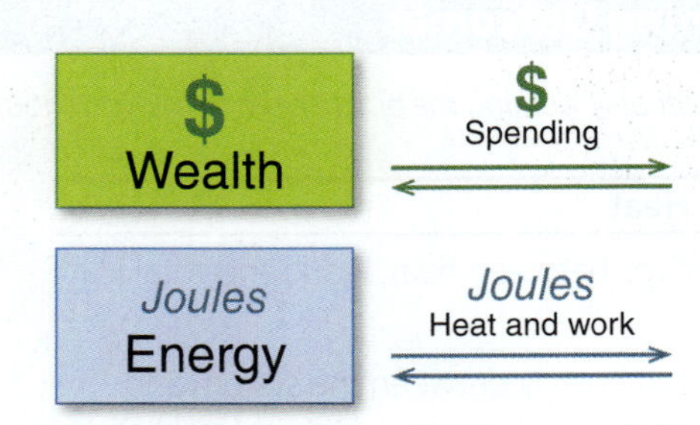

Figure 8.3 In much the same way that spending is the transfer of wealth, heat and work are the transfer of energy. We describe heat and work using the same units we use to describe energy.

it on something—say, a hot dog and a Coke. If I'm hungry, I may transfer my money to the hot dog guy. Now the guy who sold me the hot dog has the wealth (but I have lunch).

In the same way that spending transfers wealth from one person to another, heat and work transfer energy from one object or substance to another.

To describe wealth, we use some form of money, such as dollars. If we want to describe spending, it makes sense to describe it in the same way—using dollars. Similarly, we describe heat and work in the same units that we use to describe energy.

Energy changes take place through heat or work.

Units of Energy

There are several common units for energy (**Table 8.1**). The standard unit of energy is the **joule (J)**. Other common units of measurement include the *British thermal unit* (BTU), the *kilowatt hour* (kWh), and the *calorie* (c); see **Figure 8.4**. In nutritional circles, a kilocalorie is often referred to as a *Calorie*, where the letter *C* is sometimes capitalized. Most scientists prefer to avoid ambiguity, so they use the term *kilocalorie* rather than the capital *C* nutritional calorie.

A *joule* is defined as the energy required to accelerate a mass of 1 kg at a rate of 1 m/s^2 over a distance of 1 m.

TABLE 8.1 Common Units of Energy

1 joule = 1 kg·m^2/s^2
1 calorie = 4.184 J
1,000 calories = 1 kcal = 1 Calorie
1 British thermal unit (BTU) = 1,055 J
1 kilowatt hour (kWh) = 3.6 × 10^6 J

Example 8.1 Converting between Energy Units

A candy bar contains 212.0 kilocalories of stored chemical energy. What is this energy in joules and in kilojoules?

We first convert from kilocalories to calories; then we multiply by a conversion factor relating calories and joules, using the relationship given in Table 8.1. This gives us a final answer of 887,000 J, or 887.0 kJ.

$$212.0\ \cancel{\text{kcal}} \times \frac{1{,}000\ \cancel{\text{cal}}}{1\ \cancel{\text{kcal}}} \times \frac{4.184\ \text{J}}{1\ \cancel{\text{cal}}} = 887{,}000\ \text{J} = 887.0\ \text{kJ}$$

TRY IT

1. One gram of gasoline releases 47.3 kJ of potential energy when it is burned. Convert this value to calories.

Check It
Watch Explanation

Courtesy Enerco Group, Inc./Mr. Heater

a

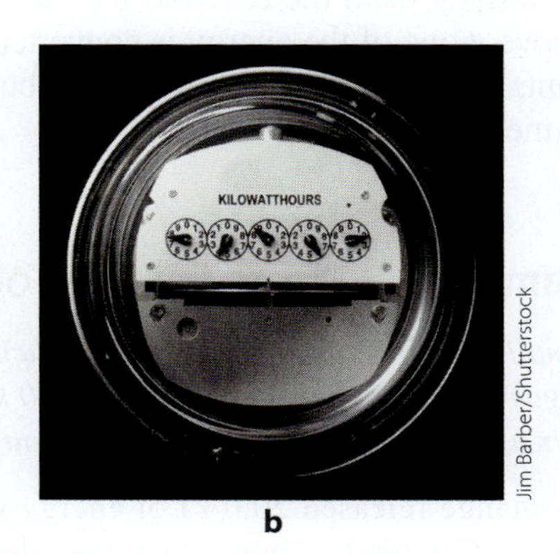

Jim Barber/Shutterstock

b

PHOTO RF/Science Source

c

Figure 8.4 Energy measurements use many different units. (a) A company advertises the energy its space heaters release in BTUs. (b) Your electric meter often measures energy in kilowatt hours. (c) Food products often report energy content in Calories (meaning kilocalories).

The two forms of energy transfer are heat and work.

Heat and Work in Chemical Changes

The energy changes that accompany a physical or chemical change take place in the forms of heat and work. For example, consider a piston in a gasoline engine (**Figure 8.5**). As the gasoline burns, it releases energy. This energy causes the gas in the piston to expand, which pushes the cylinder up and turns the engine. The energy released has performed work on the piston.

Explore **Figure 8.5**

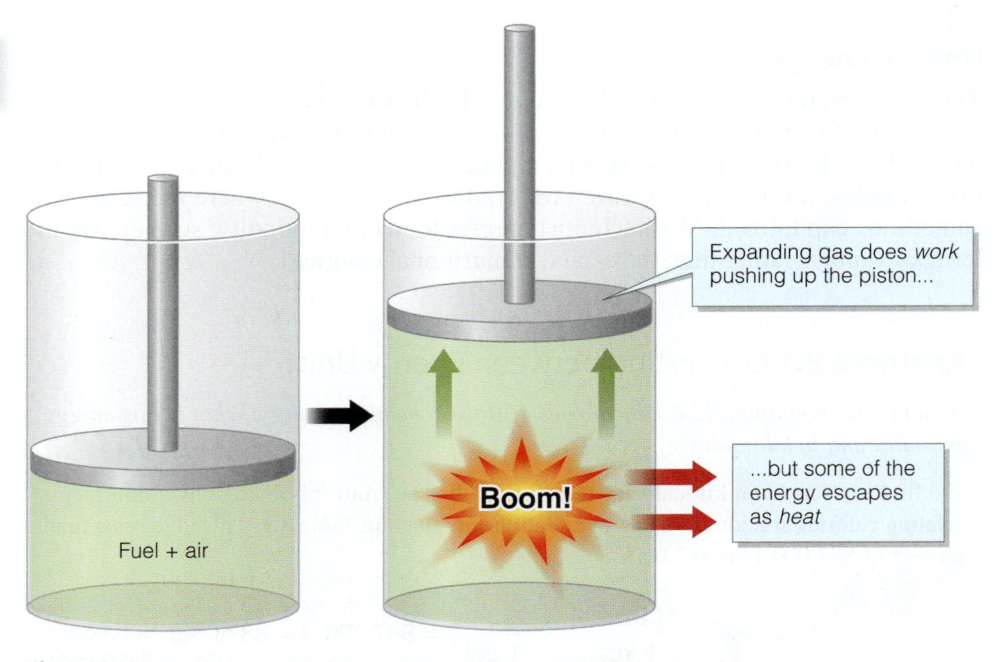
Expanding gas does *work* pushing up the piston...
Boom!
...but some of the energy escapes as *heat*
Fuel + air

Figure 8.5 As the fuel mixture in a piston explodes, it produces heat and work.

Scientists use the capital Greek letter delta (Δ) to indicate a change. *E* means energy, and Δ*E* means a change in energy.

However, not all of the energy transfers to the piston as work. Part of the energy escapes to the surroundings as heat. The total energy released from the gasoline (ΔE) is the sum of the heat (q) and the work (w):

$$\Delta E = q + w$$

So, here is a question: What if we welded the piston shut and then ignited the gasoline? (Don't try this at home.) Assuming the piston chamber doesn't explode, all of the energy from the reaction would be released as heat. Because the piston does not move, none of the energy is converted into motion—that is, no work is done. The amounts of heat and work can change, but the total amount of energy released will be the same.

Ed Aldridge/Shutterstock

An engine like this one converts the potential energy in gasoline into work and heat. Car designers try to make engines run as efficiently as possible—that is, they try to maximize the amount of energy transferred as work and minimize the amount of energy lost as heat.

Example 8.2 Measuring the Total Energy Change

A small sample of propane burns, producing carbon dioxide and water vapor. As the hot gas mixture expands, it releases 20.0 kJ of heat and does 31.0 kJ of work pushing against a piston. What is the total amount of energy released in this reaction?

The change released 20.0 kJ of energy in the form of heat and 31.0 kJ in the form of work. The total amount of energy released is the sum of the heat and the work, or 51.0 kJ.

$$\text{Energy released} = q + w = 20.0\text{ kJ} + 31.0\text{ kJ} = 51.0\text{ kJ}$$

Endothermic and Exothermic Changes

Let's examine the reaction inside the piston from a chemical perspective. When gasoline reacts with oxygen, it produces carbon dioxide and water. It also releases a large amount of energy. We describe this reaction as an **exothermic change**—meaning it releases heat energy. In fact, we sometimes include "energy" with the reaction products:

$$2\ C_8H_{18}\ (l) + 25\ O_2\ (g) \rightarrow 16\ CO_2\ (g) + 18\ H_2O\ (g) + \text{energy}$$

Conversely, an **endothermic change** absorbs energy. When describing an endothermic change, we sometimes show energy as a reactant. For example, when an ice cube melts, it absorbs energy to convert the solid ice to liquid. We can express this as an equation:

$$\text{Energy} + H_2O\ (s) \rightarrow H_2O\ (l)$$

When describing energy changes, we must be very careful to define the system that is changing. In thermochemistry, the **system** is the part of the universe being studied. The term **surroundings** refers to the rest of the universe. So, when we say that burning fuel is an exothermic process, we mean that energy is released from the system (the fuel) to the surroundings. In the example of the melting ice cube, the ice is the system that absorbs heat from the surroundings.

System: The part of the universe being studied.

So, if someone puts an ice pack on an injured ankle, is the flow of heat endothermic or exothermic? To answer that question, we must first define the system and the surroundings. We know that heat energy flows from the ankle into the ice, so the change is exothermic for the ankle, but it is endothermic for the ice (**Figure 8.6**).

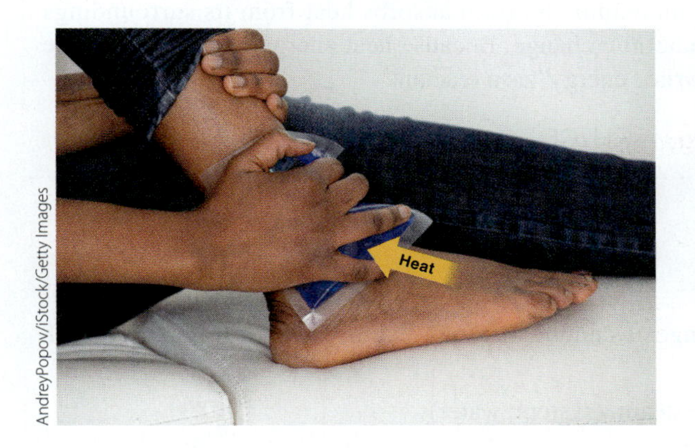

AndreyPopov/iStock/Getty Images

Figure 8.6 An ice pack is often the first treatment for a sprained ankle. The heat energy flows out of the ankle and into the ice. This change is exothermic from the perspective of the ankle, but endothermic from the perspective of the ice.

When describing energy changes, we must be careful to show the *direction* of energy flow from the perspective of the system we are studying. In an exothermic reaction, the system loses heat energy, and so the energy change for that system is negative. On the other hand, if a system gains heat (an endothermic reaction), the sign of the energy change is positive.

Endothermic: $+q$

Exothermic: $-q$

Some say the world will end in fire;
Some say, in ice.
From what I've tasted of desire
I hold with those who favor fire.

But if it had to perish twice
I think I know enough of hate
To say that for destruction ice
Is also great and would suffice.
–Robert Frost

© ch_ch/depositphotos.com

Katvic/Shutterstock

Similarly, if a system does work on its surroundings (for example, when the gas pushes the piston up), the sign of w for that system is negative. If the surroundings do work on the system (that is, increase the energy of the system), then the sign of w is positive (**Figure 8.7**).

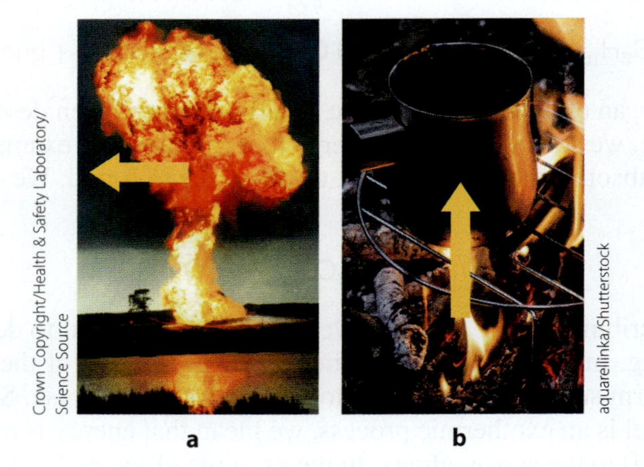

Figure 8.7 The signs of q and w depend on the system being studied. (a) As gasoline explodes, it releases heat to the surroundings and pushes back the air around it. If gasoline is the system, both q and w are negative. (b) Heat travels from the campfire to the coffeepot. The heat change, q, is positive for the water in the pot but negative for the wood in the fire.

Example 8.3 Expressing Energy Changes in Chemical Equations

When ammonium chloride dissolves, it absorbs heat from the surroundings. Write a chemical expression for this change, including energy where appropriate.

In this case, the system (the ammonium chloride) absorbs heat from its surroundings as it dissolves. This is an endothermic change. Because heat energy is required for the reaction to take place, we can write "energy" as a reactant.

$$\text{Energy} + NH_4Cl\,(s) \rightarrow NH_4Cl\,(aq)$$

TRY IT

2. Describe each of these changes as endothermic or exothermic, using the frame of reference given:
   - **a.** Water boils, converting to steam. (system: water)
   - **b.** An acid and base react, releasing heat. (system: acid and base molecules)
   - **c.** A log burns in a fireplace. (system: the log)
3. Freezing is an exothermic process. When a liquid solidifies, it releases heat to its surroundings. Write an equation for the freezing of water, and include energy in the chemical equation where appropriate.

The law of conservation of energy is also called the first law of thermodynamics.

The Law of Conservation of Energy

In each example we've discussed so far, energy moved between a system and its surroundings. This transfer of energy leads us to a key idea, commonly called the **law of conservation of energy**:

Energy cannot be created or destroyed.

In a chemical or physical change, energy passes from one system to another, but the total energy of the universe remains constant. The energy that flows out of the system flows into the surroundings, and vice versa. Mathematically, we can say that the energy change of the system (ΔE_{system}) is equal to the energy change of the surroundings ($\Delta E_{\text{surroundings}}$) but is opposite in sign:

$$(\Delta E_{\text{system}}) = -(\Delta E_{\text{surroundings}})$$

As an analogy, think of the classic board game Monopoly® (**Figure 8.8**): At the beginning of the game, you pass out play money to all players. As the game progresses, the players buy and sell property. If you land on someone else's property, you pay them rent. Your wealth goes down, and the other player's wealth goes up by the same amount. *But the total amount of money in the game does not change.* At the end of the game, the game box contains the same amount of money as when you started.

txking/Shutterstock

Figure 8.8 In the board game Monopoly®, players may gain or lose money, but the total money in the game does not change. Similarly, chemical systems may gain or lose energy, but this does not change the total energy of the universe.

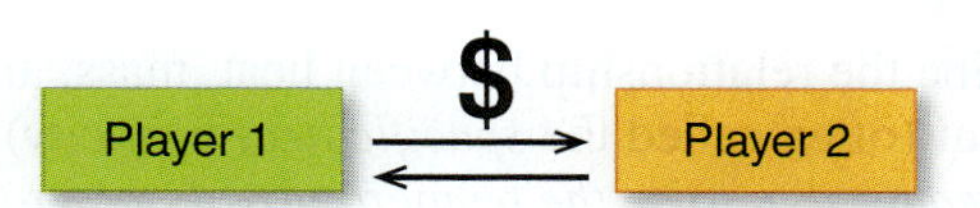

It's the same thing with the conservation of energy: Energy can be passed back and forth between the system and its surroundings, but the total amount of energy remains constant.

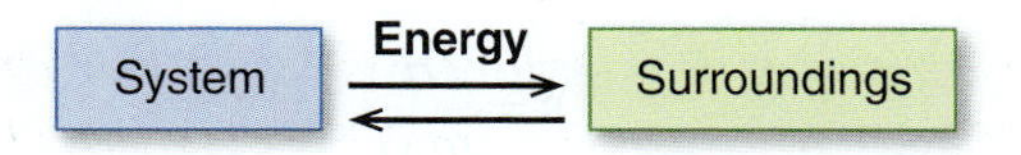

Example 8.4 Using the Law of Conservation of Energy

A chemical reaction releases 200 J of heat energy to its surroundings. Write this change of energy for the system (the chemical reaction) and for the surroundings.

Because the system loses energy, we write this change as –200 J. Based on the law of conservation of energy, the energy that the system loses is gained by the surroundings. Therefore, the change in energy for the surroundings is +200 J. Notice that these values are equal in magnitude but opposite in sign.

In Chapter 16, we'll explore nuclear reactions—and we'll see that the law of conservation of energy is not a complete description of the universe. But for nonnuclear chemical and physical changes, this law holds true.

TRY IT

4. As a sample of butane gas burns, it releases 55.0 J of heat and performs 23.0 J of work on its surroundings. What is the total change in energy for the chemical system? What is the total change in energy for the surroundings? Include the correct signs in your answers.

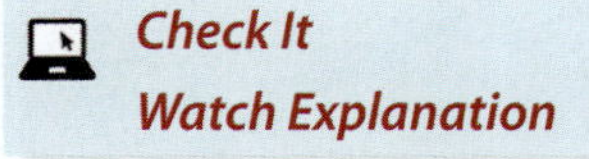

8.2 Heat Energy and Temperature

We commonly use the words *heat* and *temperature* to describe the world around us. As we explore the effects of energy changes, it's important to understand the difference between these terms. *Heat* refers to the *total* kinetic energy transferred from one substance or object to another. **Temperature** is a measure of the *average* kinetic energy of the particles in a substance. For example, suppose a spoon of boiling water

Tobias1900/Shutterstock

Figure 8.9 The water in the spoon and the water in the pot are the same temperature, but because the pot has more water, it has a larger amount of heat energy available to cook the meal.

and a pot of boiling water (**Figure 8.9**) both have the same temperature—in other words, the same average kinetic energy. Which one can we use to cook a meal? The pot, of course: because it holds a larger amount of water, there is a larger amount of heat energy available to cook the food.

Specific Heat and Heat Capacity

The relationship between heat energy and temperature depends on the substance.

The precise relationship between heat and temperature is unique to each substance. For example, if we add 1,000 calories of heat energy to one kilogram of water, the temperature increases by just one degree Celsius. However, if we add the same amount of heat energy to one kilogram of iron, the temperature increases nearly 10 times as much. While the reasons for this are not obvious, the relationships between heat, temperature, and the composition of different materials are well known and very important.

The specific heat capacity relates heat, mass, and temperature change.

Scientists describe the relationship between heat, mass, and temperature using the term **specific heat** (often called the *specific heat capacity*). The specific heat is *the amount of heat required to raise the temperature of one gram of material by one degree Celsius*. Written mathematically, this relationship can be expressed as

$$\text{Specific heat} = \frac{\text{heat}}{(\text{mass}) \times (\text{change in temperature})}$$

or

$$s = \frac{q}{m\Delta T}$$

or, in rearranged form,

$$q = ms\Delta T$$

where q is the amount of heat, m is the mass, s is the specific heat, and ΔT is the change in temperature. The specific heats of several common substances are given in **Table 8.2**. Example 8.5 is based on this equation.

One calorie (4.184 J) is the amount of heat required to raise one gram of water by one degree Celsius. That is, the specific heat of water = 1 cal/g · °C, or 4.184 J/g · °C.

California CPA/Moment Open/Getty Images

Compare the specific heat of water to the other materials in Table 8.2. Do you notice how much larger the value is for water? This subtle fact is essential to human life: Although our bodies require a very narrow temperature range (about 80–110 °F) to stay alive, we are able to survive in much colder or somewhat warmer settings. This is partly because the specific heat of water helps to regulate our body temperatures, making us less susceptible to short-term temperature extremes.

TABLE 8.2 Specific Heats for Several Materials

	Material	Specific Heat (J/g · °C)
Gas	Air (dry)	1.01
Liquid	Water (liquid)	4.184
	Ethanol	2.597
	Oil (petroleum)	1.74
	Gasoline	2.2
Solid	Glass (quartz)	0.70
	Concrete	0.880
	Ice	2.10
	Sand	0.799
	Aluminum	0.897
	Chromium	0.449
	Gold	0.129
	Iron	0.449
	Lead	0.130
	Nickel	0.444
	Zinc	0.388
	Steel	0.50

Example 8.5 Using Specific Heat in Calculations

How many kilojoules of heat are required to raise the temperature of 120.0 grams of water by 5.0 °C?

Using the relationship described earlier, we can answer this question. After rounding to two significant digits, we obtain an answer of 2,500 J, or 2.5 kJ.

$$q = ms\Delta T = (120.0\ \cancel{g})\left(4.184\ \frac{J}{\cancel{g}\cdot\cancel{°C}}\right)(5.0\ \cancel{°C}) = 2{,}500\ J = 2.5\ kJ$$

Sometimes scientists and engineers need to determine heat–temperature relationships for objects containing many substances or for objects whose mass is difficult to measure. In these situations a simpler measure is used, called the *heat capacity*. **Heat capacity** (represented by the letter *C*) is defined as the amount of heat required to raise the temperature of an object, regardless of its mass.

$$\text{Heat capacity} = \frac{\text{Heat}}{\text{change in temperature}}$$

or

$$C = \frac{q}{\Delta T}$$

Heat capacity is a useful tool in many situations. For example, suppose a team of workers at a chemical plant needs to raise the temperature of a steel reactor filled with aqueous sodium hydroxide. Calculating the specific heats of all the different components of the reactor is too complicated. Instead, they use the heat capacity of the whole unit (the reactor vessel and the solution) to predict the temperature changes. Example 8.6 illustrates a calculation of this type.

Heat capacity calculations do not involve mass.

Example 8.6 Using Heat Capacity in Calculations

When filled with water, a large reaction vessel in a chemical plant has a heat capacity of 5.41×10^5 kJ/°C. How many kJ of heat are required to heat this entire vessel from 25.0 °C to 48.2 °C?

From the initial and final temperature values, we can determine the change in temperature, ΔT, for this reaction:

$$\Delta T = T_{\text{final}} - T_{\text{initial}} = 48.2\ °C - 25.0\ °C = 23.2\ °C$$

Rearranging the equation for heat capacity and substituting the numerical values, we find this change requires 1.26×10^7 kJ of heat.

$$q = C\Delta T = \left(5.41 \times 10^5\ \frac{kJ}{\cancel{°C}}\right)(23.2\ \cancel{°C}) = 1.26 \times 10^7\ kJ$$

TRY IT

5. If 425 joules of heat flow into a 100.0-gram block of chromium, how much will the temperature of the block rise? The specific heat of chromium is 0.449 J/g · °C.
6. When filled with water, a reaction vessel in a manufacturing plant has a heat capacity of 28,400 kJ/°C. If 1.41×10^5 kJ of heat are added, by how much will the temperature of the vessel change?

Check It

Watch Explanation

Figure 8.10 A simple coffee cup calorimeter.

Calorimetry: Measuring Heat Flow

Scientists and engineers often need to measure how much heat is transferred from one source to another. Experiments that measure heat flow are called **calorimetry** experiments. The two most common calorimetry techniques are *coffee cup calorimetry* and *bomb calorimetry*.

Coffee Cup Calorimetry

Coffee cup calorimetry is exactly what it sounds like. In this method, water is placed in an insulated cup (a Styrofoam coffee cup works well) that is covered with a cork lid and a thermometer (**Figure 8.10**). The system to be studied is placed inside the coffee cup, and the reaction or change is allowed to take place. Because Styrofoam is a very good insulator, nearly all of the heat involved in the change is absorbed or released by the water. By measuring the temperature change for the water, we can determine the amount of heat that the system gained or lost.

For example, if we wanted to measure the specific heat of a block of iron, we would heat the block, and then place it in the calorimeter. When the hot iron is placed in the cool water, heat is transferred from the iron to the water. According to the law of conservation of energy, the total energy inside the cup is conserved. This means that the metal loses the same amount of heat that the water gains. That is:

$$q_{water} = -q_{metal}$$

In this equation, the minus sign shows that the metal is losing energy. This type of experiment is shown in **Figure 8.11**. Because $q = \text{ms}\Delta T$, we can say that

$$m_w s_w \Delta T_w = -m_m s_m \Delta T_m$$

where the subscript "m" refers to the metal, and the subscript "w" refers to the water. We can then rearrange this equation to find the specific heat for the metal.

$$s_m = -\frac{m_w s_w \Delta T_w}{m_m \Delta T_m}$$

Example 8.7 illustrates this type of problem.

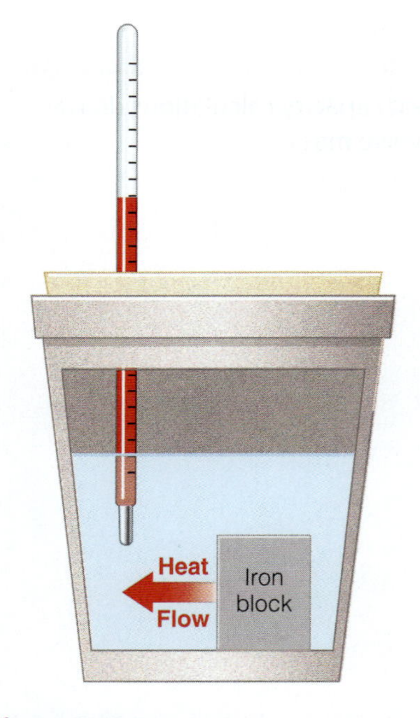

Figure 8.11 The specific heat of a metal can be measured using coffee cup calorimetry. When a hot metal sample is placed in water, the water temperature rises. Using the mass, specific heat, and temperature change of water, we can determine the amount of heat absorbed by the water. This quantity is equal in magnitude to the amount of heat released by the metal.

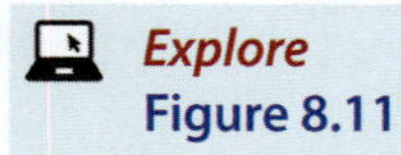

Explore Figure 8.11

Example 8.7 Finding the Specific Heat Using Coffee Cup Calorimetry

To find the specific heat of nickel, a chemist heats a 26.0-g sample to 100.0 °C, and then places it into a coffee cup calorimeter containing 52.1 g of water at an initial temperature of 20.0 °C. After some time, both the metal and water reach an equal temperature of 24.0 °C. What is the specific heat of the metal?

To solve this problem, we first need to find the change in temperature of the metal and the water:

$$\Delta T_{metal} = T_{final} - T_{initial} = 24.0\ °\text{C} - 100.0\ °\text{C} = -76.0\ °\text{C}$$
$$\Delta T_{water} = T_{final} - T_{initial} = 24.0\ °\text{C} - 20.0\ °\text{C} = 4.0\ °\text{C}$$

Next, we use the equation for finding specific heat. The specific heat of water (s_w) is equal to 4.184 J/g·°C. Using the rearranged form of the equation, we can find the specific heat of the metal.

$$s_m = -\frac{m_w s_w \Delta T_w}{m_m \Delta T_m} = -\frac{(52.1\ \cancel{g})\left(4.184\ \frac{J}{\cancel{g} \cdot \cancel{°C}}\right)(4.0\ \cancel{°C})}{(26.0\ g)(-76.0\ °C)}$$

$$s_m = 0.44\ \text{J/g} \cdot °\text{C}$$

TRY IT

7. To find the specific heat of an unknown metal, a chemist heats a 50.0-g block of the metal to 100.0 °C. She then places it in a coffee cup calorimeter containing 50.0 g of water at a temperature of 24.8 °C. After the metal is placed in the water, both the metal and water reach an equal temperature of 51.2 °C. What is the specific heat of the material?

Coffee cup calorimetry is used to measure the gain or loss of heat energy for many physical and chemical changes. For example, many "instant cold packs" contain ammonium chloride. When this compound dissolves in water, it absorbs heat from its surroundings. We can write the endothermic reaction like this:

$$\text{Heat} + NH_4Cl\ (s) \rightarrow NH_4Cl\ (aq)$$

How much does the temperature drop when we dissolve ammonium chloride in water? Of course, this depends on how much NH_4Cl and how much water we use. But we can measure the amount of heat absorbed by NH_4Cl by doing a coffee cup calorimetry experiment: Fill a coffee cup with a known amount of water. Add a known amount of NH_4Cl, quickly cover the calorimeter, and measure the change in temperature that results. The amount of heat lost from the water (the aqueous solution) will be equal to the amount of heat gained by the solid:

$$q_{\text{solid}} = -q_{\text{aq}}$$

Because we know that $q = ms\Delta T$, this equation becomes

$$q_{\text{solid}} = -m_{\text{aq}}s_{\text{aq}}\Delta T_{\text{aq}}$$

Example 8.8 illustrates this type of calculation.

Example 8.8 Finding the Heat of a Reaction Using Coffee Cup Calorimetry

A 10.4-gram sample of NH_4Cl was combined with 100.0 grams of water in a coffee cup calorimeter, causing the water temperature to decrease by 6.20 °C. Based on this result, how much heat energy was required to dissolve the sample of NH_4Cl? Calculate the heat of solution for NH_4Cl in kJ/mol.

In this example, ammonium chloride absorbs heat energy as it dissolves. The amount of heat absorbed by the solid is equal to the amount of heat lost from the aqueous solution. We can solve this using the specific heat equation:

$$q_{\text{solid}} = -m_{\text{aq}}s_{\text{aq}}\Delta T_{\text{aq}}$$

The mass of the resulting solution (m_{aq}) includes both the water and the dissolved ammonium chloride (110.4 g). Because it is mostly water, the solution will have a specific heat nearly identical to that of water (4.184 J/g · °C). To solve, we substitute these values into the equation:

$$q_{\text{solid}} = -(110.4\ \cancel{g})\left(4.184\frac{\text{J}}{\cancel{g}\cdot\cancel{°C}}\right)(-6.20\ \cancel{°C}) = 2{,}860\ \text{J} = 2.86\ \text{kJ}$$

Finally, we are asked to calculate the heat of solution in kJ/mol. We begin by converting 10.4 grams of NH_4Cl from grams to moles:

$$10.4 \text{ g } \cancel{NH_4Cl} \times \frac{1 \text{ mole } NH_4Cl}{53.49 \text{ g } \cancel{NH_4Cl}} = 0.194 \text{ moles } NH_4Cl$$

To calculate the heat of the solution in kJ/mol, we divide the heat absorbed by the number of moles of NH_4Cl.

$$\text{Heat of solution} = \frac{2.86 \text{ kJ}}{0.194 \text{ moles } NH_4Cl} = 14.7 \text{ kJ/mol}$$

Check It
Watch Explanation

TRY IT

8. A chemist fills a coffee cup calorimeter with 50.0 g of water. He then adds a sample of KOH (5.21 g) and stirs the mixture until all of the KOH dissolves. As he mixes the KOH, the temperature rises from 24.0 °C to 46.4 °C. Based on this result, how much heat energy does the KOH release? Calculate the heat of solution for KOH in kcal/mol.

Bomb Calorimetry

Bomb calorimetry is not what it sounds like. Bomb calorimetry measures the heats of reaction for fuels and other high-energy chemical changes. In bomb calorimetry, the test substance is placed in a heavy steel container. Outside this container is a second container that is filled with water and equipped with a thermometer (**Figure 8.12**). A pair of ignition wires detonates the sample. The gases produced cannot expand, so all of the energy is released as heat into the calorimeter.

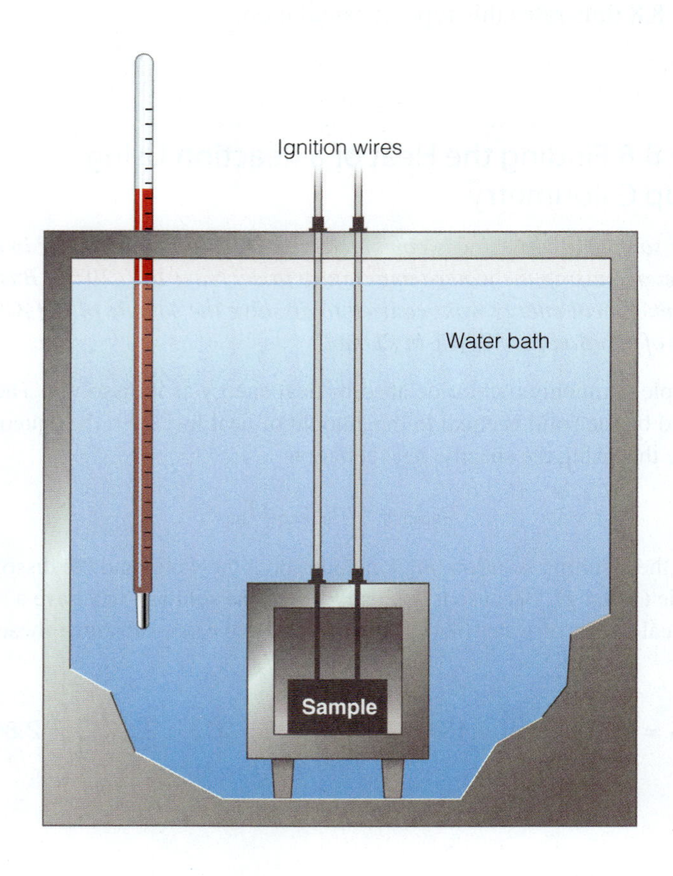

Figure 8.12 A bomb calorimeter contains a sealed sample chamber that is surrounded by a water bath. When the sample detonates, heat flows from the sample chamber into the water bath.

Explore
Figure 8.12

Using the heat capacity (joules/°C) for the calorimeter, the amount of energy produced in the reaction can be determined.

We can use bomb calorimetry to answer the important questions we asked at the beginning of this chapter: How well does ethanol work as a fuel? Does a gallon of ethanol provide the same energy as a gallon of gasoline?

To test this question, we would place a small sample of ethanol and excess oxygen inside the bomb calorimeter. We would then ignite the sample inside the container and measure how much the temperature of the calorimeter increased. Using this temperature change, we could determine the joules of energy per gram of ethanol. We could repeat the experiment with gasoline and then compare the results of the two experiments. Look at the data in Example 8.9 and in question 9, and see how the results compare.

In bomb calorimetry, a reaction takes place inside a sealed container.

In coffee cup calorimetry, the reaction takes place at a *constant pressure*.

In bomb calorimetry, the reaction takes place at a *constant volume*.

Bomb calorimetry can measure the energy content of foods as well as fuels. For example, a milkshake like this releases about 380 nutritional Calories (1,600 kJ) when it reacts with oxygen.

Heather Winters/Photodisc/Getty Images

Example 8.9 Finding the Heat of a Reaction Using Bomb Calorimetry

A chemist places a 20.0-g sample of ethanol inside a bomb calorimeter with a known heat capacity of 28.72 kJ/°C. When the ethanol ignites, the temperature of the calorimeter rises from 22.04 °C to 42.74 °C. How much heat did the ethanol release? Calculate the energy released in kilojoules per gram of ethanol.

From the initial and final temperatures, we see that the calorimeter undergoes a temperature change, ΔT, of 20.70 °C. Using the heat capacity of the bomb calorimeter, we can determine the heat released from this reaction:

$$q = C\Delta T = \left(28.72\,\frac{\text{kJ}}{\cancel{^\circ\text{C}}}\right)(20.70\,\cancel{^\circ\text{C}}) = 594.5\ \text{kJ}$$

We can also express this result in kilojoules per gram of ethanol:

$$\frac{\text{heat released}}{\text{grams fuel}} = \frac{594.5\ \text{kJ}}{20.0\ \text{g}} = 29.7\ \text{kJ/g}$$

Question 9 that follows repeats this problem for gasoline. I encourage you to try this practice problem and then compare the results for the two fuels.

TRY IT

9. When a chemist repeated the experiment in Example 8.9 using a 20.0-g sample of gasoline, the temperature rose from 22.02 °C to 54.41 °C. Calculate the energy released by gasoline in kilojoules per gram. How does this compare to the energy released by ethanol in Example 8.9?

Check It
Watch Explanation

8.3 Heat Energy and Chemical Reactions

Calorimetry provides scientists and engineers with valuable information about the amount of energy that is absorbed or produced in chemical reactions. Scientists and engineers rely on this type of information for many applications, from understanding how reactions take place to designing efficient automobile engines to creating safe manufacturing environments.

The energy absorbed or released in a reaction is an *extensive property*—that is, it relies on how much matter is involved (**Figure 8.13**). Because of this, we express energy changes as a function of the amount of substances involved in the reaction.

© itsajoop/depositphotos.com

Patrick Orton/Getty Images

Figure 8.13 The energy released from a chemical reaction depends on the amount of matter that reacts. A small campfire releases much less heat than a massive forest fire.

TABLE 8.3 Fuel Values for Common Combustion Fuels

Fuel	Fuel Value (kJ/g)
Methane	55.5
Natural gas	54.0
Propane	50.3
Butane	49.5
Gasoline	46.5
Anthracite coal	34.6
Ethanol	29.7
Wood (oak)	18.9

Data from *CRC Handbook of Chemistry and Physics*, 92nd ed. (Boca Raton, FL: CRC Press, 2011).

Fuel Value

Earlier, we looked at how chemists use bomb calorimetry to measure the energy stored in fuels such as gasoline and ethanol. The **fuel value** of a substance is the amount of energy that can be produced by its combustion. We often express the fuel value in terms of energy per unit mass (for example, kilojoules/gram). **Table 8.3** gives the fuel value of several common fuels. The higher the fuel value, the more energy it releases when it burns.

Example 8.10 Using Fuel Values

Based on the data in Table 8.3, how much energy (in kJ) would be released by the combustion of 2.50 kilograms of coal? How much energy would be released by the combustion of 2.50 kilograms of oak?

To answer these questions, we first convert the mass of each species from kilograms to grams. Then we use the fuel value to convert from grams to kilojoules of energy. Rounding to three significant digits, we obtain values of 86,500 kJ of energy from coal and 47,300 kJ of energy from oak.

$$2.50 \text{ kg coal} \times \frac{1{,}000 \text{ g}}{1 \text{ kg}} \times \frac{34.6 \text{ kJ}}{\text{g coal}} = 86{,}500 \text{ kJ energy from coal}$$

$$2.50 \text{ kg oak} \times \frac{1{,}000 \text{ g}}{1 \text{ kg}} \times \frac{18.9 \text{ kJ}}{\text{g coal}} = 47{,}300 \text{ kJ energy from oak}$$

Check It
Watch Explanation

TRY IT

10. Based on Table 8.3, which releases more energy, the combustion of 1.0 kg of natural gas or the combustion of 3.0 kg of anthracite coal?

Reaction Enthalpy

One of the most important measures of energy change is the **reaction enthalpy**, represented by the symbol ΔH_{rxn}. The reaction enthalpy is the amount of heat energy that is absorbed or released in a chemical reaction at constant pressure.

Chemists often write ΔH_{rxn} along with a chemical reaction. For example, consider the combustion of ethanol:

$$C_2H_6O\ (l) + 3\ O_2\ (g) \rightarrow 2\ CO_2\ (g) + 3\ H_2O\ (g) \qquad \Delta H_{rxn} = -1{,}368\ \text{kJ}$$

Notice that the sign of ΔH_{rxn} is negative. This means that heat is released from the system (the reaction). This is an exothermic reaction.

ΔH_{rxn} is negative for an exothermic reaction.

The reaction enthalpy relates the heat change to the number of moles of reactants and products. The equation above means that each mole of C_2H_6O releases 1,368 kJ of heat when it burns. The ΔH_{rxn} is related to the number of moles of each reactant or product by the coefficients in the balanced equation. For this equation, we can write the following conversion factors:

ΔH_{rxn} is the heat energy change per number of moles in the balanced equation.

$$\frac{-1{,}368\ \text{kJ}}{1\ \text{mol}\ C_2H_6O} \text{ or } \frac{-1{,}368\ \text{kJ}}{3\ \text{mol}\ O_2} \text{ or } \frac{-1{,}368\ \text{kJ}}{2\ \text{mol}\ CO_2} \text{ or } \frac{-1{,}368\ \text{kJ}}{3\ \text{mol}\ H_2O}$$

Practice
Energy Stoichiometry
Use this interactive to practice reaction enthalpy calculations.

The enthalpy of a reaction allows us to relate the heat changes in a reaction to the amount of a substance that reacts. Examples 8.11 and 8.12 illustrate this important calculation.

Example 8.11 Predicting Changes in Heat Energy Using the Enthalpy of a Reaction

The enthalpy of reaction for the combustion of ethanol (C_2H_6O) is −1,368 kJ/mol, as in the following equation. How much heat will be released by the combustion of 789.0 g of ethanol?

$$C_2H_6O\ (l) + 3\ O_2\ (g) \rightarrow 2\ CO_2\ (g) + 3\ H_2O\ (g) \qquad \Delta H_{rxn} = -1{,}368\ \text{kJ}$$

In this problem, we're given the number of grams of ethanol. The ΔH_{rxn} relates the energy per mole of ethanol. To solve this problem, we must first convert to moles of ethanol and then to kilojoules of energy:

$$\text{Grams } C_2H_6O \Rightarrow \text{Moles } C_2H_6O \Rightarrow \text{Kilojoules of energy}$$

To convert from grams to moles, we use the molar mass of C_2H_6O (46.08 g/mol). We then use the conversion factor that relates ΔH_{rxn} to moles of C_2H_6O:

$$789.0\ \cancel{\text{g}\ C_2H_6O} \times \frac{1\ \cancel{\text{mol}\ C_2H_6O}}{46.08\ \cancel{\text{g}\ C_2H_6O}} \times \frac{-1{,}368\ \text{kJ}}{1\ \cancel{\text{mol}\ C_2H_6O}} = -2.342 \times 10^4\ \text{kJ}$$

The negative value in this answer means that this amount of heat is released to the surroundings as this reaction occurs.

Let's return to the analogy of wealth and spending for a moment. Right now, what is your total net worth? That is, what is the value of all of your possessions? It's hard to answer this question exactly. But if I asked how much you spent on lunch today, you probably could tell me much more precisely. It's hard to measure total wealth, but easy to measure a change in wealth.

The same is true with energy changes. We rarely know the total energy of a system, with all the different forms of potential energy. However, we can easily measure energy *changes* that take place. These energy changes drive the physical and chemical properties that we can observe in the world around us.

Example 8.12 Predicting Changes in Heat Energy Using the Enthalpy of a Reaction

Many manufacturers produce hydrogen gas from methane gas, as in the following reaction. This reaction is endothermic, with $\Delta H_{rxn} = 206.1$ kJ. How much heat energy is required to produce 1.00 kg of hydrogen gas?

$$CH_4\,(g) + H_2O\,(g) \rightarrow CO\,(g) + 3\,H_2O\,(g) \qquad \Delta H_{rxn} = 206.1\text{ kJ}$$

To solve this problem, we must convert from the mass of hydrogen to moles of hydrogen and then finally to kilojoules of energy:

$$\text{Kilograms } H_2 \Rightarrow \text{Grams } H_2 \Rightarrow \text{Moles } H_2 \Rightarrow \text{Kilojoules of energy}$$

As before, we use the molar mass of H2 to convert from grams to moles and then use the conversion factor from the balanced equation to convert from moles to kilojoules.

$$1.00\text{ kg } H_2 \times \frac{1{,}000\text{ g}}{1\text{ kg}} \times \frac{1\text{ mol } H_2}{2.02\text{ g } H_2} \times \frac{206.1\text{ kJ}}{3\text{ mol } H_2} = 3.40 \times 10^4\text{ kJ}$$

Notice that the final conversion factor related the energy in kilojoules to 3 moles of H2. When writing a conversion factor involving ΔH_{rxn}, the coefficients from the balanced equation indicate the number of moles present.

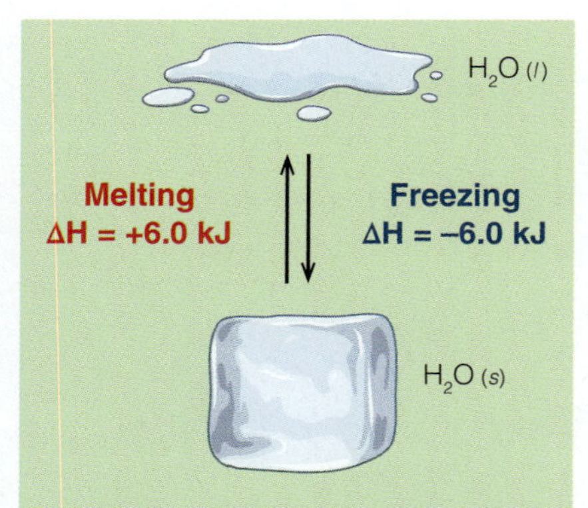

Notice that the energy absorbed when ice melts is the opposite of the energy released when ice freezes.

Physical changes, such as a phase change or the formation of an aqueous solution, also involve enthalpy changes (ΔH). For example, these two equations describe the enthalpy changes for melting and freezing water:

$$\text{Melting:} \quad H_2O\,(s) \rightarrow H_2O\,(l) \qquad \Delta H = 6.0\text{ kJ}$$
$$\text{Freezing:} \quad H_2O\,(l) \rightarrow H_2O\,(s) \qquad \Delta H = -6.0\text{ kJ}$$

Similarly, we can describe the ΔH for the formation of a solution:

$$KOH\,(s) \rightarrow KOH\,(aq) \qquad \Delta H = -57.7\text{ kJ}$$
$$NH_4Cl\,(s) \rightarrow NH_4Cl\,(aq) \qquad \Delta H = 14.0\text{ kJ}$$

Example 8.13 illustrates the use of enthalpies to describe physical changes.

Example 8.13 Using Enthalpy to Describe a Physical Change

Dissolving sodium nitrate in water is an endothermic process. What is the enthalpy change when 200.0 grams of sodium nitrate dissolves in water?

$$NaNO_3\,(s) \rightarrow NaNO_3\,(aq) \qquad \Delta H = 21.7\text{ kJ}$$

As before, we solve this problem by converting from grams to moles and then using the ΔH as a conversion factor.

$$200.0\text{ g } NaNO_3 \times \frac{1\text{ mol } NaNO_3}{85.00\text{ g } NaNO_3} \times \frac{21.7\text{ kJ}}{1\text{ mol } NaNO_3} = 51.1\text{ kJ}$$

TRY IT

Check It
Watch Explanation

11. Acetylene (C_2H_2) reacts with hydrogen gas in the combination reaction shown here. How much heat energy is released if 501.6 grams of H_2 react with excess acetylene?

$$C_2H_2\,(g) + 2\,H_2\,(g) \rightarrow C_2H_6\,(g) \qquad \Delta H_{rxn} = -311.5 \text{ kJ}$$

12. A company needs 5.02×10^3 kJ of heat energy to power a manufacturing process. The enthalpy of reaction for the combustion of propane is 2,202 kJ/mol. How many kilograms of propane are needed to produce this amount of heat energy?

$$C_3H_8\,(s) + 5\,O_2\,(g) \rightarrow 3\,CO_2\,(g) + 4\,H_2O\,(g) \qquad \Delta H_{rxn} = -2{,}202 \text{ kJ}$$

Capstone Question

Capstone Video

While working for a food science company, you need to analyze the energy content of a sugar substitute with the molecular formula $C_5H_{12}O_5$. As part of this process, you conduct a series of bomb calorimetry experiments. In each experiment, you place a sample of the sweetener in the calorimeter, and then measure the temperature change when the sweetener undergoes combustion. The heat capacity of the calorimeter is 4.371 kJ/°C. (a) Write a balanced equation for the combustion of this compound. (b) Using the data in the table below, calculate the heat energy produced in each reaction. (c) Report the fuel value for this sweetener in kJ/g, in kcal/g, and in kJ/mol. Report your answers to the correct number of significant digits.

Sample	Mass	Temperature Change (calorimeter)	$q_{reaction}$
1	1.018 g	7.11 °C	
2	0.989 g	6.86 °C	
3	0.954 g	6.67 °C	
4	1.053 g	7.29 °C	
5	1.104 g	7.69 °C	

SUMMARY

In this chapter, we've explored the relationships between energy and change. When a physical or chemical change takes place, energy passes from one location or system to another. Energy transfer takes place in the form of work or of heat.

When describing energy changes, it is important to define the system under investigation. In an endothermic reaction, a system absorbs heat energy from its surroundings. In an exothermic reaction, a system releases energy to its surroundings. The law of conservation of energy states that energy is not created or destroyed but moves from one form to another.

Heat and temperature are closely related. Heat energy refers to the *total* kinetic energy transferred from one substance or object to another. Temperature is a measure of the *average* kinetic energy of the molecules in a substance. The temperature change brought about by the gain or loss of heat energy is unique to each substance. For any substance, the specific heat (also called the *specific heat capacity*) is the amount of heat required to raise one gram of a substance by one degree Celsius. A similar term is the *heat capacity*—this is the amount of heat required to raise the temperature of any object by one degree Celsius. Heat capacity is useful for larger, complex objects where temperature changes are important.

Scientists measure heat changes using a technique called *calorimetry*. Two common forms of this technique are coffee cup calorimetry and bomb calorimetry. These studies typically involve an insulated container that holds water and the system being studied. As the physical or chemical change takes place, the temperature of the water changes. Using the specific heat of water and the heat capacity of the insulated container, scientists can measure the heat absorbed or released by the system.

The amount of energy absorbed or released in a chemical reaction is an extensive property, meaning it depends on the amount of the substances involved. Two common measurements of reaction energy are fuel value and reaction enthalpy. Fuel value gives the amount of energy released by different fuels in combustion reactions. Reaction enthalpy, ΔH_{rxn}, gives the relationship between the heat change in a constant-pressure reaction and the amount of each substance (in moles) involved in the reaction. Enthalpy changes are also useful for describing physical changes, such as phase changes and the formation of solutions.

Gasoline or Ethanol—Which Fuel Is Better?

At the beginning of this chapter, we considered a question: Which makes a better fuel, ethanol or gasoline? Using the concepts in this chapter, we can measure and compare the energy available from the combustion of a gallon of ethanol and from a gallon of gasoline.

Gasoline is a mixture of compounds made up primarily of carbon and hydrogen. A major component of gasoline is octane (C_8H_{18}). Using calorimetry experiments, scientists have carefully measured the reaction enthalpies for the combustion of ethanol and of octane. Let's compare the ΔH_{rxn} for the combustion of one mole of these two compounds:

$$\text{Ethanol: } C_2H_6O + \frac{7}{2}\,O_2 \rightarrow 2\,CO_2 + 3\,H_2O \qquad \Delta H_{rxn} = -1{,}368 \text{ kJ}$$

$$\text{Octane: } C_8H_{18} + \frac{25}{2}\,O_2 \rightarrow 8\,CO_2 + 9\,H_2O \qquad \Delta H_{rxn} = -5{,}470 \text{ kJ}$$

Notice that we used fractional coefficients for oxygen. While this is not the conventional format for a balanced equation, we write it this way so the ΔH for each reaction is the energy released by one mole of fuel.

If we divide each ΔH_{rxn} by the molar mass and take the absolute value, we can obtain the fuel value of these compounds in kilojoules per gram:

$$\text{Fuel value of ethanol (kJ/g)} = \left|\frac{-1{,}367 \text{ kJ/mol}}{46.08 \text{ g/mol}}\right| = 29.69 \text{ kJ/g}$$

$$\text{Fuel value of octane (kJ/g)} = \left|\frac{-5{,}470 \text{ kJ/mol}}{114.26 \text{ g/mol}}\right| = 47.87 \text{ kJ/g}$$

Finally, we can use the densities of ethanol and octane to convert this result from kJ/gram to kJ/gallon:

$$\text{Fuel value of ethanol} = \frac{29.69 \text{ kJ}}{\text{g}} \times \frac{0.789 \text{ g}}{\text{mL}} \times \frac{3{,}785 \text{ mL}}{\text{gallon}} = 88{,}700 \text{ kJ/gallon}$$

$$\text{Fuel value of octane} = \frac{47.87 \text{ kJ}}{\text{g}} \times \frac{0.703 \text{ g}}{\text{mL}} \times \frac{3{,}785 \text{ mL}}{\text{gallon}} = 127{,}000 \text{ kJ/gallon}$$

This means that a gallon of octane contains 43% more chemical potential energy than a gallon of ethanol. Because of this, vehicles that use ethanol require more gallons of fuel than those that use gasoline. The U.S. government website fueleconomy.gov reports that E15 mixtures (gasoline containing about 15% ethanol) get 4–5% fewer miles per gallon than fuel containing 100% gasoline.

Key Terms

8.1 Energy, Work, and Heat

thermodynamics The scientific field that deals with energy and temperature changes.

energy The ability to do work.

potential energy Energy that is stored.

kinetic energy The energy of motion; the faster an object is moving, the greater kinetic energy it has.

heat energy A form of kinetic energy involving the kinetic energy of the particles within a substance; heat energy specifically refers to the total kinetic energy transferred from one substance to another.

work The transfer of energy from one form to another.

joule (J) The standard unit of energy; $1\ J = 1\ kg \cdot m^2/s^2$.

exothermic change A physical or chemical change that releases energy to the surroundings.

endothermic change A physical or chemical change that absorbs energy from the surroundings.

system In thermodynamics, the part of the universe being studied.

surroundings In thermodynamics, everything that exists around the system being studied.

law of conservation of energy In a chemical or physical change, the total energy of the universe remains constant.

8.2 Heat Energy and Temperature

temperature A measure of the average kinetic energy of the molecules in a substance.

specific heat The amount of heat required to raise the temperature of one gram of a substance by one degree Celsius; sometimes called the *specific heat capacity*.

heat capacity The amount of heat required to raise the temperature of a given object.

calorimetry An experimental technique used to measure heat changes.

coffee cup calorimetry A technique for measuring heat changes that uses an insulated container (such as a Styrofoam coffee cup) to measure heat changes.

bomb calorimetry A technique for measuring heat changes using a sealed container; commonly used to measure high-energy reactions.

8.3 Heat Energy and Chemical Reactions

fuel value The amount of heat energy that can be released by a combustion reaction of a certain substance.

reaction enthalpy (ΔH_{rxn}) The amount of heat energy that is absorbed or released in a chemical reaction at constant pressure.

Additional Problems

8.1 Energy, Work, and Heat

13. How are energy, work, and heat related?

14. What is the difference between potential energy and kinetic energy?

15. Each of these objects contains potential energy. How may the energy be released?

a. an anvil positioned at the top of a cliff
b. a plate of spaghetti
c. a tightly pulled bowstring
d. a stick of dynamite

16. Each of these things contains potential energy. How may the energy be released?

a. a combustible gas such as acetylene
b. a high-calorie lunch
c. an apple hanging from a branch
d. water in a dam-formed lake

17. Refer to Table 8.1 to complete these energy conversions:

a. Convert 12.8 kilowatt hours (kWh) to joules (J).
b. Convert 259.3 kilocalories to joules.
c. Convert 300 kWh to British thermal units (BTUs).

18. Refer to Table 8.1 to complete these energy conversions:

a. Convert 14.8 kilojoules to kilocalories.
b. Convert 25,000 BTU to kWh.
c. Convert 28.3 kWh to J.

19. A recent-model refrigerator is estimated to use 722 kWh each year. What is this energy consumption in BTUs? What is it in kJ?

20. In 2012, the average U.S. home used 10,837 kWh. What is this energy consumption in BTUs? What is it in kJ?

21. An average banana has about 105 kilocalories of energy content. What is this value in joules?

22. A jelly donut has a food value of about 925 kilojoules. What is this value in kilocalories?

23. What are the two ways in which energy transfers take place?

24. How are heat energy and kinetic energy related?

25. Consider the following chemical reaction. Is this change endothermic or exothermic? Which has the higher potential energy, the starting materials or the products?

$$CH_4\ (g) + 2\ O_2\ (g) \rightarrow CO_2\ (g) + 2\ H_2O\ (g) + \text{heat energy}$$

26. Consider the following chemical reaction. Is this change endothermic or exothermic? Which has the higher potential energy, the starting materials or the products?

$$\text{heat energy} + Cu(OH)_2\ (s) \rightarrow CuO\ (s) + H_2O\ (g)$$

27. Identify each of these reactions as endothermic or exothermic:

a. heat + $H_2O\ (l) \rightarrow H_2O\ (g)$
b. $H_2O\ (g) \rightarrow H_2O\ (l)$ + heat
c. $K_2CO_3\ (s) \rightarrow K_2CO_3\ (aq)$ + heat
d. heat + $NH_4Cl\ (s) \rightarrow NH_4Cl\ (aq)$

28. Identify each of these reactions as endothermic or exothermic:

a. $C_3H_8 + 5\ O_2 \rightarrow 3\ CO_2 + 4\ H_2O$ + energy
b. energy + $Zn(OH)_2 \rightarrow ZnO + H_2O$
c. $KOH\ (s) \rightarrow KOH\ (aq)$ + heat
d. $NaOH\ (aq) + HCl\ (aq) \rightarrow NaCl\ (aq) + H_2O\ (l)$ + heat

29. Write a balanced equation for each of these reactions. Include heat with the starting materials or products as appropriate.

a. Liquid water freezes, forming ice (exothermic).
b. Magnesium and oxygen combine to form magnesium oxide (exothermic).
c. Calcium hydroxide forms calcium oxide plus water (endothermic).

30. Write a balanced equation for each of these reactions. Include heat with the starting materials or products as appropriate.

a. Carbon dioxide changes from a solid to a gas (endothermic).
b. Solid carbon reacts with oxygen gas to form carbon dioxide gas (exothermic).
c. Potassium (solid) reacts with chlorine gas to form solid potassium chloride (exothermic).

31. Describe each of these changes as endothermic or exothermic, using the frame of reference given:

a. The water in a lake releases heat to the air, forming ice. (system: water)
b. The wax in a candle burns. (system: candle)
c. You warm your hands by a fire. (system: your hands)

32. Describe each of these changes as endothermic or exothermic, using the frame of reference given:

a. A grenade detonates, creating a fireball. (system: the grenade)
b. A stove heats water to boiling. (system: the stove)
c. A stove heats water to boiling. (system: the water)

33. When many ionic compounds dissolve, they release heat. For example, when calcium sulfate mixes with water, the temperature of the water may rise by 10 °C. If calcium sulfate is the system, what makes up the surroundings? Is this reaction endothermic or exothermic?

34. Ammonium chloride is commonly used in instant ice packs. When this compound dissolves, the temperature of the water quickly drops. If ammonium chloride is the system, what makes up the surroundings? Is this reaction endothermic or exothermic?

35. On a cold December day in Indiana, a boy foolishly responds to a dare and touches a flagpole with his tongue. The cold flagpole quickly freezes the moisture on his tongue, which then sticks to the pole. Describe this change as endothermic or exothermic:

a. if the boy is the system
b. if the flagpole is the system

36. A man places some hamburgers on a charcoal grill. As the charcoal burns, the heat cooks the burgers. Describe this change as endothermic or exothermic:

a. if the charcoal is the system
b. if the burgers are the system

37. Find the total change in energy for these situations:

a. A gas absorbs 20 kJ of heat and has 30 kJ of work done to it.
b. A gas releases 30 kJ of heat and does 45 kJ of work on its surroundings.

38. Find the total change in energy for these situations:

a. A gas releases 53.2 kJ of heat and does 23.7 kJ of work on its surroundings.
b. A gas absorbs 15.7 kJ of heat and has 42.3 kJ of work done to it.

39. Inside a small engine, the combustion of gasoline releases 47.0 kJ of energy. If the sample does 15.0 kJ of work on the engine, how much energy is lost as heat?

40. A sample of ethanol reacts, releasing 42.1 kJ of energy. If 25.0 kJ are released as heat, how much work is done on the surroundings?

41. A military group tests a high-energy chemical explosive underground. As the explosive is detonated, the ground around it shakes. The potential energy has been released from the explosive, but where has it gone? Has the total energy of the universe changed?

42. Miners extract coal from the mountains of West Virginia. The coal is shipped to a power plant in New York, where it is burned to produce electricity. What is the relationship between the chemical energy stored in the coal, the heat released, and the electricity produced?

43. A bottle of frozen water is dropped into a cooler filled with water. As the water inside the bottle melts, it absorbs 182,000 joules of heat energy from the surrounding water. What happens to the temperature of the water as a result? If the ice inside the bottle is the system, is this change endothermic or exothermic? Use this table to complete your answer.

	System: Ice Bottle	Surroundings: Water in Cooler
Change	+82,000 J	
Result	Ice melts, temperature increases	
Heat Absorbed or Released?		

44. A blacksmith immerses a red-hot iron rod in a large bucket of water. As the iron cools, it releases 138 kJ of heat energy into the surrounding water. What happens to the temperature of the water as a result? If the iron is the system, is this change endothermic or exothermic? Use this table to complete your answer.

	System: Iron Rod	Surroundings: Water in Bucket
Change	−138 kj	
Result	Iron cools	
Heat Absorbed or Released?		

8.2 Heat Energy and Temperature

45. What is the difference between heat and temperature?

46. A pot of water is heated on a stove. How does the motion of the molecules in the liquid change as the temperature increases?

47. What is the difference between an object's specific heat and its heat capacity?

48. A bomb calorimeter contains a steel casing, insulation, and water surrounding the inner bomb. When measuring temperature changes, does it make more sense to use the specific heat for this calorimeter or the heat capacity?

49. Zinc has a specific heat of 0.39 J/g · °C while iron has a specific heat of 0.45 J/g · °C. If a 100-g sample of each metal is cooled from 100 °C to room temperature (25 °C), which one releases more heat energy?

50. The temperature of a spoonful of water increases by one degree for each 20 J of energy added. The temperature of a bathtub full of water increases one degree for each 7.5×10^4 J of energy added. Which has the greater heat capacity, the water in the spoon or the water in the bathtub? Which (if either) has the greater specific heat?

51. An engineer tests the thermal properties of a metal alloy. Using a 50.0-g sample, she finds that adding 485 J of heat energy to the alloy causes a temperature change of 4.10 °C. What is the specific heat of this alloy?

52. A chemist tests the properties of a 230.0-g sample of an unknown metal. He finds that 1,217 J of heat increases the temperature of the metal by 5.88 °C. What is the specific heat of this metal?

53. Aluminum has a specific heat of 0.91 J/g·°C. If a 4,000-g sheet of aluminum absorbs 28,000 joules of energy, how much will the temperature increase?

54. Tin has a specific heat of 0.21 J/g·°C. If a 3.2-kg sheet of tin absorbs 8.4 kJ of energy, how much will the temperature increase?

55. Gold has a specific heat of 0.031 cal/g·°C. How many calories of heat are required to raise the temperature of a 100.0-g gold bar by 15 °C?

56. Iron has a specific heat of 0.108 cal/g·°C. If a 2.1-kg iron bar cools from a temperature of 180 °C to a temperature of 25 °C, how much heat does it release?

57. An industrial reaction vessel is found to undergo a change in temperature of 0.061 °C for each kilojoule of energy absorbed. What is the heat capacity for this vessel in kJ/°C?

58. A boiler system requires 15,000 kJ to heat it from 25 °C to 200 °C. What is the heat capacity for this system in kJ/°C?

59. A bomb calorimeter is a device used to measure heat changes. One model of bomb calorimeter has a heat capacity of 14.5 kJ/°C. How much heat would be required to raise the temperature of this calorimeter from 25.3 °C to 33.7 °C?

60. When filled with water, a large reaction vessel in a chemical plant has a heat capacity of 540,000 kJ/°C. How many kJ of heat are needed to heat this entire vessel from 25 °C to 65 °C? How much energy will this require in kilowatt hours?

61. An industrial reaction vessel has a heat capacity of 6.13×10^3 kJ/°C. If 7,702 MJ of heat is added to the system, by how much will the temperature change?

62. An insulated reaction vessel has a heat capacity of 105.32 kJ/°C and an initial temperature of 24.3 °C. If 1,000 kJ of heat energy is added to the system, what temperature will the reaction vessel reach?

63. A house is built with a granite countertop. The heat capacity of the countertop is 158.2 kJ/°C. A hot pan of water is placed on the countertop, and 8,000 J of heat energy is transferred into the countertop. By how much does the temperature of the countertop change?

64. A house is built with a concrete floor to help keep the house cool. How many kJ of heat would be required to raise the temperature of a concrete slab having a mass of 15,000 kg from 65 °F to 75 °F? The specific heat of concrete is 0.880 kJ/kg · °C. How much energy will this require in kilowatt hours?

65. How is coffee cup calorimetry different from bomb calorimetry?

66. In bomb calorimetry, the calorimetry is sealed shut so that no work can be done by the reaction process. In light of this, how does the heat released in a bomb calorimetry experiment compare to the total energy change?

67. A student places a block of hot metal into a coffee cup calorimeter containing 150.0 g of water. The water temperature rises from 23.7 °C to 32.1 °C. How much heat (in calories) did the water absorb? How much heat did the metal lose?

68. A student places a block of hot metal into a coffee cup calorimeter containing 171.0 g of water. The water temperature rises from 23.3 °C to 42.5 °C. How much heat (in calories) did the water absorb? How much heat did the metal lose?

69. In a coffee cup calorimetry experiment, a block of hot metal is placed in a calorimeter containing 100.0 g of water. The water temperature rises from 23.1 °C to 31.5 °C. Calculate q_{metal} and q_{water} for this experiment.

70. In a calorimetry experiment, a reaction takes place in 100.0 g of an aqueous solution. The solution temperature drops from 22.8 °C to 15.5 °C. Calculate $q_{reaction}$ and $q_{solution}$ for this experiment. Assume the solution has a specific heat identical to that of pure water.

71. A chemist conducts a calorimetry experiment on an unknown metal. The mass of the sample is 24.4 g. The chemist heats the metal to 100 °C and then places it in a coffee cup calorimeter containing 108.5 g of water with an initial temperature of 21.3 °C. After some time, both the metal and water reach an equal temperature of 25.0 °C.

a. How much heat was absorbed by the water?
b. How much heat was released by the metal?
c. What is the specific heat of the metal?
d. Based on **Table 8.4**, is the unknown metal most likely aluminum, lead, nickel, or tin?

72. A chemist conducts a calorimetry experiment on an unknown metal. The mass of the sample is 52.4 g. The chemist heats the metal to 100 °C and then places it in a coffee cup calorimeter containing 95.4 g of water. The temperature of the water rises from 21.9 °C to 23.2 °C.

a. How much heat was absorbed by the water?
b. How much heat was released by the metal?
c. What is the specific heat of the metal?
d. Based on Table 8.4, is the unknown metal most likely aluminum, lead, nickel, or tin?

TABLE 8.4 Specific Heats for Several Metals

Material	Specific Heat (cal/g · °C)
Aluminum	0.220
Iron	0.108
Titanium	0.125
Lead	0.031
Nickel	0.104
Tin	0.054

73. A chemist dissolved an 11.6-g sample of KOH in 100.0 grams of water in a coffee cup calorimeter. When she did so, the water temperature increased by 25.5 °C. Based on this, how much heat energy was required to dissolve the sample of KOH? Calculate the heat of solution for KOH in kJ/mol. (Assume the specific heat of the solution is 4.184 J/g · °C.)

74. When a 20.8-g sample of CsBr was combined with 115.0 grams of water in a coffee cup calorimeter, the water temperature decreased by 4.47 °C. Based on this, how much heat energy was released when CsBr was dissolved? Calculate the heat of solution for CsBr in kJ/mol. (Assume the specific heat of the solution is 4.184 J/g · °C.)

75. A sample of jet fuel is tested in a bomb calorimeter. Before detonation, the temperature of the calorimeter is 22.31 °C. If the fuel releases 558.1 kJ of heat, and the heat capacity of the calorimeter is 13.12 kJ/°C, determine the final temperature of the calorimeter.

76. A sample of charcoal is incinerated in a bomb calorimeter. Before detonation, the temperature of the calorimeter is 22.56 °C. If the fuel releases 33.58 kJ of heat, and the heat capacity of the calorimeter is 13.12 kJ/°C, determine the final temperature of the calorimeter.

77. A 1.0-g sample of a candy bar was placed in a bomb calorimeter with a heat capacity of 3.54 kcal/°C. The sample was completely burned, causing the temperature of the calorimeter to rise by 1.36 °C. How many kilocalories of energy were stored in the candy bar?

78. A 2.0-g sample of a glazed donut was placed in a bomb calorimeter with a heat capacity of 3.54 kcal/°C. The sample was completely burned, causing the temperature of the calorimeter to rise by 2.02 °C. How many kilocalories of energy were stored in the donut sample?

8.3 Heat Energy and Chemical Reactions

79. A synthetic fuel blend is tested by bomb calorimetry. It is found that an 80.10-g sample releases 28,096 kJ of heat. What is the fuel value of this blend in kJ/g?

80. A 15.0-g sample of coal is tested by bomb calorimetry. Upon combustion, the sample releases 334.5 kJ of heat. What is the fuel value of this blend in kJ/g?

81. Using data from Table 8.3 in the text, calculate the energy released by the combustion of each of these compounds:

a. 100.0 g of methane
b. 15.0 mL of gasoline (density of gasoline = 0.70 g/mL)

82. Using data from Table 8.3 in the text, calculate the energy released by the combustion of each of these compounds:

a. 10.0 kg of propane
b. 18.0-mL sample of ethanol (density of ethanol = 0.79 g/mL)

83. What mass of propane must be burned to produce 30,000 BTUs of heat energy? (Refer to Tables 8.1 and 8.3 for the proper conversion factors and fuel values. Report your answer to three significant digits.)

84. What mass of methane must be burned to produce 15,000 BTUs of heat energy? (Refer to Tables 8.1 and 8.3 for the proper conversion factors and fuel values. Report your answer to three significant digits.)

85. Recently scientists have explored the use of crop residues as a renewable fuel. In one such test, a 21.05-g sample of dried cornstalks was placed inside a bomb calorimeter (**Figure 8.14**) with a known heat capacity of 28.72 kJ/°C. When the stalks were burned, the temperature of the calorimeter rose from 24.10 °C to 36.93 °C. How much heat was released by the combustion of the cornstalks? What is the fuel value of the cornstalks in kJ/g?

86. Recently scientists have explored the use of switchgrass as a renewable fuel. In one such test, a 53.18-g sample of dried switchgrass (**Figure 8.15**) was placed inside a bomb calorimeter with a known heat capacity of 30.31 kJ/°C. When the grass was burned, the temperature of the calorimeter rose from 24.35 °C to 55.95 °C. How much heat was released by the combustion of the switchgrass? What is the fuel value of the grass in kJ/g?

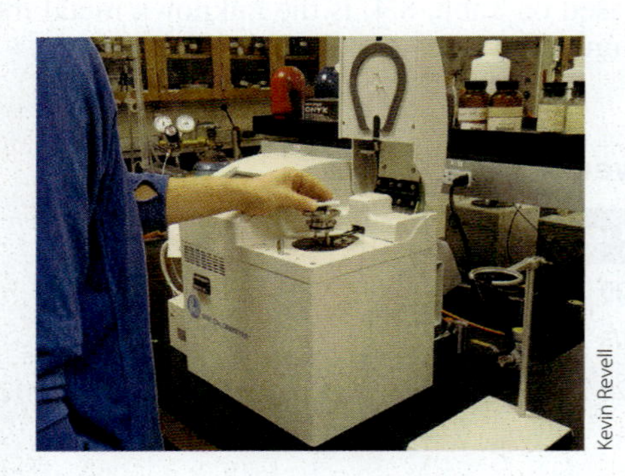
Kevin Revell

Figure 8.14 Scientists use bomb calorimeters like this one to measure the fuel value of different substances.

Stephen Ausmus/USDA/Science Source

Figure 8.15 Switchgrass is a hardy plant that grows easily in many places. Scientists have explored the use of switchgrass as a renewable fuel.

87. Using the reaction enthalpy data shown, determine the amount of heat released in each situation:

$$2\,Ca\,(s) + O_2\,(g) \rightarrow 2\,CaO\,(s) \quad \Delta H_{rxn} = -1{,}269.8\text{ kJ}$$

a. If calcium reacts with 3 moles of O_2
b. If 8 moles of calcium reacts with O_2
c. If calcium and oxygen react to form 1 mole of CaO

88. Using the reaction enthalpy data shown, determine the amount of heat released in each situation:

$$2\,K\,(s) + Cl_2\,(g) \rightarrow 2\,KCl\,(s) \quad \Delta H_{rxn} = -873.0\text{ kJ}$$

a. If potassium reacts with 5 moles of chlorine
b. If 10 moles of potassium react with chlorine
c. If potassium and chlorine react to form 0.4 moles of KCl

89. Using the reaction enthalpy data shown, determine the heat change in each situation:

$$Sn\,(s) + 2\,Cl_2\,(g) \rightarrow SnCl_4\,(s) \quad \Delta H_{rxn} = -511.3\text{ kJ}$$

a. If 23.13 g of tin reacts with chlorine gas
b. If tin reacts with 18.68 g of chlorine gas
c. If tin and chlorine react to form 11.83 g of tin(IV) chloride

90. Using the reaction enthalpy data shown, determine the heat change in each situation:

$$MgCO_3\,(s) \rightarrow MgO\,(s) + CO_2\,(g) \quad \Delta H_{rxn} = +100.7\text{ kJ}$$

a. If 18.19 g of $MgCO_3$ decomposes to form MgO and CO_2
b. If $MgCO_3$ decomposes to form 139.1 g solid MgO
c. If $MgCO_3$ decomposes to form 12.1 g CO_2 gas

91. Methane is the main component of natural gas. Using the reaction enthalpy given here, calculate the heat energy produced by the combustion of one kilogram of methane.

$$CH_4\ (g) + 2\ O_2\ (g) \rightarrow CO_2\ (g) + 2\ H_2O\ (l)$$
$$\Delta H_{rxn} = -861.3\ kJ$$

92. Calcium oxide is the major component of cement mix. Manufacturers produce this compound by the decomposition of calcium hydroxide, shown here. Using the reaction enthalpy, calculate the energy required to produce one kilogram of calcium oxide by this process.

$$Ca(OH)_2\ (s) \rightarrow H_2O\ (g) + CaO\ (s)$$
$$\Delta H_{rxn} = +108.5\ kJ$$

93. The acid-base neutralization reaction shown here is an exothermic process. How many grams of NaOH are neutralized if 218.5 kJ of heat are released in the process?

$$HBr\ (aq) + NaOH\ (aq) \rightarrow NaBr\ (aq) + H_2O\ (l)$$
$$\Delta H_{rxn} = -55.8\ kJ$$

94. The acid-base neutralization reaction shown here is an exothermic process. How many grams of HBr are neutralized if 119.0 kJ of heat are released in the process?

$$HBr\ (aq) + KOH\ (aq) \rightarrow KBr\ (aq) + H_2O\ (l)$$
$$\Delta H_{rxn} = -55.8\ kJ$$

95. Sodium hydroxide dissolves in water in an exothermic process as shown here. How much heat is released if 139.8 g of NaOH dissolve in water?

$$NaOH\ (s) \rightarrow NaOH\ (aq) \qquad \Delta H_{rxn} = -44.0\ kJ$$

96. When steam condenses to form liquid water, it releases heat to its surroundings. How much heat is released when 100.0 g of steam returns to the liquid state?

$$H_2O\ (g) \rightarrow H_2O\ (l) \qquad \Delta H_{rxn} = -40.67\ kJ$$

97. Thermite reactions, such as the reaction of iron(III) oxide with aluminum, are used to produce intense heat (**Figure 8.16**). Based on the reaction enthalpy below, how many grams of aluminum are needed to produce 1,000,000 kJ of heat by this reaction? Report your answer to three significant digits.

$$2\ Al\ (s) + Fe_2O_3\ (s) \rightarrow Al_2O_3\ (s) + 2\ Fe\ (s)$$
$$\Delta H_{rxn} = -851.5\ kJ$$

98. Butane, C_4H_{10}, is the fuel used in many handheld lighters (**Figure 8.17**). Based on the reaction enthalpy below, how many grams of butane are needed to produce 1,000 kJ of heat by this reaction? Report your answer to three significant digits.

$$2\ C_4H_{10}\ (g) + 13\ O_2\ (g) \rightarrow 8\ CO_2\ (g) + 10\ H_2O\ (l)$$
$$\Delta H_{rxn} = -5{,}755\ kJ$$

Figure 8.16 Thermite reactions produce intense heat.

Figure 8.17 Handheld lighters like this one use butane fuel.

99. Magnesium reacts with oxygen in the very exothermic reaction shown here. If 30.18 g of magnesium were reacted in a bomb calorimeter having a heat capacity of 38.33 kJ/°C, by how much would the temperature rise?

$$2\ Mg\ (s) + O_2\ (g) \rightarrow 2\ MgO\ (s)$$
$$\Delta H_{rxn} = -1{,}203.2\ kJ$$

100. A chemical manufacturer needs to neutralize a mixture containing 25.2 kg of sodium hydroxide by reacting it with excess hydrochloric acid. The chemical engineers estimate that once filled, the insulated reaction vessel will have a heat capacity of 318.5 kJ/°C. Based on this data, determine the amount of heat released by this reaction. If the neutralization starts at a temperature of 30.0 °C, how hot will the reaction vessel get?

$$HCl\ (aq) + NaOH\ (aq) \rightarrow NaCl\ (aq) + H_2O\ (l)$$
$$\Delta H_{rxn} = -55.8\ kJ$$

CHAPTER NINE

Covalent Bonding and Molecules

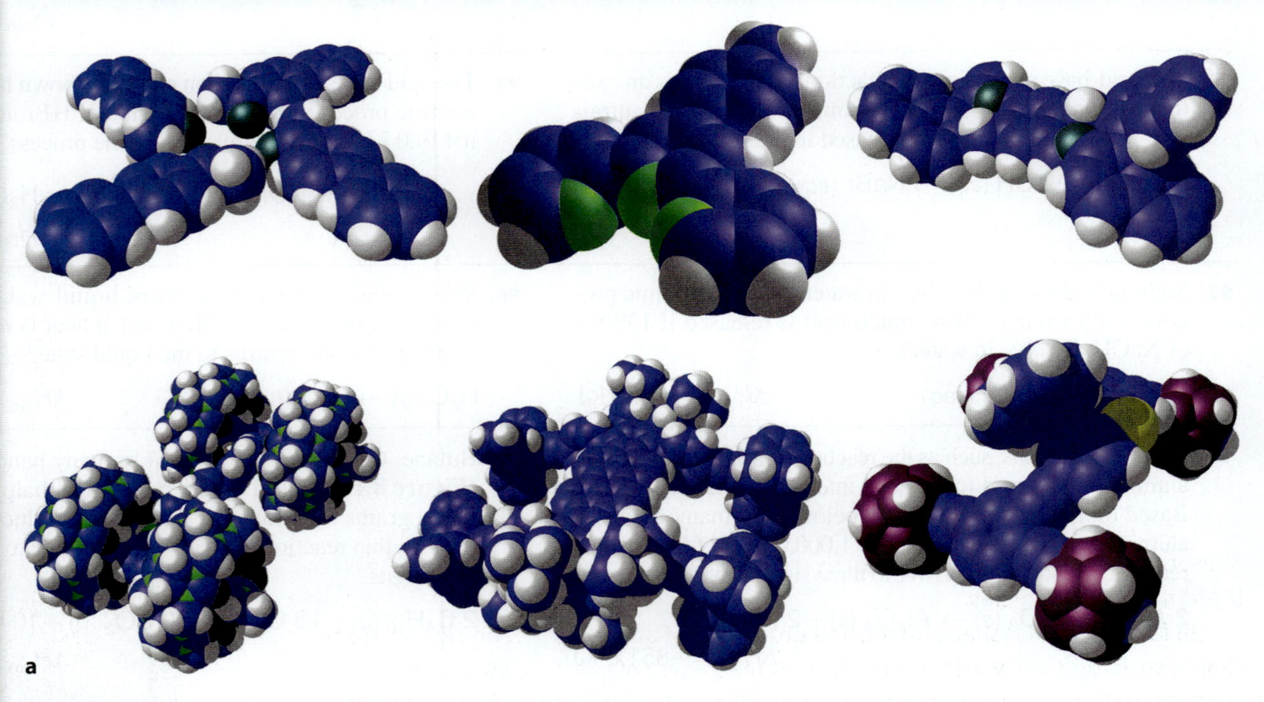

Figure 9.1 (a) The first Nanocar Race featured single-molecule cars from six different research teams. *(continued)*

The Shortest Race

In April 2017, six teams competed in the shortest race in history. The race was only 100 nanometers long—much too small to see with the naked eye. And the cars that competed were each composed of a single molecule. Scientists call these tiny vehicles *nanocars* (**Figure 9.1**).

The idea was pioneered by Dr. James Tour, a professor of chemistry and materials science at Rice University. Inspired by the way nature builds structures, Tour set out to create molecular-scale machines. In 2005, Tour and his students reported building the first nanocar, and they showed that this single molecule could roll across a smooth gold surface. In the years since, they've learned to power the cars with "motors" that convert light energy to push the car along. They've built single-molecule submarines that propel themselves through water. They've even built "nanotrucks" and "nanotrains" that transport a tiny cargo from one place to another.

Tour believes that, although it may take centuries for the technology to develop, we'll eventually use tiny machines to build large structures. "Right now, we're learning to control molecular motion," he says.

His groundbreaking ideas are catching on. In 2017, several research teams met in Toulouse, France, for the first ever Nanocar Race. The incredibly tiny course was on a smooth gold surface inside an instrument called a *scanning tunneling microscope*.

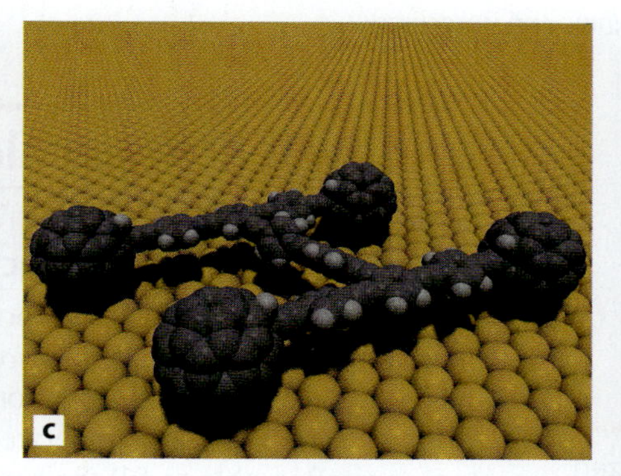

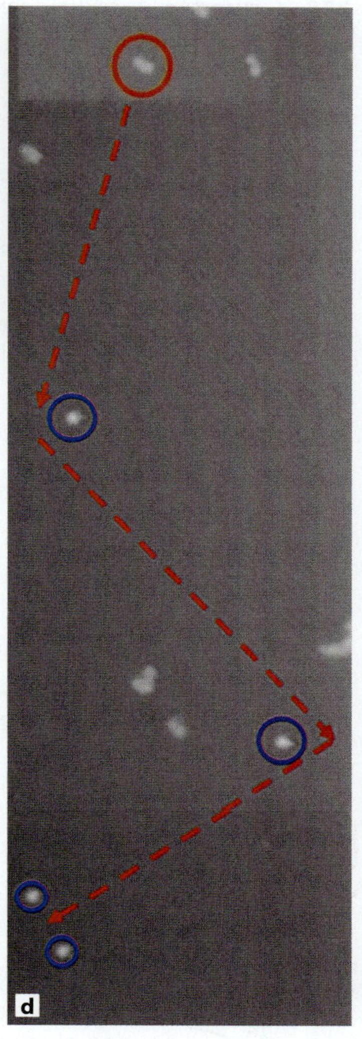

Figure 9.1 (continued) (b) Professor James Tour (right) works with students in his Rice University laboratory. (c) This image is an artist's rendition of an early nanocar on a gold surface. (d) The Nanocar from Rice University ran on a silver surface, rather than gold. It was piloted by students from the Grill research group at the University of Graz in Austria. This image shows the nanocar (circled in red) at the top. The spots circled in blue are single-atom obstacles that the car had to navigate. (a) © Technical University Dresden, Nano-Windmill company; © University of Basel, Swiss Nano Dragster; © Mana-NIMS, Nano-vehicle MANA-NIMS team; © Ohio University, Ohio Bobcat Nano-Wagon Team; © G. RAPENNE/P. ABEILHOU/CEMES-CNRS Nanomobile-Club; © Rice University, Nano-car team; All Images provided by CNRS Photo Library; (b) Jeff Fitlow / Rice University; (c) James M. Tour Group / Rice University; (d) © L. Grill (University of Graz)

On the surface, single atoms marked the obstacles and finish line. The "drivers" used the instrument to guide the cars around the obstacles and across the finish line.

So, how do these tiny machines work? The answer to this question begins with the *shape* of the molecule. When atoms combine to form covalent bonds, they produce structures with predictable shapes. Tour and his students take advantage of this phenomenon. They use chemical reactions to connect groups of atoms as if they were Lego® blocks, creating wheels, axles, and even motors for their molecular vehicles.

In fact, molecular shape affects the properties of any covalent compound, from simple structures like water to the complex machinery of living creatures. In this chapter, we'll see how atoms share electrons through covalent bonds, and look at how electron arrangement determines the shape of a molecule. We'll consider the way charges spread across molecules and ions, and preview the impacts of shape and charge distributions on the behavior of different compounds.

The concepts in this chapter are vitally important. Bonding, shape, and charge form a link between the fundamental aspects of chemistry and the properties that we see in the world around us. So fasten your nano-seatbelt; let's take a drive.

Explore Figure 9.1

Intended Learning Outcomes

After completing this chapter and working the practice problems, you should be able to:

9.1 Covalent Molecules

- Describe the electronic arrangements of covalent structures, including single, double, and triple covalent bonds, lone pairs, and filled and expanded octets.
- Draw Lewis structures for simple molecules.

9.2 Molecules and Charge

- Calculate formal charge, and relate it to the overall charge of polyatomic ions.
- Draw Lewis structures for polyatomic ions.
- Use resonance structures to describe bonding and charge distribution.

9.3 Shapes of Molecules

- Apply the VSEPR model to predict the electronic and molecular geometry for molecules having two, three, or four charge sets.

9.4 Polar Bonds and Molecules

- Describe the trends in electronegativity across the periodic table.
- Use differences in electronegativity to differentiate between covalent, polar covalent, and ionic bonds.
- Estimate molecular dipoles through the combination of polar covalent bonds and molecular shape.

9.1 Covalent Molecules

Representing Covalent Structures

We first encountered chemical bonding in Chapter 5. Remember that there are two broad classes of chemical bonds, *ionic* and *covalent*. Ionic bonds result from the attraction between oppositely charged ions. In a covalent bond, two atoms share electrons.

The octet rule: Atoms are stabilized by eight electrons in their valence level. ■

Covalent bonds occur between nonmetal atoms. By sharing electrons, these atoms are often able to fill their valence levels and satisfy the *octet rule*. Recall that the octet rule states that atoms are stabilized by completely filled *s* and *p* sublevels—this corresponds to eight electrons in the valence shell.

For example, consider chlorine gas, Cl_2. Each neutral chlorine atom has seven valence electrons. By forming a two-electron bond, both chlorine atoms fill their valence levels. In *Lewis structures*, we use a dash to represent two shared electrons:

:C̈l• + •C̈l: —(Two electrons shared)→ :C̈l••C̈l: = :C̈l—C̈l:

In a double bond, atoms share two pairs of electrons. ■

Sometimes, atoms share two pairs of electrons. This is called a covalent *double bond*. We can see an example of a covalent double bond with molecular oxygen, O_2. Each oxygen atom has six valence electrons and therefore needs two additional electrons to fulfill the octet rule.

:Ȯ• + •Ȯ: ⟶ :O::O: = :O=O:

Similarly, atoms can share three pairs of electrons to form a covalent *triple bond*. This type of bonding occurs in a molecule of nitrogen gas. Each nitrogen atom has five valence electrons and therefore needs three additional electrons to fill its valence. The nitrogen atoms form a triple bond.

:Ṅ• + •Ṅ: ⟶ :N:::N: = :N≡N:

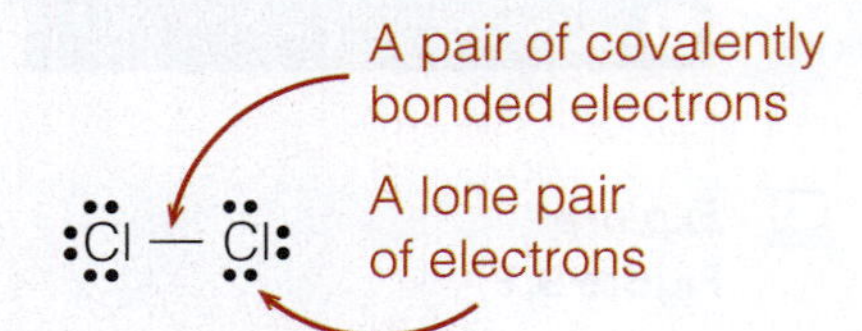

Figure 9.2 This image shows the Lewis structure for a chlorine molecule. The dash between the two atoms represents a two-electron covalent bond. We refer to pairs of unshared molecules as *lone pairs*.

In the Lewis structures just described, we also showed the unshared valence electrons. Although unshared electrons do not form bonds, they often affect the shape of a molecule.

Because electrons pair up with opposite spins, we typically represent unshared electrons in pairs. For example, in the Lewis structure of chlorine, each chlorine atom has four pairs of electrons. One pair of electrons forms the covalent bond; the other three pairs of electrons are unshared. Chemists refer to pairs of unshared valence electrons as **lone pairs** (**Figure 9.2**). We can think of an octet as four pairs of electrons.

An octet contains four pairs of electrons. ■

In most covalent molecules and polyatomic ions, all atoms follow the octet rule. For example, consider the Lewis structure of carbon dioxide, CO_2:

Ö=C=Ö

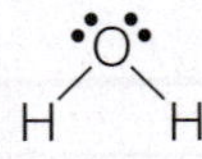

Figure 9.3 In a water molecule, the oxygen atom's valence level is filled with eight electrons, and each hydrogen atom's valence level is filled with two electrons.

In this molecule, the carbon atom forms four covalent bonds. This means there are eight valence electrons around the carbon atom. Each oxygen atom forms two covalent bonds and has two lone pairs. Each oxygen atom also has eight valence electrons. Both carbon and the two oxygen atoms fulfill the octet rule.

Figure 9.3 shows the Lewis structure for water. Notice that the oxygen atom is surrounded by eight valence electrons—a complete octet. Each hydrogen atom has only two electrons (the covalent bond). Because only two electrons are needed to fill energy level 1, each hydrogen atom in this structure also has a complete valence level.

Hydrogen needs just two electrons to fill its valence. ■

Example 9.1 Identifying the Valence Electrons in a Lewis Structure

Urea (CH_4N_2O) is a covalent compound that is a major component of urine. The Lewis structure for urea is shown below. Indicate the number of covalent bonds and lone pairs around each carbon, hydrogen, nitrogen, and oxygen atom. Do these atoms fulfill the octet rule?

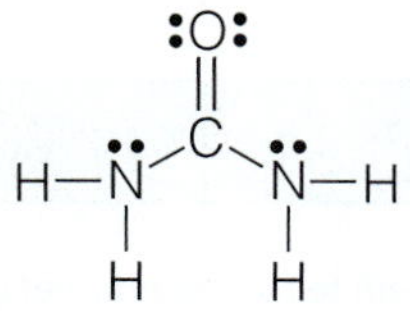

The carbon atom in this molecule has four covalent bonds (two single bonds and one double bond). This arrangement corresponds to eight valence electrons and fulfills the octet rule. The two nitrogen atoms each have three covalent bonds and one lone pair. This also adds up to eight electrons. The oxygen atom also has eight valence electrons (two covalent bonds and two lone pairs). So the carbon, nitrogen, and oxygen atoms all fulfill the octet rule.

Each hydrogen atom has only two electrons in its valence shell. Because energy level 1 holds only two electrons, the hydrogen atoms also have a filled valence level. This table summarizes these common bonding patterns.

Atom	Covalent bonds	Lone pairs	Valence electrons
C	4	0	8
N	3	1	8
O	2	2	8
H	1	0	2

TRY IT

1. The Lewis structure of hypochlorous acid, HClO, is shown here. Identify the number of covalent bonds and lone pairs around each atom. Do these atoms fulfill the octet rule?

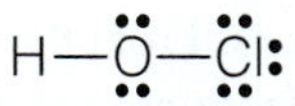

Exceptions to the Octet Rule

Although most covalent molecules follow the octet rule, some exceptions occur. For example, the element boron often forms covalent compounds in which the boron atom has an incomplete valence level (**Figure 9.4**).

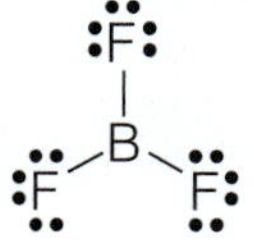

Figure 9.4 Many boron compounds have an incomplete valence level.

A more common exception occurs with larger elements. The nonmetals in rows 3–7 of the periodic table sometimes form bonds that result in an **expanded octet** containing more than eight electrons. This is possible because the *d* sublevels in the higher energy levels can accommodate extra electrons. Atoms with expanded octets often contain five or six covalent bonds.

Atoms in rows 3–7 can form *expanded octets* with more than eight electrons.

For example, phosphorus commonly forms two compounds with chlorine, PCl_3 and PCl_5 (**Figure 9.5**). The first compound, PCl_3, adheres to the octet rule. In the second compound, the phosphorus atom has an expanded octet with 10 valence electrons.

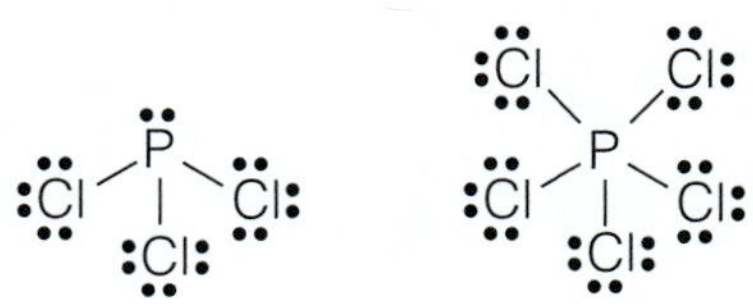

Figure 9.5 PCl_3 follows the octet rule. The phosphorus atom in PCl_5 has an expanded octet with 10 valence electrons.

Drawing Lewis Structures

To understand how covalent compounds behave, we often need to understand their structures. The following four-step process will help you draw Lewis structures from a molecular formula (**Figure 9.6**):

1. *Add up all the valence electrons.* Use the periodic table to determine the number of valence electrons in each atom. Recall that the main-group numbers

1. Sum electrons.
2. Draw framework.
3. Fill octets on outer atoms.
4. Fill octet on central atom.

Figure 9.6 Follow these steps to draw Lewis structures.

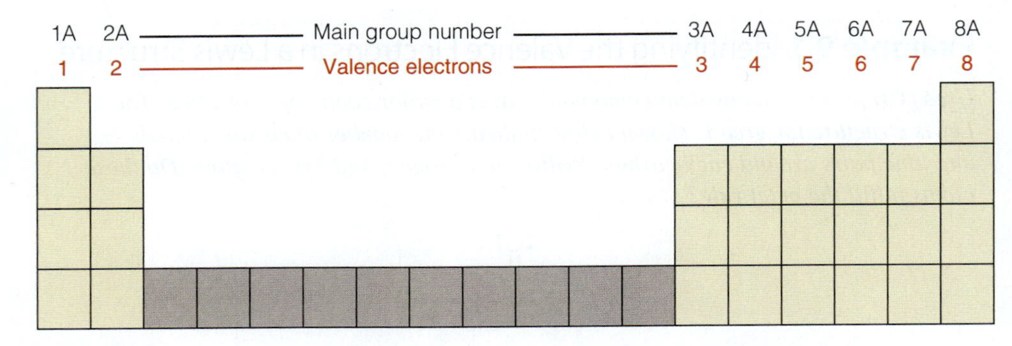

Figure 9.7 The main-group numbers 1A–8A tell us the number of valence electrons in each main-group element.

(1A–8A) tell us the number of valence electrons in each main-group element (**Figure 9.7**).

2. *Frame the structure*, using single bonds. Typically the atom that is nearer to the lower left of the periodic table is the central atom. Because it can hold only two electrons, hydrogen is never a central atom.
3. *Fill the octets of the outer atoms first.* Add lone pairs to the outer atoms to complete their valence levels.
4. *Fill the octet on the central atom.* After filling in the octets on the outer atoms, place any remaining electrons on the central atom. These electrons may be enough to fulfill the octet on the central atom (and may even result in an expanded octet). If the central atom does not have at least eight electrons, use double or triple bonds to fulfill the octet.

Examples 9.2 and 9.3 illustrate this important technique.

Example 9.2 Drawing a Lewis Structure

Draw a Lewis structure for nitrogen trichloride, NCl_3.

We can solve this problem by carefully following the four-step process:

Step 1: Add up all the valence electrons. From the periodic table, we can determine that the nitrogen atom has five valence electrons, and each chlorine atom has seven valence electrons. Based on this information, there are a total of 26 valence electrons in this structure.

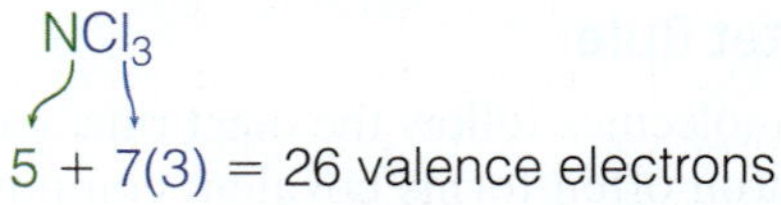

NCl_3

$5 + 7(3) = 26$ valence electrons

Step 2: Frame the structure. We draw nitrogen as the central atom because it is to the left of chlorine on the periodic table. We connect the central nitrogen atom to the outer chlorine atoms with a single bond, which we show as a dash.

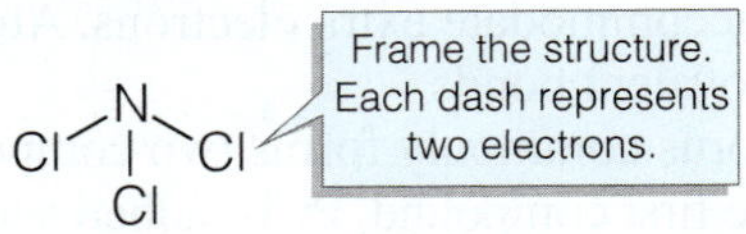

Cl–N–Cl, with a third Cl bonded below N

Frame the structure. Each dash represents two electrons.

Step 3: Fill the octets of the outer atoms first. The covalent bonds each represent two electrons. To complete the octets on the outer atoms, we must add six more electrons to each chlorine:

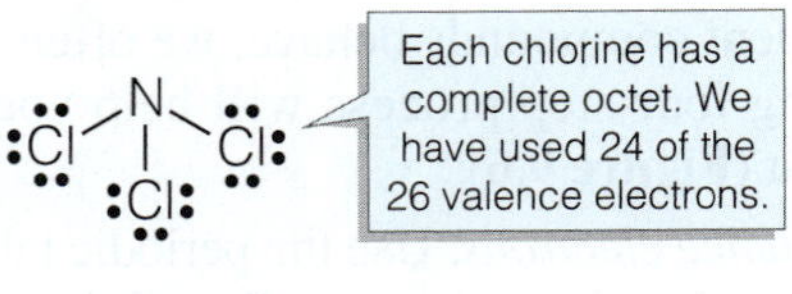

:Cl–N–Cl:, with a third :Cl: bonded below N (each Cl with three lone pairs)

Each chlorine has a complete octet. We have used 24 of the 26 valence electrons.

Step 4: Fill the octet on the central atom. We used 24 of the 26 valence electrons to complete the octets around the chlorine atoms. We draw the remaining two electrons on the central atom:

Each atom has a complete octet. The complete structure shows all 26 valence electrons.

Notice that in this Lewis structure, we have accounted for all 26 valence electrons, and each atom has a filled octet.

Example 9.3 Drawing a Lewis Structure

Formaldehyde, CH_2O, is a gas that is commonly used as a precursor to manufacture plastics. Draw the Lewis structure for a formaldehyde molecule.

As we did for Example 9.2, let's follow the four-step process for drawing this structure.

Step 1:

$$\text{Carbon} + \text{oxygen} + \text{two hydrogens} = 4 + 6 + 1(2) = 12 \text{ valence electrons}$$

Step 2: Draw carbon as the central atom, surrounded by the oxygen and hydrogen atoms. We connect the carbon to each outer atom using single bonds:

Step 3: We draw octets around the outer atoms. Because hydrogen is a row 1 element, its valence is filled by the single bond to carbon. We draw six additional electrons around the oxygen atom to complete its octet:

Step 4: Now we need to fill the octet on the central atom. However, we have used all 12 valence electrons to complete the octets of the outer atoms. Therefore, we need to share a pair of electrons between an outer atom and the central carbon atom. To accomplish this, we move two electrons down from the oxygen to form a double bond to the carbon, so that both atoms have a filled octet.

The valence around carbon is not filled...

...So we move two electrons from the oxygen to form a double bond...

...Now all of the atoms have a filled valence.

TRY IT

2. Draw Lewis structures for these molecules:

a. NH_3 **b.** SF_6 **c.** CS_2

3. Draw Lewis structures for these molecules:

a. $SiCl_4$ **b.** HCN **c.** HNO

9.2 Molecules and Charge

Polyatomic Ions and Formal Charge

In Chapter 5, we introduced *polyatomic ions* as groups of atoms with an overall charge (see Section 5.2). The distribution of these charges across the ions affects many of their chemical properties. Now that we have a better understanding of covalent bonding, let's look at the structures of eight common polyatomic ions (**Figure 9.8**). Most of these ions follow the octet rule, but two ions (sulfate and phosphate) have expanded octets.

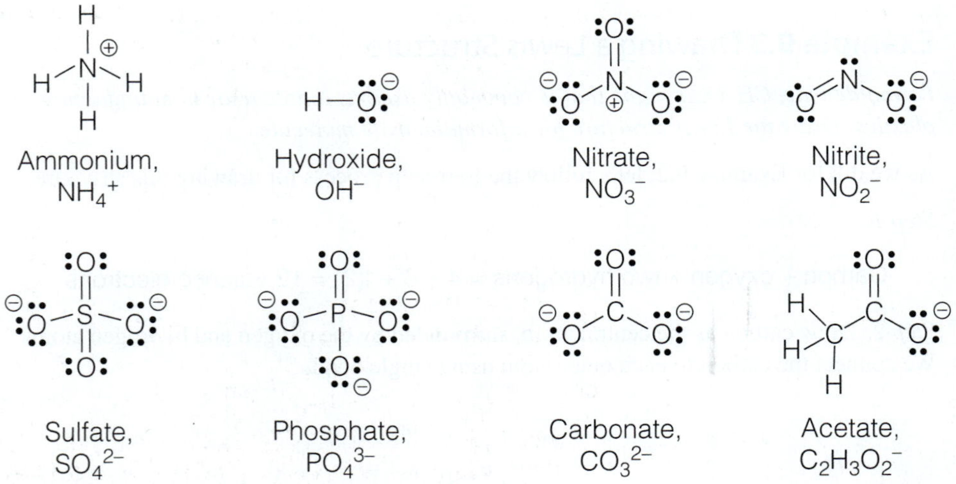

Ammonium, NH_4^+ | Hydroxide, OH^- | Nitrate, NO_3^- | Nitrite, NO_2^-

Sulfate, SO_4^{2-} | Phosphate, PO_4^{3-} | Carbonate, CO_3^{2-} | Acetate, $C_2H_3O_2^-$

Figure 9.8 The Lewis structures of several common polyatomic ions.

In Figure 9.8, notice that each polyatomic ion has at least one atom with a positive or negative charge. These charges are called **formal charges**. They allow us to locate charged sites within a molecule or polyatomic ion. Formal charges are not a perfect indicator of the charge on an atom, but they are a good tool for keeping track of the overall charge of a molecule or ion. We can calculate the formal charge on any atom using this equation:

Formal charge = Valence electrons in the neutral atom − Number of covalent bonds − Number of unshared electrons

Formal charge estimates the total electron charge around bonded atoms. Formal charge assumes that atoms share electrons evenly. Because two atoms share two electrons, each covalent bond contributes a charge equivalent to one electron (**Figure 9.9**).

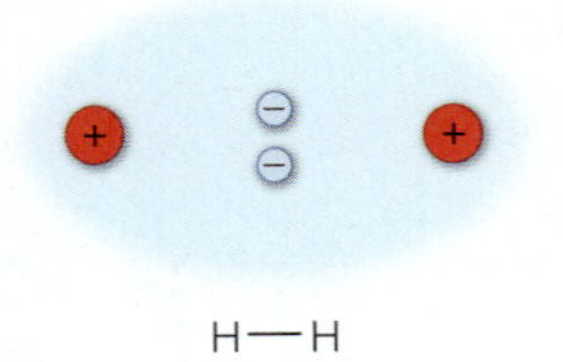

Figure 9.9 In a covalent bond, the charge from two electrons is spread over two atoms. As a result, each nucleus feels the effective charge of one electron.

For example, the hydroxide ion has a charge of –1. Based on the following Lewis structure, does the negative charge lie on the oxygen atom or on the hydrogen atom?

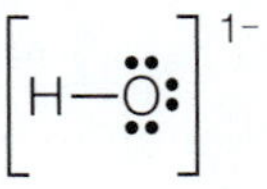

$$[\text{H—}\ddot{\text{O}}\text{:}]^{1-}$$

To answer this, we calculate the formal charge on each atom. Let's begin with the hydrogen atom: A neutral hydrogen atom has one electron. The hydrogen has one covalent bond and zero unshared electrons. Therefore, its formal charge is zero:

$$\text{Formal charge (hydrogen)} = 1 - 1 - 0 = 0$$

1 valence electron; 1 covalent bond; 0 unshared electrons

Now let's examine the formal charge on the oxygen atom: A neutral oxygen atom has six valence electrons. The oxygen in the OH^- ion has six unshared electrons plus an additional covalent bond. This means it has an excess of electrons and an overall negative charge:

$$\text{Formal charge (oxygen)} = 6 - 1 - 6 = -1$$

When we draw the Lewis structure for hydroxide, we commonly show the minus charge adjacent to the oxygen atom:

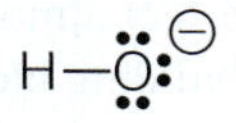

H—O: ⊖

Let's look at a second example. The thiocyanate ion, SCN^-, has an overall charge of minus one. If we draw the Lewis structure as shown, the sulfur atom has a formal charge of minus one, and the carbon and nitrogen atoms both have formal charges of zero:

:S—C≡N:

Sulfur: FC = 6−1−6 = −1

Carbon: FC = 4−4−0 = 0

Nitrogen: FC = 5−3−2 = 0

The sum of the formal charges on the atoms equals the overall charge of the molecule or ion.

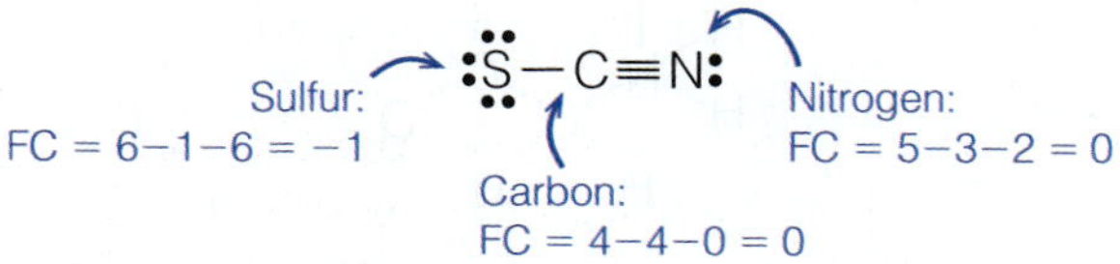

Formal charges are a powerful tool and are especially important in biological chemistry. Two elements that are critical for life, oxygen and nitrogen, frequently carry formal charges. When covalently bonded, a neutral oxygen atom has two bonds and four unshared electrons. If the oxygen atom has three bonds and two unshared electrons, it will have a formal charge of +1. On the other hand, if an oxygen atom has only one bond and six unshared electrons, it will have a formal charge of −1 (**Figure 9.10**).

Nitrogen is similar: While a neutral nitrogen atom has three bonds and two unshared electrons, it can gain one bond (giving it a formal charge of +1) or lose one bond (giving it a formal charge of −1).

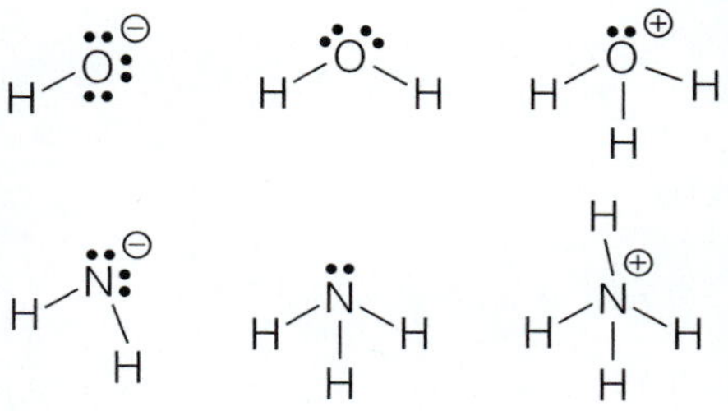

Figure 9.10 Oxygen and nitrogen both form species with different formal charges.

Example 9.4 Calculating Formal Charge

Automotive airbags contain sodium azide, NaN_3. The Lewis structure for the azide ion (without charges) is shown below. Calculate the formal charge on each atom in this structure. What is the overall charge of the azide ion?

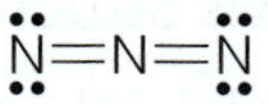

N=N=N

Because nitrogen is in group 5A, we know that a neutral nitrogen atom has five valence electrons. The central nitrogen atom has four covalent bonds and zero unshared electrons. Therefore, the formal charge of this atom is +1:

$$\text{Formal charge (central atom)} = 5 - 4 - 0 = +1$$

The two outer nitrogen atoms each have two covalent bonds and four unshared electrons. Therefore, the formal charges of these atoms are each −1:

$$\text{Formal charge (outer atoms)} = 5 - 2 - 4 = -1$$

So the atoms in this ion have formal charges of −1, +1, and −1, as shown below. Based on this, the overall charge of the ion is −1. The correct formula for the azide ion is N_3^-.

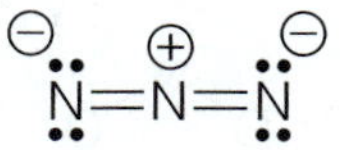

⊖ ⊕ ⊖
N=N=N

Check it
Watch Explanation

TRY IT

4. Calculate the formal charge on each atom in these structures:

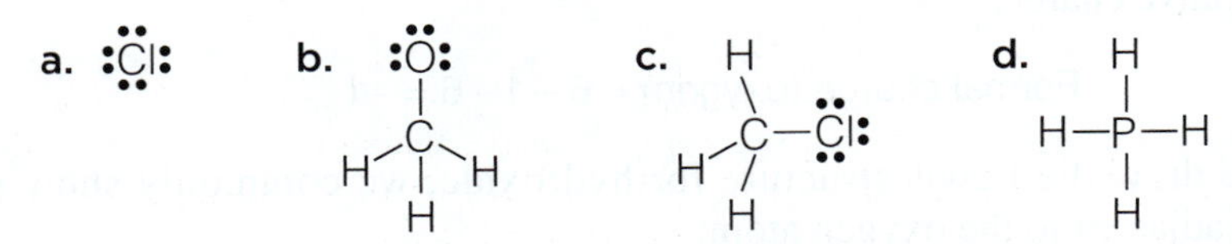

5. Proteins are large molecules that affect almost every function of the human body. Proteins are composed of smaller building blocks called amino acids. The simplest amino acid is glycine. In water, two atoms in a glycine molecule have a formal charge. In the structure of glycine shown here, all bonds are drawn, but the unbonded valence electrons are missing. Draw in the missing valence electrons, and locate both sites that have a nonzero formal charge. Calculate the formal charge at these sites.

Drawing Lewis Structures for Polyatomic Ions

Drawing Lewis structures for polyatomic ions is similar to drawing Lewis structures for neutral molecules. However, we must consider the charge of the ion when we determine the number of valence electrons present. If an ion has a negative charge, it means that extra electrons are present.

For example, how many valence electrons are present in a hydroxide ion? We know that hydroxide has the formula OH^-. The oxygen has six valence electrons, and the hydrogen has one valence electron. The charge (−1) means that one additional electron is present:

Total electrons = 6 (oxygen) + 1 (hydrogen) + 1 (charge) = 8 valence electrons

Example 9.5 illustrates this idea further.

Example 9.5 Drawing a Lewis Structure for a Polyatomic Ion

Draw a Lewis structure for the nitrite ion, NO_2^-. Show all nonzero formal charges.

We can solve this problem by carefully following the steps given earlier:

Step 1: Add up all the valence electrons. From the periodic table, we can determine that the nitrogen atom has five valence electrons and that each oxygen atom has six valence electrons. Because there is a −1 charge on the ion, we know that one additional electron is present on this ion. Therefore, this ion has a total of 18 valence electrons:

$$NO_2^-$$

5 + 6(2) + 1 = 18 valence electrons

Step 2: Frame the structure. We draw nitrogen as the central atom because it is to the left of oxygen on the periodic table:

O–N–O

Step 3: Fill the octets of the outer atoms first. Placing electrons around each oxygen atom uses 16 of the 18 available valence electrons:

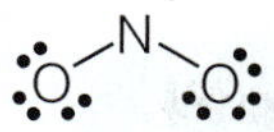

Step 4: Fill the octet on the central atom. We draw the remaining two electrons on the central atom, as shown below. This leaves the nitrogen atom two electrons short of a filled octet. To fulfill the octet, we must move two electrons from one of the oxygen atoms to form a double bond between the nitrogen and oxygen:

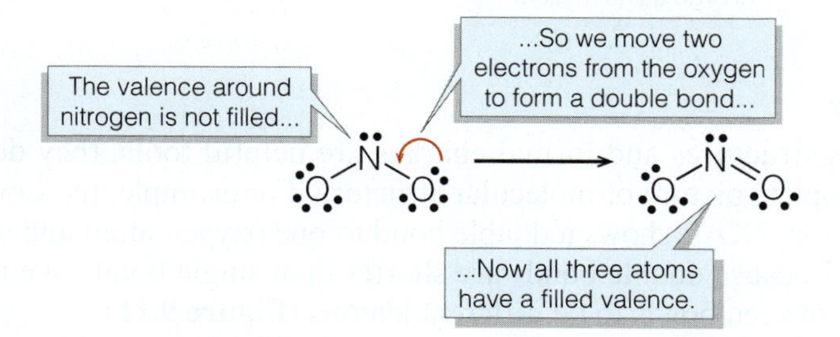
The valence around nitrogen is not filled...
...So we move two electrons from the oxygen to form a double bond...
...Now all three atoms have a filled valence.

Show formal charges. As drawn, the oxygen atom on the left has a formal charge of minus one. We indicate this with a minus charge, to give the final answer.

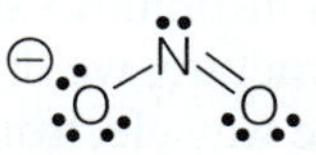

TRY IT

6. Draw a Lewis structure for the methylthiolate anion, CH_3S^-. Carbon is the central atom. Show any nonzero formal charges.

Check it
Watch Explanation

Choosing the Best Lewis Structure

Formal charges are useful for identifying the best Lewis structures when more than one structure is possible. For example, let's draw the structure of phosgene ($COCl_2$), a highly toxic gas that was used as a chemical weapon in World War I. We can draw a Lewis structure for this compound by following the steps outlined earlier. This molecule has a total of 24 valence electrons. Placing the carbon atom in the center and filling in the octets of the outer atoms, we get a structure like this:

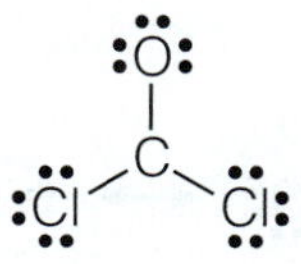

Now we need to fill the octet on the central atom. We could do this either with a double bond to the oxygen atom or a double bond to one of the chlorine atoms, as shown below. Which structure is more correct?

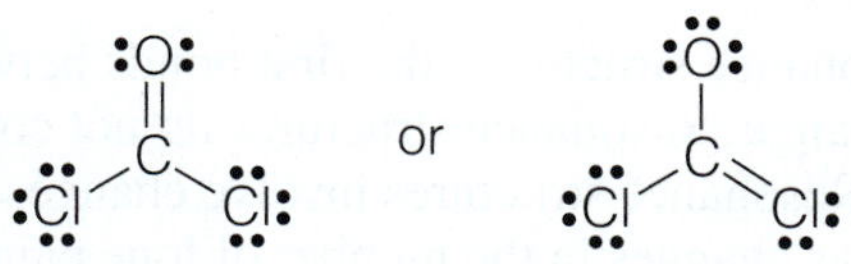

or

In general, the best Lewis structures are those that minimize the charge values or have zero values for all of the atoms. The best structure for phosgene has a double bond to the oxygen, and single bonds to both chlorines, because all atoms have a formal charge of zero:

The best Lewis structure has the smallest formal charge on each atom.

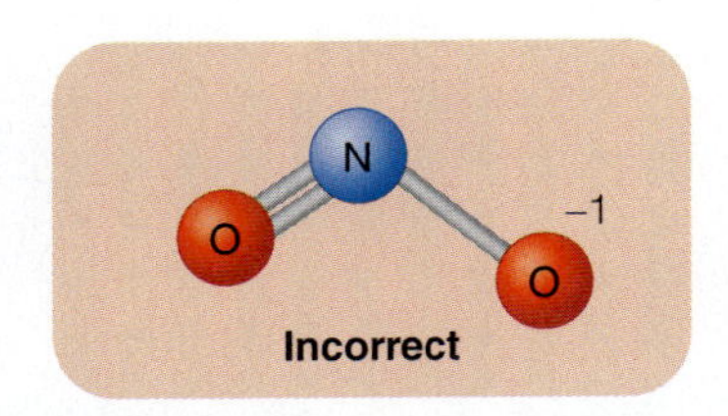

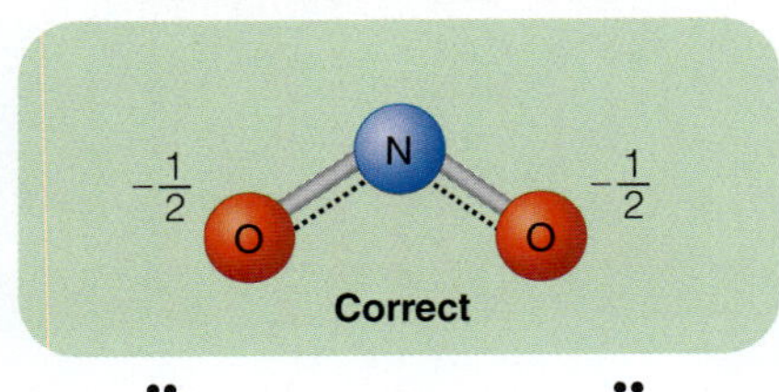

Figure 9.11 In the nitrite ion, both oxygen atoms have bonds of equal length, and the negative charge is shared evenly between the two oxygen atoms. We represent this arrangement with resonance structures.

Resonance structures involve second or third covalent bonds as well as lone pairs.

Resonance

While Lewis structures and formal charges are helpful tools, they do not always give us a complete picture of molecular structure. For example, the Lewis structure of the nitrite ion, NO_2^-, shows a double bond to one oxygen atom and a single bond to the other. Because double bonds are shorter than single bonds, we might expect the nitrogen–oxygen bonds to be different lengths (**Figure 9.11**).

In fact, both of the bonds are the same length. What's more, the negative charge is shared equally between the two oxygens, so each atom has about one-half the charge of an electron. How can we explain this?

The reality of the nitrite ion is difficult to show with a single Lewis structure. To better describe the electronic structure, we draw a set of Lewis structures, called **resonance structures**, that show the way electrons are distributed around a molecule or ion. Figure 9.11 shows two Lewis structures for the nitrite ion. Neither structure by itself is totally correct, but the *average* of the two structures conveys the truth that the negative charge is shared between the two oxygen atoms. Chemists use a double-headed arrow to indicate resonance structures (**Figure 9.12**).

Figure 9.12 Chemists indicate resonance structures using a double-headed arrow.

Generally, ions that spread charge over multiple atoms are more stable than those that do not. We can draw resonance structures for most nonmetal oxyanions (including nitrate, nitrite, phosphate, sulfate, carbonate, and acetate), and they all exhibit remarkable stability. In these resonance structures, the outer oxygen atoms share the negative charge on the ion (**Figure 9.13**).

Figure 9.13 The phosphate ion has four major resonance structures.

When drawing resonance structures, the first bonds between atoms (that is, the single bonds) do not change. Resonance structures do not completely break or form bonds between atoms. Resonance structures involve changes in the second (or third) covalent bonds as well as changes in the number of lone pairs. This concept is illustrated in Example 9.6.

Example 9.6 Evaluating Charge Distribution from Resonance Structures

The nitrate ion (NO_3^-) has three major resonance structures. Draw each structure. Based on these structures, what is the average charge on each oxygen atom?

Following the steps given in Section 9.1, we can first draw a single Lewis structure for the nitrate ion, as shown in the left structure below. In this structure, one of the three oxygen atoms contains a double bond. To draw the other resonance structures, we shift the electrons so that the double bond lies on each of the other oxygen atoms. We use a double-headed arrow between the structures to indicate that these are resonance structures:

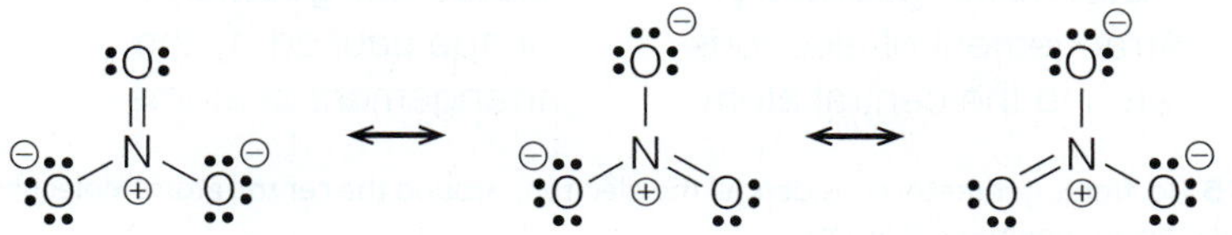

Notice in these structures that the first bonds between the atoms do not change. The only changes involve the second bond and the unshared electrons. The actual structure of the nitrate ion is an average of these three structures.

Each oxygen atom has a negative charge in two of the three resonance structures. Based on this, we can say that the charge on each oxygen atom is approximately $-\frac{2}{3}$.

Volodymyr Burdiak/Shutterstock

There is an ancient parable about several blind men who touch a different part of an elephant (the trunk, the side, the leg). Each describes the elephant differently—one as a snake, one as a wall, one as a tree. Similarly, each resonance structure is a partial description; the actual bonding is a composite of these individual pictures.

TRY IT

7. The resonance structures for the carbonate ion are shown below, including formal charges. Based on these structures, what is the average charge on each oxygen atom?

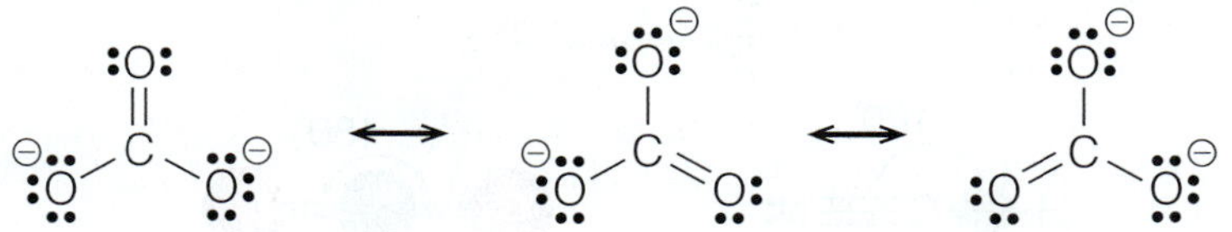

8. The formate ion has the formula CHO_2^- (C is the central atom). Draw two resonance structures for the formate ion. Include formal charges.

Check It

Watch Explanation

9.3 Shapes of Molecules

The shapes of molecules critically affect their properties. The nanocars that we discussed at the beginning of this chapter are a dramatic example of this, but shape affects many properties in ways that are more subtle. For example, when we breathe (**Figure 9.14**), our bodies expel two compounds, carbon dioxide (CO_2) and water (H_2O). Both are small, simple molecules. But CO_2 is a gas at room temperature, and water is a liquid. Why the difference? As we'll see in the pages to come, the shape of each molecule plays a big part in determining these properties.

Fortunately, we can use Lewis structures to predict the shapes of molecules. To do this, we use an approach called the **valence shell electron pair repulsion (VSEPR) model**. The VSEPR model is based on the idea that electrons repel each other, and therefore they occupy regions that are as far away from each other as possible. From this idea, a fairly limited set of possible geometries emerges.

Brilliant Eye/Shutterstock

Figure 9.14 When we breathe, we emit carbon dioxide and water. The properties of these molecules are partially determined by their shape.

We describe geometries in two ways. First is the **electronic geometry**. This is the arrangement of electrons around a central atom. The second approach, called the **molecular geometry**, describes the arrangement of *atoms* within a molecule (**Figure 9.15**). The molecular geometry (sometimes simply called the *shape*) depends on the electronic geometry, but it considers only the location of the atoms.

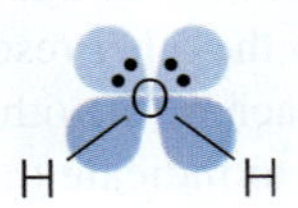

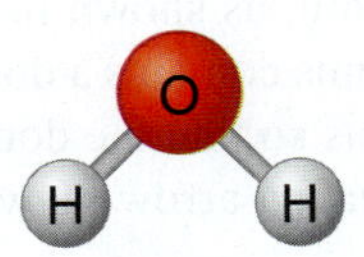

Electronic geometry
Arrangement of electrons around the central atom

Molecular geometry
Shape caused by the arrangement of atoms

Figure 9.15 Electronic geometry considers all the electrons around the central atom. Molecular geometry considers only the arrangement of atoms.

Two electron sets. Let's begin with a simple molecule, hydrogen cyanide. **Figure 9.16** shows the Lewis structure for this molecule. There are two "sets" of electrons around the central carbon atom—the set bonded to the hydrogen atom (highlighted in green) and the set bonded to the nitrogen atom (highlighted in orange). In the VSEPR model, we consider the triple bond to be one "set" of electrons because these electrons all occupy the area between the carbon and nitrogen atoms. The two electron sets (orange and green) repel each other, so they will move as far away from each other as possible—meaning they will end up on opposite sides of the carbon atom. The electronic geometry around the central carbon is **linear**; that is, the bond angle between the green set and the yellow set is 180°. As a result, the molecular geometry (the arrangement of atoms) is also linear.

In the VSEPR model, single, double, and triple bonds are all counted as one electron "set."

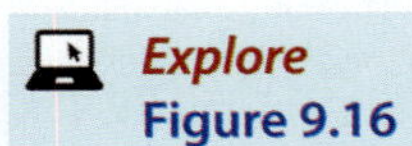

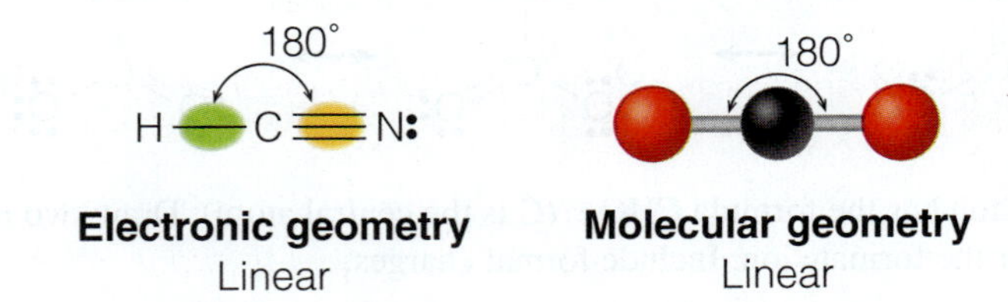

Figure 9.16 The electronic and molecular geometries of hydrogen cyanide are linear. For simplicity, this figure and the ones that follow use black for the central atom and red for all outer atoms. Additionally, these figures show single, double, and triple bonds as a single set of electrons.

Let's consider another molecule, carbon dioxide, CO_2. Which of the three geometries shown here would be the most stable?

a b c

Again, we need to consider the number of electron sets around the central atom. Like our first example, this molecule has only two sets of electrons around the central carbon atom (each double bond counts as one set), and so the most stable arrangement is the one that keeps the electrons as far away from each other as possible—that is, the geometry shown in option c. Any time the central atom has only two electron sets, the electronic and molecular geometry around that atom are both linear, with a bond angle of 180°.

Three electron sets. Next let's examine a molecule that has three electron sets around the central atom, such as formaldehyde, CH_2O (**Figure 9.17**). As before, these three electron sets repel each other, pushing as far away from each other as possible. In this case, the electrons adopt a geometry that we describe as **trigonal planar**. In a trigonal planar geometry, the molecule is flat, and each set of electrons is separated by an angle of 120°.

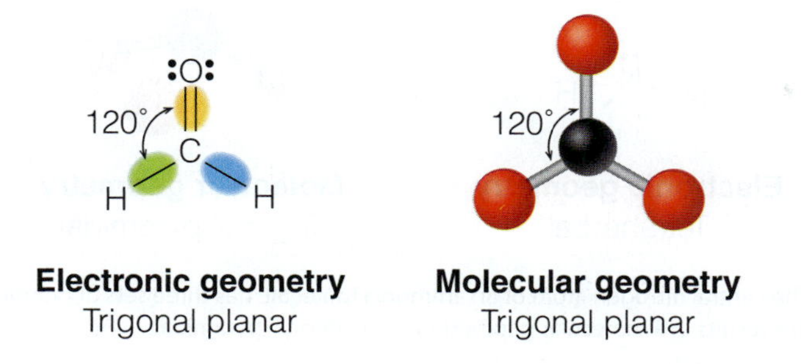

Figure 9.17 The electronic and molecular geometries of formaldehyde (CH_2O) are trigonal planar.

The nitrite ion, shown in **Figure 9.18**, also has three electron sets around the central atom: the double bond to the oxygen atom (highlighted in green), the single bond to the other oxygen atom (highlighted in blue), and the lone pair (in orange). The lone pair takes up as much or more space than a bonded set of electrons, and so we must consider it in predicting the geometry of the molecule. Three electron sets adopt a trigonal planar electronic geometry.

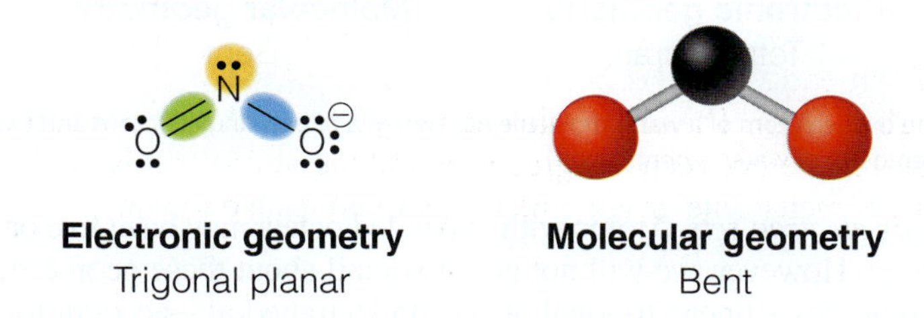

Figure 9.18 The nitrite ion has an electronic geometry (represented by the three shaded clouds) that is trigonal planar. The molecular geometry, which considers only the arrangement of atoms, is bent.

In this example, the electronic geometry and the molecular geometry are not the same. Although the electronic geometry is trigonal planar, only two of the three electron sets are connected to atoms, and so the molecular geometry is better described as *bent*.

Four electron sets. Finally, let's consider a molecule with four sets of electrons around the central atom, such as CH_4. What shape would this molecule take? You might initially guess a cross shape, like a plus sign or the letter *x*. In this geometry, the sets of electrons are separated by only 90°. By adopting a three-dimensional geometry, the electron sets create more space between themselves. In a **tetrahedral geometry**, the sets of electrons are separated by 109.5°. A tetrahedral geometry can be thought of as an X-shape in which two of the branches have been turned sideways, as shown in **Figures 9.19** and **9.20**.

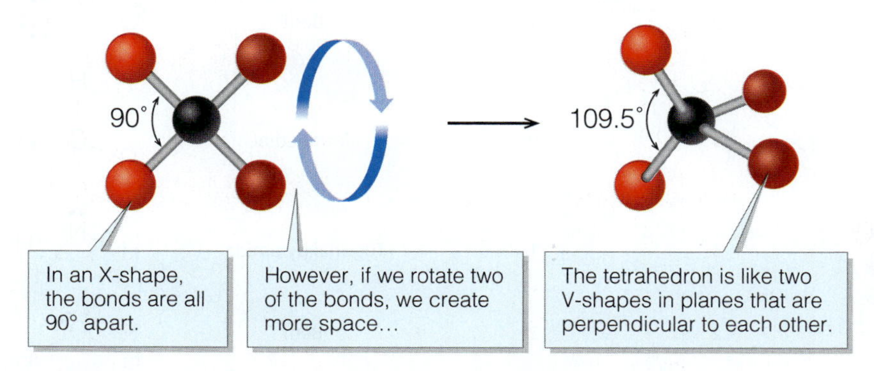

90°
109.5°
In an X-shape, the bonds are all 90° apart.
However, if we rotate two of the bonds, we create more space…
The tetrahedron is like two V-shapes in planes that are perpendicular to each other.

Figure 9.19 Moving from an X-shape to a tetrahedron provides more space between the electron sets.

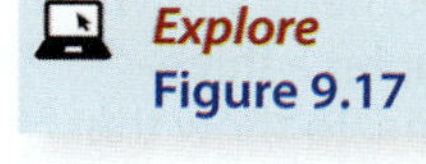

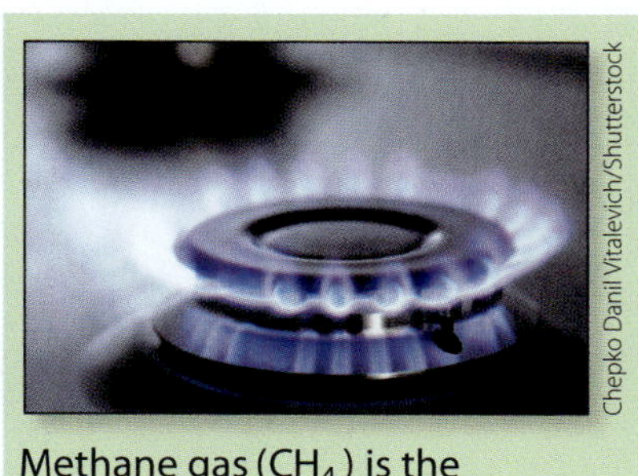

Methane gas (CH_4) is the main component of natural gas. This compound has a tetrahedral geometry. (See Figure 9.20 below.)

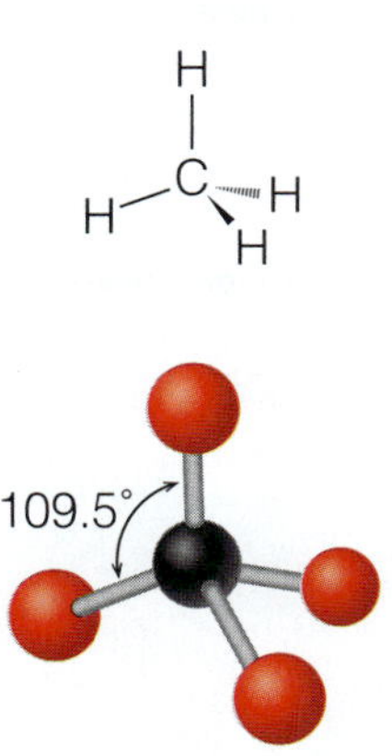

Figure 9.20 The electronic and molecular geometries of methane (CH_4) are tetrahedral.

The tetrahedral electronic geometry can also lead to more than one molecular geometry. For example, look at the structure of ammonia shown in **Figure 9.21**. Because there are four electron sets around the central atom, ammonia has a tetrahedral electronic geometry. However, if we consider only the atoms, the molecule forms a pyramid shape. We refer to the molecular geometry of ammonia as *trigonal pyramidal*.

The tetrahedral geometry shown in Figure 9.20 occupies a three-dimensional volume. To show this shape clearly, chemists sometimes draw bonds using wedges and dashes. The wedge indicates that the atom is coming toward you; the dash indicates it is going away from you.

Electronic geometry
Tetrahedral

Molecular geometry
Trigonal pyramidal

Figure 9.21 The central nitrogen atom of an ammonia molecule has three sets of bonding electrons and one lone pair. This results in a molecular geometry that is trigonal pyramidal.

The electronic geometry of a water molecule is also tetrahedral (**Figure 9.22**). Notice that the water molecule has two bonding sets and two lone pairs. The two lone pairs push the hydrogen atoms toward each other. We refer to the molecular geometry of water as *bent*.

Each electronic geometry produces one or more molecular geometries.

Electronic geometry
Tetrahedral

Molecular geometry
Bent

Figure 9.22 The oxygen atom of a water molecule has two sets of bonding electrons and two lone pairs. The molecular geometry of water is bent.

Five and six electron sets. Atoms with expanded octets may have five or six electron sets around them. However, we will not go into detail about these geometries here. The three main geometries—linear, trigonal planar, and tetrahedral—account for most of the shapes you will encounter in chemistry. These geometries are summarized in **Table 9.1**.

TABLE 9.1 Electronic and Molecular Geometries

Electron sets	Electronic geometry	Model	Bonding sets	Nonbonding sets (lone pairs)	Molecular geometry	Examples
2	Linear		2	0	Linear	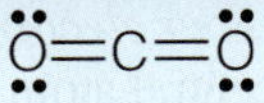 CO_2
3	Trigonal planar	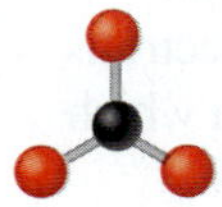	3	0	Trigonal planar	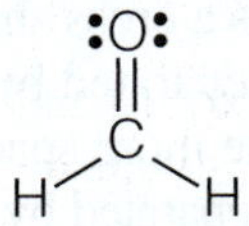 H_2CO
			2	1	Bent	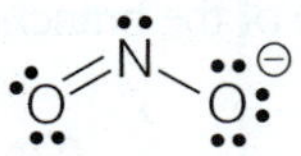 NO_2^-
4	Tetrahedral	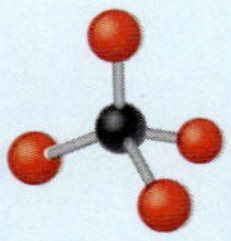	4	0	Tetrahedral	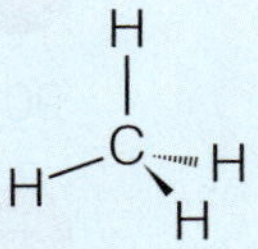 CH_4
			3	1	Trigonal pyramidal	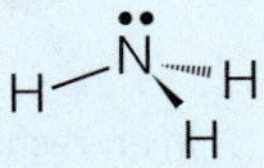 NH_3
			2	2	Bent	 H_2O

Remember that these geometries describe only the shape around *one central atom*: If a molecule contains multiple atoms along the backbone, the geometry of each atom must be considered. As an example, look at this molecule:

The atoms highlighted in green triangles have a trigonal planar electronic geometry because they are each surrounded by three sets of electrons. Those highlighted in yellow squares have a tetrahedral electronic geometry because they are surrounded by four sets of electrons. Notice that the peripheral atoms (those connected to only one other atom) are not highlighted. Only the central atoms affect the geometry of the molecule, and so we don't consider the geometries of atoms bonded to only one other atom.

Example 9.7 Determining the Molecular and Electronic Geometry of a Molecule

Dihydrogen sulfide, H_2S, is an awful-smelling gas produced by rotting eggs. Draw the Lewis structure for this compound. What is the electronic geometry of this compound? What is the molecular geometry?

To solve this problem, we must first draw a Lewis structure for H_2S. A correctly drawn structure shows that the sulfur atom has two bonds to hydrogen atoms and two lone pairs:

Because this molecule has four electron sets around the central atom, its electronic geometry is tetrahedral. However, its molecular geometry—the arrangement of its three atoms—is bent.

Example 9.8 Determining the Molecular and Electronic Geometry of a Molecule

Lithium carbonate is a simple ionic compound that is widely used to treat bipolar disorder. What is the molecular geometry of the carbonate ion?

Use the four-step technique from Section 9.1 to draw Lewis structures for ions. ■

Although this question does not explicitly ask for a Lewis structure, it is essential for solving the problem. The Lewis structure for the carbonate ion is shown here:

Based on this structure, we can see that the central carbon atom has three electron sets around it and therefore an electronic geometry that is trigonal planar. Because all three electron sets are bonded to atoms, the molecular geometry is also trigonal planar.

TRY IT

9. Draw a Lewis structure for carbon disulfide. What is the electronic geometry around the central atom?

10. Salicylic acid is found in willow bark and was used for centuries as an anti-inflammatory and fever-reducing agent. Using the structure of salicylic acid below, indicate the electronic and molecular geometries on each of the numbered atoms. The first is done as an example.

Atom	Electronic geometry	Molecular geometry
1	Trigonal planar	Trigonal planar
2		
3		
4		
5		
6		

9.4 Polar Bonds and Molecules

Electronegativity and Polar Covalent Bonds

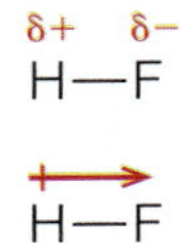

Figure 9.23 There are two ways to represent a polar covalent bond.

So far, we've described covalent bonds as electrons being "shared" between two atoms. But electrons are often not shared evenly. For example, consider the simple molecule hydrogen fluoride (**Figure 9.23**). The fluorine pulls the electrons more strongly than the hydrogen, so the electrons are, on average, closer to the fluorine than to the hydrogen. Consequently, the fluorine atom has a small negative charge, while the hydrogen atom has a small positive charge. We show this by writing $\delta+$ (read as "delta-plus") above the positive ion and $\delta-$ ("delta-minus") above the negative ion. (The lowercase Greek letter delta is commonly used in math to show a very small increment.)

Polar covalent bonds are like two people sharing a bed when one of them hogs the blanket. The one who gets a little more of the blanket is slightly warmer; the one who gets less is slightly colder. They are sharing the blanket, but not evenly.

Alternatively, we can show this polarity by drawing an arrow pointing in the direction that the electrons are pulled, with a small cross on the back of the arrow to indicate the positive end. We refer to this type of bond as a **polar covalent bond**.

How can we determine which bonds are polar covalent? To do this, we use a property called **electronegativity**, which was first described by the great American scientist Linus Pauling (**Figure 9.24**). Electronegativity is a measure of how strongly atoms pull bonded electrons. The electronegativities of the elements are shown in **Table 9.2**. This table conveys one of the most important ideas in understanding chemical bonding, and it is worth exploring in detail.

Figure 9.24 Linus Pauling (1901–1994) developed the electronegativity scale and helped to develop the concepts relating to ionic and covalent bonds. He won the Nobel Prize in Chemistry in 1954 and the Nobel Peace Prize in 1962.

TABLE 9.2 Pauling's Table of Electronegativities

							H 2.1										
Li 1.0	Be 1.5											B 2.0	C 2.5	N 3.0	O 3.5	F 4.0	
Na 0.9	Mg 1.2											Al 1.5	Si 1.8	P 2.1	S 2.5	Cl 3.0	
K 0.8	Ca 1.0	Sc 1.3	Ti 1.5	V 1.6	Cr 1.6	Mn 1.5	Fe 1.8	Co 1.8	Ni 1.8	Cu 1.9	Zn 1.7	Ga 1.6	Ge 1.8	As 2.0	Se 2.4	Br 2.8	
R 0.8	Sr 1.0	Y 1.2	Zr 1.4	Nb 1.6	Mo 1.8	Tc 1.9	Ru 2.2	Rh 2.2	Pd 2.2	Ag 1.9	Cd 1.7	In 1.7	Sn 1.8	Sb 1.9	Te 2.2	I 2.5	
Cs 0.7	Ba 0.9	La 1.1	Hf 1.3	Ta 1.5	W 1.7	Re 1.9	Os 2.2	Ir 2.2	Pt 2.2	Au 2.4	Hg 1.9	Tl 1.8	Pb 1.8	Bi 1.9	Po 2.0	At 2.2	
Fr 0.7	Ra 0.9																

Notice that electronegativity generally increases from left to right across the periodic table. This pattern is consistent with what we already know from ionic bonding: Metals tend to lose electrons easily, while nonmetals tend to pull, or gain, electrons.

Electronegativity increases as we go up a group or column on the periodic table. For example, iodine has a value of 2.5, while the elements above it—bromine, chlorine, and fluorine—have increasingly higher values. Fluorine is the most electronegative of the elements, meaning it has the strongest pull for electrons. On the lower left-hand side, cesium and francium are the least electronegative, meaning they have the weakest hold on their electrons.

If a metal atom bonds with a nonmetal atom, the electronegative nonmetal pulls the electrons away from the metal. This results in the formation of positive and negative ions, and an ionic bond.

Polar covalent bonds lie somewhere between purely ionic and purely covalent bonds. In fact, there is a continuum between purely covalent (two atoms having equal electronegativities) and purely ionic (**Figure 9.25**). We measure just how covalent or how ionic a bond is by measuring the difference between the electronegativities of the two elements (see Table 9.2).

Table 9.2 does not include the noble gases. Electronegativity is a measure of how strongly atoms pull electrons in a bond. Because noble gases generally do not form compounds, they do not have electronegativity values.

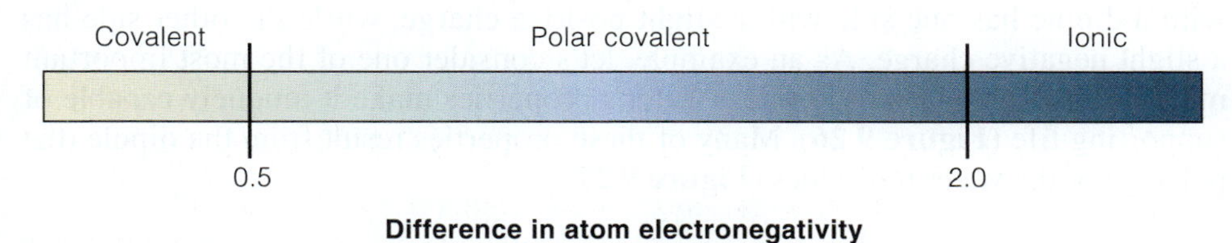

Figure 9.25 Bonds may be classified by the differences in electronegativity of the two bonded atoms.

In general, bonds are classified this way:

- *Covalent:* difference in electronegativity less than 0.5
- *Polar covalent:* difference between 0.5 and 2.0
- *Ionic:* difference greater than 2.0

Electronegativities help us understand how charge is distributed in a bond. In a polar covalent bond, the more electronegative element has a greater share of the electrons and therefore a slight negative charge. As we'll see in later sections, the polarity and charge differences described by these simple rules have a huge impact on the chemical and physical properties of substances.

Kevin Revell

Neither dog will let go of the rope, but the bigger dog has a much stronger hold on it. Another example of a polar covalent bond.

Example 9.9 Determining the Polarity of a Bond

Which bond is more polar, a C—O bond or an F—S bond? Show the direction of polarity for both bonds.

To determine the polarity of a bond, we use the table of electronegativities (Table 9.2). Carbon has an electronegativity of 2.5, while oxygen has a value of 3.5. The difference between these values is 1.0, meaning this is a polar covalent bond.

Similarly, the difference between F (4.0) and S (2.5) is 1.5. This bond is also polar covalent. Because the difference is slightly larger in this case, we can say that the F—S bond is more polar than the C—O bond.

The atom with the larger electronegativity has a slight negative charge, and the atom with the smaller electronegativity has a slight positive charge. We can show the polarities of these two bonds using either the arrow or the $\delta+/\delta-$ convention.

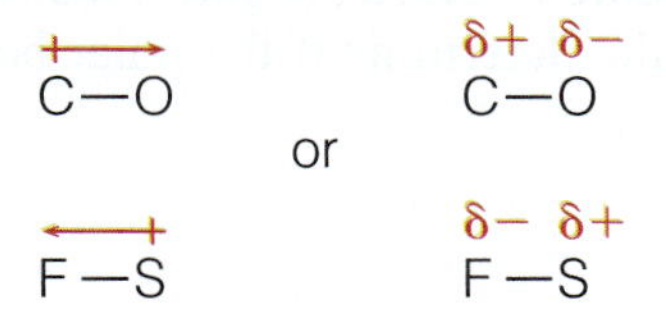

or

$\delta+$ $\delta-$
C—O

$\delta-$ $\delta+$
F—S

Check it
Watch explanation

TRY IT

11. Find the difference in electronegativity between each of these atoms, and describe the bond as covalent, polar covalent, or ionic:

a. C—O **b.** N—H **c.** Na—Cl **d.** Fe — Cl **e.** C — H

12. Use either the $\delta+$ and $\delta-$ symbols or an arrow to show the direction of bond polarity for each of these bonds:

a. O—H **b.** Zn—S **c.** C—Li

13. Which bond is more polar, carbon–oxygen or carbon–nitrogen?

Rich Carey/Shutterstock

Figure 9.26 Water's extraordinary properties depend on its shape and the polarity of its bonds.

Molecules with Dipoles

Polar covalent bonds have a tremendous impact on a molecule's properties. Molecules with polar covalent bonds often exhibit an overall polarity, called a **molecular dipole** (sometimes called a *net dipole* or simply a *dipole*). A molecule with a dipole has one side with a slight positive charge, while the other side has a slight negative charge. As an example, let's consider one of the most important molecules in the universe—water. Water's properties make it uniquely capable of supporting life (**Figure 9.26**). Many of these properties result from the dipole (the polarity) of the water molecules (**Figure 9.27**).

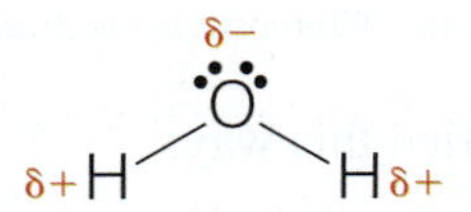

Figure 9.27 Because of its shape and the polarity of its bonds, a water molecule contains a dipole—one side of the molecule is slightly positive, while the other side is slightly negative.

Molecular dipoles result from polar covalent bonds and the shape of the molecules.

Water's dipole is a result of its polar covalent bonds and its shape. In earlier sections, we saw that water has a bent molecular geometry because of the two lone pairs on the oxygen atom. And because oxygen is more electronegative than hydrogen, the oxygen side of the molecule has a slight negative charge while the hydrogen side has a slight positive charge.

Polar bonds add together to produce a net molecular dipole.

Identifying Molecules with a Net Dipole

The polarity of a molecule depends on its shape and the polarity of its bonds. If we add together the effects of all polar covalent bonds in a molecule, we can determine the net dipole for a molecule. Often this is done graphically, by visual inspection.

To see how this works, let's compare two molecules, water and carbon dioxide. **Figure 9.28** shows the direction of the polar covalent bonds in these molecules. In the water molecule, the polar covalent bonds pull electron density toward the oxygen atom. To find the net dipole of this molecule, we add together the pull of each polar covalent bond.

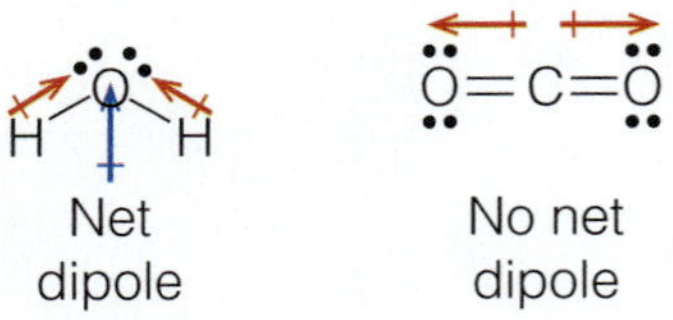

Net dipole

No net dipole

Figure 9.28 In a water molecule, the polar covalent bonds pull the electrons toward the oxygen atom, producing a net dipole. Because carbon dioxide pulls the electrons in opposite directions, the molecule does not have a net dipole—it does not have positive and negative ends.

The CO_2 molecule also contains polar covalent bonds. However, the molecule has a linear geometry, so the two polar covalent bonds pull in opposite directions. As a result, the molecule has no net dipole.

To determine whether a molecule has a net dipole, first identify any polar covalent bonds using the table of electronegativities (see Table 9.2). Then, based on the shape of the molecule, determine if the polar bonds pull together or cancel each other out. Let's consider three more examples, illustrated in **Figure 9.29**. The first molecule, CH_4, does not have a net dipole. Because carbon and hydrogen have very similar electronegativities, the bonds in CH_4 are nonpolar. The second molecule, BCl_3, contains polar bonds; but the symmetrical pull of the three chlorine atoms cancel each other out, so there is no overall dipole. In the third molecule, CH_2F_2, the two carbon–fluorine bonds are polar, creating a net dipole in the direction of the two fluorine atoms.

No net dipole   No net dipole   Net dipole

Figure 9.29 We can determine the net dipoles by adding together the pull of the polar bonds. Individual bond dipoles are shown in red; net dipoles are shown in blue.

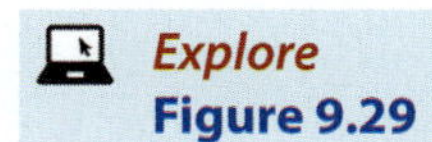

***Explore* Figure 9.29**

Example 9.10 Identifying a Molecular Dipole

Does sulfur dichloride have a net dipole? Draw a Lewis structure to support your answer.

To answer this question, we must first draw a correct Lewis structure for this compound. Using the rules in Section 9.1, we obtain the following Lewis structure:

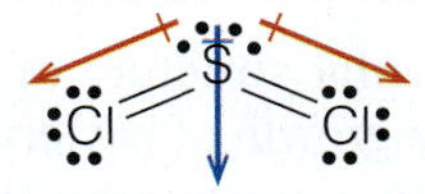

The central sulfur atom has four electron sets, resulting in a tetrahedral electronic geometry, and a bent molecular geometry. Because of the difference in electronegativity, the sulfur–chlorine bonds are polar (shown with the red arrows). And because the molecule has a bent molecular geometry, these polar bonds add together to produce a small net dipole (the blue arrow). The chlorine side of the molecule is slightly negative, while the sulfur side of the molecule is slightly positive.

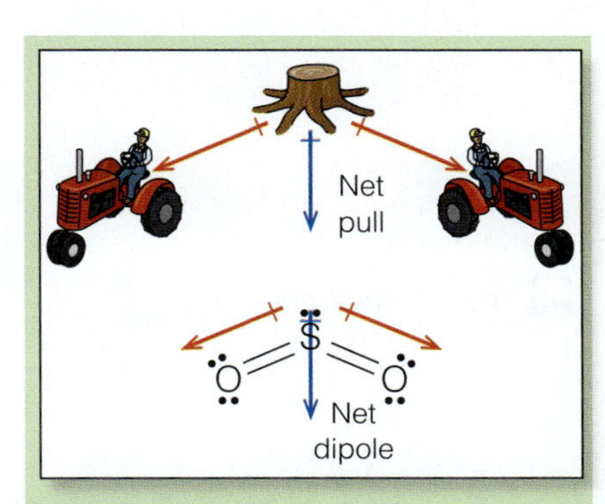

Imagine two tractors pulling at a stump in different directions. The net pull on the stump can be found by adding together the pull of the two tractors. Similarly, we find the net dipole by adding together the pull of each polar covalent bond.

TRY IT

14. Draw Lewis structures for these molecules. Determine if the molecules have a net dipole.

a. HCN   **b.** CS_2 **c.** SiF_4

Check it* *Watch Explanation

How Dipoles Affect Properties—A Preview

The shape and polarity of molecules influence their behavior. For example, let's compare the properties of carbon dioxide and water. The two molecules are about the same size, and both have polar covalent bonds. But because of their different shapes, water has a net dipole while carbon dioxide does not (see Figure 9.28).

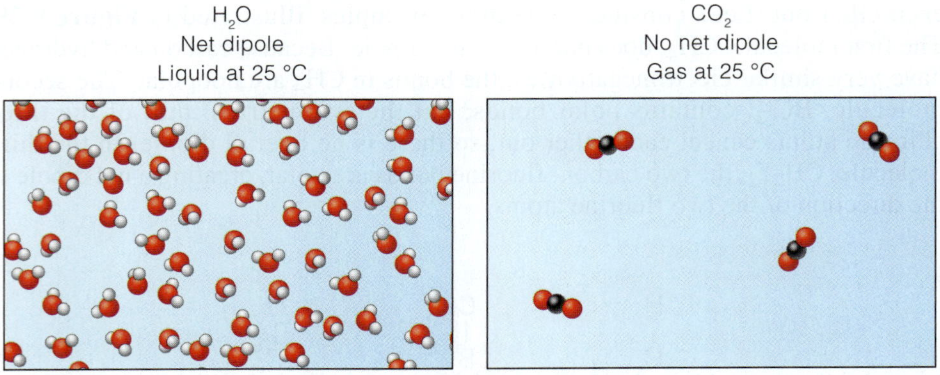

H_2O
Net dipole
Liquid at 25 °C

CO_2
No net dipole
Gas at 25 °C

Figure 9.30 Water's net dipole causes the molecules to stick closely together, making it a liquid at room temperature. Carbon dioxide has no net dipole, and it is a gas at room temperature.

Because of their dipoles, water molecules have an attraction for each other. This causes the molecules to stick closely together (**Figure 9.30**). Water is a liquid at room temperature because of the attraction between the molecules. Water is also able to dissolve many ionic compounds because charged particles (ions) are attracted to the polar water molecules.

On the other hand, carbon dioxide is a gas at room temperature. Because its molecules do not have a net dipole, they do not stick together as tightly as water molecules do.

In the chapters ahead, we will explore these ideas in much more detail and see how the shape and polarity of molecules determines how they interact with their surroundings.

Capstone Video

Capstone Question

Figure 9.31 shows the molecular structure of allicin, a pungent substance found in garlic that inhibits the growth of bacteria like *Staphylococcus aureus*.

Analyze this substance by determining the electronic and molecular geometry of each numbered atom. Use the electronegativities in Table 9.2 to identify any polar covalent bonds, and indicate the direction of polarity.

Use the structure to determine the molecular formula, molar mass, and percent by mass of sulfur in this compound. Biological studies have found that aqueous solutions containing as little as 28 μg of allicin per milliliter inhibit bacterial growth. Express this amount as the number of moles of allicin per liter of solution.

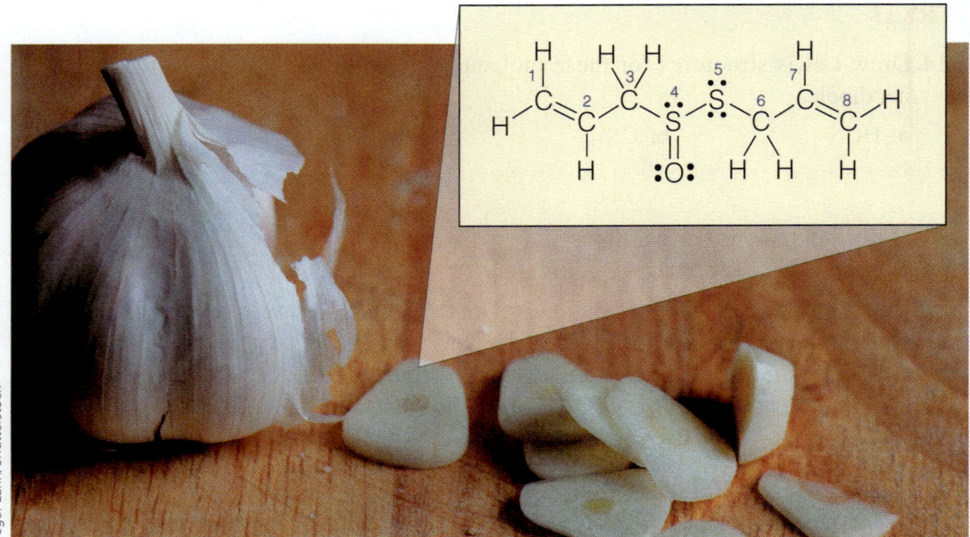

Figure 9.31 This image shows the molecular structure of Allicin, a pungent substance found in garlic.

SUMMARY

In this chapter, we've explored the bonding, shape, and polarity of covalent molecules. In covalent bonds, two nonmetal atoms share electrons. Atoms can share two electrons (a single bond), four electrons (a double bond), or six electrons (a triple bond). We commonly represent molecules with covalent bonds using Lewis structures.

In most covalent structures, atoms fulfill the octet rule by having eight electrons in their valence level. Hydrogen, a row 1 element, can accommodate only two electrons in its valence level. Atoms in rows 3 through 7 of the periodic table sometimes have expanded octets with more than eight electrons.

Polyatomic ions are groups of atoms that are covalently bonded but possess an overall charge. We often use formal charge to represent the charges on atoms in a polyatomic ion. While formal charge is not a perfect indicator of charge distribution, it does give us a tool for keeping track of the total number of electrons as well as the total charge in a molecule or ion. Some molecules and ions share electrons in a way that we cannot adequately describe with a single Lewis structure. Resonance structures are sets of Lewis structures that together describe the distribution of electrons through a molecule or ion.

The shapes of molecules are critically important to their behavior. The molecular geometry describes the arrangement of atoms in a molecule. A related term is *electronic geometry*, which is the arrangement of electrons around a bonded atom. Using the valence shell electron pair repulsion (VSEPR) model, we can predict the electronic geometries of most compounds. The three most common electronic geometries are linear, trigonal planar, and tetrahedral, corresponding to two, three, or four sets of electrons around a central atom. The molecular geometries are a function of the basic electronic geometries.

Another factor that defines a substance's behavior is the polarity of its bonds. Pauling's table of electronegativities defines how strongly an atom pulls bonded electrons. By comparing the electronegativities of two atoms, we can predict whether bonds between those atoms will be covalent, polar covalent, or ionic.

The polarities of the individual bonds in a molecule add together to produce a net dipole. If a molecule has a net dipole, one side of the molecule has a slight positive charge, while the other side has a slight negative charge. We can often estimate the direction of a net dipole by considering the shape of the molecule and the direction of the individual bond dipoles.

The type of bonds, shape, and polarity of a molecule profoundly affect its physical properties. As we'll see in upcoming chapters, properties such as melting point and boiling point depend on these important factors.

Building a Nanocar

At the beginning of this chapter, we introduced Dr. James Tour, the Rice University chemist behind the first "nanocar"—a single-molecule vehicle that can propel itself across a smooth gold surface. Now that we've developed a better understanding of molecular shape, let's take a closer look at how the Rice University team designed their molecular car (**Figure 9.32**).

To build the wheels, the Tour group needed a spherical structure. Fortunately, several common molecules have this type of shape. Their first nanocar used a molecule called a *fullerene* as the wheels. This fascinating molecule is composed of 60 carbon atoms that form a perfect sphere (Figure 9.32b). The fullerene wheels rolled nicely on the gold surface, but they created problems when Tour and his team added a light-driven motor. To avoid this issue, they had to change the wheel design. Adamantane is another nearly spherical molecule with the molecular formula $C_{10}H_{16}$. The team found that adamantane wheels also rolled smoothly on the gold surface but didn't cause the problems that the fullerene wheels did.

To build the axles, the Tour group used two carbons connected by a triple bond. Each carbon has a linear geometry, producing a straight axle (Figure 9.32c). A single bond connects the wheels to the axle. Because atoms can rotate freely around a single bond, the wheels are able to spin.

Finally, the team designed a motor. Some double bonds change shape when they absorb light (Figure 9.32d, e). The nanocar uses this change in shape to turn a paddlewheel that pushes the molecule across the surface.

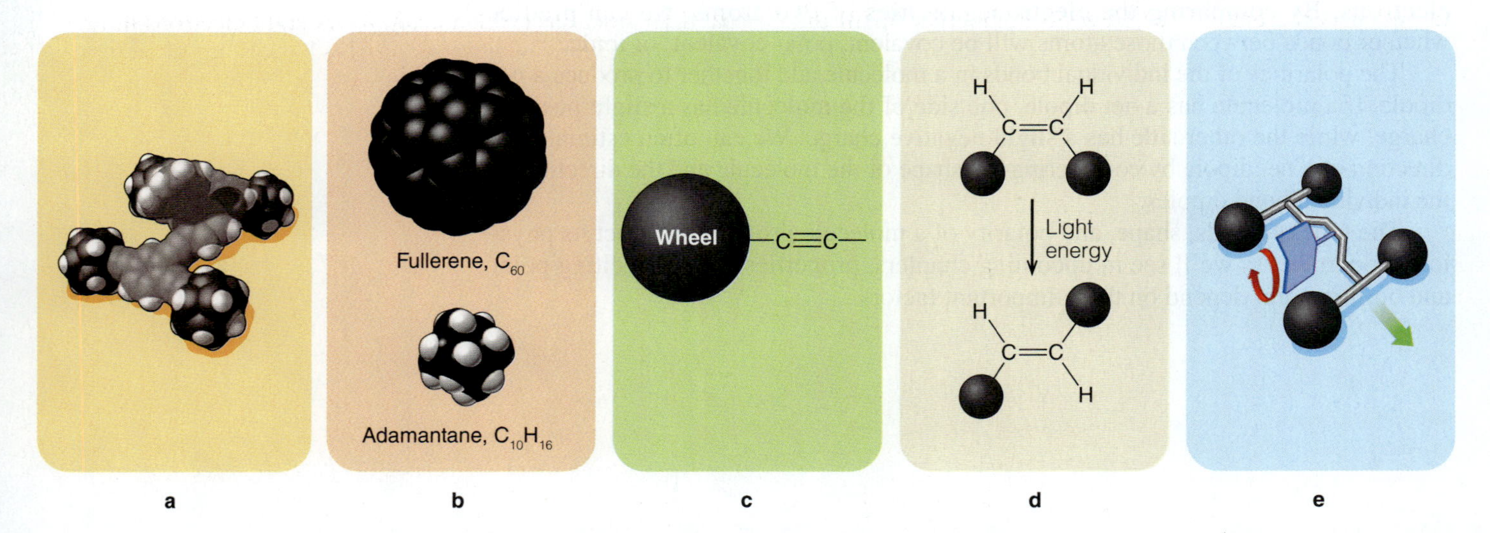

Figure 9.32 (a) The complete nanocar features four wheels, two axles, a chassis, and a paddlewheel-style motor. (b) Tour's first nanocars used fullerene as the wheels. Newer designs use adamantane. (c) The carbon–carbon triple bond produces a linear geometry that is perfect for the axles. (d) The paddlewheel motor is powered by a carbon–carbon double bond that changes shape when it absorbs light energy. (e) This image is a simplified representation of the nanocar.

Key Terms

9.1 Covalent Molecules

lone pair A pair of unshared valence electrons.

expanded octet A bonding arrangement in which an atom has 10 or 12 valence electrons; this is possible only with elements in rows 3–7 of the periodic table.

9.2 Molecules and Charge

formal charge A method of identifying charged sites on a molecule or ion. The formal charge of an atom is the number of electrons in the valence of that atom in its neutral, unbonded state, minus the number of covalent bonds, minus the number of unshared electrons.

resonance structures A set of Lewis structures that show how electrons are distributed around a molecule or ion. Resonance structures are used when a single Lewis structure cannot adequately depict the structure.

9.3 Shapes of Molecules

valence shell electron pair repulsion (VSEPR) model A way of predicting the geometry of molecules based on the number of electron sets around a central atom.

electronic geometry A description of the arrangement of electrons around a central atom.

molecular geometry A description of the arrangement of atoms within a molecule; the molecule shape.

linear A geometry in which two atoms or electron sets are separated by 180° angles.

trigonal planar A geometry in which three atoms or electron sets are separated by 120° angles.

tetrahedral A geometry in which four atoms or electron sets are separated by 109.5° angles.

9.4 Polar Bonds and Molecules

polar covalent bond A covalent bond in which the atoms do not share the electrons evenly; in this type of bond, one atom has a slight positive charge, while the other has a slight negative charge.

electronegativity A measure of how strongly atoms pull bonded electrons.

molecular dipole An overall polarity in which different sides of a molecule have slight positive and negative charges; sometimes called a *net dipole* or simply a *dipole*.

Additional Problems

9.1 Covalent Molecules

15. What is the octet rule? How do nonmetals fulfill the octet rule?

16. If an atom has a filled octet, it has eight electrons in its valence level. What two energy sublevels are filled in a complete octet?

17. How many bonded electrons are represented in this Lewis structure? How many nonbonded electrons are represented?

H—N̈=Ö:

18. How many bonded electrons are represented in this Lewis structure? How many nonbonded electrons are represented?

:O: ‖ C with H and H

19. Complete the table to indicate the number of covalent bonds, lone pairs, and valence electrons around each atom in the molecule shown. Do these atoms fulfill the octet rule?

H—Ö—Ö—H

Atom	Covalent bonds	Lone pairs	Valence electrons
H			
O			
O			
H			

20. Complete the table to indicate the number of covalent bonds, lone pairs, and valence electrons around each atom in the molecule shown. Do these atoms fulfill the octet rule?

:O: ‖ C with H and :C̈l:

Atom	Covalent bonds	Lone pairs	Valence electrons
C			
O			
Cl			
H			

21. Determine whether the central atom in each of these structures has an incomplete octet, a complete octet, or an expanded octet:

a. 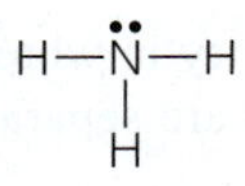 b. 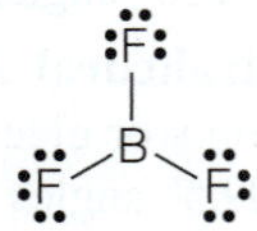 c.

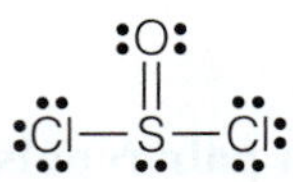

22. Determine whether the central atom in each of these structures has an incomplete octet, a complete octet, or an expanded octet:

a. 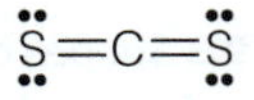 b. 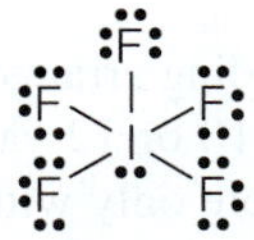 c.

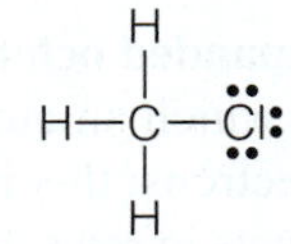

23. Which of these elements—boron, carbon, nitrogen, or oxygen—commonly forms compounds with an incomplete octet?

24. Which of these elements—nitrogen, sulfur, or iodine—never forms an expanded octet? Why not?

25. Find the total number of valence electrons in these molecules:

a. HCl b. NH_3 c. PCl_3 d. CH_3Cl

26. Find the total number of valence electrons in these molecules:

a. N_2 b. C_2H_2 c. SF_6 d. C_2H_4SO

27. Find the total number of valence electrons in these molecules:

a. PCl_5 b. HBr c. H_2SO_4 d. HNO_3

28. Find the total number of valence electrons in these molecules:

a. BBr_3 b. $HCSN$ c. C_2H_4 d. $C_6H_{12}O_6$

29. Draw Lewis structures for each of these molecules:

a. H_2 b. Cl_2 c. HCl

30. Draw Lewis structures for each of these molecules:

a. HBr b. H_2O c. CCl_4

31. Draw Lewis structures for each of these molecules:

a. N_2 b. HNO (nitrogen is the central atom) c. SCl_2

32. Draw Lewis structures for each of these molecules:

a. CO_2 b. PCl_3 c. HCN (carbon is the central atom)

33. Draw Lewis structures for each of these molecules:

a. $BrCl$ b. SiF_4 c. CS_2

34. Draw Lewis structures for each of these molecules:

a. SeF_4 b. PBr_3 c. CH_2Cl_2 (carbon is the central atom)

9.2 Molecules and Charge

35. The carbonate ion (shown here) has an overall charge of −2. Calculate the formal charge on each atom in this structure.

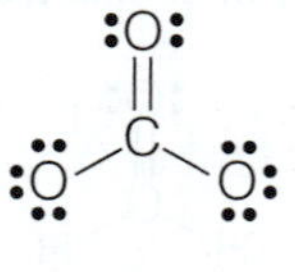

36. The sulfate ion (shown here) has an overall charge of −2. Calculate the formal charge on each atom in this structure.

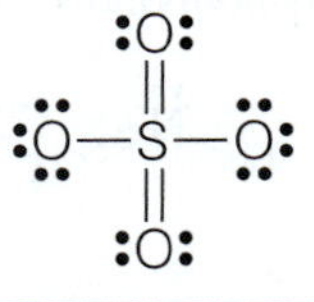

37. Determine the formal charge on the nitrogen atom in each of the following:

a. 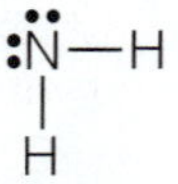 b. 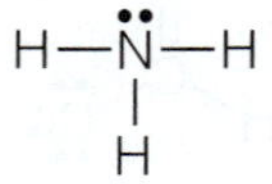 c.

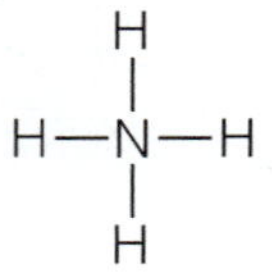

38. Determine the formal charge on the sulfur atoms in each of the following:

a. 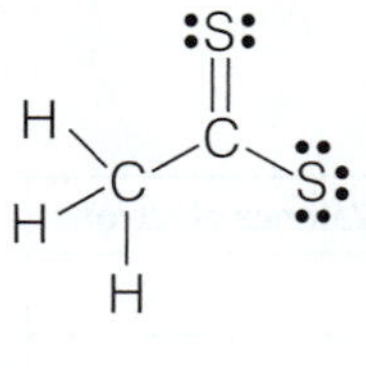 b. 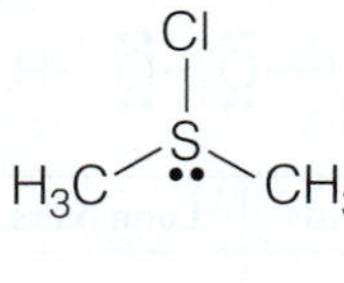 c.

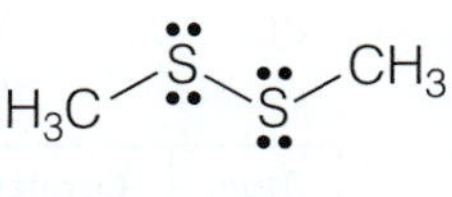

39. Identify the formal charge on each of these atoms:

a. an oxygen atom having one covalent bond and three lone pairs
b. an oxygen atom having two covalent bonds and two lone pairs
c. an oxygen atom having three covalent bonds and one lone pair

40. Identify the formal charge on each of these atoms:

a. a nitrogen atom having two covalent bonds and two lone pairs
b. a nitrogen atom having three covalent bonds and one lone pair
c. a nitrogen atom having four covalent bonds

41. In each of these Lewis structures, one atom has a nonzero formal charge. Locate that atom, and indicate the charge.

a. 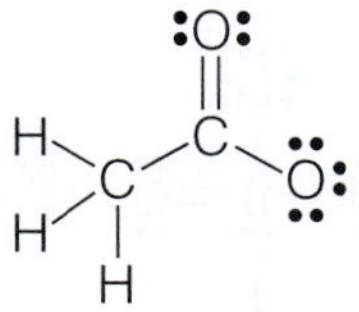 b. 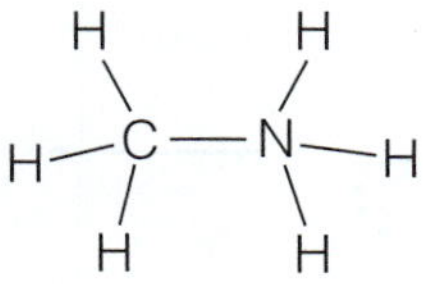 c.

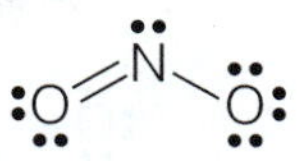

42. In each of these Lewis structures, two atoms have nonzero formal charges. Locate those atoms, and indicate the charges.

a. 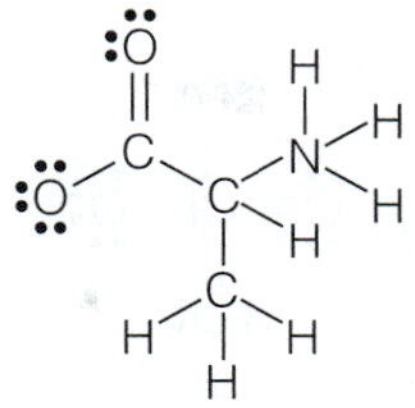 b. 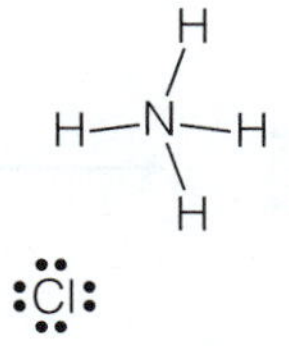 c.

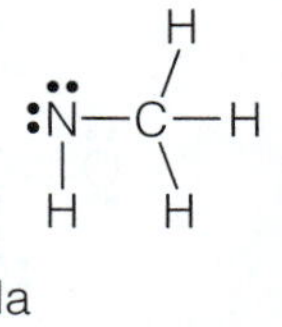

43. Find the total number of valence electrons in each of these ions:

a. NH_4^+ b. SO_3^{2-} c. BH_4^- d. OH^-

44. Find the total number of valence electrons in each of these ions:

a. H_3O^+ b. PO_4^{3-} c. IO_3^- d. N_3^-

45. Draw Lewis structures for each of these ions. Show formal charges where appropriate.

a. NH_4^+ b. NO_2^- c. NO_3^-

46. Draw Lewis structures for each of these ions. Show formal charges where appropriate.

a. BH_4^- b. IO^- c. ICl_4^-

47. Draw Lewis structures for each of these molecules. Show formal charges where appropriate.

a. SO b. SO_2 c. PF_5

48. Draw Lewis structures for each of these ions. Show formal charges where appropriate.

a. BrO^- b. CN^- c. PH_4^+

49. Draw Lewis structures for each of these oxyanions. Show formal charges where appropriate.

a. SO_4^{2-} b. SO_3^{2-} c. ClO_3^-

50. Draw Lewis structures for each of these oxyanions. Show formal charges where appropriate.

a. PO_4^{3-} b. BrO_2^- c. BrO_4^-

51. In this structure, all of the atoms have a formal charge of zero. Draw in any missing electrons to complete the Lewis structure.

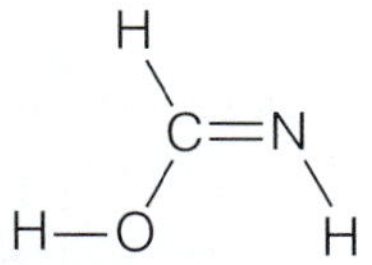

52. In this structure, the formal charges have been included, but the unshared electrons have not. Complete the Lewis structure by drawing in any additional valence electrons.

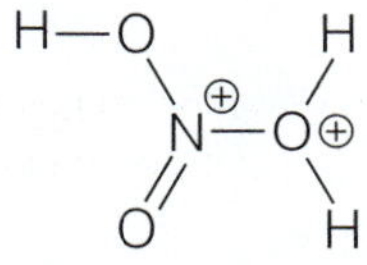

53. What are resonance structures? How do chemists represent resonance structures?

54. When drawing resonance structures, which of the following do not change? (Select all that apply.)

a. total number of electrons
b. location of the lone pairs
c. location of second and third bonds
d. location of first bonds

55. Thiocyanate, SCN^-, can be drawn in two resonance structures, shown here. Calculate the formal charge on each atom of the two forms.

56. The thioformate ion, $CHSO^-$, has two major resonance structures. The first resonance structure is shown. Of structures a–d, which one is the correct second resonance structure? For each other structure, indicate why it is not a correct resonance structure for this ion.

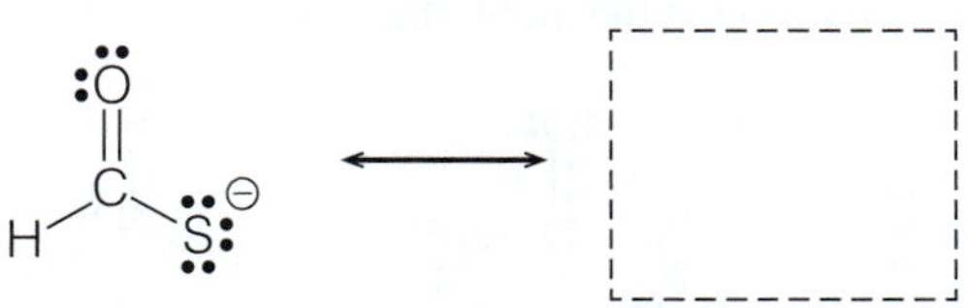

a. 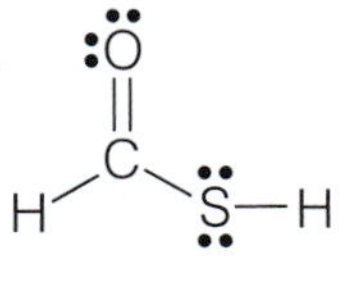 b. 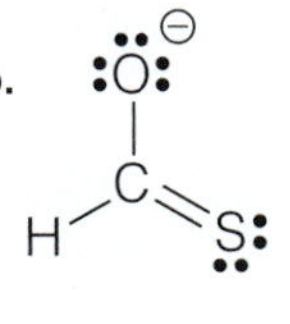 c. 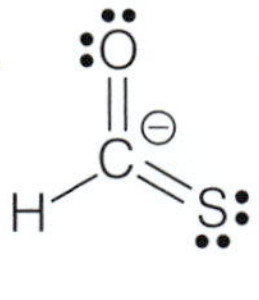 d.

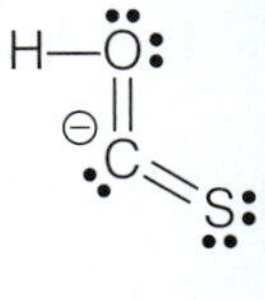

57. The resonance structures for the chlorate ion are shown here with formal charges. Based on these structures, what is the average charge on each oxygen atom?

58. The resonance structures for the sulfite ion are shown here with formal charges. Based on these structures, what is the average charge on each oxygen atom?

59. These structures show only the bonds in the bicarbonate ion, HCO_3^-. Complete the resonance structures by adding electrons and charges to show how the negative charge is spread out in this ion.

60. These structures show only the bonds in the bisulfate ion, HSO_4^-. Complete the resonance structures by adding electrons and charges to show how the negative charge is spread out in this ion.

9.3 Shapes of Molecules

61. For each of the following, indicate the electronic geometry and the bond angle that would occur around the central atom:

a. an atom with 4 charge sets
b. an atom with 2 charge sets
c. an atom with 3 charge sets

62. What is the difference between the electronic geometry and the molecular geometry around an atom?

63. For each of these molecules or ions, indicate the number of charge sets (using the VSEPR model) and give the electronic geometry around the highlighted atom:

a. S=C=S

b. NO_3^-

c. NH_4^+

64. For each of these molecules, indicate the number of charge sets (using the VSEPR model) and give the electronic geometry around the highlighted atom:

a. SiF_4

b. O=Si=O

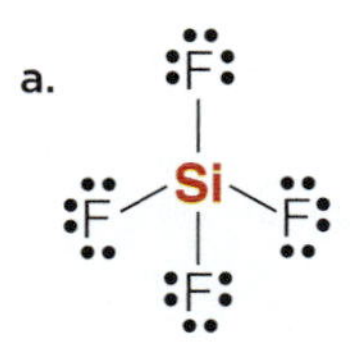

c. CH_3SH

65. For each of these molecules, indicate the number of charge sets (using the VSEPR model) and give the electronic geometry around the highlighted atom:

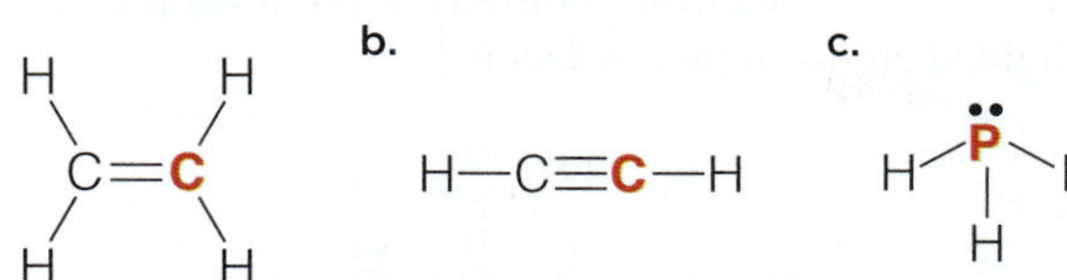

66. For each of these molecules or ions, indicate the number of charge sets (using the VSEPR model) and give the electronic geometry around the highlighted atom:

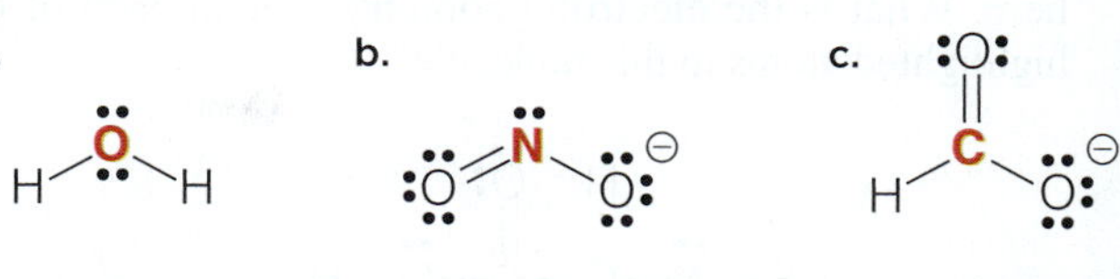

67. For each of these structures, describe the electronic and molecular geometry around the central atom:

68. For each of these structures, describe the electronic and molecular geometry around the central atom:

69. For each of these structures, describe the electronic and molecular geometry around the central atom:

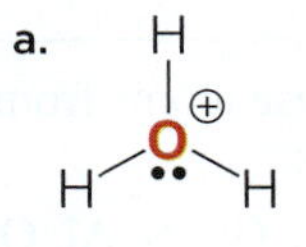

70. For each of these structures, describe the electronic and molecular geometry around the central atom:

71. Draw Lewis structures for these molecules, and describe the electronic geometry around the central atom:

a. boron trichloride, BCl_3
b. phosphorus trichloride, PCl_3
c. carbon dioxide, CO_2

72. Draw Lewis structures for these molecules, and describe the electronic geometry around the central atom:

a. carbon tetrachloride, CCl_4
b. nitrogen trichloride, NCl_3
c. carbon disulfide, CS_2

73. Draw Lewis structures for these molecules, and describe the electronic geometry around the central atom:

a. silicon tetrachloride, $SiCl_4$
b. sulfur trioxide, SO_3

74. Draw Lewis structures for these molecules, and describe the electronic geometry around the central atom:

a. hydrogen cyanide, HCN (C is the central atom)
b. hypochlorous acid, HOCl (O is the central atom)

75. Draw Lewis structures for these two ions. Identify the electronic and molecular geometry around the central atom in each ion.

a. sulfate ion
b. sulfite ion

76. There are four oxyanions of bromine: hypobromite, bromite, bromate, and perbromate. Draw a Lewis structure for each ion, and identify the electronic and molecular geometry around each bromine atom.

77. Identify the electronic geometry around each highlighted atom:

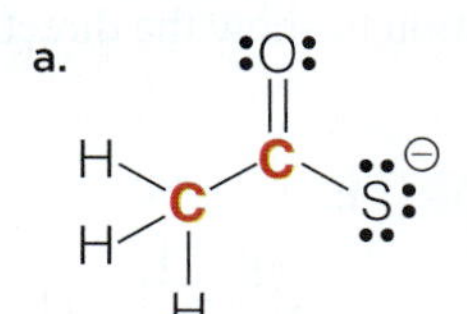

78. Identify the electronic geometry around each highlighted atom:

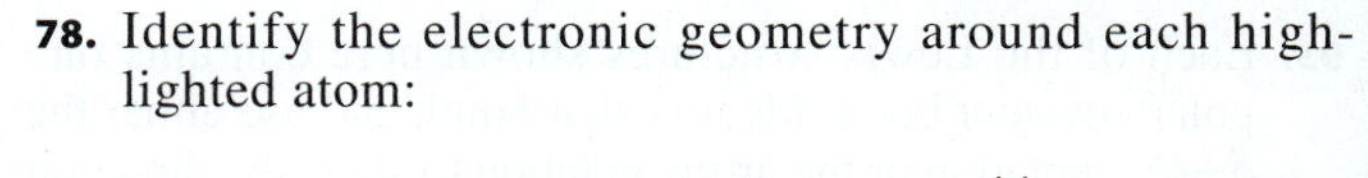

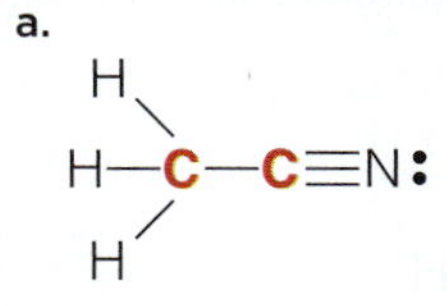

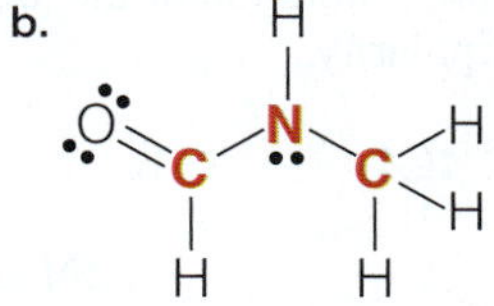

79. Amino acids are naturally occurring compounds that are the building blocks for many important biological molecules. One of the main amino acids is asparagine, shown here. What is the electron geometry around each of the highlighted atoms in this molecule?

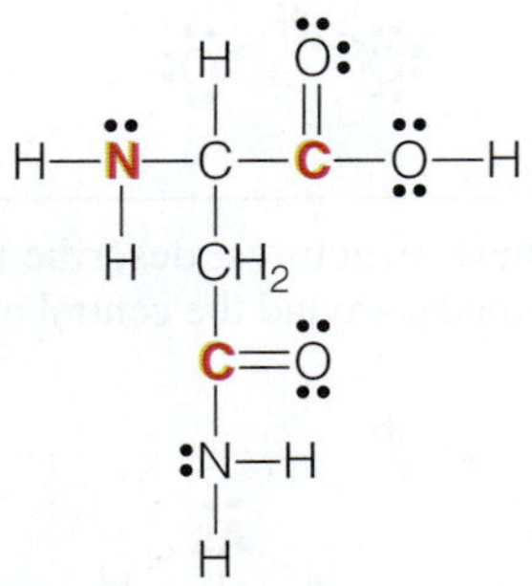

80. Amino acids are naturally occurring compounds that are the building blocks for many important biological molecules. One of the main amino acids is cysteine, shown here. What is the electron geometry around each of the highlighted atoms in this molecule?

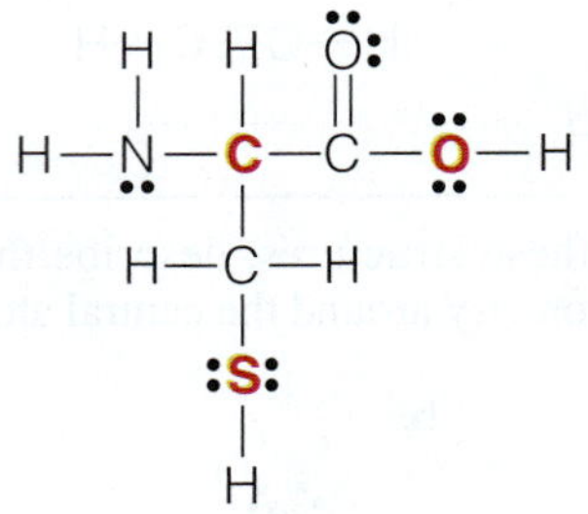

9.4 Polar Bonds and Molecules

81. Using the periodic table as a guide, select the more electronegative atom from each set:

a. Br or Cl b. Si or C c. N or O d. K or Ca

82. Using the periodic table as a guide, select the more electronegative atom from each set:

a. C or N b. N or P c. Ge or Si d. Mg or Na

83. Using the periodic table as a guide, rank these atoms from most electronegative to least electronegative:

a. Ca, O, Si, Ti b. I, Cs, Br, O c. K, In, N, F

84. Using the periodic table as a guide, rank these atoms from most electronegative to least electronegative:

a. Sc, Fe, N, F b. Po, Fr, Os, S c. Ga, N, Al, O

85. Using Pauling's table of electronegativities (Table 9.2), describe each of these bonds as covalent, polar covalent, or ionic:

a. C–P b. Ca–O c. Al–Cl d. N–H

86. Using Pauling's table of electronegativities (Table 9.2), describe each of these bonds as covalent, polar covalent, or ionic:

a. C–Cu b. Cs–Br c. P–Nd d. C–S

87. Based on Pauling's table of electronegativities (Table 9.2), identify which bond in each of these sets is more polar:

a. C–S or C–O b. Ca–Cl or Ni–Cl c. Ag–S or W–S

88. Based on Pauling's table of electronegativities (Table 9.2), identify which bond in each of these sets is less polar:

a. Se–Cl or Ti–Ge b. H–N or H–O c. Ca–N or Si–F

89. Use the $\delta+/\delta-$ notation to show the more positive and negative ends of each of these polar covalent bonds:

a. N–H b. O–H c. P–Cl

90. Use the $\delta+/\delta-$ notation to show the more positive and negative ends of each of these polar covalent bonds:

a. Br–Si b. H–Cl c. Cl–C

91. Use the arrow notation to show the more positive and negative ends of each of these polar covalent bonds:

a. Sn–Cl b. S–Ge c. I–Cl

92. Use the arrow notation to show the more positive and negative ends of each of these polar covalent bonds:

a. Al–Br b. F–Cl c. Si–O

93. Each of the Lewis structures shown here contains one polar covalent bond. Identify that bond, and use either the $\delta+/\delta-$ notation or the arrow notation to show the direction of polarity.

a. 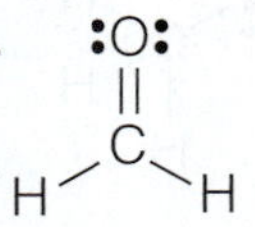 b.  c.

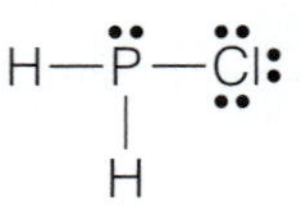

94. Each of the Lewis structures shown here contains one polar covalent bond. Identify that bond, and use either the $\delta+/\delta-$ notation or the arrow notation to show the direction of polarity.

a. 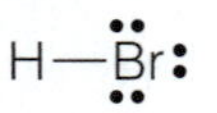 b. 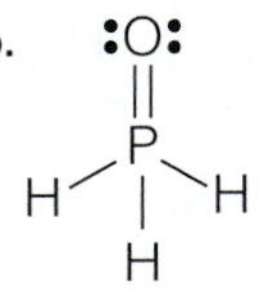 c.

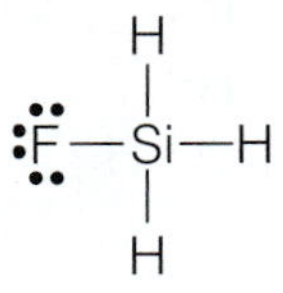

95. Using the symbol $+\!\!\longrightarrow$, show the direction of the polar bonds in each of these molecules:

a. 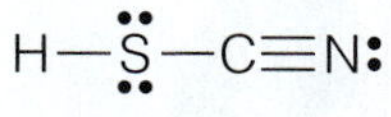 H—S—C≡N:

b. 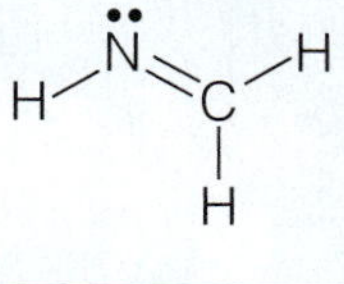 H—N=C(—H)—H

96. Using the symbol $+\!\!\longrightarrow$, show the direction of the polar bonds in each of these molecules:

a. 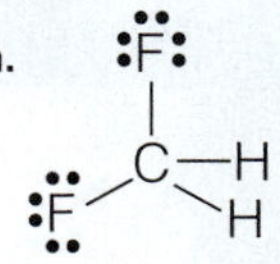 F—C(—F)(—H)—H

b. 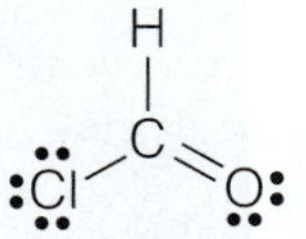 H—C(—Cl)=O

97. Each of these molecules or ions has a net dipole. Show the direction of polarity of the individual bonds and of the net dipole.

a. 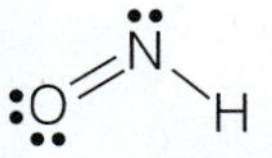 O=N—H

b. 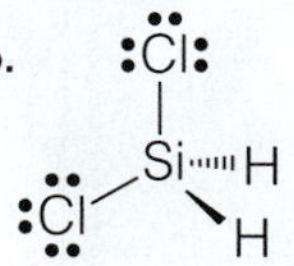 Cl—Si(—Cl)(—H)—H

c. 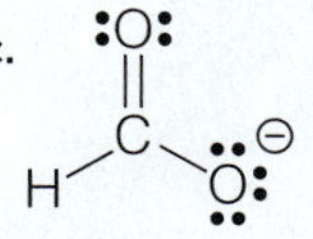 H—C(=O)—O⊖

98. Each of these molecules has a net dipole. Show the direction of polarity of the individual bonds and of the net dipole.

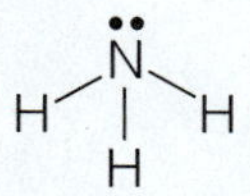

a. 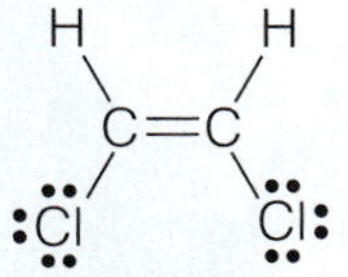 H(Cl)C=C(H)Cl

b. H—N(—H)—H

c. H—O—H

99. Draw Lewis structures for each of these molecules. Identify any polar covalent bonds, and state whether the molecule has a net dipole. Show the direction of the net dipole.

a. Cl_2 b. HCN c. PF_3

100. Draw Lewis structures for each of these molecules. Identify any polar covalent bonds, and state whether the molecule has a net dipole. Show the direction of the net dipole.

a. BCl_3 b. FCl c. SCl_2

101. Draw Lewis structures for each of these molecules. Identify any polar covalent bonds, and state whether the molecule has a net dipole. Show the direction of the net dipole.

a. CS_2 b. SF_2 c. $SOCl_2$

102. Draw Lewis structures for each of these molecules. Identify any polar covalent bonds, and state whether the molecule has a net dipole. Show the direction of the net dipole.

a. NCl_3 b. PH_2Cl c. CHF_3

Challenge Questions

103. The structure of aspirin is shown here. How many atoms in this molecule have a tetrahedral electronic geometry? How many atoms have a trigonal planar electronic geometry? How many polar covalent bonds are present? Show the direction of polarity of these bonds.

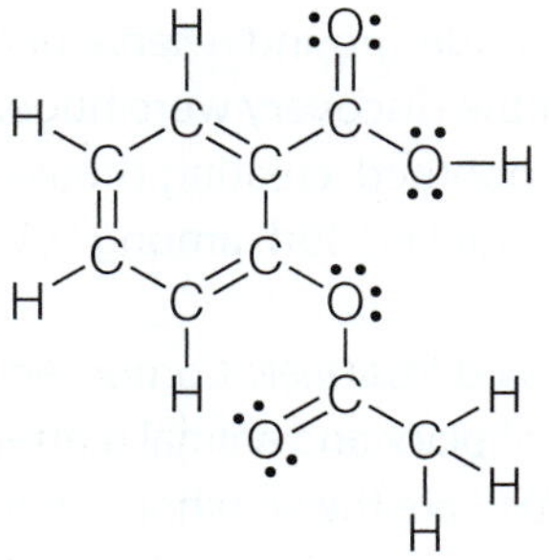

104. The structure here is levodopa, a compound used to treat Parkinson's disease. How many atoms in this molecule have a tetrahedral electronic geometry? How many atoms have a trigonal planar electronic geometry? How many polar covalent bonds are present? Show the direction of polarity of these bonds.

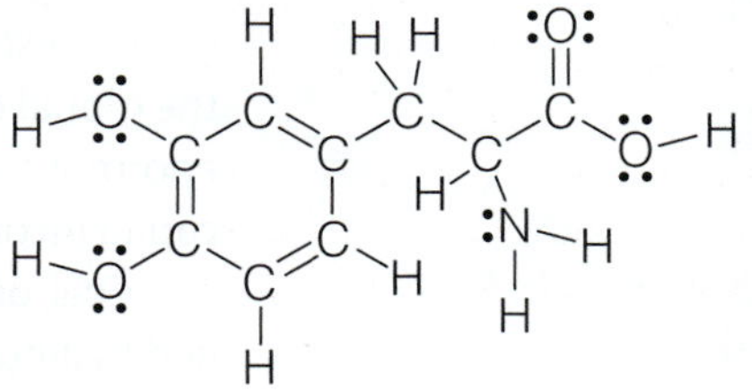

Solids, Liquids, and Gases

CHAPTER TEN

Figure 10.1 (a) Coal is the solid form of fossil fuels. It must be mined. (*continued*)

The North Dakota Boom

In 2006, geologists discovered an enormous underground reserve of oil and natural gas in western North Dakota. The impacts of the discovery were huge: While much of the global economy staggered, the Dakotas boomed, creating thousands of jobs and enormous wealth. In 2006, North Dakota was ranked 40th among U.S. states for per capita income. By 2013, it was ranked second.

Coal, oil, and natural gas are collectively called *fossil fuels*. Composed mainly of carbon and hydrogen, these fuels form by the decay of plant and animal matter over long periods of time. Along with nuclear energy, fossil fuels are the workhorse of the industrialized world: Each year, fossil fuels produce over 80% of the energy used in the United States.

Among fossil fuels, natural gas has steadily grown in importance. The surplus of oil and gas coming out of North Dakota and Montana has pushed down the cost of fuel, which in turn has reduced the cost of other commodities. Your gas bill and your grocery bill are lower because of the massive 2006 discovery. And though all fossil fuels produce carbon dioxide, natural gas burns more cleanly and efficiently than coal or oil, reducing the environmental impact of energy consumption. Natural gas is widely used for heating, and many vehicle fleets (like buses, cabs, and delivery trucks) have transitioned from liquid gasoline to natural gas.

Collecting and transporting each type of fossil fuel has its own challenges. Coal, a solid, has to be mined and then transported by ship or train. Companies extract oil (a liquid) and natural gas from wells and often transport these commodities through massive pipelines (**Figure 10.1**).

Figure 10.1 (continued) (b) Oil is the liquid form of fossil fuels. (c) An operating oil well. (d) A worker manages natural gas lines. (e) Many newer buses run on natural gas. This fuel burns more cleanly than gasoline. (a) Vyacheslav Svetlichnyy/Shutterstock; (b) Lowell Georgia/Getty Images; (c) Ed Reschke/Stone/Getty Images; (d) ZoranOrcik/Shutterstock; (e) George Rose/Getty Images

Safely handling and transporting gases can be especially challenging. Workers must follow special safety guidelines when working with natural gas. In fact, many careers, from health care to mechanical work to restaurant management, involve work with compressed gases. Understanding how gases behave will serve you well in nearly any field.

In this chapter, we'll examine the properties of matter in its three phases: solid, liquid, and gas. We begin with solids and liquids, exploring how chemical bonds affect a substance's structure and properties. Then we'll shift our attention to gases. We will see how the pressure and volume of gases vary with temperature and with the amount of gas present. We'll explore chemical processes that involve gases, such as combustion reactions or the production of beer.

At the end of this chapter, we will return to the challenge of safely storing and transporting natural gas. We'll look at new discoveries that are changing the way scientists and engineers think about this problem and opening up exciting opportunities for the future.

Intended Learning Outcomes

After completing this chapter and working the practice problems, you should be able to:

10.1 Interactions between Particles

- Describe the motion of particles in a solid, liquid, or gas.

10.2 Solids and Liquids

- Describe the bonding and arrangement of particles in ionic, metallic, molecular, polymeric, and covalent-network substances.
- Describe the different types of intermolecular forces, and relate these differences to relative melting or boiling temperatures.

10.3 Describing Gases

- Describe the key features of an ideal gas.
- Describe how to use a liquid barometer to determine pressure.

10.4 The Gas Laws

- Apply the combined gas laws to relate changes in the pressure, volume, and temperature of a gas.
- Relate the pressure, volume, number of moles, and temperature of a gas using the ideal gas law.
- Relate the temperature, volume, and pressure of a gas to atomic or molecular motion.

10.5 Diffusion and Effusion

- Describe the motion of larger and smaller gas particles at a given temperature, and apply these concepts to the principles of diffusion and effusion.

10.6 Gas Stoichiometry

- Apply the principles of stoichiometry to solve problems involving reactions of gases.

10.1 Interactions between Particles

In Chapter 1, we discussed the three common *states of matter* (also called the *phases of matter*): solid, liquid, and gas. We saw that the macroscopic (human-scale) properties are a function of their structure on the atomic or molecular level. In a solid, particles pack closely together. Each atom is held in a fixed position by the atoms around it. In a liquid, the particles are close together, but they are not held in a fixed position; particles move freely past each other. In a gas, the particles move independently of each other—they are spaced far apart and have little or no interaction with the other gas particles as they move. These concepts are summarized in **Table 10.1**.

TABLE 10.1 The States of Matter

	Atomic/Molecular Arrangement	Macroscopic Properties
Solid	Particles are close together and held in a fixed place.	Definite shape and volume
Liquid	Particles are close together, but move freely past each other.	Definite volume Adopts shape of container
Gas	Particles are far apart and have very little interaction.	Adopts shape and volume of container

© yekophotost/depositphotos.com

The melting and boiling points of a substance depend on the forces holding the particles together. The stronger the forces of attraction, the higher the phase transition temperatures.

Substances with strong forces of attraction have high melting and boiling points. ■

A transition from one state of matter to another is called a *phase change*. For example, let's look at the phase changes that take place with water: In ice, the water molecules pack tightly together (**Figure 10.2**). If we add heat (kinetic energy) to the ice, the molecules in the solid vibrate more rapidly. If we continue to add heat, the molecules gain enough energy to overcome the forces holding them in place. They break out of the solid framework and begin to move freely past each other. The ice melts—a phase change from solid to liquid. If we then heat the liquid water, the molecules move faster and faster until they begin to break out of the liquid phase and enter the gas phase. Particles in the gas phase move about freely, interacting very little with each other.

For a compound to change from solid to liquid or from liquid to gas, there must be enough kinetic energy to overcome the forces of attraction. The stronger the forces between the particles, the greater the energy required to change

Figure 10.2 This image captures water in all three phases: solid ice, liquid water, and gaseous steam. Each phase change requires heat energy to overcome the forces of attraction between the water molecules.

Explore Figure 10.2

phase. A substance with very strong forces between its particles will have a high melting point and boiling point. We can use phase transition temperatures to understand and compare the forces that hold substances together.

TRY IT

1. Methane, ammonia, and water are all covalent compounds that form molecules. Based on the melting points shown, which compound exhibits the strongest forces of attraction between molecules? Which one exhibits the weakest forces of attraction?

Substance	Formula	Melting Point
Methane	CH_4	−182 °C
Ammonia	NH_3	−78 °C
Water	H_2O	0 °C

10.2 Solids and Liquids

In solids and liquids, forces of attraction hold particles close together. These forces vary for different types of substances. In this section, we will explore the different ways that particles can be arranged within solids and liquids and learn about the properties that result from their unique structures.

Ionic Substances

We first talked about ionic compounds in Chapter 5. In these compounds, positive and negative ions pack tightly together in rigid frameworks, called *lattices* (**Figure 10.3**). Because the interactions between the ions are so strong, it takes a tremendous amount of energy to disrupt the lattices. As a result, most ionic compounds have very high melting points (**Table 10.2**).

Figure 10.3 Ionic compounds form ordered lattices of alternating positive and negative charges.

TABLE 10.2 Melting Points for Ionic Compounds

Compound	Melting Point (°C)
NaCl	801
KCl	770
$MgCl_2$	714
CaO	2,572
Al_2O_3	2,072

Metallic Substances

Elemental metals (as well as mixtures of metals, called *alloys*) usually form ordered lattices of tightly packed atoms (**Figure 10.4**). Metallic atoms have a loose hold on their outer electrons, and the electrons move easily from one atom to the next. Because of this property, metals are good conductors of electricity.

While atoms pack close together in metallic solids, the neutral metal atoms do not bind as tightly to each other as the oppositely charged ions in an ionic compound. Because of this, a metal lattice is not as rigid as an ionic lattice, and the shapes of metallic solids are easily altered. Metals are *malleable*, meaning they can be pounded into different shapes, and they are *ductile*, meaning they can be stretched into wire.

Metals have moderate to high melting points (**Table 10.3**). Metals like iron, aluminum, and copper are commonly melted and molded into useful shapes (**Figure 10.5**).

Love Silhouette/Shutterstock

Figure 10.4 Metals pack into close-fitting arrangements of atoms.

TABLE 10.3 Melting Points for Common Metals

Element	Melting Point (°C)
Lead	327
Aluminum	660
Gold	1,064
Copper	1,085
Iron	1,538

Cultura Creative (RF)/Alamy

Figure 10.5 Workers pour molten (liquid) iron into a mold.

Molecular Substances

Recall that molecules are discrete units of atoms held together by covalent bonds. Covalent bonds within a molecule are very strong, but the forces *between* the individual molecules—called **intermolecular forces**—are much weaker (**Figure 10.6**). As a result, molecular compounds usually have lower melting and boiling points than ionic compounds.

Recall that covalent bonds form between nonmetals.

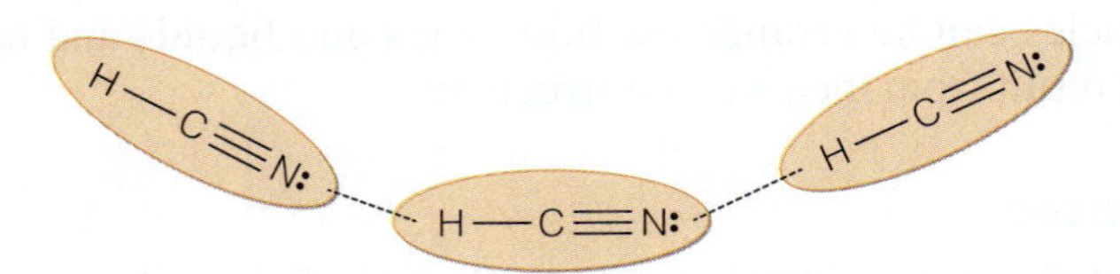

Figure 10.6 The forces between molecules (shown as dashed lines) are weaker than the covalent bonds within a molecule.

The strength of intermolecular forces varies widely, depending on the structure of the molecules. As a result, molecular materials exhibit a stunning diversity of physical properties. Broadly, intermolecular forces are divided into three key groups: *dipole–dipole interactions*, *hydrogen bonds*, and *dispersion forces*.

Intra– versus Inter–

Be careful when using the prefixes *intra–* and *inter–*. *Intra–* means "within." *Inter–* means "between." So "*intra*molecular forces" are the bonds within a molecule—strong covalent bonds. "*Inter*molecular forces" are the weaker forces between different molecules.

Dipole–Dipole Interactions

As we saw in Chapter 9, many molecules contain polar covalent bonds. Depending on their shape, molecules with polar bonds often have a *net dipole*, meaning different sides of the molecule have a slight positive and negative charge (**Figure 10.7**). If a compound has a net dipole, the molecules of that compound tend to "stick" together due to the attraction of the positive and negative poles (**Figure 10.8**). This type of interaction is called a **dipole–dipole interaction**.

Figure 10.7 Recall that we show polar covalent bonds and molecular dipoles with the δ+ and δ− symbols, or with an arrow symbol that has a cross on the positive side and the arrowhead on the negative side.

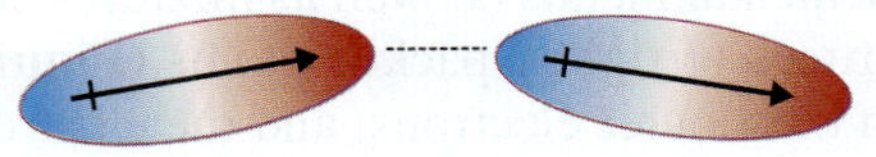

Figure 10.8 Molecules with net dipoles experience an attractive force between their positive and negative ends.

In a polar covalent bond, the more electronegative atom has a slight negative charge.

Dipole–dipole attractions affect a substance's melting and boiling points. To illustrate this idea, consider the molecules in **Figure 10.9**. The molecules in Flask A have a molecular dipole, while those in Flask B do not. Because of the forces of

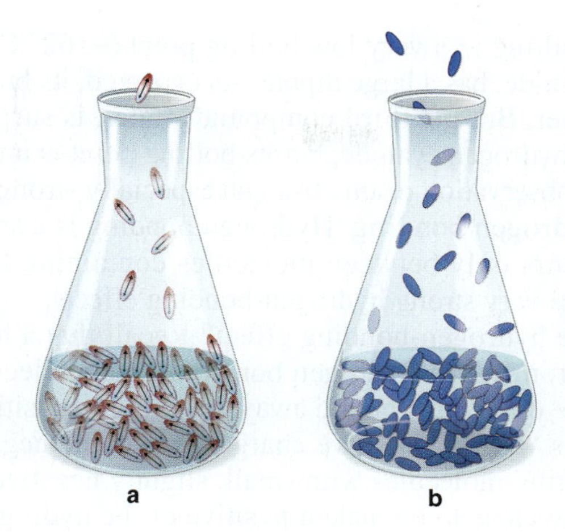

Figure 10.9 The dipoles cause the molecules in flask A to stick together more tightly, so more energy is needed for them to escape into the gas phase. As a result, molecules with a dipole tend to have higher boiling points than those that do not.

Explore **Figure 10.9**

attraction between the dipolar molecules, it takes more heat energy to pull them apart. Consequently, the dipolar molecules have a higher melting point and boiling point than the nonpolar molecules have.

An example of this effect is shown in **Table 10.4**. Carbon dioxide has no net dipole, sulfur dioxide has a small net dipole (owing to its bent shape), and acetonitrile has a large net dipole. Notice that the larger the size of the dipole, the higher the boiling point.

Polar molecules tend to have higher melting and boiling points than nonpolar molecules.

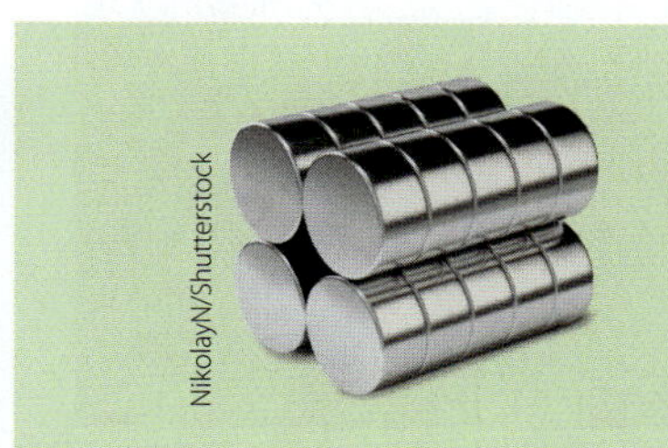

Dipole–dipole interactions are similar to the interactions between magnets—the opposite ends are attracted to each other, and they tend to stick together.

TABLE 10.4 Molecules with Larger Dipoles Exhibit Higher Boiling Points

	Carbon dioxide	Sulfur dioxide	Acetonitrile
Geometry	Linear	Bent	Linear
Dipole	Zero	Small	Large
Boiling Point	−79 °C*	−10 °C	+82 °C

*Carbon dioxide *sublimes* (transitions from solid directly to gas) at this temperature.

Melting and boiling points are a handy way to compare the strength of different intermolecular forces. However, we must be careful in our comparisons: Heavier molecules tend to have higher melting and boiling temperatures than lighter molecules, so it is best to compare compounds having a similar formula mass.

Hydrogen Bonding

Look at the three small compounds in **Table 10.5**. The first compound, methane (CH_4), has no dipole. Because of this, CH_4 molecules have very weak interactions

TABLE 10.5 A Comparison of the Properties of Methane, Hydrogen Cyanide, and Water

	Methane	Hydrogen cyanide	Water
Formula mass	16.0 u	27.0 u	18.0 u
Dipole strength*	0	2.98	1.85
Boiling point	−162 °C	+26 °C	+100 °C

*These numbers convey the relative size of each dipole.

Hydrogen bonds are stronger than other dipole–dipole interactions.

with each other, resulting in a very low boiling point (−162 °C). The second compound, hydrogen cyanide, has a large dipole: As expected, its boiling point (+26 °C) is significantly higher. But the third compound, water, is surprising: Its dipole is smaller than that of hydrogen cyanide, but its boiling point is much higher.

This surprising observation results from an especially strong type of intermolecular force called **hydrogen bonding**. Hydrogen bonding is a special dipole–dipole interaction that occurs only between molecules containing H–F, H–O, or H–N bonds. Water exhibits very strong hydrogen-bonding effects.

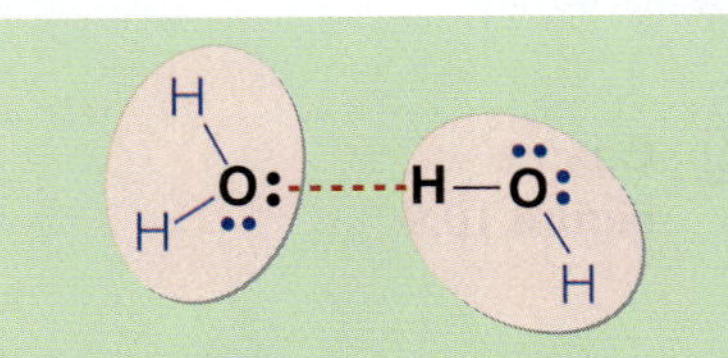

To form a hydrogen bond, the hydrogen atom must be covalently bonded to one electronegative atom (F, O, or N), but also attracted to the electrons on an F, O, or N atom in a neighboring molecule.

What causes the hydrogen-bonding effect? Recall that a hydrogen atom contains only one electron. When hydrogen bonds to a more electronegative element (F, O, or N), this one electron is pulled away, leaving the positive nucleus exposed (**Figure 10.10**). This "naked" positive charge attracts the negative poles of other molecules. Neighboring molecules with small, slightly negative atoms (F, O, or N) can get exceptionally close to the naked positive of the hydrogen atom—nearly as close as a covalent bond. The result is an unusually strong intermolecular force.

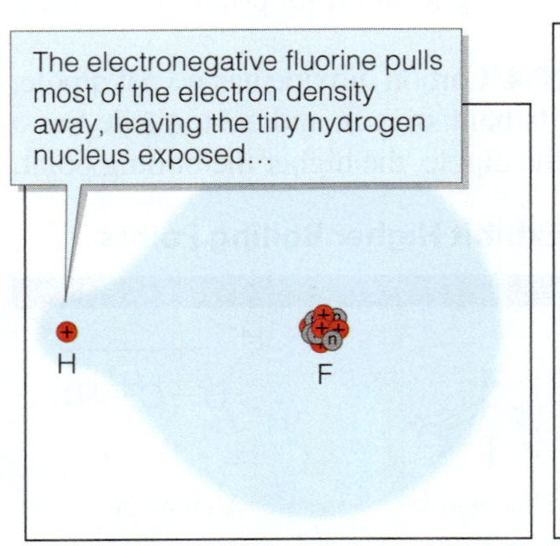

The electronegative fluorine pulls most of the electron density away, leaving the tiny hydrogen nucleus exposed...
H
F

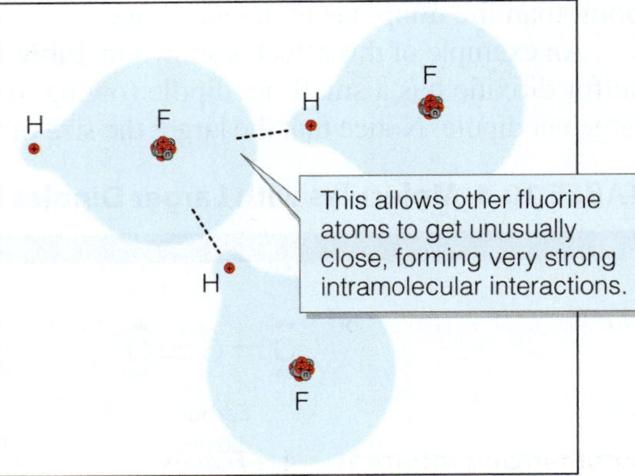

This allows other fluorine atoms to get unusually close, forming very strong intramolecular interactions.
H
F
H
F
H
F

Figure 10.10 Hydrogen bonds arise when hydrogen is bonded to the electronegative elements nitrogen, oxygen, or fluorine.

Water contains two O–H bonds, resulting in multiple hydrogen-bonding interactions that cause water molecules to stick tightly to each other (**Figure 10.11**). As a result, the melting and boiling points of water are much higher than those of other small molecules. The interactions between water molecules are strong enough that some insects are able to walk on the surface of water, supported by the forces of attraction between individual molecules. The tendency of water to form droplets in the air or beads on a waxed surface also arises from this strong attraction of water molecules for each other (**Figure 10.12**). Hydrogen bonding plays a critical role in the chemistry of life, from the fundamental way water molecules interact to the shapes of huge biological molecules like proteins and DNA.

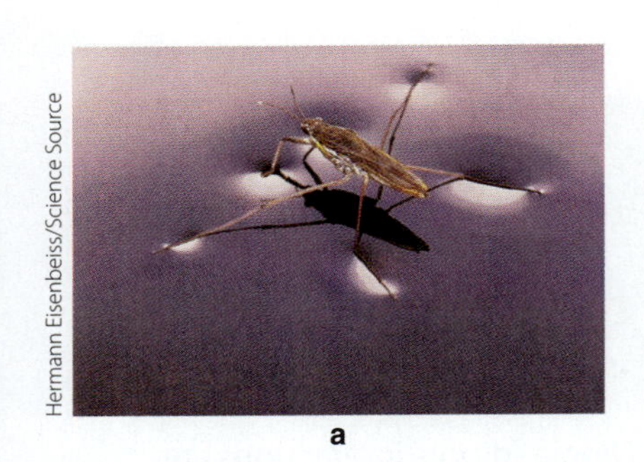

a

b

c

Figure 10.12 Hydrogen bonding causes water molecules to stick tightly together, a phenomenon called surface tension. These forces can (a) support the weight of a bug on the water, and (b) cause water to form droplets in the air or (c) on a waxed surface. London dispersion forces are the weakest intermolecular force.

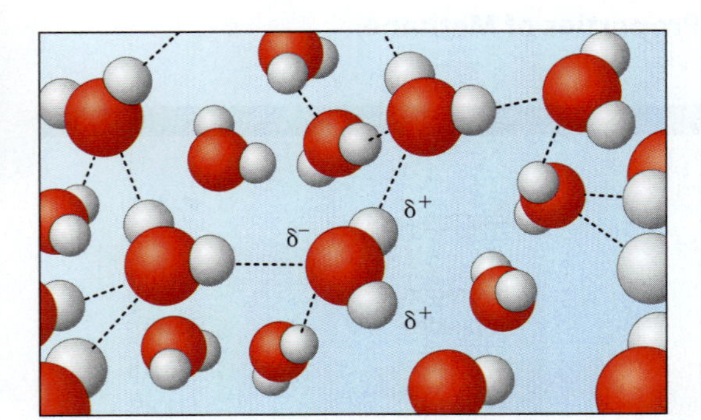

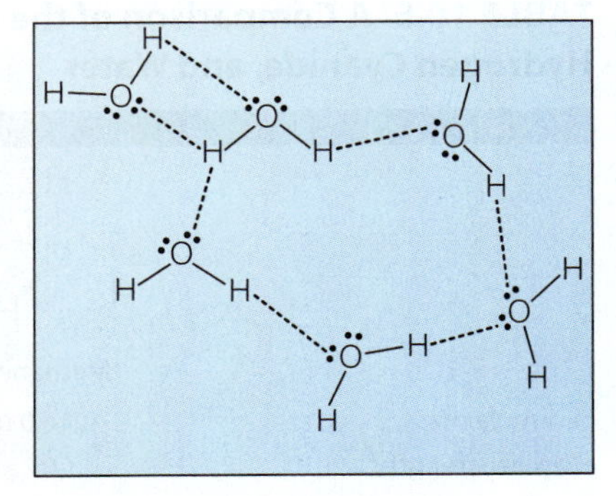

Figure 10.11 Water molecules are held tightly together by hydrogen bonds, as depicted by a space-filling model (left) and Lewis structures (right).

London Dispersion Forces

So what about covalent compounds that don't have dipoles? At first glance, we might expect there to be no attraction between them. However, this is not the case. Even molecules with no overall dipole exhibit a weak attraction for each other (**Figure 10.13**). These forces of attraction are called **London dispersion forces** (sometimes simply called *dispersion forces* or *London forces*). London dispersion forces are related to dipole–dipole forces but are much smaller in magnitude.

Even if they are shared evenly between two atoms, electrons are constantly moving. This motion produces slight, fleeting areas of positive and negative charges called *instantaneous dipoles*. These rapidly changing dipoles produce ripple effects in the surrounding molecules as well: A slight buildup of negative charge on one molecule will push the electron density away on a neighboring molecule. The result is a dipole on the secondary molecule and an instant of attraction between the two molecules (**Figure 10.14**).

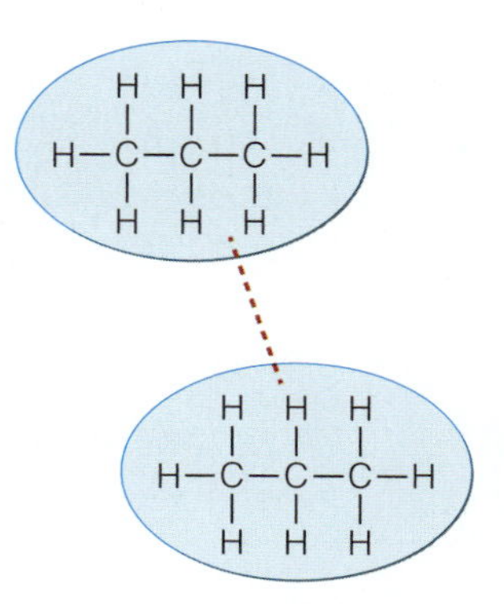

Figure 10.13 Although propane (C_3H_8) molecules do not have a net dipole, they weakly attract each other through London dispersion forces.

London dispersion forces are the weakest intermolecular force.

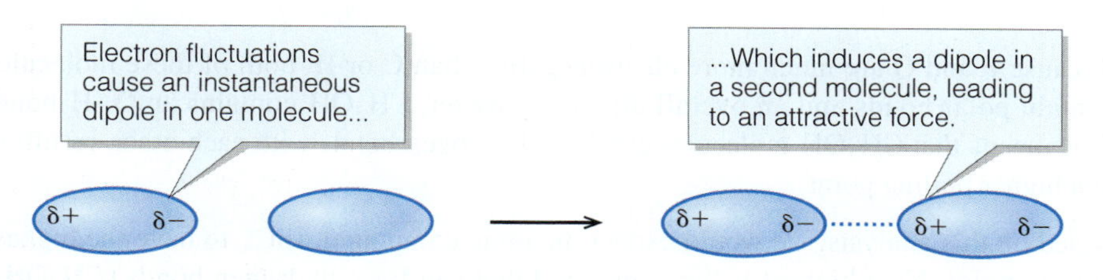

Figure 10.14 London dispersion forces are caused by the interaction of tiny instantaneous dipoles.

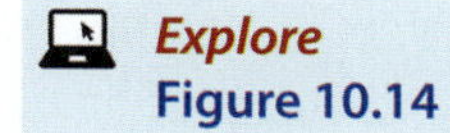

Explore Figure 10.14

Summarizing Intermolecular Forces

In summary, there are three types of intermolecular forces: London dispersion forces, dipole–dipole forces, and hydrogen bonds (**Table 10.6**). Of these, dispersion forces are the weakest—resulting in molecules with the lowest melting and boiling points. Hydrogen bonds are the strongest—resulting in molecules with the highest melting and boiling points.

TABLE 10.6 Types of Intermolecular Forces

Type	Description	Strength
Hydrogen bonding	Molecules with H–F, H–O, or H–N bonds	Strongest
Dipole–dipole forces	Molecules with net dipoles	
London dispersion forces	All covalent molecules	Weakest

The stars are always in the sky, but their faint light is often blocked by the light of the moon or the bright light of the sun. In the same way, London dispersion forces are always present—but their effects are slight in molecules that have dipole–dipole or hydrogen-bonding interactions.

Instantaneous dipoles are like ripples on the surface of a pool: On average, the surface of the pool is level, but the ripples make the water uneven. For an instant, there is more water in one part of the pool than in another. Similarly, the "ripples" caused by electron motion produce a slight attraction between neighboring molecules.

Example 10.1 Predicting the Properties of Compounds

Classify these three compounds, LiCl, CH_3F, and CH_3OH, as ionic, metallic, or covalent. Predict which compound would have the highest and lowest boiling points.

The first compound, LiCl, contains both a metal and a nonmetal. Because of this, we know that it is an ionic compound and therefore has high melting and boiling points.

The other compounds, CH_3F and CH_3OH, are composed entirely of nonmetal atoms and therefore are covalent (molecular) compounds. To predict the relative boiling points of these compounds, we need to draw Lewis structures and consider what polar bonds may be present:

Polar bond

```
      H
      |
  H — C — F:
      |
      H
```

Polar bonds

O–H bond means hydrogen bonding can occur.

```
      H
      |
  H — C — O:
      |     \
      H      H
```

Because F and O are much more electronegative than C or H, both of these molecules contain polar bonds and an overall dipole. However, CH_3OH contains an O–H bond. This means that CH_3OH molecules can form hydrogen bonds with each other, resulting in a higher boiling point.

Based on this analysis, we would expect the ionic compound, LiCl, to have the highest boiling point. Next highest is the compound that can form hydrogen bonds (CH_3OH), followed by the compound with only dipole–dipole interactions (CH_3F).

Check it
Watch Explanation

TRY IT

2. Classify the following as ionic, metallic, or covalent solids:

a. potassium nitrate **b.** phosphorus tribromide **c.** chromium

3. Which of these compounds would you expect to have the highest melting point, and why?

a. hexane, C_6H_{14} **b.** sodium fluoride, NaF

4. Based on their polarity, which of these compounds would you expect to have the highest boiling point?

a.
```
H     H
 \   /
  C=C
 /   \
H     H
```

b.
```
H
 \
  C=O:
 /
H
```

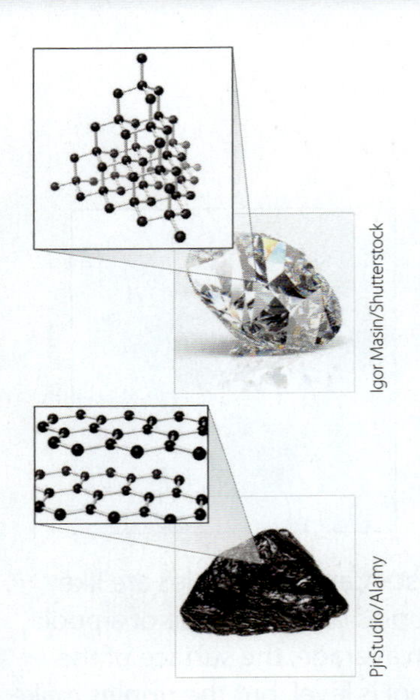

Figure 10.15 Diamond and graphite are both covalent networks of carbon. In diamond, each carbon has a tetrahedral geometry; in graphite, each carbon has a trigonal planar geometry.

Covalent Networks and Polymers

Some compounds contain covalent bonds but differ significantly from the molecular solids described earlier. For example, elemental carbon exists in two primary forms, *diamond* and *graphite*. Both of these are examples of **covalent networks**—lattices of connected covalent bonds, forming one giant molecule (**Figure 10.15**). In diamond, each carbon atom has four single bonds and a tetrahedral geometry. The single bonds repeat indefinitely in three dimensions, resulting in a rigid structure that is among the hardest and most durable substances known. In graphite, the carbon atoms each have a trigonal planar geometry. The atoms are arranged in two-dimensional "sheets" that slide easily over each other. As a result, graphite is softer and is used as the writing material in most pencils.

Polymers are compounds that contain long chains of covalently bonded atoms. There are many naturally occurring polymers, including cellulose (the structural component of wood), starch (a major component of many staple foods, such as rice and potatoes), and proteins. *Plastics* such as poly(ethylene) and poly(vinyl chloride), or PVC, are synthetic polymers (**Figure 10.16**).

Polymers are composed of small, repeating units that are bonded together. For example, **Figure 10.17** shows the structure of poly(vinyl chloride). Each small unit is a single link in the longer chain. We'll discuss polymers in more detail in Chapter 15.

Figure 10.16 Polyethylene is commonly used in disposable plastics like shrink wrap. Poly(vinyl chloride), or PVC, is commonly used in plumbing applications.

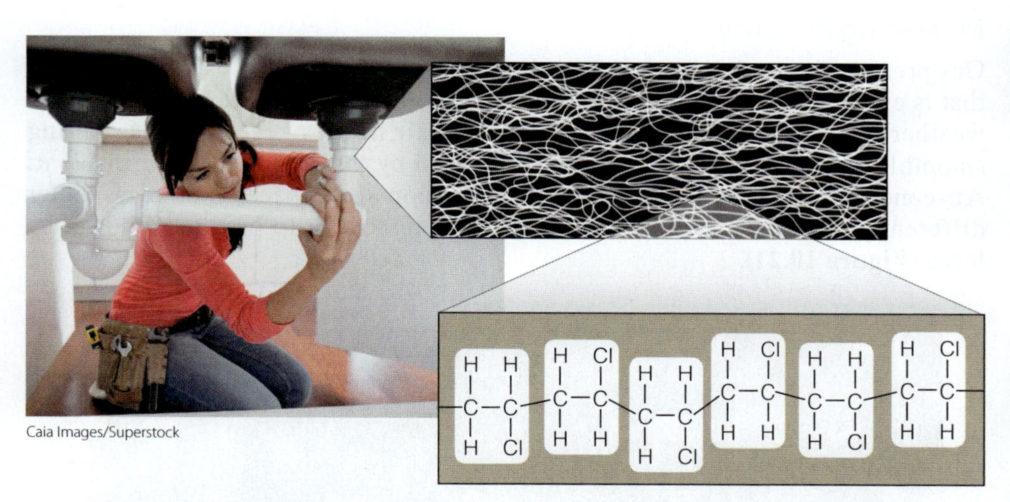

Figure 10.17 Polymers like PVC are composed of long chains of small, repeating units. Each "link" of a PVC chain has the formula C_2H_3Cl.

Besides being used as a fuel source, fossil fuels provide the starting materials to make most plastics.

Plastics or polymers are made up of long chains of covalent bonds.

10.3 Describing Gases

In the previous section, we saw how the properties of solids and liquids depend on the forces of attraction between particles. In the remaining sections of this chapter, we'll shift our attention to gases. In a gas, the particles are spaced far apart. They move freely, interacting very little with the particles around them. Gas particles travel in a straight line until they bounce off another particle or off the walls of the container (**Figure 10.18**).

When describing gases, we often assume that they are behaving as *ideal gases*. An **ideal gas** has two key properties:

1. The volume of the particles is much, much less than the volume of the container.
2. The particles have no attraction for each other. When they pass each other, they do not slow down. When they collide, they bounce off each other like billiard balls.

When gases behave this way, we can predict their behavior mathematically by using simple relationships between the temperature of the gas, the volume the gas occupies, and the pressure it exerts.

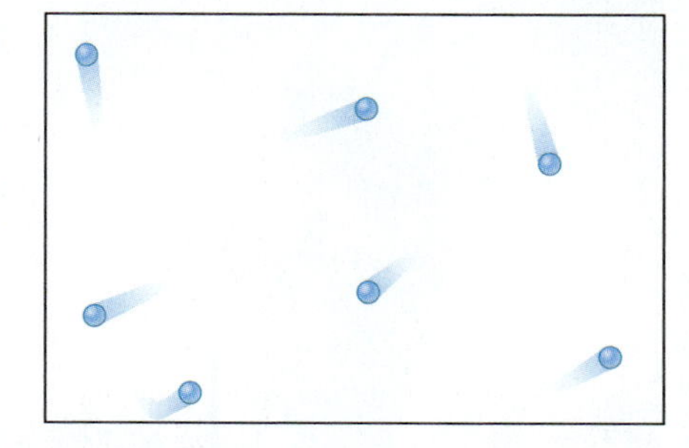

Figure 10.18 In a gas, particles move freely and have very little interaction with each other.

Explore Figure 10.18

We commonly describe gases by the pressure they exert. ■

Pressure

We commonly describe gases by the **pressure** they exert on their surroundings. We can think of pressure as the "push" that particles exert on everything around them. For example, consider the gas represented in Figure 10.18. The particles move freely around the box. As they move, they collide with the walls of the container. Collectively, these tiny collisions create a larger push. This is the pressure on the container (**Figure 10.19**).

Cathyrose Melloan/Alamy

Figure 10.19 The gas inside this rural propane tank exerts pressure on its walls.

Formally, pressure is defined as the force that is exerted divided by the area over which it is applied:

$$\text{Pressure} = \frac{\text{Force}}{\text{Area}}$$

As we will see in the following sections, the pressure caused by the gas depends on the volume, temperature, and amount of gas present.

Measuring Pressure

Gas pressure is a normal part of our lives. Earth's atmosphere exerts a pressure that is essential to life. Subtle variations in atmospheric pressure cause wind and weather patterns (**Figure 10.20**). Pressurized air in tires keeps vehicles rolling smoothly. The bounce of a basketball is caused by the compressed air inside it. Air-conditioning systems rely on the compression and expansion of gases under different pressures. Pneumatic tools use compressed air to produce a powerful force (**Figure 10.21**).

www.dennisoswald.de/Getty images

Figure 10.20 A tornado occurs when areas of high and low air pressure come together.

Photononstop/Alamy

Carolyn Franks/Alamy

Mark Hunt/Design Pics/Media Bakery

Figure 10.21 From tires to air-conditioning systems and pneumatic tools, compressed gas serves many useful purposes.

A **barometer** is a device used to measure atmospheric pressure. A classic barometer is shown in **Figure 10.22**. In this device, a long tube is filled with a liquid (usually mercury), then turned upside down in a reservoir of the liquid. Gravity pulls the liquid down, creating a vacuum in the top of the tube. At the same time, pressure from the outside air pushes the liquid up into the tube. The higher the air pressure, the higher the liquid is pushed into the tube. We measure the pressure of the atmosphere by measuring the height of the mercury inside the tube. Because of this, atmospheric pressure is often reported in **millimeters of mercury (mm Hg**; also called *torr*). At sea level, the average atmospheric pressure is 760 mm Hg, or 760 torr. This is referred to as *standard pressure*.

Why do barometers use mercury rather than a less-toxic liquid like water or oil? Because mercury is over 13 times denser than water. Atmospheric pressure is enough to push a column of mercury 760 millimeters (0.76 meters) into the air. If we used water in a barometer, the column would have to be over 33 feet (10 meters) high. Of course, there are many other designs for pressure gauges. The shapes and styles of these gauges differ based on their application (**Figure 10.23**).

For practical applications, we sometimes refer to the **gauge pressure** of a compressed gas. The gauge pressure is the difference between the compressed gas pressure and the atmospheric pressure. For example, if the gauge on an air compressor reads zero, it doesn't mean there is a vacuum (meaning no air) inside the cylinder; it means that the air pressure inside the cylinder is equal to the air outside. For the problems in this chapter, you can assume that the pressures given are absolute pressures, not gauge pressures.

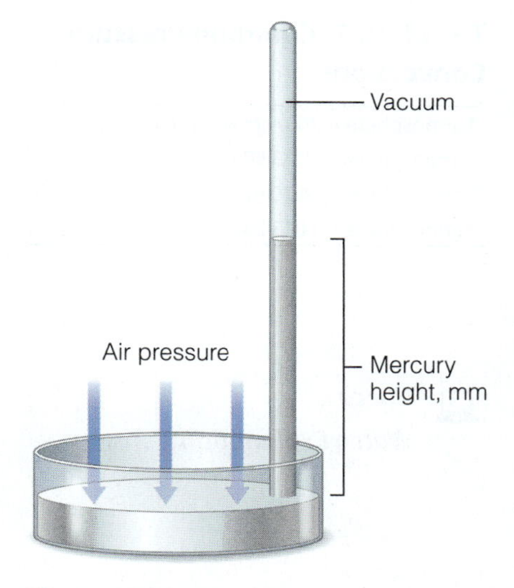

Figure 10.22 A simple barometer is used to measure the pressure of the atmosphere.

Explore Figure 10.22

1 torr = 1 mm Hg

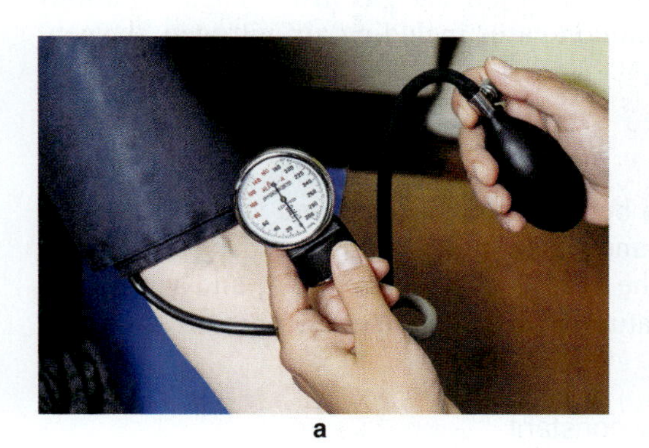
a

b

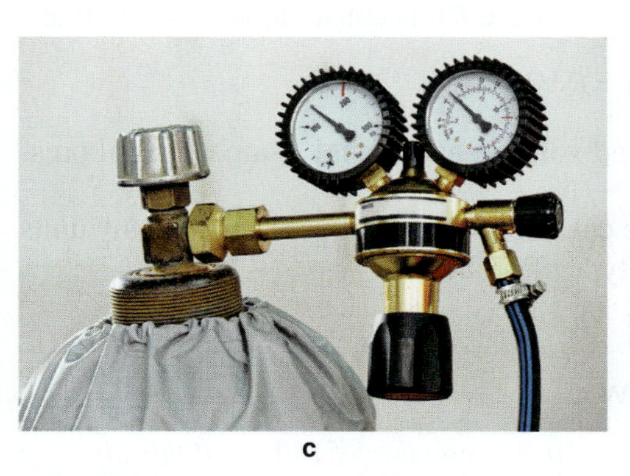
c

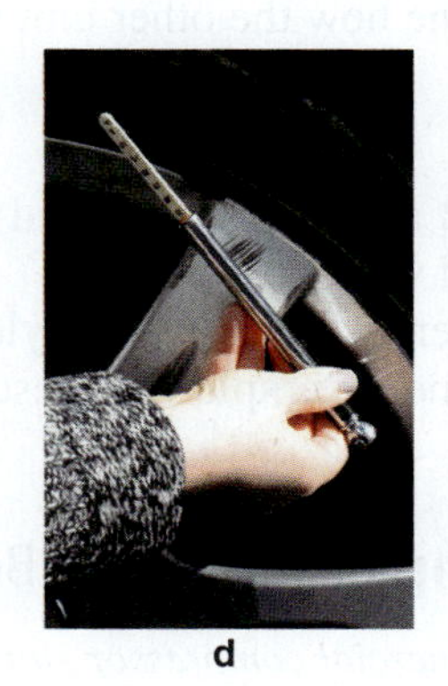
d

Figure 10.23 Other uses for pressure gauges include (a) for measuring blood pressure, (b) for an acetylene tank used in welding, (c) for a scuba diving regulator, and (d) for measuring tire pressure. (a) Fotosr52/Shutterstock; (b) Sergiy Zavgorodny/Shutterstock; (c) Dmitry Kalinovsky/Shutterstock; (d) Science Source

© Kiankhoon/depositphotos.com

When you pull the air out of a straw, you lower the air pressure inside the straw. Because the pressure outside is higher than inside, the liquid gets pushed up into the straw. A barometer works the same way.

TABLE 10.7 Common Pressure Conversions

1 atmosphere = 760 mm Hg (torr)
1 atmosphere = 101.3 kPa
1 atmosphere = 14.70 psi
1 atmosphere = 1.013 bar

We can use many different units to describe gas pressure. A common unit is the **atmosphere (atm)**, based on the standard air pressure at sea level. One atmosphere is defined as 760 mm Hg (29.92 inches Hg). In the United States, pressure is often described in terms of *pounds per square inch* (psi). A basketball has a gauge pressure of about 8 psi, while the tires on your car likely have a gauge pressure of 30–40 psi. Other common units of pressure include the *kilopascal* (kPa) and the *bar*. The relationships between these units are summarized in **Table 10.7**.

Check it
Watch Explanation

TRY IT

5. A tire has a maximum recommended pressure of 276 kPa. What is this pressure in psi?

6. Use the relationships given in Table 10.7 to complete these conversions:

a. Convert 2.4 atmospheres to psi.

b. Convert 0.892 kPa to atmospheres.

c. Convert 1,500 kPa to psi.

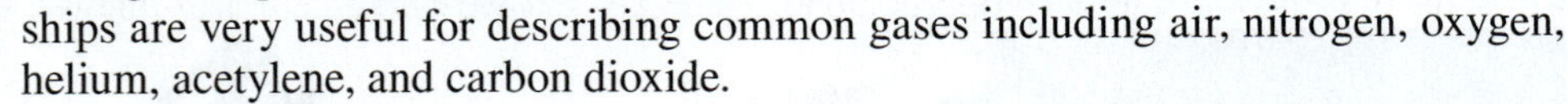

10.4 The Gas Laws

A scientific law describes an observed behavior, but it doesn't explain why the behavior occurs. The gas laws are an example of this idea.

The pressure, volume, and temperature of an ideal gas are related to each other through simple mathematical relationships called the **gas laws**. These relationships are very useful for describing common gases including air, nitrogen, oxygen, helium, acetylene, and carbon dioxide.

Boyle's Law

Boyle's law describes the relationship between the pressure and volume of a gas. According to Boyle's law, the pressure and the volume of a gas are inversely related. That is, if the volume (V) goes up, the pressure (P) goes down, and vice versa (**Figure 10.24**). As long as the temperature does not change, *the pressure times the volume of a gas is constant.*

$$PV = \text{constant}$$

Boyle's law is useful when the pressure or the volume is changing, and we wish to determine how the other unit will change. Because PV is constant, we can say that

$$P_1V_1 = P_2V_2$$

where P_1 and V_1 are the initial pressure and volume, and P_2 and V_2 are the final pressure and volume.

When working with Boyle's law, we can use any units of pressure and any units of volume. Example 10.2 describes this type of problem.

Wm. Baker/GhostWorx images/Alamy

Figure 10.24 A firefighter carries an air tank on his back. The air is compressed into a small volume, and so it has a high pressure.

Example 10.2 Using Boyle's Law

A commercial compressor stores 2.8 liters of air at a pressure of 150 psi. If this air is allowed to expand until the pressure is equal to 15 psi (just over atmospheric pressure), what volume will the air occupy?

In this example, we are looking for the volume of air after the pressure is decreased. P_1 is 150 psi, P_2 is 15 psi. V_1 is 2.8 L, and we are trying to solve for V_2. To solve this problem, we rewrite the equation above to isolate V_2 and then plug in the pressure and volume quantities to solve:

$$V_2 = \frac{P_1V_1}{P_2} = \frac{(150\ \text{psi})(2.8\text{L})}{15\ \text{psi}} = 28\ \text{L}$$

Notice that when we solved this problem, our units of pressure canceled out. It is important to make sure that the units of P_1 and P_2 are the same. The units for V_1 and V_2 will also be the same.

TRY IT

7. A balloon has a volume of 2.5 liters at a pressure of 1.0 bar. If the pressure around the balloon is decreased to 0.80 bar, what is the new volume of the balloon?

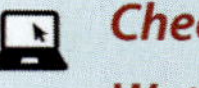

Check it
Watch Explanation

Charles's Law

Charles's law states that at constant pressure, the volume of a gas is directly proportional to its temperature. If the temperature goes up, the volume goes up. If the temperature goes down, the volume goes down. Mathematically, we can represent this by the following equation:

$$V \propto T$$

where the $\propto$ symbol means "is proportional to." Alternatively, we can write this as follows:

$$\frac{V}{T} = \text{constant}$$

Using Charles's Law to Find Absolute Zero

In the mid-nineteenth century, William Thomson (who was later granted the title Lord Kelvin) compared the plots of volume versus temperature for a number of different gas samples, similar to the graph in **Figure 10.25**. Although he could not measure gas pressures at very low temperatures, Thomson observed that if the volume continued to decrease in a linear fashion, all of the gases would reach a volume of zero at the same temperature. This temperature, –273.15 °C, is **absolute zero**—the lowest possible

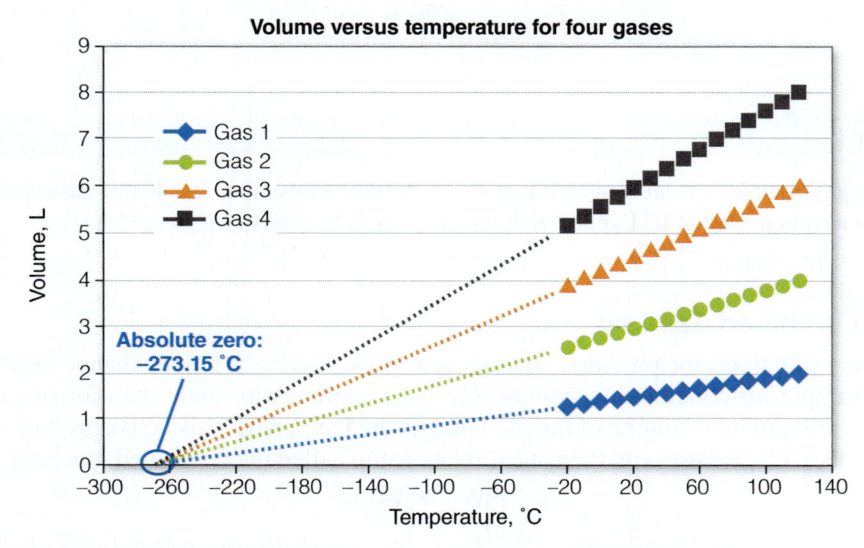

Figure 10.25 Absolute zero is calculated by extrapolating the volume–temperature relationship of a gas back to a volume of zero.

temperature. Recall that temperature is the measure of the kinetic energy of the particles in a substance. At absolute zero, the particles in a material have zero kinetic energy.

In the Kelvin temperature scale, absolute zero is given the value of 0 kelvin. A unit kelvin is the same size as a degree Celsius. We convert between the Celsius and Kelvin scales by using the following relationship:

$$\text{kelvin} = {}^\circ\text{C} + 273.15$$

Solving Volume–Temperature Problems Using Charles's Law

We commonly use Charles's law to predict how the volume of a gas will change if the temperature changes, or vice versa. Because the ratio of volume to temperature is constant, we can write Charles's law using this relationship:

$$\frac{V_1}{T_1} = \frac{V_2}{T_2}$$

where V_1 and T_1 are the initial volume and temperature, and V_2 and T_2 are the final volume and temperature.

When working a Charles's law problem, you can use any units of volume; but *you must express the temperature in kelvins*. Use of Celsius in a gas-law problem will give an incorrect answer. Example 10.3 illustrates this type of problem.

For gas-law problems, temperatures must be in kelvins.

Example 10.3 Using Charles's Law

A balloon has a volume of 3.2 liters at room temperature (25 °C). The gas inside the balloon is then heated to 100 °C. What is the new volume of the balloon?

In this example, we are looking for the volume of the balloon after the temperature has increased. V_1 is 3.2 L. Before solving for V_2, we must convert the temperatures to the Kelvin scale:

$$T_1 = 25\ ^\circ\text{C} + 273 = 298\ \text{K}$$

$$T_2 = 100\ ^\circ\text{C} + 273 = 373\ \text{K}$$

In these calculations, notice that we rounded 273.15 to 273 because we knew the Celsius temperature only to the nearest degree. Finally, we can rewrite the equation to isolate V_2, then plug in the quantities to solve.

$$V_2 = \frac{V_1 T_2}{T_1} = \frac{(3.2\ \text{L})(373\ \cancel{\text{K}})}{298\ \cancel{\text{K}}} = 4.0\ \text{L}$$

Check it
Watch Explanation

TRY IT

8. A gas occupies a volume of 15.2 L at 25 °C. At what temperature would this gas expand to a volume of 30.4 L? Express your answer in both kelvins and degrees Celsius.

The Combined Gas Law

In many situations, the pressure, volume, and temperature of a gas all change together. For example, air-conditioning systems rely on the compression and expansion of gases. As the pressure and volume of the gas change, the temperature also changes. For situations like this, we use a mathematical relationship called the **combined gas law**:

$$\frac{P_1 V_1}{T_1} = \frac{P_2 V_2}{T_2}$$

As in Boyle's and Charles's laws, we can use any units of pressure and volume, but the temperatures must be expressed in kelvins. Examples 10.4 and 10.5 illustrate the use of the combined gas law.

The combined gas law can be used to solve any Boyle's law or Charles's law problem.

Example 10.4 Using the Combined Gas Law

A gas with a temperature of 280 K, a pressure of 200 kPa, and a volume of 25.8 L is compressed to 15.8 L, causing the pressure to increase to 350 kPa. What is the temperature of the gas under the new conditions?

In this example, we're given P_1, V_1, and T_1, as well as P_2 and V_2. We're asked to solve for T_2. We can rearrange the combined gas law equation and insert the values to reach the answer.

$$T_2 = \frac{P_2V_2T_1}{P_1V_1} = \frac{(350\ \cancel{\text{kPa}})(15.8\ \cancel{\text{L}})(280\ \text{K})}{(200\ \cancel{\text{kPa}})(25.8\ \cancel{\text{L}})} = 300\ \text{K}$$

Example 10.5 Using the Combined Gas Law

Under constant-pressure conditions, a sample of hydrogen gas initially at 88 °C and 1.62 L is cooled until its final volume is 942 mL. What is its final temperature?

In this example, the pressure is constant, so $P_1 = P_2$. Because of this, it may be canceled from the equation. Also, notice that V_1 is given in liters, but V_2 is in milliliters: For the volume units to cancel, they must be the same. Therefore, let's express V_2 as 0.942 L. Converting the temperature to kelvins and substituting the values into the equation, we obtain a final answer of 210 K.

$$T_2 = \frac{\cancel{P_2}V_2T_1}{\cancel{P_1}V_1} = \frac{(0.942\ \cancel{\text{L}})(361\ \text{K})}{(1.62\ \cancel{\text{L}})} = 210\ \text{K}$$

TRY IT

9. On a cold Iowa morning when the temperature is –30 °C, a truck tire is inflated to a pressure of 45.0 psi. The truck is then driven south until the temperature reaches +33 °C. If the tire has not lost any air, what is the pressure in the tire after it warms up?

Avogadro's Law

Avogadro's law states that, if pressure and temperature are constant, the volume of a gas is proportional to the number of moles of gas present. The more gas that is present, the larger the volume it occupies. Mathematically, we can say that

$$V \propto n$$

where n is the number of moles of the gas.

The exact volume that a mole of gas occupies depends on its temperature and pressure. A temperature of 0 °C (273 K) and a pressure of 1.0 atmosphere is called *standard temperature and pressure*, or STP. At STP, one mole of gas occupies a volume of 22.4 liters (**Figure 10.26**).

Figure 10.26 At standard temperature and pressure (STP), 1 mole of gas occupies 22.4 L, the size of a large balloon.

 Check it
Watch Explanation

TRY IT

10. At STP, 1 mole of gas occupies a volume of 22.4 L. At the same temperature and pressure, what volume does 5 moles of gas occupy?

The Ideal Gas Law

The gas laws described so far allow us to relate different properties of gases. We can connect all of these together using the **ideal gas law**. This extraordinarily useful law relates the amount of gas present to its pressure, volume, and temperature. We usually write this law as

$$PV = nRT$$

Remember, in an ideal gas, the particles do not interact with each other at all. Real gases will have some interactions, but most common gases behave in a nearly ideal manner.

where P is the pressure, V is the volume, T is the temperature, n is the number of moles of gas, and R is a constant, called the *gas constant*, with a value of 0.0821 L·atm/mol·K.

When using the ideal gas law, it is important to make sure that the temperatures are expressed in kelvins and that the units of volume and pressure align with the values included in the gas constant (for example, liters and atmospheres). Examples 10.6 and 10.7 illustrate this law.

Example 10.6 Using the Ideal Gas Law

What volume does 1.00 mole of gas occupy at a temperature of 0.00 °C and a pressure of 1.00 atmosphere?

To solve this problem, we rearrange the ideal gas law equation to isolate V and then insert the values:

$$V = \frac{nRT}{P} = \frac{(100\ \text{mol})\left(0.0821 \frac{\text{L} \cdot \text{atm}}{\text{mol} \cdot \text{K}}\right)(273.15\ \text{K})}{(1.00\ \text{atm})} = 22.4\ \text{L}$$

Notice that in the solution above, all of the units cancel except for liters, and so the answer is given in liters. This problem corresponds to standard temperature and pressure (STP).

Example 10.7 Using the Ideal Gas Law

A portable oxygen tank has a volume of 2.40 L and a pressure of 243 psi at a temperature of 22 °C. How many moles of oxygen are present in this cylinder? What is the mass of the oxygen in grams?

To solve this problem, we first need to convert temperature to the Kelvin scale (22 °C = 295 K). To align our units with the gas constant, we also need to convert the pressure (243 psi = 16.5 atm). Using the ideal gas equation, we can then find the number of moles:

$$n = \frac{PV}{RT} = \frac{(16.5\ \text{atm})(2.40\ \text{L})}{\left(0.0821 \frac{\text{L} \cdot \text{atm}}{\text{mol} \cdot \text{K}}\right)(295\ \text{K})} = 1.64\ \text{mol}$$

Once we know the moles, we can then convert to grams of oxygen, as described in Chapter 7.

$$1.64\ \text{mol} \times \frac{32.00\ \text{g O}_2}{1\ \text{mol}} = 52.5\ \text{g O}_2$$

TRY IT

11. At what temperature does 1.20 moles of hydrogen gas occupy a volume of 28.1 L at a pressure of 121 kilopascals?

12. A room with a volume of 50 m^3 has a pressure of 750 torr at a temperature of 72 °F. How many moles of gas occupy the room?

Mixtures of Gases

We commonly encounter gases that contain several components. For example, air is a mixture of gases, composed of about 78% nitrogen and 21% oxygen. The remaining 1% is a mixture of argon, carbon dioxide, water vapor, and other trace components (**Figure 10.27**).

Philip and Karen Smith/Getty images

Figure 10.27 Air is a mixture of nitrogen, oxygen, argon, carbon dioxide, and smaller amounts of other gases.

Fortunately, the pressure and volume depend only on the amount of gas present, not on the identity of the gas. One mole of oxygen gas has the same pressure and volume as one mole of nitrogen gas or one mole of carbon dioxide gas.

When working with mixtures of gases, we sometimes use the **partial pressure** of the gases that are present. The partial pressure is the pressure caused by one gas in a mixture. We can add up the partial pressures to find the total pressure of the mixture. Example 10.8 illustrates this idea.

Adding up all the partial pressures gives the total pressure.

Example 10.8 Using Partial Pressures for a Mixture of Gases

If a 40.0-L cylinder is filled with 5.00 moles of nitrogen, 2.00 moles of oxygen, and 3.00 moles of carbon dioxide at a temperature of 400 K, what is the pressure inside the cylinder?

We could take two approaches in solving this problem. The first is to calculate the pressure that is produced by each of the individual gases. The partial pressures of nitrogen, oxygen, and carbon dioxide (labeled as P_{N_2}, P_{O_2}, and P_{CO_2}) are calculated as shown here:

$$P_{N_2} = \frac{nRT}{V} = \frac{(5.00\ \text{mol})\left(0.0821\frac{\text{L}\cdot\text{atm}}{\text{mol}\cdot\text{K}}\right)(400\ \text{K})}{(40.0\ \text{L})} = 4.11\ \text{atm}$$

$$P_{O_2} = \frac{nRT}{V} = \frac{(2.00\ \text{mol})\left(0.0821\frac{\text{L}\cdot\text{atm}}{\text{mol}\cdot\text{K}}\right)(400\ \text{K})}{(40.0\ \text{L})} = 1.64\ \text{atm}$$

$$P_{CO_2} = \frac{nRT}{V} = \frac{(3.00\ \text{mol})\left(0.0821\frac{\text{L}\cdot\text{atm}}{\text{mol}\cdot\text{K}}\right)(400\ \text{K})}{(40.0\ \text{L})} = 2.46\ \text{atm}$$

Once we've done this, we can find the total pressure by adding together the partial pressures:

$$P_{total} = P_{N_2} + P_{O_2} + P_{CO_2} = 4.11 \text{ atm} + 1.64 \text{ atm} + 2.46 \text{ atm} = 8.21 \text{ atm}$$

The second approach to this problem simplifies the calculation. Because the total pressure depends on the number of moles of gas (not the identity of the gas), we can find the total number of moles present and then calculate the pressure based on this combined value.

$$5.00 \text{ moles } N_2 + 2.00 \text{ moles } O_2 + 3.00 \text{ moles } CO_2 = 10.00 \text{ moles total}$$

$$P_{total} = \frac{nRT}{V} = \frac{(10.00 \cancel{\text{mol}})\left(0.0821 \frac{\cancel{\text{L}} \cdot \text{atm}}{\cancel{\text{mol}} \cdot \cancel{\text{K}}}\right)(400 \cancel{\text{K}})}{(40.0 \cancel{\text{L}})} = 8.21 \text{ atm}$$

Check it

Watch explanation

TRY IT

13. An air sample contains 0.3% water vapor. If the total air pressure is 765 torr, what is the partial pressure due to the water?

14. A 24-liter cylinder contains 150.0 grams of NH_3 gas and 600.0 grams of argon gas at a temperature of 300 K. What is the pressure inside the cylinder?

15. While walking along the ocean, you take a deep breath. If your lung capacity is 5.0 liters, the temperature is 87 °F, and the pressure is 765 torr, how many moles of air can you take in? Since air is composed of about 21% oxygen (by mole percent), how many grams of oxygen have you taken in?

A Molecular View of the Gas Laws

In the preceding sections, we used mathematical relationships to describe how gases behave. By thinking about gases on the atomic or molecular level, we can begin to understand why these mathematical relationships are so, and we can also predict the behavior of gases in other situations. Let's ask three questions about the relationship between the particles in a container and the pressure they exert:

Pressure increases with temperature.

1. *How does the pressure change if the temperature of the gas increases?* At higher temperatures, the particles move faster. Faster-moving particles strike their surroundings with more force, exerting more pressure. On the other hand, if the temperature decreases, the pressure drops. This is why tires are sometimes partially flat on cold mornings (**Figure 10.28**).

Keith Beaty/Getty Images

Figure 10.28 What happens to the pressure in bike tires on a cold morning? Why is this so? What happens to the pressure if we pump more gas particles into the tire?

2. *What happens to the pressure of a gas if the volume increases?* If the volume of a container increases, the particles have to travel farther before reaching the walls of the container. As a result, the particles collide with the container less frequently. An increase in volume produces a decrease in pressure.
3. *What happens to the pressure of a gas if the number of particles in a container is increased?* If you've ever pumped up a tire, you know that increasing the amount of gas present increases the pressure. On a molecular level, more gas particles mean the particles will strike the sides of the container more often, creating more pressure. The dependence of pressure on temperature, number of particles, and volume of the container is shown in **Figure 10.29**.

Increasing the amount of gas increases the pressure.

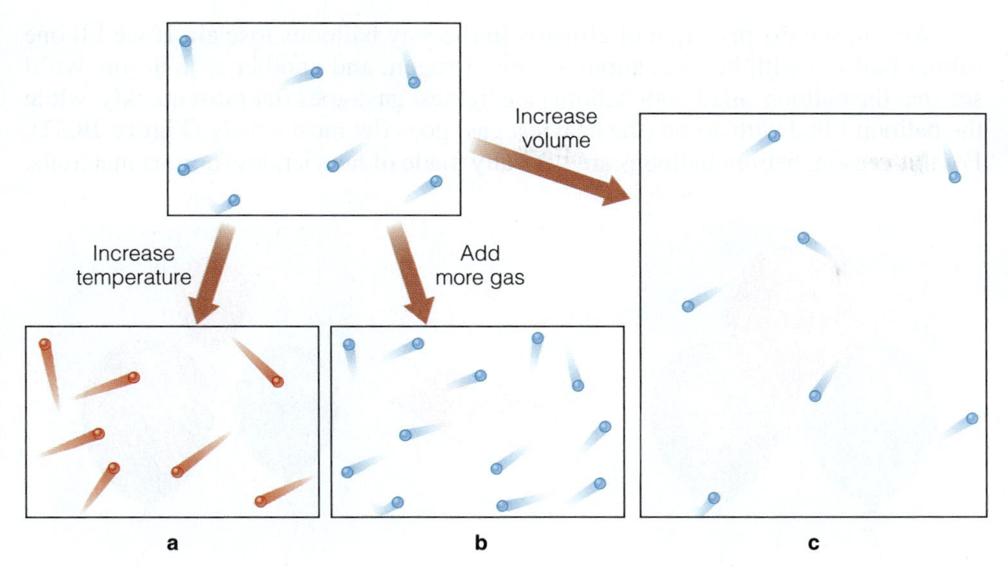

Figure 10.29 How does the pressure change as the temperature, volume, and number of particles change? (a) Increasing temperature increases pressure. (b) Increasing the number of particles increases the pressure. (c) Increasing the volume decreases the pressure.

***Explore* Figure 10.29**

TRY IT

16. How do these changes affect the pressure of a gas?

a. decreasing the temperature

b. decreasing the volume

c. decreasing the amount of gas

10.5 Diffusion and Effusion

Diffusion is the spread of particles through random motion. Diffusion can refer to either the liquid or the gaseous state, but the principle is the same: Particles randomly move, and as they do, they slowly spread from areas of higher concentration to lower concentration. For example, gases emitted from a smokestack diffuse out into the atmosphere (**Figure 10.30**). Because lighter particles move more quickly than heavier particles, lighter gases diffuse more quickly than heavier gases.

Effusion is the process of a gas escaping from a container. Like diffusion, effusion depends on the velocity of the gas particles. For example, **Figure 10.31** shows two containers, one filled with helium, one filled with air. Each container has a single small hole out of which the gases can escape. The smaller helium particles move faster and therefore collide with the container walls more frequently. As a result, they are more likely to encounter the opening, and they leak out of the container more quickly.

Figure 10.30 Gases spread from areas of high concentration to low concentration. This process is called *diffusion*.

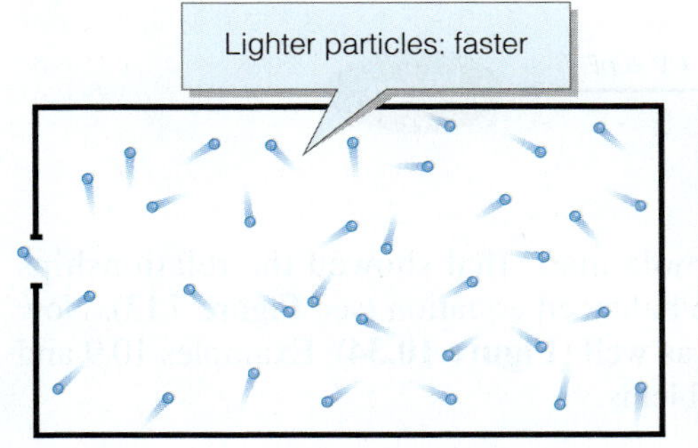

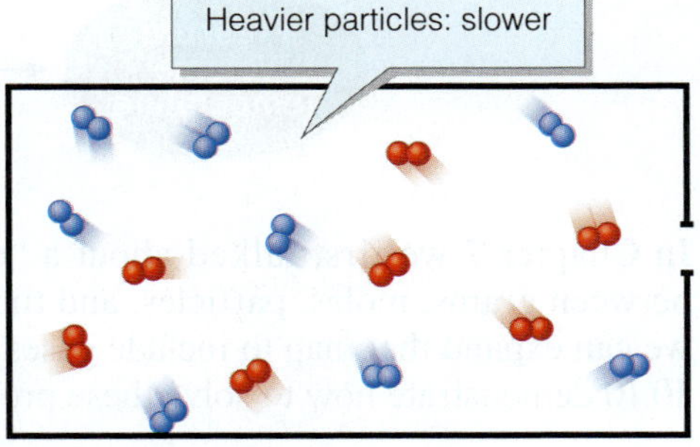

Figure 10.31 Because they move faster, lighter gases escape from a container more quickly.

***Explore* Figure 10.31**

Lighter gases diffuse and effuse faster.

We can see the principle of effusion in the way balloons lose air. If we fill one rubber balloon with helium, another with nitrogen, and another with argon, we'll see that the balloon filled with helium (the lightest gas) goes flat most quickly, while the balloon filled with argon (the heaviest gas) goes flat most slowly (**Figure 10.32**). For this reason, helium balloons are typically made of heavier, less porous materials.

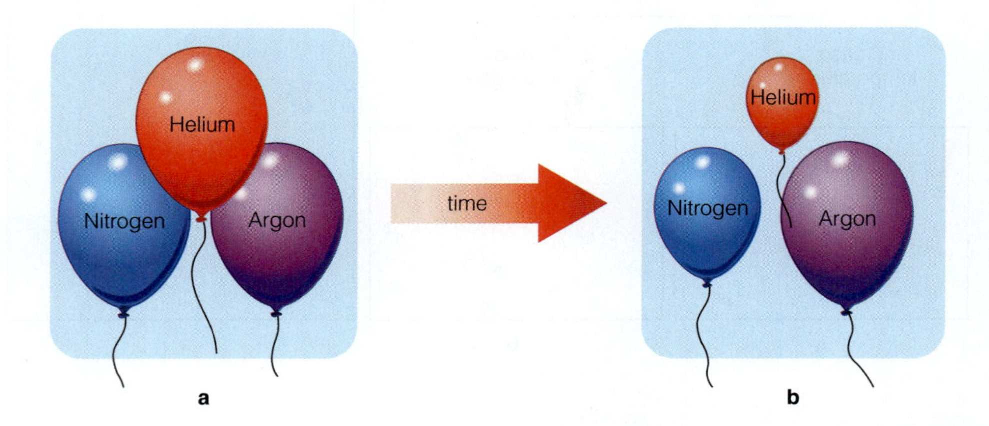

Figure 10.32 (a) When balloons are filled with different gases, (b) the one containing the lightest gas goes flat most quickly.

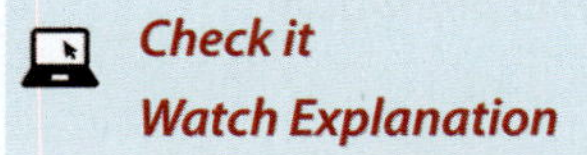

TRY IT

17. Four containers of gas are opened to the air at the same time. One contains carbon dioxide (CO_2), one contains methane (CH_4), one contains propane (C_3H_8), and one contains carbon disulfide (CS_2). Which of these gases would mix with the air most quickly? Which would mix most slowly?

10.6 Gas Stoichiometry

Sudden changes in gas pressure can have explosive results. These changes often arise when gases form or are consumed in a chemical reaction. To safely conduct reactions involving gases, we must know how much gas is produced or consumed. For example, the process of fermentation, used in making beer and bread (**Figure 10.33**), involves the reaction of glucose to form ethanol and carbon dioxide gas:

$$C_6H_{12}O_6\,(s) \rightarrow 2\ C_2H_6O\,(l) + 2\ CO_2\,(g)$$

In this fermentation reaction, how much carbon dioxide is produced for each kilogram of glucose that reacts?

Scott Eisen/Bloomberg via Getty images

Figure 10.33 These copper boiling and fermentation tanks are from the Samuel Adams Brewhouse in Boston, Massachusetts. Because the fermentation reaction produces carbon dioxide gas, brewers must monitor the pressure of gases inside the tanks.

To answer this question, we must combine the rules of stoichiometry (covered in Chapter 7) with the gas laws. At the center of these two concepts is the mole. Using the ideal gas law, we can relate *P*, *V*, and *T* to the number of moles of a gas:

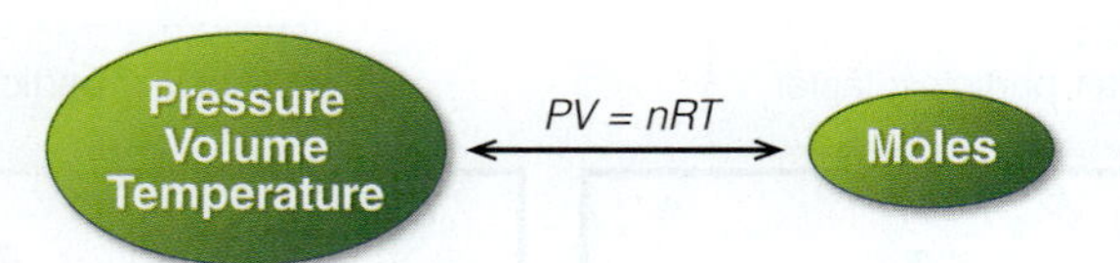

In Chapter 7 we first talked about a "mole map" that showed the relationships between grams, moles, particles, and the balanced equation (see Figure 7.13). Now we can expand that map to include gases as well (**Figure 10.34**). Examples 10.9 and 10.10 demonstrate how to solve these problems.

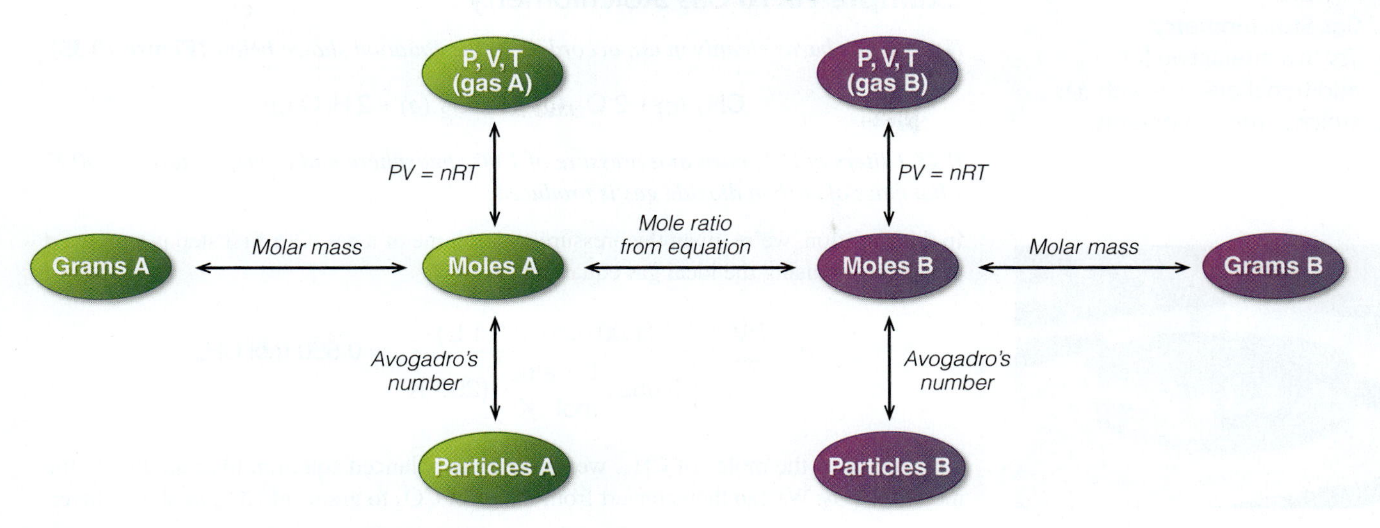

Figure 10.34 The mole is the central hub for conversion between units.

Example 10.9 Gas Stoichiometry

In the fermentation of glucose, how many moles of carbon dioxide are produced for each kilogram of glucose that reacts? If the reaction takes place in a sealed container and the gas occupies a volume of 8.10 liters at a temperature of 21 °C, find the pressure of the carbon dioxide gas inside the container.

$$C_6H_{12}O_6\,(s) \rightarrow 2\,C_2H_6O\,(l) + 2\,CO_2\,(g)$$

To solve this problem, we first relate grams of glucose to moles of carbon dioxide:

$$\text{Grams } C_6H_{12}O_6 \rightarrow \text{Moles } C_6H_{12}O_6 \rightarrow \text{Moles } CO_2$$

We begin with 1 kg (1,000 g) of glucose:

$$1{,}000\ \cancel{g\ C_6H_{12}O_6} \times \frac{1\ \cancel{mole\ C_6H_{12}O_6}}{180.18\ \cancel{g\ C_6H_{12}O_6}} \times \frac{2\ \text{moles } CO_2}{1\ \cancel{mole\ C_6H_{12}O_6}} = 11.10\ \text{moles } CO_2$$

Once we've found the moles of CO_2, we use the ideal gas equation to find the pressure. Be sure to convert the temperature to kelvins and then solve:

$$P = \frac{nRT}{V} = \frac{(11.10\ \cancel{mol}\ CO_2)\left(0.0821\frac{\cancel{L}\cdot \text{atm}}{\cancel{mol}\cdot \cancel{K}}\right)(294\ \cancel{K})}{(8.10\ \cancel{L})} = 33.1\ \text{atm } CO_2$$

A pressure of 33.1 atm is a dangerous explosion risk. To prevent this risk, brewers must be careful to release some of the CO_2 formed and to monitor the pressure inside fermentation vessels.

In this example, we first used the balanced equation to solve for moles; then we used the ideal gas law to convert from moles to pressure. In other problems, we may be given the pressure and volume of a gas and asked to relate it to another reagent. In this case, we first convert to moles using the ideal gas law and then solve the stoichiometry problem. Example 10.10 illustrates this process.

Practice
Gas Stoichiometry
Try this interactive for additional practice with gas stoichiometry problems.

Example 10.10 Gas Stoichiometry

*Natural gas burns cleanly in air, according to the equation shown below (***Figure 10.35***):*

$$CH_4\ (g) + 2\ O_2\ (g) \rightarrow CO_2\ (g) + 2\ H_2O\ (g)$$

If 13.1 liters of CH_4 burn at a pressure of 1.00 atmosphere and a temperature of 290 K, what mass of carbon dioxide gas is produced?

In this question, we're given the pressure and volume of a gas. Our first step is to convert this to moles, using the ideal gas equation:

$$n = \frac{PV}{RT} = \frac{(1.00\ \cancel{\text{atm}})(13.1\ \cancel{\text{L}})}{\left(0.0821 \frac{\cancel{\text{L}} \cdot \cancel{\text{atm}}}{\text{mol} \cdot \cancel{\text{K}}}\right)(290\ \cancel{\text{K}})} = 0.550\ \text{mol CH}_4$$

Vitaliy Maselko/Alamy

Figure 10.35 The characteristic blue flame of a natural gas stove.

Once we have the moles of CH_4, we can use the balanced equation to relate this to the moles of CO_2. We can then convert from moles of CO_2 to grams of CO_2, as shown here:

$$0.550\ \cancel{\text{moles CH}_4} \times \frac{1\ \cancel{\text{mole CO}_2}}{1\ \cancel{\text{mole CH}_4}} \times \frac{44.01\ \text{g CO}_2}{1\ \cancel{\text{mole CO}_2}} = 24.2\ \text{g CO}_2$$

Because these problems involve multiple steps, it will take some time before you become proficient at them. I encourage you to try the following problems.

Check it
Watch Explanation

TRY IT

18. In carbonated soft drinks, the carbon dioxide reacts with water to produce carbonic acid. This gives the drinks a pungent flavor that is balanced with a large amount of sugar. Over time, carbonic acid converts back into carbon dioxide and water, as shown in the equation below. What mass of H_2CO_3 would be required to produce a volume of 2.0 liters of CO_2 at a pressure of 1.0 atm and a temperature of 298 K?

$$H_2CO_3\ (aq) \rightarrow H_2O\ (l) + CO_2\ (g)$$

19. An airbag contains a solid fuel, which must react very rapidly to produce a large amount of gas. Airbags typically use several reactions in tandem, one of which is the decomposition of sodium azide to give sodium and nitrogen:

$$2\ NaN_3\ (s) \rightarrow 2\ Na\ (s) + 3\ N_2\ (g)$$

In this reaction, what volume of gas would be produced by the decomposition of 15 grams of NaN_3 if the temperature is 350 K and the pressure is 1.2 atm?

20. Acetylene torches burn at very high temperatures, using this reaction:

$$2\ C_2H_2\ (g) + 5\ O_2\ (g) \rightarrow 4\ CO_2\ (g) + 2\ H_2O\ (g)$$

At standard temperature and pressure, what volume of oxygen gas is required to completely react with a 4.8-liter cylinder of acetylene having an absolute pressure of 80 psi and a temperature of 20 °C?

Capstone Question

At the beginning of this chapter, we introduced the fossil fuels—coal, oil, and natural gas. Many fleet vehicles use compressed natural gas as an alternative to liquid gasoline (**Figure 10.36**). Natural gas is mainly composed of methane, CH_4. This equation shows the combustion of methane:

$$CH_4\,(g) + 2\,O_2\,(g) \rightarrow 2\,H_2O\,(g) + CO_2\,(g)$$

Using the table below, draw the Lewis structure for each substance in this reaction, and determine the electronic and molecular geometries for CH_4, CO_2, and H_2O. Compare the types of intermolecular forces in liquid CH_4 and liquid H_2O. Which substance has the stronger intermolecular forces? How does this impact their boiling points?

Methane burns with a fuel value of 55.5 kJ/g. If a 1.00-L cylinder containing pure methane gas at a pressure of 5,000 psi at 298 K is completely burned, how much energy will be released? What mass of CO_2 will form in this reaction?

Substance	Mass (g/mol)	Boiling Point	Lewis Structure	Electronic Geometry	Molecular Geometry
CH_4	16.0	−162 °C			
O_2	32.0	−183 °C			
CO_2	44.0	—*			
H_2O	18.0	100 °C			

*At standard atmospheric pressure, CO_2 sublimes (converts from solid to gas) at −79 °C.

Capstone Video

Gado Reportage/Alamy

Figure 10.36 Many fleet vehicles like this delivery truck run on compressed natural gas.

SUMMARY

In this chapter, we've explored the structure and properties of solids, liquids, and gases. In a solid, the particles pack closely together in fixed positions. In a liquid, the particles are close together but are not held in a fixed position. In a gas, the particles are far apart and have little or no interaction with the particles around them. The melting point and boiling point of a substance depend on how strongly the particles are held together.

The properties of a substance depend on its composition and structure. Ionic compounds form rigid lattices with each particle strongly attracted to those around it. Ionic compounds have very high melting and boiling points. In metallic substances, the atoms pack closely together and share electrons loosely between multiple atoms. This structure creates strong interactions, but not the rigid framework of ionic compounds.

Molecular compounds contain discrete groups of atoms, called molecules, that are connected by covalent bonds. The physical properties of molecular compounds depend on the interactions between molecules. Three major types of intermolecular forces are London dispersion forces, dipole–dipole interactions, and hydrogen bonds. Dispersion forces are the weakest interactions, while hydrogen bonds are the strongest. Molecular compounds typically have lower melting and boiling points than ionic compounds.

Some substances, such as diamond and graphite, form extensive networks of covalent bonds rather than small molecules. Polymers are compounds that contain long chains of covalent bonds. Plastics are synthetic polymers.

Gases have neither fixed shape nor fixed volume. In an ideal gas, the particles do not interact with each other and occupy a negligible fraction of the volume of the container.

Gases are typically measured by their pressure. A barometer is a device that measures atmospheric pressure. Pressure is measured in several different units, including atmospheres, torr (also called mm Hg), pounds per square inch (psi), bars, and kilopascals (kPa).

Boyle's law states that at constant temperatures, the product of the pressure and volume of an ideal gas are constant. This means that as pressure increases, the volume decreases, and vice versa.

Charles's law states that the volume of a gas is directly proportional to its temperature at constant pressure. By extrapolating the volume of a gas down to zero, we can identify the lowest possible temperature, called absolute zero. At this temperature, particles have no kinetic energy. The Kelvin temperature scale measures the absolute temperature of a substance: Absolute zero is given the value of 0 kelvin. When working with the gas laws, we must convert temperatures to the Kelvin scale.

The combined gas law integrates Charles's and Boyle's laws into a single law that relates the pressure, volume, and temperature of a gas. This law is most commonly expressed as

$$\frac{P_1V_1}{T_1} = \frac{P_2V_2}{T_2}$$

where P, V, and T denote the pressure, volume, and temperature, and the subscripts denote initial and final conditions.

Avogadro's law states that at constant pressure, the volume of a gas is proportional to the number of moles present. At standard temperature and pressure (STP), one mole of gas occupies a volume of 22.4 liters.

The ideal gas law describes the relationship of the P, V, and T to the number of moles of gas present. The relationship is given by the expression

$$PV = nRT$$

where n is the number of moles of gas, and the gas constant, R, has a value of 0.0821 L·atm/mol·K.

In a mixture of gases, each component contributes to the overall pressure of the system. The total pressure of the system is simply the sum of the partial pressures of the gases present.

We can understand the relationships described in the gas laws by thinking about gases on a molecular level. Because the pressure inside a container arises from the collisions of gases with the walls of the container, an increase in pressure occurs as the amount of gas increases (more particles striking the walls), as temperature increases (faster-moving particles strike the walls more frequently and with more energy), or as the volume decreases (the particles strike the walls more frequently).

Diffusion is the spread of particles through random motion, such as the motion of gas particles. Because lighter gas particles move more quickly than heavier particles, lighter gas particles diffuse more quickly. A related term, *effusion*, refers to the escape of gas from a container. Because smaller particles move faster, they also effuse more quickly than heavier particles.

As we saw in Chapter 7, we can use a balanced equation to predict the grams and moles that will be consumed or produced in a reaction. By combining stoichiometry calculations with the ideal gas law, we can predict changes in pressure and volume that accompany reactions involving gases.

Rethinking Gas Storage

Compressed gases are a part of modern life. From fountain drinks to air conditioners and from mechanical work to health care, compressed gases are all around us. But compressed gases can be hard to handle. Because of their high pressures, gas cylinders require heavy steel walls. They can be difficult to transport, and they are potentially dangerous. If a gas cylinder ruptures, the pressure inside the container can fire the cylinder like a missile.

Omar Yaghi is a chemist at the University of California, Berkeley (**Figure 10.37**). For the past two decades, he has worked on a class of compounds called *metal-organic frameworks*, or MOFs, that hold the potential to change the way people store gases.

Originally from Jordan, Yaghi arrived in the United States when he was 15. Although he spoke very little English, he enrolled in classes at a community college near Albany, New York. His abilities and hard work paid off, and he eventually earned a Ph.D. in chemistry from the University of Illinois. As he describes it, "I was drawn to the beauty of chemistry. I didn't necessarily want to change the world—I was driven by curiosity and passion for science."

The MOF compounds that Yaghi creates *are* beautiful, but also very useful. Like a sponge or a honeycomb, these substances contain many tiny pores—except they are much smaller. Normally, gases move freely around a container, interacting very little with their surroundings. A MOF is able to trap gases within the tiny pores (Figure 10.37e). As a result, there are fewer free particles, and the pressure of the gas decreases. If a gas is placed in a cylinder that contains MOFs, the pressure is lower than if the gas were placed in an empty cylinder!

MOF technology has created a surge of interest in gas storage. For example, several companies are developing MOFs for the fuel tanks of vehicles powered by natural gas. Vehicles using this technology can safely store more fuel, so they have to be refueled less frequently.

Figure 10.37 (a) The transport and storage of gas requires heavy steel containers. (b) Omar Yaghi is a chemist at UC Berkeley who studies gas storage. (c) A sponge is an example of a porous material. The empty spaces in these structures can trap other materials. (d) Yaghi's compounds, called MOFs, contain tiny pores. (e) These pores can trap gas particles, reducing the number of free particles and decreasing the pressure in the container. This image depicts CH_4 molecules trapped in a MOF. (f) Dr. Yaghi sits behind the wheel of a vehicle operating on a MOF methane fuel tank.
(a) DWD-photo/Alamy; (b) Omar M. Yaghi Research group at University of California Berkeley; (c) ©Ksena32/depositphotos.com; (d) Omar M. Yaghi Research group at University of California Berkeley; (e) Omar M. Yaghi Research group at University of California Berkeley; (f) Omar M. Yaghi Research group at University of California Berkeley

a

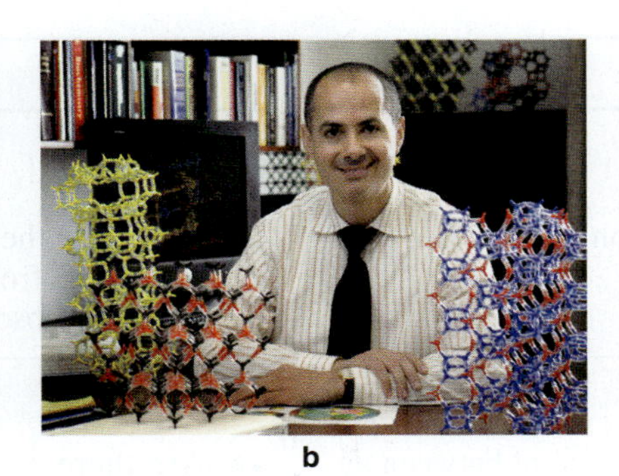
b

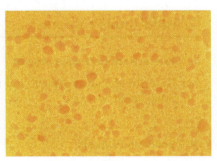
c

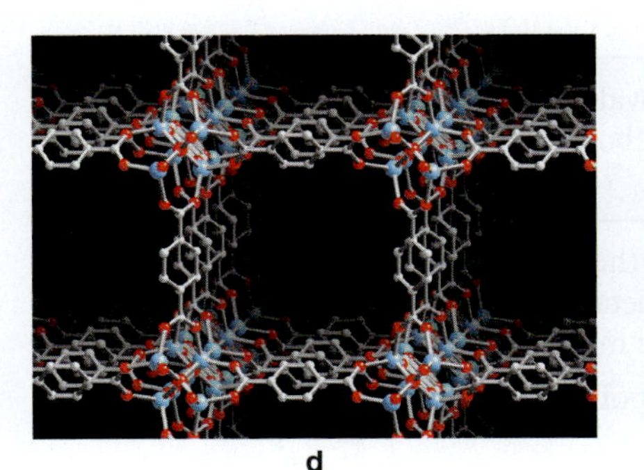
d

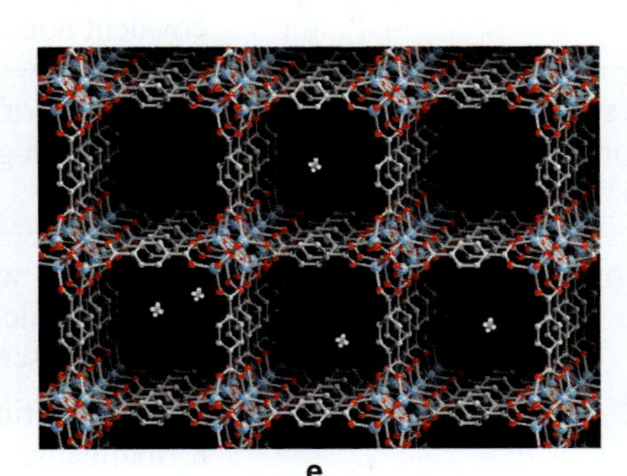
e

f

Key Terms

10.2 Solids and Liquids

intermolecular forces The forces of attraction or repulsion that take place between molecules.

dipole–dipole interaction An intermolecular force between two molecules containing net dipoles.

hydrogen bond An unusually strong dipole–dipole interaction that occurs between molecules containing H–F, H–O, or H–N bonds.

London dispersion forces Intermolecular forces that result from fluctuations in charge density called *instantaneous dipoles*.

covalent networks Long two- or three-dimensional sequences of covalent bonds, resulting in very large single molecules.

polymer A compound composed of long chains of covalently bonded atoms.

10.3 Describing Gases

ideal gas A gas in which the volume of the particles is much less than the volume of the container, and in which the particles have no attraction for each other.

pressure The force that an object exerts divided by the area over which it is applied; for gases, pressure describes the force that gases exert on their surroundings.

barometer A device used to measure atmospheric pressure.

millimeters of mercury (mm Hg) A measure of gas pressure; this unit originates from the height to which atmospheric pressure can push a column of mercury in a barometer; 1 mm Hg = 1 torr.

gauge pressure The difference between a compressed gas pressure and atmospheric pressure.

atmosphere (atm) A unit of gas pressure; 1 atm = 760 mm Hg.

10.4 The Gas Laws

gas laws Mathematical relationships between the pressure, volume, and temperature of gases.

Boyle's law The pressure and volume of an ideal gas are inversely related; the product of PV is constant at constant temperature.

Charles's law The volume of an ideal gas is directly proportional to its temperature; the relationship between V and T is constant at constant pressure.

absolute zero The lowest possible temperature, corresponding to 0 K or −273.15 °C; at this temperature, the particles in a substance have zero kinetic energy.

combined gas law A combination of Boyle's law and Charles's law; it states that for an ideal gas, the quantity PV/T is constant; usually expressed by the equation $P_1V_1/T_1 = P_2V_2/T_2$, where the subscripts 1 and 2 denote two different conditions.

Avogadro's law If pressure and temperature are constant, the volume of a gas is proportional to the number of moles of gas present.

ideal gas law The relationship between pressure, volume, temperature, and the number of moles of an ideal gas; typically expressed in the form $PV = nRT$, where R is the gas constant.

partial pressure The pressure caused by one gas in a mixture.

10.5 Diffusion and Effusion

diffusion The spread of particles through random motion; lighter gases diffuse more quickly than heavier gases.

effusion The process of a gas escaping from a container; lighter gases effuse more quickly than heavier gases.

Additional Problems

10.1 Interactions between Particles

21. Describe the arrangement and motion of particles in a solid, liquid, and gas.

22. How does the motion of particles change as a substance transitions from liquid to gas? In this transition, does the substance absorb or release heat energy?

10.2 Solids and Liquids

23. How is the arrangement of particles different between an ionic and a metallic solid?

24. Are there examples in which an ionic solid contains covalent bonds?

25. What are the *intra*molecular forces in a molecular solid? What types of *inter*molecular forces can exist in a covalent solid?

26. What types of covalent bonds are necessary for an intermolecular hydrogen bond to form?

27. Describe each of the following as ionic, metallic, or molecular solids:

a. calcium fluoride
b. glucose, $C_6H_{12}O_6$
c. bronze, an alloy of copper and tin
d. table salt, NaCl

28. Determine whether the following properties broadly describe ionic, metallic, or molecular solids. Some properties may describe more than one group.

a. high melting point
b. malleable
c. low boiling point
d. loosely shared electrons within a network of atoms

29. These four compounds have very similar formula masses. Classify them as ionic or covalent. Predict which of the compounds would have the highest and lowest boiling points.

a. LiF b. H_2O c. N_2 d. HCl

30. These four compounds have very similar formula masses. Classify them as ionic or covalent. Predict which of the compounds would have the highest and lowest boiling points.

a. Li_2S b. HF c. CH_3F d. O_2

31. The Lewis structure of propane, C_3H_8, is shown here. What is the strongest type of intermolecular force in liquid propane?

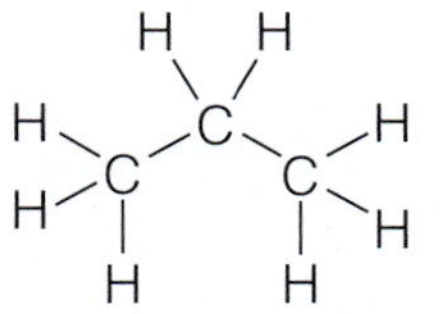

32. The Lewis structure of methylamine, CH_3NH_2, is shown here. What is the strongest type of intermolecular force in liquid methylamine?

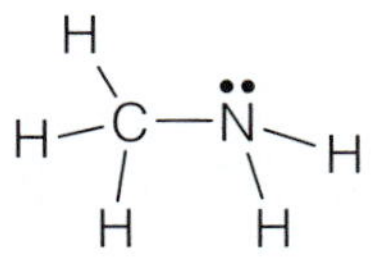

33. The Lewis structure of formaldehyde is shown here. What is the strongest type of intermolecular force in liquid formaldehyde? Sketch two molecules, using a dashed line (as in Figure 10.6) to show the positive region of one molecule interacting with the negative region of another molecule.

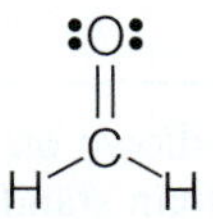

34. The Lewis structure for ammonia, NH_3, is shown here. What is the strongest type of intermolecular force in liquid ammonia? Sketch two molecules, using a dashed line (as in Figure 10.6) to show the positive region of one molecule interacting with the negative region of another molecule.

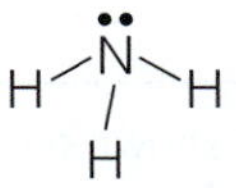

35. Draw Lewis structures for each of the following molecules. Determine whether each molecule has a net dipole, and identify the strongest intermolecular force that would act between molecules of each pure substance.

a. CH_4 b. CH_2F_2 c. HNO (N is the central atom)

36. Draw Lewis structures for each of the following molecules. Determine whether each molecule has a net dipole, and identify the strongest intermolecular force that would act between molecules of each pure substance.

a. HCN b. HOCl (O is the central atom) c. CS_2

37. Methanol (CH_3OH) forms hydrogen bonds. Sketch Lewis structures for two methanol molecules, and use dashed lines to show the hydrogen bonds that can form.

38. Methanol (CH_3OH) dissolves easily in water because hydrogen bonds can form between these molecules. Sketch a methanol molecule and a water molecule, and show two possible hydrogen bonding interactions between the two mole cules.

39. What are polymers? How is a polymer different from a molecular solid?

40. How is a covalent network different from a molecular solid?

10.3 Describing Gases

41. Describe the motion of particles in a gas.

42. What are the two criteria for a gas to be considered an ideal gas?

43. Describe how the pressure on a container arises from the particles inside and outside the container.

44. What happens if the pressure inside a sealed container is much greater than the pressure outside the container? What happens if the pressure outside a container is much greater than the pressure inside?

45. When drinking through a straw, you are able to control the height of the liquid inside the straw by changing the pressure inside your mouth, as shown in **Figure 10.38**. What happens if the pressure in your mouth is lower than the air pressure outside? What happens if the pressure in your mouth is higher than the air pressure outside?

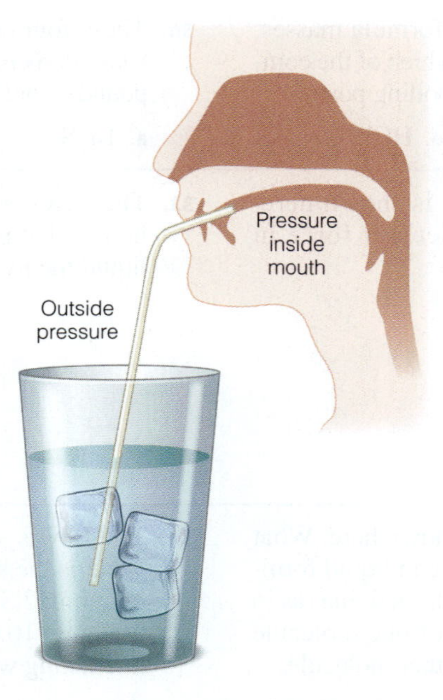

Figure 10.38 A straw is like a barometer.

46. Ignoring the fact that humans can't survive in a vacuum, is it possible to drink through a straw in the vacuum of space? Why or why not?

47. Measure the pressure shown on the barometer. Is this pressure higher or lower than standard atmospheric pressure? (1 inch = 25.4 mm)

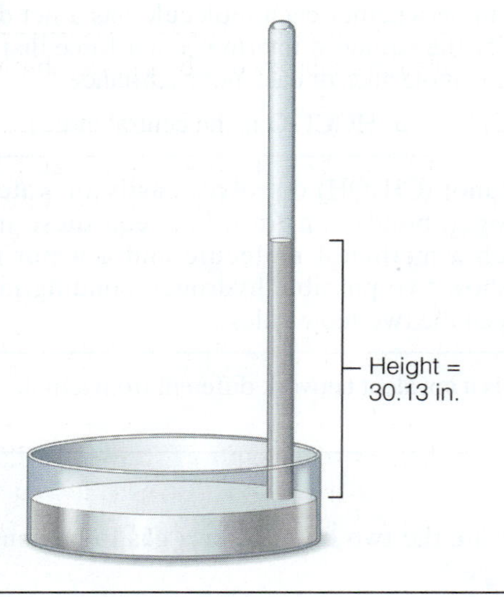

48. Measure the pressure shown on the barometer. Is this pressure higher or lower than standard atmospheric pressure? (1 inch = 25.4 mm)

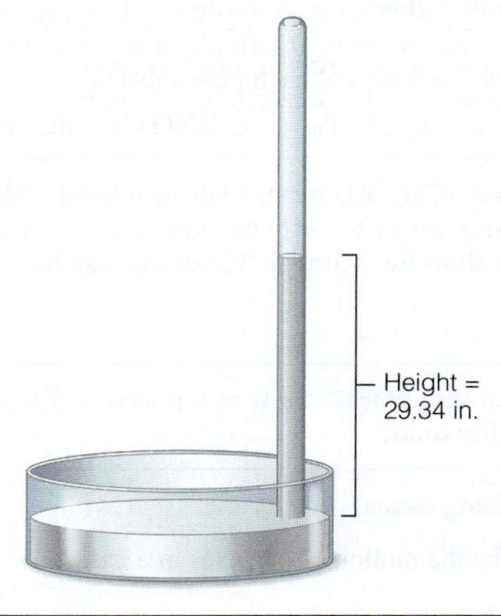

49. In August 1992, Hurricane Andrew slammed into the Miami area, causing 23 deaths and destroying over 25,000 homes. When it made landfall, the pressure in the eye of the storm was 922 millibars. What is this pressure in torr? Is this pressure higher or lower than standard atmospheric pressure?

50. In November 2013, Typhoon Haiyan struck the Philippine islands, causing over 6,000 deaths. At one point, the pressure in the eye of the storm was measured at 26.43 inches of mercury. What is this pressure in torr? What is this pressure in millibars? Is this pressure higher or lower than standard atmospheric pressure?

51. Express standard atmospheric pressure in millibars, torr, and inches of mercury.

52. Express standard atmospheric pressure in mm Hg, kilopascals, and bars.

53. Convert the following pressures:

a. Express 698 mm Hg in millibars.
b. Express 3.2 atm in torr.
c. Express 1.42 bars in psi.

54. Convert the following pressures:

a. Express 1.2 atmospheres in bars and millibars.
b. Express 32.41 psi in torr.
c. Express 23.29 inches of mercury in atmospheres.

55. Gauge pressure refers to the difference in pressure between a compressed gas and atmospheric pressure. If the atmospheric pressure is 15.0 psi, and you inflate your bicycle tire to a gauge pressure of 110.0 psi, what is the absolute pressure in the tires?

56. Gauge pressure refers to the difference in pressure between a compressed gas and atmospheric pressure. If the atmospheric pressure is 15.0 psi, and you inflate your football to a gauge pressure of 10.0 psi, what is the absolute pressure in the football?

10.4 The Gas Laws

57. How does the pressure change if the volume of a gas decreases? How does it change if the temperature decreases?

58. If the volume of a gas increases, how does the pressure change? How does the pressure change if the temperature of a gas increases?

59. A gas with an initial pressure of 1.0 atmosphere and a volume of 4.3 liters is compressed to a pressure of 5.1 atmospheres. What is the new volume of the gas?

60. A gas with an initial pressure of 780 torr and a volume of 150 liters is compressed to a volume of 32 liters. What is the pressure of the compressed gas?

61. A balloon with a pressure of 15.0 psi has a volume of 1.24 liters. If the pressure drops, the balloon will expand. What will be the volume of the balloon if the pressure decreases to 10.4 psi?

62. A pressurized gas is allowed to expand. If the gas originally occupied a volume of 420 mL at a pressure of 300 psi, what volume will it occupy if pressure is lowered to 150 psi?

63. A balloon with a volume of 1.41 liters at a temperature of 300.0 K is heated to 350.0 K. What is the new volume of the balloon?

64. A balloon with a volume of 4,320 cm^3 at a temperature of 25 °C is heated to 95 °C. What is the new volume of the balloon?

65. An ideal gas is allowed to cool at a constant pressure. The gas occupies a volume of 20.0 L at a temperature of 273.15 K (0 °C). At what temperature will it occupy a volume of 10.0 L?

66. A gas occupies a volume of 800.0 mL at a temperature of 25 °C. At what temperature would the gas occupy only half this volume?

67. What is absolute zero? Describe absolute zero with regard to the motion and kinetic energy of molecules.

68. Is it possible for a substance to be a gas at absolute zero?

69. A gas has a pressure of 900.0 millibars at a temperature of 30.0 °C. If the volume is unchanged but the temperature is increased to 80.0 °C, what is the new pressure of the gas?

70. A gas occupying a volume of 43.0 liters has a pressure of 850 millibars at a temperature of 35 °C. If the volume is unchanged, but the temperature is increased to 120 °C, what is the new pressure of the gas?

71. A gas cylinder at a temperature of 0 °C has a pressure of 22.4 psi. At what temperature would the pressure inside the cylinder increase to 35.0 psi?

72. A tank of gas at a temperature of 0 °C has a pressure of 415 psi. At what temperature would the pressure inside the tank increase to 600 psi?

73. A gas inside a piston initially has a pressure of 1.2 bars at a temperature of 25 °C and a volume of 20.0 mL. If the temperature is increased to 120 °C, and the piston expands to a volume of 80 mL, what is the new pressure inside the piston?

74. A gas inside a piston occupies a volume of 220 cm^3 at a pressure of 1.03 bars and a temperature of 100 °C. The gas is then cooled to room temperature (25.0 °C), and the volume of gas drops to 180 cm^3. What is the pressure of the gas under these new conditions?

75. A gas inside a balloon has a temperature of 293 K and a volume of 2.40 liters. The gas is cooled to 273 K, and pressure decreases from 790 mm Hg to 750 mm Hg. What is the new volume of the balloon? Report your answer to three significant digits.

76. In air-conditioning systems, compressed gases are allowed to expand, and this expansion results in cooling. A gas with a volume of 5.00 mL at a pressure of 8.0 bars at a temperature of 40 °C is allowed to expand to a volume of 15.0 mL at a pressure of 2.0 bars. What is the temperature of the gas after it expands?

77. What is the volume of two moles of gas, calculated at standard temperature and pressure?

78. What is the volume of 3.5 moles of gas, calculated at standard temperature and pressure?

79. At STP, how many moles of gas occupy a room with a volume of 5.0 m^3? (1 m^3 = 1,000 L)

80. At STP, how many moles of gas occupy a tank with a volume of 1.2 m^3? (1 m^3 = 1,000 L)

81. What is the pressure of 2.31 moles of gas at a temperature of 400.0 K and a volume of 3.5 liters?

82. What is the pressure of 12.5 moles of gas at a temperature of 360.0 K and a volume of 5.02 liters?

83. At what temperature does 4.0 moles of gas under a pressure of 1.0 atmosphere occupy a volume of 120.3 liters?

84. At what temperature does 1.3 moles of gas under a pressure of 1.2 atmospheres occupy a volume of 108.4 L?

85. A gas cylinder contains 80.0 grams of helium gas, occupying a volume of 20.0 liters at a temperature of 280 K. What is the pressure inside the cylinder?

86. What is the pressure of 14.0 moles of argon gas at a temperature of 0 °C and a volume of 800 mL?

87. What volume is occupied by 2 moles of nitrogen gas at a temperature of 73 °F and a pressure of 238.0 psi?

88. What volume would be required to store 150 moles of argon gas at 70 °F with a maximum pressure of 350.0 psi?

89. A gas cylinder contains 80.0 grams of carbon dioxide gas, occupying a volume of 20.0 liters at a temperature of 280 K. What is the pressure inside the cylinder?

90. How many grams of helium gas can a 12.0-liter cylinder at a temperature of 30 °C contain before the pressure exceeds 180.0 psi?

91. A 1.0-L cylinder of carbon dioxide has a pressure of 8.0 atmospheres and a temperature of 373 K. How many CO_2 molecules are in this cylinder?

92. What is the pressure caused by 1.5×10^{21} helium atoms occupying a volume of 0.0023 L at a temperature of 398 K?

93. A gas cylinder contains an unknown gas. It is found that a 1.90-gram sample of the gas occupies a volume of 2.30 liters at a pressure of 1.0 atmosphere and a temperature of 298 K.

a. How many moles are in the 1.90-gram sample?
b. What is the formula mass of this gas, in grams per mole?
c. Based on the formula mass, identify the gas as one of the following: helium, neon, argon, nitrogen, oxygen, or carbon dioxide.

94. A gas cylinder at a temperature of 298 K contains an unknown gas. When 48.0 grams of the gas are released from the cylinder, the pressure of the gas drops by 53.9 psi (3.67 atmospheres). How many moles of gas were released from the tank? What is the formula mass of this gas, in grams per mole? Identify the gas as one of the following: helium, neon, argon, nitrogen, oxygen, or carbon dioxide.

95. What does the term *partial pressure* mean?

96. A gas mixture contains 50% oxygen, 30% nitrogen, and 20% carbon dioxide by mole ratio. If the total pressure is 1.0 atmosphere, what is the partial pressure of oxygen, nitrogen, and carbon dioxide in the sample?

97. A gas mixture contains 40% methane and 60% oxygen by mole ratio. If the total pressure is 45 psi, what is the partial pressure of methane and oxygen in the sample?

98. A cylinder of gas contains 100.0 moles of nitrogen gas and 25.0 moles of oxygen gas. The overall pressure of the cylinder is 214.2 psi. Calculate the partial pressure of oxygen in the cylinder.

99. A 24.0-liter gas cylinder contains 8.0 moles of nitrogen gas and 2.0 moles of oxygen gas. If the temperature inside the cylinder is 0 °C, what is the total pressure inside the cylinder?

100. A balloon is filled with 10.9 grams of N_2, 9.6 grams of O_2, 2.2 grams of CO_2, and 0.04 grams of Ar. The pressure inside the balloon is 1.05 bars, and the temperature is 70 °F. What is the volume of the balloon?

101. An air sample contains 0.5% water vapor. If the total air pressure is 750 torr, what is the partial pressure due to the water vapor?

102. Imagine you are sitting in a room with a volume of 25,000 liters and a temperature of 72 °F. The total air pressure is 750 mm Hg, and the partial pressure from oxygen is 21% of the total pressure. How many grams of oxygen are in the room?

103. A mixture of nitrogen gas (N_2) and oxygen gas (O_2) has a pressure of 1.0 atmosphere at a temperature of 72 °F. At this temperature, which type of molecules is moving with more velocity?

104. The pressure that a gas exerts on a container is determined by the collisions the gas makes with the inside walls of the container. How would these collisions change (in frequency or magnitude) under the following conditions?

a. Keep the temperature constant and decrease the volume.
b. Decrease the temperature and decrease the volume.
c. Increase the temperature and decrease the volume.
d. Increase both the temperature and the volume.

10.5 Diffusion and Effusion

105. What is the difference between diffusion and effusion?

106. If you have a balloon filled with neon and a balloon filled with argon, which one will deflate the fastest? How do you know?

10.6 Gas Stoichiometry

107. The reaction of potassium metal with water produces hydrogen gas, as in the equation below. How many moles of hydrogen gas can be produced from the reaction of 5 moles of potassium? At STP, how many liters of hydrogen gas can be produced?

$$2\ K\ (s) + 2\ H_2O\ (aq) \rightarrow 2\ KOH\ (aq) + H_2\ (g)$$

108. Hydrogen and oxygen react explosively to form water, as in the reaction below. If a balloon containing 1.5 liters of hydrogen gas at 25 °C and a pressure of 1.0 atmosphere reacts with excess oxygen, how many grams of water can be produced?

$$2\ H_2\ (g) + O_2\ (g) \rightarrow 2\ H_2O\ (g)$$

109. Propane gas (C_3H_8) reacts with oxygen according to this balanced equation:

$$C_3H_8\ (g) + 5\ O_2\ (g) \rightarrow 3\ CO_2\ (g) + 4\ H_2O\ (g)$$

a. How many moles of water can be formed from 15.0 moles of propane?
b. How many moles of carbon dioxide can be formed from 15.0 moles of propane?
c. At STP, what volume of CO_2 would be produced from 15.0 moles of propane?
d. Are more moles of gas produced or consumed in this reaction?

110. The Born–Haber process is used to manufacture ammonia (NH_3) from nitrogen gas and hydrogen gas:

$$3\ H_2\ (g) + N_2\ (g) \rightarrow 2\ NH_3\ (g)$$

In this reaction,

a. How many moles of nitrogen are needed to react with 15 moles of hydrogen?
b. How many moles of ammonia can be produced from 15 moles of hydrogen?
c. At a temperature of 800 K and a pressure of 4.00 atm, how many liters of ammonia can be produced from 15.0 moles of hydrogen?
d. If this reaction proceeded to the right in a sealed container, would the pressure inside the container increase or decrease? How do you know?

111. Natural gas, CH_4, burns in oxygen as shown in the reaction below. A 1.0-liter cylinder containing CH_4 with a pressure of 2.1 atmospheres at a temperature of 298 K is completely used to power a portable stove.

$$CH_4\ (g) + 2\ O_2\ (g) \rightarrow CO_2\ (g) + 2\ H_2O\ (g)$$

a. How many moles of CH_4 were in the cylinder?
b. How many moles of oxygen gas were necessary to react with this amount of CH_4?
c. How many moles of water were produced in this reaction?
d. How many grams of carbon dioxide were produced in this reaction?

112. When a beverage is carbonated, the carbon dioxide reacts with water to produce carbonic acid, which gives soft drinks a bitterness that is balanced with a large amount of sugar. Over time, carbonic acid converts back into carbon dioxide and water:

$$H_2CO_3\ (aq) \rightarrow H_2O\ (l) + CO_2\ (g)$$

In this reaction, what volume of CO_2 can be produced from 2.0 grams of H_2CO_3 at standard temperature and pressure? (Standard temperature and pressure is 273 K and 1.0 atmosphere.)

113. Glucose fermentation takes place through the following reaction:

$$C_6H_{12}O_6\ (aq) \rightarrow 2\ C_2H_6O\ (aq) + 2\ CO_2\ (g)$$

If 2.05 grams of $C_6H_{12}O_6$ were placed in a sealed container having a temperature of 200 °C and a volume of 5.0 liters, and if this reaction went to completion, what would be the pressure from the carbon dioxide inside the container?

114. Ammonia, NH_3, is a colorless gas with a pungent odor. Its applications range from cleaning supplies to fertilizer, and it is a building block for the production of many industrial chemicals and consumer products. For example, the first step in preparing nitric acid involves the reaction of ammonia gas with oxygen gas to produce nitrogen monoxide and water, as shown in this reaction:

$$4\ NH_3\ (g) + 5\ O_2\ (g) \rightarrow 4\ NO\ (g) + 6\ H_2O\ (g)$$

a. If this reaction takes place in a sealed container, and the temperature inside the container is kept constant, will the pressure inside the container increase or decrease? How do you know?
b. If 170 grams of NH_3 react in this way, how many grams of H_2O can be produced?
c. If 170 grams of NH_3 react in this way, how many liters of H_2O can be produced, given a temperature of 373 K and a pressure of 1.00 atmosphere?

115. Octane, a component of gasoline, burns according to this equation:

$$2\ C_8H_{18}\ (l) + 25\ O_2\ (g) \rightarrow 16\ CO_2\ (g) + 18\ H_2O\ (g)$$

Octane has a formula mass of 114.26 g/mol and a density of 0.703 kg/L. At standard temperature and pressure, what volume of CO_2 is produced by the combustion of one liter of octane?

116. During photosynthesis, plants absorb sunlight and use this energy to convert carbon dioxide and water into simple sugars and oxygen. The process is exactly the reverse of combustion and is described by this equation:

$$6\ CO_2\ (g) + 6\ H_2O\ (l) \rightarrow C_6H_{12}O_6\ (s) + 6\ O_2\ (g)$$

At a temperature of 25 °C and a pressure of 1.00 atm, how many liters of carbon dioxide gas are consumed by the production of 1.00 kg of $C_6H_{12}O_6$?

Challenge Questions

117. Propane gas reacts with oxygen according to this balanced equation:

$$C_3H_8\ (g) + 5\ O_2\ (g) \rightarrow 3\ CO_2\ (g) + 4\ H_2O\ (g)$$

Assuming an air sample contains 21% oxygen (by volume), in which of the following mixtures will the propane and oxygen almost completely react, leaving almost no excess of either gas?

a. a mixture of 50% air and 50% propane
b. a mixture of 74% air and 26% propane
c. a mixture of 82% air and 18% propane
d. a mixture of 96% air and 4% propane

118. Octane, a component of gasoline, burns according to this equation:

$$2\ C_8H_{18}\ (l) + 25\ O_2\ (g) \rightarrow 16\ CO_2\ (g) + 18\ H_2O\ (g)$$

In an engine cycle, 0.030 g of C_8H_{18} is mixed with 0.30 L of air at a pressure of 3.2 atmospheres and a temperature of 450 K inside a piston. The spark plug detonates the gasoline, causing the explosive reaction to take place. What mass of CO_2 and what mass of H_2O can be produced in this engine cycle? Assume the air contains 21% oxygen. In what two ways does this reaction increase the pressure and push the piston?

Solutions

CHAPTER ELEVEN

Figure 11.1 (a) Producing a great cup of coffee requires careful attention to detail. *(continued)*

The Perfect Cup of Coffee

What makes a really good cup of coffee?

For Aaron Blanco, this is more than just a question — it's an obsession. Aaron is the founder of the Brown Coffee Company in San Antonio, Texas. For the past decade, he has explored every facet of the coffee-making process, from growing coffee plants to roasting and grinding the beans to brewing the perfect cup. He makes amazing coffee (**Figure 11.1**).

Aaron started his business in 2005. He dreamed of opening a coffee shop but lacked the funds to get it started. So instead, he bought a small oven and began selling fresh-roasted beans to local restaurants. Roasting creates tiny pores in the bean, so water can reach and dissolve the molecules inside. It also breaks complex molecules into simpler, tastier structures. But it has to be done just right: Too little heating leaves the flavors trapped; too much heating destroys them. By blending the scientific method with the nuances of taste, Aaron explored how the heating temperature and time affect the flavor of coffee.

After several years of roasting, he finally opened his first shop. Preparing coffee for customers turned his attention to the grinding and brewing processes: How finely should the beans be ground? How does the water temperature affect the flavor? What volume of water is needed for each gram of coffee grounds? How should the water and grounds be mixed?

As his business has grown, Aaron has returned to the beginning — to the bean itself. Farmers grow two species of the tropical coffee plant: *Coffea robusta* grows at

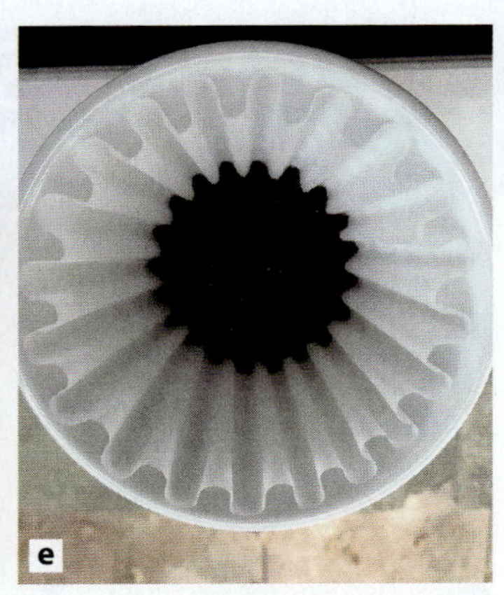

Figure 11.1 (continued) (b) Aaron Blanco works with his coffee roaster. (c) Ripe coffee cherries on a tree in El Salvador. (d) Coffee beans drying in the sun. (e) Freshly ground coffee, ready to brew. (a) Lux Superich/EyeEm/Getty Images; (b–e) Courtesy of the Brown Coffee Company

any altitude and is well suited to massive farms and mechanized harvests. Robusta coffee is bitter, but cheap. Its more delicate cousin, *Coffea arabica*, produces a milder, tastier coffee—but it grows well only at high altitudes. Aaron frequently travels to the highlands of Africa and Central America where arabica is grown. He meets with farmers to discuss the soil, local weather, and subtle differences in farming techniques.

Aaron's passion and hard work have paid off. In 2015, Aaron opened his third coffee shop, and his coffee was recognized as the "Best Coffee in San Antonio." Food Network star Alton Brown called it "The best cup of coffee I've ever had in my life." Aaron's travels in search of the best coffee beans were even described in the documentary *Coffee Hunting: Kenya*. As he puts it, "It took ten years to be an overnight success."

So what goes into a perfect cup of coffee? Aaron will tell you there is still a lot he doesn't know. Coffee is a complex mixture of hundreds of components dissolved in water. But while we may never fully understand coffee, we can begin by understanding how solutions behave. How do substances dissolve in water? How do we measure the amount of a substance that is in a solution? Why do different compounds exhibit different solubilities? How are the properties of solutions different from the properties of pure liquids?

In this chapter, we'll explore these questions. We'll describe the composition of solutions and see how different substances behave when dissolved in water. We'll explore chemical reactions that occur in solution and apply the principles of stoichiometry to reactions in solution. Pour a cup of coffee, and let's get started.

Intended Learning Outcomes

After completing this chapter and working the practice problems, you should be able to:

11.1 Describing Concentration

- Calculate solution concentrations by percent, parts per million, parts per billion, and molarity.
- Convert between moles, volume, and molarity.
- Quantitatively describe the preparation of solutions of a known molarity.

11.2 Electrolyte Solutions

- Describe the behavior of electrolytes and nonelectrolytes in aqueous solution.
- Determine the molar concentrations of ions in solution.
- Qualitatively describe the changes in freezing point, boiling point, and osmotic pressure as a function of solute concentration.

11.3 Reactions in Solution—a Review and a Preview

- Describe reactions that take place in aqueous solution using molecular and ionic equations.

11.4 Solution Stoichiometry

- Apply the principles of stoichiometry to solve problems involving solutions.

11.1 Describing Concentration

When sugar is added to coffee, the sugar *dissolves* (**Figure 11.2**). That is, the sugar particles disperse throughout the coffee, forming a homogeneous mixture. This type of mixture is called a **solution**. The substance that dissolves (for example, the sugar) is the **solute**. The liquid in which the solute dissolves is the **solvent**.

The amount of solute that is present in a solution is called its **concentration**. If a solution contains a small amount of solute, we say it is *dilute*. If it contains a large amount of solute, we say it is *concentrated*. A solution that contains the maximum amount of dissolved solute is *saturated*.

There are several different ways to describe concentration, and they each have different applications. In this section, we will look at three common ways of describing concentration: by percent, by fraction, or by molarity.

fotogarfoingles/Deposit Photos

Figure 11.2 Sugar dissolves in hot coffee. In a solution, the solute particles disperse evenly throughout the solvent.

Concentration by Percent

Percent by Mass and Volume

We frequently describe solutions by the percentage of solute present. One way to do this is **percent by mass**:

$$\text{Mass \%} = \frac{\text{mass of solute}}{\text{mass of solution}} \times 100\%$$

More commonly, we express concentration as **percent by volume**:

$$\text{Volume \%} = \frac{\text{volume of solute}}{\text{volume of solution}} \times 100\%$$

We often use percent by volume when both the solute and solvent are liquids. For example, instructions for cleaning a stove might call for a 5% (v/v) solution of bleach in water. The abbreviation (v/v) indicates that both the solute and the solution are measured by volume. In this solution, the solute accounts for 5% of the total volume of the solution. We could prepare this solution by adding 5 mL of bleach to enough water to make 100 mL of solution.

vvoe/Shutterstock

Cleaning solutions are often prepared based on volume percent.

Example 11.1 Using Percent Concentration

A cleaning solution contains 2% (v/v) ammonia in water. What volume of ammonia is needed to prepare 10 liters of this solution?

To solve this problem, we substitute the volume of solution and the target percentage into the percent by volume equation.

$$2\% = \frac{\text{volume of solute}}{\text{10 L solution}} \times 100\%$$

$$0.02 = \frac{\text{volume of solute}}{\text{10 L solution}}$$

$$\text{Volume of solute} = 0.2\text{ L or }200\text{ mL}$$

TRY IT

1. A chemist prepares a solution by adding 1.0 kg of salt to 5.0 kg of water. What is the percent by mass of salt in this solution?
2. A common light beer contains 4.2% (v/v) alcohol. What is the volume of alcohol (in ounces and in milliliters) in a 16-ounce bottle of beer?

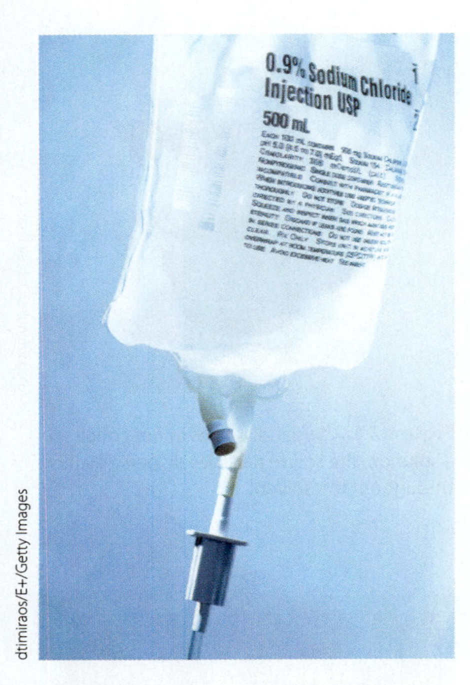

dtimiraos/E+/Getty Images

Figure 11.3 Normal saline solution is used in hospital IV packs. It has a concentration of 0.9% (m/v).

For aqueous solutions, mass/volume % is the number of grams of solute per 100 mL.

Mass/Volume Percent

When the solute is a solid, we often express concentration by **mass/volume percent**:

$$\text{Mass/volume \%} = \frac{\text{mass of solute}}{\text{volume of solution}} \times 100\%$$

We write this as "% (m/v)," and we frequently use units of g/mL. For example, normal saline solution (used for intravenous fluids) is 0.9% (m/v). This means that 0.9 grams of salt are present per 100 mL of solution (**Figure 11.3**).

Seawater has a higher salt concentration than normal saline. Typically, seawater has a salt concentration of about 3.5% (m/v). This is still far from saturated: In the Middle East, the Dead Sea has a salt concentration of approximately 34% (m/v); see **Figure 11.4**.

Image Professionals GmbH/Alamy

Figure 11.4 These salt deposits lie along the Dead Sea, which has a salt concentration of about 34% (m/v).

Example 11.2 Using Percent Concentration

Normal saline solution is 0.90% NaCl (g/mL). How many grams of sodium chloride are needed to prepare 0.50 L of normal saline solution?

Recall that a percentage is a fraction of 100. Therefore, if a solution is 0.90% (g/mL), this means it contains 0.90 g per 100 mL, or 9.0 grams per liter:

$$0.90\%\ (\text{g/mL}) = \frac{0.90\ \text{g}}{100\ \text{mL}} = \frac{9.0\ \text{g}}{\text{L}}$$

To find the total mass of NaCl needed, we multiply the volume of solution that we need by the ratio of grams to volume. This gives us a final answer of 4.5 g of NaCl.

$$0.50\,\cancel{\text{L}} = \frac{9.0\ \text{g NaCl}}{\cancel{\text{L}}} = 4.5\ \text{g NaCl}$$

A different approach to this problem is to set up a ratio.

$$\frac{9.0\ \text{g NaCl}}{1\ \text{L}} = \frac{x}{0.50\ \text{L}}$$

$$x = 4.5\ \text{g NaCl}$$

TRY IT

3. What mass of potassium chloride is present in 100 mL of a solution that is 15.0% KCl (m/v)? How much more water would you need to add to dilute this solution to 3.0% (m/v)?

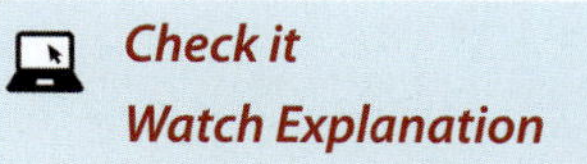

Check it
Watch Explanation

Very Dilute Solutions: ppm and ppb

Clean water is vitally important. Toxic substances in the water supply can cause widespread health issues or have drastic effects on ecosystems, even when they are present in very dilute quantities. For example, pharmaceutical compounds, cosmetics, pesticides, and components of plastic containers often find their way into rivers, streams, and reservoirs, endangering the wildlife and humans who drink from them (**Figure 11.5**). This is why governmental agencies and watchdog groups closely monitor the levels of pollutants in lakes and streams. **Table 11.1** summarizes the maximum allowable levels of several chemical contaminants in drinking water.

Bommanna Loganathan

Figure 11.5 Water monitoring helps protect the health of people and ecosystems. This chemist is collecting a water sample from a stream to analyze for pollutants.

TABLE 11.1 Contaminants Monitored in U.S. Drinking Water

Contaminant	Source	Health Effects	Maximum Concentration Level in Drinking Water (mg/L)
Polychlorinated biphenyls (PCBs)	Runoff from landfills	Reproductive and nervous system difficulties; increased risk of cancer	5×10^{-4}
Dioxin (2,3,7,8-TCDD)	Emissions from waste incineration; discharge from chemical factories	Reproductive difficulties; increased risk of cancer	3×10^{-8}
Nitrate	Runoff from fertilizer use; leaching from septic tanks, sewage, erosion of natural deposits	Serious illness for infants below the age of six months	10
Mercury	Erosion of natural deposits, discharge from refineries and factories; runoff from landfills and croplands	Kidney damage	2×10^{-3}
Atrazine	Runoff from herbicides	Cardiovascular and reproductive problems	3×10^{-3}

Source: Data from the U.S. Environmental Protection Agency (www.epa.gov).

We often express very dilute concentrations in **parts per million (ppm)** or **parts per billion (ppb)**. For example, if a pond contains 1 gram of nitrate for every 1,000,000 grams of solution, we describe the concentration as 1 part per million, or 1 ppm. Mathematically, we calculate the parts per million using the following relationship:

$$\text{Concentration (ppm)} = \frac{\text{mass of solute}}{\text{mass of solution}} \times 10^6$$

Even smaller concentrations are expressed using parts per billion, or ppb. The relationship is similar:

$$\text{Concentration (ppb)} = \frac{\text{mass of solute}}{\text{mass of solution}} \times 10^9$$

For a dilute aqueous solution,
1 mg/L = 1 ppm
1 μg/L = 1 ppb

When calculating ppm or ppb, any unit of mass may be used, but the units for the solute and the solution must be the same. By definition, 1 liter of pure water at 4 °C has a mass of exactly 1 kg. Under these conditions, 1 ppm means 1 *milligram* of solute per liter of solution. Similarly, 1 ppb means 1 *microgram* of solute per liter of solution. Example 11.3 demonstrates this principle.

Because 1 μg = 10^{-6} g, we can write 45 μg as 45×10^{-6} g. While this is not proper scientific notation, it is a simple way to convert from μg to g.

Example 11.3 Concentrations in ppm and ppb

Phthalates are a class of compounds that increase the flexibility and durability of plastics. These compounds slowly leach out of plastic products and may contaminate waterways. In a recent study, scientists found that samples from a European waste-water treatment plant contained about 45 μg of phthalates per liter of solution. What is the concentration of this compound in ppm? In ppb? (Assume the solution has the same density as pure water at 4 °C.)

To solve this problem, let's express both the mass of solute and the mass of solution in grams: 45 μg = 45×10^{-6} g, and one liter of water has a mass of 1 kg, or 1,000 g:

$$\text{Concentration (ppm)} = \frac{45 \times 10^{-6}\ \cancel{g}}{1{,}000\ \cancel{g}} \times 10^6 = 0.045 \text{ ppm}$$

$$\text{Concentration (ppb)} = \frac{45 \times 10^{-6}\ \cancel{g}}{1{,}000\ \cancel{g}} \times 10^9 = 45 \text{ ppb}$$

So 45 μg of solute per liter of aqueous solution is the same as 45 ppb.

Check it
Watch Explanation

TRY IT

4. In 1999, scientists reported that groundwater from a waste site near Atlantic City, New Jersey, had concentrations of diazepam (Valium®) as high as 40 ppb. At this concentration, how many milligrams of diazepam were present in a 100-liter sample of this wastewater?

5. A 0.500-kg water sample contains 14.3 μg of contaminant. What is the concentration of the contaminant in ppb? What is the concentration in % g/mL?

Molarity

Concentrations based on the mass or volume of solute are useful for many common measurements. However, when describing chemical reactions, we are frequently concerned with the number of *moles* of a substance that are consumed or produced.

In these cases, it is helpful to express concentration in a way that allows us to quickly convert to moles. For this reason, chemists often describe concentration in terms of **molarity (M)**, which is the moles of solute per liter of solution:

$$\text{Molarity (M)} = \frac{\text{moles of solute}}{\text{liters of solution}}$$

We commonly write this in an abbreviated form:

$$M = \frac{\text{moles}}{V}$$

where V is the volume in liters.

The molarity equation also allows us to find the number of moles present, given the concentration and the volume. Because molarity is equal to moles over volume, we can rearrange the equation to solve for moles:

$$\text{moles} = MV$$

Examples 11.4 and 11.5 illustrate these ideas.

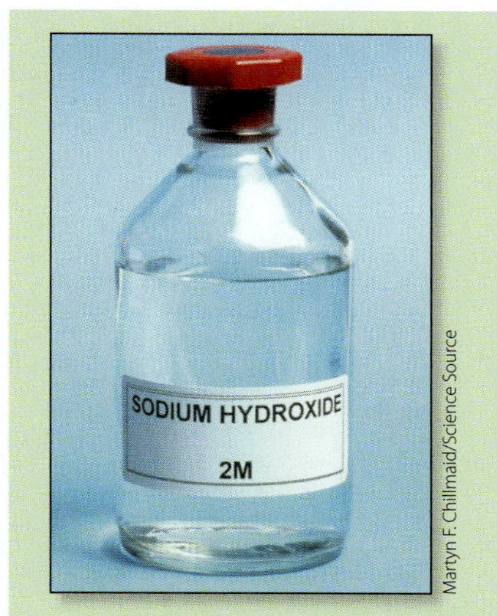

For many solutions, such as acids and bases, it is important to label both the ingredients and the concentration.

Example 11.4 Calculating Molarity

An aqueous solution with a volume of 9.41 L contains 412.3 grams of dissolved potassium chloride. What is the molarity of this solution?

To solve this problem, we must first convert the amount of KCl from grams to moles. The molar mass of KCl is 74.55 g/mol, so we can write:

$$412.3\ \cancel{\text{g KCl}} \times \frac{1\ \text{mol KCl}}{74.55\ \cancel{\text{g KCl}}} = 5.531\ \text{moles KCl}$$

We can then find the molarity.

$$\text{Molarity} = \frac{\text{moles of solute}}{\text{liters of solution}} = \frac{5.531\ \text{mol KCl}}{9.41\ \text{L}} = 0.588\ \text{M KCl}$$

Example 11.5 Finding Moles from Molarity and Volume

How many moles of HCl are in a 250-mL sample of 6.0 M HCl?

To answer this question, we first convert the volume to liters (250 mL = 0.250 L). We then use the relationship given earlier to solve:

$$\text{Moles} = MV = \left(6.0\frac{\text{mol}}{\cancel{\text{L}}}\right)(0.250\ \cancel{\text{L}}) = 1.5\ \text{moles}$$

Molarity is an important and useful tool for describing concentration. Before moving on, I encourage you to attempt the following problems.

TRY IT

6. A 2.4-liter solution contains 23.9 grams of calcium chloride. What is the molarity of calcium chloride in this solution?

7. How many moles of silver nitrate are present in 15.0 mL of a 1.2-M *aq*. $AgNO_3$ solution?

Check it
Watch Explanation

Preparing Solutions of Known Molarity

Here's a question: If we dissolve 1.0 mole of sugar in 1.0 L of water, do we have a 1.0-M solution? Not exactly. Because the sugar also takes up space, the volume of the solution after mixing is a little more than 1.0 L. The solution is almost 1.0 M, but not exactly. So, how do we prepare solutions whose molarity is precisely known? Chemists routinely use this three-step procedure (**Figure 11.6**):

1. Measure out the desired amount of solute, and add it to a volumetric flask (these flasks have long, slender necks for precisely measuring volume).
2. Partially fill the flask with solvent, and mix until the solute dissolves completely.
3. Dilute the solution, slowly adding solvent until the correct volume is reached.

By following this procedure, we know the amount of solute and the volume of solution are both measured precisely.

Explore Figure 11.6

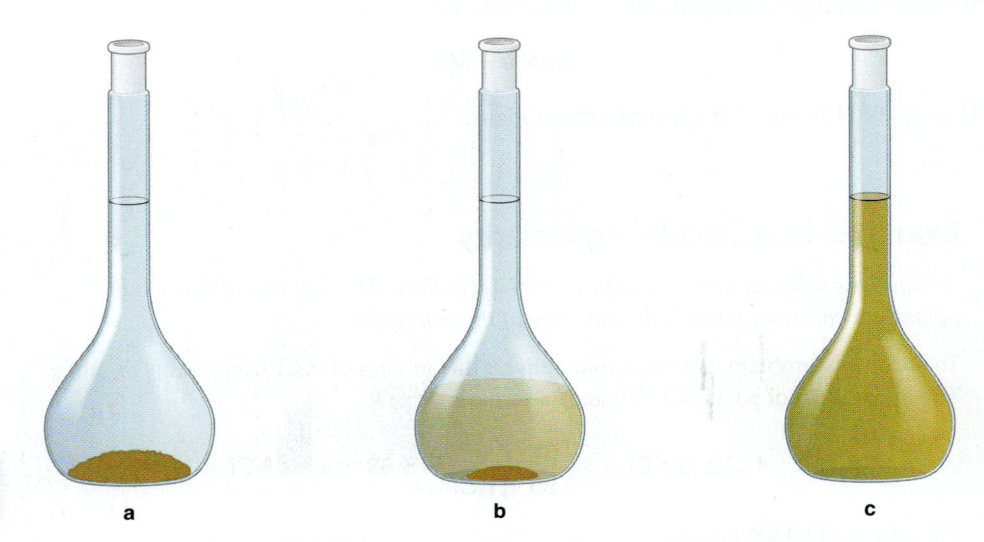

Figure 11.6 To prepare a solution with a precisely known concentration: (a) Measure out the correct amount of solute; (b) partially fill the volumetric flask, and mix until the solid dissolves; and (c) slowly dilute the solution to the correct volume.

Check it
Watch Explanation

TRY IT

8. While working in a forensics lab, you need to prepare 2.00 L of a 1.40-M *aq.* sodium hydroxide solution. What mass of sodium hydroxide is required to do this? Describe how you would prepare the solution.

PHOTO RESEARCHERS/Science Source

Figure 11.7 Two glasses of orange drink. Both glasses contain the same amount of drink mix, but the one on the right has been diluted by adding more water.

Explore Figure 11.7

Preparing Dilute Solutions

Many laboratories store common chemicals, such as acids and bases, as concentrated solutions. While this saves space, chemists often need more dilute solutions for routine use. For example, if we need a solution of 1.0 M *aq.* HCl, how do we prepare it from a concentrated 12.0-M solution?

Chemists prepare dilute solutions by mixing the concentrated solution with additional solvent (**Figure 11.7**). Doing this increases the volume of the solution, but the moles of solute *do not change*. That is, the number of moles of solute in the initial (concentrated) solution is equal to the number of moles in the final (dilute) solution:

$$\text{moles}_{\text{initial}} = \text{moles}_{\text{final}}$$

Because the moles of a solute are equal to the molarity times the volume, we can rewrite this expression based on the molarity and volume of the initial and final solutions:

$$M_iV_i = M_fV_f$$

where M_i and V_i are the initial molarity and volume, and M_f and V_f are the final molarity and volume.

MV = moles

Using this relationship, we can prepare dilute solutions of known concentration, as illustrated in Examples 11.6 and 11.7.

Example 11.6 Preparing Dilute Solutions

You have 25.0 mL of a 10.0-M solution of potassium hydroxide. To this, you add enough water to make 500.0 mL of a dilute solution. What is the molarity of the final solution?

To solve this problem, we rearrange the molarity and volume equation to find the final molarity (M_f), then substitute the values in the question:

$$M_f = \frac{M_iV_i}{V_f} = \frac{(10.0\ \text{M})(25.0\ \cancel{\text{mL}})}{500.0\ \cancel{\text{mL}}} = 0.500\ \text{M KOH}$$

Notice that, although we would need to convert the volume from milliliters to liters in order to solve for the moles present, we did not have to convert the volume units to solve this problem because the units cancel each other out.

Turtle Rock Scientific/Science Source

When diluting acid solutions, it is good practice to add small amounts of acid slowly to water, rather than adding water to the acid. This method prevents the solution from overheating and splattering hot, concentrated acid.

Example 11.7 Preparing Dilute Solutions

Starting from a 12.0-M concentrated HCl solution, you need to prepare 180 mL of 1.0-M aq. HCl. What volume of concentrated HCl do you need?

In this example, we need to solve for the initial volume of the concentrated solution, V_i. Rearranging the relationship to find V_i, we get

$$V_i = \frac{M_fV_f}{M_i} = \frac{(1.0\ \cancel{\text{M}})(180\ \text{mL})}{12.0\ \cancel{\text{M}}} = 15\ \text{mL conc. HCl}$$

Notice again that units of molarity cancel out, and it was not necessary to change the volume from milliliters to liters. We used milliliters for the volume of V_f and were left with milliliters for the volume of V_i.

Using Square Brackets to Represent Concentration

Sometimes chemists use square brackets to indicate the concentration of a solute. For example, to represent the concentration of H_2SO_4 in a sample, we might write $[H_2SO_4]$, meaning "concentration of sulfuric acid." If we wish to say the concentration of sodium nitrate is 1.3 molar, we write this as $[NaNO_3] = 1.3$ M.

TRY IT

9. You have 30.0 mL of a 7.80-M solution of $MgBr_2$ that you wish to dilute until $[MgBr_2] = 0.50$ M. To what volume must you dilute the sample?

Check it

Watch Explanation

11.2 Electrolyte Solutions

To review electrolyte solutions, see Sections 5.6 and 6.5.

In previous chapters, we discussed solutions of ionic compounds in water. Recall that when ionic compounds dissolve in water, they *dissociate*—that is, they separate into positive and negative ions, as shown in **Figure 11.8**. Because these compounds increase the ability of water to conduct electricity, they are commonly called **electrolyte solutions**.

Explore Figure 11.8

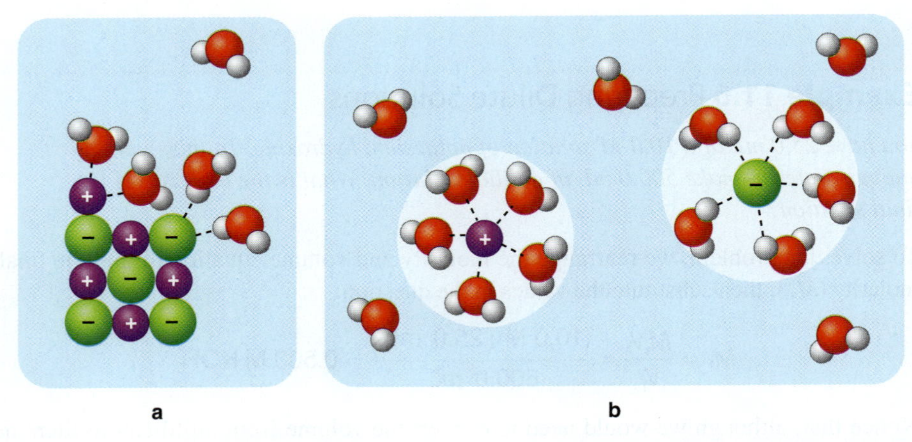

Figure 11.8 (a) When an ionic solid dissolves, the water molecules pull the ions away from the solid and into solution. (b) In solution, the positive ions are surrounded by the negative ends of the water molecules, while the negative ions are surrounded by the positive ends of the water molecules.

Covalent compounds generally do not ionize in water. An exception to this occurs with *acids*, compounds that contain weak polar covalent bonds to hydrogen. In water, these bonds break to produce an H^+ cation and a corresponding anion.

Sometimes we show the process of dissociation or ionization using *ionic equations*. For example, solid KBr and gaseous hydrogen bromide both produce ions when dissolved in water. We can write these as either molecular equations or ionic equations:

Molecular equations:
$$KBr\ (s) \rightarrow KBr\ (aq)$$
$$HBr\ (g) \rightarrow HBr\ (aq)$$

Ionic equations:
$$KBr\ (s) \rightarrow K^+\ (aq) + Br^-\ (aq)$$
$$HBr\ (g) \rightarrow H^+\ (aq) + Br^-\ (aq)$$

Electrolyte Concentrations

Suppose one mole of magnesium chloride, $MgCl_2$, is dissolved in enough water to make one liter of solution. The concentration of $MgCl_2$ is 1.0 M. However, magnesium chloride is an ionic compound, so it dissociates when dissolved in water:

$$MgCl_2\ (s) \rightarrow Mg^{2+}\ (aq) + 2Cl^-\ (aq)$$

$$\underset{MgCl_2}{1\ \text{mole/L}} \rightarrow \underset{Mg^{2+}}{1\ \text{mole/L}} + \underset{Cl^-}{2\ \text{moles/L}}$$

Notice that when the solid dissociates, two moles of chloride ions form for every one mole of magnesium ions. This means that the concentration of chloride ions in the solution is twice as much as that of magnesium. We could describe this solution as having a 1.0-M concentration of $MgCl_2$, or, we could describe it as having a 1.0-M concentration of Mg^{2+} ions and a 2.0-M concentration of Cl^- ions. Both descriptions are correct.

Let's try a second example: What is the concentration of lithium ions and sulfate ions in a 3.5-M solution of lithium sulfate? Again, we must recognize that the ionic compound dissociates in solution:

$$LiSO_4\ (s) \rightarrow 2\ Li^+\ (aq) + SO_4^{2-}\ (aq)$$

Notice that we get two moles of lithium ions for each mole of lithium sulfate. As a result, the concentration of lithium ions is 2×3.5 M, or 7.0 M. Because there is only one mole of sulfate in each mole of lithium sulfate, the concentration of sulfate ions is 3.5 M.

© duskbabe/depositphotos.com

A grilled cheese sandwich is made of two slices of bread and one slice of cheese. If everyone at the table has one sandwich, you can also say that everyone has one slice of cheese and two slices of bread.

Example 11.8 Determining Ion Concentrations

Write an ionic equation showing the ions formed when ammonium phosphate dissolves in water. What is the concentration of ammonium and phosphate ions in a 0.928-M ammonium phosphate solution?

Ammonium phosphate has the formula $(NH_4)_3PO_4$. When dissolved in water, this compound dissociates, producing three ammonium ions and one phosphate ion from each unit of the compound:

$$(NH_4)_3PO_4\,(s) \rightarrow 3\ NH_4^+\ (aq) + PO_4^{3-}\ (aq)$$

Because three ammonium ions are produced for each unit of compound, the ammonium ion concentration is three times the concentration of the original compound:

$$[NH_4^+] = 3 \times 0.928\ M = 2.78\ M$$

One unit of phosphate is formed for each unit of compound, so the phosphate ion concentration is the same as that of the original compound.

$$[PO_4^{3-}] = 1 \times 0.928\ M = 0.928\ M$$

TRY IT

10. Find the concentration of each of the dissociated ions in these solutions:

a. 5.0 M *aq.* $AlBr_3$ **b.** 2.4 M *aq.* K_2CO_3 **c.** 7.3 M *aq.* NH_4Cl

11. If 0.50 L of 1.0-M *aq.* $NaNO_3$ solution is combined with 2.40 L of 3.2-M *aq.* $Fe(NO_3)_2$ solution, how many moles of nitrate are in the combined solution? What is the molarity of nitrate in the combined solution?

Check it
Watch Explanation

Colligative Properties

The **colligative properties** are properties of a solution that do not depend on the *type* of particles that are dissolved, but rather on the *number* of dissolved particles in the solution. There are three common and important colligative properties: freezing point depression, boiling point elevation, and osmotic pressure.

Freezing Point Depression

In colder climates, salt trucks are a common sight in winter: These trucks spread salt on roads to melt ice and snow (**Figure 11.9**). This works because a saltwater solution has a lower freezing point than pure water. In general, the presence of a solute (or a contaminant) lowers the freezing point of the liquid. This property of aqueous solutions is called **freezing point depression**.

Anton J. Geisser/AGE Fotostock

Figure 11.9 A salt truck deposits salt on a road during a snowstorm. The salt lowers the freezing temperature of water, preventing the roads from icing over.

Colligative properties depend on the number of dissolved particles in solution. ■

How much the temperature is lowered depends on the total concentration of dissolved particles. The more dissolved particles, the lower the freezing temperature.

Figure 11.10 shows the freezing point of water containing sugar (sucrose), sodium chloride, and calcium chloride. For each compound, the freezing point of the solution drops as more solute is added. However, the change in freezing point depends on the solute: It drops the most for calcium chloride and the least for sucrose. Why is this?

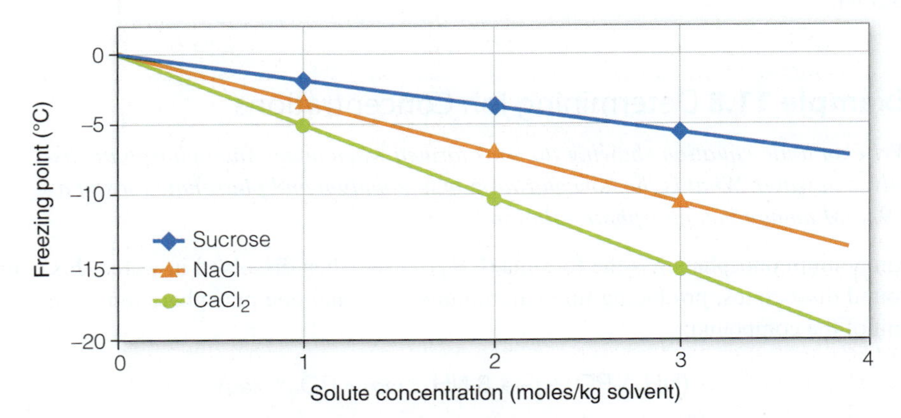

Figure 11.10 A comparison of the freezing points of sucrose, sodium chloride, and calcium chloride solutions at different concentrations.

The freezing point is the same thing as the melting point—at this temperature, solid and liquid interconvert.

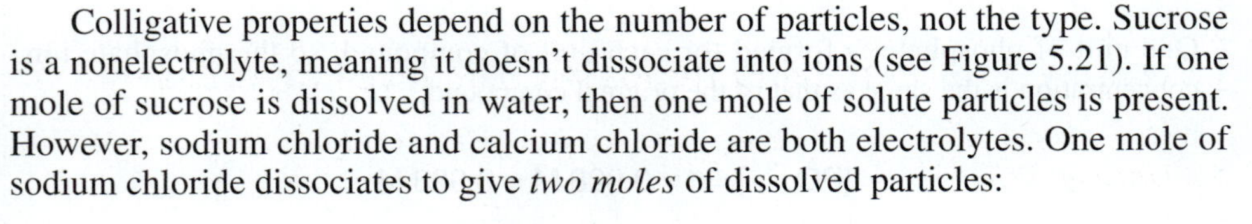

Colligative properties depend on the number of particles, not the type. Sucrose is a nonelectrolyte, meaning it doesn't dissociate into ions (see Figure 5.21). If one mole of sucrose is dissolved in water, then one mole of solute particles is present. However, sodium chloride and calcium chloride are both electrolytes. One mole of sodium chloride dissociates to give *two moles* of dissolved particles:

$$NaCl\ (s) \rightarrow Na^{+}\ (aq) + Cl^{-}\ (aq)$$

$$\underset{NaCl}{1\ \text{mole}} \rightarrow \underset{Na^{+}}{1\ \text{mole}} + \underset{Cl^{-}}{1\ \text{mole}}$$

Similarly, one mole of calcium chloride dissociates to give *three moles* of dissolved particles:

$$CaCl_2\ (s) \rightarrow Ca^{2+}\ (aq) + 2Cl^{-}\ (aq)$$

$$\underset{CaCl_2}{1\ \text{mole}} \rightarrow \underset{Ca^{2+}}{1\ \text{mole}} + \underset{Cl^{-}}{2\ \text{moles}}$$

When making pasta, many people add a little salt to the water. Dissolved solids do raise the boiling temperature of water, but a tiny amount of salt is not enough to make an appreciable difference.

Because calcium chloride dissociates into more particles, it drops the freezing point of water more than sucrose or sodium chloride does at the same concentration. Calcium chloride drops the freezing point more effectively than sodium chloride, and it is usually the salt of choice for melting ice on roadways.

Boiling Point Elevation

The presence of dissolved solids also raises the boiling temperature of a liquid. This property is commonly called **boiling point elevation**. Like freezing point depression, this property depends on the number (not the type) of solute particles present.

One important example of both freezing point depression and boiling point elevation is the use of mixtures as engine coolant (**Figure 11.11**). Automobiles must have liquid coolant to carry excess heat away from the engine, but the use of pure

Figure 11.11 A mixture of ethylene glycol and water is used as engine coolant. This mixture has a lower freezing point and a higher boiling point than pure water.

water in the coolant lines is risky: Because water expands when it freezes, water lines may burst in cold weather. To prevent this, ethylene glycol (commonly called *anti-freeze*) is mixed with water. A 40% (by mass) solution of ethylene glycol in water has a freezing point of –45 °C. The ethylene glycol/water mix also has a higher boiling temperature than pure water. This lowers the gas pressure inside the coolant lines and allows the engine to run hotter without letting the coolant escape as easily.

Example 11.9 Identifying Trends in Colligative Properties

Which solution has a lower freezing point, a 3.0-M solution of Na_2CO_3 or a 2.0-M solution of $AlCl_3$?

To solve this problem, we must identify the total number of moles of ions present in each solution. For each liter of the solution, 3.0 moles of sodium carbonate dissociates to give a total of 9.0 moles of ions:

$$Na_2CO_3\,(aq) \rightarrow 2\,Na^+\,(aq) + CO_3^{2-}\,(aq)$$

$$3\text{ moles/L} \rightarrow 6\text{ moles/L} + 3\text{ moles/L}$$

(total ion concentration: 9 moles/L)

Similarly, each liter of 2.0-M $AlCl_3$ solution dissociates to give a total of 8.0 moles of ions:

$$AlCl_3\,(aq) \rightarrow Al^{3+}\,(aq) + 3\,Cl^-\,(aq)$$

$$2\text{ moles/L} \rightarrow 2\text{ moles/L} + 6\text{ moles/L}$$

(total ion concentration: 8 moles/L)

Because the Na_2CO_3 solution has a greater concentration of dissolved ions, it has a lower freezing point than the $AlCl_3$ solution.

TRY IT

12. Which of these solutions has the lowest freezing point? Which one has the highest freezing point?

a. 2.0 M *aq.* $MgSO_4$ **b.** 2.0 M *aq.* K_2CO_3 **c.** 3.2 M *aq.* NH_4Cl

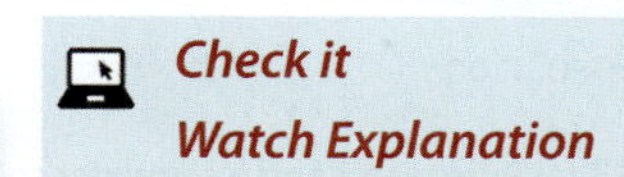

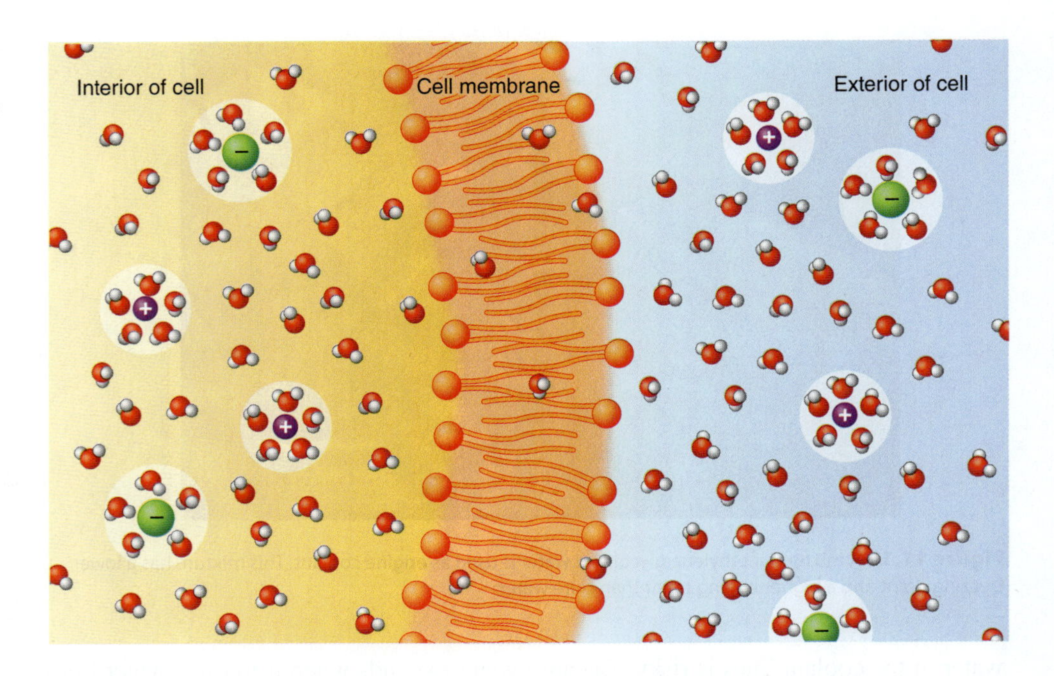

Figure 11.12 Water molecules are able to move across a semipermeable membrane (such as a cell membrane), but dissolved ions (like sodium and chloride) cannot.

Osmotic Pressure

Cell membranes are composed of a thin double layer of molecules called a *lipid bilayer*. These membranes enclose the cell, holding its contents in place. The lipid bilayer is a *semipermeable membrane*: This means that water can pass back and forth through the membrane, but most molecules and ions cannot (**Figure 11.12**).

When the solutions on either side of the membrane have the same solute concentration, water flows equally through the membrane in both directions. However, if one side of the membrane has a higher solute concentration, the water will migrate toward the more concentrated side. This tendency of water to move toward the more concentrated solution is called osmotic pressure.

We can show this property in a laboratory by placing pure water and a saltwater solution on either side of a synthetic membrane (**Figure 11.13**). Over time, water will migrate toward the more concentrated side, causing the volume of the saltwater solution to increase.

Why does water do this? To help you understand this property, here's an analogy: Imagine two towns with equal populations (**Figure 11.14**). Both towns are nice, and people like to live close to work. People with jobs stay put, while those without jobs may move back and forth between towns.

If the towns are the same size, and if each town has the same number of jobs, then the same number of people would move in each direction between Towns A and B. But what if a new factory opens in Town A? Suddenly, more people have a job in Town A, and so fewer of them will migrate to Town B, while people in Town B are still moving to Town A. The result is that the population of Town A increases, and the population of Town B decreases.

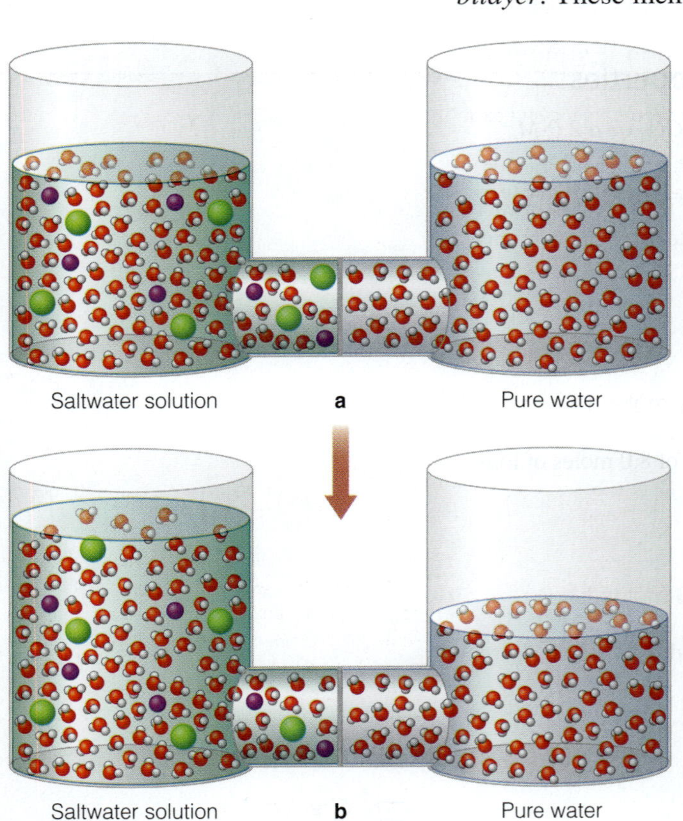

Figure 11.13 (a) Pure water and a saltwater solution are separated by a semipermeable membrane. (b) The water molecules will flow through the membrane toward the solute.

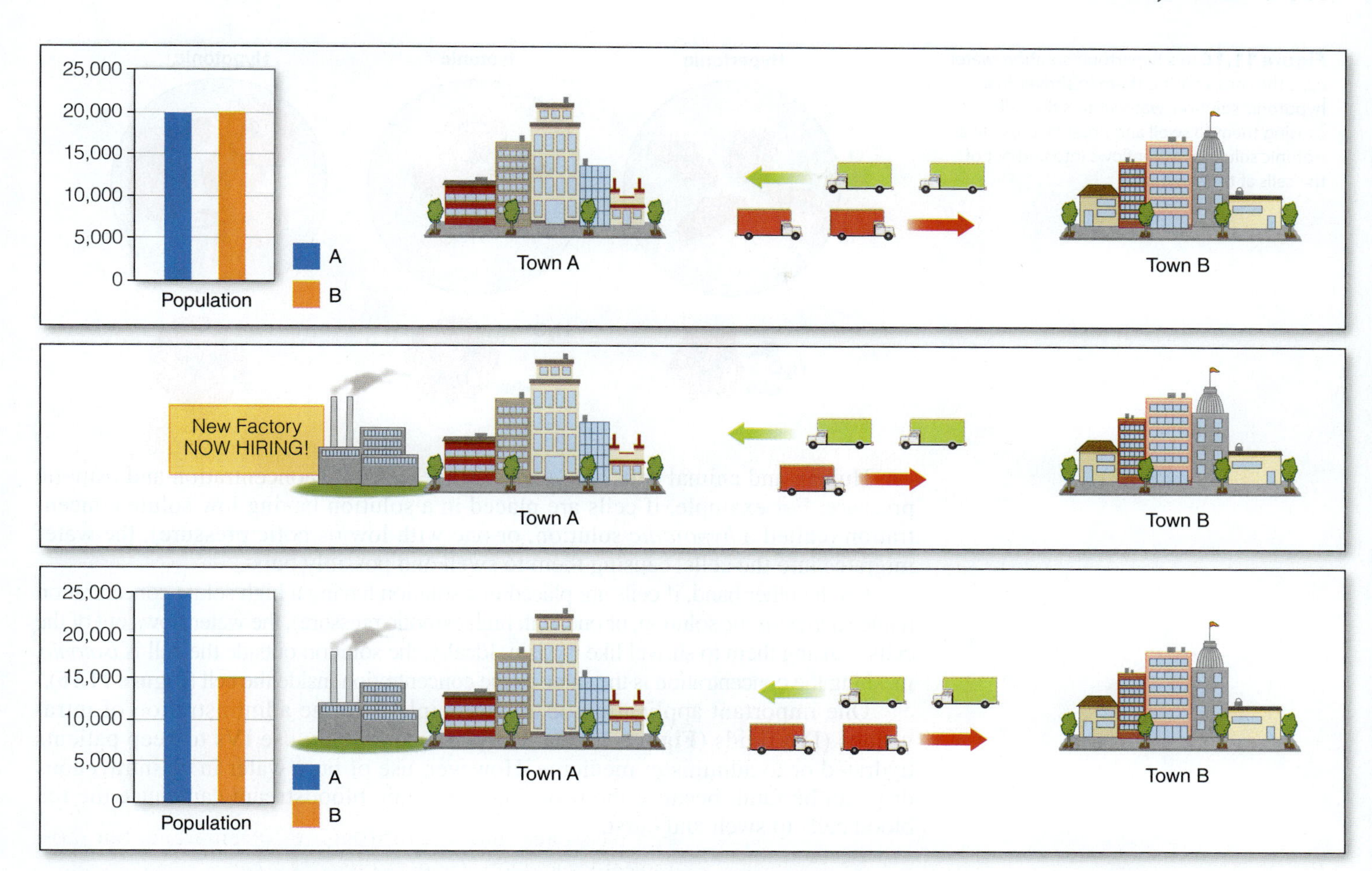

Figure 11.14 We can think about osmotic pressure in terms of two towns whose people are moving back and forth between them. Top: The two towns have equal populations, and people freely move between them. Middle: A new factory opens in Town A. Suddenly, fewer people leave Town A, but people continue to move in from Town B. Bottom: As a result, the population of Town A grows while that of Town B diminishes, until the number of people moving to and from each town balances.

Just as jobs keep people in a town, solute ions form strong interactions with water molecules, preventing them from moving across the membrane. The more concentrated a solution is, the fewer "jobless" water molecules migrate across the membrane. As a result, if two solutions are divided by a membrane, there will always be a net migration of water toward the solution with more ions—that is, toward the higher concentration (**Figure 11.15**).

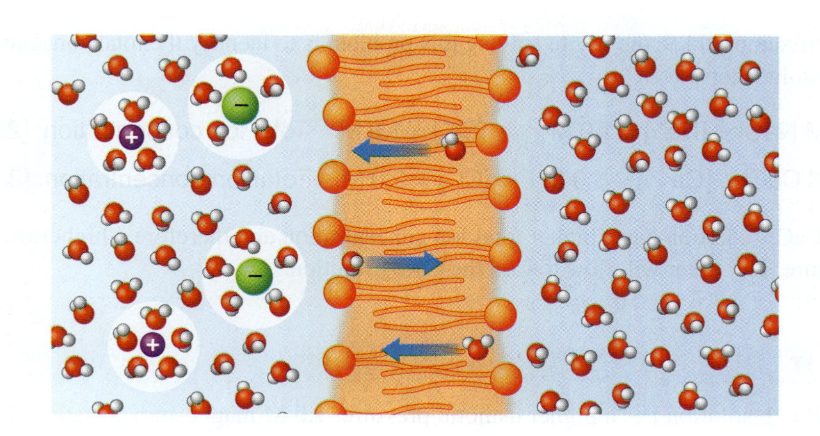

Figure 11.15 Because water aggregates around ions (shown as green and violet spheres), there are fewer "free" water molecules on the left-hand side. As a result, the net migration of water will be from right to left, approaching a balance of free water molecules on each side.

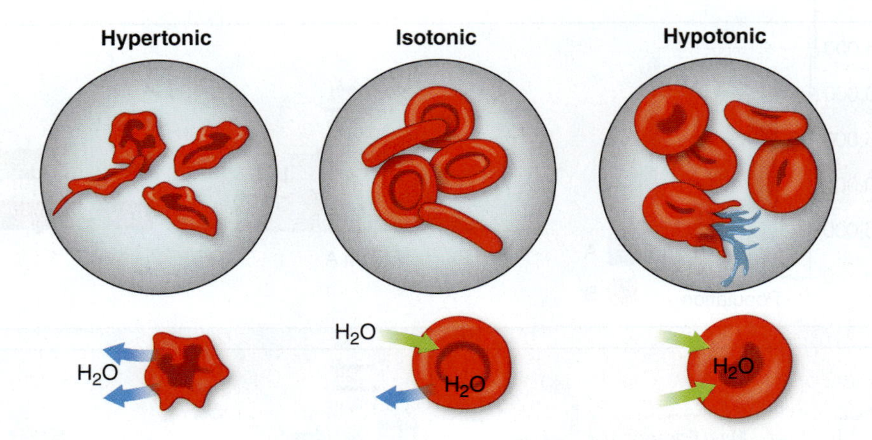

Figure 11.16 In a hypertonic solution, water exits the cells, causing them to shrivel. In a hypotonic solution, water enters the cells, causing them to swell and possibly burst. In an isotonic solution, water flows into and out of the cells at the same rate.

Human and animal cells are very sensitive to solute concentration and osmotic pressure. For example, if cells are placed in a solution having low solute concentration (called a *hypotonic* solution, or one with low osmotic pressure), the water migrates into the cells, causing them to swell and possibly burst.

On the other hand, if cells are placed in a solution having a high solute concentration (called a *hypertonic* solution, or one with high osmotic pressure), the water flows out of the cells, causing them to shrivel like a raisin. Ideally, the solution outside the cell is *isotonic*, meaning the concentration is the same as the concentration inside the cell (**Figure 11.16**).

One important application of this principle is in the administration of intravenous (IV) fluids (**Figure 11.17**). Hospitals commonly use IVs to keep patients hydrated or to administer medicine. However, use of pure water in an intravenous drip can be fatal, because the resulting hypotonic bloodstream can cause the red blood cells to swell and burst.

Figure 11.17 Normal saline solution is isotonic with the typical osmotic pressure of the cells.

Example 11.10 Osmotic Pressure

A 1.0-M aq. NaCl solution and a 1.0-M aq. $CaCl_2$ *solution are separated by a semipermeable membrane, as shown. Which solution has a higher osmotic pressure? Will the water flow toward the NaCl solution or the* $CaCl_2$ *solution?*

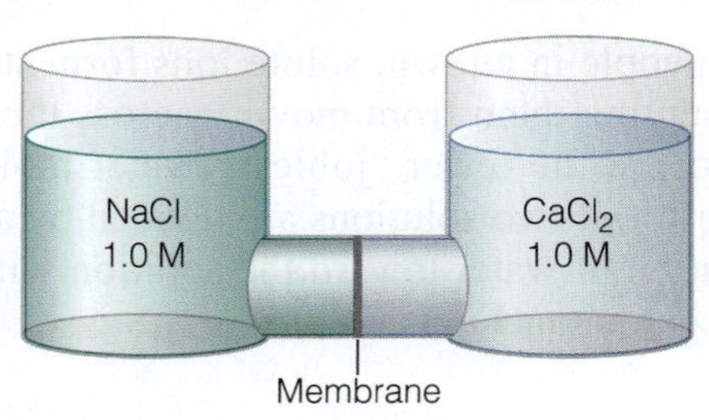

As in Example 11.9, the key to solving this problem is to identify the total concentration of dissolved ions.

1.0 M NaCl: $[Na^+] = 1.0$ M $[Cl^-] = 1.0$ M. Total ion concentration: [2.0 M]

1.0 M $CaCl_2$: $[Ca^{2+}] = 1.0$ M $[Cl^-] = 2.0$ M. Total ion concentration: [3.0 M]

The $CaCl_2$ solution has a higher total ion concentration and therefore a higher osmotic pressure. The water will flow toward the $CaCl_2$ solution.

Check it
Watch Explanation

TRY IT

13. Which solution has a higher osmotic pressure, 1.0 M magnesium acetate or 2.0 M sodium acetate?

11.3 Reactions in Solution—a Review and a Preview

In Chapter 6, we looked at some common types of reactions that take place in aqueous solution. Before moving on, let's briefly review those reaction types and preview another type of reaction (metal displacement reactions) that also takes place in solution.

Macmillan Learning

Figure 11.18 When silver and chloride ions are combined, a white precipitate forms.

Precipitation Reactions

In *precipitation reactions*, two solutions are combined, and a solid product is formed. For example, aqueous silver nitrate reacts with aqueous potassium chloride to produce silver chloride as a white solid (**Figure 11.18**):

$$AgNO_3\,(aq) + KCl\,(aq) \rightarrow AgCl\,(s) + KNO_3\,(aq)$$

The driving force for this reaction is the formation of the precipitate. We can also express this reaction using an ionic equation:

$$Ag^+\,(aq) + NO_3^-\,(aq) + K^+\,(aq) + Cl^-\,(aq) \rightarrow AgCl\,(s) + K^+\,(aq) + NO_3^-\,(aq)$$

In the equation above, the precipitating ions are shown in red. The other ions, potassium and nitrate, are not changed. These are called *spectator ions*. Net ionic equations do not include spectator ions—they show only the ions directly involved in the reaction. Here is the net ionic equation for this reaction:

$$Ag^+\,(aq) + Cl^-\,(aq) \rightarrow AgCl\,(s)$$

Ion solubility (see Table 6.3) determines which precipitates form.

Precipitation reactions depend on the solubility of the products formed. If two ions can combine to produce a product that is insoluble in water, a precipitation reaction will take place. The solubilities of many ionic compounds are given in Chapter 6 (see Table 6.3 and Figure 6.18). These are useful references for determining whether a precipitation reaction will take place.

Acid-Base Neutralization Reactions

In an *acid-base neutralization reaction*, an acid combines with a hydroxide base to produce water and an ionic compound (called a *salt*). For example, hydrochloric acid reacts with magnesium hydroxide to produce water plus magnesium chloride:

$$2\,HCl\,(aq) + Mg(OH)_2\,(aq) \rightarrow 2\,H_2O\,(l) + MgCl_2\,(aq)$$

The driving force of a neutralization reaction is the formation of water. Neutralization and precipitation reactions are both examples of double-displacement reactions. We will discuss other reactions of acids in Chapter 12.

Macmillan Learning

a

Macmillan Learning

b

Figure 11.19 (a) When elemental copper is placed in a silver nitrate solution, (b) a metal displacement reaction occurs, causing elemental silver to coat the copper surface.

Metal Displacement Reactions

In a **metal displacement reaction**, a solution containing the ion of one metal reacts with the elemental form of another metal. For example, copper metal reacts with silver nitrate to produce copper nitrate and elemental silver (**Figure 11.19**):

Molecular equation: $Cu\,(s) + 2\,AgNO_3\,(aq) \rightarrow Cu(NO_3)_2\,(aq) + 2\,Ag\,(s)$

Net ionic equation: $Cu\,(s) + 2\,Ag^+\,(aq) \rightarrow Cu^{2+}\,(aq) + 2\,Ag\,(s)$

These reactions are examples of single-displacement reactions. These reactions will be described in more detail in Chapter 14.

Check it
Watch Explanation

TRY IT

14. Predict the products of the following reactions. Refer to Table 6.3 and Figure 6.18 as needed.

$$Pb(NO_3)_2\ (aq) + MgCl_2\ (aq) \rightarrow$$

$$H_3PO_4\ (aq) + 3\ NaOH\ (aq) \rightarrow$$

11.4 Solution Stoichiometry

In Chapter 7, we used balanced equations to answer the question, "How much?" The first step in most stoichiometry problems is to convert the amounts given to moles. When working with solutions, we commonly know the concentration and the volume of a solution. From this, we can determine the moles of solute present:

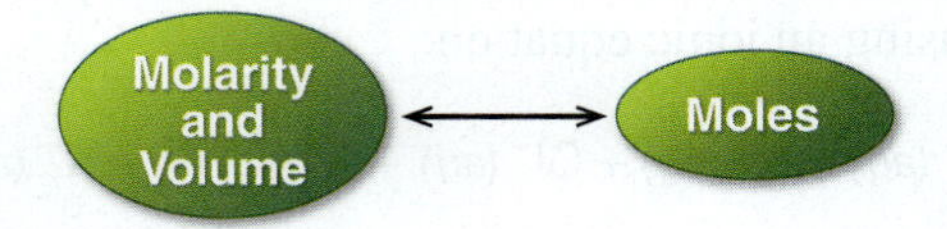

Once we have done this, we can predict the number of moles of any other species consumed or produced in the chemical reaction. Let's begin by looking at Example 11.11.

Example 11.11 Solution Stoichiometry

Silver and chloride ions react as shown in the following net ionic equation. Based on this reaction, how many grams of AgCl precipitate can be produced if 10.0 mL of solution containing 0.0023 M chloride ions reacts with excess Ag^+?

$$Ag^+\ (aq) + Cl^-\ (aq) \rightarrow AgCl\ (s)$$

This question gives us the volume and the molarity of the chloride ion. To solve the problem, we must first find the number of moles of chloride ion present:

$$\text{Volume (chloride)} = 10.0\ \text{mL} = 0.0100\ \text{L}$$

$$\text{Moles (chloride)} = MV = \left(0.0023\ \frac{\text{mol}}{\cancel{\text{L}}}\right)(0.0100\ \cancel{\text{L}}) = 2.3 \times 10^{-5}\ \text{moles}\ Cl^-$$

Once we have found the moles of Cl^-, we can relate this to the moles of AgCl, and then to the grams of AgCl, using the balanced equation.

$$2.3 \times 10^{-5}\ \cancel{\text{mol}\ Cl^-} \times \frac{1\ \cancel{\text{mole AgCl}}}{1\ \cancel{\text{mole}\ Cl^-}} \times \frac{143.32\ \text{g AgCl}}{1\ \cancel{\text{mole AgCl}}} = 3.3 \times 10^{-3}\ \text{g AgCl}$$

Check it
Watch Explanation

TRY IT

15. Calcium and carbonate ions combine in a precipitation reaction, as shown in this net ionic equation. If 25.0 mL of a solution containing 0.0015 M Ca^{2+} reacts with excess carbonate, how many grams of $CaCO_3$ will form?

$$Ca^{2+}\ (aq) + CO_3^{\ 2-}\ (aq) \rightarrow CaCO_3\ (s)$$

Gravimetric Analysis

To determine the concentration of ions in solution, chemists often use a technique called **gravimetric analysis.** In this technique, the mass of a precipitate is used to determine the concentration of a reactant.

For example, suppose a chemist needs to find the concentration of iodide ions in a solution. How can she do this? Recall that iodide ions react with lead(II) ions to produce lead(II) iodide as an insoluble precipitate:

$$Pb^{2+}\ (aq) + 2\ I^-\ (aq) \rightarrow PbI_2\ (s)$$

The chemist therefore takes a small sample of the unknown I^- solution and combines it with a second solution containing an excess of Pb^{2+}. A precipitate forms. She collects the precipitate by filtration and then allows it to dry. From the mass of the precipitate, she can determine the number of moles of iodide present, and therefore the concentration of the original solution. This idea is illustrated in **Figure 11.20** and Example 11.12.

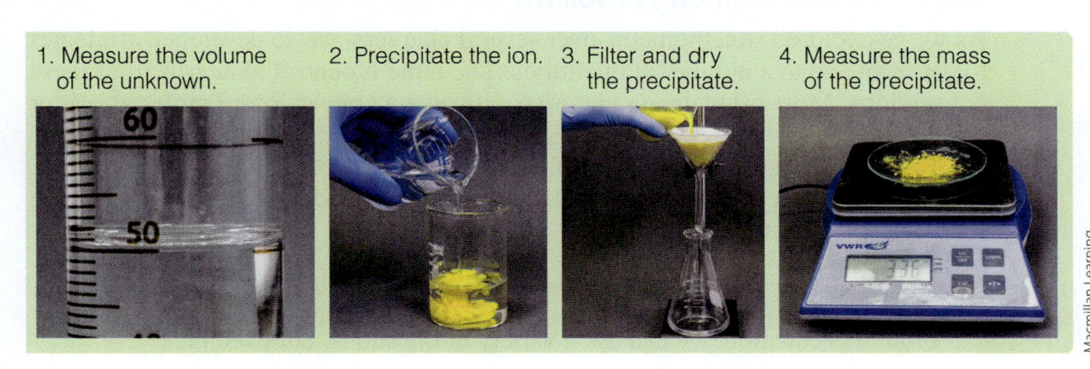

Figure 11.20 Gravimetric analysis follows a four-step procedure.

Example 11.12 Gravimetric Analysis

The concentration of sulfate ion (SO_4^{2-}) in a solution may be measured using a precipitation reaction with barium chloride. The net ionic equation for this reaction is

$$Ba^{2+}\ (aq) + SO_4^{2-}\ (aq) \rightarrow BaSO_4\ (s)$$

You need to measure the sulfate concentration in a large drum containing aqueous sodium sulfate. From the drum, you measure a 0.200-L sample and react it with excess barium chloride. The reaction produces $BaSO_4$ as a precipitate, which you carefully filter and dry. The $BaSO_4$ precipitate has a mass of 0.309 g. What is the concentration (in molarity) of sulfate ion in the drum?

This question gives us the mass of $BaSO_4$ and asks us to find the concentration of SO_4^{2-} in the drum. To do this, we must convert from grams of $BaSO_4$ to moles of $BaSO_4$, then relate that to moles of SO_4^{2-}, and then solve for the concentration:

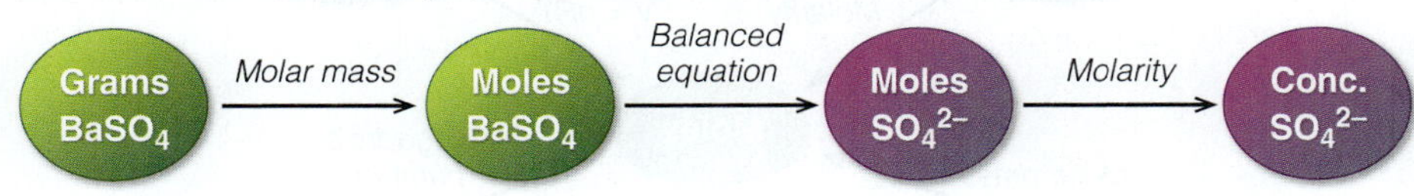

First, convert grams of $BaSO_4$ to moles of $BaSO_4$ and then to moles of SO_4^{2-}:

$$0.309\ \cancel{\text{g BaSO}_4} \times \frac{1\ \cancel{\text{mole BaSO}_4}}{233.39\ \cancel{\text{g BaSO}_4}} \times \frac{1\ \text{mole SO}_4^{2-}}{1\ \cancel{\text{mole BaSO}_4}} = 1.32 \times 10^{-3}\ \text{moles SO}_4^{2-}$$

This is the number of moles of sulfate present in the 0.200-L sample, so we can calculate the molarity.

$$\text{Concentration (molarity)} = \frac{1.32 \times 10^{-3}\ \text{mol}}{0.200\ \text{L}} = 6.60 \times 10^{-3}\ \text{M}$$

Check it
Watch Explanation

TRY IT

16. The presence of lead(II) in a water sample can be determined by precipitating the lead(II) ions with a solution of potassium chromate, K_2CrO_4. The reaction takes place according to this equation:

$$Pb^{2+}\ (aq) + K_2CrO_4\ (aq) \rightarrow PbCrO_4\ (s) + 2\ K^+\ (aq)$$

You have been assigned to find the concentration of lead(II) in the waste stream of a manufacturing process. You therefore react a 1.00-L sample from the waste stream with an excess of K_2CrO_4. The resulting precipitate has a mass of 34.1 mg. What is the concentration of Pb^{2+} in the wastewater?

Advanced Stoichiometry Problems

As we've explored stoichiometry over several chapters, we've developed strategies to get from and to a number of quantities. The mole is central to nearly all of these conversions. Most problems in chemistry involve a conversion to moles at one or more steps.

Figure 11.21 shows a map for solving a wide range of problems. For example, if solid sodium bicarbonate ($NaHCO_3$, baking soda) is added to aqueous HCl, the following reaction takes place:

$$NaHCO_3\ (s) + HCl\ (aq) \rightarrow NaCl\ (aq) + H_2O\ (l) + CO_2\ (g)$$

From this equation, we can ask a variety of questions. For example:

1. Starting from the moles of $NaHCO_3$, how many moles of H_2O can form?
2. Starting from the molarity and volume of HCl, how many grams of CO_2 can form?
3. What volume of 1.0 M HCl is required to react with 2.5 grams of $NaHCO_3$?
4. If this reaction produces 30.5 grams of NaCl at standard temperature and pressure, what volume of CO_2 gas forms?
5. If 200 mL of H_2O forms, how many moles of HCl are consumed?

Each of these questions can be solved using the map in Figure 11.21. There are many other possibilities—can you think of a few? How would you solve them? Example 11.13 gives a similar problem.

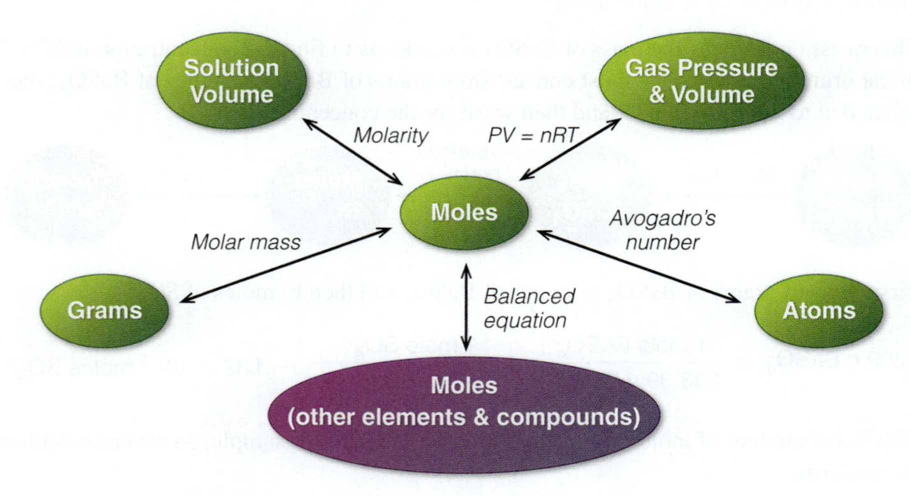

Figure 11.21 The mole is the intermediate for most calculations in chemistry.

Example 11.13 Advanced Stoichiometry Problems

Calcium metal reacts with aqueous hydrobromic acid to produce calcium bromide and hydrogen gas, as shown in the equation below. What volume of 2.50-M aq. HBr is required to produce 200.0 L of hydrogen gas at a temperature of 298 K and a pressure of 1.20 atm?

$$Ca\,(s) + 2\,HBr\,(aq) \rightarrow CaBr_2\,(aq) + H_2\,(g)$$

We are given the volume of H_2 gas and then asked to relate this to the volume of the aqueous HBr solution. Based on the mole map in Figure 11.21, we see that our pathway for solving this problem must be

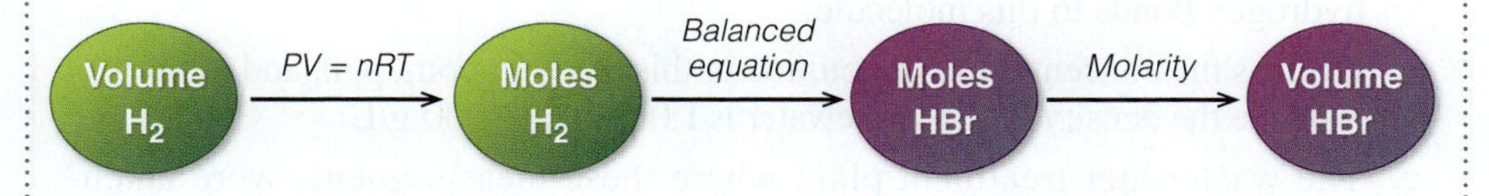

The first step requires the ideal gas law, which we covered in Chapter 10:

$$n = \frac{PV}{RT} = \frac{(1.20\ \text{atm})(200.0\ \text{L}\ H_2)}{\left(0.0821\frac{\text{L}\cdot\text{atm}}{\text{mol}\cdot\text{K}}\right)(298\ \text{K})} = 9.81\ \text{mol}\ H_2$$

Following the pathway outlined above, we are able to relate this value to the moles of HBr and reach a final answer of 7.85 L *aq.* HBr.

$$9.81\ \text{moles}\ H_2 \times \frac{2\ \text{moles HBr}}{1\ \text{mole}\ H_2} \times \frac{1\ \text{L}}{2.50\ \text{mole HBr}} = 7.85\ \text{L}\ aq.\ \text{HBr solution}$$

TRY IT

Check it
Watch Explanation

17. Ammonium phosphate reacts with barium chloride to form a white precipitate, as shown here:

$$2\,(NH_4)_3PO_4\,(aq) + 3\,BaCl_2\,(aq) \rightarrow 6\,NH_4Cl\,(aq) + Ba_3(PO_4)_2\,(s)$$

a. How many moles of ammonium phosphate are needed to completely react with 96.2 mL of 2.10-M *aq. $BaCl_2$*?

b. How many grams of barium phosphate could be produced in this reaction?

18. Aqueous hydrochloric acid reacts with zinc metal according to this reaction:

$$2\,HCl\,(aq) + Zn\,(s) \rightarrow ZnCl_2\,(aq) + H_2\,(g)$$

a. If 15.0 mL of 3.4 M *aq.* HCl reacts with excess zinc,

b. How many moles of $ZnCl_2$ are produced?

c. How many grams of H_2 gas are produced?

d. If the reaction takes place at standard temperature and pressure (273 K, 1.0 atm), what volume of H_2 gas is produced?

Capstone Video

Capstone Question

Scientists can learn a great deal about a community's drug use (both legal and illegal) by analyzing the contaminants in wastewater. In a recent study in New York State, chemists measured the concentration of cotinine, a derivative of nicotine that is produced in the liver and excreted in the urine of tobacco users (**Figure 11.22**). The chemists found the average cotinine concentration in the wastewater stream was 1.43 μg/L.

a. Cotinine is slightly soluble in water. Using the structure in Figure 11.22, identify any polar covalent bonds that are present in this substance. Sketch several water molecules around the structure to show how water forms hydrogen bonds to this molecule.

b. Express the concentration of cotinine in this study in ppm, ppb, and molarity. Assume the density of the wastewater is 1.0 g/mL (1,000 g/L).

c. The wastewater treatment plant where these measurements were taken serves a community of 100,000 people and processes a volume of 83,000 m^3 of wastewater per day. Based on this, what mass of cotinine enters the treatment plant each day? Report your answer in grams.

d. If 17% of the nicotine that humans consume is excreted as cotinine, estimate the total daily nicotine consumption for this community. If the nicotine dose in a single cigarette is 1.6 mg, estimate the number of doses consumed per day in this community.

Jonutis/Shutterstock

a

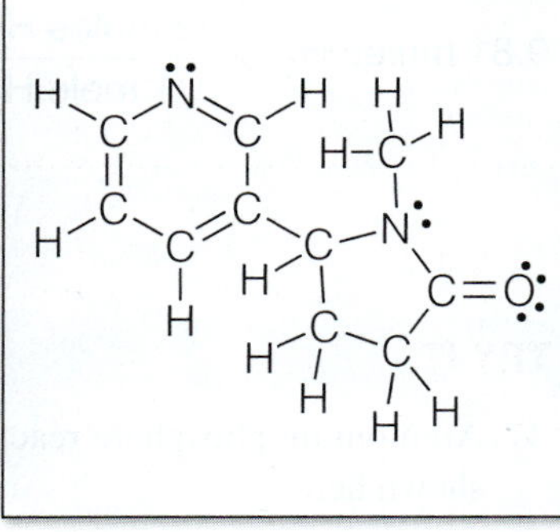

b

Figure 11.22 (a) Wastewater treatment plants remove solids and dissolved impurities from city sewage. Scientists can gain many public health insights by analyzing the contaminants coming into plants like this. (b) This is the structure of cotinine, a derivative of nicotine found in the urine of tobacco users.

SUMMARY

A key characteristic of any solution is its concentration; that is, the amount of solute present in the solution. Concentration is measured in a variety of units, including percentage, parts per million (ppm), parts per billion (ppb), and molarity. Molarity is defined as the moles of solute per liter of solution; it is particularly useful because it allows us to easily relate volume to moles.

When ionic compounds dissolve in water, the compounds dissociate into positive and negative ions. Because aqueous ionic solutions conduct electricity, water-soluble ionic compounds are sometimes called *electrolytes*. When electrolytes dissociate, the concentration of the individual ions may be higher than the overall concentration of the ionic compound. Most covalent compounds are nonelectrolytes; they do not dissociate in water.

Colligative properties are those properties of solutions that depend only on the quantity, not the identity, of solute present. Colligative properties include melting point depression, boiling point elevation, and osmotic pressure.

A variety of chemical reactions involve aqueous solutions. They include precipitation, acid-base neutralization, and metal displacement reactions.

We can extend the principles of stoichiometry to solve problems pertaining to solutions. Given the volume and molarity of a solution, we can determine the number of moles present in a solution or in a reaction. As noted in previous chapters, we can relate moles of any reactant or product to those of any other component in the reaction. From moles, we can convert to grams, atoms, pressure/volume measurements for gases, and others.

Amazing Coffee — The Importance of Concentration

By Aaron Blanco, President of Brown Coffee Co., San Antonio, Texas

What you measure improves. This is true in any endeavor in life, and it is evident in the world of specialty coffee.

For decades, most Americans bought their coffee already ground and in a can. Grab a spoon, pile some — or a lot — of coffee into the filter; add some water from the faucet. Close the hatch, flip the switch, and *voilà*!

"Coffee."

"Joe."

"Just a regular cuppa."

Most coffee served in restaurants and diners was equally uninspired. However, over the last decade or so, coffee has experienced a renaissance as people take a more science-based approach to the art of coffee brewing. At its core is the intersection of flavor and intensity, and how that is measured. What does a particular coffee from a particular place taste like? How intense is the flavor? These qualities can be thought of in terms of two parameters, *strength* and *extraction yield* (**Figure 11.23**).

Let's tackle strength first. Coffee strength is measured in total dissolved solids (TDS). This is typically expressed as a percent (m/v), or grams of solute per mL of solution. The TDS is normally between 0.8% and 1.6%. The higher the TDS, the "stronger" a cup of coffee is. But strength doesn't taste like anything. It is merely a measurement of intensity. You can detect something lovely in a cup that is very faint, very powerful, or somewhere in between.

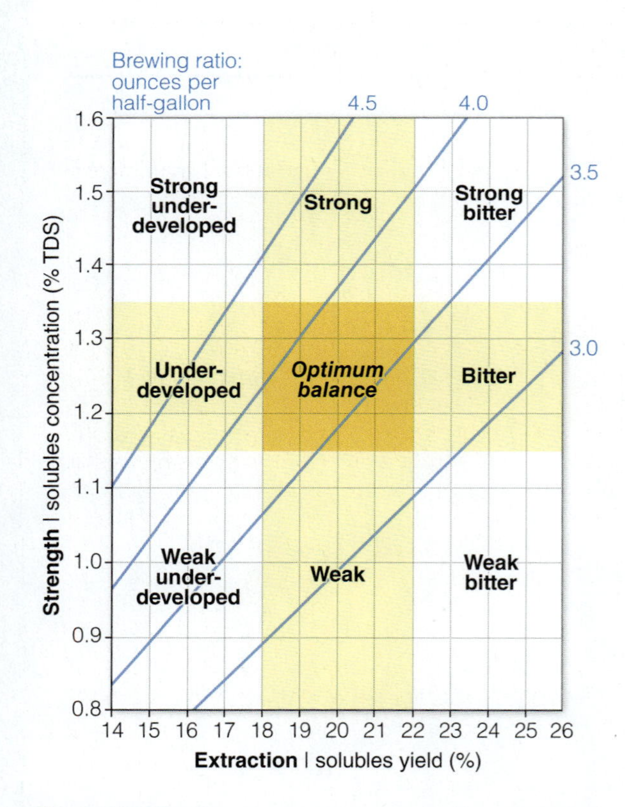

Figure 11.23 The strength (total dissolved solids, or TDS) and flavor (extraction yeast) both affect the quality of coffee.
Data from CoffeeChemistry.com

The second parameter, extraction yield, affects the flavor of the coffee. About 30% of the mass of a coffee bean is soluble in water. However, some compounds are more soluble than others. If too little of the material is extracted, then the most soluble compounds predominate, and the flavor is simple and bland. As the extraction yield increases, the flavor becomes more complex. However, if too much material is extracted, the coffee may take on a bitter flavor.

A number of factors affect both TDS and extraction yield. They include such things as the size and uniformity of the grounds, the temperature of the water used in brewing (hotter water is a more efficient solvent), external agitation (stirring), freshness, and so on.

Historically, most American tastes have fallen between 1.15% and 1.35% TDS and between 18% and 22% extraction yield (see Figure 11.23). But tastes change. This chart, developed in the 1950s, was made in an era when lower-quality coffees were the norm. People then used less than optimal grinding and brewing equipment and often stale coffee from a giant can.

I personally prefer coffees at 1.4%–1.45% TDS and 22%–23% extraction yield for regular filter coffee. Espresso is much stronger, with 10.5%–11% TDS and 24%–25% extraction yield. I expect that as technology helps us isolate and improve each variable, we'll start to see coffees regularly pushing 26%–28% extraction yields. But only time will tell if tastes will follow science, or if old coffee-drinking habits will die hard.

Key Terms

11.1 Describing Concentration

solution A homogeneous mixture; a solution is usually a liquid.

solute A substance that is dissolved in a solution.

solvent The major component of a solution.

concentration The amount of solute present in a solution.

percent by mass A measure of concentration; the mass of solute divided by the total mass of the solution, expressed as a percentage.

percent by volume A measure of concentration; the volume of solute divided by the total volume of the solution, expressed as a percentage.

mass/volume percent A measure of concentration; the mass of solute divided by the total volume of the solution, expressed as a percentage; often measured in grams of solute per milliliter of solution.

parts per million (ppm) A measure of concentration used for dilute solutions; 1 ppm = 1 g of solute per 1,000,000 grams of solution, or 1 mg of solute per liter of solution.

parts per billion (ppb) A measure of concentration used for very dilute solutions; 1 ppb = 1 g of solute per 1,000,000,000 grams of solution, or 1 μg of solute per liter of solution.

molarity (M) A measure of concentration defined as the moles of solute per liter of solution.

11.2 Electrolyte Solutions

electrolyte solutions An aqueous solution containing dissociated ions; this type of solution conducts electricity more effectively than pure water.

colligative properties Properties that depend on the number of particles dissolved in the solution, but not on the type of particles dissolved.

freezing point depression A colligative property of water; the presence of solute in an aqueous solution lowers the freezing point below that of pure water.

boiling point elevation A colligative property of water; the presence of solute in an aqueous solution raises the boiling point above that of pure water.

osmotic pressure The tendency of water to move toward regions of greater solute concentrations; an imbalance in concentration inside and outside of a living cell can cause the cell to swell or shrink.

11.3 Reactions in Solution—a Review and a Preview

metal displacement reaction A reaction in which the ion of one metal reacts with the elemental form of another metal.

11.4 Solution Stoichiometry

gravimetric analysis A technique that uses the mass of a precipitate to determine the concentration of a reactant.

Additional Problems

11.1 Describing Concentration

19. A solution contains 0.0025 grams of potassium nitrate dissolved in 2.0 liters of water. At 20 °C, one liter of water can dissolve over 300 grams of this compound. Identify the solvent and the solute in this solution. Is this solution best described as dilute, concentrated, or saturated?

20. What is the difference between a concentrated solution and a saturated solution?

21. A calcium chloride solution contains 40.0 grams of calcium chloride dissolved in 100 grams of water. What is the percent by mass of calcium chloride in this solution?

22. You wish to prepare a solution that is 12% sodium chloride by mass. How much sodium chloride is necessary to prepare 2.0 kg of this solution?

23. A cleaning solution calls for a 20% (v/v) mixture of cleaner in water. How much cleaner and water are necessary to prepare 4.0 liters of the solution?

24. Antifreeze is a mixture of ethylene glycol and water. What volume of ethylene glycol would be needed to prepare 8.0 liters of antifreeze that is 20% (v/v) ethylene glycol?

25. Wine typically contains about 12% (v/v) alcohol. What is the volume of alcohol in a 200-mL glass of wine?

26. Which of these drinks has the greatest alcohol content?

a. two 12-ounce glasses of beer, which is 4.5% alcohol (v/v)
b. one 8-ounce glass of wine, which is 12% alcohol (v/v)
c. one 1.5-ounce shot of whiskey, which is 40% alcohol (v/v)

27. Calculate the percent by mass (g/mL) in these solutions:

a. an aqueous solution containing 52.0 g of potassium carbonate in a solution of 1,000 mL
b. a saturated sugar solution with a volume of 1.0 liter, containing 2.0 kg of dissolved sugar
c. a solution containing 28.4 grams of salt dissolved in a 150-mL solution

28. How many grams of sodium nitrate are in 40 gallons of an 8.0% (g/mL) sodium nitrate solution?

29. You need to prepare 4.0 liters of an aqueous solution that is 5% (m/v or g/mL) sodium chloride. What mass of sodium chloride do you need to prepare this solution?

30. You need to prepare 15.0 liters of an aqueous solution that is 0.4% (g/mL) ammonium carbonate. What mass of ammonium carbonate do you need to prepare this solution?

31. Express each of these concentrations in parts per million (ppm):

a. 5.0 milligrams per kilogram of water
b. 5.0 micrograms per kilogram of water
c. 5.0 nanograms per kilogram of water

32. Express each of these concentrations in parts per billion (ppb):

a. 8.0 milligrams per kilogram of water
b. 8.0 micrograms per kilogram of water
c. 8.0 nanograms per kilogram of water

33. The water from a river is found to contain 10 ppm of dissolved oxygen. Assuming the density of the water is 1.0 g/mL, how many milligrams of oxygen are in a liter of the river water?

34. A lake is found to contain 8.0 ppm of dissolved nitrates. Assuming the density of the water is 1.0 g/mL, what mass of nitrates are dissolved in one liter of the lake water?

35. Waterways in the Mississippi Valley region of the United States commonly contain 0.03 mg of dissolved phosphorus per liter of water. Assuming the density of the water is 1.0 g/mL, what is this concentration in ppm?

36. Triazines are compounds commonly used as herbicides, which may leach into waterways. A 1.0-L sample of lake water is found to contain 8.0 micrograms of triazines. What is the concentration of triazines in ppm?

37. Fluoride is commonly added to drinking water to promote bone health. While a very small amount of fluoride can be beneficial, larger amounts can be harmful or deadly. Don't swallow your toothpaste. The U.S. Environmental Protection Agency (EPA) currently sets the maximum concentration level (MCL) of fluoride in drinking water at 4.0 mg/L. What is this concentration in parts per million? What is this concentration in % (g/mL)?

38. Arsenic is a highly toxic metal that enters the water supply through erosion of natural deposits or as waste from agricultural and industrial applications. The EPA sets the maximum concentration limit of arsenic in drinking water as 0.010 mg/L. Given the density of water as 1.0 g/mL, what is the maximum concentration of arsenic in ppm? In ppb?

39. Polychlorinated biphenyls (PCBs) were commonly used as plastic additives, hydraulic fluids, and lubricating oils through the mid-twentieth century, until they were banned because of their toxic effects. A recent study around the Savannah River in Georgia showed PCB levels of 5 ng per gram of sediment. What is this concentration in ppm? What is it in ppb?

40. Mercury is a highly toxic metal that can cause nerve damage. The EPA sets the maximum concentration level of mercury in drinking water as 2.0×10^{-6} grams per liter of water. What is this concentration level in ppm?

41. Tetrachloroethylene is sometimes discharged into the water supply from factories and dry cleaners. The EPA considers the maximum concentration level (MCL) of tetrachloroethylene in drinking water to be 5 ng/L. What is this concentration in ppm? What is this concentration in ppb?

42. A river pollutant is found to have a concentration of 0.0025% m/v. If the density of the river water is 1.03 g/mL, what is the concentration of this pollutant in ppm?

43. Calculate the molarity of each of these solutions:

a. a 2.20-liter solution containing 48.3 grams of potassium iodide
b. a 4.9-liter solution containing 12.1 grams of dissolved carbon dioxide
c. a 500-mL solution containing 93.1 grams of magnesium acetate

44. Calculate the molarity of each of these solutions:

a. a 55.0-liter solution containing 2.3 kg of dissolved sugar ($C_{12}H_{22}O_{11}$)
b. a solution with a volume of 892 milliliters containing 1.27 grams of dissolved sodium fluoride
c. a solution containing 2.53 g of ammonium acetate with a volume of 170.2 mL

45. Find the following values:

a. moles of NaOH in 2.0 L of a 5.0-M NaOH solution
b. moles of KCl in an 800-mL solution of 1.5-M KCl
c. moles of Na_2CO_3 in 300 μL of a 1.0-M Na_2CO_3 solution

46. Find each of the following:

a. moles of barium hydroxide in a 400-mL sample of 0.47-M $Ba(OH)_2$
b. grams of K_3PO_4 in a 350-mL sample of a 1.4-M potassium phosphate solution
c. kilograms of $BeCl_2$ in a completely filled, 45-gallon drum containing 5.0 M *aq.* $BeCl_2$

47. Using a 1.0-M aqueous solution of KOH, a chemist wishes to add 14.30 moles of potassium hydroxide to an acidic solution. What volume of KOH solution does she need to add?

48. What volume (in milliliters) of 3.5-M *aq.* HCl contains 0.025 moles of hydrochloric acid?

49. While working in a pharmaceutical laboratory, you need to prepare 3.0 L of a 1.00-M NaCl solution. What mass of NaCl would be required to prepare this solution? How would you go about preparing the solution?

50. You've just started a new job in the laboratory of the Food and Drug Administration. Your supervisor asks you to prepare a 1.50-M solution of aqueous potassium permanganate, $KMnO_4$. What mass of $KMnO_4$ will you need to prepare 500 mL of this solution? Describe how you would prepare the solution.

51. A 12.0-M solution of HCl is diluted from an initial volume of 25 mL to a final volume of 800 mL. What is the new concentration of the HCl solution?

52. A 150-mL sample of a solution of 4.80 M $FeCl_2$ is diluted with enough water to raise the volume to 3.40 L. What is the new concentration of this solution? How many moles of $FeCl_2$ are present in this solution?

53. You need to prepare 100 mL of 0.100-M *aq.* $CuSO_4$. Describe how you would prepare this solution from a stock solution of 5.0-M *aq.* $CuSO_4$.

54. You need to prepare 300 mL of 0.100-M *aq.* sodium acetate. Describe how you would prepare this solution.

a. from solid sodium acetate
b. from a stock solution of 6.0-M *aq.* $NaC_2H_3O_2$

11.2 Electrolyte Solutions

55. Indicate whether each of the following is or is not an electrolyte:

a. KCl
b. $FeBr_3$
c. C_2HO
d. CH_5N

56. Indicate whether each of the following is or is not an electrolyte:

a. an ionic compound that dissolves readily in water
b. a nonionic compound that dissolves readily in water
c. a solution of magnesium chloride
d. a solution of acetone, a nonacidic covalent compound

57. Sketch the dissociation of a crystal of potassium bromide in water.

58. Sketch the dissociation of a crystal of sodium nitrate in water.

59. Rewrite these expressions as ionic equations:

a. $Ba(OH)_2\ (s) \rightarrow Ba(OH)_2\ (aq)$
b. $(NH_4)_2SO_4\ (s) \rightarrow (NH_4)_2SO_4\ (aq)$

60. Rewrite these expressions as ionic equations:

a. $Mg(NO_3)_2\ (s) \rightarrow Mg(NO_3)_2\ (aq)$
b. $(NH_4)_2CO_3\ (s) \rightarrow (NH_4)_2CO_3\ (aq)$

61. Write ionic equations to show the dissociation of these ionic solids:

a. $MgBr_2$ b. $AlCl_3$ c. $Ca(NO_3)_2$

62. Write ionic equations to show the dissociation of these ionic solids:

a. CsBr b. $(NH_4)_3PO_4$ c. K_3PO_4

63. Find each of the following:

a. $[OH^-]$ in 1 M *aq.* NaOH
b. $[Cl^-]$ in 2 M *aq.* $MgCl_2$
c. $[Br^-]$ in 3 M *aq.* $FeBr_3$

64. Find the concentration of sodium ions in each of these solutions:

a. 0.60 M *aq.* NaCl
b. 0.60 M *aq.* Na_2CO_3
c. 0.60 M *aq.* Na_3PO_4

65. Find the concentration of all ions in each of these solutions:

a. 2.0 M *aq.* Li_2S
b. 0.7 M *aq.* Li_3PO_4
c. 1.3×10^{-4} M *aq.* $Mg(NO_2)_2$

66. Find the concentration of all ions in each of these solutions:

a. 3.0 M *aq.* LiCl
b. 0.70 M *aq.* $Fe(NO_3)_2$
c. 1.7×10^{-4} M *aq.* NH_4Cl

67. How many moles of chloride ion are in a 400.0-mL solution of 2.30-M *aq.* $MgCl_2$?

68. How many moles of sulfate ion are in a 350.0-mL solution of 1.20-M *aq.* $(NH_4)_2SO_4$?

69. A 300-mL solution of 1.3-M *aq.* NaCl is mixed with 500 mL of 2.5-M *aq.* $CaCl_2$. How many moles of chloride ion are in the combined solution? What is the molarity of chloride in the combined solution?

70. A 2.10-L solution of 1.0-M *aq.* HBr is mixed with 1.25 L of 3.0-M *aq.* $BeBr_2$. How many moles of bromide ion are in the combined solution? What is the molarity of bromide in the combined solution?

71. If 400 mL of 1.2-M *aq.* $MgSO_4$ is mixed with 550 mL of 2.7-M *aq.* K_2SO_4, what are the concentrations of magnesium, potassium, and sulfate ions in the resulting solution?

72. If 0.040 L of 12.0-M KNO_3 is mixed with 90 mL of 2.5 M $Ca(NO_3)_2$, what is the concentration of potassium, calcium, and nitrate in the resulting solution?

73. What is a colligative property?

74. If additional solute is added to a solution:

a. How does the freezing point change?
b. How does the melting point change?
c. How does the boiling point change?
d. How does the osmotic pressure change?

75. Which of these aqueous solutions has a lower freezing temperature: a 1.0-M solution of $MgCl_2$ or a 1.0-M solution of KBr? Explain your answer.

76. Which of these aqueous solutions has a higher boiling temperature: a 2.0-M solution of CsBr or a 2.0-M solution of $Ca(NO_2)_2$? Explain your answer.

77. Between a solution of 2.0 M *aq.* $NaNO_3$ and 1.5 M *aq.* $AlCl_3$:

a. Which would have a lower freezing point?
b. Which would have a higher boiling point?
c. Which would have a higher osmotic pressure?

78. Between a solution of 1.1 M *aq.* K_2CO_3 and 1.5 M *aq.* CH_4O (a nonelectrolyte):

a. Which would have a lower freezing point?
b. Which would have a higher boiling point?
c. Which would have a higher osmotic pressure?

79. A 3.0-M *aq.* NaCl solution and a 1.0-M *aq.* Na_2SO_4 solution are separated by a semipermeable membrane, as shown. Which solution has a higher osmotic pressure? Will the water flow toward the NaCl solution or the Na_2SO_4 solution?

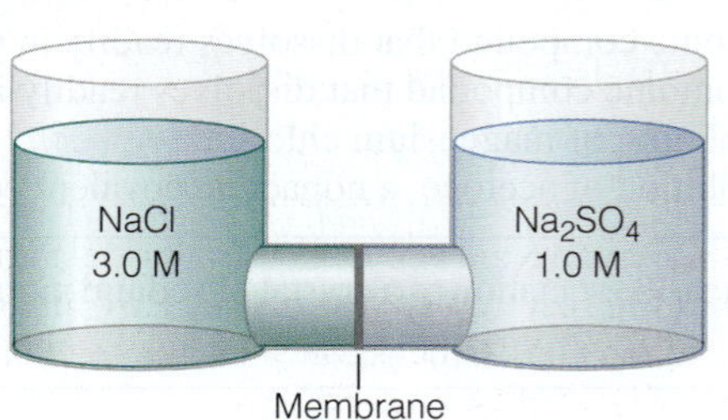

80. A chemist prepares two aqueous solutions, each with a volume of 1.00 L. The first solution contains 51.35 g of ammonium sulfate. The second solution contains 73.25 g of sodium chloride. What is the total ion concentration in each solution? Which solution has a higher osmotic pressure?

81. Normal saline is a solution of aqueous sodium chloride with a concentration of 0.90% (g/mL). Express this concentration in molarity; then indicate whether the osmotic pressure of each of these solutions is hypotonic, isotonic, or hypertonic compared to normal saline:

a. 1.0 M *aq.* KCl
b. 0.5 M *aq.* K_2SO_4
c. 4.3 M *aq.* NH_4Cl
d. 0.30 M *aq.* sugar

82. In the figure shown, water molecules migrate from the area of lower solute concentration (right) to the area of higher solute concentration (left). Do water molecules "know" which way to travel? If not, how can we explain the migration of water molecules toward a higher solute concentration?

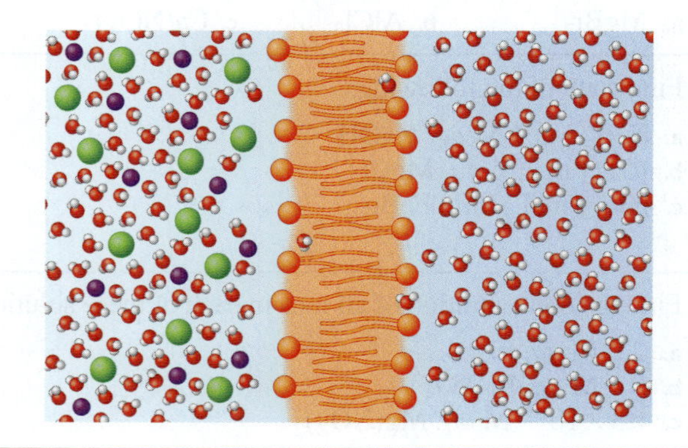

11.3 Reactions in Solution—a Review and a Preview

83. What is the driving force in a precipitation reaction?

84. What is the driving force in a neutralization reaction?

85. Write this precipitation reaction as a complete ionic equation and then as a net ionic equation:

$$CaBr_2\,(aq) + Pb(NO_3)_2\,(aq) \rightarrow PbBr_2\,(s) + Ca(NO_3)_2\,(aq)$$

86. Write this neutralization reaction as a complete ionic equation and then as a net ionic equation:

$$Ca(OH)_2\,(aq) + 2\,HBr\,(aq) \rightarrow CaBr_2\,(aq) + 2\,H_2O\,(l)$$

87. Identify each of these reactions as precipitation, neutralization, or metal displacement:

a. $Mg\ (s) + NiCl_2\ (aq) \rightarrow Ni\ (s) + MgCl_2\ (aq)$
b. $Mg(OH)_2\ (aq) + H_2SO_4\ (aq) \rightarrow MgSO_4\ (aq) + 2\ H_2O\ (l)$
c. $MgCl_2\ (aq) + Pb(C_2H_3O_2)_2\ (aq) \rightarrow Mg(C_2H_3O_2)_2\ (aq) + PbCl_2\ (s)$

88. Identify each of these reactions as precipitation, neutralization, or metal displacement:

a. $Ba(OH)_2\ (aq) + FeCl_2\ (s) \rightarrow BaCl_2\ (aq) + Fe(OH)_2\ (s)$
b. $Ba(OH)_2\ (aq) + 2\ HCl\ (aq) \rightarrow BaCl_2\ (aq) + 2\ H_2O\ (l)$
c. $3\ Ba\ (s) + 2\ Al(NO_3)_3\ (aq) \rightarrow 3\ Ba(NO_3)_2\ (aq) + 2\ Al\ (s)$

89. These reactions take place in aqueous solutions. Complete and balance each reaction.

a. $FeCl_3\ (aq) + ZnSO_4\ (aq) \rightarrow$
b. $HBr\ (aq) + LiOH\ (aq) \rightarrow$

90. These reactions take place in aqueous solutions. Complete and balance each reaction.

a. $KOH\ (aq) + H_2SO_4\ (aq) \rightarrow$
b. $Ba(OH)_2\ (aq) + CuBr_2\ (aq) \rightarrow$

91. Write a balanced equation to show the precipitation reaction that occurs when aqueous silver chlorate is combined with aqueous magnesium chloride.

92. Write a balanced equation to show the neutralization reaction that occurs between nitric acid and aqueous barium hydroxide.

11.4 Solution Stoichiometry

93. Magnesium nitrate dissociates in water, as shown in this equation. How many moles of nitrate ion are in a 2.47-liter solution where $[Mg(NO_3)_2] = 1.93\ M$?

$$Mg(NO_3)_2\ (s) \rightarrow Mg^{2+}\ (aq) + 2\ NO_3^-\ (aq)$$

94. Potassium phosphate dissociates in water, as shown in this equation. How many moles of potassium ion are present in 25.0 mL of a 1.30-M *aq.* potassium phosphate solution?

$$K_3PO_4\ (s) \rightarrow 3\ K^+\ (aq) + PO_4^{3-}\ (aq)$$

95. Aluminum reacts with nitric acid, as in the equation below. How many moles of aluminum are required to completely react with 100 mL of 12.0-M HNO_3?

$$2\ Al\ (s) + 6\ HNO_3\ (aq) \rightarrow 2\ Al(NO_3)_3\ (aq) + 3\ H_2\ (g)$$

96. Carbonic acid decomposes according to the equation below. How many moles of CO_2 can be produced from 100 mL of a 1.3-M *aq.* H_2CO_3 solution?

$$H_2CO_3\ (aq) \rightarrow H_2O\ (l) + CO_2\ (g)$$

97. Tin metal reacts with hydrobromic acid according to the following equation. How many moles of H_2 gas can be produced if 500 mL of a 1.3-M HBr solution is added to excess tin?

$$Sn\ (s) + 2\ HBr\ (aq) \rightarrow SnBr_2\ (aq) + H_2\ (g)$$

98. Potassium hydroxide and iron(II) chloride combine in the precipitation reaction shown below. How many moles of iron(II) hydroxide can be produced in the reaction of 0.520 L of 3.4-M KOH with excess $FeCl_2$?

$$2\ KOH\ (aq) + FeCl_2\ (aq) \rightarrow 2\ KCl\ (aq) + Fe(OH)_2\ (s)$$

99. Sodium reacts violently with water according to this equation:

$$2\ Na\ (s) + 2\ H_2O\ (l) \rightarrow 2\ NaOH\ (aq) + H_2\ (g)$$

If 0.50 grams of sodium metal is added to 2.0 liters of water, how many moles of NaOH are produced? Assuming the volume of the resulting solution is also 2.0 liters, what is the molarity of NaOH in the resulting solution?

100. Zinc reacts with hydrochloric acid according to this equation:

$$Zn\ (s) + 2\ HCl\ (aq) \rightarrow ZnCl_2\ (aq) + H_2\ (g)$$

If 0.50 grams of zinc metal is added to an excess of aqueous HCl, how many moles of $ZnCl_2$ are produced? If the volume of the resulting solution is 0.750 liters, what is the molarity of $ZnCl_2$ in the resulting solution?

101. Zinc metal reacts with aqueous copper(II) chloride according to this equation:

$$Zn\ (s) + CuCl_2\ (aq) \rightarrow ZnCl_2\ (aq) + Cu\ (s)$$

In this reaction, what mass of copper metal can be produced from the reaction of 500 mL of 1.20-M *aq.* $CuCl_2$ with excess zinc?

102. Magnesium metal reacts with aqueous silver nitrate according to this equation:

$$Mg\ (s) + 2\ AgNO_3\ (aq) \rightarrow Mg(NO_3)_2\ (aq) + 2\ Ag\ (s)$$

In this reaction, what mass of silver metal can be produced from the reaction of 691 mL of 1.30-M *aq.* $AgNO_3$ with excess magnesium?

103. The concentration of bromide ion may be determined by gravimetric analysis, using this reaction:

$$Ag^+\ (aq) + Br^-\ (aq) \rightarrow AgBr\ (s)$$

A 0.500-L aqueous solution containing bromide was reacted with excess Ag^+. If the AgBr precipitate formed in this reaction has a mass of 0.0035 g, how many moles of Br^- were in the solution? What was the molarity of Br^- in the solution?

104. The concentration of mercury(II) ion in water was determined by gravimetric analysis, using this reaction:

$$Hg^{2+}\ (aq) + H_2S\ (aq) \rightarrow HgS\ (s) + 2\ H^+\ (aq)$$

A chemist combined a 400.0-mL sample of aqueous solution with excess H_2S, resulting in the formation of 12.09 mg of mercury sulfide precipitate. How many moles of mercury were in the original sample? What was the molar concentration of Hg^{2+} in the original sample?

105. Concentration of calcium in a sample may be determined by precipitation using the chromate ion, CrO_4^{2-}:

$$Ca^{2+}\ (aq) + CrO_4^{2-}\ (aq) \rightarrow CaCrO_4\ (s)$$

A chemist combined 0.250 L of an unknown calcium solution with an excess of ammonium chromate. This resulted in the precipitation of calcium chromate. The mass of the precipitate was 314.1 mg. What was the molar concentration of Ca^{2+} in the original sample?

106. The concentration of lead(II) ion in water may be determined by gravimetric analysis, using this reaction:

$$Pb^{2+}\ (aq) + SO_4^{2-}\ (aq) \rightarrow PbSO_4\ (s)$$

A chemist treated an unknown lead(II) solution with a volume of 0.150 L with an excess of sodium sulfate solution. The mass of the resulting precipitate was found to be 0.00215 grams. What was the mass of lead(II) in the original sample? What was the molar concentration of the original sample?

Challenge Questions

107. Hydrochloric acid reacts with potassium hydroxide according to this equation:

$$HCl\ (aq) + KOH\ (aq) \rightarrow H_2O\ (l) + K_2SO_4\ (aq)$$

What volume of 3.34-M KOH is required to exactly react with 0.25 liters of 1.35-M HCl?

108. Iron reacts with hydrochloric acid to produce aqueous iron(II) chloride, as shown in this equation:

$$Fe\ (s) + 2\ HCl\ (aq) \rightarrow H_2\ (g) + FeCl_2\ (aq)$$

If 20.2 grams of Fe react with 100.0 mL of 2.1-M *aq.* HCl, how many moles of H_2 will be produced?

109. Mussel shells are composed primarily of calcium carbonate, produced by the reaction of calcium ions and dissolved CO_2 in water. We can describe the overall reaction this way:

$$Ca^{2+}\ (aq) + CO_2\ (aq) + H_2O\ (l) \rightarrow CaCO_3\ (s) + 2H^+\ (aq)$$

In 2017, scientists observed a dramatic increase in the mussel population in Kentucky Lake. Between 2012 and 2018, the levels of calcium ions in Kentucky Lake increased from 16.0 ppm to 40.0 ppm. Scientists suspect that the use of calcium chloride to de-ice roads is a major source for this increase.

a. Is calcium chloride an electrolyte?

b. Create a sketch showing one formula unit of $CaCl_2$ dissolved in water. Your sketch should show 10–15 water molecules and depict the way the polar water molecules interact with the calcium and chloride ions.

c. What is the mass of calcium ions per liter in a 40.0-ppm aqueous solution?

d. Based on the equation above, what is the maximum amount of $CaCO_3$ that could form in a body of water containing 1.0×10^7 L of water with a calcium concentration of 40.0 ppm?

Acids and Bases

Figure 12.1 (a) The leaves of the coca tree are the source of cocaine. *(continued)*

CHAPTER TWELVE

Cocaine: Ruin and Recovery

Cocaine is a naturally occurring stimulant, produced in the leaves of the coca tree that grows in Central and South America (**Figure 12.1**). Just as you or I might drink coffee for a midday boost, the inhabitants of this region chew coca leaves to provide energy for long workdays. The tradition dates back thousands of years.

But in the late 1800s, pharmaceutical companies began exploring the region in search of new medicines. Drawing from native traditions, they extracted cocaine from the coca leaves. They found that treating the compound with hydrochloric acid produced a white crystalline powder that is easy to handle and ship and is stable for long periods of time.

At first the dangerous properties of this substance were not well understood. Cocaine was marketed as a stimulant in the United States and Europe. In Sir Arthur Conan Doyle's famous Sherlock Holmes stories, the brilliant detective used cocaine occasionally. Cocaine was even a chief flavoring for the original Coca-Cola® soft drink. (Although cocaine was removed from the formula in 1904, other extracts from coca leaves are still used in cola flavoring.)

Cocaine abuse was rare until the 1960s and 1970s, when the rock 'n' roll era saw a surge in its use—largely in affluent circles. Then, in the mid-1980s, drug dealers began converting the powder cocaine to a different form, called "crack." This form, which chemists refer to as the *free-base* form, was devastatingly addictive. In the years that followed, untold numbers of lives were destroyed by crack cocaine use.

Dr. Susan Hurley has seen this damage firsthand. In her work as a substance abuse counselor, she helps people overcome their addictions. "Crack is amazing to me because it is so highly addictive," she says. "The first time people use it, it takes over their lives. Crack addicts will use their entire paycheck on the substance instead of paying bills, buying groceries, or tending to their family."

Both powder and crack cocaine produce a euphoric high, but the intensity and duration of the high is very different. The powder produces a milder high that may

b

Figure 12.1 (continued) (b) A Bolivian woman chews on coca leaves. (c) An early advertisement for Coca-Cola® praises its ability to boost energy. The manufacturer removed cocaine from the soft drink in 1904. (d) The "crack" form of cocaine usually exists as small, hard pieces. (e) Dr. Susan Hurley is substance abuse counselor in Georgia. (a) Bram Smits/Shutterstock; (b) Jorge Bernal/AFP/Getty Images; (c) Granger, NYC – All rights reserved; (d) DEA/Science Source; (e) Courtesy of Dr. Susan Hurley

last for several hours. As Dr. Hurley observes, "Powder cocaine users can be quite functional, partying, working, doing surgery, etc." On the other hand, crack users experience an intense high that lasts for only a few minutes. The euphoric high (followed by a miserable low) leads crack users to binge on the drug. Addicts will go for days without eating or sleeping, consuming as much as they can until their money or supply runs out — or until they pass out from hunger and exhaustion.

Treatment strategies for recovering users are also different: Powder cocaine is more common in affluent circles, and because its effects are less debilitating, many addicts enter treatment with a support network of family and friends. But crack is devastating. Many of the crack addicts Dr. Hurley works with have lost jobs, families, and homes. Yet, they are not without hope: Through the vitally important work of counselors like Dr. Hurley, addicts of all types can find new opportunities and discover a life free of addiction.

So, what is the difference between powder and crack cocaine? And why do they have such drastically different properties? To understand the differences between the two drug forms, we must first understand two related types of compounds, called *acids* and *bases*. In this chapter we'll explore the structures and properties of acids and bases. These compounds are all around us, and they are vital to our lives. As we'll see, the chemical differences between acids and bases (like the two forms of cocaine) are slight, but they drastically affect how the compounds behave. Subtle changes can have huge impacts.

Intended Learning Outcomes

After completing this chapter and working the practice problems, you should be able to:

12.1 Introduction to Acids and Bases

- Define acids and bases using both the Arrhenius and the Brønsted-Lowry definitions.
- Describe the process of acid ionization and base dissociation in aqueous solution.

12.2 Acid-Base Equilibrium Reactions

- Explain the difference between strong and weak acids.
- Identify the acid, base, conjugate acid, and conjugate base in a chemical equilibrium.

12.3 Reactions Involving Acids and Bases

- Predict the reactions of metals with acid to form metal ions and hydrogen gas.
- Predict the products from acid-base neutralization reactions.

12.4 Acid and Base Concentration

- Explain how the addition of acid or base to an aqueous solution affects the concentration of H^+ and OH^- in an aqueous solution.
- Relate the values of pH, pOH, $[H^+]$, and $[OH^-]$ in an aqueous solution.

12.5 Measuring Acid and Base Concentration

- Identify common indicators for acid and base solutions.
- Apply data from a titration experiment to find the concentration of an unknown acid or base.

12.6 Buffers and Biological pH

- Describe how buffers are able to stabilize the pH of solutions.

12.1 Introduction to Acids and Bases

Think for a moment about the term *acidic*. You might picture powerfully corrosive materials that can disintegrate clothes, skin, or even metal. On the other hand, people often describe foods like soda, oranges, and barbecue sauce as acidic. What's the difference between the nightmarishly corrosive materials and the harmless flavorings of common foods? As we will see in the coming sections, the behavior of an acid depends on two factors: its strength and its concentration. We will carefully examine both of these.

Bases are essentially the opposite of acids. Like acids, bases can be very strong (like drain cleaner) or quite mild (like baking soda). Many household products have acidic or basic properties (**Figure 12.2**). Bases and acids neutralize each other, negating their powerful corrosive properties and producing water and simple ionic compounds.

In this chapter we'll explore the chemical properties of acids and bases, and describe them in terms of their strength and concentration. We'll look at reactions of acids and bases, and discover how we can use neutralization reactions in laboratory analysis. We'll begin to see that acid-base behavior is all around us: in every breath, in every drink, literally in every fiber of our bodies.

Courtesy of Kevin Revell

Figure 12.2 Many common household items have acidic or basic properties.

The Arrhenius Definition

Let's begin with a classic definition for acids and bases, called the **Arrhenius definition**:

- **Acids** are compounds that produce H^+ ions in water.
- **Bases** are compounds that produce OH^- ions in water.

Recall from earlier chapters that compounds that separate into ions in an aqueous solution are called *electrolytes*. Acids and bases fall into this category. **Table 12.1** presents the names and formulas of some common acids and bases.

TABLE 12.1 Common Acids and Bases

Formula	Name
Acids	
HF	Hydrofluoric acid
HCl	Hydrochloric acid
HBr	Hydrobromic acid
HI	Hydroiodic acid
H_2CO_3	Carbonic acid
HNO_3	Nitric acid
HNO_2	Nitrous acid
H_2SO_4	Sulfuric acid
H_3PO_4	Phosphoric acid
$HC_2H_3O_2$	Acetic acid
Bases	
NaOH	Sodium hydroxide
KOH	Potassium hydroxide
$Mg(OH)_2$	Magnesium hydroxide
$Ca(OH)_2$	Calcium hydroxide

In most acids, a hydrogen atom is bonded to a strongly electronegative element, like oxygen or one of the halogens. In aqueous solution, these compounds *ionize* to produce an H^+ cation and a corresponding anion. For example, hydrochloric acid ionizes to form H^+ and Cl^-:

$$HCl\ (aq) \rightarrow H^+\ (aq) + Cl^-\ (aq)$$

In an aqueous solution, the H^+ ions are stabilized by water molecules. We can represent this in two ways: Sometimes, we show the H^+ as an unattached cation with strong (but transient) forces of attraction to several water molecules (**Figure 12.3**). Other times, we show the H^+ attached to a single water molecule to form H_3O^+, called a **hydronium ion**:

$$HCl\ (aq) + H_2O\ (l) \rightarrow H_3O^+\ (aq) + Cl^-\ (aq)$$

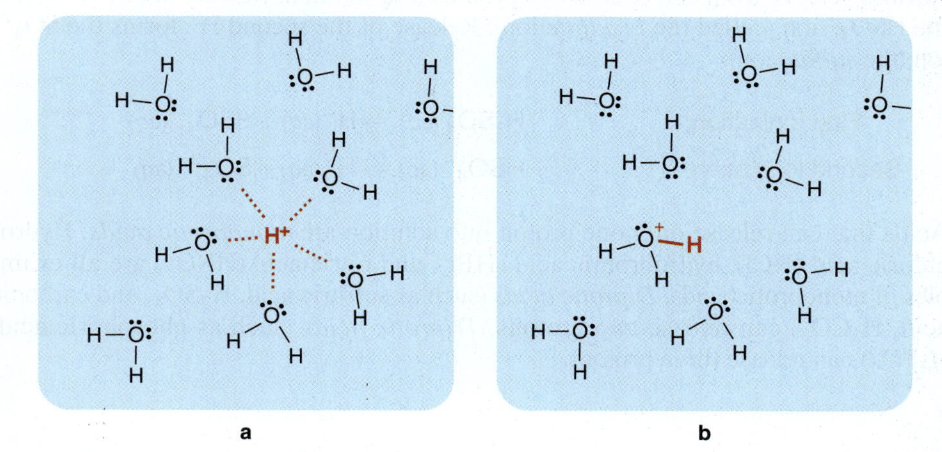

Figure 12.3 When an acid dissolves in water, the H^+ can be represented (a) by itself (H^+) or (b) bonded to a water molecule (H_3O^+). Both symbols are correct and used routinely.

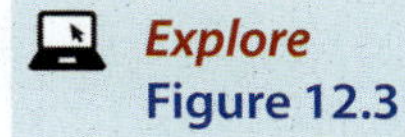

Explore Figure 12.3

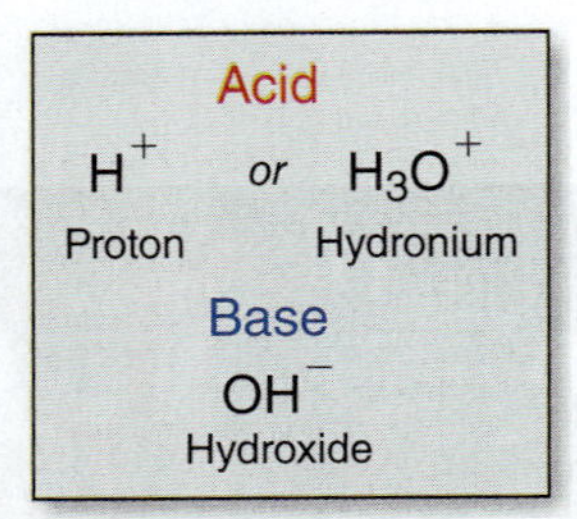

Figure 12.4 Acidic and basic solutions involve these ions.

The two forms of the acid equation mean the same thing: In aqueous solution, the H^+ separates from the chloride ion and is stabilized by the water molecules around it. Because an H^+ cation is composed of only one proton, it is sometimes referred to simply as a *proton*, and the gain or loss of an H^+ is called a *protonation* or a *deprotonation*. For example, we could say that the H_2O molecule is "protonated" to form the hydronium ion (**Figure 12.4**).

An aqueous H^+ ion is also represented as H_3O^+. ■

A second example of an acid is nitric acid, HNO_3. Again, acids release H^+ in aqueous solution. We can show this in two ways:

Water not included: $HNO_3\ (aq) \rightarrow H^+\ (aq) + NO_3^-\ (aq)$

Water included: $HNO_3\ (aq) + H_2O\ (l) \rightarrow H_3O^+\ (aq) + NO_3^-\ (aq)$

When discussing acids and bases, a *proton* refers to an H^+ ion. ■

Bases dissociate in water to form hydroxide (OH^-) ions. Most common bases are metal hydroxides, which dissociate to form the metal cation and the hydroxide anion. For example, potassium hydroxide and magnesium hydroxide dissociate in water, as shown here:

$$KOH\ (s) \rightarrow K^+\ (aq) + OH^-\ (aq)$$

$$Mg(OH)_2\ (s) \rightarrow Mg^{2+}\ (aq) + 2\ OH^-\ (aq)$$

There are many other common bases of this type, such as sodium hydroxide (NaOH), lithium hydroxide (LiOH), and calcium hydroxide [$Ca(OH)_2$].

Check it
Watch Explanation

TRY IT

1. Complete the following equations to show the ions these acids would form in aqueous solution:

   $HBr\ (aq) + H_2O\ (l) \rightarrow$

   $HClO_4\ (aq) + H_2O\ (l) \rightarrow$

2. Write an ionic equation to show the ions that form when calcium hydroxide is dissolved in water.

Coffee and tea are both mildly acidic, due to a class of compounds called *tannins*. These compounds also cause the dark coloring of these beverages.

Polyprotic Acids

Polyprotic acids can release more than one H^+ into aqueous solution. For example, sulfuric acid, H_2SO_4, can release two protons into solution. Release of one H^+ forms the HSO_4^- ion (called the *bisulfate* ion). Release of the second H^+ forms the SO_4^{2-} ion (the *sulfate* ion):

First ionization: $H_2SO_4\ (aq) \rightarrow H^+\ (aq) + HSO_4^-\ (aq)$

Second ionization: $HSO_4^-\ (aq) \rightarrow H^+\ (aq) + SO_4^{2-}\ (aq)$

Acids that can release only one proton into solution are *monoprotic acids*. Hydrochloric acid (HCl), hydrobromic acid (HBr), and nitric acid (HNO_3) are all examples of monoprotic acids. *Diprotic acids* (such as sulfuric acid, H_2SO_4, and carbonic acid, H_2CO_3) can release two protons. *Triprotic acids* (such as phosphoric acid, H_3PO_4) can release three protons.

TRY IT

3. Complete the second and third reactions to show the dissociation of phosphoric acid to lose its first, second, and third protons.

First ionization: $H_3PO_4\ (aq) \rightarrow H^+\ (aq) + H_2PO_4^-\ (aq)$

Second ionization: $H_2PO_4^-\ (aq) \rightarrow$

Third ionization:

Check it
Watch Explanation

The Brønsted-Lowry Definition

While the Arrhenius description of acids and bases is useful, it is limited to reactions that occur in water. But compounds that behave as acids or bases in water undergo very similar reactions in other solvents. For example, consider the two reactions below. What do they have in common?

$$HCl + H_2O \rightarrow H_3O^+ + Cl^-$$
$$HCl + H_3N \rightarrow H_4N^+ + Cl^-$$

In both examples, the HCl molecule transfers an H^+ to the other species to produce a cation (hydronium or ammonium) and a chloride anion. We describe HCl as an acid in the first reaction, so why not in the second? The Arrhenius definition is too narrow. A broader description of acids and bases is the **Brønsted-Lowry definition**. According to this definition,

- **Acids** are compounds that donate H^+ ions.
- **Bases** are compounds that accept H^+ ions.

In the two reactions, HCl is the acid—it donates the H^+ to the base. Water and ammonia (NH_3) are the bases—they receive the H^+ from the acid. These reactions are shown structurally in **Figure 12.5**.

Acids are proton donors; bases are proton acceptors.

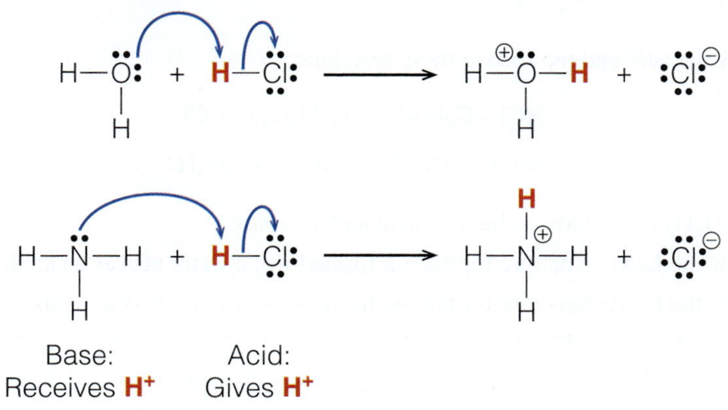

Base: Receives H^+

Acid: Gives H^+

Figure 12.5 HCl reacts similarly with H_2O and with NH_3. The blue arrows show the electron motion: The oxygen and nitrogen atoms donate electrons to form a new covalent bond with H^+. When this happens, the electrons in the H–Cl bond move to the chlorine atom, producing a chloride ion.

Let's look at another example of a Brønsted-Lowry acid-base reaction. In the reaction below, ammonia and water combine in an acid-base reaction. Which compound behaves as the acid? Which one behaves as the base?

$$NH_3\ (aq) + H_2O\ (l) \rightarrow NH_4^+\ (aq) + OH^-\ (aq)$$

In this example, the NH_3 gains an H^+ to form NH_4^+. That is, NH_3 is the base. On the other hand, H_2O loses an H^+ to form OH^-. Water behaves as the acid in this

Charles Brutlag/Shutterstock

Ammonia, NH_3, is a weakly basic substance that is commonly used as a fertilizer.

reaction. Also notice that in this reaction, ammonia combines with water to produce a hydroxide ion. Substances that are basic in the Brønsted-Lowry sense (H^+ acceptors) also behave as bases in the Arrhenius sense (produce hydroxide in aqueous solutions). **Figure 12.6** provides a summary of the acid-base definitions.

Acid
- Produces H^+ or H_3O^+ in water
- Proton (H^+) donor

Base
- Produces OH^- in water
- Proton (H^+) acceptor

Figure 12.6 Acid-base definitions.

Example 12.1 Identifying Acids and Bases

Identify the acid and base in the following reaction:

$$H_2S\ (aq) + H_2O\ (l) \rightarrow HS^-\ (aq) + H_3O^+\ (aq)$$

This reaction does not contain any of the familiar acids and bases in Table 12.1, but we can still identify the acid and base from the changes that take place. Notice that as we go from reactants to products, the H_2O gains an H^+ to become H_3O^+. This means that H_2O is the proton acceptor—the base. Similarly, as we go from reactants to products, H_2S loses a hydrogen ion and a charge of +1 to become HS^-. This means that H_2S is the proton donor—the acid.

Check It
Watch Explanation

TRY IT

4. Identify the acid and the base in these reactions:

$$HCl + C_2H_6O \rightarrow C_2H_6OH^+ + Cl^-$$

$$HNO_2 + CH_5N \rightarrow NO_2^- + CH_6N^+$$

5. Water (H_2O) can behave as both an acid and as a base.
   - **a.** Draw the Lewis structure for the ion formed when water acts as an acid.
   - **b.** Draw the Lewis structure for the ion formed when water acts as a base.

TABLE 12.2 Common Strong Acids

Formula	Name
HCl	Hydrochloric acid
HBr	Hydrobromic acid
HI	Hydroiodic acid
HNO_3	Nitric acid
H_2SO_4	Sulfuric acid
$HClO_4$	Perchloric acid
$HClO_3$	Chloric acid

12.2 Acid-Base Equilibrium Reactions

Strong acids completely ionize in water. The examples of acids given earlier (HCl and HNO_3) are both strong acids. In fact there are only seven common strong acids, as shown in **Table 12.2**.

Most acids only partially ionize in water—these are **weak acids**. For example, hydrofluoric acid (HF) is a weak acid. In solution, only about 2% of HF breaks into ions; the remaining 98% remains intact.

$$HF\ (aq) + H_2O\ (l) \rightleftharpoons H_3O^+\ (aq) + F^-\ (aq)$$

98%   2%

The ionization of weak acids is an example of an **equilibrium reaction**. This type of reaction occurs in both the forward and backward directions. Hydrofluoric acid can ionize to form H_3O^+ and F^-, but the ions can also combine to re-form HF. To indicate this movement, we use an *equilibrium arrow* ($\rightleftharpoons$), which shows the reaction going in both directions. The concept of equilibrium will be covered further in Chapter 13.

Let's consider the forward and backward reactions of this equilibrium in more detail (**Figure 12.7**). In the forward reaction (read from left to right), the HF donates the H^+ to a water molecule. This means that HF is the acid and H_2O is the base. What about the reverse reaction? Going from right to left, the hydronium (H_3O^+) gives the H^+ back to the fluoride ion. In the reverse reaction, H_3O^+ is the acid and F^- is the base. When describing an acid-base equilibrium, we refer to the acid and base on the right-hand side of the chemical equation as the **conjugate acid** and the **conjugate base**.

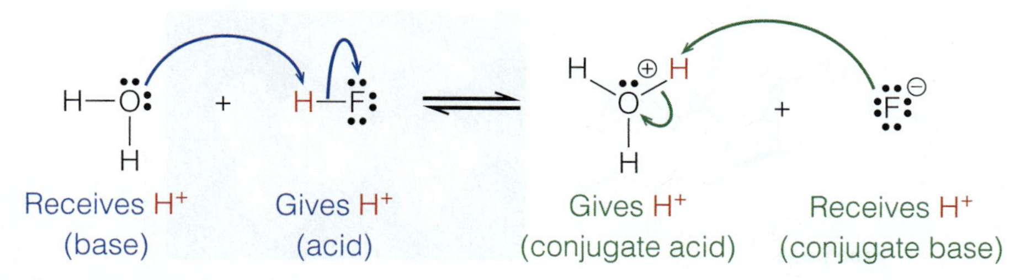

Figure 12.7 In the forward reaction (blue arrows), HF is the acid and H_2O is the base. In the reverse reaction (green arrows), H_3O^+ is the acid and F^- is the base.

Let's look at another example: Trichloroacetic acid ($HC_2Cl_3O_2$) is a weak acid. In water, the ionization of this acid is written as follows:

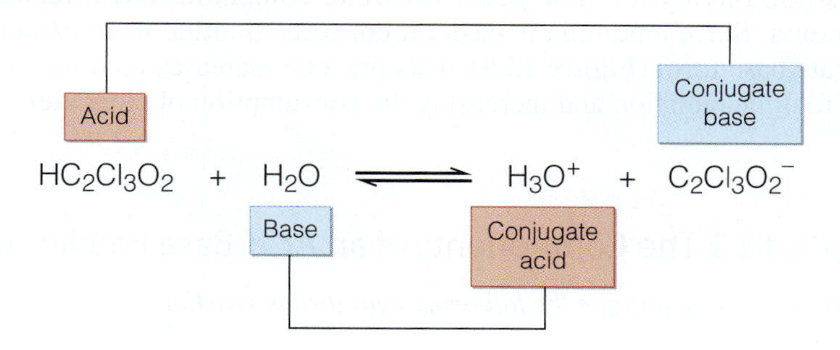

In the forward direction, $HC_2Cl_3O_2$ donates the H^+. It is the acid. In the reverse direction, $C_2Cl_3O_2^-$ accepts the H^+; it is the conjugate base. Similarly, H_2O is the base in the forward direction, but H_3O^+ is the acid in the reverse direction.

Notice that the only difference between the acid and conjugate base is the H^+. *The acid form has the H^+; the base form does not.* Because of this, we refer to the acid and conjugate base as an *acid-base pair*. The base and conjugate acid are also a pair—the only difference between them is the H^+.

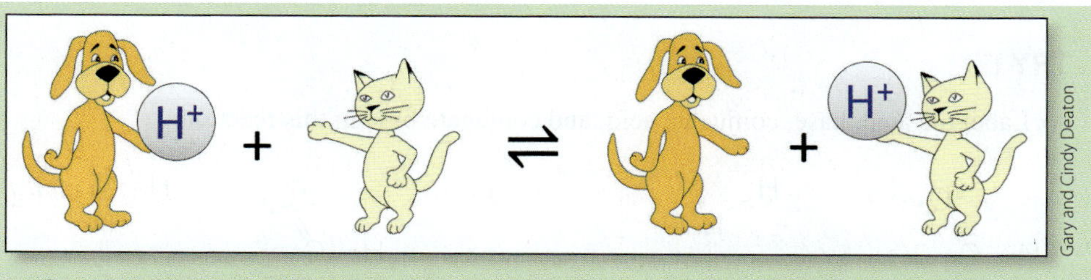

In this acid-base equilibrium, the dog is the acid. Can you identify the conjugate-base dog? Of course. A conjugate base is identical to the acid, but without the H^+.

The free-base form of cocaine has much lower melting and boiling points than the ionic acid form. This is why it can be smoked rather than snorted. When cocaine enters the body through the lungs, it reaches the brain much more quickly, leading to a faster and more intense high.

Many drug molecules (both legal and illegal) can undergo acid-base reactions. For example, drug producers commonly ship cocaine in the acid form, called the *acid salt* or *powder* form (**Figure 12.8**). In the 1980s, drug dealers began treating acid cocaine with base. This acid-base reaction converts cocaine to the conjugate base (also called the *free-base* form), commonly known as *crack*.

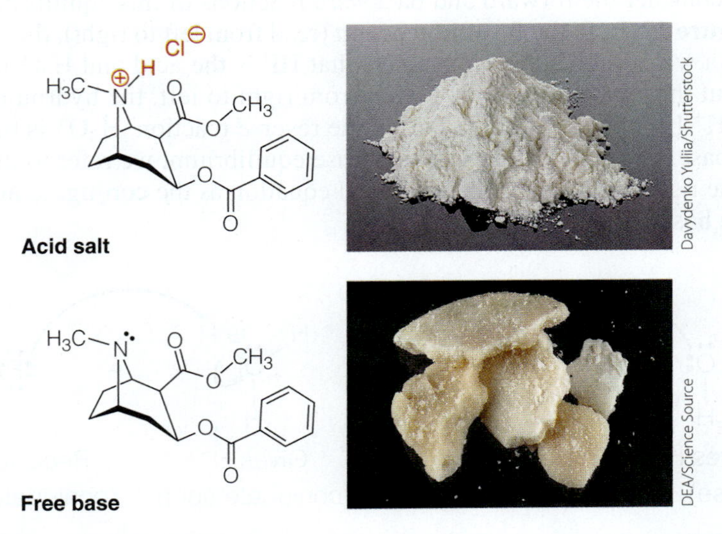

Figure 12.8 Powder and crack cocaine are an acid-conjugate base pair. Notice that the difference between them is the presence of an H^+ (and a counter ion, the Cl^-). The line structures are a simplified way to show carbon atoms. To learn more about line structures, check out Section 15.3.

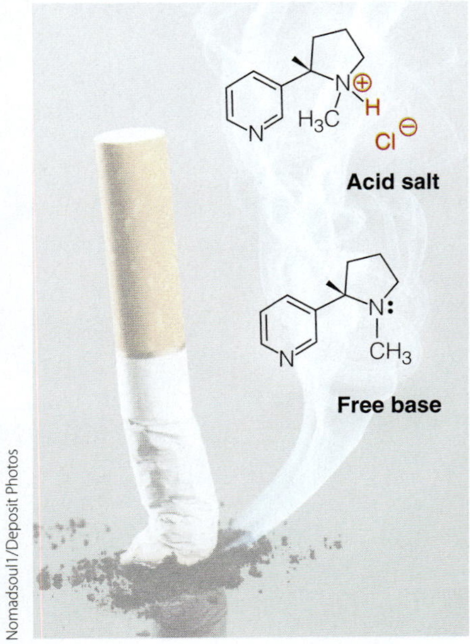

It was not a new trick. For years, cigarette companies have treated tobacco with ammonia. Since ammonia is basic, it converts nicotine from its acid form to its conjugate base form (**Figure 12.9**). This practice increases nicotine intake when smoking, fueling addiction and increasing the consumption of cigarettes.

Acid salt

Free base

Nomadsoul1/Deposit Photos

Figure 12.9 Like cocaine, nicotine exists as an acid salt or a free base.

Example 12.2 The Components of an Acid-Base Equilibrium

Identify the acid-base pairs in the following equilibrium reaction:

$$HBr + CH_4O \rightleftharpoons CH_5O^+ + Br^-$$

The key to identifying acid-base pairs is to realize that they are the same species, with or without an H^+. In the forward reaction, notice that HBr donates an H^+, leaving the Br^- alone. Therefore, HBr is the acid, and Br^- is the conjugate base. By extension, CH_4O is the base. Its conjugate acid is the same species, but with an H^+ attached. Therefore, CH_4O and CH_5O^+ are a base/conjugate acid pair.

Check It
Watch Explanation

TRY IT

6. Label the acid, base, conjugate acid, and conjugate base in this reaction:

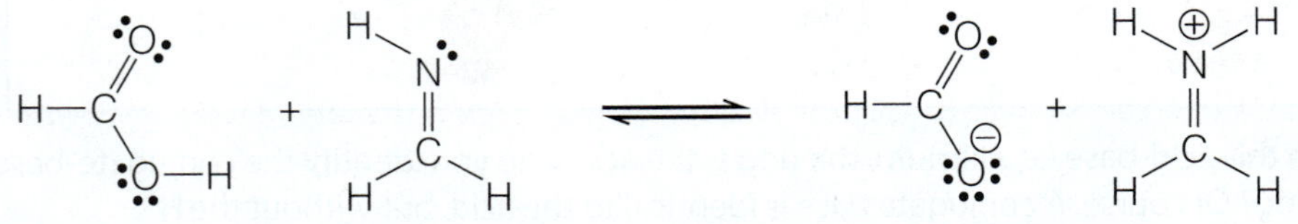

Are Conjugate Bases Basic?

Let's take one more look at the reaction of HF with water:

$$HF\ (aq) + H_2O\ (l) \rightleftharpoons H_3O^+\ (aq) + F^-\ (aq)$$

In this equilibrium, we say that HF is the acid, and F^- is the conjugate base. Does this mean that F^- is basic? Actually, yes. In a solution of fluoride ions, the ions act as a weak base, removing H^+ from water to produce hydroxide:

$$F^-\ (aq) + H_2O\ (l) \rightleftharpoons OH^-\ (aq) + HF\ (aq)$$

Because we wrote the F^- on the left-hand side of this equation, F^- is the base, and HF is the conjugate acid.

Let's look at another example: Hypochlorous acid, HClO, is a weak acid. This means that its conjugate, ClO^-, is basic. We can write two reactions, showing how both the acid and the conjugate base react with water:

Acidic reaction of HClO: $HClO\ (aq) + H_2O\ (l) \rightleftharpoons H_3O^+\ (aq) + ClO^-\ (aq)$

Basic reaction of ClO^-: $ClO^-\ (aq) + H_2O\ (l) \rightleftharpoons OH^-\ (aq) + HClO\ (aq)$

As a general rule, *the conjugates of weak acids are basic, and the conjugates of weak bases are acidic.*

The conjugates of weak acids are basic.

The conjugates of weak bases are acidic.

TRY IT

7. Carbonic acid, H_2CO_3, is weakly acidic. Write two equilibria equations, one showing the acidic reaction of H_2CO_3 in water and the other showing the basic reaction of HCO_3^- in water.

8. Hypobromite, BrO^-, is a weak base. Write two equilibria equations, one showing the reaction of BrO^- in water and the other showing the reaction of its conjugate acid in water.

Check It
Watch Explanation

12.3 Reactions Involving Acids and Bases

Neutralization Reactions

Acid-base **neutralization reactions** were introduced in Chapter 6. In these reactions, an acid and a hydroxide base combine to produce water and an ionic compound, commonly referred to as a *salt*:

Neutralization reactions were introduced in Section 6.5. See that section for other examples.

$$\text{Acid} + \text{Base} \rightarrow \text{Water} + \text{Salt}$$

For example, hydrofluoric acid reacts with sodium hydroxide:

$$HF\ (aq) + NaOH\ (aq) \rightarrow H_2O\ (l) + NaF\ (aq)$$

The H^+ from the acid combines with the OH^- from the base, forming water. The remaining ions end up together as NaF. Similarly, perchloric acid reacts with potassium hydroxide:

$$HClO_4\ (aq) + KOH\ (aq) \rightarrow H_2O\ (l) + KClO_4\ (aq)$$

Again, the H^+ from the acid reacts with the OH^- from the hydroxide base to form water. The remaining ions combine to produce a salt.

What drives a neutralization reaction to take place? To understand this question, it's helpful to rewrite the reaction as a complete ionic equation:

$$H^+\ (aq) + ClO_4^-\ (aq) + K^+\ (aq) + OH^-\ (aq) \rightarrow H_2O\ (l) + K^+\ (aq) + ClO_4^-\ (aq)$$

The formation of water is the driving force for neutralization reactions.

Notice that the K^+ and the ClO_4^- ions are unchanged in this reaction. They are spectator ions. The net ionic reaction for neutralization is the formation of water from H^+ and OH^-. The formation of water drives neutralization reactions.

$$H^+ (aq) + OH^- (aq) \rightarrow H_2O (l)$$

Example 12.3 Predicting the Products from Acid-Base Neutralization Reactions

Write a balanced equation to show the reaction of aq. nitric acid with aq. calcium hydroxide.

In this reaction, the H^+ from the acid and the OH^- from the base combine to produce water. The remaining ions, Ca^{2+} and NO_3^-, combine to produce an ionic compound, $Ca(NO_3)_2$.

$$HNO_3 (aq) + Ca(OH)_2 (aq) \rightarrow H_2O (l) + Ca(NO_3)_2 (aq) \quad \text{not balanced}$$

A simple way to balance neutralization reactions is to balance the number of H^+ and OH^- ions that combine. Because $Ca(OH)_2$ contains two hydroxide ions, two hydrogen ions (and therefore two HNO_3 molecules) are required to react with it. This produces two water molecules and leads us to the balanced form.

$$2\ HNO_3 (aq) + Ca(OH)_2 (aq) \rightarrow 2\ H_2O (l) + Ca(NO_3)_2 (aq) \quad \text{balanced}$$

Check It

Watch Explanation

TRY IT

9. Predict the products for these neutralization reactions. Make sure the equations are balanced.

$$HBr (aq) + NaOH (aq) \rightarrow$$

$$HNO_2 (aq) + Al(OH)_3 (aq) \rightarrow$$

$$H_2SO_4 (aq) + Mg(OH)_2 (aq) \rightarrow$$

10. Write a complete ionic equation and a net ionic equation for this neutralization reaction:

$$Ba(OH)_2 (aq) + 2\ HNO_3 (aq) \rightarrow Ba(NO_3)_2 (aq) + 2\ H_2O (l)$$

Reactions of Acids with Metal

Most metals react with acid to produce metal cations and hydrogen gas (**Figure 12.10**). For example, iron metal reacts with hydrobromic acid to produce iron(II) bromide and hydrogen gas. We can write this as either a molecular equation or as a net ionic equation:

Molecular equation: $Fe (s) + 2\ HBr (aq) \rightarrow FeBr_2 (aq) + H_2 (g)$

Net ionic equation: $Fe (s) + 2\ H^+ (aq) \rightarrow Fe^{2+} (aq) + H_2 (g)$

The reactions of metals with acid are single-displacement reactions. The reactants begin with one element (metal) and produce another element (hydrogen).

Similarly, aluminum reacts with nitric acid to produce aluminum nitrate and hydrogen gas:

Molecular equation: $2\ Al (s) + 6\ HNO_3 (aq) \rightarrow 2\ Al(NO_3)_3 (aq) + 3\ H_2 (g)$

Net ionic equation: $2\ Al (s) + 6\ H^+ (aq) \rightarrow 2\ Al^{3+} (aq) + 3\ H_2 (g)$

Precious metals (silver, gold, platinum) do not react with acid. In fact, the reason these metals are "precious" is that they are rare, beautiful, and durable. They are not easily destroyed or chemically changed the way more common metals are.

De Agostini/M. Carrieri/Universal Images Group/Newscom

The gold in this Incan figurine has endured for centuries. Interestingly, the lump on his left cheek is thought to represent a wad of coca leaves.

$$Mg\ (s) + 2\ HCl\ (aq) \rightarrow MgCl_2\ (aq) + H_2\ (g)$$

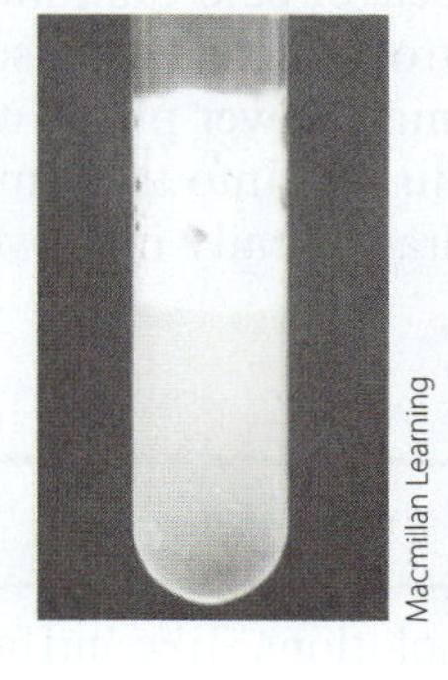

Macmillan Learning

$$Zn\ (s) + 2\ HCl\ (aq) \rightarrow ZnCl_2\ (aq) + H_2\ (g)$$

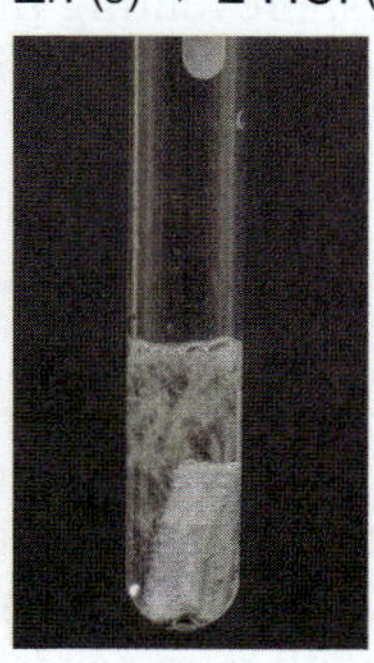

Macmillan Learning

$$Au\ (s) + 2\ HCl\ (s) \rightarrow \text{No reaction}$$

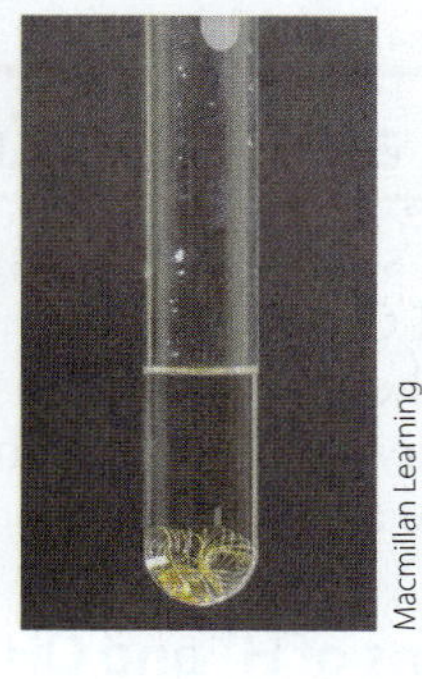

Macmillan Learning

Figure 12.10 Most metals react with acid to produce metal cations and hydrogen gas. Precious metals like gold, silver, and platinum do not.

Explore Figure 12.10

TRY IT

11. Potassium, calcium, and aluminum all react with acid. Complete and balance these reactions to show the products that form:

$K\ (s) + HBr\ (aq) \rightarrow$

$Ca\ (s) + H_2SO_4\ (aq) \rightarrow$

$Al\ (s) + HNO_3\ (aq) \rightarrow$

Check It

Watch Explanation

Formation of Acids from Nonmetal Oxides

As their name implies, **nonmetal oxides** are compounds or ions that contain a nonmetal covalently bonded to one or more oxygen atoms. Many nonmetal oxides react with water to form acids. For example, carbon dioxide reversibly reacts with water to form carbonic acid:

$$CO_2\ (g) + H_2O\ (l) \rightleftharpoons H_2CO_3\ (aq)$$

In carbonated beverages, like beer and soda, the dissolved CO_2 forms carbonic acid through this reaction (**Figure 12.11**). As we'll see in Section 12.6, this reaction also occurs as carbon dioxide is dissolved in the bloodstream—a process that helps regulate the acid-base balance in the blood.

Jesus Ayala/Getty Images

Figure 12.11 Dissolved CO_2 in soda forms carbonic acid, giving soda its pungent quality.

Nonmetal oxides react with water to form acids.

Sulfur oxides and nitrogen oxides also react with water to form acidic compounds. For example, sulfur trioxide reacts to form sulfuric acid, while dinitrogen pentoxide reacts to form nitric acid:

$$SO_3\ (g) + H_2O\ (l) \rightarrow H_2SO_4\ (aq)$$

$$N_2O_5\ (g) + H_2O\ (l) \rightarrow 2\ HNO_3\ (aq)$$

Fletcher & Baylis/Science Source

Figure 12.12 The erosion of this statue was caused by acid rain.

These reactions are important because sulfur and nitrogen are common in fossil fuels, such as coal and oil. Burning these "dirty fuels" can produce sulfur oxides (sometimes called SO_x emissions) and nitrogen oxides (called NO_x emissions). These oxides combine with moisture in the air, leading to the effect known as **acid rain**. In heavily industrialized areas, acid rain accelerates erosion and is harmful to plant and animal life (**Figure 12.12**). To minimize the effects of acid rain, most first-world nations have developed strategies for removing nitrogen oxides and sulfur oxides from fossil fuel emissions. For example, coal-burning power plants use "scrubbers" to trap SO_x and NO_x gases, preventing their emission into the atmosphere. While these technologies are not perfect, they have dramatically improved air quality over the past half-century.

12.4 Acid and Base Concentration

Strongly acidic solutions can dissolve lead. Weakly acidic solutions, like buffalo wing sauce and Coca-Cola, taste great during football season. What is the difference between the two? How can we measure the strength of acids and bases? To answer these questions, we must begin with pure water and the two species, H^+ and OH^-, that define acids and bases.

Concentrations of H^+ and OH^- in Aqueous Solutions

Pure water always contains a tiny amount of both H^+ and OH^-. In its liquid form, water molecules constantly collide with each other. In a small fraction of these collisions, an H^+ moves from one water molecule to another:

$$H_2O\ (l) + H_2O\ (l) \rightleftharpoons H_3O^+\ (aq) + OH^-\ (aq)$$

Cheng Wei/Shutterstock

In wet cement, calcium oxide reacts with water to produce calcium hydroxide, $Ca(OH)_2$. Wet cement is basic.

This process is called the **self-ionization** reaction of water. For simplicity, we sometimes show it as a single water molecule breaking into two ions:

$$H_2O\ (l) \rightleftharpoons H^+\ (aq) + OH^-\ (aq)$$

In pure water, one in every 560,000,000 water molecules is ionized. This means that the concentrations of H^+ and OH^- in pure water are both equal to 1.0×10^{-7} M. Mathematically, we say that in pure water,

$$[H^+][OH^-] = 1.0 \times 10^{-14}$$

Recall that $[H^+]$ means "concentration of H^+."

What happens if we add acid to pure water? Suddenly, the concentration of H^+ jumps up. This excess H^+ reacts with the small amount of OH^- that is present, and $[OH^-]$ decreases. On the other hand, if we add excess base, the opposite occurs: The $[OH^-]$ jumps up, and $[H^+]$ decreases. The concentrations of OH^- and H^+ are like a seesaw: If one goes up, the other goes down (**Figure 12.13**).

Interestingly, even if $[H^+]$ and $[OH^-]$ change, *the product of the two concentrations remains constant*. For liquid water near 25 °C, $[H^+]$ times $[OH^-]$ always equals 1.0×10^{-14}.

$$[H^+][OH^-] = 1.0 \times 10^{-14}$$

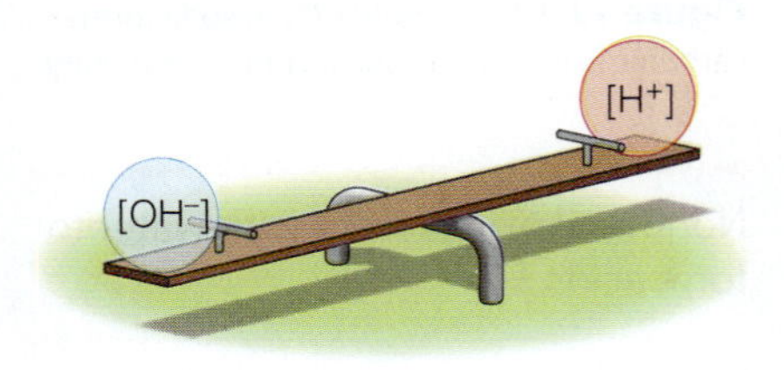

Figure 12.13 The $[H^+]$ and $[OH^-]$ are like two sides of a seesaw: If one goes up, the other goes down by the same factor.

Math Review: Exponential and Scientific Notation

The values of $[H^+]$ and $[OH^-]$ are usually expressed in scientific or in *exponential notation*. We're going to use these notations quite a bit in the coming sections, so a quick review may be helpful.

Scientific notation and exponential notation are similar in that they both involve 10 raised to some power. The difference is that scientific notation uses a *coefficient* times a multiplier (10 raised to an integer power), whereas exponential notation raises 10 to some non-integer power and does not use a coefficient (**Figure 12.14**). The value of the exponent in exponential notation is always within one unit of properly written scientific notation.

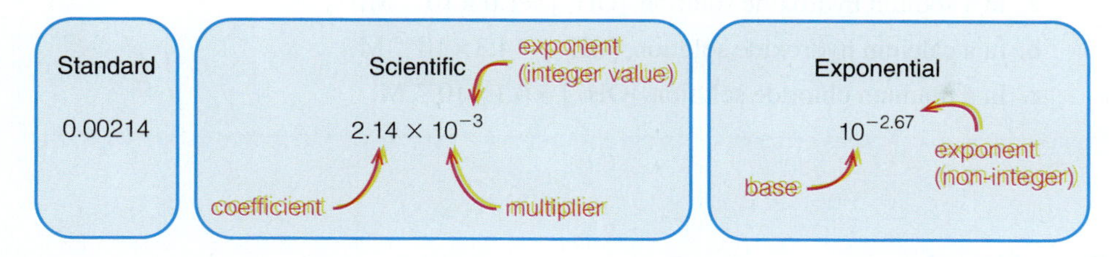

Figure 12.14 Scientific notation uses a coefficient, a multiplier, and an integer exponent. In exponential notation, there is no coefficient, and the exponent may be a non-integer. The three forms shown here represent the same value.

Remember that when multiplying exponents, we always add the exponents together. Mathematically, this is represented by the following generic equation:

$$x^m x^n = x^{(m+n)}$$

For example, we could find the product of 5^2 and 5^7:

$$(5^2)(5^7) = 5^9$$

Because $[H^+][OH^-] = 10^{-14}$, this means that if we express the concentration as exponents of 10, the two exponents will always add up to equal –14.

In water, $[H^+][OH^-] = 10^{-14}$.

Connecting $[OH^-]$ and $[H^+]$

Table 12.3 shows the H^+ and OH^- concentrations in solutions ranging from strongly acidic to strongly basic. Notice that the product of $[H^+]$ and $[OH^-]$ always equals 1.0×10^{-14}. At the neutral point, the concentration of both species is 1.0×10^{-7} M. The $[H^+]$ is greater than this in acidic solutions, but less in basic solutions.

TABLE 12.3 $[H^+]$ and $[OH^-]$ in Solutions Ranging from Strongly Acidic to Strongly Basic

	$[H^+]$ (mol/L)	$[OH^-]$ (mol/L)
Strongly acidic	1×10^{0}	1×10^{-14}
	1×10^{-1}	1×10^{-13}
	1×10^{-3}	1×10^{-11}
Weakly acidic	1×10^{-5}	1×10^{-9}
Neutral	1×10^{-7}	1×10^{-7}
Weakly basic	1×10^{-9}	1×10^{-5}
	1×10^{-11}	1×10^{-3}
	1×10^{-13}	1×10^{-1}
Strongly basic	1×10^{-14}	1×10^{0}

Example 12.4 Relating $[H^+]$ and $[OH^-]$

The concentration of H^+ in an aqueous solution is 1.0×10^{-6} M. What is the concentration of hydroxide? Is this solution acidic or basic?

We begin with the relationship between $[H^+]$ and $[OH^-]$ and then substitute in the value of $[H^+]$:

$$[H^+][OH^-] = 1.0 \times 10^{-14}$$

$$(1.0 \times 10^{-6})[OH^-] = 1.0 \times 10^{-14}$$

Rearranging this equation to solve, we find that $[OH^-] = 1.0 \times 10^{-8}$ M:

$$[OH^-] = \frac{1.0 \times 10^{-14}}{1.0 \times 10^{-6}} = 1.0 \times 10^{-8} \text{ M}$$

In this example, $[H^+] = 1.0 \times 10^{-6}$ M, which is a larger amount than in neutral water (remember that in neutral water, $[H^+] = [OH^-] = 1.0 \times 10^{-7}$ M). Therefore, this solution is acidic.

Check It
Watch Explanation

TRY IT

12. For each of these solutions, find the H^+ concentration. Label the solution as acidic, basic, or neutral.

a. In a sodium hydroxide solution, $[OH^-] = 1.0 \times 10^{-4}$ M.

b. In a calcium hydroxide solution, $[OH^-] = 4.3 \times 10^{-7}$ M.

c. In a titanium chloride solution, $[OH^-] = 6.1 \times 10^{-9}$ M.

The pH Scale

As we saw in Table 12.3, the acidity or basicity of a solution depends on the concentration of H^+ or OH^-. These concentrations vary widely but usually range from about 1.0 (that is, 1.0×10^0) all the way down to 0.00000000000001 (1×10^{-14}) molar. To describe these vastly different concentrations, chemists use the **pH scale**. The pH of a solution is defined as the negative log of the H^+ concentration:

$$pH = -\log[H^+]$$

The log of a number is defined as the exponent, when a number is expressed in exponential notation with a base of 10 (**Table 12.4**). For example, the log of 10^{18} is 18. The log of $10^{-2.3}$ is –2.3. What is the log of 100? Because $100 = 10^2$, the log of 100 = 2.

TABLE 12.4 Logarithmic Notations

Value	Exponential Notation	Log
1,000.	10^3	3
100.	10^2	2
10.	10^1	1
1.	10^0	0
0.1	10^{-1}	−1
0.01	10^{-2}	−2
0.001	10^{-3}	−3

The pH scale reverses the sign of the log of H^+ concentrations. This is because $[H^+]$ is generally small, and so the exponent is negative. Taking the negative log allows us to work with positive numbers:

- If $[H^+] = 10^{-7}$ M, the pH is 7.
- If $[H^+] = 10^{-13}$ M, the pH is 13.

For a neutral aqueous solution, $[H^+] = [OH^-] = 10^{-7}$ M. Based on this, *the neutral point on the pH scale is 7*. Acidic solutions have pH values lower than 7, and basic solutions have pH values higher than 7 (**Table 12.5**).

A pH of 7 is neutral. Lower pH values are acidic; higher pH values are basic.

It is also possible to measure the pOH, defined as the negative log of the hydroxide ion concentration. Because the product of $[H^+]$ and $[OH^-] = 10^{-14}$, we can take the negative log of both sides of the equation, which gives us another helpful relationship:

$$pH + pOH = 14$$

TABLE 12.5 The pH and pOH Values Always Add Up to 14

	$[H^+]$ (mol/L)	pH	$[OH^-]$ (mol/L)	pOH
Strongly acidic	1×10^0	0	1×10^{-14}	14
	1×10^{-1}	1	1×10^{-13}	13
	1×10^{-3}	3	1×10^{-11}	11
Weakly acidic	1×10^{-5}	5	1×10^{-9}	9
Neutral	1×10^{-7}	7	1×10^{-7}	7
Weakly basic	1×10^{-9}	9	1×10^{-5}	5
	1×10^{-11}	11	1×10^{-3}	3
	1×10^{-13}	13	1×10^{-1}	1
Strongly basic	1×10^{-14}	14	1×10^0	0

However, pOH is seldom used. Because we can describe both acidic and basic solutions using pH, we do not need to use pOH in most applications. Before moving on, I encourage you to look at the following examples of conversions between pH, pOH, and the ion concentrations in aqueous solutions.

Example 12.5 Finding pH from $[H^+]$

A chemist measures the acidity of a solution and finds that $[H^+] = 0.01$ M. What is the pH of this solution? Is the solution acidic or basic?

To solve this problem, we first convert the concentration to scientific notation and then to exponential notation:

$$[H^+] = \underbrace{0.01\ \text{M}}_{\text{standard}} = \underbrace{1.0 \times 10^{-2}\ \text{M}}_{\text{scientific}} = \underbrace{10^{-2}\ \text{M}}_{\text{exponential}}$$

Now we can convert from exponential notation to pH:

$$\text{pH} = -\log[H^+] = -\log(10^{-2}) = 2$$

Because the pH is less than 7, the solution is acidic.

Example 12.6 Finding pH from $[H^+]$

A solution of hydrochloric acid in water is found to have $[H^+] = 0.0030$ M. What is the pH of this solution?

Estimate: Before using a calculator, let's use the definition of the log to find the approximate value: We know that 0.0030 M is slightly greater than 0.001, or 10^{-3}. It's also less than 0.01, or 10^{-2}. In light of this, the exponent will have a value between −2 and −3. Therefore, the pH will be somewhere between 2 and 3 (**Figure 12.15**).

Being able to estimate the log value is very helpful: Quickly approximating the concentration allows you to rule out incorrect answers, and it is a fast way to check that you've done the problem correctly. It's also fun to do at parties, if things get slow.

Calculator: To calculate the pH, we first find the log[0.0030] and then take the negative of that value:

$$\text{pH} = -\log[H^+] = -\log(0.0030) = 2.52$$

One of the most common problems students encounter with pH problems is entering the numbers into the calculator correctly. Different calculators require slightly different ways to enter data for these problems; I encourage you to try each of the sample problems in this section on your specific calculator to make sure you get answers that agree with those listed here.

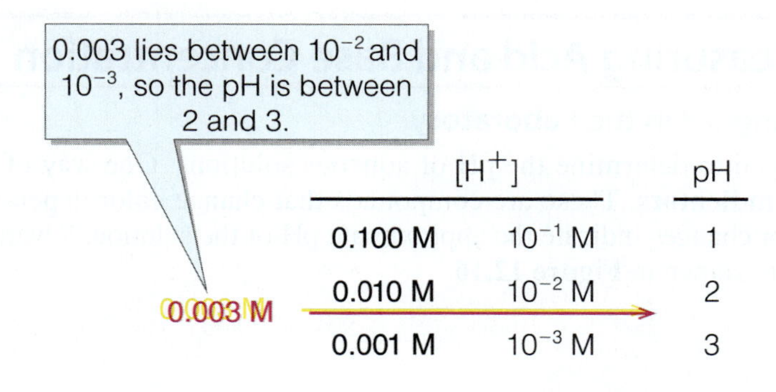

Figure 12.15 The pH tells how many spaces lie between the decimal point and the first nonzero digit. With practice, you can approximate pH values without a calculator.

Check It
Watch Explanation

TRY IT

13. If the $[H^+]$ of a solution is 10^{-4} M, what is the pH? Is the solution acidic or basic?

14. The pH of a solution is 9.3. Is the solution acidic or basic? What is the pOH of this solution?

Example 12.7 Finding $[H^+]$ and $[OH^-]$ from pH

Carbonated soft drinks are fairly acidic and typically have a pH of about 3. What are $[H^+]$ and $[OH^-]$ at this pH?

Because the pH = 3, this means that $[H^+] = 10^{-3}$ M, or 0.001 M.

Next we find $[OH^-]$: We know that $[H^+][OH^-] = 10^{-14}$, so we can say that

$$[OH^-] = \frac{10^{-14}}{[H^+]} = \frac{10^{-14}}{10^{-3}} = 10^{-11} \text{ M}$$

We could also solve this problem using the pOH. Since the pH = 3, the pOH must equal 11. Therefore, $[OH^-] = 10^{-11}$ M.

Example 12.8 Finding $[H^+]$ from pH

What is the $[H^+]$ in an aqueous solution with a pH of 9.43?

Again, using the definition of pH, $[H^+] = 10^{-9.43}$ M. We typically enter this in the calculator as 10^−9.43, or as the inverse log of −9.43. This calculation should give an answer of

$$10^{-9.43} = 3.7 \times 10^{-10} \text{ M}$$

In answering this question, notice that we've gone from exponential notation ($10^{-9.43}$) to scientific notation (3.7×10^{10}).

TRY IT

15. If the hydronium concentration of a solution is 0.00001 M, what is the pH? What is the pOH?

16. A solution of baking soda in water has a pH of 8.3. What is the $[H^+]$ in this solution?

17. A solution has an H^+ concentration of 0.041 M. Is it strongly acidic or strongly basic? Without using a calculator, estimate the pH of this solution. Then check your answer on a calculator.

12.5 Measuring Acid and Base Concentration

Determining pH in the Laboratory

We often need to determine the pH of aqueous solutions. One way of doing this is with **pH indicators**. These are compounds that change color depending on the pH; the color changes indicate the approximate pH of the solution. Several common indicators are shown in **Figure 12.16**.

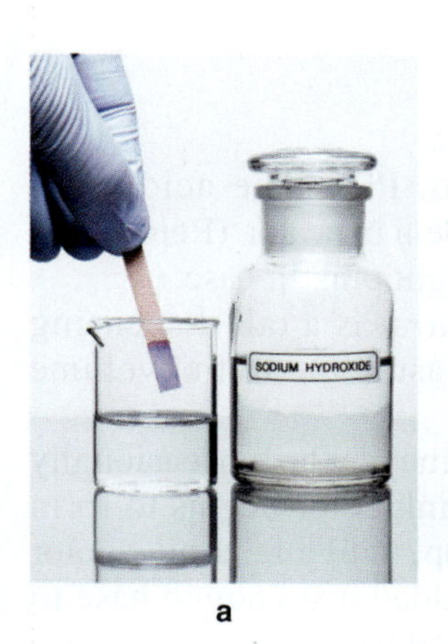

a

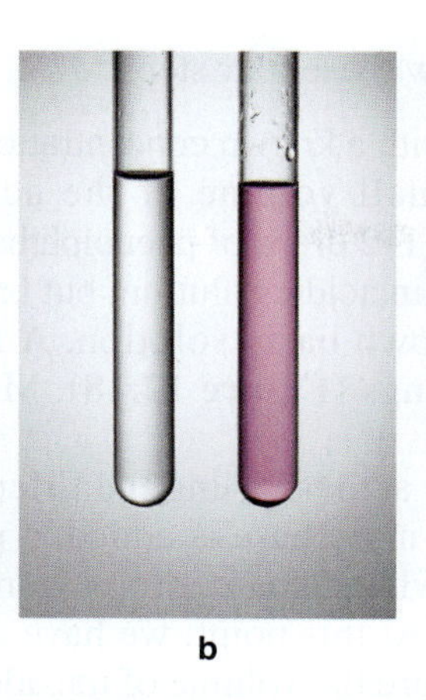
b

c

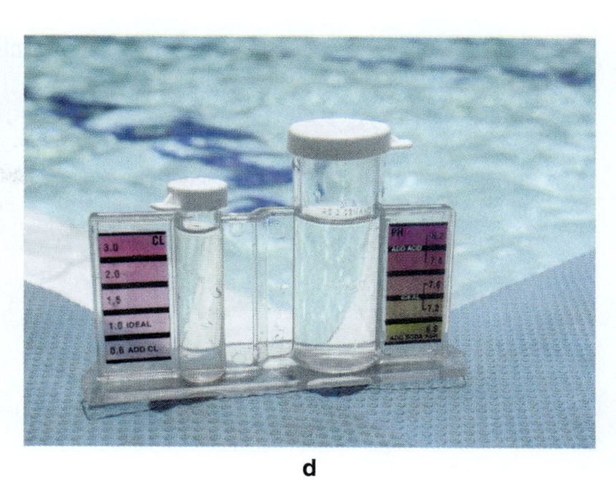
d

Figure 12.16 pH indicators change colors depending on the pH of the solution. (a) Litmus turns blue in base. (b) Phenolphthalein is colorless in acid, but pink in base. (c) pH paper contains a blend of indicators, giving a spectrum of colors that chemists use to approximately measure pH. (d) Pool test kits also use pH indicators. (a) Turtle Rock Scientific/Science Source; (b) Turtle Rock Scientific/Science Source; (c) ©1998 Larry Stepanowicz/FundamentalPhotographs, NYC; (d) Luisecheverriurrea/depositphotos.com

Explore Figure 12.16

One of the oldest and most common indicators is **litmus**. Litmus turns blue in the presence of base, and red in the presence of acid. Chemists sometimes test for acids or bases using *litmus paper*, small strips of paper that are pretreated with litmus. The paper comes in red or blue, and its color changes or stays the same depending on the pH.

A slightly more advanced form of litmus paper is **pH paper**. This paper contains a blend of indicators and produces a series of colors ranging from deep red to deep blue. Rather than simply indicate whether the solution is acidic or basic, pH paper allows us to approximate the pH to the nearest integer.

A third common indicator is a compound called **phenolphthalein**. In acidic solution, phenolphthalein is colorless. In basic solution, it is bright pink.

For more precise pH measurements, chemists use an analytical instrument called a *pH meter* (**Figure 12.17**).

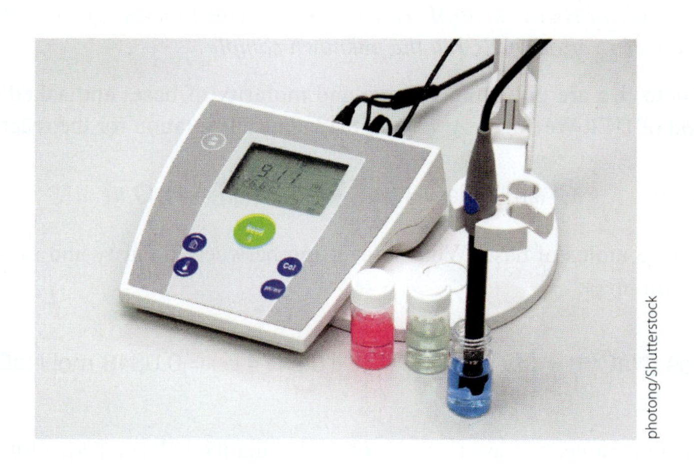
photong/Shutterstock

Figure 12.17 A pH meter measures the pH of a solution more precisely than simple chemical indicators.

Acid-Base Titrations

Titration is an analytical technique used to precisely determine the concentration of an acid or base by measuring the volume required for a neutralization to occur. For example, to determine the concentration of a solution of hydrochloric acid, we could neutralize it with sodium hydroxide:

$$HCl\ (aq) + NaOH\ (aq) \rightarrow NaCl\ (aq) + H_2O\ (l)$$

Titrations are used to determine the molarity of an unknown acid or base.

To perform the titration, we follow these four steps:

1. Prepare a basic solution with a known concentration.
2. Precisely measure a small volume of the acid. Place the acid in an Erlenmeyer flask, and add 1–2 drops of phenolphthalein indicator. (Remember, this indicator is colorless in acidic solution, but bright pink in base.)
3. Fill a buret with the known basic solution. A *buret* is a tall dispensing tube with volume markings (**Figure 12.18**). Measure the initial volume in the buret.
4. Begin adding base to the acid, swirling the Erlenmeyer flask occasionally to mix the solutions. As more base is added, a pink color begins to form and then disappears. Slowly add base, drop by drop, until a faint pink color remains (**Figure 12.19**). At this point, we have added just enough base to neutralize the acid. Measure the volume of the added base.

Loekiepix/Shutterstock

Figure 12.18 Titrations use a long tube with volume markings called a buret.

Once the titration is complete, we can calculate the concentration of the unknown solution. We begin by using the volume and molarity of the base to find the moles of base. Using the balanced equation, we relate the moles of base to the moles of acid. From the moles and volume of the acid, we can determine its molarity:

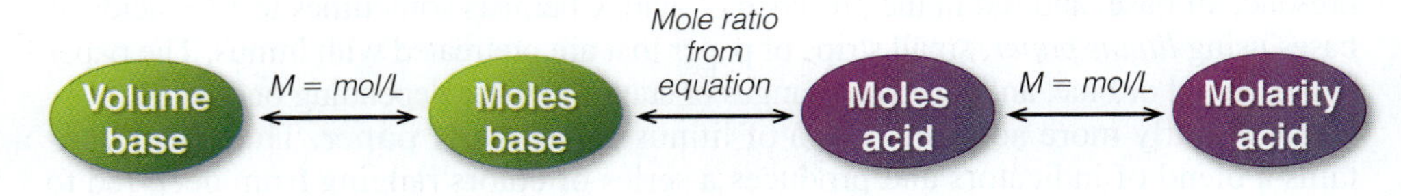

The examples that follow illustrate this process.

Example 12.9 Titration of an Unknown Acid

A 25.0-mL sample of HCl is neutralized by reaction with 17.4 mL of 2.00-M NaOH. What is the concentration of HCl in the unknown sample?

In this problem, we are given the volume and molarity of base, and asked to find the concentration of HCl. We begin by writing the balanced equation for the reaction:

$$HCl\ (aq) + NaOH\ (aq) \rightarrow NaCl\ (aq) + H_2O\ (l)$$

We first find the moles of base. When converting between molarity and moles, be sure the volume is in liters:

$$\text{Moles (NaOH)} = MV = \left(2.00\,\frac{\text{mol}}{\cancel{\text{L}}}\right)(0.0174\,\cancel{\text{L}}) = 0.0348\ \text{mol NaOH}$$

We then relate the moles of base to moles of acid using the balanced equation:

$$0.0348\ \cancel{\text{mol NaOH}} \times \frac{1\ \text{mol HCl}}{1\ \cancel{\text{mol NaOH}}} = 0.0348\ \text{mol HCl}$$

Finally, we divide the moles of acid by the volume of the acid solution (in liters) to find its molarity:

$$\frac{0.0348\ \text{mol HCl}}{0.0250\ \text{L}} = 1.39\ \text{M HCl}$$

Courtesy of Kevin Revell

Figure 12.19 An acid-base titration. Base is slowly added to the solution until it turns pink.

Explore Figure 12.19

Example 12.10 Titration of an Unknown Base

A 30.0-mL sample of aq. $Ba(OH)_2$ is neutralized with an HCl solution having a concentration of 0.400 M. If 22.6 mL of aq. HCl are required to complete the neutralization, what is the concentration of the $Ba(OH)_2$ solution?

Again, we begin with a balanced equation for the neutralization reaction:

$$2\ HCl\ (aq) + Ba(OH)_2\ (aq) \rightarrow BaCl_2\ (aq) + 2\ H_2O\ (l)$$

In this example, we know the volume and molarity of the acid, and we need to find the concentration of the base. Our strategy is the same as in the previous example, except that we start with the acid and find the base:

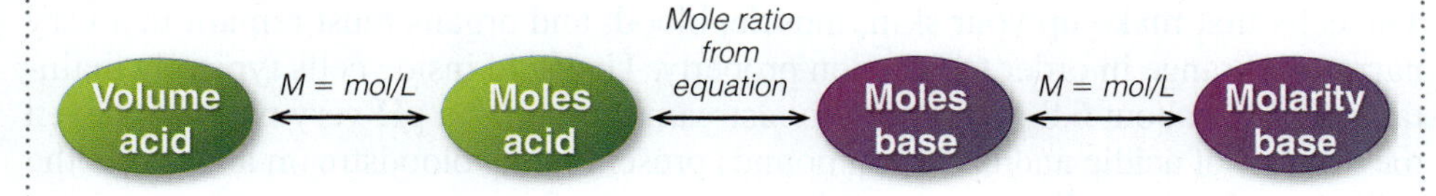

$$\text{Moles (HCl)} = MV = \left(0.400\frac{\text{mol}}{\cancel{\text{L}}}\right)(0.0226\ \cancel{\text{L}}) = 0.00904\ \text{mol HCl}$$

$$0.00904\ \cancel{\text{mol HCl}} \times \frac{1\ \text{mol Ba(OH)}_2}{2\ \cancel{\text{mol HCl}}} = 0.00452\ \text{mol Ba(OH)}_2$$

$$\frac{0.00452\ \text{mol Ba(OH)}_2}{0.0300\ \text{L}} = 0.151\ \text{M Ba(OH)}_2$$

TRY IT

18. A solution containing nitric acid, HNO_3, is analyzed by titration. A volume of 19.5 mL of 1.50-M NaOH solution is required to neutralize 50.0 mL of the acid. What is the original concentration of the nitric acid solution?

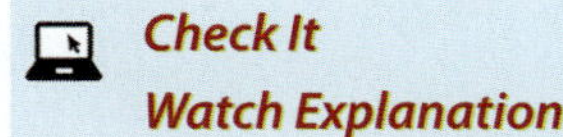

Example 12.11 Predicting the Volume in a Neutralization Reaction

A chemist needs to completely neutralize 0.600 L of a 3.0 M aq. H_2SO_4 solution. What volume of 0.80 M aq. KOH is necessary to do this? If the chemist adds 5.0 L of the aq. KOH to the acid solution, will the resulting solution be acidic, basic, or neutral?

To solve this problem, we begin by writing a balanced equation for the neutralization:

$$H_2SO_4\ (aq) + 2\ KOH\ (aq) \rightarrow K_2SO_4\ (aq) + 2\ H_2O\ (l)$$

Using the volume and molarity of the H_2SO_4, we can find the moles of acid:

$$\text{Moles (H}_2\text{SO}_4) = MV = \left(3.0\frac{\text{mol}}{\cancel{\text{L}}}\right)(0.600\ \cancel{\text{L}}) = 1.8\ \text{mol H}_2\text{SO}_4$$

We then use the balanced equation to find the moles of KOH needed. Finally, we use the molarity of the KOH solution to convert from moles to volume:

$$1.8\ \cancel{\text{mol H}_2\text{SO}_4} \times \frac{2\ \cancel{\text{mol KOH}}}{1\ \cancel{\text{mol H}_2\text{SO}_4}} \times \frac{1\ \text{L solution}}{0.80\ \cancel{\text{mol KOH}}} = 4.5\ \text{L solution}$$

So 4.5 L of the basic solution are needed to neutralize the acid. If the chemist adds 5.0 L of base, the base will be the excess reagent, and the final solution will be basic.

Check It
Watch Explanation

TRY IT

19. What volume of 2.4-M *aq*. NaOH is required to completely neutralize 200 mL of 6.0-M *aq*. HBr?

20. A drop of phenolphthalein is added to a solution of 1.0-M H_2SO_4 with a volume of 10.0 mL. The solution is mixed with 8.2 mL of 0.70-M NaOH. After the reaction is complete, is the solution acidic, basic, or neutral? What color is the indicator?

12.6 Buffers and Biological pH

The cells that make up your skin, muscle, blood, and organs must remain in a very narrow pH range in order to function properly. The fluid inside cells typically maintains a pH of about 6.8, while the bloodstream maintains a pH very near 7.4. Given the number of acidic and basic compounds present in the bloodstream and inside the cells, how do our bodies remain within these ranges?

Our bodies use **buffer** systems to keep the pH at healthy levels. A buffer is a solution containing a mixture of acidic and basic components that resists changes in pH. In the absence of a buffer, adding a small amount of acid or base can drastically affect the pH of a solution. However, in a buffered solution, adding the same amount of acid or base causes a much smaller pH change (**Figure 12.20**).

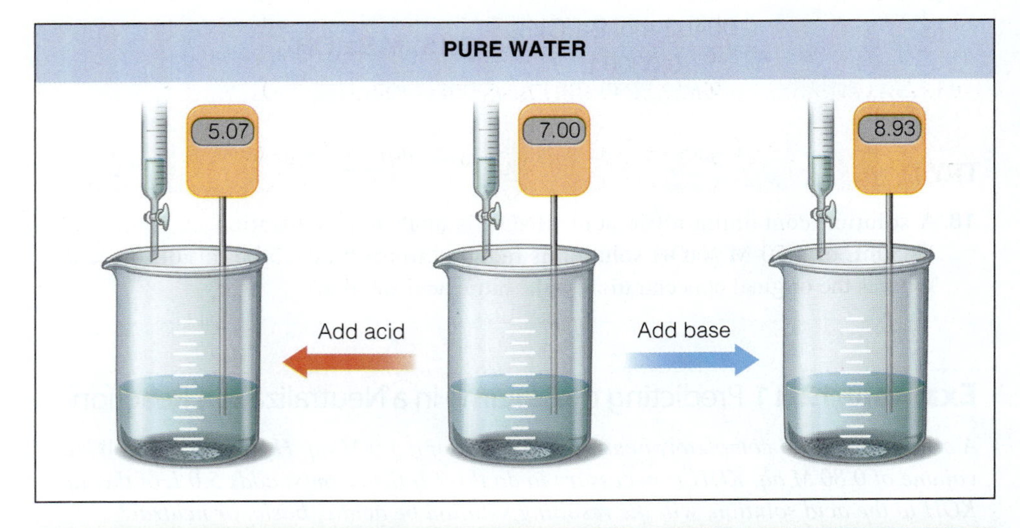

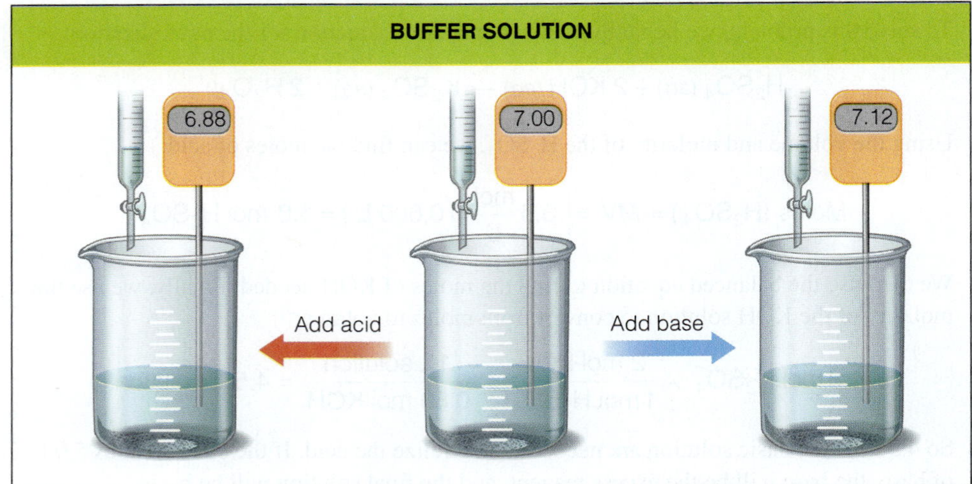

Figure 12.20 The pH of water changes drastically when an acid or base is added. Buffer solutions resist changes in pH.

How do buffer solutions work? Buffer solutions contain one of the following two combinations:

- a mixture of a weak acid and its conjugate base, or
- a mixture of a weak base and its conjugate acid

Both species must be present in large quantities. For example, animal cells contain a buffer consisting of $H_2PO_4^-$ (a weak acid) and HPO_4^{2-} (its conjugate base). If acid is added to the cells, the H^+ reacts with the conjugate base. If base is added to the cells, the OH^- is consumed by the acid. The ratio of $H_2PO_4^-$ to HPO_4^{2-} will change, but the amount of H^+ and OH^- present changes only slightly. This is shown in **Figure 12.21**.

Chemically, we represent this buffer using the equilibrium shown here:

$$H_2PO_4^- + H_2O \rightleftharpoons H_3O^+ + HPO_4^{2-}$$

The acid consumes excess OH^-

The conjugate base consumes excess H^+

A buffer solution can be thought of as two guards watching opposite directions. One component guards against an influx of acid, the other guards against an influx of base.

The bloodstream and the fluids between cells use a different buffer system, based on carbonic acid and its conjugate base, the bicarbonate ion. When carbon dioxide dissolves in the bloodstream, it reacts with water to form carbonic acid. This acid partially ionizes to form the bicarbonate ion (HCO_3^-):

$$\underset{\text{carbonic acid}}{H_2CO_3\ (aq)} + H_2O\ (l) \rightleftharpoons H_3O^+\ (aq) + \underset{\text{bicarbonate ion}}{HCO_3^-\ (aq)}$$

This buffer typically maintains blood pH in the range of 7.35 to 7.45.

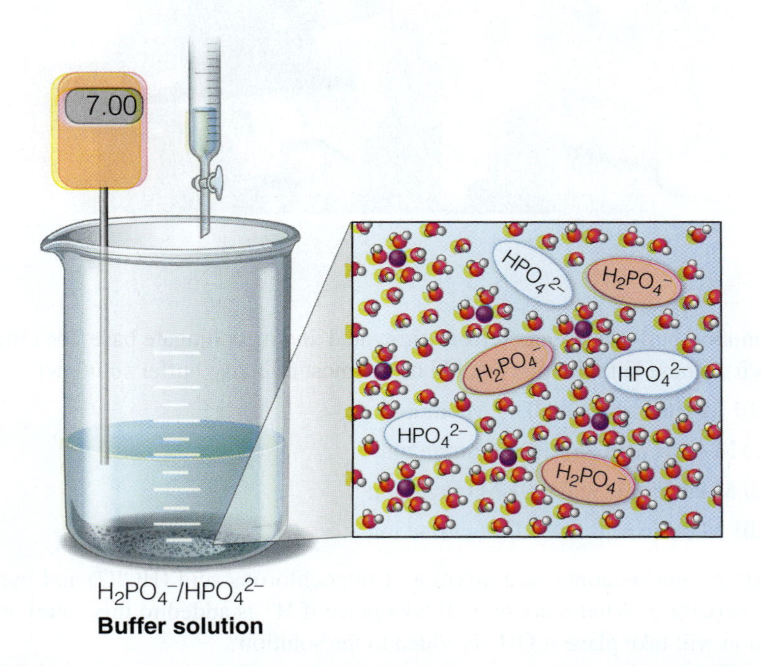

Figure 12.21 The phosphate buffer is composed of $H_2PO_4^-$ (a weak acid, in red) and HPO_4^{2-} (its conjugate base, in blue). If base is added, the acid neutralizes it. If acid is added, the base consumes it.

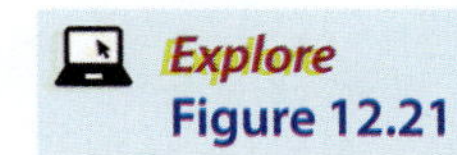

JamesBrey/Getty Images

It's hard to sneak up on two guard dogs facing in opposite directions. Either way you approach, one of the dogs will see you. Similarly, with a buffer solution, the acid guards against excess base, and the conjugate base guards against excess acid.

Certain conditions can cause blood pH to fluctuate. For example, hyperventilation (**Figure 12.22**) is a condition in which a person breathes abnormally fast, taking in more oxygen than the body needs. This condition is usually brought on by extreme stress or panic, and it causes dissolved CO_2 levels in the bloodstream to drop. This in turn reduces the amount of H_2CO_3 present in the bloodstream, causing the pH of the blood to quickly rise. While this situation may cause other symptoms, it can usually be brought into balance after normal breathing is restored.

It is possible to overload a buffer. For example, if we add a large amount of acid to a phosphate buffer, eventually all of the base will be consumed. At that point, the pH of the solution will change sharply.

Buffers can be adjusted to specific pH values. The pH of a buffer solution depends on the specific combination of acid and conjugate base, and on the concentration of each component present in solution.

Leonid Iastremskyi/Pixel-shot/Alamy

Figure 12.22 Hyperventilation is often brought on by stress and causes a rapid rise in blood pH.

Check It
Watch Explanation

TRY IT

21. A common buffer is prepared from citric acid and its conjugate base, the citrate ion. Which of these combinations would be the most effective buffer solution?

a. 0.01 M citric acid and 0.01 M citrate ion

b. 1.0 M citric acid and 1.0 M citrate ion

c. 1.0 M citric acid and 0.01 M citrate ion

d. 0.01 M citric acid and 1.0 M citrate ion

22. A buffer solution contains a mixture of hypochlorous acid (HClO) and hypochlorite ion (ClO^-). What reaction will take place if H^+ is added to this solution? What reaction will take place if OH^- is added to the solution?

Capstone Question

Vinegar is an aqueous solution of acetic acid, a monoprotic, weak acid with the formula $HC_2H_3O_2$. The structure for this acid is shown in **Figure 12.23**. The hydrogen attached to the oxygen atom is the acidic hydrogen. Vinegar ionizes as shown here:

$$HC_2H_3O_2\ (aq) \rightleftharpoons H^+\ (aq) + C_2H_3O_2^-\ (aq)$$

a. Draw the Lewis structure for the conjugate base of acetic acid. Include formal charge where appropriate. What is the name of this ion?

b. Both acetic acid and its conjugate base dissolve easily in water. Sketch water molecules around both species, using dashed lines to show where hydrogen bonds can form between water and the dissolved species.

c. Distilled white vinegar has a concentration of 5.0% (m/v), or 5.0 g of acetic acid per 100 mL solution. What is the molarity of acetic acid in this solution?

d. Using a pH meter, a chemist determines that the pH of a 5.0% vinegar solution is 2.42. What is the molarity of H^+ in this solution?

e. Based on your answers in parts c and d, what percentage of acetic acid is ionized in an aqueous solution?

Capstone Video

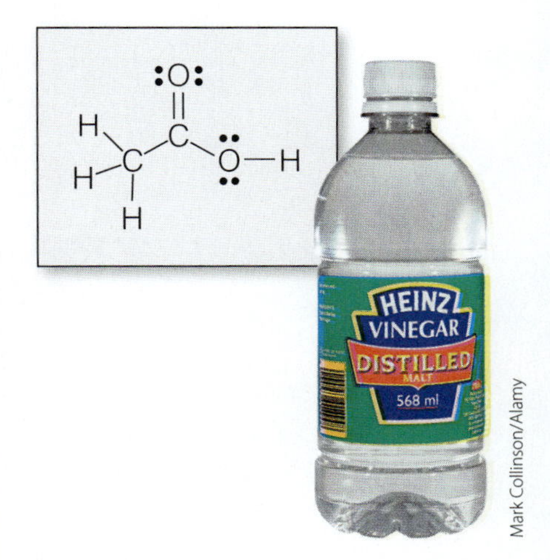

Figure 12.23 Vinegar is an aqueous solution of acetic acid, commonly used in cooking. Standard distilled white vinegar is usually sold as a 5% (m/v) solution.

SUMMARY

Acids and bases are all around us, and they are very important to our air, water, and living systems. In the Arrhenius definition, acids produce H^+ ions in water, which may also be represented by the hydronium ion, H_3O^+. Bases produce hydroxide (OH^-) ions in water.

The Brønsted-Lowry definitions for acids and bases are broader and enable us to describe reactions outside of an aqueous solution. A Brønsted-Lowry acid is any compound that gives up an H^+ (called a *proton*) in a chemical reaction. Conversely, a Brønsted-Lowry base is any compound that accepts an H^+.

There are seven common strong acids; nearly all other acids are weak acids. Strong acids completely ionize in water, but weak acids only partly ionize. Weak acids and bases exchange protons (that is, H^+ ions) in equilibrium reactions. Equilibrium reactions occur in both the forward and reverse directions and are shown by two arrows pointing in opposite directions ($\rightleftharpoons$).

On the left-hand side of an acid-base equilibrium reaction, the proton donor is the acid, and the proton acceptor is the base. On the right-hand side of the equation, the proton donor and acceptor are called the *conjugate acid* and the *conjugate base*. A compound that gives a proton (the acid) will gain the proton back in the reverse reaction. As a result, the acid and the conjugate base are the same species, differing only by one H^+.

Acids and bases react with each other in neutralization reactions. Acids react with hydroxide bases to form water and ionic compounds, called *salts*. Acids also react with many metals, producing metal cations and hydrogen gas. Nonmetal oxides often react with water to produce acidic compounds.

Pure water undergoes a self-ionization reaction to produce a small amount of H^+ and OH^- ions. These ions are always present, even in pure water. At 25 °C, the product of the molar H^+ and OH^- concentrations equals 10^{-14}. That is, $[H^+][OH^-] = 1.0 \times 10^{-14}$.

Acidic solutions have a higher concentration of H^+ than of OH^-. In basic solutions, the opposite is true. The concentration of H^+ in an acidic or basic solution is measured using the pH scale. The pH is the negative log of the H^+ concentration. In neutral solutions, $[H^+] = [OH^-] = 1.0 \times 10^{-7}$ M, and so the pH of a neutral solution is 7. Acidic solutions have a lower pH, and basic solutions have a higher pH.

Indicators are compounds that change color depending on the pH of their surroundings. Two common indicators are litmus (which turns blue in basic solution and red in acidic solution) or phenolphthalein (which is colorless in acid and bright pink in base). Combinations of indicators are used to make pH paper. Instruments called *pH meters* measure pH more precisely than simple indicators.

Titration is an analytical technique for measuring the concentration of an acidic or basic solution. In this technique, chemists use an indicator to determine the exact amount of base needed to react with an unknown acid (or vice versa).

Living systems use buffers to maintain the pH of cells and intercellular fluid (such as blood) within a narrow pH range. Buffers use a weak acid and its conjugate base (or a weak base and its conjugate acid) in high concentrations. In a buffer, the acidic component will react with any excess base, and the basic component will react with any excess acid. The body uses a mixture of $H_2PO_4^-$ and HPO_4^{2-} in the cells, and it uses a mixture of H_2CO_3 and HCO_3^- in the bloodstream.

The War on Drugs, Then and Now

In 1986, the United States was in a panic over crack cocaine. Across the country, countless lives were devastated by the powerfully addictive drug. The "crack epidemic" produced a spike in theft, violent crime, and gang activity. Things seemed to be spiraling out of control.

In this context, Congress passed the Anti-Drug Abuse Act of 1986. The bill included mandatory sentencing for possession and trafficking of drugs, including cocaine and marijuana, and imposed especially harsh sentences for crack (**Figure 12.24**).

The law mandated a 5-year minimum prison sentence for the possession of 500 grams of powder cocaine—and the same minimum for just 5 grams of crack. In the years that followed, the 100:1 ratio was widely criticized as unfair. In 1995, the U.S. Sentencing Commission reported that this gap was "a primary cause of the growing disparity between sentences for black and white federal defendants." In addition to racial inequities, the sentencing guidelines disproportionately targeted low-level offenders. Due to the stability of the powder form, major drug traffickers nearly always import cocaine as the acid salt. Local dealers converted the cocaine to the crack form, but these small-time dealers faced harsher penalties.

In the decades following the Anti-Drug Abuse Act, the problems inherent in the law became clear, and policy makers struggled to formulate more just sentencing guidelines. Finally, in 2010, the Fair Sentencing Act was passed by a unanimous vote in the Senate and signed into law by the president. This law narrowed the gap in cocaine sentencing and eliminated minimum sentencing for low-level offenders.

Lucy Nicholson/Reuters/Newscom

Figure 12.24 The Anti-Drug Abuse Act of 1986 led to a spike in incarceration rates.

Key Terms

12.1 Introduction to Acids and Bases

Arrhenius definition Describes an acid as a compound that produces H^+ or H_3O^+ ions in water, and a base as a compound that produces OH^- ions in water.

acid A compound that produces H^+ or H_3O^+ ions in water (Arrhenius definition); a compound that donates H^+ ions (Brønsted-Lowry definition).

base A compound that produces OH^- ions in water (Arrhenius definition); a compound that accepts H^+ ions (Brønsted-Lowry definition).

hydronium ion H_3O^+, the ion formed in acidic aqueous solutions.

polyprotic acids Acids that can release more than one H^+ into aqueous solution.

Brønsted-Lowry definition Describes an acid as a compound that donates H^+ ions, and a base as a compound that accepts H^+ ions.

12.2 Acid-Base Equilibrium Reactions

strong acid An acid that completely ionizes in water.

weak acid An acid that only partially ionizes in water.

equilibrium reaction A reaction that occurs in both the forward and backward directions.

conjugate acid The acid on the right-hand side of a chemical equation in an acid-base equilibrium; the acid formed when a base reacts with H^+.

conjugate base The base on the right-hand side of a chemical equation in an acid-base equilibrium; the base formed when an acid releases an H^+.

12.3 Reactions Involving Acids and Bases

neutralization reaction A reaction in which an acid and a base combine to produce water and an ionic compound (called a *salt*).

nonmetal oxide A compound or ion containing a nonmetal covalently bonded to one or more oxygen atoms.

acid rain The effect observed when nonmetal oxides combine with moisture in the atmosphere to produce acidic rainfall.

12.4 Acid and Base Concentration

self-ionization A process that occurs in water when a water molecule fragments to produce H^+ and OH^- ions.

pH scale A scale that indicates the relative concentration of acid or base in an aqueous solution; pH is defined as the negative log of the H^+ concentration.

12.5 Measuring Acid and Base Concentration

pH indicator A compound that changes color depending on the pH; the color changes indicate the approximate pH of the solution.

litmus A pH indicator that turns blue in the presence of base and red in the presence of acid.

pH paper Paper containing a blend of indicators that can be used to estimate pH based on color.

phenolphthalein A pH indicator that is bright pink in base, but colorless in acid.

titration An analytical technique that can precisely measure the concentration of an acid or base by measuring the volume required for a neutralization to occur.

12.6 Buffers and Biological pH

buffer A solution containing a mixture of acidic and basic components; it resists changes in pH.

Additional Problems

12.1 Introduction to Acids and Bases

23. Complete the following acid dissociation reactions:

a. $HBr\ (aq) \rightarrow$ __________ $+ Br^-\ (aq)$
b. $HBr\ (aq) + H_2O\ (l) \rightarrow$ __________ $+ Br^-\ (aq)$
c. $H_2SO_4\ (aq) \rightarrow$ __________ $+ HSO_4^-\ (aq)$
d. $H_2CO_3\ (aq) + H_2O\ (l) \rightarrow H_3O^+\ (aq) +$ __________

24. Complete the following acid dissociation reactions:

a. $HNO_3\ (aq) \rightarrow$ __________ $+ NO_3^-\ (aq)$
b. $HNO_3\ (aq) + H_2O\ (l) \rightarrow$ __________ $+ NO_3^-\ (aq)$
c. $HClO_4\ (aq) \rightarrow$ __________ $+ ClO_4^-\ (aq)$
d. $H_3PO_4\ (aq) \rightarrow H^+\ (aq) +$ __________

25. Complete the following base dissociation reactions:

a. $NaOH\ (s) \rightarrow Na^+\ (aq) +$ __________
b. $Ca(OH)_2\ (s) \rightarrow$ __________ $+ 2\ OH^-\ (aq)$

26. Complete the following base dissociation reactions:

a. $CsOH\ (s) \rightarrow$ __________ $+$ __________
b. __________ $\rightarrow Ba^{2+}\ (aq) + 2\ OH^-\ (aq)$

27. Write ionic equations to show the dissociation of these acids in water:

a. HCl b. HNO_3 c. $HClO_4$

28. Write ionic equations to show the dissociation of these acids in water:

a. HBr b. HI c. H_2SO_4

29. Write ionic equations to show the dissociation of these bases in water:

a. $NaOH$ b. $Mg(OH)_2$ c. $CsOH$

30. Write ionic equations to show the dissociation of these bases in water:

a. KOH b. $Ca(OH)_2$ c. $Ba(OH)_2$

31. Identify these acids as monoprotic, diprotic, or triprotic:

a. H_2CO_3 b. HCl c. $HClO_4$ d. H_2SO_4

32. Identify these acids as monoprotic, diprotic, or triprotic:

a. H_3PO_4 b. HBr c. HNO_3 d. $H_2C_2O_4$

33. What is the difference between the Arrhenius definition of an acid and base and the Brønsted-Lowry definition?

34. Is it possible to be an acid or base in the Arrhenius sense without being an acid or base in the Brønsted-Lowry sense? Is it possible to be a Brønsted-Lowry acid or base without being an Arrhenius acid or base?

35. Draw a sketch showing the dissociation of HBr in water. Explain why it is reasonable to show the dissociation as forming H^+ *(aq)* or H_3O^+ *(aq)*.

36. According to the Arrhenius definition, a base is anything that produces hydroxide (OH^-) in aqueous solution. Although ammonia (NH_3) does not dissociate to form an OH^-, it is still an Arrhenius base. Why is this so? Write a chemical equation to support your answer.

37. Identify the acid and the base in these reactions:

a. $KOH\ (aq) + HF\ (aq) \rightarrow KF\ (aq) + H_2O\ (l)$
b. $HBr\ (aq) + H_2O\ (l) \rightarrow H_3O^+\ (aq) + Br^-\ (aq)$
c. $H_3N\ (aq) + H_2SO_4\ (aq) \rightarrow H_4N^+\ (aq) + HSO_4^-\ (aq)$

38. Identify the acid and the base in these reactions:

a. $H_2O\ (l) + HCl\ (aq) \rightarrow H_3O^+\ (aq) + Cl^-\ (aq)$
b. $NH_4^+\ (aq) + H_2O\ (l) \rightarrow NH_3\ (aq) + H_3O^+\ (aq)$
c. $HNO_3\ (aq) + CH_3NH_2\ (aq) \rightarrow NO_3^-\ (aq) + CH_3NH_3^+\ (aq)$

12.2 Acid-Base Equilibrium Reactions

39. What is the difference between a strong acid and a weak acid?

40. Consider the following two acid-base equations. Why is one arrow used for the first expression, but two arrows for the second? What do the different arrows mean?

$$HNO_3\ (aq) + H_2O\ (l) \rightarrow H_3O^+\ (aq) + NO_3^-\ (aq)$$

$$HNO_2\ (aq) + H_2O\ (l) \rightleftharpoons H_3O^+\ (aq) + NO_2^-\ (aq)$$

41. Write the molecular formulas for each of these acids. Which one is not a strong acid?

a. hydrochloric acid b. hydrobromic acid
c. sulfuric acid d. phosphoric acid

42. Write the molecular formulas for each of these acids. Which of these is not a strong acid?

a. hydroiodic acid b. nitric acid
c. carbonic acid d. perchloric acid

43. Identify these compounds as strong or weak acids:

a. H_3PO_4 b. H_2CO_3 c. HSO_4^- d. HNO_3

44. Identify these compounds as strong or weak acids:

a. HCl b. HF c. $HClO_4$ d. HCN

45. What is an equilibrium?

46. Which of the following are examples of an equilibrium?

a. water flowing between the deep end and shallow end of a pool
b. people moving back and forth between two towns
c. water flowing over a waterfall
d. oxygen gas flowing back and forth between different rooms in a house
e. a candle burning to produce carbon dioxide and water

47. Ammonia (NH_3) can behave as both an acid and a base.

a. Draw the Lewis structure for the conjugate acid of ammonia.
b. Draw the Lewis structure for the conjugate base of ammonia.

48. Methyl alcohol has the molecular formula CH_4O. This compound can behave as both an acid and a base.

a. Draw the Lewis structure for the conjugate acid of methyl alcohol.
b. Draw the Lewis structure for the conjugate base of methyl alcohol.

49. In the following acid-base equilibria of weak acids in water, label the acid (A), the base (B), the conjugate acid (CA), and the conjugate base (CB):

a. $HClO_2\ (aq) + H_2O\ (l) \rightleftharpoons H_3O^+\ (aq) + ClO_2^-\ (aq)$
b. $H_2CO_3\ (aq) + H_2O\ (l) \rightleftharpoons H_3O^+\ (aq) + HCO_3^-\ (aq)$
c. $H_2O\ (l) + CH_3NH_3^+\ (aq) \rightleftharpoons CH_3NH_2\ (aq) + H_3O^+\ (aq)$

50. In the following acid-base equilibria, label the acid (A), the base (B), the conjugate acid (CA), and the conjugate base (CB):

a. $CH_3CO_2H\ (aq) + H_2O\ (l) \rightleftharpoons H_3O^+\ (aq) + CH_3CO_2^-\ (aq)$
b. $H_2O\ (l) + H_2O\ (l) \rightleftharpoons H_3O^+\ (aq) + OH^-\ (aq)$
c. $HNO_2\ (aq) + HCl\ (aq) \rightleftharpoons H_2NO_2^+\ (aq) + Cl^-\ (aq)$

51. Label the acid (A), base (B), conjugate acid (CA), and conjugate base (CB) in these acid-base equilibria:

a. $HCl\ (g) + CH_3OH\ (g) \rightleftharpoons CH_3OH_2^+\ (g) + Cl^-\ (g)$

b. $CH_3OH + PBr_3 \rightleftharpoons CH_3O^- + HPBr_3^+$

c. $H_2S\ (g) + HF\ (aq) \rightleftharpoons H_3S^+\ (aq) + F^-\ (aq)$

52. Label the acid (A), base (B), conjugate acid (CA), and conjugate base (CB) in these acid-base equilibria:

a. $H_3CN + HF \rightleftharpoons H_4CN^+ + F^-$

b. $C_2H_6O\ (l) + HCl\ (aq) \rightleftharpoons C_2H_7O^+\ (aq) + Cl^-\ (aq)$

c. $CH_3O^-\ (aq) + H_2O\ (l) \rightleftharpoons CH_3OH\ (aq) + OH^-\ (aq)$

53. HF is a weak acid. Fill in the missing species in these equations to show the acidic behavior of HF and the basic behavior of F^-.

$HF\ (aq) + H_2O\ (l) \rightleftharpoons$ ________

$F^-\ (aq) + H_2O\ (l) \rightleftharpoons$ ________

54. HCN is a weak acid. Fill in the missing species in these equations to show the acidic behavior of HCN and the basic behavior of CN^-.

________ $(aq) + H_2O\ (l) \rightleftharpoons H_3O^+\ (aq) + CN^-\ (aq)$

________ (aq) + ________ $(l) \rightleftharpoons OH^-\ (aq) + HCN\ (aq)$

55. Complete these acid-base reactions to show the missing species:

$H_2CO_3\ (aq) + H_2O\ (l) \rightleftharpoons H_3O^+\ (aq)$ + ________ (aq)

$HNO_2\ (aq) + H_2O\ (l) \rightleftharpoons$ ________ + ________

$NO_2^-\ (aq) + H_2O\ (l) \rightleftharpoons$ ________ + ________

56. Complete these acid-base reactions to show the missing species:

$HClO_3\ (aq) + H_2O\ (l) \rightleftharpoons H_3O^+\ (aq)$ + ________ (aq)

$HF\ (aq) + H_2O\ (aq) \rightleftharpoons$ ________ + ________

$N_3^-\ (aq) + H_2O\ (l) \rightleftharpoons$ ________ + ________

57. Complete these acid-base reactions to show the missing species:

$H_3PO_4\ (aq) + H_2O\ (l) \rightleftharpoons H_3O^+\ (aq)$ + ________ (aq)

$H_2PO_4^{2-}\ (aq) + H_2O\ (l) \rightleftharpoons H_3O^+\ (aq)$ ________ (aq)

________ $+ H_2O\ (l) \rightleftharpoons H_3O^+\ (aq) + PO_4^{3-}\ (aq)$

58. Complete these acid-base reactions to show the missing species:

$H_2SO_3\ (aq) + H_2O\ (l) \rightleftharpoons H_3O^+\ (aq)$ + ________ (aq)

________ $+ H_2O\ (l) \rightleftharpoons H_3O^+\ (aq) + SO_3^{2-}\ (aq)$

59. Hypochlorous acid (HClO) is a weak acid. The conjugate base of this acid is the hypochlorite ion (ClO^-).

a. Write a balanced equation showing the reaction of HClO with water.

b. Write a balanced equation showing the reaction of ClO^- with water.

60. Methylamine (CH_3NH_2) is a weakly basic compound. The conjugate acid of this compound is $CH_3NH_3^+$.

a. Write a balanced equation showing the reaction of CH_3NH_2 with water.

b. Write a balanced equation showing the reaction of $CH_3NH_3^+$ with water.

61. Sometimes acid-base reactions show only a single product. For example, the compound methylamine (CH_5N) reacts with HCl to produce a single solid product. What is the conjugate acid and the conjugate base in this reaction? (*Hint*: Write the products as ions rather than as a neutral compound.)

$HCl\ (g) + CH_5N\ (g) \rightarrow CH_6NCl\ (s)$

62. Sometimes acid-base reactions show only a single product. For example, in the gas phase, ammonia reacts with hydrofluoric acid to form solid ammonium fluoride. What is the conjugate acid and the conjugate base in this reaction? (*Hint*: Write this as an ionic equation.)

$NH_3\ (g) + HF\ (g) \rightarrow NH_4F\ (s)$

63. Water can act as both an acid and as a base. Draw the conjugate acid and the conjugate base of a water molecule.

64. Ammonia, NH_3, can act as both an acid and as a base. Draw the conjugate acid and the conjugate base of an ammonia molecule.

65. Trichloroacetic acid ($HC_2Cl_3O_2$) is a weak acid that is used to treat warts and remove tattoos. Label the acid (A), base (B), conjugate acid (CA), and conjugate base (CB) in the acid-base equilibrium of trichloroacetic acid shown below.

$Cl_3C{-}C({=}O){-}OH + H_2O \rightleftharpoons Cl_3C{-}C({=}O){-}O^- + H_3O^+$

66. Sodium naproxide is the active component in the anti-pain and anti-inflammatory compound Aleve®. In aqueous solution, this compound undergoes the acid-base equilibrium reaction shown. Label the acid (A), base (B), conjugate acid (CA), and conjugate base (CB) for this reaction.

(sodium naproxide: CH_3O–naphthalene–$CH(CH_3)COO^{\ominus}\ Na^{\oplus}$) + $H_2O \rightleftharpoons Na^{\oplus}\ OH^{\ominus}$ + (CH_3O–naphthalene–$CH(CH_3)COOH$)

12.3 Reactions Involving Acids and Bases

67. Nonmetal oxides often react with water to form acids. Which of these compounds are nonmetal oxides?

a. CO_2 b. BaO c. NH_3 d. SO_3

68. Complete these reactions to show the formation of carbonic acid and sulfuric acid from water and a nonmetal oxide:

________ $+ H_2O\ (l) \rightleftharpoons H_2CO_3\ (aq)$

________ $+ H_2O\ (l) \rightleftharpoons H_2SO_4\ (aq)$

69. Each of the following metals reacts with hydrochloric acid to form the metal chloride and hydrogen gas. Complete and balance the reactions to show the formation of these products.

a. $Sn\ (s) + HCl\ (aq) \rightarrow$
b. $Li\ (s) + HCl\ (aq) \rightarrow$
c. $Fe\ (s) + HCl\ (aq) \rightarrow$
d. $Al\ (s) + HCl\ (aq) \rightarrow$

70. Each of the following metals reacts with nitric acid to form the metal chloride and hydrogen gas. Complete and balance the reactions to show the formation of these products.

a. $Zn\ (s) + HNO_3\ (aq) \rightarrow$
b. $Pb\ (s) + HNO_3\ (aq) \rightarrow$
c. $Na\ (s) + HNO_3\ (aq) \rightarrow$
d. $Al\ (s) + HNO_3\ (aq) \rightarrow$

71. Complete and balance these reactions. One of the metals does not react.

a. $Be\ (s) + HF\ (aq) \rightarrow$
b. $Sr\ (s) + H_2SO_4\ (aq) \rightarrow$
c. $Au\ (s) + HNO_3\ (aq) \rightarrow$

72. Complete and balance these reactions. One of the metals does not react.

a. $Ni\ (s) + HNO_3\ (aq) \rightarrow$
b. $Pt\ (s) + HClO_4\ (aq) \rightarrow$
c. $Ba\ (s) + HBr\ (aq) \rightarrow$

73. Show the products from these acid-base neutralization reactions:

a. $HCl + NaOH \rightarrow$
b. $2\ HBr + Mg(OH)_2 \rightarrow$
c. $3\ HNO_3 + Al(OH)_3 \rightarrow$

74. Show the products from these acid-base neutralization reactions:

a. $HBr + KOH \rightarrow$
b. $H_2SO_4 + 2\ NaOH \rightarrow$
c. $4\ HNO_3 + Ti(OH)_4 \rightarrow$

75. Complete and balance these acid-base neutralization reactions:

a. $HCl\ (aq) + Ca(OH)_2\ (s) \rightarrow$
b. $Mg(OH)_2\ (aq) + H_2SO_4\ (aq) \rightarrow$
c. $Al(OH)_3\ (s) + HF\ (aq) \rightarrow$

76. Complete and balance these acid-base neutralization reactions:

a. $CsOH\ (aq) + HBr\ (aq) \rightarrow$
b. $HBr\ (aq) + Fe(OH)_2\ (s) \rightarrow$
c. $H_3PO_4\ (aq) + KOH\ (aq) \rightarrow$

77. Write a complete ionic equation and a net ionic equation for this neutralization reaction:

$$Mg(OH)_2\ (aq) + 2\ HNO_3\ (aq) \rightarrow Mg(NO_3)_2\ (aq) + 2\ H_2O\ (l)$$

78. Write a complete ionic equation and a net ionic equation for this neutralization reaction:

$$H_2SO_4\ (aq) + 2\ KOH\ (aq) \rightarrow K_2SO_4\ (aq) + 2\ H_2O\ (l)$$

79. Citric acid has the formula $H_3C_6H_5O_7$. It is a triprotic acid, meaning it can lose three H^+ ions, and the citrate anion has a charge of –3 ($C_6H_5O_7^{3-}$). Write a balanced equation showing the reaction of citric acid with sodium hydroxide to produce sodium citrate and water.

80. Phosphoric acid has the formula H_3PO_4. It is a triprotic acid, meaning it can lose three H^+ ions, and the phosphate anion has a charge of –3 (PO_4^{3-}). Write a balanced equation showing the reaction of phosphoric acid with potassium hydroxide to produce potassium phosphate and water.

81. Nonmetal oxides often react with water to form acids. Write a balanced equation showing the reaction of gaseous sulfur trioxide with water to form aqueous sulfuric acid.

82. Write a balanced equation showing the reaction of carbon dioxide with water to produce aqueous carbonic acid.

12.4 Acid and Base Concentration

83. The self-ionization reaction of water is shown below. In this reaction, one water molecule acts as an acid; the other acts as a base. Label the conjugate acid and conjugate base in this reaction.

$$H_2O\ (l) + H_2O\ (l) \rightleftharpoons H_3O^+\ (aq) + OH^-\ (aq)$$

acid base

84. How do the H^+ and OH^- concentrations change

a. when an acid is added to water?
b. when a base is added to water?

85. Find the concentration of H^+ ions in each of these situations:

a. a strongly acidic solution where $[OH^-] = 1.0 \times 10^{-12}$ M
b. a weakly acidic solution where $[OH^-] = 5.23 \times 10^{-9}$ M
c. a basic solution where $[OH^-] = 4.28 \times 10^{-5}$ M

86. Find $[OH^-]$ in each of these situations:

a. a strongly acidic solution where $[H^+] = 1.0 \times 10^{-2}$ M
b. a neutral solution where $[H^+] = 1.0 \times 10^{-7}$ M
c. a basic solution where $[H^+] = 4.28 \times 10^{-11}$ M

87. Calculate the pH for each of the following. Identify whether the solution is acidic, basic, or neutral.

a. $[H^+] = 1.0 \times 10^{-3}$ M
b. $[H^+] = 1.0 \times 10^{-7}$ M
c. $[H^+] = 5.23 \times 10^{-11}$ M

88. Calculate the pH for each of the following. Identify whether the solution is acidic, basic, or neutral.

a. $[H^+] = 1.0 \times 10^{-8}$ M
b. $[H^+] = 2.5 \times 10^{-12}$ M
c. $[H^+] = 6.91 \times 10^{-5}$ M

89. Calculate the pH for each of the following. Identify whether the solution is acidic, basic, or neutral.

a. $[OH^-] = 1.0 \times 10^{-7}$ M
b. $[OH^-] = 1.0 \times 10^{-9}$ M
c. $[OH^-] = 1.26 \times 10^{-2}$ M

90. Calculate the pH for each of the following. Identify whether the solution is acidic, basic, or neutral.

a. $[OH^-] = 1.0 \times 10^{-3}$ M
b. $[OH^-] = 1.0 \times 10^{-7}$ M
c. $[OH^-] = 8.21 \times 10^{-11}$ M

91. Complete this table to show the pH and pOH of each solution:

	$[H^+]$	$[OH^-]$	pH	pOH
Solution A	1.0×10^{-5}		5	
Solution B		1.0×10^{-3}		3
Solution C	1.0×10^{-9}			
Solution D			8	

92. Complete this table to show the pH and pOH of each solution:

	$[H^+]$	$[OH^-]$	pH	pOH
Solution A	1.0×10^{-8}		8	
Solution B		1.0×10^{-3}	11	
Solution C	1.0×10^{-10}			
Solution D				6

93. Identify each solution as acidic, basic, or neutral:

a. an aqueous solution with a pH of 7.0
b. an aqueous bleach solution with a pH of 8.2
c. a solution of monosodium phosphate with a pH of 6.1
d. a solution with $[H^+] = 1.5 \times 10^{-5}$ M
e. a solution with $[H^+] = 5.2 \times 10^{-10}$ M

94. Identify each solution as acidic, basic, or neutral:

a. an aqueous solution where the hydroxide ion concentration is greater than the hydronium ion concentration
b. a solution of vinegar in water, with a pH of 5.2
c. a solution with $[OH^-] = 5.2 \times 10^{-2}$ M
d. a solution where the $-\log[H^+] = 7.0$
e. a solution with a pOH of 4.0

95. Find the concentration of hydronium ions in each of these solutions:

a. a solution with a pH of 4.2
b. a solution with a pH of 11.3
c. a solution with a pOH of 6.7
d. a solution with a pOH of 3.66

96. Find the concentration of hydroxide ions in each of these solutions:

a. a solution with a pOH of 4.7
b. a solution with a pOH of 10.3
c. a solution with a pH of 6.2
d. a solution with a pH of 13.4

97. In a dilute sodium bicarbonate solution, $[H^+] = 1.3 \times 10^{-8}$ M. Without using a calculator, what integer value will this pH be near? How can you tell? Estimate the pH and then use a calculator to verify your answer.

98. A solution has a hydronium ion concentration of 4.1×10^{-11} M. Without using a calculator, what integer value will this pH be near? How can you tell? Estimate the pH and then use a calculator to verify your answer.

99. Which has a lower pH, a 0.010-M solution of HCl, or a 0.010-M solution of HF? How do you know?

100. Which has a lower pH, a 1.0×10^{-4} M solution of HCl, or a 1.0×10^{-4} M solution of H_2SO_4? How do you know?

12.5 Measuring Acid and Base Concentration

101. Between litmus paper, pH paper, and a pH meter, which gives the most precise reading of pH? Which gives the least precise reading?

102. Describe the color of litmus and the color of phenolphthalein in both acidic and basic solutions.

103. Describe the color of the indicator in each of these solutions:

a. litmus at a pH of 3.0
b. litmus at a pH of 9.3
c. phenolphthalein in a solution where $[H^+] = 1.5 \times 10^{-4}$ M
d. phenolphthalein in a solution where $[H^+] = 1.5 \times 10^{-10}$ M

104. Describe the color of the indicator in each of these solutions:

a. phenolphthalein at a pH of 12.1
b. phenolphthalein at a pH of 9.3
c. litmus in a solution with a pOH of 3.4
d. litmus in a solution where $[OH^-] = 1.0 \times 10^{-2}$ M

105. While working in an analytical laboratory, you are asked to find the molarity of a sodium hydroxide solution. Describe how you would do this using a titration.

106. Your research laboratory uses a sample of hydrochloric acid and obtains unexpected results. Your research advisor asks you to measure the concentration of this solution using a titration. Describe how you would go about this process.

107. In a titration experiment, a 25.0-mL nitric acid solution is neutralized with 28.3 mL of 1.50-M aqueous sodium hydroxide. What is the molarity of the nitric acid solution?

108. In a titration experiment, 85.3 mL of 2.00-M *aq.* KOH is required to neutralize 20.0 mL of a hydrochloric acid solution. What is the molarity of the HCl solution?

109. In a titration experiment, two drops of phenolphthalein are added to 25.0 mL of *aq.* NaOH solution, causing it to turn bright pink. A solution of 3.05-M hydrochloric acid is added to the solution. It is found that 5.2 mL of the acid solution is required to turn the solution colorless. What is the concentration of the original base solution?

110. A solution of cesium hydroxide is analyzed by titration. It is found that 0.0824 L of 1.50-M aqueous nitric acid is required to neutralize 0.0250 L of the base solution. What is the concentration of the original base solution?

111. In a titration experiment, 12.42 mL of a 0.105-M *aq.* sodium hydroxide solution is required to neutralize a 5.00-mL sample of sulfuric acid. Write a balanced equation for this neutralization reaction. What is the concentration of the sulfuric acid solution?

112. In a titration experiment, 15.3 mL of a 0.010-M hydrochloric acid solution is required to neutralize a 0.500-L solution of barium hydroxide. Write a balanced equation for this neutralization reaction. What is the concentration of the barium hydroxide solution?

113. What volume of 5.0-M *aq.* NaOH would be required to completely neutralize 0.100 liters of 5.32-M *aq.* HI solution?

114. What volume of 1.4-M *aq.* KOH would be required to completely neutralize 0.500 liters of 4.23-M *aq.* H_2SO_4 solution?

115. Describe the color of the indicator in each of these solutions:

a. litmus in an acidic solution
b. phenolphthalein in a solution where $[OH^-] > [H_3O^+]$
c. phenolphthalein in a solution where $[H^+] = 0.0043$ M
d. litmus in a solution with a pH of 3

116. Describe the color of the indicator in each of these solutions:

a. litmus in a basic solution
b. phenolphthalein in a solution where $[OH^-] < [H_3O^+]$
c. phenolphthalein in a solution where $[H^+] = 3.2 \times 10^{-11}$ M
d. litmus in a solution with a pH of 9.4

117. A drop of phenolphthalein is added to a solution of 0.042-M HBr with a volume of 10.0 mL. The solution is mixed with 11.7 mL of 0.70-M NaOH. After the reaction is complete, is the solution acidic, basic, or neutral? What color is the indicator?

118. If 4.0 mL of 0.014-M *aq.* HBr is reacted with 1.0 mL of 0.010-M $Ba(OH)_2$, will the solution be acidic, basic, or neutral? How many moles of the excess product will remain?

12.6 Buffers and Biological pH

119. What are buffers?

120. What are the two components of a buffer solution? How do these components neutralize excess acid and base?

121. What buffer system is found inside animal cells? What are the acid and (conjugate) base species in this buffer?

122. What buffer system is used to regulate the pH of the bloodstream? What are the acid and (conjugate) base species in this buffer?

123. A mixture of acetic acid and sodium acetate makes an effective buffer. Here is the equilibrium equation for these two species:

$$HC_2H_3O_2\ (aq) + H_2O\ (l) \rightleftharpoons C_2H_3O_2^-\ (aq) + H_3O^+\ (aq)$$

a. What reaction would take place if acid were added to this buffer?
b. What reaction would take place if base were added to this buffer?

124. The carbonic acid/bicarbonate buffer that is found in the bloodstream is described by this equation:

$$H_2CO_3\ (aq) + H_2O\ (l) \rightleftharpoons HCO_3^-\ (aq) + H_3O^+\ (aq)$$

a. What reaction would take place if acid were added to this buffer?
b. What reaction would take place if base were added to this buffer?

Challenge Questions

125. If 100.0 mL of 1.21-M HCl is reacted with 50.0 mL of 1.06-M NaOH, how many moles of HCl solution will remain unreacted? What will be the H^+ concentration in the resulting solution? What will be the pH of the resulting solution?

126. Albuterol is commonly used in breathing treatments. Albuterol can be converted from the free-base form to the acid salt form by reacting it with HCl, as shown in the reaction below. What volume of 0.100-M HCl would be required to completely react with 500.0 grams of albuterol?

HOCH2 / HO / OH / N–C(CH3)3 / H + HCl → HOCH2 / HO / OH / ⊕ H / N–C(CH3)3 / H / Cl⊖

Albuterol (base)
Chemical formula: $C_{13}H_{21}NO_3$
Molar mass: 239.34 g/mol

Albuterol (HCl salt)

CHAPTER THIRTEEN

Reaction Rates and Equilibrium

Figure 13.1 (a) Today there are far more people in the world than ever before. (*continued*)

The Haber-Bosch Process

Before 1800, the world population was less than 1 billion. But the dawn of the nineteenth century was a turning point: Advances in science and medicine reduced infant mortality rates and increased life spans. More people were being born, and fewer were dying young. The population began to rise. Today the world has eight times as many people, and the number continues to climb (**Figure 13.1**).

For many, the population jump was a nightmare scenario: How could the world grow enough food to keep up with the population? Would there be mass starvation? Would the wars of the future be clashes over food?

Fortunately, this scenario has not come to pass, largely due to a technological breakthrough that changed agriculture and multiplied the world's food-production capability. Today it is estimated that a third of the world's population has food because of this technology. The breakthrough, called the *Haber-Bosch process*, enabled mass production of nitrogen fertilizer.

Nitrogen is the fourth most common element in your body (after carbon, hydrogen, and oxygen). Nitrogen atoms are in every bit of muscle tissue and every fragment of DNA. We get nitrogen from our diet, which means that ultimately it comes from plant life. Despite the abundance of nitrogen in the atmosphere, N_2 gas is unreactive, and plants can't absorb it directly. Rather, they draw nitrogen from the soil, where tiny but vital microbes perform a process called *nitrogen fixation* — converting atmospheric nitrogen into usable ammonia (or its conjugate acid, the ammonium ion). The problem is, microbes produce a limited amount of "fixed" nitrogen, and the amount of nitrogen in the soil limits how well crops can grow.

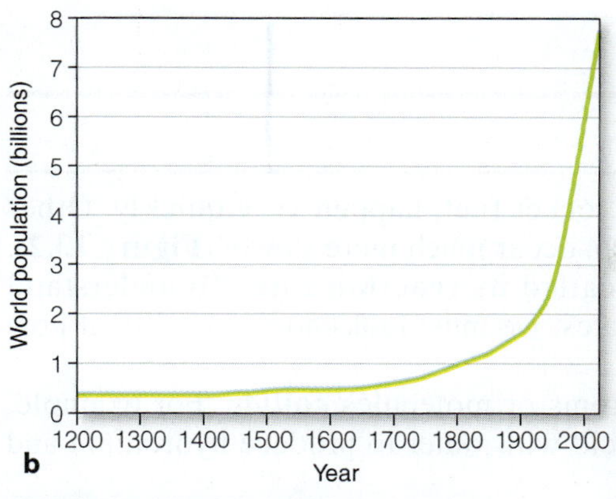

Figure 13.1 (continued) (b) This graph shows the dramatic jump in the world's population since 1800. (c) Ammonia fertilizer enhances crop growth. (d) Farmers today produce much more food per acre than ever before. (e) Fritz Haber developed the process used to produce ammonia. (a) Thomas La Mela/Shutterstock; (c) Doug Martin/Science Source; (d) Courtesy of Kevin Revell; (e): ullstein bild/Getty Images

In the early 1900s, a German scientist named Fritz Haber tackled the problem of producing ammonia from nitrogen gas. In theory, the reaction should be easy — combine nitrogen with hydrogen to form ammonia:

$$N_2\,(g) + 3\,H_2\,(g) \rightarrow 2\,NH_3\,(g)$$

In practice, it was much harder. Even though ammonia is a stable product, the reaction doesn't occur at room temperature. And when heated, the reaction produces only a tiny amount of ammonia, leaving mostly unreacted nitrogen and hydrogen. Why wouldn't this reaction work?

Haber analyzed the changes in *energy* that controlled this reaction. Further, he recognized that this was an *equilibrium* — a reaction that occurred in both the forward and reverse directions. By analyzing these two closely related concepts, energy and equilibrium, he developed a new process for producing large amounts of ammonia.

Haber's process changed the face of agriculture. Farmers today can grow four times more food per acre than was possible in 1900, making them better able to deal with the needs of a growing population. It has been estimated that 80% of the nitrogen in our bodies was once ammonia, produced through the Haber-Bosch process.

In this chapter, we're going to look at the concepts that were essential to unlocking the nitrogen problem. We'll see how energy changes drive chemical reactions to take place and determine how quickly these changes occur. Building on this concept, we'll turn our attention to equilibrium reactions. We'll describe equilibria conceptually as well as mathematically. Finally, we'll see how Haber was able to overcome the nitrogen problem and impact the world so profoundly.

Intended Learning Outcomes

After completing this chapter and working the practice problems, you should be able to:

13.1 Reaction Rates

- Describe the effects of concentration and temperature on reaction rates.
- Describe the energy changes that accompany chemical reactions using a reaction diagram.
- Describe the relationship between activation energy and rate of reaction.

13.2 Equilibrium Reactions

- Describe equilibrium in terms of the rates of opposite reactions.
- Given a reaction energy diagram, predict whether an equilibrium will favor the reactants or the products.

13.3 Equilibrium Expressions

- Write equilibrium expressions for equilibria involving solutions, solids, and gases.
- Relate the magnitude of the equilibrium constant, *K*, to equilibrium concentrations.

13.4 Le Chatelier's Principle

- Apply Le Chatelier's principle to describe how the addition or removal of reactants or products affects an equilibrium.
- Describe the effect of temperature and pressure changes on an equilibrium.

Figure 13.2 Different chemical reactions occur at different rates: Some are very fast while others take place much more slowly.

13.1 Reaction Rates

Some reactions, like the explosion of rocket fuel, happen very quickly. Other reactions, like the gradual rusting of iron, occur much more slowly (**Figure 13.2**). The speed of a chemical reaction is called its **reaction rate**. To understand why reactions take place at different rates, we must look more carefully at how they occur.

Chemical reactions occur when atoms or molecules collide. For example, consider the reaction of hydrofluoric acid with water to produce hydronium and fluoride ions:

$$HF\,(aq) + H_2O\,(l) \rightarrow F^-\,(aq) + H_3O^+\,(aq)$$

For this reaction to happen, the HF and H_2O molecules must collide with each other. However, not every collision produces a reaction. The molecules must collide with enough energy and at an orientation that allows the rearrangement to take place (**Figure 13.3**).

Three major factors that affect how collisions occur and the rates of chemical reactions are concentration, temperature, and energy changes. We'll discuss each of these in the sections that follow.

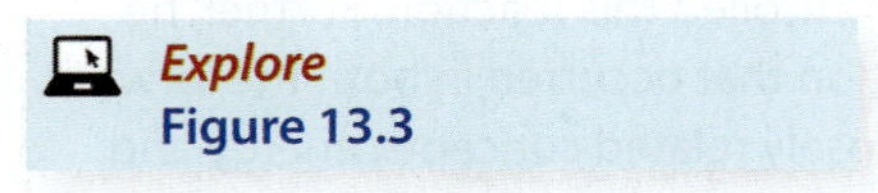

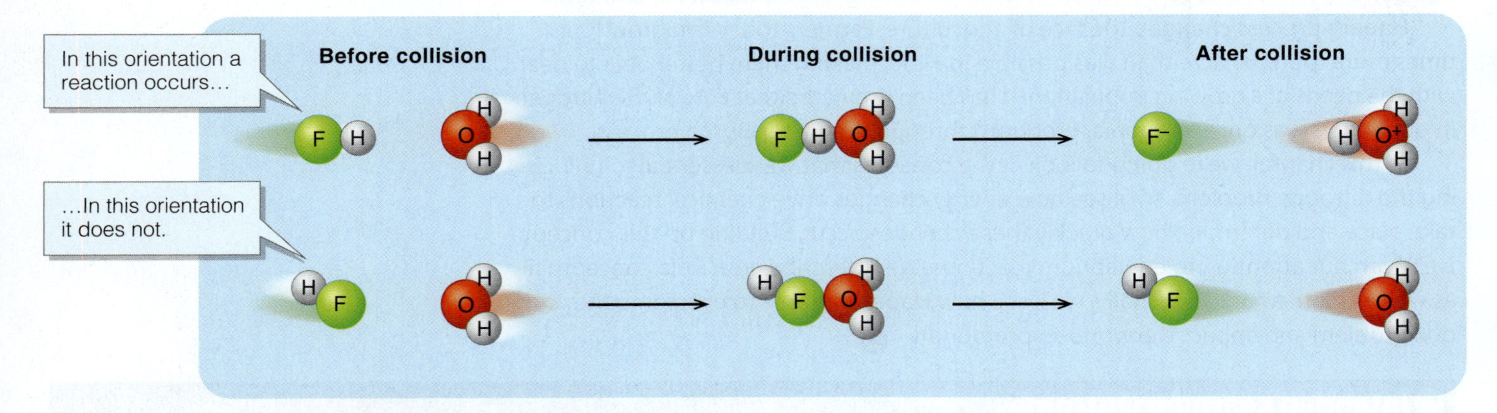

Figure 13.3 Chemical reactions occur when molecules collide. For a reaction to take place, the molecules must be oriented correctly relative to each other.

Reactions occur when atoms or molecules collide.

When a bat hits a baseball, the result depends on both the angle and the energy at which the bat and ball collide. In much the same way, the angle and energy of molecular collisions determine the outcome. Not every hit is a home run, and not every collision results in a reaction.

How Concentration and Temperature Affect Reaction Rates

Because chemical reactions require the collision of molecules, the rate of reactions depends in part on how often collisions occur. When molecules collide more frequently, reactions occur at a faster rate.

One way to increase the number of collisions is to increase the concentration of reactants. We can see this experimentally by reacting a metal (such as zinc) with hydrochloric acid (**Figure 13.4**). The dilute acid reacts slowly with the zinc, gradually forming a few bubbles of hydrogen gas. By contrast, the concentrated acid reacts violently, quickly producing a large volume of hydrogen gas.

Increasing concentration increases the reaction rate.

A second way to increase the rate of a reaction is to increase the temperature. Recall that at higher temperatures, molecules move more quickly. Whether the reaction takes place in a solid, liquid, or gas, this means that the molecules collide more frequently (**Figure 13.5**). It also means that when particles do collide, they do so more forcefully, making the molecules more likely to rearrange.

Increasing the temperature increases the reaction rate.

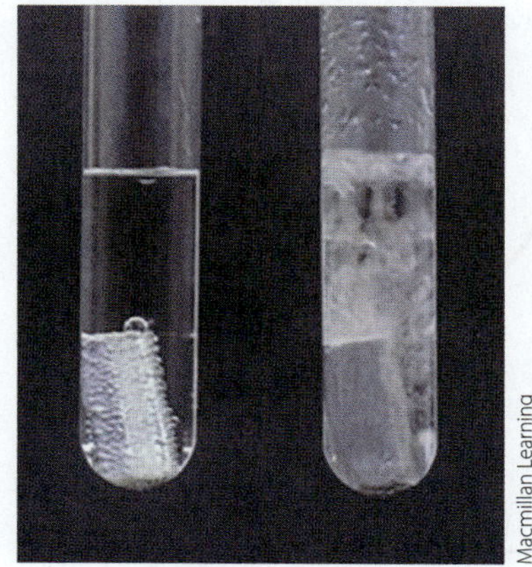

Figure 13.4 Increasing the concentration increases the reaction rate. Zinc reacts slowly with 1 M HCl but quickly with 6 M HCl.

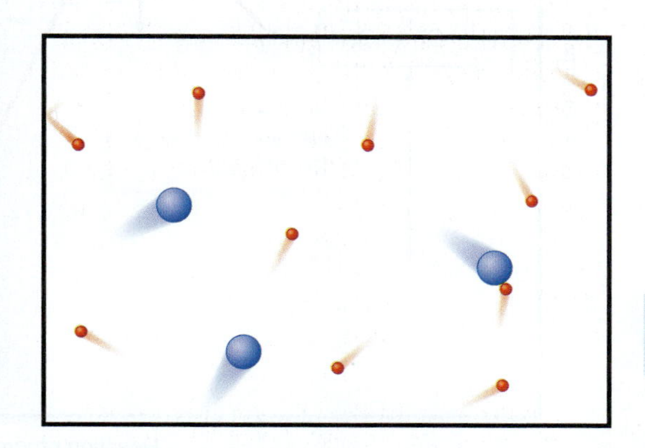

Figure 13.5 Increasing the temperature increases the velocity of the molecules in a gas. The faster the molecules move, the more often they collide.

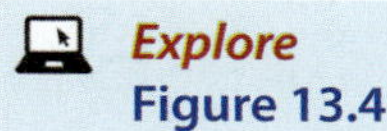

Figure 13.5

How Changes in Energy Affect Reaction Rates

Energy changes also affect reaction rates. Recall that chemical reactions can be *endothermic* (absorbing heat energy from their surroundings) or *exothermic* (releasing heat energy to their surroundings). At first glance, the relationship between energy change and reaction rate may seem unclear. For example, many exothermic reactions happen very quickly, but others are much slower (**Figure 13.6**). How can we explain this?

a

b

Figure 13.6 Reaction rates vary for different compounds. (a) The reaction between sodium and chlorine is exothermic. The two elements react immediately and violently. (b) The reaction of wood with oxygen is also exothermic, but the reaction takes place very slowly unless heat is added first.

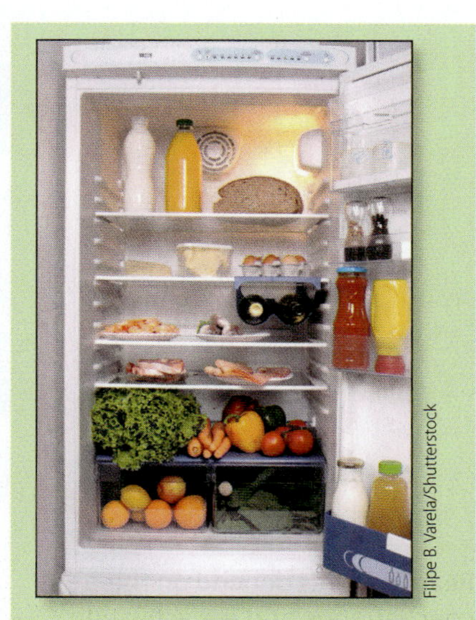

Keeping food in the refrigerator slows down the chemical reactions that cause food to spoil.

Describing Energy Changes in Chemistry: Reaction Energy Diagrams

To understand the relationships between energy changes and reaction rates, it's helpful to map the energetic factors that make reactions happen (or prevent them from happening). This type of map, called a **reaction energy diagram**, shows the energetic changes that accompany a chemical reaction.

Remember your first date? Or your first breakup? Even if it gets you to a more stable place, making or breaking bonds always involves an uncomfortable, high-energy transition.

For example, **Figure 13.7** shows a reaction energy diagram for the reaction of hydrochloric acid with hydroxide:

$$HCl\ (aq) + OH^- \ (aq) \rightarrow Cl^- \ (aq) + H_2O\ (l) \qquad \Delta H^\circ = -55.8 \text{ kJ/mol}$$

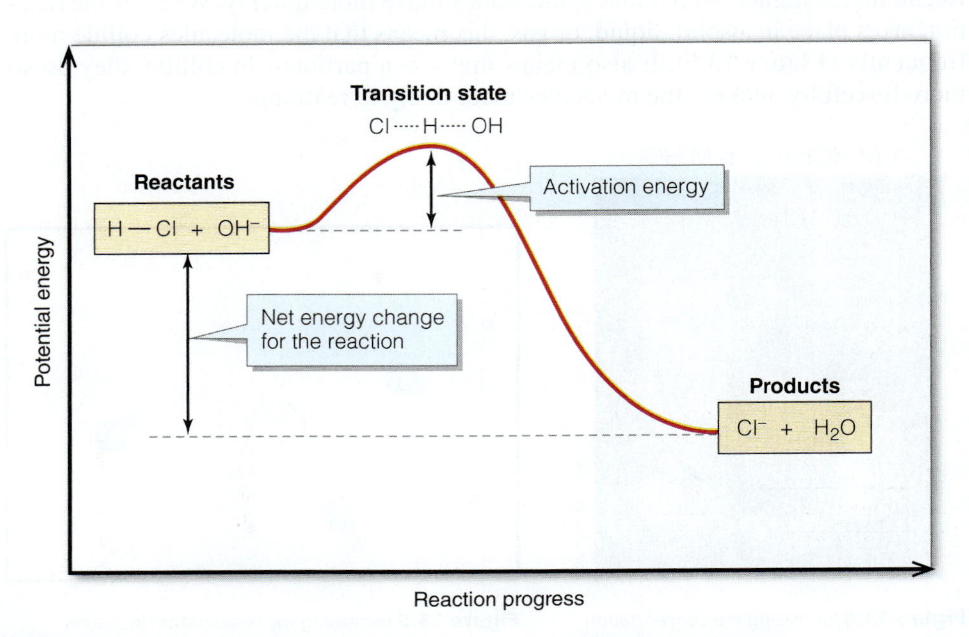

Figure 13.7 A reaction energy diagram shows energy changes as a reaction progresses.

Explore Figure 13.7

We know that the reaction is exothermic, as indicated by the negative enthalpy value. But does this reaction also *require* energy for the changes to occur? Let's think carefully about the reaction that takes place. For the hydroxide to take the hydrogen away from the chloride, there must be some instant when the hydrogen ion is pulled between the OH and the Cl, but it is not fully bonded to either. This instant, called the **transition state**, is higher in energy than either the reactants or the products.

$$\underset{\text{Reactants}}{Cl{-}H + {}^{\ominus}OH} \longrightarrow \underset{\text{Transition state}}{[Cl{\cdots}H{\cdots}OH]^{\ddagger}} \longrightarrow \underset{\text{Products}}{Cl^{\ominus} + H_2O}$$

The transition state is the highest-energy arrangement of atoms that occurs during a chemical reaction. This arrangement lasts only an instant, but it determines the energy "hill" that a reaction must get over. This energy barrier is called the **activation energy**. Along with concentration, the activation energy determines how quickly a reaction occurs. Even if a reaction will form a more stable product, if the activation energy is too high, the reaction will not take place.

Reaction rates depend on the concentration and the activation energy. ■

This idea also explains why many reactions require energy to get started. For example, the combustion of charcoal with oxygen is exothermic, but it doesn't happen until you add a match—and probably some lighter fluid (**Figure 13.8**). The match provides the activation energy, in the form of heat, to initiate the reaction. After that, the heat released by the combustion enables the reaction to continue.

As a rule of thumb, reactions with an activation energy of less than about 40 kJ/mol (or 10 kcal/mol) occur at room temperature. Reactions with a higher activation energy require heating before they take place.

Figure 13.8 The match provides enough heat to get over the activation energy and initiate the exothermic reaction.

Catalysts

A **catalyst** is a species that is not part of the balanced equation but causes the reaction to go more quickly. A catalyst stabilizes the transition state in the reaction. This *reduces the activation energy* required to go from reactant to product. As a result, reactions are able to go more quickly (**Figure 13.9**).

Catalysts are critically important in a wide variety of settings. In living creatures, compounds called *enzymes* act as catalysts to make complex reactions proceed quickly and with minimal energy cost. Manufacturers use catalysts to reduce the time, energy, and cost necessary to produce goods.

In Chapter 15, we'll discuss a specific catalyst used to make everything from medicines to plastics.

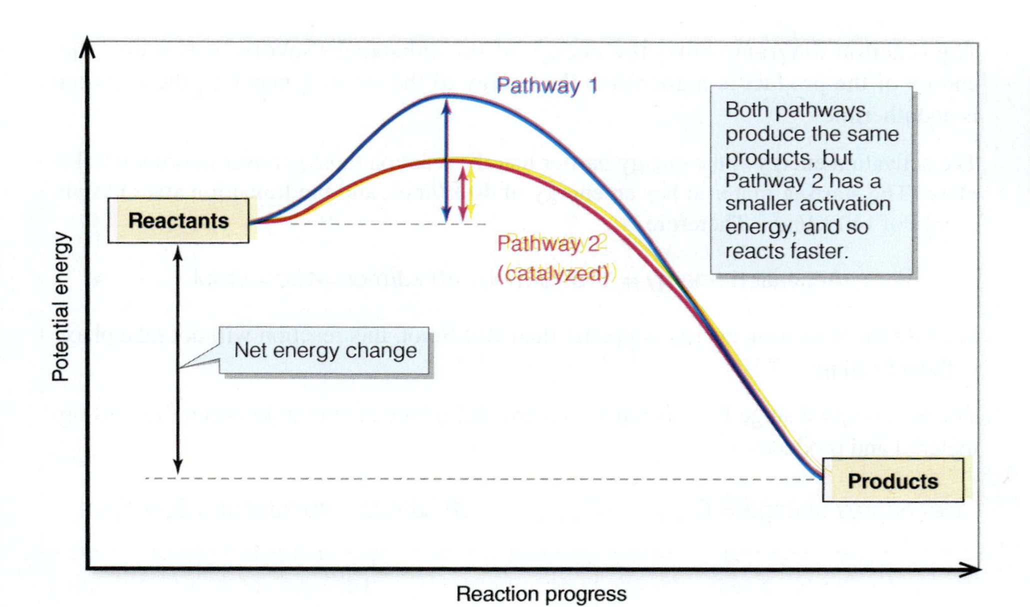

Figure 13.9 A catalyst reduces the activation energy, causing the reaction to occur more quickly.

Have you ever wanted to start a business? If so, you've probably thought about the initial investment. For example, to open a restaurant, you first need to build or rent a building, secure permits, hire staff, and develop a menu. The cost to do this is the activation energy for getting the business started. Regardless of how lucrative the business could be, if the resources are not available to get over the activation barrier, the business cannot get started. In the same way, even exothermic reactions don't take place if the activation energy is too high.

Example 13.1 Interpreting Reaction Energy Diagrams

Consider a simple reaction where compound A is converted to compound B. Based on the energy diagram shown, is this reaction endothermic or exothermic? Find both the activation energy and the net energy change for this reaction.

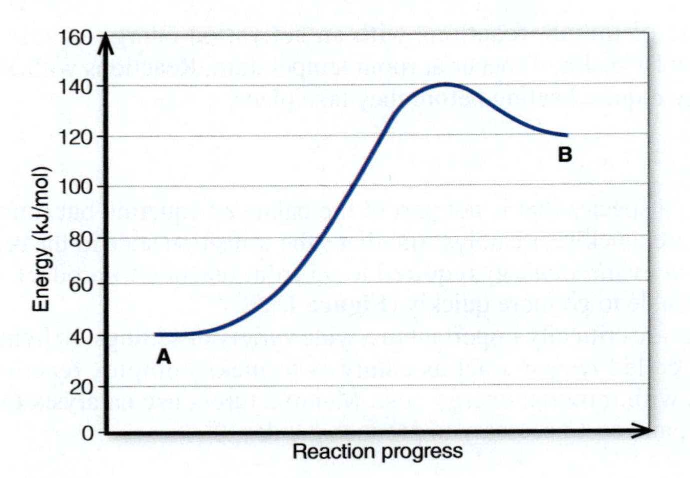

The reaction diagram shows the energy of the substances involved. Because the energy of the product is higher than the energy of the starting material, the reaction is endothermic.

The activation energy is the energy barrier that the reaction must get over in order to take place. The starting material has an energy of 40 kJ/mol, and the transition state has an energy of 140 kJ/mol. Therefore,

$$\text{Activation energy} = 140 \text{ kJ/mol} - 40 \text{ kJ/mol} = 100 \text{ kJ/mol}$$

Because the activation energy is greater than 40 kJ/mol, this reaction will not take place without heating.

The net energy change for this reaction is the difference in energy between the starting material and product.

$$\text{Net energy change} = E_{\text{product}} - E_{\text{reactant}} = 120 \text{ kJ/mol} - 40 \text{ kJ/mol} = 80 \text{ kJ/mol}$$

Check It
Watch Explanation

TRY IT

1. Compare the two reactions diagrammed below. What is the net energy change for each reaction? Which one is more exothermic? Assuming the concentrations and temperatures are the same, which reaction will occur more quickly?

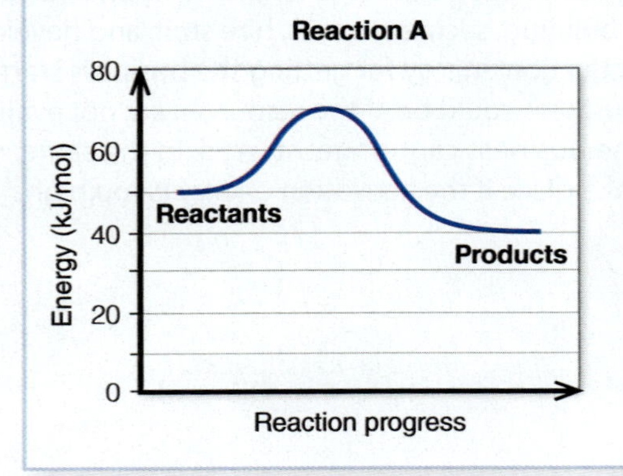

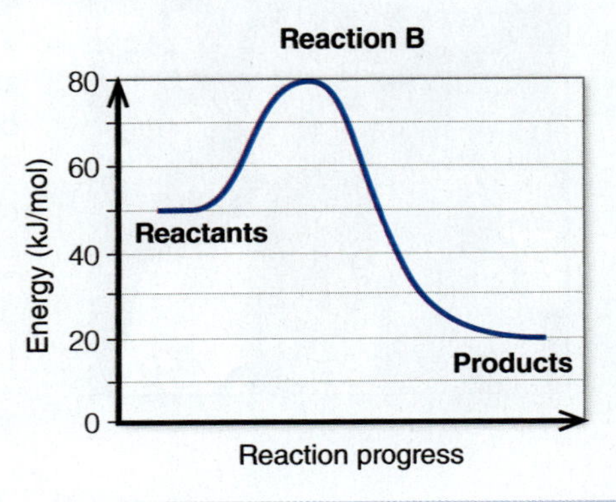

13.2 Equilibrium Reactions

Many reactions are *reversible*—that is, they can occur in both the forward and the reverse directions. For example, consider how hydrofluoric acid reacts in water:

$$HF(aq) \rightleftharpoons H^+(aq) + F^-(aq)$$

The reaction can go in the forward direction (the acid ionizes to form H^+ and F^-), but it can also go in the backward direction (the ions combine to reform HF). This type of back-and-forth reaction is called an **equilibrium reaction**. In a chemical equation, we use a double arrow ($\rightleftharpoons$) to show that the reaction proceeds in both directions.

At equilibrium, the forward and reverse reactions occur at the same rate, so the concentrations of reactants and products do not change.

Consider what happens if pure hydrogen fluoride (HF) dissolves in water: For the first instant, there is a large concentration of HF but essentially no H^+ or F^- ions. But as the HF molecules begin to ionize, the concentration of H^+ and F^- ions increases (**Figure 13.10**).

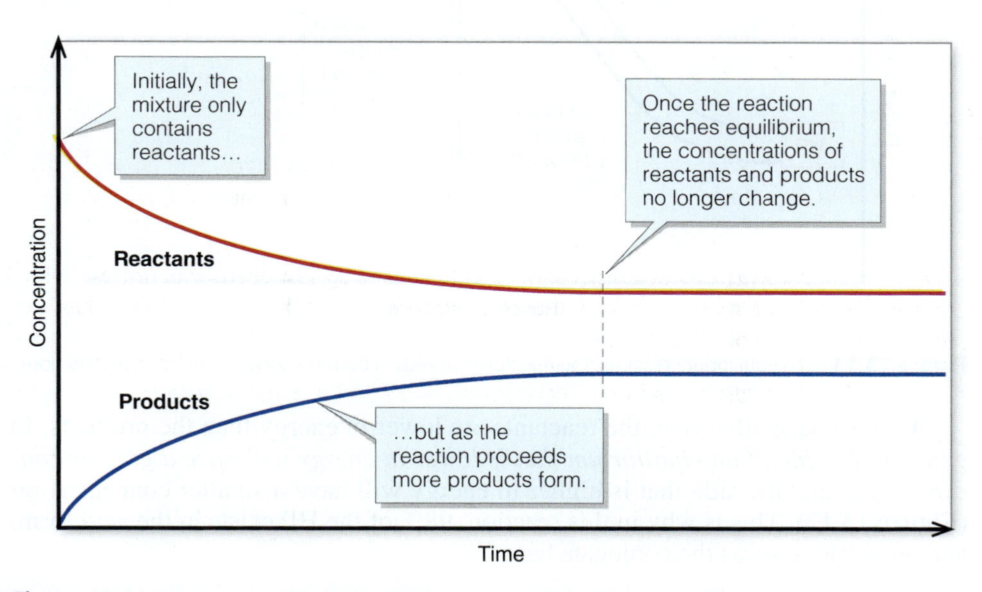

Figure 13.10 This graph shows how the concentrations of reactants and products change in an equilibrium reaction. Once a reaction reaches equilibrium, the concentrations of reactants and products no longer change.

As their concentrations increase, these ions begin to encounter each other in solution, and they react to form HF. Although HF molecules continue to ionize, the reverse reaction is also taking place. Soon the amount of each species stabilizes. Both the forward and reverse reactions are taking place, but because they are happening at the same rate, the amounts of each component are not changing. This situation is called **equilibrium**.

At equilibrium, the rates of the forward and reverse reactions are the same.

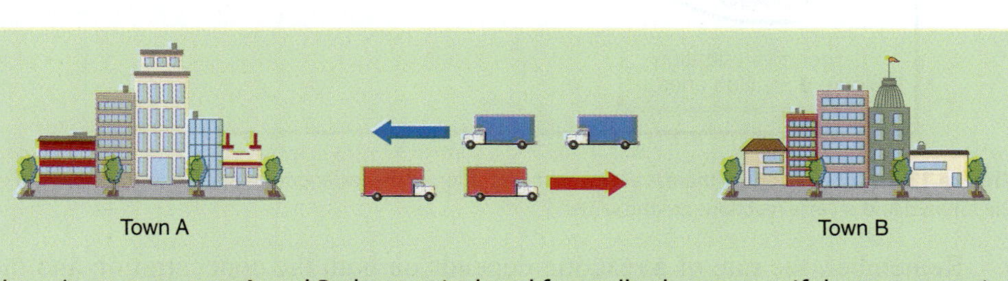

Imagine two towns, A and B, that are isolated from all other towns. If the towns are in equilibrium, the number of people who are moving in each direction is the same, so the overall populations of the towns don't change.

Some equilibrium reactions favor the reactants, while others favor the products. For example, recall that hydrofluoric acid, HF, is a weak acid. This means that at equilibrium, there is much more reactant (HF) than products (H^+ and F^-):

$$HF\,(aq) \rightleftharpoons H^+\,(aq) + F^-\,(aq)$$

98% 2%

We can analyze equilibrium reactions like this one by using reaction energy diagrams (**Figure 13.11**). In this reaction, the activation energy hill is small enough that the reactants can get over the hill to form products, but the products can also get over the hill in the reverse direction.

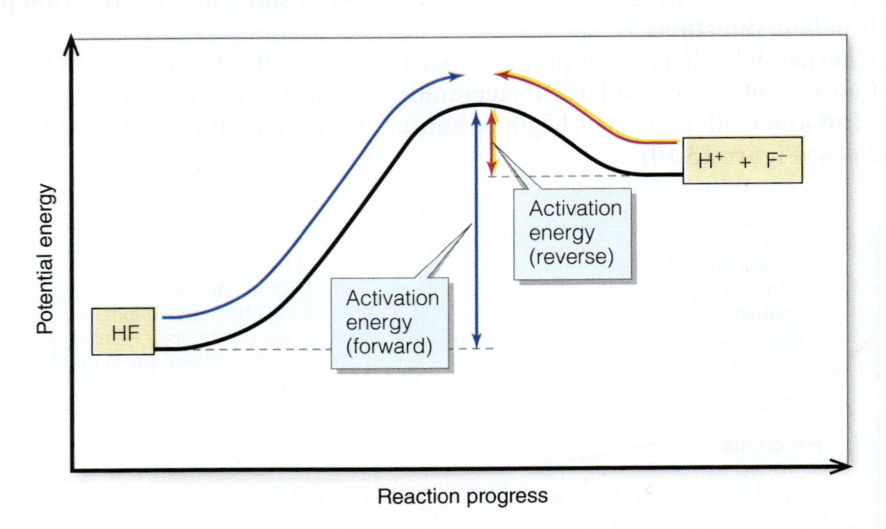

Figure 13.11 In an equilibrium reaction, the reaction proceeds in both the forward and reverse directions.

In this energy diagram, the reactants are lower in energy than the products. In general, *the side of an equilibrium that is lower in energy will have a greater concentration*, and the side that is higher in energy will have a smaller concentration (**Figure 13.12**). This is why in this reaction, 98% of the HF exists in the acid form, and only 2% exists as the conjugate base (F^-).

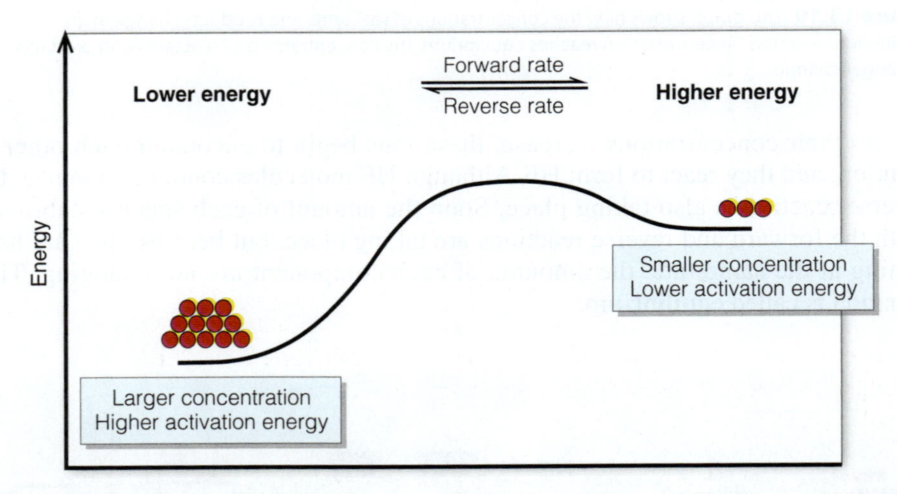

Figure 13.12 At equilibrium there is a larger concentration of the lower energy species, but the rates of the forward and reverse reactions are the same.

In general, equilibria favor the lower-energy species.

Remember, the rate of a reaction depends on both the concentration and the activation energy. The low-energy side of an equilibrium has a greater concentration but a higher activation energy. The high-energy side has a smaller concentration but a lower activation energy. Together, the balance of concentration and activation energy causes the forward and reverse reactions to occur at the same rate.

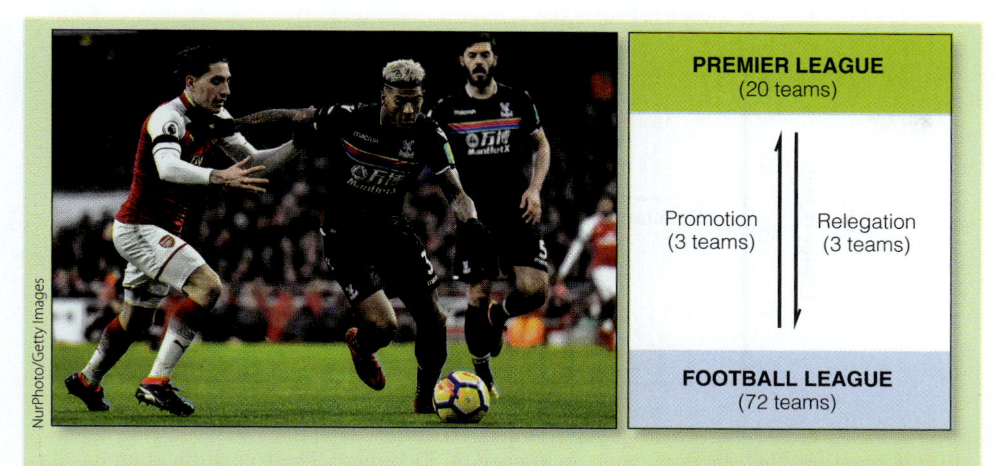

In the English football system, the top 20 teams play in the Premier League. Below this league is the 72-team Football League. After each season, the top three teams in the lower league are promoted to the higher league, and the bottom three teams in the higher league are relegated to the lower league. This is an example of an equilibrium: Because the rates in both directions are the same, the number of teams in each league doesn't change. And much like the Premier League, which has fewer teams, the higher energy levels in a chemical equilibrium have a lower concentration of species.

Example 13.2 Energy Diagrams and the Position of an Equilibrium

The equilibrium reaction and energy diagram given below represent the ionization of acetic acid, $HC_2H_3O_2$. At equilibrium, is there a higher concentration of reactants (left) or products (right) in solution?

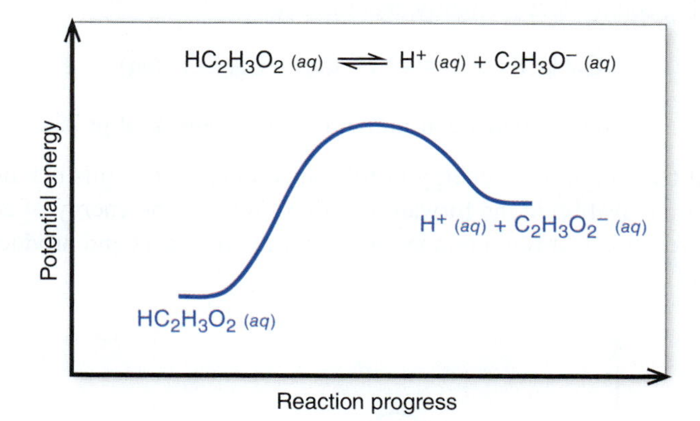

Notice in the figure that the potential energy is lower for the reactants than for the products. Because of this, the equilibrium favors the reactants.

Example 13.3 Finding Activation Energies in the Forward and Reverse Directions

The energy diagram for an equilibrium reaction is shown here. What is the activation energy for the forward reaction? What is the activation energy for the reverse reaction? Will this equilibrium favor the reactants or the products?

80
60
40
20
0
Energy (kJ/mol)
Reactants
Products
Reaction progress

In this reaction, the reactants have an energy of 50 kJ/mol. The transition state has an energy of 60 kJ/mol. This means that in the forward reaction, the activation energy is 10 kJ/mol:

$$\text{Activation energy (forward)} = \text{Energy (transition state)} - \text{Energy (reactants)}$$
$$= 60 \text{ kJ/mol} - 50 \text{ kJ/mol} = 10 \text{ kJ/mol}$$

The products have an energy of 20 kJ/mol. Again, the transition state has an energy of 60 kJ/mol. In the reverse reaction, the activation energy is 40 kJ/mol:

$$\text{Activation energy (reverse)} = \text{Energy (transition state)} - \text{Energy (products)}$$
$$= 60 \text{ kJ/mol} - 20 \text{ kJ/mol} = 40 \text{ kJ/mol}$$

Because the products are lower in energy than the reactants, there will be a higher concentration of products than reactants at equilibrium.

Check It
Watch Explanation

TRY IT

2. The following equilibrium is slightly endothermic:

$$C_6H_6OH\,(aq) \rightleftharpoons H^+\,(aq) + C_6H_5O^-\,(aq)$$

At equilibrium, is there a higher concentration of reactants or of products?

3. A chemical reaction has the energy profile shown below. For this reaction, what is the energy of activation in the forward reaction? What is the energy of activation in the reverse reaction? At room temperature, will the reactants and products reach an equilibrium?

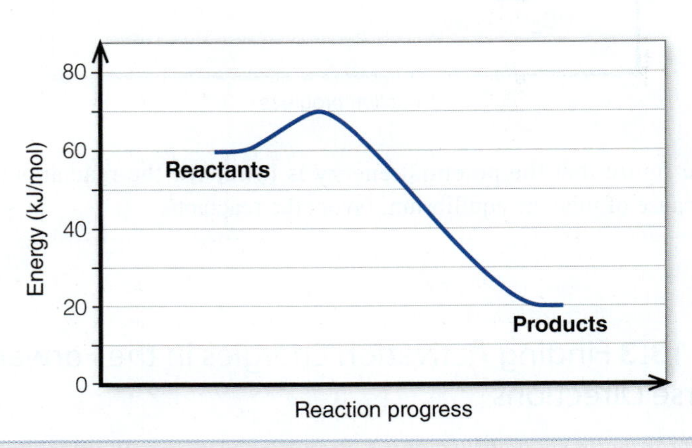

13.3 Equilibrium Expressions

Some equilibria strongly favor the products, while others strongly favor the reactants. For example, hydrochloric acid (HCl) is a strong acid that almost completely ionizes in aqueous solution. On the other hand, HF is a weak acid; only about 2% of HF ionizes in aqueous solution.

$$\underset{0.0001\%}{HCl} \rightleftharpoons \underset{99.9999\%}{H^+ + Cl^-}$$

$$\underset{98\%}{HF} \rightleftharpoons \underset{2\%}{H^+ + F^-}$$

When describing equilibria, chemists often need to describe how far to the left or the right an equilibrium lies. To do this, we write an equation called an **equilibrium expression**, which describes the balance between the reactants and products in an equilibrium. To see how equilibrium expressions work, let's begin with a generic chemical equation, where A and B are reactants, C and D are products, and the lowercase letters are the coefficients for each species.

$$aA + bB \rightleftharpoons cC + dD$$

For this equation, we write the equilibrium expression as follows:

$$K = \frac{[C]^c[D]^d}{[A]^a[B]^b}$$

[A] means "concentration of A."

Recall that the square brackets around each species mean the concentration of that species. For solutions, we express this as molarity (moles/L). In this equation, K is the **equilibrium constant** for the reaction. The value of K is very important *because it tells us whether the equilibrium favors the reactants (left-hand side) or the products (right-hand side).*

Put simply, equilibrium expressions show the concentrations of the products over the reactants:

$$K = \frac{\text{products}}{\text{reactants}}$$

For example, the ionization of HCl strongly favors the products, so the equilibrium constant has a large value. (In fact, this equilibrium favors the products so completely that it is normally written with just a single arrow.) On the other hand, the K value for the HF equilibrium is small, since the equilibrium favors the starting materials (**Figure 13.13**). The equilibrium constant does not have units.

Examples 13.4 and 13.5 illustrate the use of equilibrium expressions.

If K is large, the equilibrium favors the products. If K is small, the equilibrium favors the reactants.

HF ⇌ H+ + F−
98% 2%
K = [H+][F−] / [HF]
Numerator is much smaller than the denominator, so
K is very small.
HCl ⇌ H+ + Cl−
0.0001% 99.9999%
K = [H+][Cl−] / [HCl]
Numerator is much bigger than the denominator, so
K is very large.

Figure 13.13 Equilibria that favor the products have very large K values; those that favor the reactants have small K values.

Far to the left:
K is small

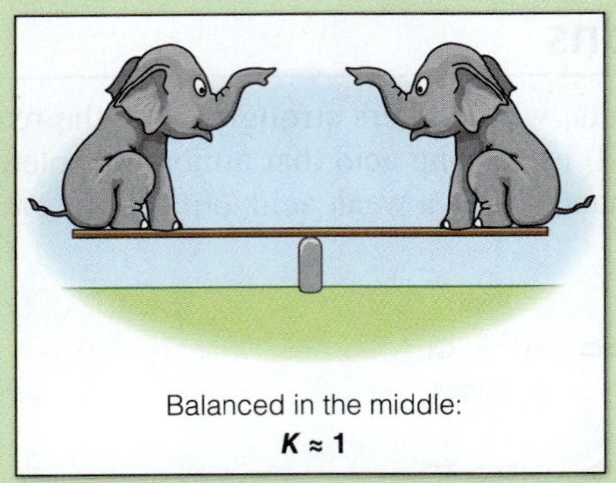
Balanced in the middle:
$K \approx 1$

Far to the right:
K is large

The equilibrium constant shows the balance between the left and the right side of an equation.

Example 13.4 Writing an Equilibrium Expression

When heated, molecular bromine breaks to form individual bromine atoms. In the reverse reaction, these atoms recombine to form molecular bromine. Write an equilibrium expression for this reaction:

$$Br_2 \rightleftharpoons 2\ Br$$

When writing an equilibrium expression, we place the products in the numerator and the reactants in the denominator. The concentration of each species is raised to the power of its coefficient. We therefore write the equilibrium expression as follows:

$$K = \frac{[Br]^2}{[Br_2]}$$

In this example, we don't know the value of the equilibrium constant, K. To find it, we must know the concentrations of each component in the equilibrium. Example 13.5 illustrates this process.

Example 13.5 Finding the Value of the Equilibrium Constant

Nitrous acid ionizes in water, as shown in the equilibrium reaction below. At equilibrium, an aqueous nitrous acid solution has these concentrations: $[H^+] = 0.0140\ M$; $[NO_2^-] = 0.0140\ M$; and $[HNO_2] = 0.493\ M$. What is the value of K for this equilibrium? Does this equilibrium lie to the right or the left?

$$HNO_2\ (aq) \rightleftharpoons H^+\ (aq) + NO_2^-\ (aq)$$

First, notice that the equilibrium concentration of HNO_2 is much higher than the concentrations of the products. Therefore, the equilibrium lies to the left, and we can expect the value of K to be small.

To find the value of K, we set up the equilibrium expression and solve:

$$K = \frac{[H^+][NO_2^-]}{[HNO_2]} = \frac{(0.0140\ M)(0.0140\ M)}{0.493\ M} = 3.98 \times 10^{-4}$$

Indeed, we find a very small value of K, indicating that the equilibrium lies far to the left. Notice that the value of K does not have units.

TRY IT

4. Write an equilibrium expression for these reactions:

$$AlCl_3\ (aq) + Cl^-\ (aq) \rightleftharpoons AlCl_4^-\ (aq)$$

$$Zn^{2+}\ (aq) + 4\ NH_3\ (aq) \rightleftharpoons [Zn(NH_3)_4]^{2+}\ (aq)$$

5. The ionization of acetic acid is shown in the equilibrium below. In water, a solution of acetic acid is found to have these concentrations: $[H^+] = 0.0040$ M; $[C_2H_3O_2^-] = 0.0040$ M; and $[HC_2H_3O_2] = 0.889$ M. What is the value of K for this equilibrium? Is this a strong or a weak acid?

$$HC_2H_3O_2\ (aq) \rightleftharpoons H^+\ (aq) + C_2H_3O_2^-\ (aq)$$

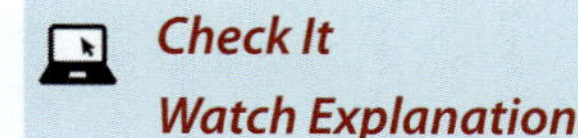

Check It
Watch Explanation

Equilibrium Expressions Involving Solvents

In some situations, special rules apply for writing equilibrium expressions. The three most important of these situations occur when an equilibrium involves a solvent, a solid, or gases.

When a chemical equilibrium involves a pure liquid or a solvent, this species is not included in the equilibrium expression. Because there is typically so much more of the solvent than there is of the other species, the concentration of the solvent doesn't change significantly. For this reason, we *do not include the solvent in the equilibrium expression.* The concentration of the solvent is considered to be part of the equilibrium constant, K. For example, we can write the ionization of HF in water in two different ways:

$$HF\ (aq) \rightleftharpoons H^+\ (aq) + F^-\ (aq) \qquad K = \frac{[H^+][F^-]}{[HF]}$$

or

$$HF\ (aq) + H_2O\ (l) \rightleftharpoons H_3O^+\ (aq) + F^-\ (aq) \qquad K = \frac{[H_3O^+][F^-]}{[HF]}$$

Because water is the solvent, we don't include the concentration of water in the expression; so, no matter which way we decide to write the expression, the value of K is the same. This type of expression is particularly important for describing the behavior of acids and bases.

Solvents are not included in equilibrium expressions. ■

Example 13.6 Writing Equilibrium Expressions When a Solvent Is Involved

When ammonia dissolves in water, it produces small amounts of hydroxide and ammonium ions through the equilibrium shown below. Write an equilibrium expression for this reaction:

$$NH_3\ (aq) + H_2O\ (l) \rightleftharpoons OH^-\ (aq) + NH_4^+\ (aq)$$

Because water is the solvent in this reaction, it is not included in the equilibrium expression.

$$K = \frac{[OH^-][NH_4^+]}{[NH_3]}$$

Example 13.7 Finding Concentrations of an Aqueous Acid

Hydrocyanic acid dissolves in water to produce H_3O^+ and CN^-, as in the equilibrium below. The equilibrium constant for this reaction is 4.9×10^{-10}. If $[H_3O^+]$ and $[CN^-]$ both equal 7.3×10^{-6} M, what is the concentration of HCN in the solution?

$$HCN\ (aq) + H_2O\ (l) \rightleftharpoons H_3O^+\ (aq) + CN^-\ (aq)$$

As before, we do not include water in the equilibrium expression:

$$K = \frac{[H_3O^+][CN^-]}{[HCN]}$$

We know the values of K, $[H_3O^+]$, and $[CN^-]$. We therefore rearrange the equation to solve for [HCN] and then substitute the values given above. From this, we find the concentration of HCN.

$$[HCN] = \frac{[H_3O^+][CN^-]}{K} = \frac{(7.3 \times 10^{-6}\ M)(7.3 \times 10^{-6}\ M)}{4.9 \times 10^{-10}} = 0.11\ M$$

Check It
Watch Explanation

TRY IT

6. The following equilibria take place in water. Write equilibrium expressions for each one.

a. $CrO_4\ (aq) + H_2O\ (l) \rightleftharpoons HCrO_4^+\ (aq) + OH^-\ (aq)$

b. $2\ H_2O\ (l) \rightleftharpoons H_3O^+\ (aq) + OH^-\ (aq)$

c. $NH_4^+\ (aq) + H_2O\ (l) \rightleftharpoons NH_3\ (aq) + H_3O^+\ (aq)$

7. Buffer solutions contain a mixture of a weak acid and their conjugate base. A chemist prepares an acetate buffer solution where $[HC_2H_3O_2] = 0.100$ M and $[C_2H_3O_2^-] = 0.200$ M at equilibrium. Based on the equilibrium data below, what is the concentration of H_3O^+ in this solution?

$$HC_2H_3O_2\ (aq) + H_2O\ (l) \rightleftharpoons H_3O^+\ (aq) + C_2H_3O_2^-\ (aq) \qquad K = 1.8 \times 10^{-5}$$

Equilibrium Expressions Involving Solids

Equilibria involving solids are very common. For example, think about making a pitcher of sweet tea. A modest amount of sugar dissolves in the tea. But if we add too much sugar, we reach a saturation point where no more sugar can dissolve. When this happens, the undissolved sugar settles in the bottom of the pitcher (**Figure 13.14**).

In this delicious situation, the sugar reaches an equilibrium with the aqueous solution. Some sugar is in the solid form, and some is dissolved. The individual sugar molecules are able to move in and out of solution:

$$C_{12}H_{22}O_{11}\ (s) \rightleftharpoons C_{12}H_{22}O_{11}\ (aq)$$

Courtesy of Kevin Revell

Figure 13.14 Once a pitcher of sweet tea is saturated, adding more sugar does not change the sweetness. The amount of solid present does not affect the equilibrium.

Solids are not included in equilibrium expressions.

Think about this for a moment: Once the solution is saturated, can we make the tea sweeter by adding more sugar? No. If we add more sugar, it simply falls to the bottom of the pitcher. The presence of additional solid sugar does not change the concentration of dissolved sugar.

This leads us to an important general rule: When equilibria involve solids, the amount of solid present doesn't affect the concentration of the other components. Because of this, *materials in the solid phase are not included in equilibrium expressions.*

Solubility Products

We often encounter equilibria involving solids in the laboratory. For example, suppose a chemist mixes solid calcium hydroxide with water. Some of the solid dissolves, but not all of it. Once the solution reaches equilibrium, the excess solid sits on the bottom of the container (**Figure 13.15**). We can write this equilibrium as follows:

$$Ca(OH)_2\ (s) \rightleftharpoons Ca^{2+}\ (aq) + 2\ OH^-\ (aq)$$

How would we write the equilibrium expression for this equation? Since $Ca(OH)_2$ is a solid, we do not include it in the expression; instead, we put a value of one in the denominator:

$$K = \frac{[\text{products}]}{[\text{reactants}]} = \frac{[Ca^{2+}][OH^-]^2}{1} = [Ca^{2+}][OH^-]^2$$

In this expression, the equilibrium constant, K, tells us the amount of calcium and hydroxide ions that can dissolve in solution. The equilibrium constant for the solution of an ionic compound is called a **solubility product**. These constants are often written as K_{sp}.

As with other equilibrium expressions, a small K_{sp} value means the equilibrium lies far to the left. The smaller the K_{sp} value, the less soluble the compound is. The solubility products for several common compounds are shown in **Table 13.1**.

Figure 13.15 This saturated solution has undissolved solid at the bottom.

In Chapter 6 we used solubility rules to broadly classify compounds as "soluble" and "insoluble." While these rules are helpful, many compounds are very slightly soluble. The solubility product allows us to predict the composition of solutions much more precisely than if we use the broad solubility rules.

TABLE 13.1 Solubility Products for Some Ionic Compounds at 25 °C

Silver Halides	K_{sp}	Lead(II) Halides	K_{sp}
AgCl	1.77×10^{-10}	$PbCl_2$	1.70×10^{-5}
AgBr	5.35×10^{-13}	$PbBr_2$	6.60×10^{-6}
AgI	8.52×10^{-17}	PbI_2	9.8×10^{-9}
Hydroxides	K_{sp}	**Carbonates**	K_{sp}
$Ca(OH)_2$	5.02×10^{-6}	$BaCO_3$	2.58×10^{-9}
$Mg(OH)_2$	5.61×10^{-12}	$FeCO_3$	3.13×10^{-11}
$Fe(OH)_2$	4.87×10^{-17}	$MgCO_3$	6.82×10^{-6}
$Fe(OH)_3$	2.79×10^{-39}	$ZnCO_3$	1.46×10^{-10}

Solubility products are normally used only for slightly soluble compounds. More soluble compounds are typically described by their solubility in moles/liter or in grams/liter.

Example 13.8 Determining the Solubility Product

In a saturated solution of manganese(II) carbonate at 25 °C, both $[Mn^{2+}]$ and $[CO_3^{2-}]$ are equal to 4.37×10^{-6} M. Write an equilibrium expression and calculate the value of K_{sp} for this compound.

We begin by writing the solubility equilibrium for this compound:

$$MnCO_3\ (s) \rightleftharpoons Mn^{2+}\ (aq) + CO_3^{2-}\ (aq)$$

Based on this, we can write the following equilibrium expression:

$$K_{sp} = [Mn^{2+}][CO_3^{2-}]$$

Substituting in the concentrations for each ion, we obtain a K_{sp} value of 1.91×10^{-11}.

$$K_{sp} = [Mn^{2+}][CO_3^{2-}] = (4.37 \times 10^{-6}\ M)(4.37 \times 10^{-6}\ M) = 1.91 \times 10^{-11}$$

Example 13.9 Determining the Concentration from the Solubility Product

The solubility product of barium sulfate is 1.08×10^{-10}. If barium sulfate is dissolved in pure water, what is the concentration of barium ions?

We begin by writing the solubility equilibrium for this compound:

$$BaSO_4\,(s) \rightleftharpoons Ba^{2+}\,(aq) + SO_4^{2-}\,(aq)$$

When it dissociates, each unit of barium sulfate produces one barium ion and one sulfate ion. Because of this, the concentrations of the two ions must be the same. If we call the unknown concentration *x*, we can say that

$$x = [Ba^{2+}] = [SO_4^{2-}]$$

Now let's substitute *x* for the concentrations of the two ions, and substitute the value of K_{sp} given in the problem. Taking the square root of both sides, we find that $x = 1.04 \times 10^{-5}$. Therefore, $[Ba^{2+}] = 1.04 \times 10^{-5}$ M.

$$K_{sp} = [Ba^{2+}][SO_4^{2-}]$$

$$1.08 \times 10^{-10} = x^2$$

$$\sqrt{1.08 \times 10^{-10}} = x$$

$$1.04 \times 10^{-5}\text{ M} = x = [Ba^{2+}]$$

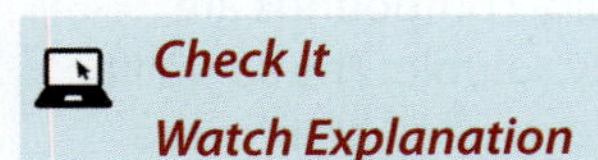

TRY IT

8. Write an equilibrium expression for each of these solubility equilibria:

   a. $NaCl\,(s) \rightleftharpoons Na^+\,(aq) + Cl^-\,(aq)$

   b. $MgBr_2\,(s) \rightleftharpoons Mg^{2+}\,(aq) + 2\,Br^-\,(aq)$

   c. $Fe(OH)_3\,(s) \rightleftharpoons Fe^{3+}\,(aq) + 3\,OH^-\,(aq)$

9. The solubility products for three compounds are shown below. Based on the K_{sp} values given, which compound is most soluble? Which compound is least soluble?

   a. $BaSO_4\,(s) \rightleftharpoons Ba^{2+}\,(aq) + SO_4^{2-}\,(aq)$ $\quad K_{sp} = 1.1 \times 10^{-10}$

   b. $LiF\,(s) \rightleftharpoons Li^+\,(aq) + F^-\,(aq)$ $\quad K_{sp} = 3.8 \times 10^{-3}$

   c. $Ni(OH)_2\,(s) \rightleftharpoons Ni^{2+}\,(aq) + 2\,OH^-\,(aq)$ $\quad K_{sp} = 2.0 \times 10^{-15}$

Equilibrium Expressions Involving Gases

Many equilibria involve gases. Equilibrium expressions for gases are identical to those for aqueous solutions, except that we use the partial pressures of the gases rather than molar concentrations. For example, at high temperatures, Br_2 and Cl_2 gases react to form BrCl. We write the equilibrium and the equilibrium expression for this reaction as shown below, where P_{Br_2}, P_{Cl_2}, and P_{BrCl} represent the partial pressure from each gas:

$$Br_2\,(g) + Cl_2\,(g) \rightleftharpoons 2\,BrCl\,(g) \qquad K = \frac{P_{BrCl}^{\,2}}{P_{Br_2}P_{Cl_2}}$$

Example 13.10 Finding Gas Pressures Using an Equilibrium Expression

A common method of manufacturing hydrogen gas involves heating a mixture of methane and steam to produce hydrogen and carbon monoxide, as shown in the equilibrium below. Given the following equilibrium data, what is the partial pressure of CO in an equilibrium mixture if $P_{CH_4} = 1.0$ *atm,* $P_{H_2O} = 1.0$ *atm, and* $P_{H_2} = 0.10$ *atm?*

$$CH_4\ (g) + H_2O\ (g) \rightleftharpoons 3\ H_2\ (g) + CO\ (g) \qquad K = 1.5 \times 10^{-6}$$

To solve this problem, we write the equilibrium expression using partial pressures:

$$K = \frac{P_{H_2}{}^3 P_{CO}}{P_{CH_4} P_{H_2O}}$$

To solve for the pressure of CO, we rearrange the equation and substitute the values given above. Under these conditions, we find that 1.5×10^{-3} atm of CO are present.

$$P_{CO} = \frac{K P_{CH_4} P_{H_2O}}{P_{H_2}{}^3} = \frac{(1.5 \times 10^{-6})(1.0\ \text{atm})(1.0\ \text{atm})}{(0.10\ \text{atm})^3} = 1.5 \times 10^{-3}\ \text{atm}$$

TRY IT

10. Ammonia is an important fertilizer, cleaning agent, and raw material for many nitrogen-containing products. Ammonia is typically prepared through the gas-phase reaction of nitrogen and hydrogen.

$$N_2\ (g) + 3\ H_2\ (g) \rightleftharpoons 2\ NH_3\ (g)$$

At 600 °C and high pressure, the equilibrium constant, *K*, for this process is 1.5×10^{-3}. If a gas mixture at equilibrium contains H_2 with a partial pressure of 10.0 atmospheres and N_2 with a partial pressure of 83.0 atmospheres, find the partial pressure of ammonia in the mixture.

Check It

Watch Explanation

13.4 Le Chatelier's Principle

At the beginning of this chapter, we introduced ammonia (NH_3), a compound of vital importance to the world's food supply. Above about 400 °C, ammonia forms by the reversible combination of nitrogen and hydrogen gas:

$$N_2\ (g) + 3\ H_2\ (g) \rightleftharpoons 2\ NH_2\ (g)$$

However, there is a problem. At higher temperatures, the equilibrium for this reaction strongly favors the starting materials. This raises a crucial question: Is it possible to control an equilibrium reaction? Can we drive an equilibrium to one side or another?

Fortunately, it is possible to do this. To understand how, let's begin with another simple equilibrium: Imagine two pools, connected by a pipe running along the bottom (**Figure 13.16**). Because water can pass back and forth between the pools, it is an equilibrium system. What happens if we add more water to the pool on the left? Water flows to the right, causing the level in the right pool to rise, until the pools level off. What happens if we remove water from the left pool? The water will flow toward the left pool, and again the pools will level off. You intuitively know that at equilibrium, the height of the water in both pools will always be the same.

Explore Figure 13.16

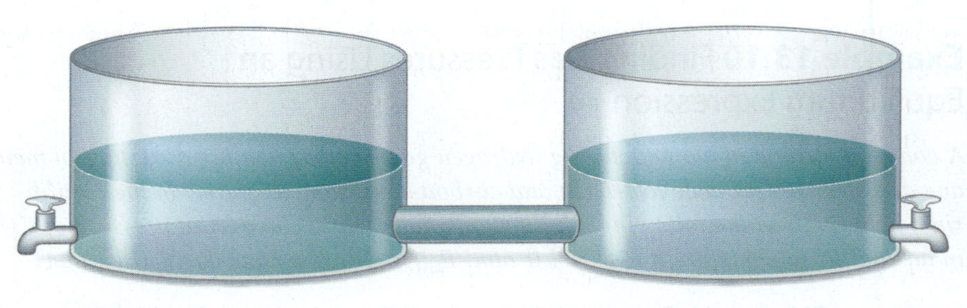

Figure 13.16 What happens if we add water to one of the pools? What if we remove water?

The same idea applies to chemical systems: *When some change takes place in concentration, temperature, pressure, or other key factor, the equilibrium shifts to minimize that change, and a new equilibrium is established.* This idea is called **Le Chatelier's principle**. In the sections that follow, we'll explore this important—concept in detail.

Equilibrium and Concentration

Changes in concentration can push the equilibrium to the right or to the left. For example, consider the equilibrium of HF with water:

$$HF\ (aq) + H_2O\ (l) \rightleftharpoons H_3O^+\ (aq) + F^-\ (aq)$$

What happens if we add more HF to this mixture? The additional HF reacts with the water, forming more F^-, until the levels of HF and F^- again reach equilibrium. Much like the pool analogy, increasing the concentration on one side also increases the concentration on the other side. Similarly, decreasing the concentration also affects both sides.

Another useful example is the evaporation of water. At room temperature, water interconverts between the liquid phase and the gas phase:

$$H_2O\ (l) \rightleftharpoons H_2O\ (g)$$

Imagine a sealed container, partially filled with water (**Figure 13.17**). Inside the container are water molecules in both the liquid and gas phases. The water molecules

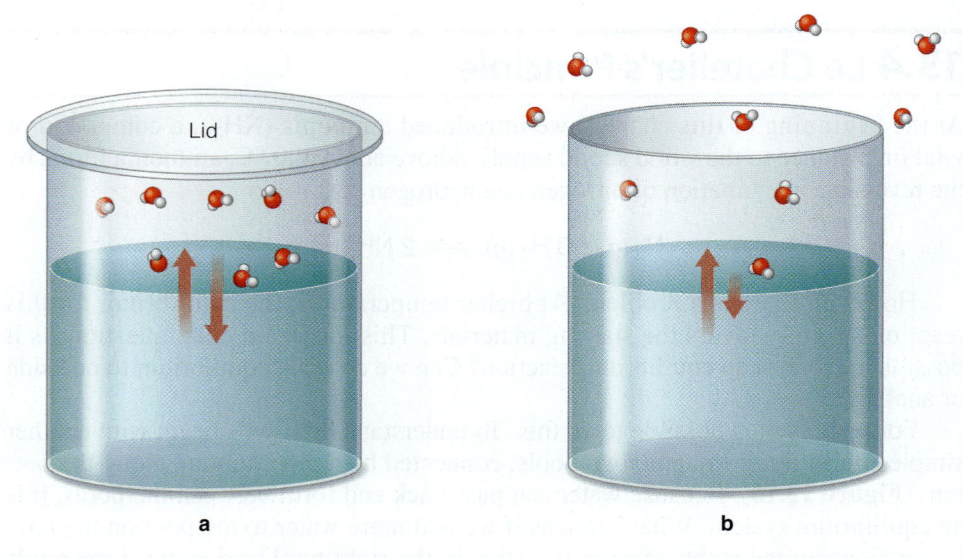

Figure 13.17 Water molecules can move back and forth between the liquid and gas phases. (a) At equilibrium, the same number of molecules move from liquid to gas as from gas to liquid. (b) If the lid is removed, the water vapors are pulled away from the liquid. The system can't reach equilibrium, and eventually all of the water evaporates.

are constantly moving. Some molecules transition from liquid to gas while others move from gas to liquid. At equilibrium, the number of molecules moving in either direction is the same.

But what if we take the cap off the container? This allows the molecules in the vapor phase to drift away from the liquid, and the system never reaches equilibrium. As you might expect, the water will continue to evaporate until there is no liquid left.

As a general rule, if we add something to an equilibrium, it pushes the equilibrium toward the other side. On the other hand, if we remove something, it pulls the equilibrium toward the side it was removed from. The example of Le Chatelier's swimming pool, shown again in **Figure 13.18**, is a good way to understand and remember this concept.

Adding to an equilibrium pushes it toward the other side.

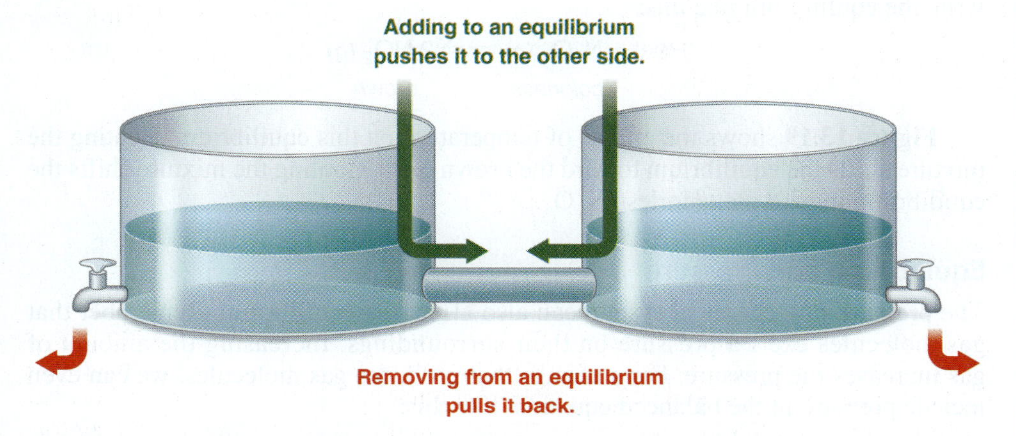
Adding to an equilibrium pushes it to the other side.
Removing from an equilibrium pulls it back.

Figure 13.18 Le Chatelier's principle says that adding to an equilibrium pushes it to the other side, but removing from an equilibrium pulls it back.

Example 13.11 Predicting Equilibrium Changes Using Le Chatelier's Principle

A reaction vessel has the following compounds present in equilibrium:

$$CH_3CO_2H + CH_3OH \rightleftharpoons CH_3CO_2CH_3 + H_2O$$

a. *Would adding water shift this equilibrium to the left or to the right?*

b. *What would be the effect of adding CH_3OH to the mixture?*

Any time we add something to an equilibrium, the reaction shifts toward the other side until balance is restored. Adding water would push the equilibrium away from water in the equation—that is, it would push the equilibrium to the left. On the other hand, adding CH_3OH would push the equilibrium toward the right.

TRY IT

11. Carbonated beverages (such as soda pop) are packaged under pressurized carbon dioxide. The CO_2 reacts with water in an equilibrium to produce carbonic acid (H_2CO_3), a weak acid that gives these drinks their sharp flavor. After the soda pop is opened, the CO_2 escapes. How does removing the CO_2 affect the concentration of H_2CO_3 in the soda?

$$CO_2\,(g) + H_2O\,(l) \rightleftharpoons H_2CO_3\,(aq)$$

Check It
Watch Explanation

Heating shifts an equilibrium toward the higher-energy side.

Equilibrium and Temperature

Le Chatelier's principle also applies to temperature changes. For example, we could write the evaporation of water like this:

$$\text{Heat} + H_2O\ (l) \rightleftharpoons H_2O\ (g)$$

The evaporation of water is an endothermic change—heat is being added in. Heating the water drives the equilibrium to the right, toward the gas. On the other hand, removing heat (cooling the system down) makes the equilibrium shift to the left, toward the liquid.

Another example of this principle is the gas-phase equilibrium reaction between N_2O_4 and NO_2. N_2O_4 is colorless. However, in the gas phase, this compound forms an equilibrium with NO_2, which is brown. The reaction is endothermic. We can write the equilibrium like this:

$$\text{Heat} + \underset{\textit{colorless}}{N_2O_4\ (g)} \rightleftharpoons \underset{\textit{brown}}{2\ NO_2\ (g)}$$

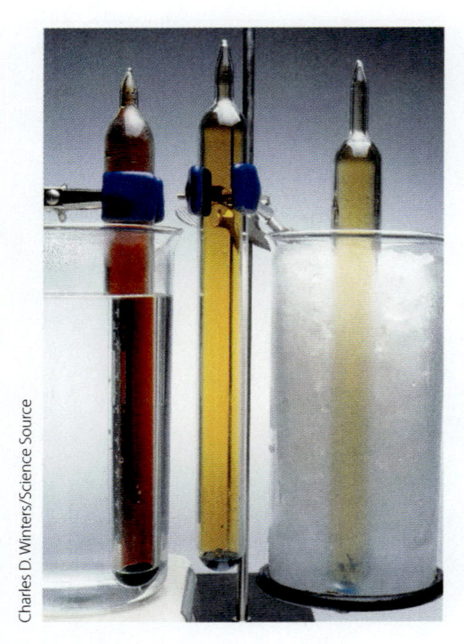

Charles D. Winters/Science Source

Figure 13.19 N_2O_4 and NO_2 exist in equilibrium. The three containers shown hold the same amount of these materials. Heating the mixture shifts the equilibrium toward the brown NO_2. Cooling the mixture shifts the equilibrium toward the colorless N_2O_4.

Explore Figure 13.19

Figure 13.19 shows the effects of temperature on this equilibrium. Heating the mixture shifts the equilibrium toward the brown NO_2. Cooling the mixture shifts the equilibrium toward the colorless N_2O_4.

Equilibrium and Pressure

The pressure of a chemical system can also affect the equilibrium. Remember that gas molecules exert a pressure on their surroundings. Increasing the amount of gas increases the pressure. For a change that produces gas molecules, we can even include pressure in the balanced equation, like this:

$$NH_3\ (l) \rightleftharpoons NH_3\ (g) + \text{pressure}$$

Again, Le Chatelier's principle applies. Adding pressure shifts the equilibrium toward the liquid. Removing pressure shifts the equilibrium toward the gas. This idea has many practical applications. For example, ammonia manufacturers use high pressure to force the ammonia into liquid form, making it easier to store and transport (**Figure 13.20**).

Fritz Haber used this important concept to drive the synthesis of ammonia. Recall that ammonia forms in an equilibrium with nitrogen and hydrogen:

$$\underbrace{N_2\ (g) + 3\ H_2\ (g)}_{\text{4 moles of gas}} \rightleftharpoons \underbrace{2\ NH_3\ (g)}_{\text{2 moles of gas}}$$

High pressure shifts equilibria toward the side with fewer moles of gas.

The left-hand side of the equation contains four moles of gas, while the right-hand side contains only two moles of gas. Haber realized that high pressures shifted the equilibrium toward the side with fewer moles of gas—that is, toward the ammonia. As a general rule, *increasing the pressure on a chemical system shifts the equilibrium toward the side with fewer moles of gas.*

DHuss/Getty Images

Figure 13.20 Under high pressure, ammonia is a liquid. Decreasing the pressure shifts the equilibrium toward the gas phase.

The Ammonia Problem and the Haber-Bosch Process

$$N_2\,(g) + 3\,H_2\,(g) \rightleftharpoons 2\,NH_3\,(g) + \text{heat}$$

The synthesis of ammonia from nitrogen posed a significant challenge. Although the reaction is exothermic, the activation energy is too high for the reaction to take place at room temperature. Heating the mixture allows the reaction to take place, but it shifts the equilibrium toward the starting material so that very little product forms at high temperature.

Fritz Haber used three techniques to solve this problem. First, he used an iron catalyst to lower the activation energy. Second, he ran the reaction at high pressure. Because there are fewer moles of gas in the products than in the reactants, high pressure pushes the equilibrium toward the product. Third, he removed ammonia from the mixture as it formed to pull the reaction toward the formation of products. Today the manufacture of ammonia relies on all three of these techniques.

Example 13.12 Predicting Equilibrium Changes Using Le Chatelier's Principle

Nitrosyl chloride, NOCl, is a toxic gas that forms in equilibrium with nitrogen monoxide and chlorine gas, as shown in this equation:

$$2\,NO\,(g) + Cl_2\,(g) \rightleftharpoons 2\,NOCl\,(g) \qquad \Delta H = -75.54 \text{ kJ/mol}$$

Based on this information, determine whether the following changes would shift the equilibrium to the left or to the right:

a. *adding Cl_2 to the mixture*

b. *increasing the temperature*

c. *increasing the pressure*

Any time we add something to an equilibrium, it pushes the equilibrium toward the other side. Adding Cl_2 will shift the equilibrium to the right, and so the amount of NOCl will also increase.

Notice that the ΔH value for this reaction is negative. This means that the reaction is exothermic. In other words, we could say that heat is a product for this reaction:

$$2\,NO\,(g) + Cl_2\,(g) \rightleftharpoons 2\,NOCl\,(g) + \text{heat}$$

Therefore, increasing the temperature for this reaction should drive the equilibrium to the left.

Finally, notice that the reactants are three moles of gas while the products are only two. Increasing the pressure will shift the equilibrium toward the products because there are fewer moles of gas on that side.

TRY IT

12. The Haber-Bosch process for making ammonia from nitrogen and hydrogen was introduced at the beginning of this chapter. Indicate whether each of the following changes would increase or decrease the amount of ammonia produced:

$$N_2\,(g) + 3\,H_2\,(g) \rightleftharpoons 2\,NH_3\,(g) + \text{heat}$$

a. adding more N_2 to the mixture

b. removing NH_3 from the mixture

c. adding NH_3 to the mixture

d. adding a catalyst

e. increasing the reaction temperature from 300 °C to 400 °C

Check It

Watch Explanation

Capstone Video

TEK IMAGE/SCIENCE PHOTO LIBRARY/Getty Images

Figure 13.21 Biologists use buffer solutions to study the reactions that take place in living systems.

Capstone Question

When biologists study the chemical reactions that occur in living systems, they often use buffer solutions to mimic the pH and osmotic pressure inside a cell (**Figure 13.21**). A common buffer solution consists of dihydrogen phosphate ($H_2PO_4^-$) and its conjugate base. This acid-base pair undergoes the equilibrium reaction shown here:

$$H_2PO_4^- (aq) + H_2O (l) \rightleftharpoons H_3O^+ (aq) + HPO_4^{2-} (aq) \qquad K = 6.31 \times 10^{-8}$$

To study the effects of a new antibacterial compound, a biologist prepares a buffer in which $[H_2PO_4^-]$ and $[HPO_4^{2-}]$ both equal 0.200 M. Write an equilibrium expression for this equation, then solve for $[H_3O^+]$ and find the pH.

To make the pH consistent with the fluid inside human cells, the biologist then adds a small amount of sodium hydroxide, which converts some of the acid into its conjugate base. After this reaction, the new concentrations are $[H_2PO_4^-] = 0.155$ M and $[HPO_4^{2-}] = 0.245$ M. What are the new $[H_3O^+]$ and pH values after the addition of base? Use the table below to summarize this information.

Did the addition of sodium hydroxide shift the equilibrium toward the $H_2PO_4^-$ or toward the HPO_4^{2-}? Describe this shift in equilibrium using Le Chatelier's principle.

	$[H_2PO_4^-]$	$[HPO_4^{2-}]$	$[H_3O^+]$	pH
Before adding base	0.200 M	0.200 M		
After adding base	0.155 M	0.245 M		

SUMMARY

In this chapter we've worked through many important ideas connecting energy, reaction rates, and equilibrium reactions. The rate of a reaction depends on the energy and frequency of molecular collisions. Reaction rates increase as the concentration and temperature increase.

Reaction rates also depend on the specific energetic changes that accompany a reaction. We commonly map these energy changes using reaction energy diagrams. These diagrams typically show the energy barrier, called the activation energy, that substances must overcome for a reaction to occur. The rate of a reaction depends on the activation energy. Catalysts are materials that lower the activation energy for a reaction, enabling the reaction to happen more quickly and at lower temperature.

Equilibrium reactions can proceed in both the forward and reverse directions. When a system is in equilibrium, the forward reaction and the reverse reaction take place at the same rate.

We use equilibrium expressions to describe equilibrium reactions. The equilibrium constant, K, describes how far to the left or right the equilibrium lies. A large K indicates the products are favored; a small K means the reactants are favored. When writing equilibrium expressions, solvents, pure liquids, and solids are not included. Equilibrium expressions may be written using concentrations, or for gas-phase reactions, using partial pressures.

Le Chatelier's principle states that when some change in concentration, temperature, pressure, or other key factor takes place, the equilibrium shifts to minimize that change, and a new equilibrium is established. Chemists commonly use Le Chatelier's principle to drive an equilibrium reaction toward the formation of desired products.

Miracles and Monstrosities: The Brutal Ironies of Fritz Haber

By nature, humans have the capacity for both goodness and evil. Most of us live between the two extremes: We are neither saints nor sociopaths — our lives are colored with love, hard work, and creative impulses, but marred with the wreckage of past mistakes and smeared with selfish and destructive desires. As St. Paul wrote in his letter to the early Christian church, "The good that I love I don't do, but the evil that I hate — that is exactly what I do!"

Seldom is the contrast between dark and bright as stark as it was in the life of Fritz Haber (**Figure 13.22**). His process for ammonia production profoundly increased the world's food supply and earned him the 1918 Nobel Prize in chemistry. But there's more to the story.

Haber was at the peak of his career when World War I broke out in Europe. A patriotic German, Haber eagerly applied his scientific ability to the war effort. It was the era of trench warfare, and Germany was locked in prolonged and costly trench battles. Haber proposed the use of gases to clear out enemy trenches. Working closely with the German army, he supervised the implementation of chlorine gas on the battlefield in April 1915 (**Figure 13.23**). His wife, who was opposed to his work with chemical weapons, committed suicide a month later. Despite this, Haber continued his efforts in chemical warfare, developing and testing several other gases for use on the battlefield.

After World War I ended, Haber resumed his work as director of the Kaiser Wilhelm Institute for Physical Chemistry and Electrochemistry in Berlin. He worked closely with great scientists in an era of extraordinary development. He helped develop a model for understanding the stability of crystal lattices. He was a good friend and neighbor of Albert Einstein.

Like Einstein, Haber was Jewish, and the rise of the Nazi party in the 1930s brought an end to his tenure. In 1933 he fled Germany; three months later, he died. In a final, horrible irony, one of the gas mixtures he had developed as an insecticide was modified and used in the murders of millions of Jews in the concentration camps, including several of Haber's close relatives.

How should history remember Fritz Haber? He is the father of chemical warfare, one of the horrors of the modern era. Yet his nitrogen fixation process has helped feed billions of people and is among the most important scientific findings of all time. His legacy is a testament to the strengths and weaknesses of human nature, and a reminder that the power of science must be coupled with wisdom and decency.

Emilio Segrè Visual Archives/American Institute of Physics/Science Source

Figure 13.22 This photo shows Fritz Haber (left) with Albert Einstein around 1914. Haber unlocked the secret of ammonia production, but he is also remembered as the father of chemical weapons.

Hulton Deutsch/Getty Images

Figure 13.23 These World War I German soldiers wore gas masks to protect themselves from exposure to chemical weapons.

Key Terms

13.1 Reaction Rates

reaction rate The speed at which a chemical reaction takes place.

reaction energy diagram A depiction of the energetic changes that accompany a chemical reaction.

transition state The highest-energy arrangement of atoms that occurs during a chemical reaction.

activation energy The energy barrier for a reaction; this energy determines how quickly a reaction occurs.

catalyst A species that is not part of a balanced equation but causes a reaction to go more quickly.

13.2 Equilibrium Reactions

equilibrium reaction A reaction that occurs in both the forward and reverse directions.

equilibrium The state in which forward and reverse reactions take place at the same rate, so the concentrations of reactants and products do not change.

13.3 Equilibrium Expressions

equilibrium expression An equation that describes the balance between reactants and products in an equilibrium.

equilibrium constant (K) The ratio of products to reactants when a reaction is at equilibrium.

solubility product (K_{sp}) The equilibrium constant for the solution of a slightly soluble ionic compound.

13.4 Le Chatelier's Principle

Le Chatelier's principle When some change takes place in concentration, temperature, pressure, or other key factor, the equilibrium shifts to minimize that change, and a new equilibrium is established.

Additional Problems

13.1 Reaction Rates

13. How do molecular collisions relate to the rates of chemical reactions?

14. Do all molecular collisions result in a chemical reaction? What other factors play a role in determining whether a reaction takes place?

15. What are two steps a chemist can take to make a reaction go more slowly?

16. At temperatures above about 1300 °C, nitrogen and oxygen gas react quickly to form nitrogen oxides, such as nitrogen monoxide. A simplified reaction scheme can be written this way:

$$N_2\,(g) + O_2\,(g) \rightarrow 2\,NO\,(g)$$

Why does this reaction occur at high temperature, but not at room temperature? How does the elevated temperature affect the collisions that occur between gas molecules?

17. A chemist wishes to convert iron into iron(II) bromide using this reaction:

$$Fe\,(s) + 2\,HBr\,(aq) \rightarrow FeBr_2\,(aq) + H_2\,(g)$$

She mixes small pieces of iron with 0.2 M aqueous HBr at room temperature and discovers that the reaction proceeds slowly. List two ways she could speed up the reaction.

18. A biologist monitoring fish populations collects tissue samples for analysis. However, the tissue samples decompose as they react with oxygen in the air, producing foul-smelling by-products. How can he slow down or stop this reaction from occurring?

19. What is a reaction energy diagram? What information does this type of diagram provide?

20. What is the difference between the activation energy and the net energy change for a reaction?

21. What is a transition state? How can you identify a transition state on an energy diagram?

22. What energetic factor determines the rate of a reaction?

23. A single-step reaction has an activation energy of +80.0 kJ/mol and a net energy change of +30.0 kJ/mol. Create an energy diagram for this reaction.

24. A single-step reaction has a transition state with an energy change of +5.0 kJ/mol and a net energy change of −18.0 kJ/mol. Create an energy diagram for this reaction.

25. A single-step reaction has an activation energy of +20.0 kJ/mol and a net energy change of +5.0 kJ/mol. Is this reaction endothermic or exothermic? Will the reaction occur at room temperature?

26. A single-step reaction has an activation energy of +7.5 kJ/mol and a net energy change of −30.2 kJ/mol. Is this reaction endothermic or exothermic? Will the reaction occur at room temperature?

27. Describe each of the following reactions as endothermic or exothermic. Also determine the activation energy and the net energy change for each reaction.

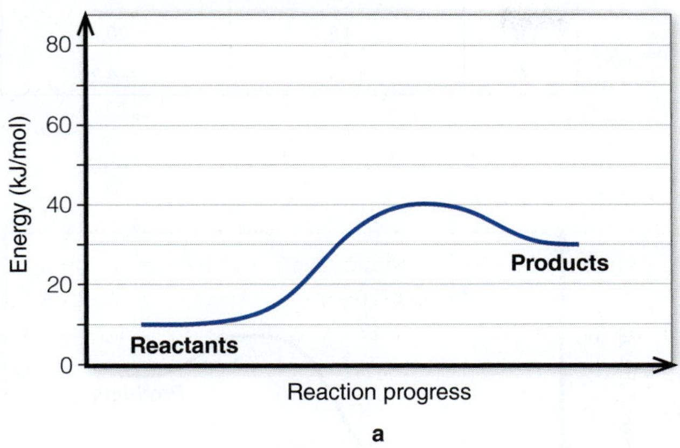

a

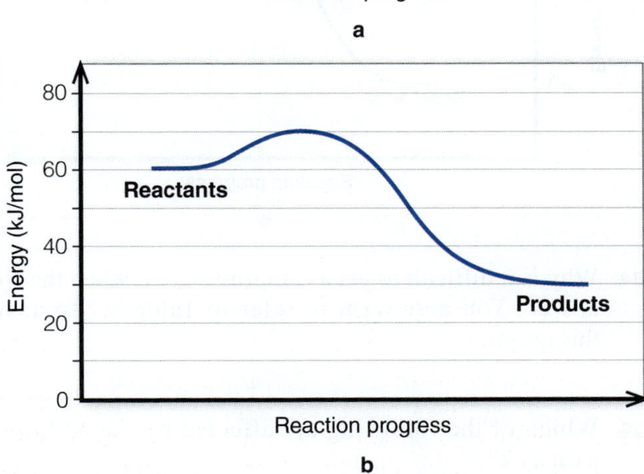

b

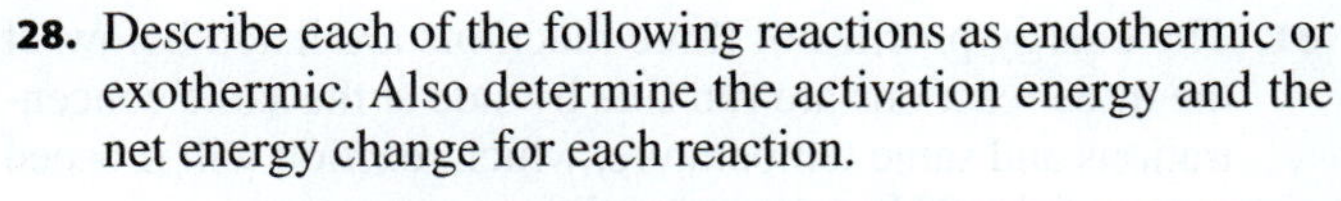

28. Describe each of the following reactions as endothermic or exothermic. Also determine the activation energy and the net energy change for each reaction.

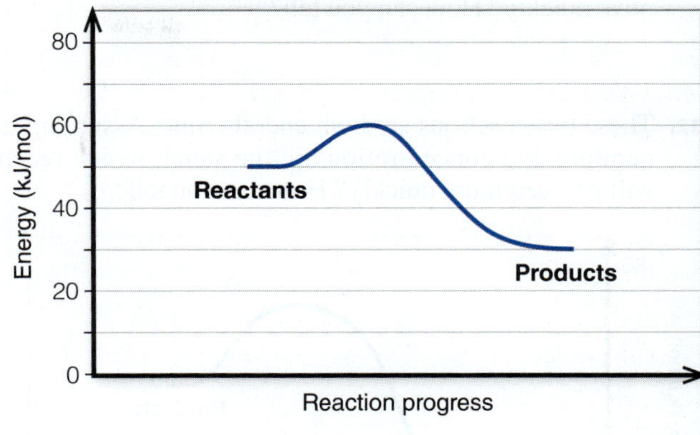

a

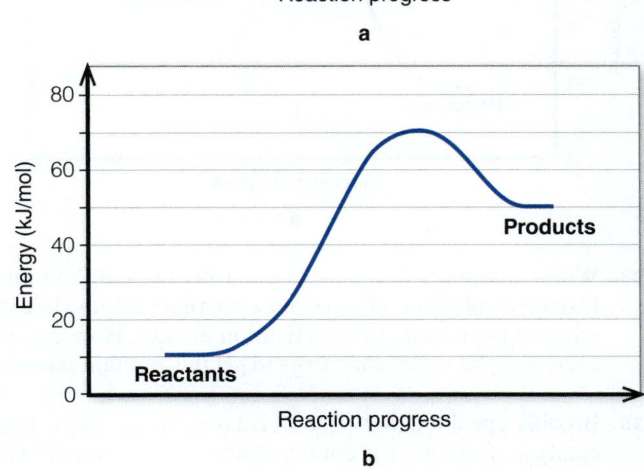

b

29. For each reaction diagram, determine the activation energy and the net energy change:

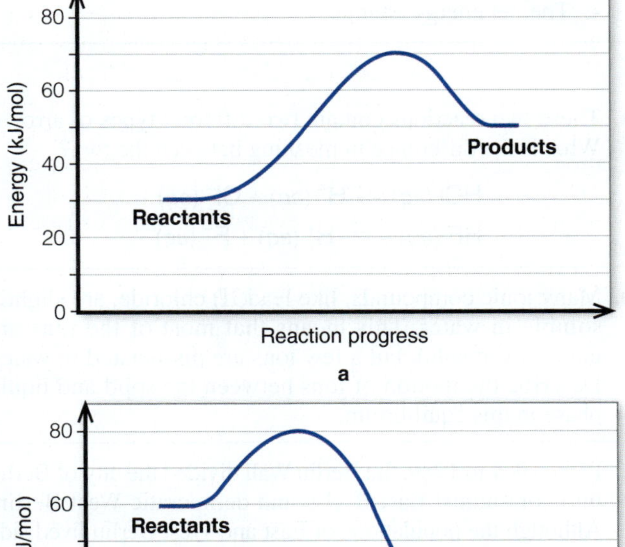

a

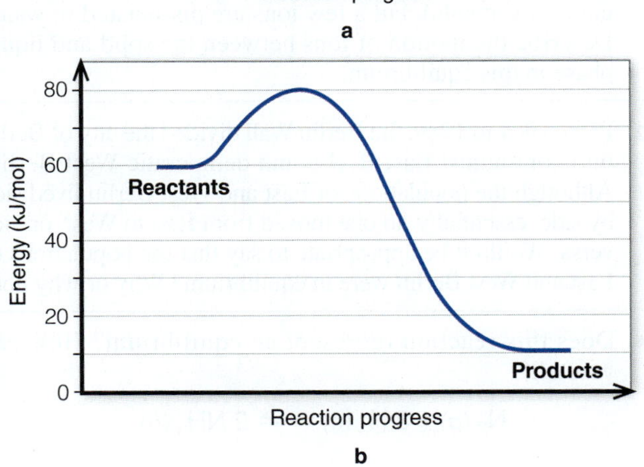

b

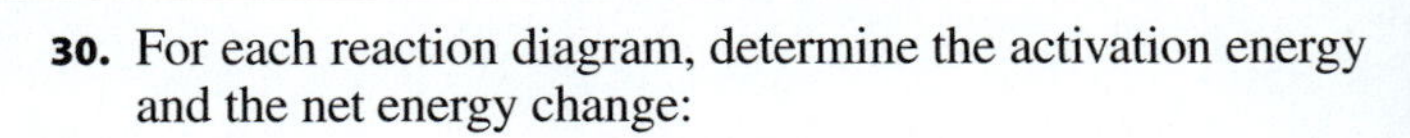

30. For each reaction diagram, determine the activation energy and the net energy change:

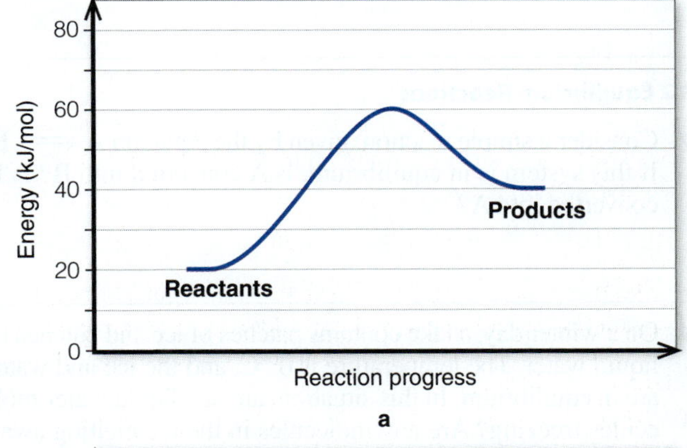

a

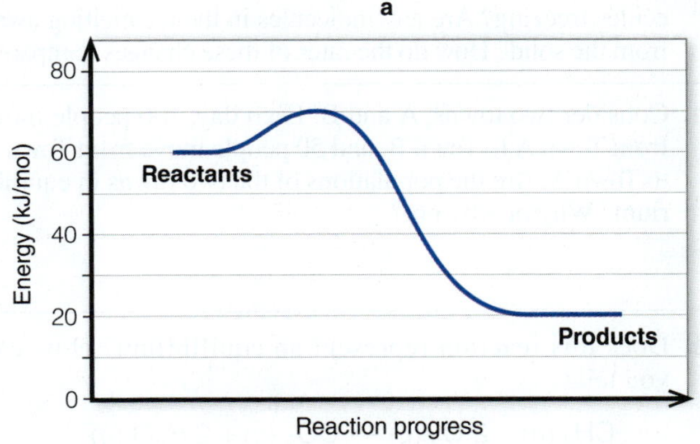

b

31. The energy profiles of three reactions are listed below. If the three reactions are run side by side at the same concentrations and same temperature, which reaction will proceed most quickly? How can you tell?

Reaction	Activation Energy (kilocalories/mole)	Net Energy Change (kilocalories/mole)
A	2.8	–8.4
B	4.8	–20.7
C	10.4	+4.2

32. These two reactions are both endothermic. Assuming temperature and concentration are the same, which reaction will proceed more quickly? How can you tell?

a

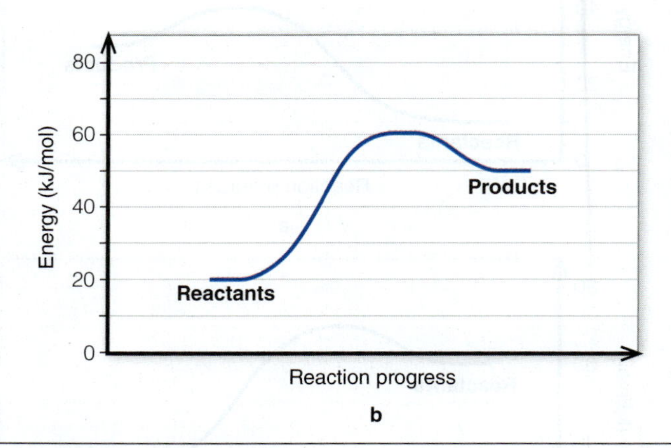

b

33. While camping, you need to get a fire started. You know that the combustion of wood is exothermic, but it won't start without overcoming the activation energy. How can you overcome the activation energy to get this reaction started?

34. Why is it difficult to get a campfire started when the wood is wet? You may want to refer to Table 8.2 to answer this question.

35. Broadly speaking, how does a catalyst work? How does a catalyst affect the net energy change of a reaction? How does it affect the activation energy?

36. Which of the following are affected by the addition of a catalyst?

a. The energy of the reactants
b. The energy of the products
c. The energy of the transition state
d. The activation energy
e. The net energy change

13.2 Equilibrium Reactions

37. Consider a simple reaction, given by the equation $A \rightleftharpoons B$. If this system is in equilibrium, is A converted into B? Is B converted into A?

38. These two reactions contain two different types of arrows. What is the difference in meaning between the two?

$$HCl\ (aq) \rightarrow H^+\ (aq) + Cl^-\ (aq)$$
$$HF\ (aq) \rightleftharpoons H^+\ (aq) + F^-\ (aq)$$

39. On a winter day, a lake contains patches of ice and patches of liquid water. The temperature is 0° C, and the ice and water are in equilibrium. In this situation, are any liquid water molecules freezing? Are any molecules in the ice melting away from the solid? How do the rates of these changes compare?

40. Many ionic compounds, like lead(II) chloride, are slightly soluble in water. This means that most of the ions are undissolved solid, but a few ions are dissociated in water. Describe the motion of ions between the solid and liquid phase in this equilibrium.

41. Consider two towns, A and B. Each day, 100 people move from Town A to Town B, and 50 people move from Town B to Town A. Are the populations of the two towns in equilibrium? Why or why not?

42. From 1961 to 1989, the Berlin Wall divided the city of Berlin into communist East Berlin and democratic West Berlin. Although the populations of East and West Berlin lived side by side, essentially no one moved from East to West, or vice versa. Would it be appropriate to say that the populations of East and West Berlin were in equilibrium? Why or why not?

43. Does this reaction represent an equilibrium? How can you tell?

$$CH_4\ (g) + 2\ O_2\ (g) \rightarrow CO_2\ (g) + 2\ H_2O\ (g)$$

44. Does this reaction represent an equilibrium? How can you tell?

$$N_2\ (g) + 3\ H_2\ (g) \rightleftharpoons 2\ NH_3\ (g)$$

45. The following reaction is endothermic. At equilibrium, is there a higher concentration of reactants or products?

$$N_2O_4\,(g) \rightleftharpoons 2\,NO_2\,(g)$$

46. The following reaction is slightly exothermic. At equilibrium, is there a higher concentration of reactants or products?

$$2\,HI\,(g) \rightleftharpoons H_2\,(g) + I_2\,(g)$$

47. Trifluoroacetic acid, $HC_2F_3O_2$, ionizes in water. This reaction is described by the equilibrium reaction and energy diagram shown. The energy levels of the starting materials and products are very nearly the same. Describe the balance of starting materials and products at equilibrium.

$$HC_2F_3O_2\,(aq) + H_2O\,(l) \rightleftharpoons H_3O^+\,(aq) + C_2F_3O_2^-\,(aq)$$

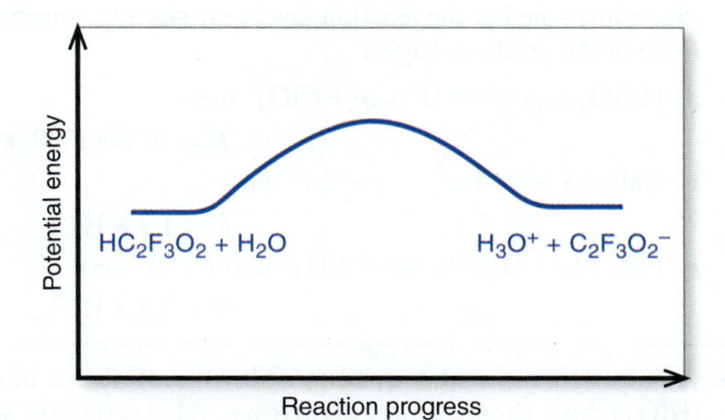

48. Boric acid, $B(OH)_3$, undergoes an unusual equilibrium in water. The equilibrium and reaction diagram are shown. At equilibrium, will there be a higher concentration of starting materials (left) or products (right) in solution?

$$B(OH)_3\,(aq) + H_2O\,(l) \rightleftharpoons B(OH)_4^-\,(aq) + H^+\,(aq)$$

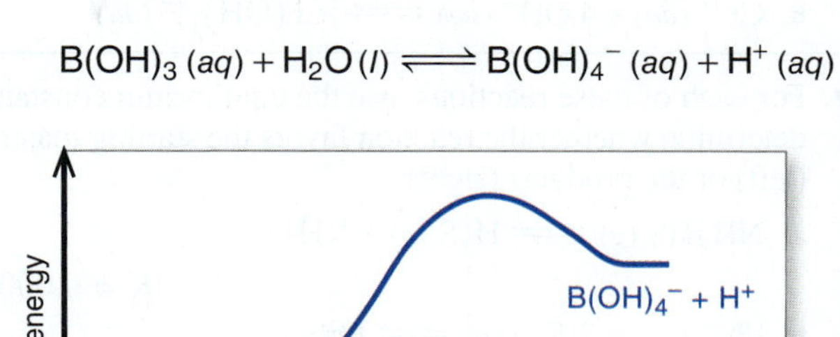
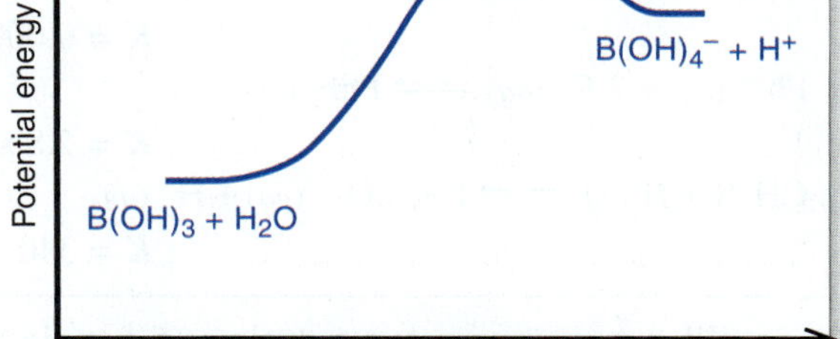

49. Consider this reaction diagram. What is the activation energy of this reaction in the forward direction? What is the activation energy in the reverse direction?

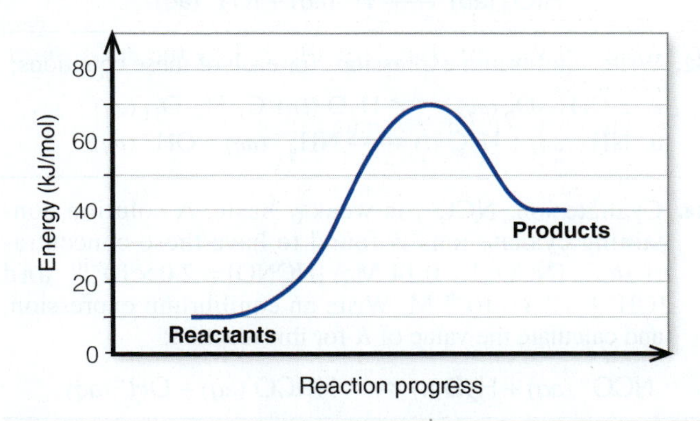

50. Consider this reaction diagram. What is the activation energy of this reaction in the forward direction? What is the activation energy in the reverse direction?

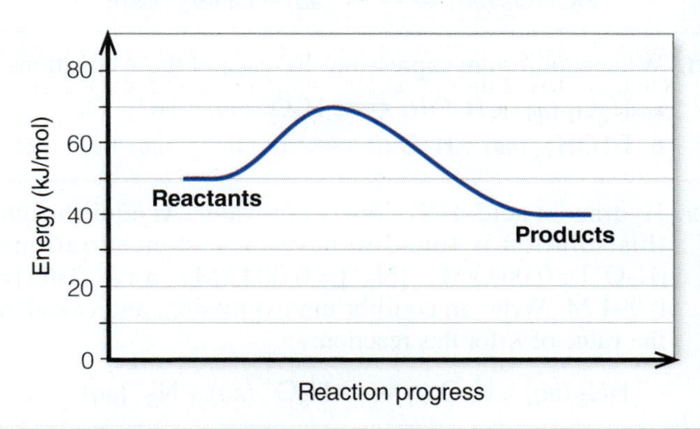

51. An equilibrium between starting materials and products can proceed through an uncatalyzed pathway (1) or a catalyzed pathway (2), as shown in the energy diagram. What effect does the catalyst have on the equilibrium?

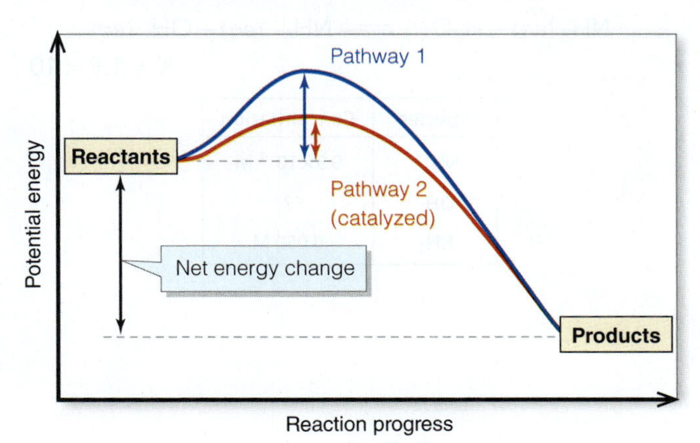

52. Indicate whether each of these factors will affect whether an equilibrium lies to the right or to the left:

a. net energy change for a reaction
b. presence of a catalyst
c. activation energy of a reaction

13.3 Equilibrium Expressions

53. Write equilibrium expressions for each of these equations:

a. $A + 2B \rightleftharpoons C$

b. $3A + B \rightleftharpoons C + 2D$

54. Write equilibrium expressions for each of these equations:

a. $A + B \rightleftharpoons 2C$

b. $2A \rightleftharpoons B$

55. Write equilibrium expressions for each of these equations:

a. $HNO_2\ (aq) \rightleftharpoons H^+\ (aq) + NO_2^-\ (aq)$

b. $Cr^{3+}\ (aq) + 4\,OH^-\ (aq) \rightleftharpoons [Cr(OH)_4]^-\ (aq)$

56. Write equilibrium expressions for each of these equations:

a. $HBrO_2\ (aq) \rightleftharpoons H^+\ (aq) + BrO_2^-\ (aq)$

b. $Ag^+\ (aq) + 2CN^-\ (aq) \rightleftharpoons [Ag(CN)_2]^-\ (aq)$

57. For each of these reactions, use the equilibrium constant to determine whether the reaction favors the starting materials (left) or the products (right):

a. $NH_4HS\ (g) \rightleftharpoons H_2S\ (g) + NH_3\ (g)$ $K = 0.00012$

b. $Pb^{2+}\ (aq) + 2\ F^-\ (aq) \rightleftharpoons PbF_2\ (s)$ $K = 2.8 \times 10^8$

c. $CH_3SO_3H\ (aq) \rightleftharpoons CH_3SO_3^-\ (aq) + H^+\ (aq)$ $K = 340$

58. For each of these reactions, use the equilibrium constant to determine whether the reaction favors the starting materials (left) or the products (right):

a. $HClO_4\ (aq) \rightleftharpoons H^+\ (aq) + ClO_4^-\ (aq)$ $K = 10{,}000{,}000{,}000$

b. $BaF_2\ (s) \rightleftharpoons Ba^{2+}\ (aq) + 2\ F^-\ (aq)$ $K = 1.7 \times 10^{-6}$

c. $NH_3\ (aq) + H_2O\ (l) \rightleftharpoons NH_4^+\ (aq) + OH^-\ (aq)$ $K = 1.8 \times 10^{-5}$

59. At equilibrium, an aqueous solution of formic acid ($HCHO_2$) has these concentrations: $[H^+] = 0.0031$ M; $[CHO_2^-] = 0.0031$ M; and $[HCHO_2] = 0.0535$ M. What is the value of K for this equilibrium? Does the equilibrium lie to the right or to the left?

$$HCHO_2\ (aq) \rightleftharpoons H^+\ (aq) + CHO_2^-\ (aq)$$

60. At equilibrium, an aqueous solution of iodic acid (HIO_3) has these concentrations: $[H^+] = 0.0203$ M; $[IO_3] = 0.0203$ M; and $[HIO_3] = 0.0024$ M. What is the value of K for this equilibrium? Does the equilibrium lie to the right or to the left?

$$HIO_3\ (aq) \rightleftharpoons H^+\ (aq) + IO_3^-\ (aq)$$

61. Write equilibrium expressions for each of these equations:

a. $H_2O_2\ (aq) + H_2O\ (l) \rightleftharpoons H_3O^+\ (aq) + HO_2^-\ (aq)$

b. $B(OH)_3\ (aq) + H_2O\ (l) \rightleftharpoons B(OH)_4^-\ (aq) + H^+\ (aq)$

62. Write equilibrium expressions for each of these equations:

a. $2\ C_6H_{12}O_6\ (aq) \rightleftharpoons H_2O\ (l) + C_{12}H_{22}O_{11}\ (aq)$

b. $NH_3\ (aq) + H_2O\ (l) \rightleftharpoons NH_4^+\ (aq) + OH^-\ (aq)$

63. Hydrazoic acid, HN_3, ionizes in water. At equilibrium, this solution is found to have these concentrations: $[H_3O^+] = 0.0043$ M; $[N_3^-] = 0.0043$ M; and $[HN_3] = 0.994$ M. Write an equilibrium expression, and calculate the value of K for this reaction:

$$HN_3\ (aq) + H_2O\ (l) \rightleftharpoons H_3O^+\ (aq) + N_3^-\ (aq)$$

64. Cyanate ion, NCO^-, is weakly basic. A solution containing cyanate ions is found to have these concentrations: $[NCO^-] = 0.14$ M; $[HCNO] = 2.0 \times 10^{-16}$ and $[OH^-] = 2.0 \times 10^{-6}$ M. Write an equilibrium expression, and calculate the value of K for this reaction:

$$NCO^-\ (aq) + H_2O\ (l) \rightleftharpoons HNCO\ (aq) + OH^-\ (aq)$$

65. Hydrofluoric acid ionizes in water as shown below. The equilibrium constant for this reaction is 6.8×10^{-14}. Given the values shown in the table below, find the concentration of HF in this solution:

$$HF\ (aq) \rightleftharpoons H^+\ (aq) + F^-\ (aq) \qquad K = 6.8 \times 10^{-4}$$

Species	Concentration
H^+	1.0×10^{-5} M
F^-	1.0×10^{-5} M
HF	?

66. The equilibrium constant for the acid-base reaction of ammonia in water (shown below) is 1.8×10^{-5} M. Given the values shown in the table below, find the concentration of OH^- in this solution:

$$NH_3\ (aq) + H_2O\ (l) \rightleftharpoons NH_4^+\ (aq) + OH^-\ (aq) \qquad K = 1.8 \times 10^{-5}$$

Species	Concentration
NH_4^+	9.5×10^{-4} M
OH^-	?
NH_3	0.050 M

67. Nitrous acid reacts with water as shown in the equilibrium below. The equilibrium constant for this reaction is 4.5×10^{-4}. A chemist prepares an aqueous solution using both nitrous acid and potassium nitrite, so that $[HNO_2] = 0.050$ M and $[NO_2^-] = 0.035$ M. What is the concentration of H_3O^+ in this solution?

$$HNO_2\,(aq) + H_2O\,(l) \rightleftharpoons H_3O^+\,(aq) + NO_2^-\,(aq)$$

68. Hypochlorous acid reacts with water as shown in the equilibrium below. The equilibrium constant for this reaction is 3.0×10^{-8}. A chemist prepares an aqueous solution using both hypochlorous acid and sodium hypochlorate. If $[HClO] = 0.100$ M and $[ClO^-] = 0.0025$ M, what is the concentration of H_3O^+ in this solution?

$$HClO\,(aq) + H_2O\,(l) \rightleftharpoons H_3O^+\,(aq) + ClO^-\,(aq)$$

69. Write equilibrium expressions for each of the following:

a. $PbI_2\,(s) \rightleftharpoons Pb^{2+}\,(aq) + 2\,I^-\,(aq)$
b. $Fe^{3+}\,(aq) + 3\,OH^-\,(aq) \rightleftharpoons Fe(OH)_3\,(s)$

70. Write equilibrium expressions for each of the following:

a. $Ca(OH)_2\,(s) \rightleftharpoons Ca^{2+}\,(aq) + 2\,OH^-\,(aq)$
b. $Ag^+\,(aq) + Cl^-\,(aq) \rightleftharpoons AgCl\,(s)$

71. What is a solubility product? What does a small solubility product indicate?

72. Solubility products are usually reported at a given temperature, such as 25 °C. How do you think the solubility product changes as the temperature increases?

73. In a saturated solution of cadmium carbonate at 25 °C, both $[Cd^{2+}]$ and $[CO_3^{2-}] = 1.0 \times 10^{-6}$. Write an equilibrium expression and calculate the value of K_{sp} for this compound.

74. In a saturated solution of barium fluoride, $[Ba^{2+}] = 3.58 \times 10^{-3}$ M and $[F^-] = 7.16 \times 10^{-3}$ M. Write an equilibrium expression and calculate the value of K_{sp} for this compound.

75. The solubility product of silver chloride at 25 °C is 1.77×10^{-10}. If silver chloride is mixed with pure water, what is the concentration of silver ions in solution?

76. The solubility product of iron(II) carbonate is 3.13×10^{-11} at 25 °C. If a chemist mixes iron(II) carbonate with pure water, what will be the equilibrium concentration of iron(II) ions?

77. Using the data in Table 13.1, find the maximum concentration of Pb^{2+} and I^- in solution when PbI_2 is mixed with pure water at 25 °C.

78. Using the data in Table 13.1, find the maximum concentration of Pb^{2+} and Cl^- in solution when $PbCl_2$ is mixed with pure water at 25 °C.

79. Using the partial pressures of each gas, write equilibrium expressions for these gas-phase equilibria:

a. $N_2O_4\,(g) \rightleftharpoons 2\,NO_2\,(g)$
b. $4\,HCl\,(g) + O_2\,(g) \rightleftharpoons 2\,Cl_2\,(g) + 2\,H_2O\,(g)$

80. Using the partial pressures of each gas, write equilibrium expressions for these gas-phase equilibria:

a. $PCl_3\,(g) + Cl_2\,(g) \rightleftharpoons PCl_5\,(g)$
b. $H_2\,(g) + I_2\,(g) \rightleftharpoons 2\,HI\,(g)$

81. At 25 °C, NO_2 and N_2O_4 exist in the equilibrium shown below. Find the equilibrium pressure of NO_2 if $P_{NO_2} = 1.5$ atm.

$$2\,NO_2\,(g) \rightleftharpoons N_2O_4\,(g) \qquad K = 7.0$$

82. At 25 °C, gaseous ClF exists in equilibrium with elemental chlorine and fluorine, as shown below. If $P_{ClF} = 1.3$ atm, find the pressure of the two elemental gases. Assume $P_{Cl_2} = P_{F_2}$.

$$2\,ClF\,(g) \rightleftharpoons Cl_2\,(g) + F_2\,(g) \qquad K = 2.9 \times 10^{-11}$$

13.4 Le Chatelier's Principle

83. What is Le Chatelier's principle?

84. Consider a U-shaped tube filled with mercury, as shown here. If additional mercury were added to the left-hand side of the tube, would the mercury inside the tube move to the left, to the right, or show no change?

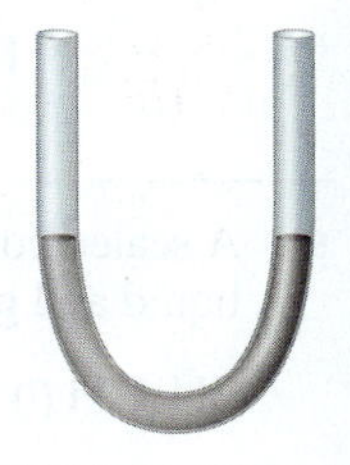

85. For the equilibrium $A + B \rightleftharpoons C$, indicate whether each of the changes given will shift the equilibrium to the right or to the left:

a. adding compound A to the mixture
b. removing compound B from the mixture
c. adding compound C to the mixture
d. removing compound C from the mixture

86. For the equilibrium $D + E \rightleftharpoons F + G$, indicate whether each of the changes given will shift the equilibrium to the right or to the left:

a. removing compound D from the mixture
b. adding compound E to the mixture
c. adding compound F to the mixture
d. removing compound G from the mixture

87. At high temperatures, bromine and chlorine gas can rearrange to form BrCl, as shown here:

$$Br_2\,(g) + Cl_2\,(g) \rightleftharpoons 2\,BrCl\,(g)$$

How would the following changes affect this equilibrium?

a. adding Br_2 to the mixture
b. removing Cl_2 from the mixture
c. removing BrCl from the mixture
d. adding a catalyst

88. In aqueous solution, simple sugars (called monosaccharides) can couple together to form disaccharides:

$$2\,C_6H_{12}O_6\,(aq) \rightleftharpoons H_2O\,(l) + C_{12}H_{22}O_{11}\,(aq)$$

simple sugar — disaccharide

How does the addition of water affect this equilibrium? How does the removal of water affect this equilibrium?

89. Methanol, CH_3OH, reacts reversibly with hydroiodic acid, as shown here:

$$CH_3OH + HI \rightleftharpoons CH_3I + H_2O$$

Based on Le Chatelier's principle, will more CH_3I be produced if water is the solvent, or if methanol is the solvent?

90. Methyl acetate is a compound commonly used in fingernail polish. When combined with water, it slowly decomposes according to the equilibrium below. If you wanted to decompose methyl acetate, which solvent would be a better choice, water or methanol? Why is this so?

$$C_3H_6O_2 + H_2O \rightleftharpoons C_2H_3O_2H + CH_3OH$$

methyl acetate — acetic acid — methanol

91. Hydroxylamine (NH_2OH) is weakly basic, as shown in the equilibrium below. Would the addition of more OH^- shift this equilibrium toward the left or the right?

$$NH_2OH\,(aq) + H_2O\,(l) \rightleftharpoons NH_3OH^+\,(aq) + OH^-\,(aq)$$

$$K = 6.6 \times 10^{-9}$$

92. Lead(II) chloride is slightly soluble in water, as shown in the equilibrium below. If more chloride ion is added to the solution, will $[Pb^{2+}]$ increase or decrease?

$$PbCl_2\,(s) \rightleftharpoons Pb^{2+}\,(aq) + 2\,Cl^-\,(aq)$$

$$K = 5.9 \times 10^{-5}$$

93. At equilibrium, water molecules move back and forth between the liquid and gas phase, as shown below. If more heat is added, does the equilibrium shift to the left or the right? What if heat is removed?

$$\text{Heat} + H_2O\,(l) \rightleftharpoons H_2O\,(g)$$

94. Barium hydroxide is only slightly soluble in water at room temperature. Based on the equilibrium below, how does the solubility change if the solution is heated?

$$\text{Heat} + Ba(OH)_2\,(s) \rightleftharpoons Ba^{2+}\,(aq) + 2\,OH^-\,(aq)$$

95. Consider the equilibrium and reaction enthalpy shown below. Would heating this mixture shift the equilibrium to the left or to the right?

$$NH_4^+\,(aq) + OH^-\,(aq) \rightleftharpoons NH_3\,(aq) + H_2O\,(l)$$

$$\Delta H^\circ = -2.8 \text{ kJ/mol}$$

96. Consider the equilibrium and reaction enthalpy shown below. Would heating this mixture shift the equilibrium to the left or to the right?

$$AgCl\,(s) \rightleftharpoons Ag^+\,(aq) + Cl^-\,(aq)$$

$$\Delta H^\circ = +65.1 \text{ kJ/mol}$$

97. When heated, PCl_3 and Cl_2 combine to form PCl_5 in equilibrium, as shown below. Would increasing the pressure on this system shift the equilibrium to the left or to the right?

$$PCl_3\,(g) + Cl_2\,(g) \rightleftharpoons PCl_5\,(g)$$

98. When heated, hydrogen and iodine gas combine to form hydrogen iodide in equilibrium. Would increasing the pressure on this system shift the equilibrium to the left, to the right, or cause no change?

$$H_2\,(g) + I_2\,(g) \rightleftharpoons 2\,HI\,(g)$$

99. A sealed container of methanol (CH_3OH) contains both the liquid and gas in equilibrium:

$$CH_3OH\,(l) \rightleftharpoons CH_3OH\,(g)$$

$$\Delta H^\circ = +37.6.1 \text{ kJ/mol}$$

Indicate whether these changes would shift the equilibrium to the left or to the right:

a. opening the container so that CH_3OH (*g*) can escape
b. increasing the pressure in the container
c. increasing the temperature

100. Silver chloride is a slightly soluble compound. It dissociates to form ions as shown in this equilibrium:

$$AgCl\,(s) \rightleftharpoons Ag^+\,(aq) + Cl^-\,(aq)$$

$$\Delta H^\circ = +65.1 \text{ kJ/mol}$$

Indicate whether these changes would shift the equilibrium to the left or to the right:

a. adding Cl^- ions
b. removing Ag^+ ions
c. increasing the temperature

Challenge Questions

101. A chemist prepares a 1.00-M solution of hypochlorous acid. This weak acid ionizes in solution, as shown here:

$$HClO\ (aq) \rightleftharpoons H^{+}\ (aq) + ClO^{-}\ (aq)$$

$$K = 1.8 \times 10^{-8}$$

To determine the pH of weakly acidic solutions, chemists often construct a table like the one below to summarize the initial concentration of each species, the change in concentration of each species, and then the equilibrium concentration. In this table, the change in concentration is given by the unknown *x*. Because *K* is so small, we can assume that the concentration of HClO does not change significantly. Construct an equilibrium expression based on this table, and find the concentrations of H^{+} and ClO^{-} in this solution. Finally, calculate the pH of this solution.

	[HClO]	[H^{+}]	[ClO^{-}]
Initial	1.00 M	–	–
Change	$-x$	$+x$	$+x$
Equilibrium	~ 1.00 M	x	x

102. Buffer solutions resist changes in pH when acid or base is added to them. A common buffer solution is composed of acetic acid and sodium acetate. This acid-conjugate base pair undergoes the equilibrium reaction shown here:

$$HC_2H_3O_2\ (aq) + H_2O\ (l) \rightleftharpoons H_3O^{+}\ (aq) + C_2H_3O_2^{-}\ (aq)$$

$$K = 1.75 \times 10^{-5}$$

A chemist prepares a buffer in which [$HC_2H_3O_2$] and [$C_2H_3O_2$] both equal 0.200 M. She then adds a base, which converts some of the acetic acid into acetate ion, and raises the pH of the solution from 4.76 to a final value of 5.20. Using the table below as a guideline, set up and solve an equilibrium expression to find the concentrations of $HC_2H_3O_2$ and $C_2H_3O_2^{-}$ in the ending solution.

	[HClO]	[ClO^{-}]
Initial	0.200	0.200
Change	$-x$	$+x$
Equilibrium	$0.200 - x$	$0.200 + x$

CHAPTER FOURTEEN

Oxidation-Reduction Reactions

Figure 14.1 (a) Batteries are a common and essential part of modern life. (*continued*)

Volta's Marvel

The year 1800 changed the world. An Italian scientist, Alessandro Volta, was working to understand the nature of electricity. As part of this work, he began to study electrical interactions between different metals. Volta discovered that when he placed plates of different metals — such as zinc and copper — on either side of a piece of cardboard soaked in dilute sulfuric acid or saltwater, he could produce an electric current. Volta called his invention a "voltaic pile." Today we know it as an *electrochemical cell*, or more simply as a *battery* (**Figure 14.1**).

Volta's invention was simple to reproduce — you can do it at your kitchen table — and many scientists quickly followed his lead. A few weeks after Volta announced his discovery, two English scientists, William Nicholson and Anthony Carlisle, built their own voltaic pile and used the electric current to separate water into hydrogen and oxygen. Other scientists improved on Volta's design, producing long-lasting batteries that resemble many batteries we still use today.

Within just a few years, another English scientist, Humphry Davy, built a massive battery at the Royal Institution in London. Through experiments with this battery, he discovered the common elements sodium, potassium, barium, calcium, magnesium, and boron.

The Volta battery gave scientists a controlled source of electricity and a way to study the behavior of electricity and charged particles more carefully than they had ever done before. These advances led to many critical inventions, including Michael

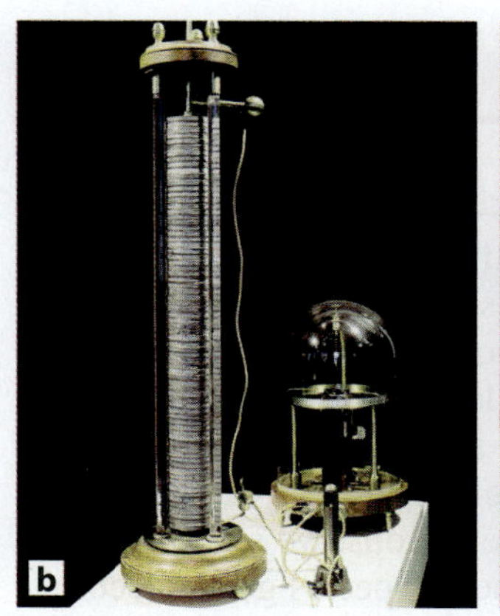

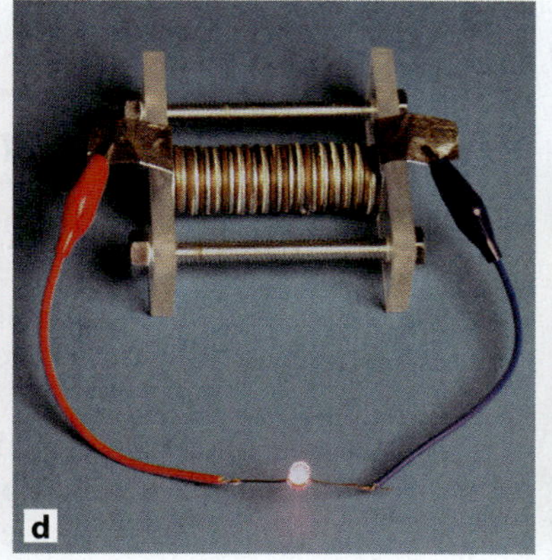

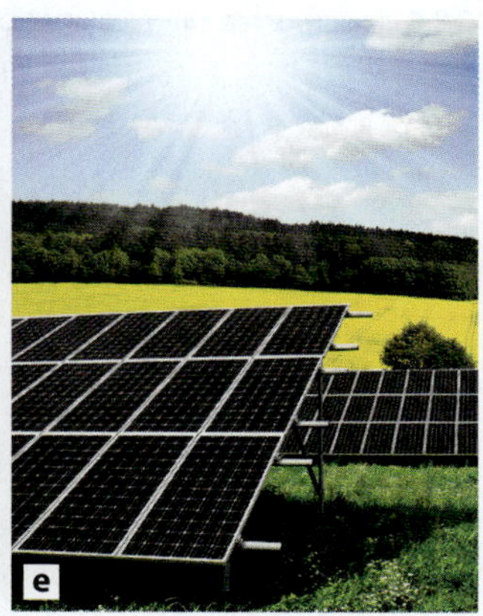

Figure 14.1 (continued) (b) The first battery, the voltaic pile, consisted of plates of copper and zinc separated by cardboard soaked in dilute acid. (c) In this print, based on a painting by Giuseppe Bertini, Volta shows his battery to the French emperor Napoleon Bonaparte in 1801. (d) A battery like Volta's can be made at home using aluminum foil, zinc washers, copper pennies, and thick paper or cardboard that has been soaked in saltwater or vinegar. The electrical potential energy can be measured with an electrical meter or simply with a small light bulb. (e) Both the chemistry of life and modern power applications are driven by oxidation-reduction reactions. (a) Gudella/iStock/Getty Images; (b) BeBa/Iberfoto/The Image Works; (c) akg-images/Newscom; (d) Ted Kinsman/Science Source; (e) © venvac/depositphotos.com

Faraday's electrical generator and Thomas Edison's light bulb. Ultimately, Volta played a critical role in ushering in the modern era.

Today, batteries are everywhere. Advanced batteries power our phones, computers, power tools, and even cars. But how do they actually work?

The electrical current produced by a battery arises from a type of chemical reaction called an *oxidation-reduction reaction*, in which electrons are transferred from one substance to another. Oxidation-reduction reactions power the chemistry of life: Plants use this type of reaction to store the energy of the sun. With every breath, our bodies use oxidation-reduction reactions to convert fuel into energy. These reactions also power the world around us: From fossil fuels to solar cells and batteries, we rely on oxidation-reduction reactions to provide energy for nearly everything we do.

In this chapter, we'll explore oxidation-reduction processes in detail. We'll examine different types of oxidation-reduction reactions and consider different ways of describing them. We'll also survey some of the applications that arise from these reactions. We'll see how batteries are built and explore some other useful and important applications of oxidation-reduction chemistry. Let's get started.

Intended Learning Outcomes

After completing this chapter and working the practice problems, you should be able to:

14.1 Oxidation and Reduction

- Determine oxidation numbers for atoms in ions and molecules.
- Identify species that are oxidized or reduced in a chemical reaction.

14.2 Types of Redox Reactions

- Describe the oxidation-reduction reactions that occur between metals and nonmetals.
- Predict the products of combustion reactions.
- Use the activity series to predict the products of metal displacement reactions.
- Predict the products of reactions of metals with water or with acid.

14.3 Half-Reactions and Batteries

- Describe the transfer of electrons in a chemical change using half-reactions.
- Describe the key components of electrochemical cells.

14.4 Balancing Redox Equations

- Balance redox equations by balancing charge and by balancing half-reactions.

14.5 Other Applications of Redox Reactions

- Describe how electroplating is used to coat surfaces.
- Describe the basic properties of a hydrogen fuel cell.

14.1 Oxidation and Reduction

The invention of the battery changed the world. Volta's discovery kicked off the field of **electrochemistry**—the study of chemical processes that involve the movement of electrons. To understand electrochemistry, we must begin with a class of reactions called **oxidation-reduction reactions** (often simply called *redox reactions*).

Electrons are negative, so if an atom gains electrons, its charge is *reduced*.

We looked briefly at oxidation-reduction reactions in Chapter 6. In this type of reaction, one species loses electrons, while another species gains electrons. The loss of electrons is called **oxidation**. When a species loses electrons, we say it has been *oxidized*. The gain of electrons is called **reduction**. When a species gains electrons, we say it has been *reduced*.

Some of the simplest redox reactions involve metals and nonmetals. For example, magnesium reacts with oxygen to produce magnesium oxide:

$$2\,Mg\,(s) + O_2\,(g) \rightarrow 2\,MgO\,(s)$$

In this reaction, each magnesium atom loses two electrons to form Mg^{2+}. Each oxygen atom gains two electrons to form O^{2-}. Magnesium is oxidized, while oxygen is reduced.

$$2\,\dot{Mg}\cdot + :\ddot{O}=\ddot{O}: \longrightarrow 2\,Mg^{2+} + 2\,:\ddot{O}:^{2-}$$

Each Mg loses two electrons.

Each O gains two electrons.

An oxidizing agent takes electrons; a reducing agent loses electrons.

The oxygen causes the magnesium to be oxidized in this reaction. A species that pulls electrons away from other atoms is sometimes called an *oxidizing agent*. Similarly, a species that loses electrons is sometimes called a *reducing agent*.

Oxidation Numbers

Oxidation numbers are a bookkeeping tool to help us track the electron changes.

When we are dealing with simple monatomic ions, the gain or loss of electrons is easy to determine. However, many redox reactions involve polyatomic ions or neutral molecules, and the change that takes place on each atom can be harder to measure. For example, consider the reaction of sodium metal with water, written as an ionic equation:

$$2\,Na\,(s) + 2\,H_2O\,(l) \rightarrow 2\,Na^+\,(aq) + 2\,OH^-\,(aq) + H_2\,(g)$$

In this reaction, the sodium is oxidized (it loses electrons). But which atom gains the electrons? At first glance, this question may be difficult to answer.

Scientists use **oxidation numbers** to keep track of electron changes in redox reactions. Oxidation numbers do not necessarily reflect the actual charge on atoms. Rather, they are a bookkeeping tool to ensure that the overall electronic changes are accounted for. We use the following rules to assign oxidation numbers to atoms:

1. The oxidation number of any species in its elemental form is zero.
2. For monatomic ions, the oxidation number is simply the charge of the ion.
3. For polyatomic ions, the oxidation numbers of the atoms must add up to the overall charge of the ion. For neutral compounds, the oxidation numbers of the atoms must add up to zero (**Figure 14.2**).
4. For atoms in molecules or polyatomic ions (**Table 14.1**):
   a. The oxidation number of fluorine is –1.
   b. The oxidation number of oxygen is –2.
   c. The lower halogens (Cl, Br, I) have an oxidation number of –1, unless they are bonded to oxygen or fluorine.
   d. If hydrogen is bonded to a nonmetal, its oxidation number is +1.
   e. If hydrogen is bonded to a metal, its oxidation number is –1.

−2 O, H +1, H +1

Oxidation numbers

Overall charge: 0

Figure 14.2 The oxidation numbers for a compound or polyatomic ion are added up to find the overall charge.

TABLE 14.1 Oxidation Numbers for Atoms in Molecules or Polyatomic Ions

Atom	Oxidation Number
F	–1
O	–2
Cl, Br, I	–1*
H	+1 or –1

*unless bonded to O or F

To see how oxidation numbers work, let's analyze the reaction of sodium metal with water. We begin by finding the oxidation number for each atom present. Na and H_2 are both elemental forms, so their oxidation numbers are zero. For the monatomic sodium ion, Na^+, its oxidation number equals its charge (+1). For H_2O and OH^-, we assign oxygen a value of –2 and hydrogen a value of +1:

The oxidation number for elemental forms is zero.

For a neutral compound or polyatomic ion, the oxidation numbers must add up to the total charge.

$$\overset{0}{2\,Na\,(s)} + \overset{+1\,-2}{2\,H_2O\,(l)} \longrightarrow \overset{+1}{2\,Na^+\,(aq)} + \overset{-2+1}{2\,OH^-\,(aq)} + \overset{0}{H_2\,(g)}$$

(+1 and 0 are printed above Na^+ and H_2 respectively.)

Na is oxidized... ...and H is reduced.

The oxidation number of sodium changes from 0 to +1. This means that sodium has lost electrons—it has been oxidized. Notice that the oxidation number of hydrogen changes from +1 (in water) to 0 (in H_2). The hydrogen atoms have gained electrons—they have been reduced.

It takes some practice to get the hang of assigning oxidation numbers. I encourage you to work through Examples 14.1 and 14.2 as well as the sample problems that follow them.

Example 14.1 Identifying Oxidation Numbers in Molecules and Polyatomic Ions

Identify the oxidation number for sulfur in the following three species: S^{2-}, $SOCl_2$, HSO_4^-.

For monatomic ions, the oxidation number is simply the charge: The oxidation number for sulfur in S^{2-} is –2.

$SOCl_2$ is a neutral compound, so the oxidation numbers must total zero. Based on the rules in Table 14.1, oxygen has an oxidation number of –2. The chlorine atoms are bonded to sulfur in this compound, and each chlorine atom has an oxidation number of –1. This means the oxidation number of sulfur must be +4. It may help you to set this up as an algebra equation, where x is the oxidation number for sulfur:

$$S + O + 2\,Cl$$
$$x + (-2) + 2\,(-1) = 0$$
$$x = +4$$

The hydrogen sulfate ion, HSO_4^-, has an overall charge of –1, so the oxidation numbers must all add up to –1. Based on the rules given above, the oxidation number of oxygen is –2, and hydrogen is +1. This means that the oxidation number of sulfur must be +6. Again, we can set this up as an algebra equation.

$$H + S + 4\,O$$
$$+1 + x + 4\,(-2) = -1$$
$$x = +6$$

Example 14.2 Identifying Oxidation Numbers in Reactions

Find the oxidation number of each atom in the following reaction. Identify the species that is oxidized or reduced.

$$CH_4\,(g) + 2\,O_2\,(g) \rightarrow CO_2\,(g) + 2\,H_2O\,(g)$$

When a substance reacts with oxygen, its oxidation number increases. The substance has been oxidized.

Based on the rules for oxidation numbers, the oxygen in O_2 (the elemental form of oxygen) has an oxidation number of zero. In the compounds CO_2 and H_2O, oxygen has a value of −2. Each hydrogen in CH_4 and in H_2O has a value of +1. The only remaining atom is carbon: In CH_4, the carbon must have a value of −4. In CO_2, carbon must have a value of +4:

$$\overset{-4\;\;4(+1)}{CH_4}\,(g) + \overset{2(0)}{2\,O_2}\,(g) \longrightarrow \overset{+4\;\;2(-2)}{CO_2}\,(g) + \overset{2(+1)\;\;-2}{2\,H_2O}\,(g)$$

In this reaction, O is reduced while C is oxidized.

Check It

Watch Explanation

TRY IT

1. Identify the oxidation number of sulfur in each of these species:
   a. S_8 **b.** H_2S **c.** H_2S_2 **d.** SO_3^{2-}
2. Identify the atoms that are oxidized and reduced in each of these reactions:
   a. $4\,Fe\,(s) + 3\,O_2\,(g) \rightarrow 2\,Fe_2O_3\,(s)$
   b. $Mg\,(s) + Br_2\,(g) \rightarrow MgBr_2\,(s)$
   c. $2\,HCl\,(aq) + Zn\,(s) \rightarrow ZnCl_2\,(aq) + H_2\,(g)$
   d. $8\,H^+\,(aq) + MnO_4^-\,(aq) + 5\,Fe^{2+}\,(aq) \rightarrow 5\,Fe^{3+}\,(aq) + Mn^{2+}\,(aq) + 4\,H_2O\,(l)$

M&M Valledor/agefotostock/AGE Fotostock

Figure 14.3 Rust is the compound iron(III) oxide, formed by the redox reaction of iron metal with oxygen gas.

14.2 Types of Redox Reactions

Redox reactions take many different forms. In this section, we'll briefly review some of the reactions you've seen before, and then we'll look at several new types of redox reactions.

Reactions of Metals with Nonmetals

The reactions between metals and nonmetals were covered in Chapter 6. Recall that when metals and nonmetals react, they form ionic compounds (**Figure 14.3**). In these reactions, the metals are oxidized to form cations, and the nonmetals are reduced to form anions. The following three reactions illustrate this type of change:

$$4\,Fe\,(s) + 3\,O_2\,(g) \rightarrow 2\,Fe_2O_3\,(s)$$

$$Zn\,(s) + Cl_2\,(g) \rightarrow ZnCl_2\,(s)$$

$$2\,Ag\,(s) + F_2\,(g) \rightarrow 2\,AgF\,(s)$$

Albert Russ/Shutterstock

Figure 14.4 Minerals are ionic compounds found in rock formations. Many minerals are metal oxides. This image shows the mineral cassiterite, which is composed of tin(IV) oxide.

Combustion Reactions

Combustion reactions are reactions that involve molecular oxygen. Most elements react with oxygen to form *oxides* (**Figure 14.4**). In these reactions the oxygen is reduced, and the other element is oxidized.

$$2\,Zn\,(s) + O_2\,(g) \rightarrow 2\,ZnO\,(s)$$

$$Si\,(s) + O_2\,(g) \rightarrow SiO_2\,(s)$$

$$2\,S\,(s) + 3\,O_2\,(g) \rightarrow 2\,SO_3\,(g)$$

Compounds also react with oxygen. Some of the most important combustion reactions involve hydrocarbons—compounds composed of carbon and hydrogen (**Figure 14.5**). Hydrocarbons react in combustion reactions to produce two compounds, carbon dioxide and water. For example, the following two reactions describe the combustion of acetylene (C_2H_2) and propane (C_3H_8):

$$2\,C_2H_2\,(g) + 5\,O_2\,(g) \rightarrow 4\,CO_2\,(g) + 2\,H_2O\,(g)$$

$$C_3H_8\,(g) + 5\,O_2\,(g) \rightarrow 3\,CO_2\,(g) + 4\,H_2O\,(g)$$

For more on combustion reactions, see Section 6.4.

Ian Plant/Solent News/REX/Shutterstock

Animals (including humans) breathe in oxygen gas, which combines with carbon and hydrogen to produce carbon dioxide and water as well as energy. Overall this process, called *cellular respiration*, is the same as a combustion reaction.

Alina Vasilescu/Shutterstock

Figure 14.5 The white-hot flame of an acetylene torch results from the reaction of acetylene with oxygen.

Metal Displacement Reactions

In a **metal displacement reaction**, an elemental metal reacts with an ionic metal. For example, consider the reaction of aqueous copper(II) sulfate with zinc metal, as shown in **Figure 14.6**. Aqueous solutions of copper(II) ions have a blue color. However, when we drop a bar of zinc into the solution, we observe several changes: The blue color of the solution becomes fainter. The zinc rod begins to dissolve, and deposits of copper metal accumulate in the bottom of the beaker and on the zinc rod.

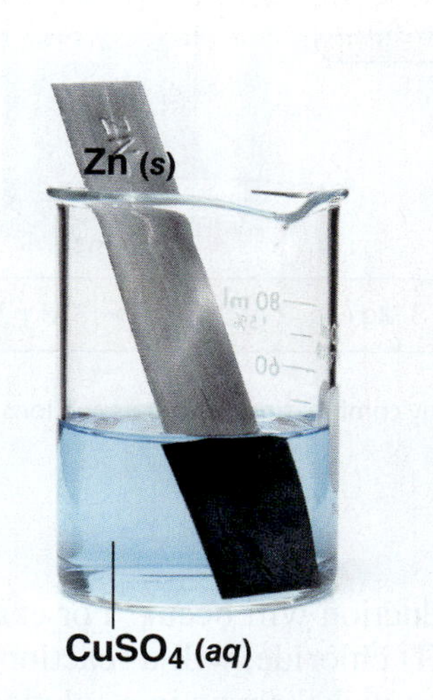

GIPhotoStock/Science Source

Figure 14.6 The reaction of zinc metal with aqueous copper(II) ions is a metal displacement reaction.

Explore
Figure 14.6

In this reaction, the zinc metal is oxidized—it loses electrons to become Zn^{2+} ions. At the same time, the copper(II) ion is reduced—it gains electrons to become elemental copper. The anion in solution, sulfate, is a spectator ion. We can describe this single-displacement reaction as a net ionic equation or as a complete molecular equation:

Net ionic equation: $Zn\,(s) + Cu^{2+}\,(aq) \rightarrow Zn^{2+}\,(aq) + Cu\,(s)$

Molecular equation: $Zn\,(s) + CuSO_4\,(aq) \rightarrow ZnSO_4\,(aq) + Cu\,(s)$

Metal displacement reactions are single-displacement reactions. ■

There are many reactions of this type (**Figure 14.7**). For example, elemental calcium reacts with iron(II) ions to produce calcium ions and elemental iron:

Net ionic equation: $Ca\ (s) + Fe^{2+}\ (aq) \rightarrow Ca^{2+}\ (aq) + Fe\ (s)$

Molecular equation: $Ca\ (s) + FeCl_2\ (aq) \rightarrow CaCl_2\ (aq) + Fe\ (s)$

Similarly, elemental magnesium reacts with aluminum ions to produce magnesium ions and elemental aluminum:

Net ionic equation: $2\ Al^{3+}\ (aq) + 3\ Mg\ (s) \rightarrow 3\ Mg^{2+}\ (aq) + 2\ Al\ (s)$

Molecular equation: $2\ Al(NO_3)_3\ (aq) + 3\ Mg\ (s) \rightarrow 3\ Mg(NO_3)_2\ (aq) + 2\ Al\ (s)$

Explore Figure 14.7

Figure 14.7 Metal displacement reactions occur for many combinations of metals and ions.

The Activity Series

So, how do we know when an oxidation reduction will occur? For example, if we place an iron nail into a solution of nickel(II) chloride, will a reaction take place? Will the opposite reaction occur if we place a nickel screw in a solution of iron(II) chloride? To answer these questions, we could run two side-by-side experiments (**Figure 14.8**). If we did this, we would find that iron reacts with nickel ions, but the opposite reaction does not occur:

$$Fe\ (s) + Ni^{2+}\ (aq) \rightarrow Fe^{2+}\ (aq) + Ni\ (s)$$

$$Fe^{2+}\ (aq) + Ni\ (s) \rightarrow \text{no reaction}$$

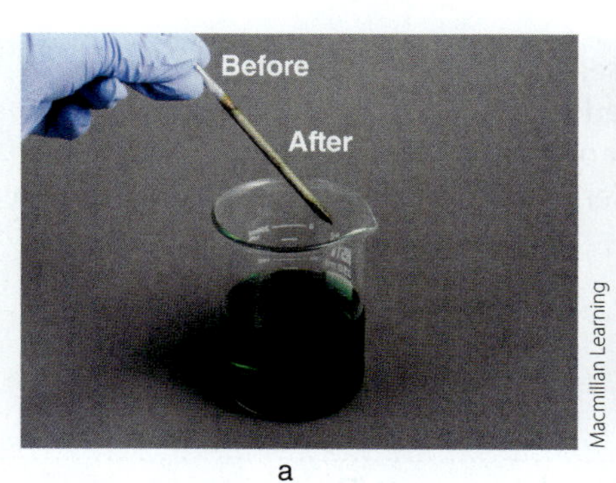

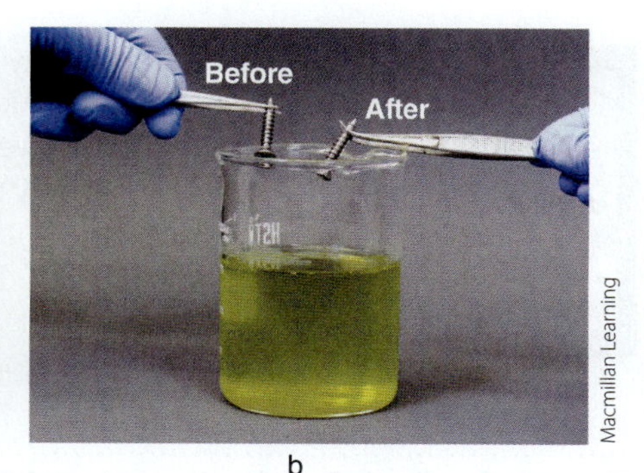

Figure 14.8 (a) An iron nail reacts in a nickel(II) chloride solution, but (b) a nickel screw does not react in an iron(II) chloride solution.

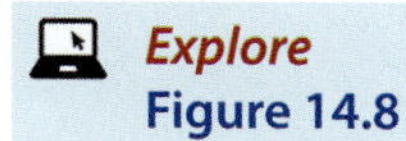
Explore Figure 14.8

Because of the profound importance of metals in so many applications, information about metal reactions like these is readily available. A helpful tool for predicting metal displacement reactions is the **activity series** (**Table 14.2**). The activity series ranks the reducing powers of metals. The alkali and alkaline earth metals appear at the top of the table. These metals release their electrons very easily. They are the most reactive metals and the strongest reducing agents. Metals at the bottom of the series are the least reactive—it takes much more energy to remove electrons from these atoms.

For example, let's compare the information on the activity series with our observations about the reactions of iron and nickel in Figure 14.8. Iron is higher on the activity series than nickel. This means that iron reduces Ni^{2+}, but nickel does not reduce Fe^{2+}. **Figure 14.9** shows this concept, and Example 14.3 describes it further.

TABLE 14.2 The Activity Series

(Reducing strength increases upward.)

Element	Half-Reaction
Lithium	$Li \rightarrow Li^{+} + e^{-}$
Potassium	$K \rightarrow K^{+} + e^{-}$
Barium	$Ba \rightarrow Ba^{2+} + 2\ e^{-}$
Calcium	$Ca \rightarrow Ca^{2+} + 2\ e^{-}$
Magnesium	$Mg \rightarrow Mg^{2+} + 2\ e^{-}$
Aluminum	$Al \rightarrow Al^{3+} + 3\ e^{-}$
Zinc	$Zn \rightarrow Zn^{2+} + 2\ e^{-}$
Chromium	$Cr \rightarrow Cr^{3+} + 3\ e^{-}$
Iron	$Fe \rightarrow Fe^{2+} + 2\ e^{-}$
Cobalt	$Co \rightarrow Co^{2+} + 2\ e^{-}$
Nickel	$Ni \rightarrow Ni^{2+} + 2\ e^{-}$
Tin	$Sn \rightarrow Sn^{2+} + 2\ e^{-}$
Lead	$Pb \rightarrow Pb^{2+} + 2\ e^{-}$
Hydrogen	$H_2 \rightarrow 2\ H^{+} + 2\ e^{-}$
Copper	$Cu \rightarrow Cu^{2+} + 2\ e^{-}$
Silver	$Ag \rightarrow Ag^{+} + e^{-}$
Platinum	$Pt \rightarrow Pt^{2+} + 2\ e^{-}$
Gold	$Au \rightarrow Au^{+} + e^{-}$

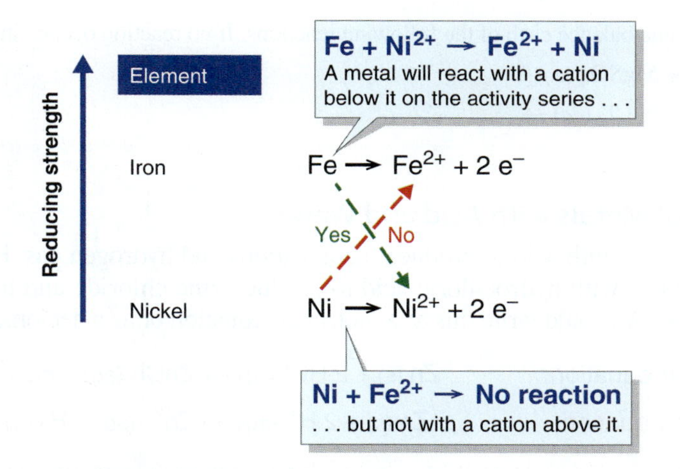

Figure 14.9 A metal can react with a cation below it on the activity series.

The metals at the bottom of the activity series are the precious metals. Elements like gold and platinum are valued not just for their exquisite beauty, but for their durability. Unlike metals such as iron, which will oxidize (rust) over time, metals like gold and platinum retain their elemental form and beauty.

The activity series is like college football rankings: Higher-ranked teams are expected to beat teams ranked lower on the list. Similarly, a metal higher on the activity series will reduce a cation that is lower on the series.

© dehooks/depositphotos.com

Example 14.3 Predicting Metal Displacement Reactions

Based on the activity series in Table 14.2, which of the following metals will react with an aqueous solution of nickel(II) chloride: gold, tin, or zinc? Write a balanced molecular equation for the reaction that occurs.

By examining Table 14.2, we can see that Zn lies above Ni^{2+} on the activity series, while Sn and Au both lie below it. Based on this ranking, Zn is the only one of these elements that will react with Ni^{2+}. We can write a balanced molecular equation for this reaction.

$$Zn\ (s) + NiCl_2\ (aq) \rightarrow ZnCl_2\ (aq) + Ni\ (s) \qquad \text{balanced}$$

Check It
Watch Explanation

TRY IT

3. Based on the activity series, which of these elements can reduce iron(II) to elemental iron?

a. calcium **b.** magnesium **c.** chromium **d.** silver

4. Complete and balance each of the following reactions. If no reaction occurs, indicate this.

a. $Zn\ (s) + MgSO_4\ (aq) \rightarrow$

b. $Co\ (s) + NiCl_2\ (aq) \rightarrow$

Charles D. Winters/Science Source

Figure 14.10 Zinc metal reacts with hydrochloric acid to produce zinc chloride and hydrogen gas. The bubbles are the hydrogen gas produced in this reaction.

Explore
Figure 14.10

Reactions of Metals with Acid and Water

Most metals react with acid to produce metal cations and hydrogen gas. For example, zinc metal reacts with hydrochloric acid to produce zinc chloride and hydrogen gas (**Figure 14.10**). We could write this as a molecular equation or as a net ionic equation:

Molecular equation: $Zn\ (s) + 2\ HCl\ (aq) \rightarrow ZnCl_2\ (aq) + H_2\ (g)$

Net ionic equation: $Zn\ (s) + 2\ H^+\ (aq) \rightarrow Zn^{2+}\ (aq) + H_2\ (g)$

Similarly, magnesium reacts with nitric acid to produce magnesium nitrate and hydrogen gas:

Molecular equation: $Mg\ (s) + 2\ HNO_3\ (aq) \rightarrow Mg(NO_3)_2\ (aq) + H_2\ (g)$

Net ionic equation: $Mg\ (s) + 2\ H^+\ (aq) \rightarrow Mg^{2+}\ (aq) + H_2\ (g)$

Not all metals react with acid. The precious metals (silver, gold, platinum) do not react with acid. Based on these observations, chemists have fit H^+ into the activity series in Table 14.2. The metals above H^+ on the activity series react with acids, and those below it do not.

Precious metals do not react with acid.

The most active metals react with water. These reactions are similar to the reactions with acid in that the metal is oxidized to form the cation, while H^+ is reduced to form hydrogen gas. For example, sodium and potassium both react violently with water to produce a metal hydroxide and hydrogen gas (**Figure 14.11**).

$$2\ Na\ (s) + 2\ H_2O\ (l) \rightarrow 2\ NaOH\ (aq) + H_2\ (g)$$

$$2\ K\ (s) + 2\ H_2O\ (l) \rightarrow 2\ KOH\ (aq) + H_2\ (g)$$

Philip Evans/Getty Images

a

Andrew Lambert Photography/Science Source

b

Figure 14.11 (a) Alkali metals like sodium and potassium react violently with water. (b) To prevent reaction with water, these elements are commonly stored under a layer of oil.

Explore **Figure 14.11**

As a general rule, alkali metals react violently with water. Alkaline earth metals also react with water, but the reactions take place more slowly. Some alkaline earth metals, like calcium, react slowly with water at room temperature. Others, like magnesium, require heating in order for the reaction with water to take place. These trends are summarized in **Table 14.3**.

Because of their extreme reactivity, sodium, potassium, and the other alkali metals must be carefully stored to keep them from coming in contact with moisture. Chemists frequently store these metals under oil to prevent exposing them to water or air.

Alkali metals react violently with water; alkaline earth metals react slowly with water.

TABLE 14.3 Reactivity of Metals with Acid and Water

	React Violently with Water	React with Water or Steam	React with Acid
Alkali metals (Li, Na, K)	✓	✓	✓
Alkaline earth metals (Mg, Ca, Ba)		✓	✓
Most transition metals* (Fe, Co, Ni)			✓
Precious metals (Cu, Au, Ag, Pt)			

*The most reactive transition metals, such as Zn, also react with steam.

Example 14.4 Predicting an Acid-Metal Reaction from the Activity Series

Will aluminum metal react with hydrochloric acid? If so, write a balanced molecular equation to describe this reaction.

We see from Table 14.2 that Al is higher than H^+ on the activity series, so we know that this reaction can occur. Aluminum forms a +3 cation, and so aluminum chloride has the formula $AlCl_3$. After balancing the equation, we arrive at the following solution.

$$2\ Al\ (s) + 6\ HCl\ (aq) \rightarrow 2\ AlCl_3\ (aq) + 3\ H_2\ (g) \qquad \text{balanced}$$

Check It
Watch Explanation

TRY IT

5. Using the activity series (Table 14.2), indicate whether these metals will react with acid:
   a. gold **b.** nickel **c.** platinum **d.** chromium

6. Use the information in Tables 14.2 and 14.3 to complete and balance each of the following reactions. Also indicate if no reaction occurs.
   a. Zn (s) + HBr $(aq) \rightarrow$ **b.** Co (s) + HNO_3 $(aq) \rightarrow$
   c. Ag (s) + HCl $(aq) \rightarrow$ **d.** Li (s) + H_2O $(l) \rightarrow$

14.3 Half-Reactions and Batteries

Half-Reactions

Oxidation-reduction reactions take place in two distinct parts. On the one hand, an atom or ion gains electrons (reduction). On the other, an atom or ion loses electrons (oxidation). We sometimes write these two separate processes as **half-reactions**. Half-reactions show the reduction without indicating where the electrons came from, or they show the oxidation without indicating where the electrons went.

Half-reactions use e^- to represent electrons in either the reactants or the products.

For example, let's look again at the reaction of zinc metal and copper(II) ions. We can write this process as two half-reactions. In one half-reaction, zinc is oxidized. In the other, copper(II) is reduced. In the reactions below, the symbol e^- represents an electron:

Oxidation half-reaction: $Zn\ (s) \rightarrow Zn^{2+}\ (aq) + 2\ e^-$

Reduction half-reaction: $2\ e^- + Cu^{2+}\ (aq) \rightarrow Cu\ (s)$

Net ionic equation: $\cancel{2\ e^-} + Cu^{2+}\ (aq) + Zn\ (s) \rightarrow Cu\ (s) + Zn^{2+}\ (aq) + \cancel{2\ e^-}$

Notice that the two electrons released by zinc are consumed by copper(II). We can add the two half-reactions together to produce the net ionic equation. Because the two electrons appear on the reactant and product sides, we can cancel them out.

The activity series (see Table 14.2) is actually a series of half-reactions. We can use the activity series to write the oxidation and reduction half-reactions for a chemical change. For example, consider the reaction of nickel metal with tin(II) ions. As **Figure 14.12** shows, nickel is above tin on the activity series. We write the half-reaction for the oxidation of nickel the same way it appears on the activity series:

Oxidation half-reaction: $Ni\ (s) \rightarrow Ni^{2+}\ (aq) + 2\ e^-$

The other half-reaction that takes place is the reduction of Sn^{2+}. Because the activity series shows oxidations, we must reverse the reactants and products to show the reaction as a reduction:

Reduction half-reaction: $2\ e^- + Sn^{2+}\ (aq) \rightarrow Sn\ (s)$

As a general rule, when writing half-reactions from the activity series, the upper half-reaction proceeds in the forward direction; the lower half-reaction will go in the reverse direction. In some instances, we must balance the half-reactions to produce a correctly balanced equation. We'll look at this process in more detail in Section 14.4.

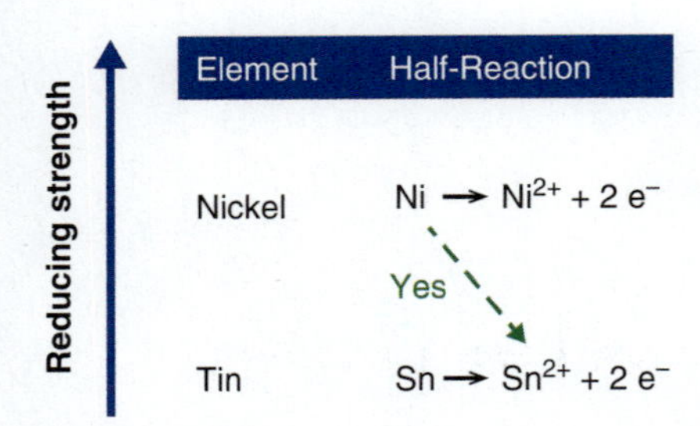

Reducing strength
Element
Half-Reaction
Nickel
Ni → Ni2+ + 2 e−
Yes
Tin
Sn → Sn2+ + 2 e−

Figure 14.12 Nickel falls above tin on the activity series. When the two react, nickel is oxidized while Sn^{2+} is reduced.

Example 14.5 Writing Half-Reactions

Write a net ionic equation for the redox reaction shown below. Then show the oxidation and reduction half-reactions that take place.

$$SnCl_2\ (aq) + Fe\ (s) \rightarrow FeCl_2\ (aq) + Sn\ (s)$$

We begin by writing the complete ionic equation:

$$Sn^{2+}\ (aq) + 2\ Cl^-\ (aq) + Fe\ (s) \rightarrow Fe^{2+}\ (aq) + 2\ Cl^-\ (aq) + Sn\ (s)$$

Chloride is a spectator ion, so we remove it from the equation to give the net ionic equation:

$$Sn^{2+}\ (aq) + Fe\ (s) \rightarrow Fe^{2+}\ (aq) + Sn\ (s)$$

Next, we break the reaction into oxidation and reduction half-reactions. Iron is oxidized from Fe to Fe^{2+}, meaning it loses two electrons. Tin is reduced from Sn^{2+} to Sn, so it gains two electrons. We can write the two half-reactions as follows:

Oxidation: $Fe\ (s) \rightarrow Fe^{2+}\ (aq) + 2\ e^-$

Reduction: $2\ e^- + Sn^{2+}\ (aq) \rightarrow Sn\ (s)$

Adding the two half-reactions together results in the net ionic equation. Note also where these two reactions are on the activity series. Iron falls higher on the activity series. It is oxidized in this reaction. Tin falls lower on the activity series, so it is reduced.

When two football players collide, the player exerting the stronger force continues to move forward, while the player exerting the weaker force is knocked backward. Similarly, the half-reaction for the stronger reducing metal on the activity series proceeds in the forward direction (oxidation). The half-reaction for the weaker metal gets reversed (reduction).

AP Photo/Kathy Willens

Example 14.6 Writing Half-Reactions

Show the oxidation and reduction half-reactions that take place when zinc reacts with nitric acid to produce zinc nitrate and hydrogen gas.

To begin, we write a balanced equation for this reaction. Zinc forms a +2 cation, so the formula for zinc nitrate is $Zn(NO_3)_2$. This leads us to the balanced equation:

$$Zn\ (s) + 2\ HNO_3\ (aq) \rightarrow Zn(NO_3)_2\ (aq) + H_2\ (g)$$

Writing the dissociated ions and removing the spectator ions (NO_3^-) gives us the net ionic equation:

$$Zn\ (s) + 2\ H^+\ (aq) \rightarrow Zn^{2+}\ (aq) + H_2\ (g)$$

Zinc is oxidized from Zn to Zn^{2+}, meaning it loses two electrons. The H^+ ions are reduced: Adding two electrons to two H^+ ions results in H_2 gas, as shown below.

Oxidation: $Zn\ (s) \rightarrow Zn^{2+}\ (aq) + 2\ e^-$

Reduction: $2\ e^- + 2\ H^+\ (aq) \rightarrow H_2\ (g)$

TRY IT

7. Write half-reactions for each of these chemical changes:

a. $Ca\ (s) + Cl_2\ (aq) \rightarrow Ca^{2+}\ (aq) + 2\ Cl^-\ (aq)$

b. $Al\ (s) + Cr(NO_3)_3\ (aq) \rightarrow Al(NO_3)_3\ (aq) + Cr\ (s)$

8. Write two half-reactions to describe the reaction of calcium with hydrochloric acid to produce calcium chloride and hydrogen gas.

Check It

Watch Explanation

Figure 14.13 Batteries are powered by oxidation-reduction reactions in which half-reactions occur at separate sites. The flow of electrons between the two sites produces an electric current.

Oxidation half-reaction (release of electrons)

Reduction half-reaction (gain of electrons)

Batteries

Batteries (also called *electrochemical cells*) are powered by chemical reactions in which oxidation occurs in one location, and reduction occurs at another. For the reaction to take place, the electrons must travel from the oxidation site to the reduction site, resulting in an electric current (**Figure 14.13**). We use half-reactions to describe the changes taking place at each site.

Figure 14.14 shows a classic electrochemical cell. The cell consists of two chambers (in this case, two beakers). One beaker contains a strip of copper immersed in a solution of copper(II) sulfate. The other beaker contains a strip of zinc immersed in a solution of zinc sulfate. If we connect these two strips with a conducting wire, the result is an electric current. We call the copper and zinc strips **electrodes**. An electrode is a site where an oxidation or reduction half-reaction takes place. The electrode where oxidation occurs is the *anode*. The electrode where reduction occurs is the *cathode*.

Reduction occurs at the cathode; oxidation occurs at the anode. ■

The current arises from a redox reaction at the two electrodes. On the zinc electrode (the anode), an oxidation takes place: The zinc atoms lose two electrons to form Zn^{2+} ions. These electrons travel through the conducting wire toward the copper electrode (the cathode). At the copper electrode, Cu^{2+} ions gain two electrons to form elemental copper.

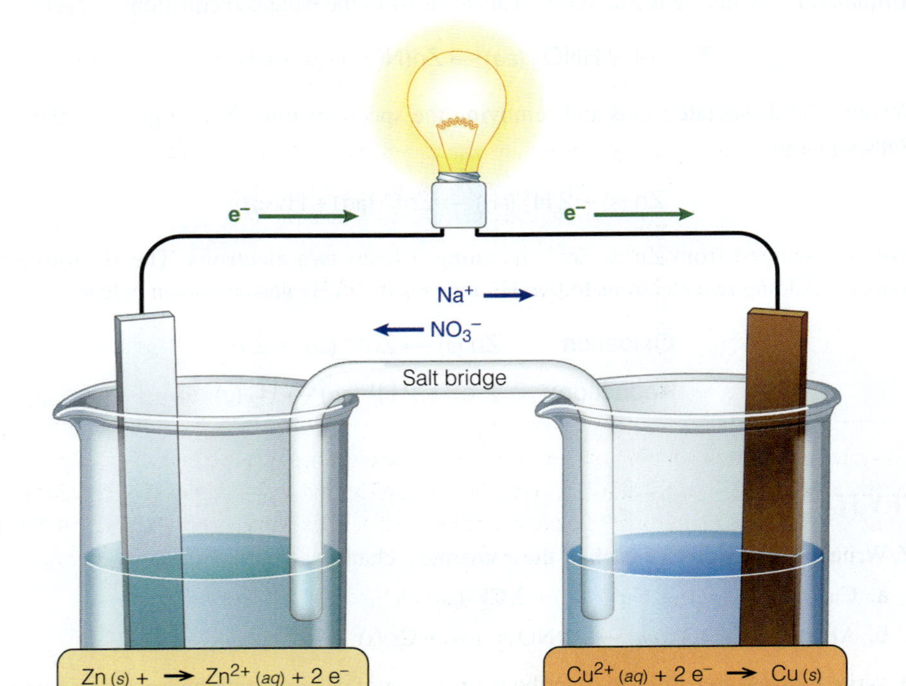

Figure 14.14 In an electrochemical cell, an oxidation takes place in one chamber while the reduction takes place in another. The transfer of electrons between the two half-reactions produces an electric current. The salt bridge balances the charges between the two solutions.

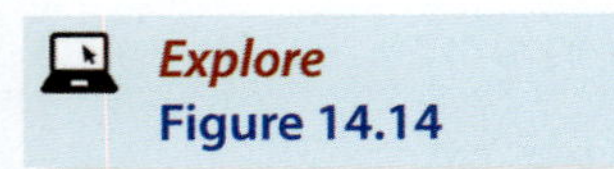

Although the oxidation and reduction half-reactions take place in separate containers, they must occur together. For the reaction to happen, the electrons must move from one atom or ion to another.

Notice that the electrochemical cell in Figure 14.14 also contains a U-shaped tube, called a *salt bridge*. This bridge is essential to the function of the cell. A salt bridge contains an ionic compound like sodium nitrate. As the electrons move from the anode to the cathode, the anode cell builds up a positive charge, while the cathode builds up a negative charge. The salt bridge provides spectator ions that can balance these charges. For example, if the bridge contains sodium nitrate, the sodium cations will migrate toward the negative charges, while the nitrate anions migrate toward the positive charges. If a salt bridge is not present, the buildup of charge on the electrodes limits the flow of electrons. The salt bridge must be present for the cell to work.

The amount of energy stored in a battery depends on the specific chemical reactions that take place at the cathode and the anode. Scientists describe the amount of electrical potential energy as the *voltage*. The unit of electrical potential energy is the *volt* (V).

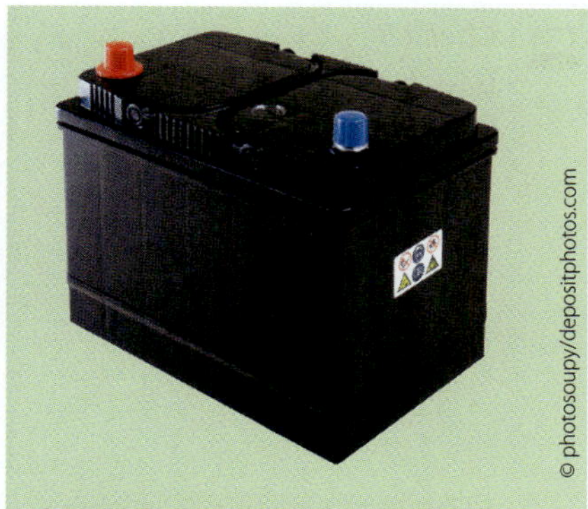

Although batteries vary widely in shape, size, and potential energy, they all rely on oxidation-reduction reactions to produce electric currents.

A salt bridge balances the charges between the cathode and anode.

Example 14.7 Predicting the Direction of Electron Flow in an Electrochemical Cell

A chemist assembles an electrochemical cell as shown. One side of the cell contains an iron electrode immersed in aqueous iron(II) nitrate solution. The other side of the cell contains a magnesium electrode immersed in aqueous magnesium nitrate solution. Using the activity series, identify the electrode where oxidation takes place (the anode) and the electrode where reduction takes place (the cathode). Write half-reactions to describe the change that takes place at each electrode.

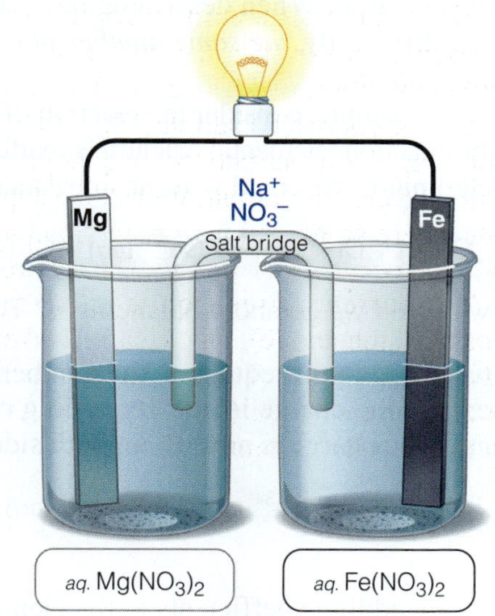

Because magnesium lies above iron on the activity series, we know that Mg reacts with Fe^{2+}:

Element	Half-Reaction
Magnesium	$Mg \rightarrow Mg^{2+} + 2\ e^-$
	Yes ✕ No
Iron	$Fe \rightarrow Fe^{2+} + 2\ e^-$

Reducing strength (increases upward)

This means that oxidation takes place at the magnesium electrode, while reduction takes place at the iron electrode. We can obtain the two half-reactions from the activity series. The reaction that is higher on the list proceeds as an oxidation reaction (that is, as written). For the reaction that is lower on the list, we reverse the reactants and products to show the reduction half-reaction.

Oxidation: $Mg\ (s) \rightarrow Mg^{2+}\ (aq) + 2\ e^-$

Reduction: $2\ e^- + Fe^{2+}\ (aq) \rightarrow Fe\ (s)$

We also can describe this process by the net ionic reaction:

$$Fe^{2+}\ (aq) + Mg\ (s) \rightarrow Fe\ (s) + Mg^{2+}\ (aq)$$

Because oxidation takes place at the anode, the magnesium strip is the anode, and the iron strip is the cathode.

Check It
Watch Explanation

TRY IT

9. A battery contains a zinc anode immersed in an aqueous zinc chloride solution and a platinum cathode immersed in an aqueous platinum(II) chloride solution. What half-reactions take place at the anode and cathode of this cell?

10. An electrochemical cell contains two chambers. The first consists of a nickel electrode immersed in aqueous nickel(II) sulfate. The second consists of a zinc electrode immersed in aqueous zinc sulfate. If these two chambers are connected by a salt bridge and a conducting wire, in which direction will the electrons flow? Write half-reactions and a net ionic equation to describe this chemical reaction.

14.4 Balancing Redox Equations

For redox equations, we must balance both the atoms and the charge. ■

Sometimes we need to balance a redox equation. As with other balanced equations, this means that the total number of each type of atom must be equal on both sides of the equation. When describing redox reactions, we must also balance the *charge*. Stated differently, *the same number of electrons lost in the oxidation will be gained in the reduction.*

For example, consider the reaction of calcium metal with aqueous chromium(III): In this reaction, elemental calcium is oxidized to Ca^{2+} while Cr^{3+} is reduced to elemental chromium. We could write an unbalanced equation to describe this change:

$$Ca\ (s) + Cr^{3+}\ (aq) \rightarrow Ca^{2+}\ (aq) + Cr\ (s) \qquad \text{not balanced (charge not equal)}$$

How should we go about balancing it? Two approaches are discussed below.

Approach 1. In this equation, the number and type of the atoms are the same on both sides, but the charge is not. By adding coefficients to the ions, we can balance the charge (the charge is now +6 on both sides):

$$Ca\ (s) + 2\ Cr^{3+}\ (aq) \rightarrow 3\ Ca^{2+}\ (aq) + Cr\ (s) \qquad \text{not balanced (charge equal, but not atoms)}$$

Then, by adding coefficients to the elemental forms, we ensure that both the charge and the atoms/ions are balanced.

$$3\ Ca\ (s) + 2\ Cr^{3+}\ (aq) \rightarrow 3\ Ca^{2+}\ (aq) + 2\ Cr\ (s) \qquad \text{balanced}$$

Approach 2. Another way to tackle this problem is to make sure that the same number of electrons are released in the oxidation and acquired in the reduction. We begin by writing two unbalanced half-reactions:

Oxidation: $Ca \rightarrow Ca^{2+} + 2\ e^-$

Reduction: $3\ e^- + Cr^{3+} \rightarrow Cr$

The oxidation releases two electrons, but the reduction requires three. For an oxidation-reduction reaction to balance, the electrons lost in the oxidation must equal the electrons gained in the reduction. To make the electrons equal, we multiply each species in the oxidation by a factor of three and each species in the reduction by a factor of two. Now six electrons are lost in the oxidation, and six electrons are gained in the reduction:

Oxidation: $(Ca \rightarrow Ca^{2+} + 2\ e^-) \times 3 \longrightarrow 3\ Ca \rightarrow 3\ Ca^{2+} + 6\ e^-$

Reduction: $(3\ e^- + Cr^{3+} \rightarrow Cr) \times 2 \longrightarrow 6\ e^- + 2\ Cr^{3+} \rightarrow 2\ Cr$

The number of electrons in the oxidation and reduction must be equal.

Finally, we add the oxidation and reduction half-reactions together to obtain the net ionic reaction. Because the number of electrons present on the left and right sides of the equation is the same, we can cancel them out, leaving the correctly balanced net ionic equation:

Oxidation half-reaction: $3\,Ca \rightarrow 3\,Ca^{2+} + 6\,e^-$

Reduction half-reaction: $6\,e^- + 2\,Cr^{3+} \rightarrow 2\,Cr$

Net ionic equation: $\cancel{6e^-} + 3\,Ca\,(s) + 2\,Cr^{3+}\,(aq) \rightarrow 3\,Ca^{2+}\,(aq) + 2\,Cr\,(s) + \cancel{6e^-}$

Example 14.8 provides another problem of this type.

Example 14.8 Balancing a Redox Equation

Silver metal undergoes a single-displacement reaction with platinum(II) nitrate to form silver nitrate and elemental platinum. Write a balanced net ionic equation for this single-displacement reaction.

Let's solve this problem using both of the approaches discussed above. First, let's do it by balancing the atoms and the charge in the net ionic equation. Nitrate is a spectator ion, so it doesn't need to be included. This leaves the following species involved with this reaction:

$$Ag + Pt^{2+} \rightarrow Ag^+ + Pt \qquad \text{not balanced}$$

In this case, we can place a coefficient (2) in front of the Ag^+ to balance the charge. We then put a 2 in front of elemental Ag to balance the number and type of atoms. Finally, we insert phase symbols to give the complete answer:

$$2\,Ag\,(s) + Pt^{2+}\,(aq) \rightarrow 2\,Ag^+\,(aq) + Pt\,(s) \qquad \text{balanced}$$

The alternative approach is to balance the electron flow in the oxidation and reduction. We begin by writing two half-reactions:

Oxidation: $Ag \rightarrow Ag^+ + e^-$

Reduction: $2\,e^- + Pt^{2+} \rightarrow Pt$

To balance the electrons in each half-reaction, we multiply each species in the oxidation by two. We then combine the two half-reactions. There are two electrons on both the reactant and product side, so we can cancel them out, leaving a balanced equation.

Oxidation half-reaction: $2\,Ag \rightarrow 2\,Ag^+ + 2\,e^-$

Reduction half-reaction: $2\,e^- + Pt^{2+} \rightarrow Pt$

Net ionic equation: $\cancel{2e^-} + 2\,Ag\,(s) + Pt^{2+}\,(aq) \rightarrow 2\,Ag^+\,(aq) + Pt\,(s) + \cancel{2e^-}$

TRY IT

11. Elemental barium reacts with a solution of cobalt(II) chloride to produce barium chloride and elemental cobalt. Write half-reactions to show the oxidation and reduction components of this change.

12. Elemental aluminum reacts with a solution of iron(II) chloride to produce aluminum chloride and elemental iron. Describe this change using half-reactions, using a net ionic equation, and using a molecular equation.

Check It
Watch Explanation

fact/Mauritius/Superstock

Figure 14.15 The brilliant chrome surfaces on this motorcycle are produced by electroplating each of the steel parts. The parts are coated with thin layers of copper, nickel, and chromium.

14.5 Other Applications of Redox Reactions

Electroplating

Electroplating is a technique that produces a thin layer of a metal such as gold, silver, copper, or chromium on the outside surface of another material. For example, "chrome" motorcycle parts are typically composed of steel, which is protected and beautified by three thin outer layers: first copper, then nickel, and finally chromium (**Figure 14.15**).

Electroplating is actually very similar to what happens in an electrochemical cell. In an electrochemical cell, a redox reaction creates an electric current. In electroplating, the reverse occurs: An electric current drives a redox reaction to take place. **Figure 14.16** illustrates the process of applying a nickel coating to an iron pipe. The pipe is immersed in an aqueous solution containing dissolved nickel(II) ions. A nickel electrode is also immersed in the solution. The two electrodes (the pipe and the nickel) are then connected to a battery. The electrical potential energy of the battery drives a redox reaction to take place. The reduction occurs on the surface of the pipe. As nickel(II) ions are reduced to elemental nickel, they adhere to the metal surface. On the other electrode, nickel atoms are oxidized to form nickel(II) ions, which dissolve in the solution and replenish the supply.

Photo Researchers /Science Source

Historically, pennies were made almost completely of copper. However, because of the rising cost of copper, in 1983 the U.S. Mint began electroplating zinc pennies with a thin copper surface.

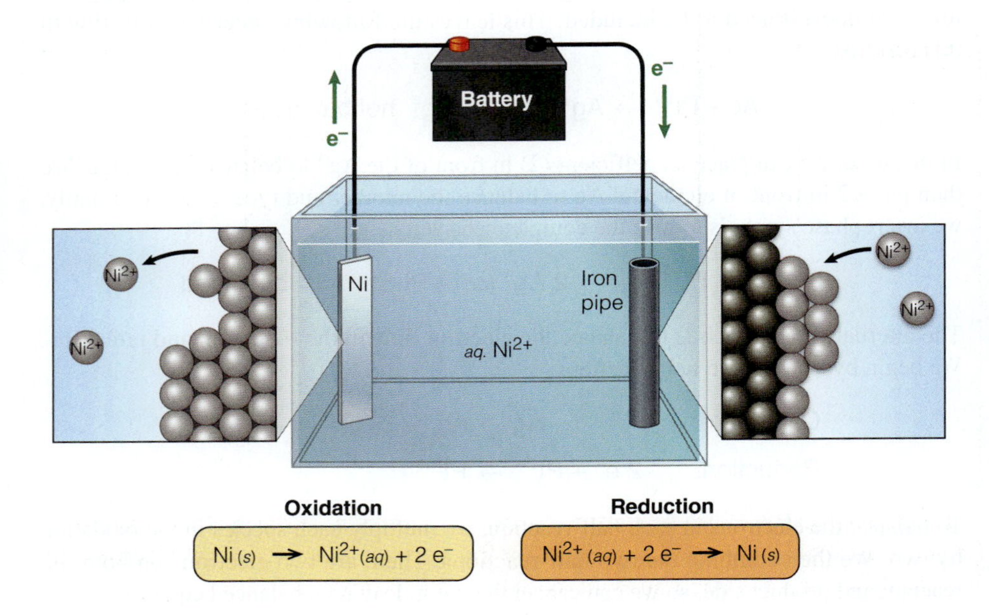

Figure 14.16 Electroplating uses electrical potential energy (voltage) to drive a redox reaction to take place. In this nickel electroplating apparatus, nickel is oxidized from the electrode at left and reduced on the electrode at right, forming a thin layer of nickel.

De Agostini Picture Library/AGE Fotostock

Following Volta's invention of the battery, the English scientist Humphry Davy discovered many of the main-group metals using a technique similar to that used in electroplating. By immersing both electrodes into a solution containing ionic compounds, he was able to reduce the metal ions to their elemental form.

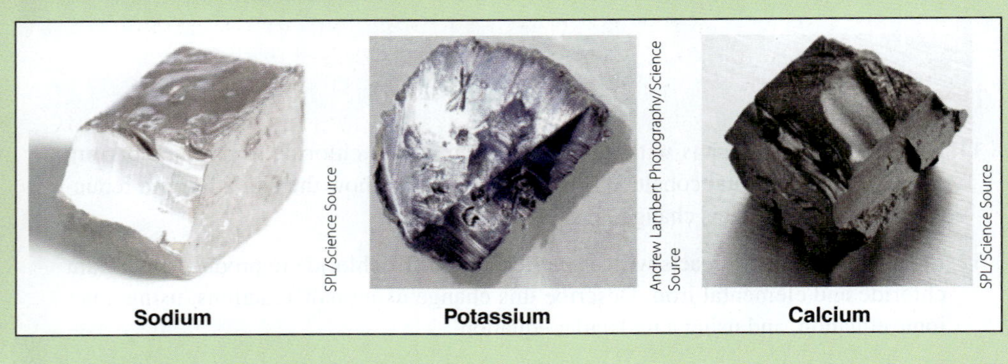

Fuel Cells

In the past few decades, **fuel cells** have emerged as an exciting new approach to storing and releasing energy. Fuel cells convert the potential energy of a combustion reaction directly into electrical energy by separating the oxidation and reduction half-reactions (**Figure 14.17**). The most common fuel cells involve the very exothermic reaction of hydrogen with oxygen:

$$2\,H_2\,(g) + O_2\,(g) \rightarrow 2\,H_2O\,(l)$$

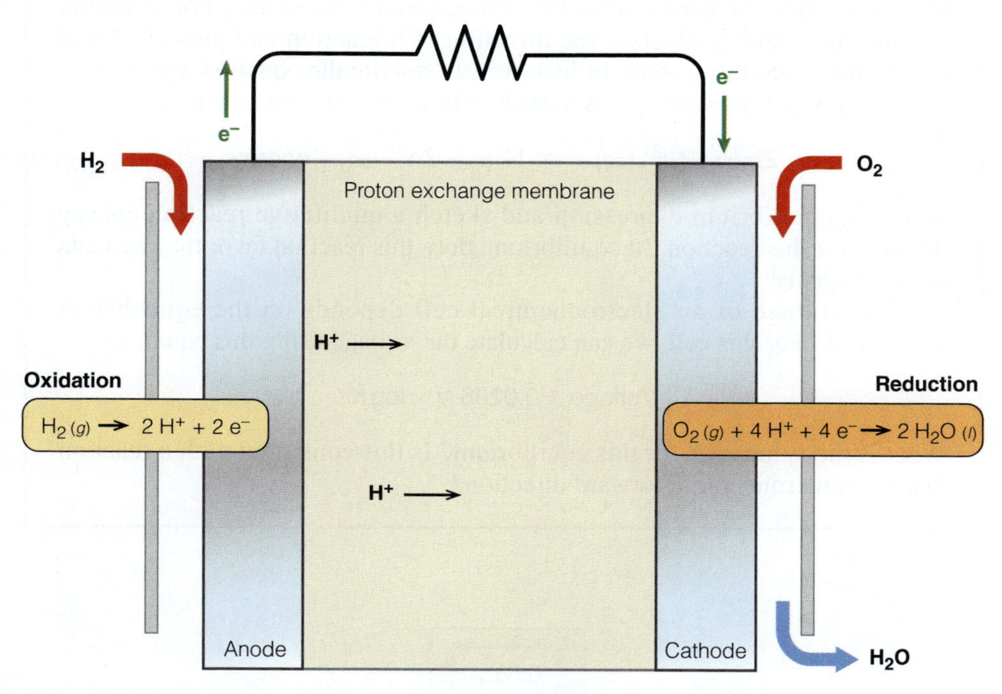

Figure 14.17 A hydrogen fuel cell divides the combustion of hydrogen into two half-reactions and uses the transfer of electrons to produce an electric current.

Recall from acid-base chemistry that we commonly refer to H^+ ions as *protons*.

A hydrogen fuel cell consists of two electrodes, separated by a substance called a *proton exchange membrane*. H^+ ions flow through the membrane, but electrons do not. At the anode, H_2 undergoes an oxidation to H^+ ions:

$$\text{Oxidation:} \quad H_2\,(g) \rightarrow 2\,H^+ + 2\,e^-$$

At the cathode, oxygen is reduced, reacting with electrons and H^+ ions to form water:

$$\text{Reduction:} \quad O_2\,(g) + 4\,H^+ + 4\,e^- \rightarrow 2\,H_2O\,(l)$$

Because the proton exchange membrane only allows H^+ ions to flow easily through it, the electrons must travel through an external route in order for the reaction to take place. The result is an electric current.

Fuel cells were developed by NASA for use in space travel. More recently, design improvements have made fuel cells a viable option for many power applications. In 2016, the automaker Hyundai released the first mass-produced vehicle powered by a hydrogen fuel cell (**Figure 14.18**).

Alex Gallardo/Reuters/Newscom

Figure 14.18 In 2016, Hyundai launched the Tucson Fuel Cell, the first mass-produced vehicle powered using a hydrogen fuel cell.

Capstone Video

Capstone Question

Figure 14.19 shows an electrochemical cell composed of a zinc electrode in a 1.0 M *aq*. Zn^{2+} solution, and a nickel electrode in a 1.0 M *aq*. Ni^{2+} solution. The electrical potential energy (voltage) of this cell is 0.53 V. When describing an electrochemical cell like this one, we often describe the reaction as only going forward:

$$Zn\ (s) + Ni^{2+}\ (aq) \rightarrow Ni\ (s) + Zn^{2+}\ (aq)$$

However, many oxidation-reduction processes are reversible. For example, recharging a battery reverses the direction of a reaction and puts electrical energy back into the system. In fact, we can rewrite the equation above as an equilibrium and show energy as a product from the forward reaction:

$$Zn\ (s) + Ni^{2+}\ (aq) \rightleftharpoons Ni\ (s) + Zn^{2+}\ (aq) + \text{energy}$$

Write an equilibrium expression and sketch a qualitative reaction energy diagram for this reaction. At equilibrium, does this reaction favor the reactants or the products?

The voltage of an electrochemical cell depends on the equilibrium constant, *K*. For this cell, we can calculate the voltage using this equation:

$$\text{Voltage} = 0.0296\ \text{V} \times \log K$$

What is the value of *K* for this equilibrium? Is this consistent with a reaction that is exothermic in the forward direction?

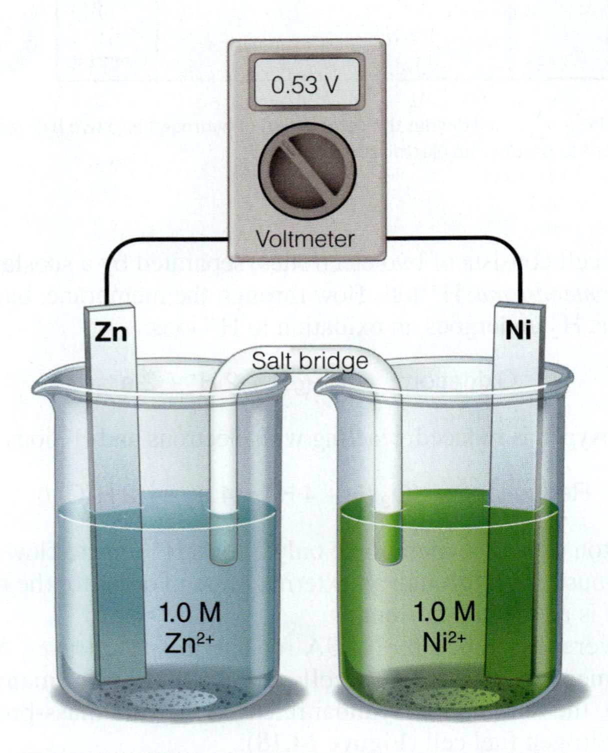

Figure 14.19 This electrochemical cell contains zinc and nickel electrodes immersed in 1.0 M Zn^{2+} and Ni^{2+} solutions.

SUMMARY

Oxidation-reduction (or redox) reactions involve the gain and loss of electrons. An atom, molecule, or ion that loses electrons is oxidized; a species that gains electrons is reduced. Oxidation-reduction reactions are central to the field of electrochemistry, the study of chemical processes that involve the movement of electrons.

There are several different classes of redox reactions. Metals and nonmetals combine in redox reactions to produce ionic compounds. Metals and nonmetals alike undergo combustion reactions, in which an element or compound reacts with oxygen to produce new compounds called oxides. These reactions often release a great deal of energy as heat or light. Hydrocarbons (compounds composed of hydrogen and carbon) are an important class of compounds because of their use as fossil fuels. Combustion reactions of hydrocarbons produce two products, carbon dioxide and water.

Metal displacement reactions are another common type of redox reaction. In these reactions, an elemental metal reacts with a metal ion, typically in aqueous solution. The elemental metal is oxidized to the ion while the ion is reduced to the elemental form. These reactions are examples of single-displacement reactions. Some metals are more active or more strongly reducing than others. The activity series is a ranking of metals by their reducing power; it allows us to predict which combinations of metals and ions will react.

Most metals react with acids to produce metal ions and hydrogen gas. The most active metals undergo a similar reaction with water.

Sometimes, it is helpful to write redox reactions as oxidation and reduction half-reactions. To balance a redox reaction, the number of electrons lost in the oxidation half-reaction must equal the number gained in the reduction half-reaction. Stated differently, a balanced redox reaction must not only balance the number and type of atom present, it must also balance the overall charge.

Batteries (also called electrochemical cells) are devices that produce electric current through a redox reaction in which the two half-reactions occur at separate sites. Simple electrochemical cells contain two electrodes connected by a salt bridge.

Batteries are just one of many important applications in the field of electrochemistry. A related technology is electroplating, in which electric potential energy is used to coat a material with a thin layer of a metal such as chromium, nickel, or gold.

Another important application of electrochemistry is the fuel cell. These devices convert the energy of a combustion reaction directly into electrical energy by separating the oxidation and reduction half-reactions. Fuel cells hold the potential to significantly impact the transportation industry in the future.

Charging Ahead: Batteries Today and Tomorrow

Figure 14.20 Chainsaws must be portable but also very powerful. This chainsaw runs on a lithium-ion battery.

Chainsaws are amazing tools (**Figure 14.20**). With a good saw, you can turn a fallen tree into a stack of firewood in an hour. But they can also be difficult to work with. They are heavy and dangerous. The small gasoline engines in most chainsaws require a mixture of gas and oil, and they are notoriously hard to start. But thanks to recent advances in battery technology, all of that is poised to change.

To power a chainsaw, a battery must be very strong but also lightweight. It must last for a good length of time, and it must be rechargeable. For years, batteries simply couldn't meet these requirements. But a new generation of batteries called *lithium-ion batteries* has changed the market for chainsaws, as well as for other products.

Lithium-ion batteries have been in use for over a decade in laptop computers and other portable electronic devices. Like Volta's original voltaic pile, these batteries rely on oxidation-reduction reactions. But innovative new materials and designs have maximized their performance. And unlike early batteries, lithium-ion batteries are rechargeable.

Figure 14.21 illustrates the design of these remarkable devices. In this type of battery, electrons are stored on graphite sheets. Lithium ions are able to move freely between the sheets, balancing the overall charge. When the two electrodes are connected, electrons flow through the circuit from the graphite sheets to a metal oxide surface, such as $Mn_2O_4^{2-}$. The result is a net reduction:

$$Mn_2O_4^{-} + e^{-} \rightarrow Mn_2O_4^{2-}$$

While the electrons flow through the circuit, the lithium ions migrate through a porous membrane and into the $Mn_2O_4^{2-}$ framework. This balances the overall charge and allows the current to continue until the supply of extra electrons on the graphite sheets is depleted.

Recharging a battery pushes the electrons back from the metal oxide surface to the graphite surface. This process oxidizes the metal while reducing the graphite. Once the electrons have been pushed back to the graphite surface, the battery is recharged. Pop the battery back in, and you're ready to go.

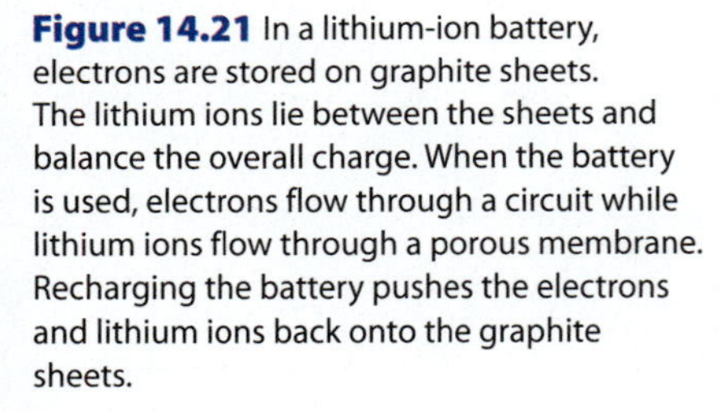

Figure 14.21 In a lithium-ion battery, electrons are stored on graphite sheets. The lithium ions lie between the sheets and balance the overall charge. When the battery is used, electrons flow through a circuit while lithium ions flow through a porous membrane. Recharging the battery pushes the electrons and lithium ions back onto the graphite sheets.

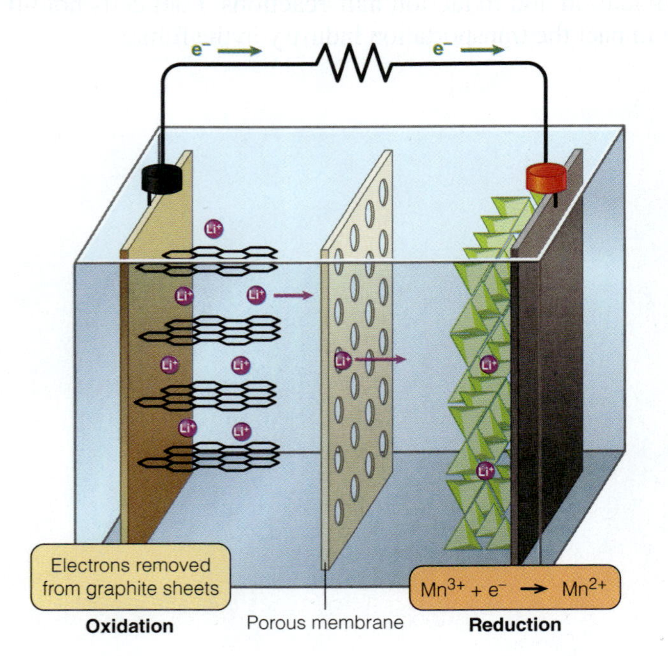

Key Terms

14.1 Oxidation and Reduction

electrochemistry The study of chemical processes that involve the movement of electrons.

oxidation-reduction reaction A chemical change in which one species loses electrons (oxidation) while another gains electrons (reduction); also called a *redox reaction.*

oxidation The loss of electrons.

reduction The gain of electrons.

oxidation number A bookkeeping tool for tracking electron changes in oxidation-reduction reactions.

14.2 Types of Redox Reactions

combustion A reaction in which oxygen gas combines with elements or compounds to produce oxide compounds.

metal displacement reaction A reaction between two metals in which one metal is oxidized to its ionic form while the other metal ion is reduced to its elemental form.

activity series A table that lists metals by their reducing powers; used to predict whether metal displacement reactions will occur.

14.3 Half-Reactions and Batteries

half-reaction A chemical equation that shows only the reduction or oxidation component of a reaction.

battery A device that produces electric current through a redox reaction in which the two half-reactions occur at separate sites; also called an *electrochemical cell.*

electrode In electrochemical cells and related devices, a site where an oxidation or reduction half-reaction takes place.

14.5 Other Applications of Redox Reactions

electroplating A technique that uses electrical potential energy (such as from a battery) to produce a thin layer of a metal such as gold, silver, copper, or chromium on the outside surface of another material.

fuel cell A device that converts the energy of a combustion reaction directly into electrical energy by separating the oxidation and reduction half-reactions.

Additional Problems

14.1 Oxidation and Reduction

13. In each of the following changes, determine whether the element or ion is oxidized or reduced:

a. A silver atom loses an electron to form an Ag^+ cation.
b. A fluorine atom gains an electron to form a fluoride ion.
c. Oxygen reacts to form oxide ions.
d. Tin reacts to form tin(IV) ions.

14. In each of the following changes, determine whether the element or ion is oxidized or reduced:

a. Elemental hydrogen forms hydrogen (H^+) ions.
b. Elemental hydrogen forms hydride (H^-) ions.
c. Sulfur reacts to form sulfide ions.
d. Magnesium reacts to form magnesium ions.

15. Why do scientists use oxidation numbers?

16. In Chapter 9, we examined the concept of formal charge. How do oxidation numbers differ from formal charges?

17. Find the oxidation number for the metal ion in each of these compounds:

a. $CoBr_2$ b. CuI c. PbO_2 d. K_2S

18. Find the oxidation number for the metal ion in each of these compounds:

a. NaBr b. $PtCl_4$ c. $NiCl_2$ d. AlF_3

19. Determine the oxidation number for each element in these compounds:

a. CuCl b. $CuCl_2$ c. SiO_2 d. H_2O

20. Determine the oxidation number for each element in these compounds:

a. $TiCl_4$ b. H_2CO c. H_2CO_3 d. $SiBr_4$

21. Determine the oxidation number for each element in these compounds:

a. $BeBr_2$ b. BBr_3 c. NH_3 d. LiH

22. Determine the oxidation number for each element in these compounds:

a. PCl_3 b. PCl_5 c. $POCl_3$ d. PH_3

23. Determine the oxidation number for each element in these ions:

a. Pb^{2+} b. Br^- c. BrO^- d. H_2N^-

24. Determine the oxidation number for each element in these ions:

a. Mg^{2+} b. ClO_2^- c. ClO_3^- d. ClO_4^-

25. Determine the oxidation number for each element in these ions:

a. Cr^{3+} b. CO_3^{2-} c. IO_3^- d. BF_4^-

26. Determine the oxidation number for each element in these ions:

a. NH_4^+ b. NO_2^- c. HCO_3^- d. CCl_3^-

27. In these reactions, identify the element that is oxidized and the element that is reduced:

a. $2\ Mg\ (s) + O_2\ (g) \rightarrow 2\ MgO\ (s)$
b. $Fe\ (s) + S\ (s) \rightarrow FeS\ (s)$
c. $Br_2\ (l) + Ca\ (s) \rightarrow CaBr_2\ (s)$

28. In these reactions, identify the element that is oxidized and the element that is reduced:

a. $2\ Cu\ (s) + O_2\ (g) \rightarrow 2\ CuO\ (s)$
b. $S\ (s) + Hg\ (l) \rightarrow HgS\ (s)$
c. $Sn\ (s) + 2\ Cl_2\ (g) \rightarrow SnCl_4\ (s)$

29. In these reactions, identify the species that is oxidized and the species that is reduced:

a. $2\ H_2\ (g) + O_2\ (g) \rightarrow 2\ H_2O\ (g)$
b. $H_2CO\ (aq) + O_2\ (g) \rightarrow H_2CO_3\ (aq)$
c. $H_2S_2\ (g) + H_2\ (g) \rightarrow 2\ H_2S\ (g)$

30. In these reactions, identify the species that is oxidized and the species that is reduced:

a. $2\ NaN_3\ (s) \rightarrow 2\ Na\ (s) + 3\ N_2\ (g)$
b. $Mg\ (s) + CuBr_2\ (aq) \rightarrow MgBr_2\ (aq) + Cu\ (s)$
c. $Ni\ (s) + 2\ HCl\ (aq) \rightarrow NiCl_2\ (aq) + H_2\ (g)$

31. In these reactions, identify the oxidizing agent:

a. $2\ Mg\ (s) + O_2\ (g) \rightarrow 2\ MgO\ (s)$
b. $CH_4\ (g) + 2\ O_2\ (g) \rightarrow CO_2\ (g) + 2\ H_2O\ (g)$
c. $Sn\ (s) + 2\ F_2\ (g) \rightarrow SnF_4\ (s)$

32. In these reactions, identify the reducing agent:

a. $H_2\ (g) + Cl_2\ (g) \rightarrow 2\ HCl\ (g)$
b. $2\ Na\ (s) + 2\ H_2O\ (l) \rightarrow 2\ NaOH\ (aq) + H_2\ (g)$
c. $Ca\ (s) + FeCl_2\ (aq) \rightarrow CaCl_2\ (aq) + Fe\ (s)$

33. In these reactions, identify the reducing agent:

a. $Ca\ (s) + 2\ H_2O\ (l) \rightarrow Ca(OH)_2\ (aq) + H_2\ (g)$
b. $F_2\ (g) + Be\ (s) \rightarrow BeF_2\ (s)$
c. $S\ (s) + O_2\ (g) + H_2O\ (g) \rightarrow H_2SO_3\ (g)$

34. In these reactions, identify the oxidizing agent:

a. $Si\ (s) + O_2\ (g) \rightarrow SiO_2\ (s)$
b. $O_2\ (g) + 4\ Ag\ (s) \rightarrow 2\ Ag_2O\ (s)$
c. $2\ H_2S\ (aq) + H_2O_2\ (aq) \rightarrow H_2S_2\ (aq) + 2\ H_2O\ (l)$

14.2 Types of Redox Reactions

35. What types of compounds are formed by the reaction of metals with nonmetals?

36. When a metal reacts with a nonmetal, which species is oxidized? Which species is reduced?

37. Write molecular equations to describe the following oxidation-reduction reactions. Include phase symbols, and identify the species that is oxidized and reduced in each reaction.

a. Solid iron metal reacts with chlorine gas to form solid iron(II) chloride.
b. Liquid mercury reacts with oxygen gas to form solid mercury(II) oxide.

38. Write molecular equations to describe the following oxidation-reduction reactions. Include phase symbols, and identify the species that is oxidized and reduced in each reaction.

a. Solid calcium metal reacts with liquid bromine to form solid calcium bromide.
b. Solid cobalt reacts with aqueous iodine to form aqueous cobalt(II) iodide.

39. Predict the products from these reactions, and balance the equations:

a. $Be\ (s) + Cl_2\ (g) \rightarrow$
b. $K\ (s) + Cl_2\ (g) \rightarrow$
c. $Co\ (s) + Cl_2\ (g) \rightarrow$
d. $Cu\ (s) + O_2\ (g) \rightarrow$

40. Predict the products from these reactions, and balance the equations:

a. $Ca\ (s) + O_2\ (g) \rightarrow$
b. $Sr\ (s) + Br_2\ (l) \rightarrow$
c. $Pb\ (s) + O_2\ (g) \rightarrow$
d. $Al\ (s) + O_2\ (g) \rightarrow$

41. What element is always involved in combustion?

42. What are hydrocarbons? What two products always result from the combustion of hydrocarbons?

43. Predict the products from these combustion reactions, and balance the equations:

a. $Mg\ (s) + O_2\ (g) \rightarrow$
b. $C_2H_4\ (g) + O_2\ (g) \rightarrow$
c. $C_4H_{10}\ (g) + O_2\ (g) \rightarrow$

44. Predict the products from these combustion reactions, and balance the equations:

a. $Zn\ (s) + O_2\ (g) \rightarrow$
b. $C_5H_{12}\ (l) + O_2\ (g) \rightarrow$
c. $C_8H_{18}\ (l) + O_2\ (g) \rightarrow$

45. Write molecular equations to describe the following chemical reactions. Include phase symbols.

a. Solid cobalt reacts with aqueous tin(II) chloride to produce aqueous cobalt(II) chloride and solid tin.
b. Aqueous gold nitrate reacts with silver metal to form solid gold metal and aqueous silver nitrate.

46. Write molecular equations to describe the following chemical reactions. Include phase symbols.

a. When barium metal is dropped into a solution of aqueous iron(II) acetate, solid iron and aqueous barium acetate are formed.
b. Aqueous nickel(II) chloride reacts with zinc metal to form aqueous zinc chloride and solid nickel.

47. Nickel(II) chloride forms a green solution when dissolved in water (**Figure 14.22**). If a strip of aluminum is placed in the solution, the green color begins to fade, and a coating can be observed on the surface of the aluminum. Why does the color fade? What is the coating on the aluminum?

Figure 14.22 Nickel(II) chloride forms a green solution in water.

48. When a copper wire is placed in a solution of silver nitrate, the wire slowly builds up a layer of silver (**Figure 14.23**). What is being oxidized in this reaction? What is being reduced? If all of the silver reacts, what ions are left in solution? Why is the solution blue?

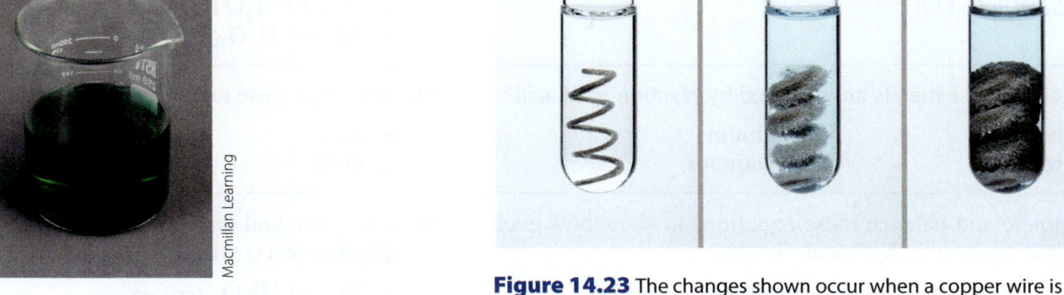

Figure 14.23 The changes shown occur when a copper wire is placed in a solution of silver nitrate.

49. Rewrite these metal displacement reactions as complete ionic and net ionic equations:

a. $Mg\ (s) + Cu(NO_3)_2\ (aq) \rightarrow Mg(NO_3)_2\ (aq) + Cu\ (s)$
b. $Al\ (s) + CrCl_3\ (aq) \rightarrow AlCl_3\ (aq) + Cr\ (s)$
c. $2\ Cr\ (s) + 3\ PtBr_2\ (aq) \rightarrow 3\ Pt\ (s) + 2\ CrBr_3\ (aq)$

50. Rewrite these metal displacement reactions as complete ionic and net ionic equations:

a. $Ni\ (s) + SnSO_4\ (aq) \rightarrow NiSO_4\ (aq) + Sn\ (s)$
b. $2\ Al\ (s) + 3\ CuSO_4\ (aq) \rightarrow 3\ Cu\ (s) + Al_2(SO_4)_3\ (aq)$
c. $Al\ (s) + 3\ AgNO_3\ (aq) \rightarrow Al(NO_3)_3\ (aq) + 3\ Ag\ (s)$

51. The following complete ionic equations represent metal displacement reactions. Rewrite them as molecular equations.

a. $Mg\ (s) + Pb^{2+}\ (aq) + 2\ NO_3^-\ (aq) \rightarrow Mg^{2+}\ (aq) + 2\ NO_3^-\ (aq) + Pb\ (s)$
b. $2\ Al\ (s) + 3\ Pt^{2+}\ (aq) + 6\ Br^-\ (aq) \rightarrow 3\ Pt\ (s) + 2\ Al^{3+}\ (aq) + 6\ Br^-\ (aq)$

52. The following complete ionic equations represent metal displacement reactions. Rewrite them as molecular equations.

a. $Fe\ (s) + 2\ Au^+\ (aq) + 2\ F^-\ (aq) \rightarrow Fe^{2+}\ (aq) + 2\ F^-\ (aq) + 2\ Au\ (s)$
b. $2\ Cr\ (s) + 3\ Ni^{2+}\ (aq) + 6\ ClO_4^-\ (aq) \rightarrow 2\ Cr^{3+}\ (aq) + 6\ ClO_4^-\ (aq) + 3\ Ni\ (s)$

53. Based on the activity series (Table 14.2), which of these metals are able to reduce copper(II) to elemental copper?

a. aluminum b. calcium
c. platinum d. gold

54. Based on the activity series (Table 14.2), which of these metals are able to reduce chromium(III) to elemental chromium?

a. lithium b. silver
c. nickel d. tin

55. Based on the activity series (Table 14.2), which of these ions can be reduced by elemental aluminum?

a. K^+ b. Mg^{2+} c. Fe^{2+} d. Pb^{2+}

56. Based on the activity series (Table 14.2), which of these ions can be reduced by elemental iron?

a. Mg^{2+} b. Ni^{2+} c. Co^{2+} d. Cu^{2+}

57. Based on the activity series (Table 14.2), determine whether these reactions would take place spontaneously:

a. $Zn^{2+}\ (aq) + Mg\ (s) \rightarrow Zn\ (s) + Mg^{2+}\ (aq)$
b. $Fe\ (s) + Sn^{2+}\ (aq) \rightarrow Sn\ (s) + Fe^{2+}\ (aq)$
c. $3\ Ag\ (s) + Al^{3+}\ (aq) \rightarrow 3\ Ag^+\ (aq) + Al\ (s)$
d. $Cr\ (s) + 3\ LiCl\ (aq) \rightarrow 3\ Li\ (s) + CrCl_3\ (aq)$

58. Based on the activity series (Table 14.2), determine whether these reactions would take place spontaneously:

a. $Ca^{2+}\ (aq) + Ni\ (s) \rightarrow Ca\ (s) + Ni^{2+}\ (aq)$
b. $Ba\ (s) + Cu^{2+}\ (aq) \rightarrow Ba^{2+}\ (aq) + Cu\ (s)$
c. $2\ Au\ (s) + Zn^{2+}\ (aq) \rightarrow 2\ Au^+\ (aq) + Zn\ (s)$
d. $Pb\ (s) + 2\ KNO_3\ (aq) \rightarrow 2\ K\ (s) + Pb(NO_3)_2\ (aq)$

59. Use the activity series (Table 14.2) to complete each of these reactions. Also indicate if no reaction occurs.

a. $Sn^{2+}\ (aq) + Ni\ (s) \rightarrow$
b. $Ni\ (s) + Fe^{2+}\ (aq) \rightarrow$
c. $Co\ (s) + Pb(C_2H_3O_2)_2\ (aq) \rightarrow$

60. Use the activity series (Table 14.2) to complete each of these reactions. Also indicate if no reaction occurs.

a. $Zn^{2+}\ (aq) + Mg\ (s) \rightarrow$
b. $Zn\ (s) + Mg^{2+}\ (aq) \rightarrow$
c. $Ba\ (s) + Cu(NO_3)_2\ (aq) \rightarrow$

61. Which of these metals are oxidized by reaction with water?

a. Li b. Na c. Ca d. Cu

62. Which of these metals are oxidized by reaction with water?

a. V b. Pt c. In d. K

63. Complete these equations to show how each element reacts with water:

a. $Na\ (s) + H_2O\ (l) \rightarrow$
b. $K\ (s) + H_2O\ (l) \rightarrow$
c. $Ca\ (s) + H_2O\ (l) \rightarrow$

64. Each of these reactions occurs slowly with liquid water, but much more quickly if the water is heated to the vapor phase (steam). Complete the reactions to show the products formed.

a. $Mg\ (s) + H_2O\ (g) \rightarrow$
b. $Zn\ (s) + H_2O\ (g) \rightarrow$
c. $Al\ (s) + H_2O\ (g) \rightarrow$

65. Which of these metals are oxidized by reaction with acid?

a. aluminum b lithium
c. nickel d. platinum

66. Which of these metals are oxidized by reaction with acid?

a. zinc b. silver
c. gold d. iron

67. Complete and balance these equations to show how each element reacts with hydrochloric acid:

a. $Mg\ (s) + HCl\ (aq) \rightarrow$
b. $Zn\ (s) + HCl\ (aq) \rightarrow$
c. $Fe\ (s) + HCl\ (aq) \rightarrow$

68. Complete and balance these equations to show how each element reacts with nitric acid:

a. $Ni\ (s) + HNO_3\ (aq) \rightarrow$
b. $Co\ (s) + HNO_3\ (aq) \rightarrow$
c. $Al\ (s) + HNO_3\ (aq) \rightarrow$

69. Write net ionic equations to show the chemical reactions that occur in the following situations. In some cases, no reaction will occur.

a. Calcium metal is combined with aqueous iron(II) ions.
b. Calcium metal is combined with aqueous acid.
c. A solution containing iron(II) ions is combined with copper metal.
d. A solution containing iron(II) ions is combined with zinc metal.

70. Write net ionic equations to show the chemical reactions that occur in the following situations. In some cases, no reaction will occur.

a. Platinum metal is combined with aqueous acid.
b. Aluminum metal is combined with aqueous acid.
c. A solution containing nickel(II) ions is combined with aluminum metal.
d. A solution containing tin(II) ions is combined with chromium metal.

14.3 Half-Reactions and Batteries

71. Write two half-reactions to describe each of these metal displacement reactions:

a. $Mg\ (s) + CoBr_2\ (aq) \rightarrow MgBr_2\ (aq) + Co\ (s)$
b. $Cu\ (s) + 2\ AuCl\ (aq) \rightarrow CuCl_2\ (aq) + 2\ Au\ (s)$

72. Write two half-reactions to describe each of these metal displacement reactions:

a. $Fe\ (s) + SnI_2\ (aq) \rightarrow FeI_2\ (aq) + Sn\ (s)$
b. $3\ Ba\ (s) + 2\ AlCl_3\ (aq) \rightarrow 3\ BaCl_2\ (aq) + 2\ Al\ (s)$

73. Using the activity series (Table 14.2) as a guide, write two half-reactions to show the reaction of iron metal with cobalt(II) ions.

74. Using the activity series (Table 14.2) as a guide, write two half-reactions to show the reaction of Zn^{2+} with Mg.

75. Write two half-reactions to describe each of these metal displacement reactions:

a. Elemental nickel reacts with silver ions to produce nickel(II) ions and elemental silver.
b. Zinc reacts with cobalt(II) chloride solution to produce zinc chloride and elemental cobalt.

76. Write two half-reactions to describe each of these metal displacement reactions:

a. Elemental aluminum reacts with tin(II) ions to produce elemental tin and aluminum ions.
b. Lead reacts with copper(II) sulfate solution to produce lead(II) sulfate and elemental copper.

77. What are the roles of the anode, the cathode, and the salt bridge in an electrochemical cell?

78. The salt bridge in an electrochemical cell contains sodium and nitrate ions. As the redox reaction takes place, which ions flow toward the cathode? Which ions flow toward the anode?

79. A chemist assembles an electrochemical cell as shown in **Figure 14.24**. One side of the cell contains a zinc electrode immersed in aqueous zinc nitrate solution. The other side of the cell contains an iron electrode immersed in aqueous iron(II) nitrate solution. In this reaction, zinc is the anode and iron is the cathode. Write equations showing the half-reactions that take place at each electrode.

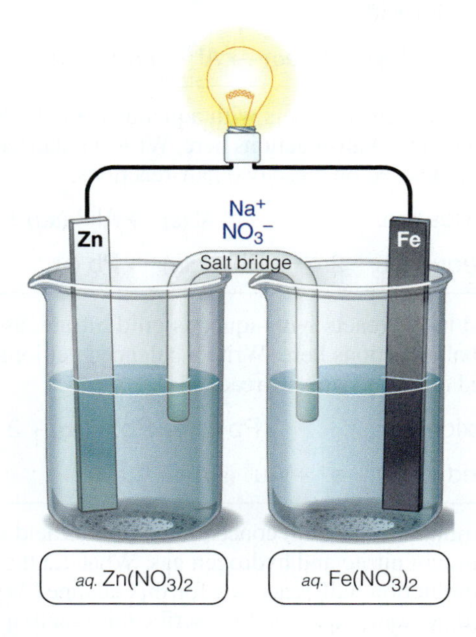

Figure 14.24 This electrochemical cell contains zinc and iron electrodes.

80. A chemist assembles an electrochemical cell as shown in **Figure 14.25**. One side of the cell contains a silver electrode immersed in aqueous silver nitrate solution. The other side of the cell contains a zinc electrode immersed in aqueous zinc nitrate solution. In this reaction, zinc is the anode and silver is the cathode. Write equations showing the half-reactions that take place at each electrode.

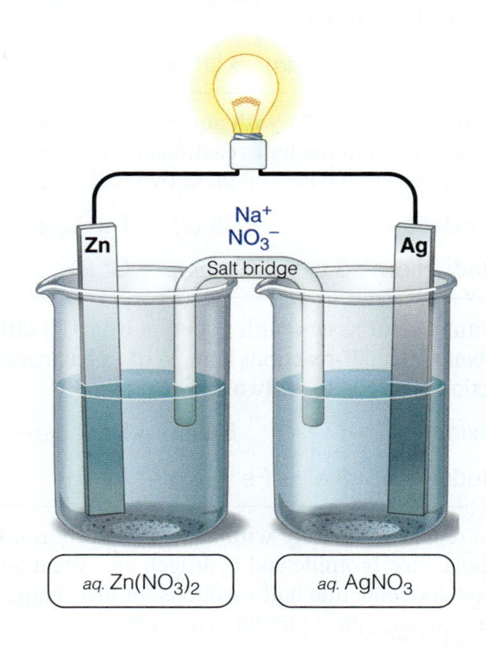

Figure 14.25 This electrochemical cell contains zinc and silver electrodes.

81. The electrochemical cell in **Figure 14.26** is called a *standard electrochemical cell*. In this type of cell, the concentration of dissolved ions is 1.0 M in each solution. If we replaced the Cu^{2+} solution with pure water, would the electrochemical cell still produce an electric current? What if we replaced the Zn^{2+} solution with pure water?

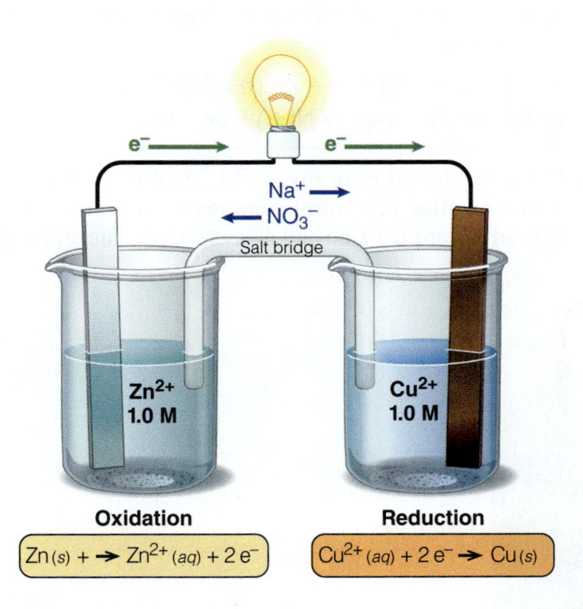

Figure 14.26 A standard electrochemical cell contains 1.0-M solutions of dissolved ions.

82. After an electrochemical cell has been in operation for some time, the appearance of the electrodes changes. For example, after a standard copper-zinc cell (such as the one shown in **Figure 14.27**) has run for a while, the zinc anode becomes much smaller, while the copper cathode becomes much larger. At the same time, the blue color of the copper(II) solution fades. How can you explain these three observations?

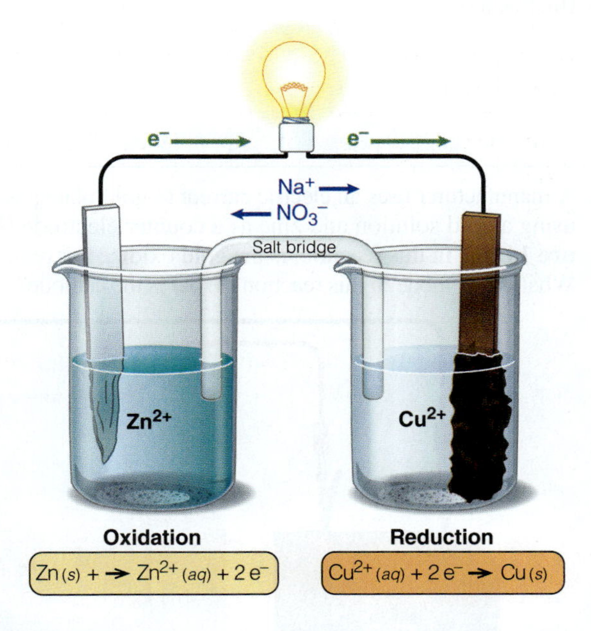

Figure 14.27 This zinc-copper electrochemical cell is almost depleted. The anode has nearly disappeared while the cathode is much larger. Why does this happen?

14.4 Balancing Redox Equations

83. Balance this ionic equation to show the reaction of aluminum metal with aqueous silver ions:

$$Al\ (s) + Ag^{+}\ (aq) \rightarrow Al^{3+}\ (aq) + Ag\ (s)$$

84. Balance this ionic equation to show the reaction of aqueous iron(II) ions with chromium metal:

$$Fe^{2+}\ (aq) + Cr\ (s) \rightarrow Cr^{3+}\ (aq) + Fe\ (s)$$

85. Balance this ionic equation to show the reaction of lithium metal with dilute acid:

$$Li\ (s) + H^{+}\ (aq) \rightarrow Li^{+}\ (aq) + H_2\ (g)$$

86. Balance this ionic equation to show the reaction of aluminum with acid:

$$Al\ (s) + H^{+}\ (aq) \rightarrow Al^{3+}\ (aq) + H_2\ (g)$$

87. Calcium metal reacts with aqueous chromium(III) chloride, as shown in the half-reactions here. Write a balanced net ionic reaction based on these two half-reactions.

Oxidation: $Ca\ (s) \rightarrow Ca^{2+}\ (aq) + 2\ e^{-}$

Reduction: $3\ e^{-} + Cr^{3+}\ (aq) \rightarrow Cr\ (s)$

88. Aluminum metal reacts with aqueous lead(II) chloride, as shown in the half-reactions here. Write a balanced net ionic reaction based on these two half-reactions.

Oxidation: $Al\ (s) \rightarrow Al^{3+}\ (aq) + 3\ e^{-}$

Reduction: $2\ e^{-} + Pb^{2+}\ (aq) \rightarrow Pb\ (s)$

89. Barium metal reacts with aqueous iron(III) chloride, as shown in the half-reactions here. Write a balanced net ionic reaction based on these two half-reactions.

Oxidation: $Ba\ (s) \rightarrow Ba^{2+}\ (aq) + 2\ e^{-}$

Reduction: $3\ e^{-} + Fe^{3+}\ (aq) \rightarrow Fe\ (s)$

90. Lead metal reacts with aqueous gold nitrate, as shown in the half-reactions here. Write a balanced net ionic reaction based on these two half-reactions.

Oxidation: $Pb\ (s) \rightarrow Pb^{2+}\ (aq) + 2\ e^{-}$

Reduction: $e^{-} + Au^{+}\ (aq) \rightarrow Au\ (s)$

91. Zinc reacts vigorously with aqueous hydrobromic acid to produce zinc bromide and hydrogen gas. What are the oxidation and reduction half-reactions for this change? Write a balanced net ionic equation to describe this reaction.

92. Aluminum reacts with concentrated nitrous acid to produce aluminum nitrite and hydrogen gas. What are the oxidation and reduction half-reactions for this change? Write a balanced net ionic equation to describe this reaction.

93. Elemental chromium reacts with aqueous lead(II) nitrate in a single-displacement reaction. What are the oxidation and reduction half-reactions for this change? Describe this change using a balanced, complete ionic reaction.

94. Aluminum metal reacts with aqueous nickel(II) bromide in a single-displacement reaction. What are the oxidation and reduction half-reactions for this change? Describe this change using a balanced, complete ionic reaction.

14.5 Other Applications of Redox Reactions

95. What is electroplating? How is electroplating different from the reactions that take place in an electrochemical cell (that is, a battery)?

96. Electroplating uses a battery to drive non-spontaneous reactions to take place. Using the activity series (Table 14.2), determine whether the following reactions would occur spontaneously or would require an electrical voltage for them to occur.

a. $Zn^{2+}\ (aq) + Mg\ (s) \rightarrow Zn\ (s) + Mg^{2+}\ (aq)$

b. $Fe\ (s) + Ni^{2+}\ (aq) \rightarrow Ni\ (s) + Fe^{2+}\ (aq)$

97. A manufacturer uses an electric current to gold-plate jewelry, using a gold solution and zinc as a counter electrode (**Figure 14.28**). In this reaction, is the gold oxidized or reduced? What is the anode in this reaction? What is the cathode?

98. Consider the setup in Figure 14.28. What would happen if the direction of current were reversed, so that electrons flowed toward the zinc and away from the nickel loop?

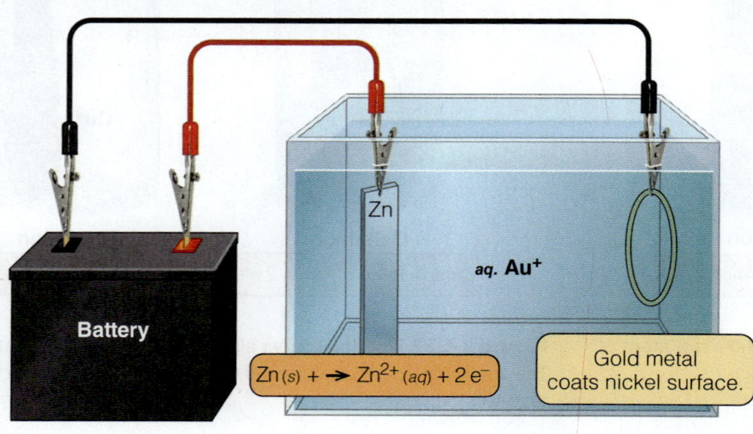

Figure 14.28 Jewelry can be gold-plated using an apparatus like the one shown here.

99. What is a fuel cell? What are the key components of a fuel cell?

100. What is the overall reaction that takes place in a hydrogen fuel cell? What half-reactions take place at the anode and at the cathode of a fuel cell?

Challenge Questions

101. Classify these oxidation-reduction reactions as synthesis, decomposition, single-displacement, or combustion reactions:

a. $Hg\ (l) + S\ (s) \rightarrow HgS\ (s)$

b. $Fe\ (s) + Cu(C_2H_3O_2)_2\ (aq) \rightarrow Cu\ (s) + Fe(C_2H_3O_2)_2\ (aq)$

c. $C_8H_{16}\ (l) + 12\ O_2\ (g) \rightarrow 8\ CO_2\ (g) + 8\ H_2O\ (g)$

102. Classify these oxidation-reduction reactions as synthesis, decomposition, single-displacement, or combustion reactions:

a. $Fe\ (s) + 2\ AgC_2H_3O_2\ (aq) \rightarrow Fe(C_2H_3O_2)_2\ (aq) + 2\ Ag\ (s)$

b. $Fe\ (s) + 2\ HCl\ (aq) \rightarrow FeCl_2\ (aq) + H_2\ (g)$

c. $2\ K\ (s) + Br_2\ (g) \rightarrow 2\ KBr\ (s)$

103. Animals convert glucose into energy through a process called *cellular respiration*. Based on the equation for cellular respiration shown here, identify the oxidation state of each atom in both the reactants and the products. In this reaction, which element is oxidized? Which element is reduced?

$$C_6H_{12}O_6\ (s) + 6\ O_2\ (g) \rightarrow 6\ CO_2\ (g) + 6\ H_2O\ (l) + \text{energy}$$

104. In Chapter 15 we will examine the process of photosynthesis, which green plants use to store energy from the sun in chemical bonds. Based on the equation for photosynthesis shown here, identify the oxidation state of each atom in both the reactants and the products. In this reaction, which element is oxidized? Which element is reduced?

$$\text{energy} + 6\ CO_2\ (g) + 6\ H_2O\ (l) \rightarrow C_6H_{12}O_6\ (s) + 6\ O_2\ (g)$$

105. Write a balanced equation showing the reaction of metallic iron with aqueous nitric acid. If 25.35 grams of iron react with 500.0 mL of 1.20 M nitric acid, how many grams of iron(II) nitrate can form?

106. What volume of 1.0-M gold nitrate is needed to completely react with 1.63 grams of zinc in a single-displacement reaction? What mass of gold could be produced by this reaction?

CHAPTER FIFTEEN

Organic Chemistry and Biomolecules

Figure 15.1 (a) Professor Robert Grubbs works in his research laboratory at Caltech. (*continued*)

Forming New Bonds: The Grubbs Catalyst

As a kid, Robert Grubbs loved putting things together. The son of a schoolteacher and a diesel mechanic, he was fascinated with how things were built. While most kids spent money on candy, he bought nails to connect pieces of scrap wood. As he grew, so did his interests in construction and mechanics. He worked with his father and on his uncles' farms, fixing cars, installing plumbing, and building houses.

In 1961, Robert left his Kentucky home and enrolled at the University of Florida as an agricultural chemistry major. He quickly found work in an animal nutrition lab. It was nasty work—analyzing the remnants from steer feces.

Fortunately, a friend worked in a chemistry lab on campus, and Robert began to help out at night. It was less smelly than his day job, and he found the work fascinating. As he later described it, "Building new molecules was even more fun than building houses." The hands-on skills he had developed as a farm kid served him well, and he excelled in the lab. Six years later, he completed his Ph.D. in chemistry from Columbia University.

Young Dr. Grubbs was fascinated with the way metal atoms interact with carbon. He explored new metal-carbon structures and unearthed new patterns of chemical reactions. As he honed his interests, he began to focus on the transition metal ruthenium. He discovered that compounds containing this metal could catalyze the formation of new carbon–carbon bonds. Instead of using metal nails to join boards, Grubbs now used metal atoms to connect carbon atoms (**Figure 15.1**).

For years—first at Michigan State and then at Caltech—Grubbs tirelessly studied these reactions that formed carbon–carbon bonds, refining his techniques and exploring

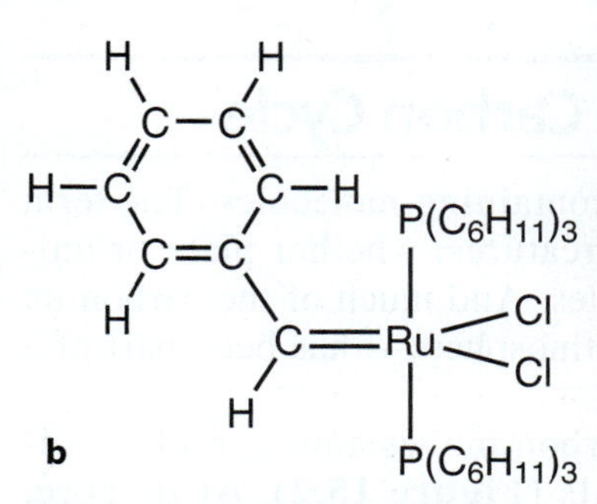

Figure 15.1 (continued) (b) The molecular structure of one of the Grubbs catalysts is shown here. (c) The durable plastic on this combine was produced using the Grubbs catalyst. (d) This factory in Indonesia produces materials for cleaning, lubricants, polymer, and personal care applications from renewable resources like palm oil, using the Grubbs catalyst. (e) Professor Grubbs received the Nobel Prize from the King of Sweden in 2005. (a) Caltech; (c) smereka/Shutterstock; (d) Courtesy of Elevance Eenewable Sciences, Inc.; (e) JONAS EKSTROMER/Getty images

new uses. Today, the ruthenium catalysts he and his students developed (commonly called "Grubbs catalysts") produce a wide variety of materials, including medicines, personal care products, automotive lubricants, and even the tough plastic on John Deere tractors. In 2005, Grubbs and others were awarded the Nobel Prize in chemistry for their development of new approaches to building carbon–carbon bonds.

Why are carbon bonds so important? As we'll see in the pages that follow, carbon provides the framework for all living organisms, both plant and animal. By bonding carbon with hydrogen, nitrogen, oxygen, and a handful of other elements, nature forms a sweeping array of compounds that range from small, simple molecules like methane gas to massive, complex structures like proteins.

And we can emulate nature. Starting from the simple carbon compounds in fossil fuels, chemists use reactions like those Grubbs developed to synthesize new compounds with innovative structures and exciting new properties.

In this chapter, we'll survey the stunning diversity of carbon-based compounds. Building on earlier chapters, we'll reexamine the way carbon and other nonmetal atoms bond. We'll explore bonding patterns in familiar molecules and observe simple reactions that take place with many different compounds. Finally, we'll look into the chemistry that takes place in living organisms and discover the elegant ways that nature builds structures, stores energy, and even stores and copies information. The chemistry of carbon — the chemistry of life — is breathtaking. Let's get started.

Intended Learning Outcomes

After completing this chapter and working the practice problems, you should be able to:

15.1 Organic Chemistry and the Carbon Cycle

- Broadly describe the exchange of carbon among the atmosphere, land, sea, and plant and animal life.

15.2 Covalent Bonding with Carbon and Other Nonmetals

- Draw neutral covalent structures for nonmetal compounds.
- Identify compounds that are isomers and draw isomers for a given molecular formula.

15.3 Drawing Covalent Structures

- Convert between Lewis, condensed, and skeletal representations for molecular structures.

15.4 Major Functional Groups

- Identify the structures and names of major functional groups.
- Describe the products of condensation reactions in the formation of ethers, esters, and amides.

15.5 Polymers and Plastics

- Qualitatively describe the structural features of synthetic polymers.

15.6 Biomolecules — An Introduction

- Identify the key structural features and functions of major classes of biomolecules, including carbohydrates, proteins, and DNA.

15.1 Organic Chemistry and the Carbon Cycle

Organic chemistry is the chemistry of carbon-containing molecules. The term *organic* refers to living organisms: All living creatures, whether plant or animal, are composed of carbon-containing molecules. And much of the carbon on Earth—whether it is found in the land, sea, or atmosphere—has been part of a plant or animal at some point in Earth's history.

The **carbon cycle** describes how Earth's carbon moves among rocks, sediments, water, atmosphere, plants, and animals (**Figure 15.2**). At its core, the carbon cycle involves two key reactions. The first is **photosynthesis**, the sequence used by green plants to harvest the energy of the sun. In photosynthesis, plants combine carbon dioxide from the atmosphere with water from the soil to produce oxygen gas and new compounds like glucose ($C_6H_{12}O_6$) that are composed of carbon, hydrogen, and oxygen. The second reaction is the reverse of the first one: Animals and plants combine oxygen with carbon compounds to produce carbon dioxide and water and release energy. This process is called **cellular respiration**.

Photosynthesis: $\text{Energy} + 6\ CO_2 + 6\ H_2O \rightarrow C_6H_{12}O_6 + 6\ O_2$

Cellular respiration: $C_6H_{12}O_6 + 6\ O_2 \rightarrow 6\ CO_2 + 6\ H_2O + \text{energy}$

Notice that the cellular respiration reaction is identical to a combustion reaction: When we say that your body "burns" calories, this is a fairly correct description. Both combustion and cellular respiration consume oxygen and produce carbon dioxide and water.

When plants and animals die, the complex carbon structures begin to decompose. The fossil fuels—coal, oil, and natural gas—form by the decay of plant and animal matter over long periods of time (**Figure 15.3**). These fuels contain a mixture of compounds that are composed mainly of carbon and hydrogen.

Figure 15.3 The black line in this rock formation is coal, which formed from the gradual decay of plant and animal matter.

Figure 15.2 Photosynthesis converts CO_2 in the atmosphere into larger carbon-based molecules in living things. In combustion or cellular respiration, the process is reversed.

Carbon is also central to aquatic life. The ocean exchanges carbon dioxide with the atmosphere. In water, carbon dioxide reacts to form carbonic acid and then bicarbonate ions:

$$CO_2\,(aq) + H_2O\,(l) \rightleftharpoons H_2CO_3\,(aq)$$

$$H_2CO_3\,(aq) \rightleftharpoons H^+\,(aq) + HCO_3^-\,(aq)$$

Many creatures produce hard calcium carbonate shells from the bicarbonate ions dissolved in the water (**Figure 15.4**). Over time, these shells accumulate on the seafloor, gradually breaking down to form sediments that are rich in carbon.

The carbon cycle is an intricate dance—a complex set of equilibria that govern Earth's composition, climate, and economic and natural resources. In this chapter we will focus on the chemistry of carbon: How does carbon bond? What kinds of carbon structures are most important to our lives? And how do the structures of carbon compounds affect their properties?

Martin Shields/Science Source

Figure 15.4 Many sea organisms create hard outer shells using calcium carbonate.

15.2 Covalent Bonding with Carbon and Other Nonmetals

In Chapter 5, we explored chemical bonding. Recall that nonmetal atoms (**Figure 15.5**) bond by "sharing" electrons through covalent bonds. Covalent bonds enable nonmetals to fill their valence level with eight electrons, thereby fulfilling the *octet rule*.

The octet rule: An atom is stabilized by having eight electrons in its valence level.

The number of covalent bonds that an atom forms depends on the number of electrons needed to complete its valence level. For example, hydrogen needs only one electron to complete its valence, so it forms one covalent bond. The halogens (F, Cl, Br, I) also need just one electron to fill their valence levels, so they also form just one bond. Oxygen and sulfur need two electrons, and so they usually form two bonds. Nitrogen and phosphorus need three electrons and therefore form three bonds. Finally, carbon needs four electrons to fulfill its valence, so it forms four covalent bonds. **Table 15.1** summarizes the bonds required in neutral atoms.

Living tissue is built primarily of four atoms: carbon, hydrogen, nitrogen, and oxygen. Using the simple bonding rules given above, these four atoms can produce an infinite combination of compounds, shapes, and properties. For example, consider the three small molecules shown in **Figure 15.6**. Each molecule has a different structure—and therefore different properties. But each molecule follows the same basic rules: four bonds to carbon, three bonds to nitrogen, two to oxygen, and one to hydrogen.

The structures in Figure 15.6 contain only single covalent bonds. However, recall that double covalent bonds (two pairs of electrons) and triple covalent bonds

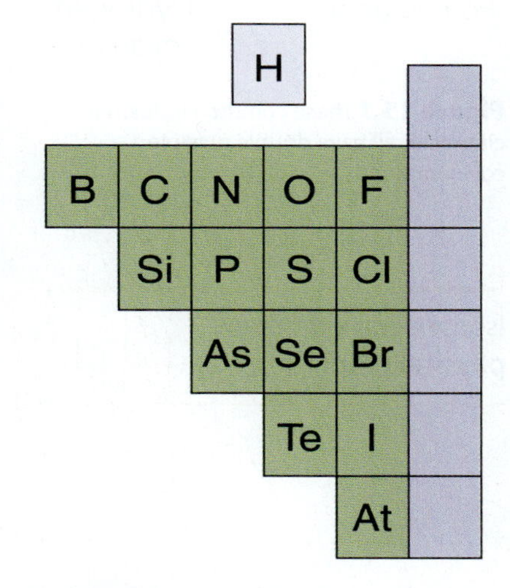

Figure 15.5 Nonmetals form covalent bonds.

TABLE 15.1 Covalent Bonds in Neutral Atoms

Atom	Valence Electrons	Electrons Needed	Covalent Bonds Formed
H	1	1	1
C	4	4	4
N	5	3	3
O	6	2	2
F	7	1	1

For more details about covalent bonding, see Chapter 9.

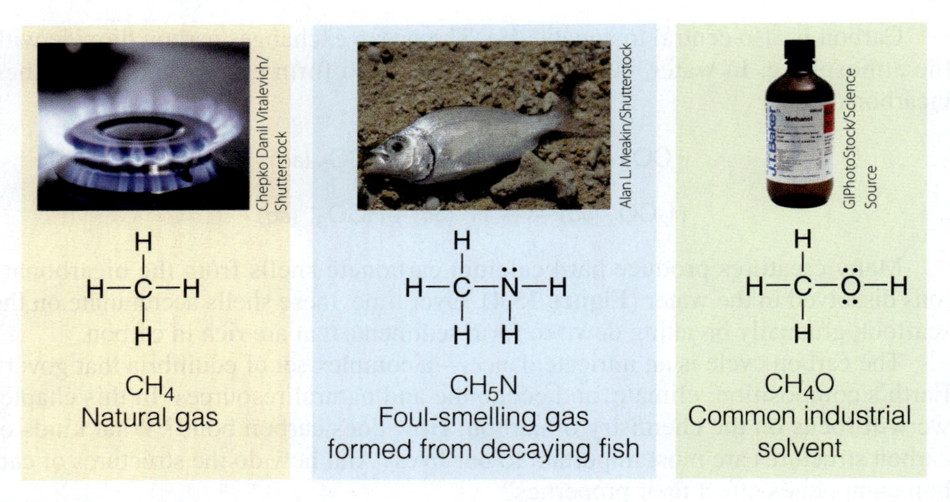

Figure 15.6 These three small molecules are composed of C, H, N, and O.

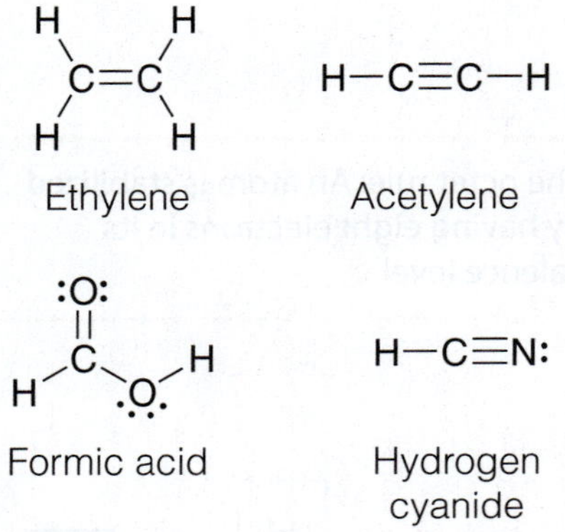

Ethylene

Acetylene

Formic acid

Hydrogen cyanide

Figure 15.7 These common industrial chemicals all have double or triple covalent bonds.

Isomers have different physical properties. ■

(three pairs of electrons) are also possible. For example, **Figure 15.7** shows four simple molecules that are commonly used as raw materials for chemical manufacturing. Notice that each of them has a double or triple covalent bond in its structure, but the fundamental bonding rules are fulfilled.

Most molecules contain many covalent bonds. As the number of atoms increases, many possible ways of combining the atoms arise. For example, a molecule with the formula C_3H_7Cl may have two different bonding arrangements:

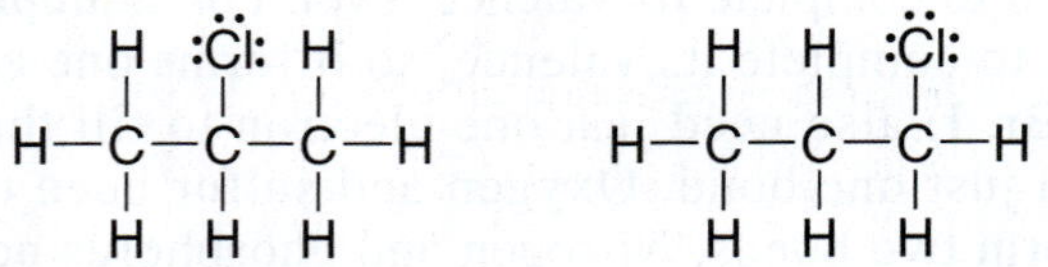

In the structure on the left, the chlorine is attached to the central carbon atom. In the structure on the right, the chlorine is attached to one of the end carbons. Compounds with the same molecular formula but different bonding sequences are called isomers. Because their structures are different, isomers have different physical properties. For example, the two C_3H_7Cl isomers have different melting and boiling points.

When distinguishing isomers, we must pay close attention to the way the atoms are connected. Molecules may be drawn in many different shapes (called *conformations*) and from many different perspectives. For example, each of the structures shown in **Figure 15.8** represents the same molecule; there are three carbon atoms in a row, and a chlorine attached to the carbon on the end. The molecules may be

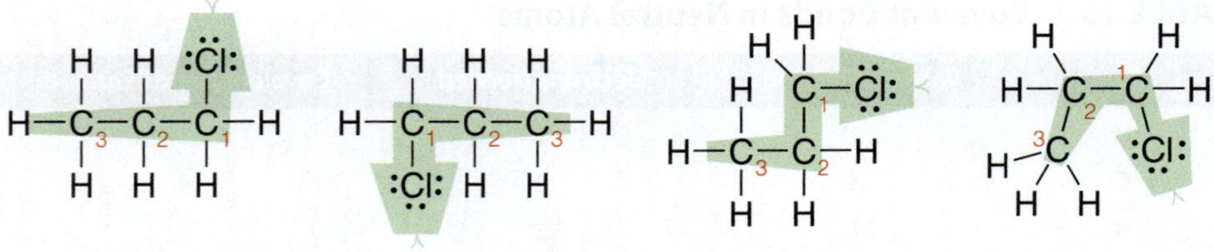

Figure 15.8 Much like a snake, many molecules coil into different shapes. These four structures all represent the same molecule.

arranged in different conformations—like a snake that coils or uncoils—but the sequence of atoms is the same. Therefore, they are not isomers; they are different conformations of the same molecule.

Because it can form four bonds, carbon forms the backbone of most organic structures. For example, consider the structure of octane, a component of gasoline (**Figure 15.9**). The structure of this molecule is based on the arrangement of the carbon atoms.

Like people, most molecules can fold into different shapes. When drawing isomers, it's important to focus on the connectivity. Molecules may be drawn in different ways but still have the same connectivity.

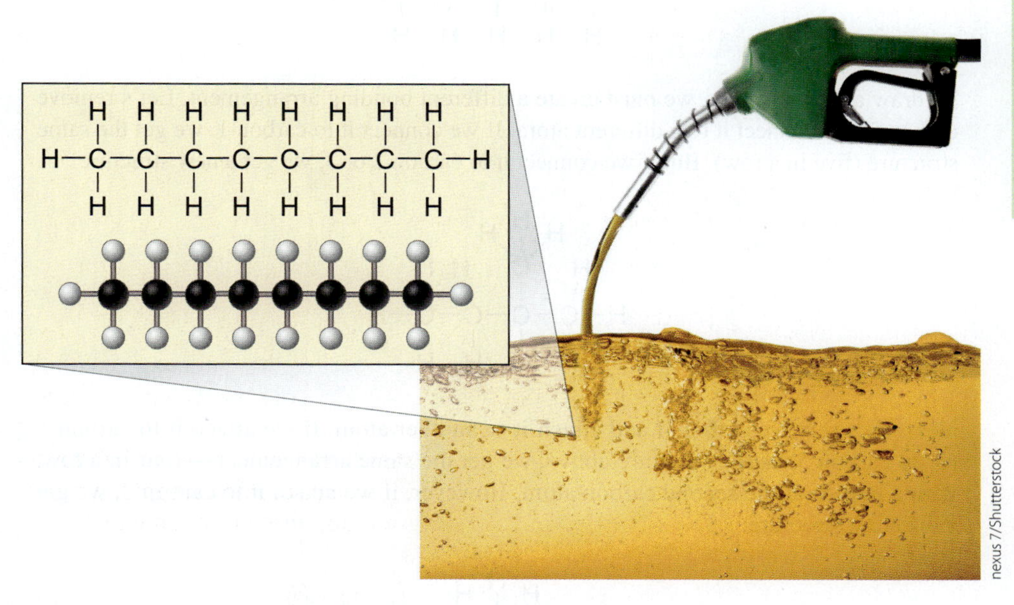

Figure 15.9 Carbon forms the backbone in the structure of octane.

Example 15.1 Drawing Isomers

There are two isomers with the molecular formula $C_2H_4F_2$. Draw Lewis structures for each of them.

Hydrogen and fluorine both form only one bond, but carbon forms four bonds. Because of this, the carbon atoms are the backbone for these molecules. We can draw a carbon-only framework like this, using lines to represent bonds to the carbon atoms:

```
   |  |
 —C—C—
   |  |
```

Each of the six peripheral spaces is occupied by a hydrogen atom or a fluorine atom. These can be arranged in two different ways: In one arrangement, each carbon atom is attached to one fluorine atom. In the other arrangement, both fluorine atoms are attached to the same carbon atom.

```
   :F̤: :F̤:          :F̤:  H
    |   |             |   |
 H—C—C—H        H—C—C—H
    |   |             |   |
    H   H            :F̤:  H
```

Example 15.2 Drawing Isomers

Draw the three isomers having the molecular formula C_5H_{12}.

The first isomer is simple: We draw five carbon atoms in a row, as shown. To help visualize the structure, it's helpful to number the carbon atoms:

```
     H   H   H   H   H
     |   |   |   |   |
 H — C1— C2— C3— C4— C5— H
     |   |   |   |   |
     H   H   H   H   H
```

To draw another isomer, we must create a different bonding arrangement. Let's remove carbon 5 and connect it to a different atom. If we connect it to carbon 1, we get the same structure (five in a row). But if we connect it to carbon 2 or 3, we get a new structure:

```
          H
        H | H
         \|/
     H    C5   H H
      \   |    | |
  H — C1— C2— C3— C4— H
      |   |    |   |
      H   H    H   H
```

Next, let's remove carbon 4 and attach it to another atom. If we attach it to carbon 1 (or to carbon 5) in the structure above, we get the same arrangement—four in a row, with a branch at the second carbon atom. However, if we attach it to carbon 2, we get a different structure:

```
           H
         H | H
          \|/
      H    C5    H
       \   |    /
   H — C1— C2— C3— H
       /   |    \
      H    C4    H
          /|\
         H H H
```

These are the only three possible combinations for these atoms. If you have drawn other structures, compare them to the ones shown here. If the structure contains the correct number of bonds to each atom, it will match one of the structures shown above, even if it is drawn in a different shape.

Check It
Watch Explanation

TRY IT

1. There are four possible isomers with the formula C_4H_9F. Draw structures for each of them.

15.3 Drawing Covalent Structures

Lewis structures are incredibly useful for describing small molecules, but they are cumbersome for larger molecules. For example, the anticancer drug Taxol® (introduced in Chapter 1) contains 113 atoms. But that is still small: Later in this chapter, we'll encounter molecules that contain thousands or even hundreds of thousands of atoms. For large molecules, we need simpler, more elegant ways to represent structures.

The first simplification is to draw Lewis structures without showing nonbonded electrons. For example, the "alcohol" in alcoholic beverages is ethyl alcohol (also called *ethanol*). **Figure 15.10** shows this molecule as a full Lewis structure and as a simplified structure in which the nonbonded valence electrons are not drawn (but understood to be present). As you continue to practice drawing molecules, you will get more comfortable with the inferred meaning in simplified structures.

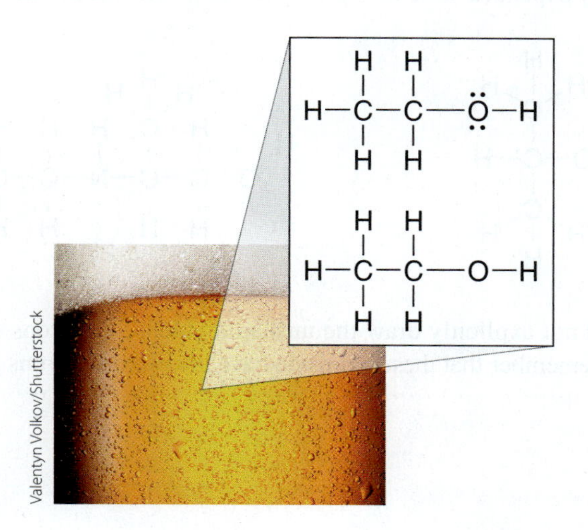

Figure 15.10 The alcohol in beer has the formula C_2H_6O. In a full Lewis structure (top), all unshared electrons are shown. In a simplified structure (bottom), the nonbonded electrons are not drawn, but inferred from the structure.

Condensed Structures

Scientists use **condensed structures** to show how atoms are connected. In condensed structures, most covalent bonds are not shown, but atoms are listed in their order of connectivity. For example, we could use condensed structures to represent the compound C_4H_{10}:

```
    H   H   H   H
    |   |   |   |
H — C — C — C — C — H   =   CH3CH2CH2CH3   or   CH3(CH2)2CH3
    |   |   |   |
    H   H   H   H
```

Lewis structure — Condensed structure

A second example is shown below. In this case we use parentheses to indicate that a larger group (the —NH_2 group) is connected to the second carbon from the left.

```
       H   H
        \ /
    H   N:  H   H
    |   |   |   |
H — C — C — C — C — H   =   CH3CH(NH2)CH2CH3
    |   |   |   |
    H   H   H   H
```

Sometimes chemists use partially condensed structures to simplify Lewis structures. In the molecule below, most C—H bonds are shown in condensed form, but all other bonds are drawn explicitly.

```
  H       CH3
   \     /
    C = C
   /     \
H3C       CH3
```

Example 15.3 Representing Structures in Different Styles

From these condensed structures, draw simplified Lewis structures that show each bond but do not show unshared electrons.

$CH_3OCH(CH_3)_2$ $HOCH_2CH(CH_3)NHCH_2CH_2OH$

We can draw fully expanded structures for these two molecules as follows:

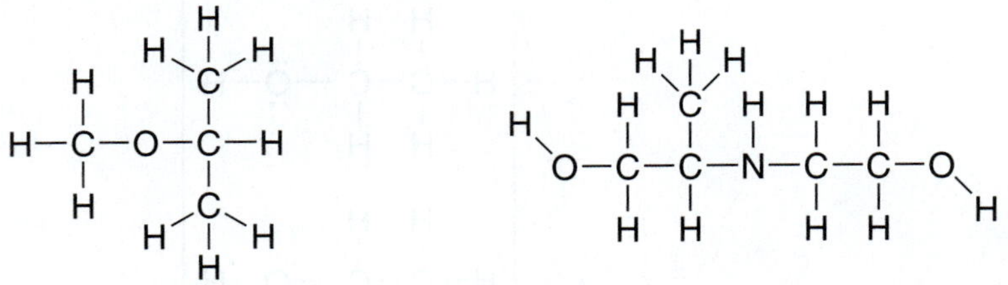

Although we did not explicitly draw the unshared valence electrons on the oxygen or nitrogen atoms, remember that these atoms do have unshared electrons present.

Check It
Watch Explanation

TRY IT

2. Write condensed representations for each of these Lewis structures:

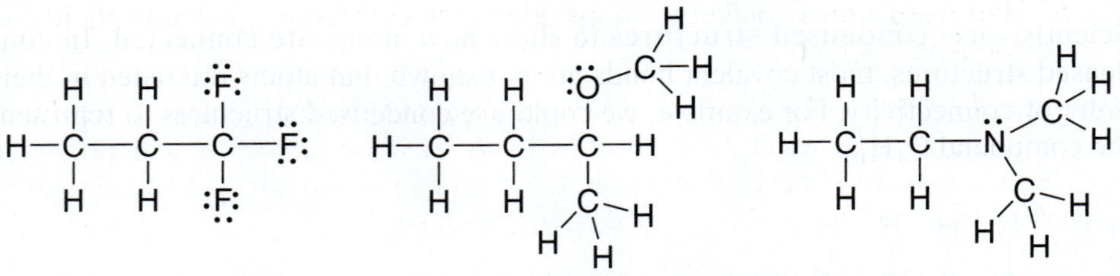

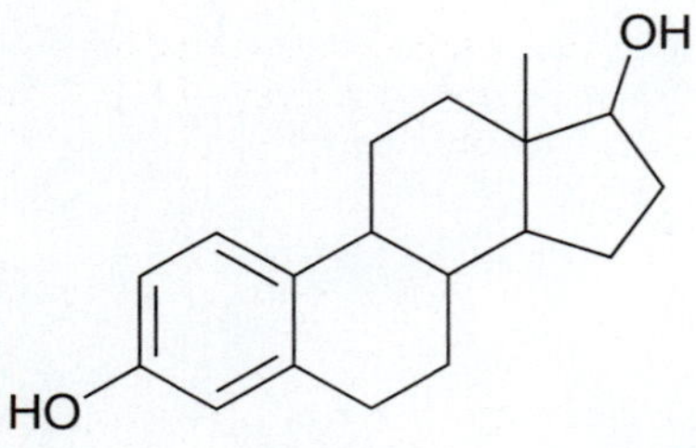

Estrogen

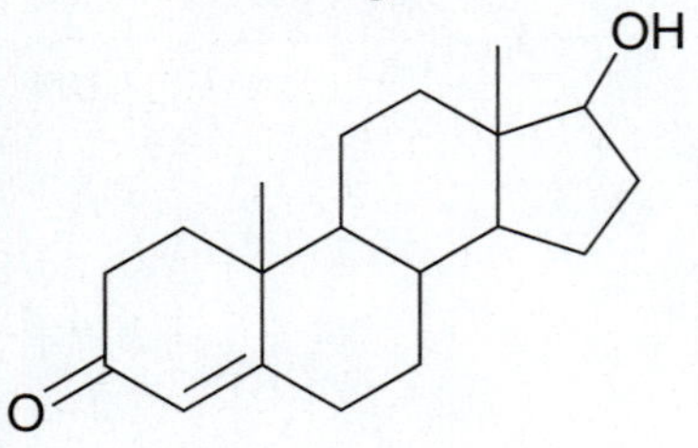

Testosterone

Figure 15.11 For complex structures, chemists often use skeletal structures. The structures above are the human female and male sex hormones. The two structures are very similar, each having four rings. The upper structure is one of several closely related structures that collectively are referred to as *estrogens*.

Skeletal Structures

While less cumbersome than Lewis structures, condensed structures are still most useful for fairly small molecules. Many molecules have larger, more complex structures. And some arrangements, such as cyclic structures (called *rings*), are hard to represent using condensed structures. In these cases, chemists use a simplified drawing technique called **skeletal structures**, as shown in **Figure 15.11**. For molecules containing only carbon and hydrogen, there are three rules for drawing skeletal structures:

1. Carbon skeletons are represented using a line structure. Each vertex (corner) of the line and each end point of the line represents a single carbon atom.
2. The hydrogen atoms bonded to carbon are not drawn, but inferred from the number of lines present.
3. Double and triple bonds are represented by double and triple lines.

For example, **Figure 15.12** shows the structures of two substances found in crude oil. Each of them contains only carbon and hydrogen. The Lewis structures are cumbersome and hard to read, but the skeletal structures clearly and elegantly depict each arrangement.

Compound name	2,4-Dimethylhexane	Toluene
Lewis structure		
Skeletal structure		

Figure 15.12 These two components of crude oil contain only carbon and hydrogen. Using skeletal structures makes it much easier to see how the atoms are connected.

If skeletal structures involve atoms other than hydrogen and carbon, two other rules apply:

4. Atoms other than carbon and hydrogen are shown with their atomic symbol.
5. Hydrogen atoms bonded to atoms other than carbon are written explicitly as condensed structures.

For example, **Figure 15.13** shows the structure of the common pain medicine acetaminophen, represented both as a full Lewis structure and as a skeletal structure. Notice that the hydrogen atoms on the oxygen and the nitrogen are shown in condensed form, but those attached to carbon are not.

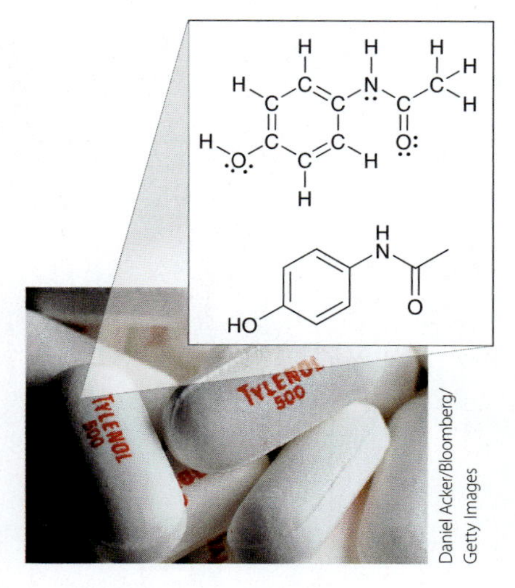

Figure 15.13 Acetaminophen is the active component of the commercial pain medicine Tylenol®.

In the classic comic strip *Calvin and Hobbes*, artist Bill Watterson used simple lines to depict characters, actions, and emotions. Similarly, line structures convey a wealth of structural meaning in a very simple way.

Example 15.4 Converting between Skeletal and Lewis Structures

The structure shown below is the commercial pain medicine ibuprofen. Redraw this molecule using a complete Lewis structure.

OH

O

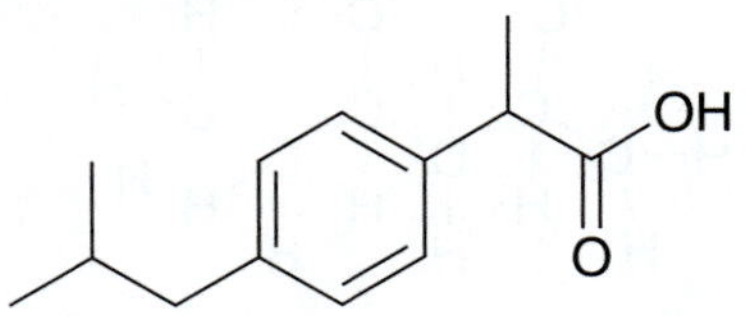

In this structure, each end point or vertex that does not have an explicit atom label represents a carbon atom. One helpful way to visualize the carbon atoms is to draw a dot on each end point. Then, because each carbon atom must have four bonds, any dot that does not have four bonds shown must have hydrogens attached. We draw in the "understood" hydrogens on each dot:

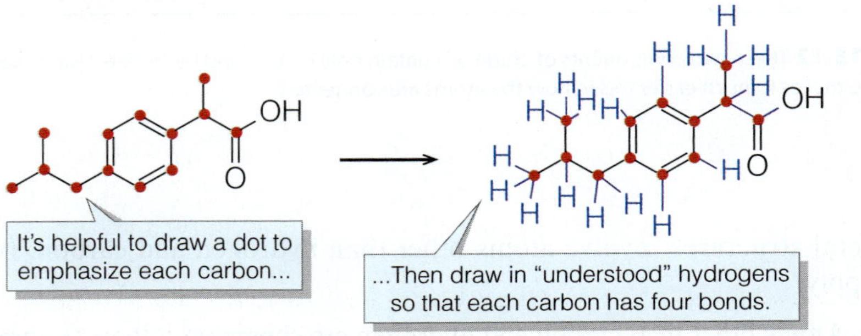

It's helpful to draw a dot to emphasize each carbon...

...Then draw in "understood" hydrogens so that each carbon has four bonds.

Finally, we can redraw our structure using atom labels for each carbon and including unshared electrons on each oxygen atom.

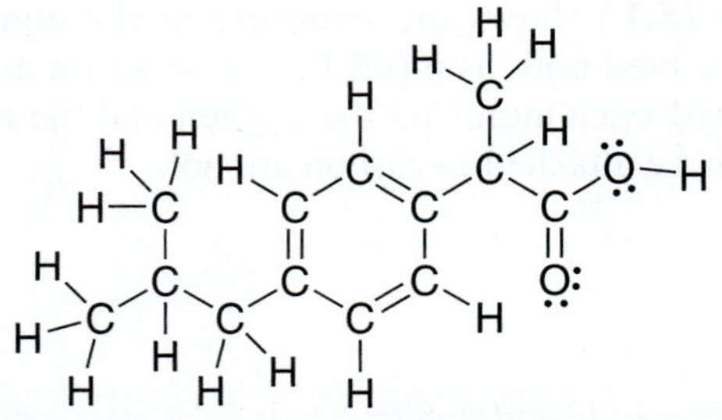

Example 15.5 Converting between Skeletal, Condensed, and Lewis Structures

Food manufacturers sometimes add propylene glycol to packaged foods to keep them moist. The condensed structure for propylene glycol is $CH_2(OH)CH(OH)CH_3$. Draw a Lewis structure and a skeletal structure for this compound.

The first part of the structure has the formula $CH_2(OH)$. This indicates that the first carbon is bonded to two hydrogens. The (OH) in parentheses means that the carbon is attached to an oxygen, which in turn is attached to another hydrogen. The second carbon is attached to one H, and also to an OH. Finally, the third carbon is connected to three hydrogens. We draw the complete Lewis structure and the corresponding skeletal structure as shown here.

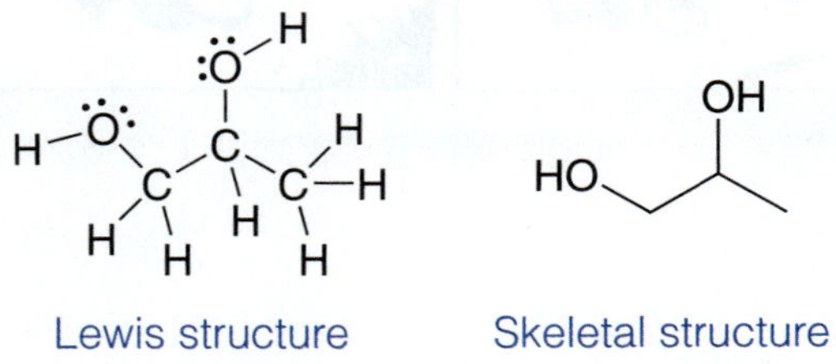

Lewis structure

Skeletal structure

TRY IT

3. Compounds called alpha-glucosidase inhibitors are used to treat diabetes. In 2015, a team of French scientists reported the preparation of a new alpha-glucosidase inhibitor using a Grubbs catalyst. Based on its skeletal structure (shown below), what is the molecular formula of this compound? Redraw this structure as a full Lewis structure.

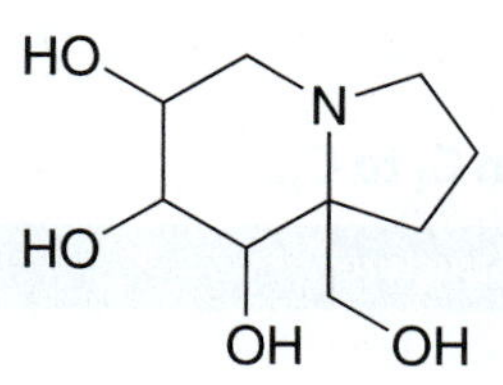

Check It
Watch Explanation

15.4 Major Functional Groups

Imagine you're driving across the United States. As you head west through rural Colorado, you see a sign for an exit one mile ahead. You're feeling hungry, so you move over into the right lane and prepare to exit. How did you know to move into the right lane?

The answer is simple: Nearly all rural exits in the United States are on the right-hand side of the road. You've driven on other roads before, so you know what to expect on this one. Similarly, if you encounter a stop sign, a traffic light, or a railroad crossing, you know what to expect: You've seen these structural features before.

In much the same way, molecules have structural features that behave in consistent ways. These structural features are called **functional groups**—small groups of atoms within a molecule that behave in a characteristic way. These behavior patterns allow us to predict the physical and chemical properties of molecules. In this section, we'll survey the common functional groups.

Hydrocarbon Functional Groups

Alkanes and Cycloalkanes

Hydrocarbons are compounds that contain only carbon and hydrogen. **Alkanes** are hydrocarbons that contain only single bonds. The simplest alkane, CH_4, is the major component of natural gas. Larger alkanes are also used as fuels. For example, butane (C_4H_{10}) is the fuel in lighters, while alkanes having six, seven, or eight carbon atoms are all components of gasoline. Larger alkanes burn more slowly: Kerosene and the wax in candles are composed of larger alkanes (**Figure 15.14**).

Alkanes contain only C and H, and all single bonds.

CSP_arvebettum/Fotosearch LBRF/AGE Fotostock

© whitestar1955/depositphotos.com

Tony Freeman/Science Source

amphaiwan/Shutterstock

Figure 15.14 Alkanes are commonly used as combustion fuels. Lighter fluid, gasoline, kerosene, and candle wax are all composed of alkanes.

Chemists name alkanes based on the number of carbons in the chain. Each alkane name uses a prefix that indicates the number of carbons. The suffix *–ane* indicates that the structure contains only single bonds. The names and structures of the straight-chain alkanes are given in **Table 15.2**. Notice from this table that each straight-chain alkane has two hydrogens per carbon atom, plus one extra hydrogen on each end of the chain. Because of this, alkanes have the general formula C_nH_{2n+2}.

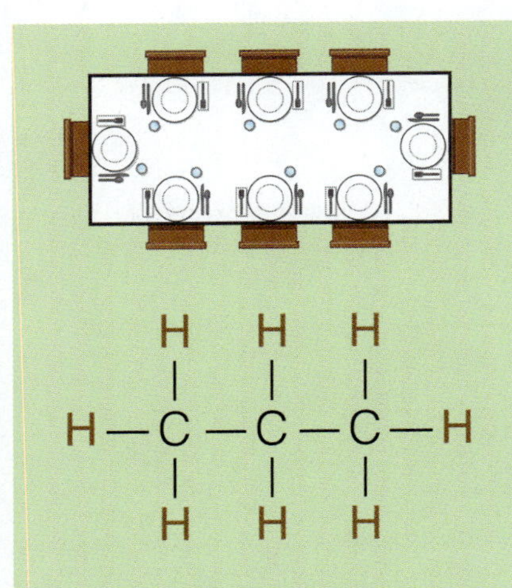

At a dinner table, seats are placed across from each other, and two additional seats are at the ends of the table. Similarly, straight-chain alkanes contain two hydrogen atoms per carbon plus two more on the ends. This structure results in the formula C_nH_{2n+2}.

TABLE 15.2 The Alkanes from C_1 to C_{10}

Formula	Lewis Structure	Condensed Structure	Name
CH_4	H H—C—H H	CH_4	Methane
C_2H_6	H H H—C—C—H H H	CH_3CH_3	Ethane
C_3H_8	H H H H—C—C—C—H H H H	$CH_3CH_2CH_3$	Propane
C_4H_{10}	H H H H H—C—C—C—C—H H H H H	$CH_3(CH_2)_2CH_3$	Butane
C_5H_{12}	H H H H H H—C—C—C—C—C—H H H H H H	$CH_3(CH_2)_3CH_3$	Pentane
C_6H_{14}	H H H H H H H—C—C—C—C—C—C—H H H H H H H	$CH_3(CH_2)_4CH_3$	Hexane
C_7H_{16}	H H H H H H H H—C—C—C—C—C—C—C—H H H H H H H H	$CH_3(CH_2)_5CH_3$	Heptane
C_8H_{18}	H H H H H H H H H—C—C—C—C—C—C—C—C—H H H H H H H H H	$CH_3(CH_2)_6CH_3$	Octane
C_9H_{20}	H H H H H H H H H H—C—C—C—C—C—C—C—C—C—H H H H H H H H H H	$CH_3(CH_2)_7CH_3$	Nonane
$C_{10}H_{22}$	H H H H H H H H H H H—C—C—C—C—C—C—C—C—C—C—H H H H H H H H H H H	$CH_3(CH_2)_8CH_3$	Decane

$CH_3CH_2CH_2CH_2CH_3$

CH_3
|
$CH_3CHCH_2CH_3$

CH_3
|
CH_3CCH_3
|
CH_3

Figure 15.15 Three isomers have the formula C_5H_{12}.

Alkanes commonly have branched structures. For instance, we saw earlier in this chapter (Example 15.2) that the formula C_5H_{12} has three isomeric forms (**Figure 15.15**). The presence of branches in the chains does not affect the number of hydrogen atoms present.

Cycloalkanes are alkanes that form a cyclic structure (commonly called a *ring*). Cycloalkanes are named based on the number of carbon atoms in the ring, but the prefix *cyclo–* is added to the alkane name to indicate that it is a cyclic structure (**Figure 15.16**).

The molecular formulas of cycloalkanes differ from those of linear alkanes. For example, compare the structures of propane and cyclopropane in **Figure 15.17**: The cyclic structure has two fewer hydrogen atoms than the straight-chain structure. Cycloalkanes have the general formula C_nH_{2n}.

Alkanes are flammable—they react quickly with oxygen—but otherwise they are fairly unreactive. Organic chemists often consider alkane and cycloalkane structures as the foundation for organic molecules, but not one of the major functional groups.

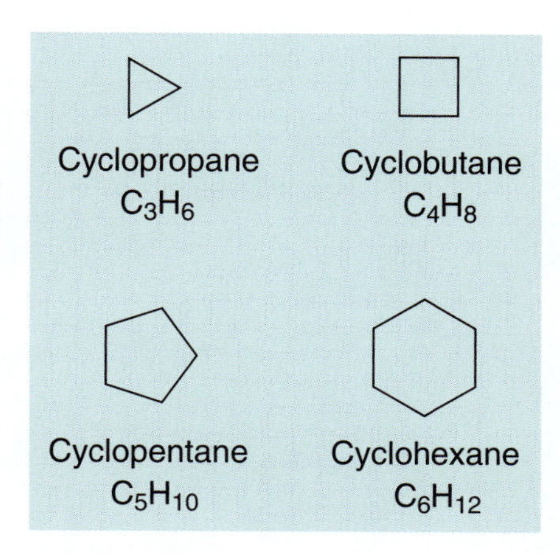

Figure 15.16 Skeletal structures of four cycloalkanes.

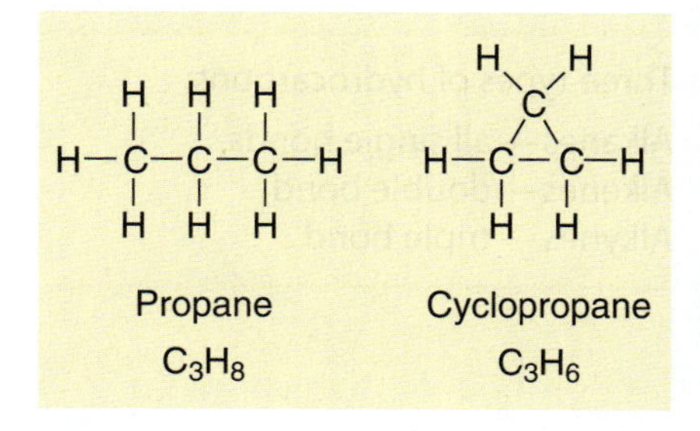

Figure 15.17 Linear alkanes have the general formula C_nH_{2n+2}. Cycloalkanes have the general formula C_nH_{2n}.

Example 15.6 Formulas for Alkanes and Cycloalkanes

Dodecane is an alkane that contains a 12-carbon chain. How many hydrogen atoms are in dodecane? How many hydrogen atoms are in cyclododecane?

For linear alkanes (those without a ring), the molecular formulas follow the general formula C_nH_{2n+2}. For dodecane, $n = 12$, so

$$2n + 2 = 2(12) + 2 = 26 \text{ hydrogen atoms}$$

Therefore, the molecular formula of dodecane is $C_{12}H_{26}$.

For cyclododecane, the formula is C_nH_{2n}, or $C_{12}H_{24}$.

TRY IT

4. What are the molecular formulas for heptane and for cycloheptane? Draw a skeletal structure for each of these molecules.

Alkenes and Alkynes

A carbon–carbon double bond is called an **alkene** functional group. Chemists also use this term for simple molecules that contain a C═C bond. The simplest alkene is ethene (also called *ethylene*), and it is shown in **Figure 15.18**. This small molecule, which is a gas at room temperature, is an important building block for making plastics (we'll discuss plastics in Section 15.5). Larger molecules also contain alkene functional groups. For example, limonene (**Figure 15.19**) is the major component of citrus oil, and it contains two alkene functional groups.

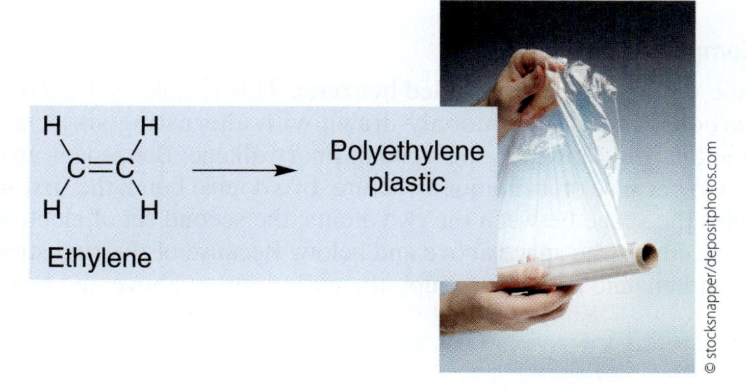

Figure 15.18 Ethylene is a precursor for polyethylene plastic.

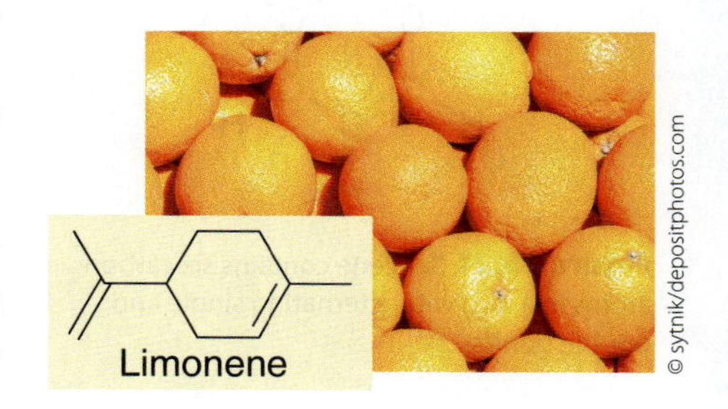

Figure 15.19 Limonene is the major component of citrus oil and is widely used in household cleaners.

$CH_3CH_2CH_2$ H

C=C

H CH_3

Chemists use *alkene* to describe the whole molecule...

...or the C=C functional group specifically.

If a molecule contains only one major functional group, the functional group name is sometimes used to describe the entire molecule. For example, *alkene* refers both to the specific functional group (the C=C bond) and to a simple molecule that contains this functional group.

Three types of hydrocarbons:
Alkanes—all single bonds
Alkenes—double bond
Alkynes—triple bond ■

Alkynes are hydrocarbons that contain a carbon–carbon triple bond. The simplest alkyne, called *acetylene* (also called *ethyne*) is a hot-burning combustion fuel used for welding (**Figure 15.20**). Acetylene is also a precursor for making plastics that conduct electricity. Although some naturally occurring compounds contain alkynes, they are much less common than alkenes.

Figure 15.20 Acetylene is used as a fuel for welding torches and as a precursor in building plastics that conduct electricity.

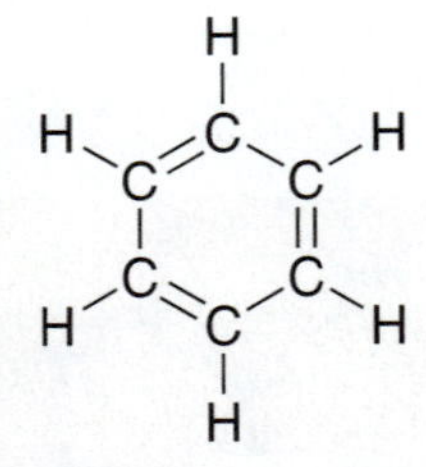

Figure 15.21 Benzene contains six carbon atoms in a ring with alternating single and double bonds.

Aromatic Compounds

The molecule in **Figure 15.21** is called **benzene**. This simple cyclic structure contains six carbons, and it is commonly drawn with alternating single and double bonds around the ring. However, this is no ordinary alkene: Benzene is an unusually stable molecule because of its unique structure. In a double bond, the first set of electrons occupies the space between the two atoms; the second set of electrons, called *pi electrons*, occupies the space above and below. Because of the ring structure, the pi electrons in benzene actually occupy the entire region above and below the six carbon atoms (**Figure 15.22**). To show this unique behavior, chemists sometimes represent benzene as two separate Lewis structures (called *resonance structures*) or as a single six-carbon skeletal structure with a circle in the middle (Figure 15.22).

For more on resonance structures, see Section 9.2.

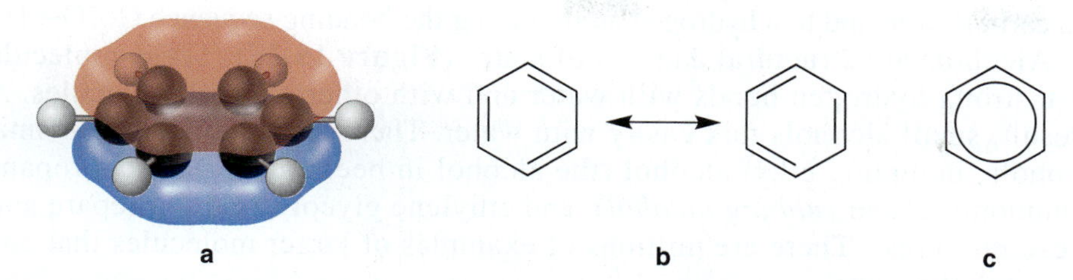

Figure 15.22 (a) In benzene, electron waves occupy the region above and below the carbon ring. Sometimes this is shown (b) as two separate structures, called resonance structures, or (c) as a single six-atom ring with a circle inside it.

Benzene is an example of an **aromatic ring**, a ring structure that contains alternating single and double bonds and is generally less reactive than simple alkenes. Although benzene contains only carbon and hydrogen, aromatic rings can contain other atoms. For example, the molecule pyridine (**Figure 15.23**) is structurally very similar to benzene and also exhibits aromatic behavior.

Figure 15.23 Pyridine is similar to benzene and is also aromatic.

Aromatic rings are present in many molecules, both naturally occurring and human-made. For example, the explosive TNT and the pain medication aspirin both contain a benzene ring (**Figure 15.24**). If you look back at Figure 15.1, you'll notice that the Grubbs catalyst also contains a benzene ring. Although benzene itself is very toxic, more complex structures that contain benzene may be perfectly safe. The benzene ring gives shape and structural stability to many types of molecules.

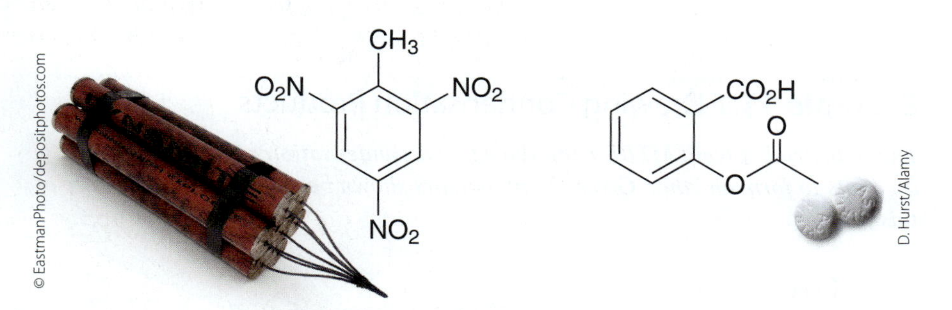

Figure 15.24 Both TNT and aspirin contain a benzene ring in their structures.

TRY IT

5. Identify the alkene, alkyne, and aromatic functional groups in these molecules:

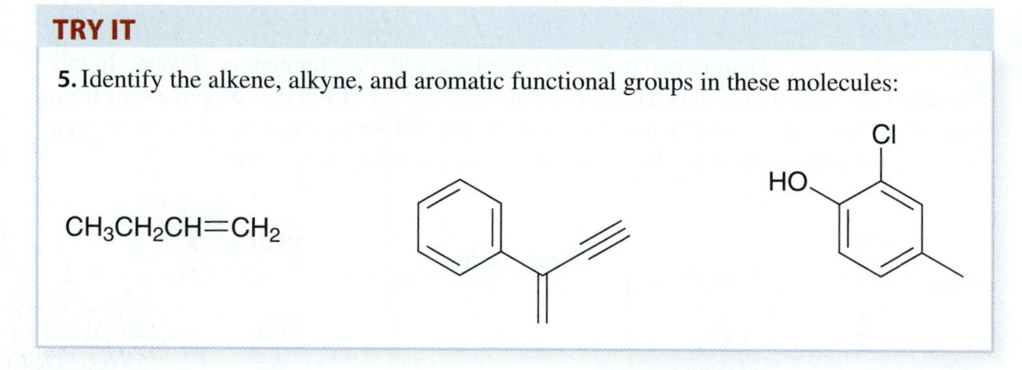

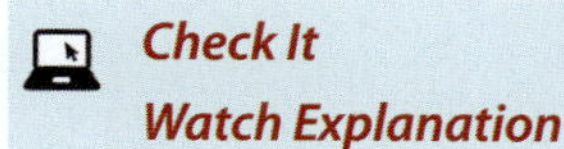

Oxygen-Containing Functional Groups

Alcohols and Ethers

Many organic molecules contain oxygen. Perhaps the most abundant functional group of all is the **alcohol** group. In an alcohol, an oxygen atom is singly bonded to a carbon atom and to a hydrogen atom, giving the bonding sequence C—O—H.

Alcohols are structural analogs of water (**Figure 15.25**). These molecules form strong hydrogen bonds with water and with other alcohol molecules. As a result, small alcohols mix easily with water. There are many common small alcohols, including ethyl alcohol (the alcohol in beer and wine), isopropanol (commonly called *rubbing alcohol*), and ethylene glycol (used to prepare antifreeze mixtures). There are millions of examples of larger molecules that contain alcohol groups.

CH_3CH_2—OH

Ethyl alcohol

H_3C—C(H)(CH_3)—OH

Isopropanol

HO—CH_2CH_2—OH

Ethylene glycol

Figure 15.25 Alcohols contain C—O—H bonds.

The **ether** functional group is composed of an oxygen atom that is singly bonded to two carbon atoms; it contains the bonding sequence C—O—C. Ethers are common in natural molecules, and they are important solvents for many manufacturing processes. Chemists commonly prepare ethers by combining two alcohols to form an ether and water. This is an example of a **condensation reaction**—a reaction in which two smaller molecules combine to produce water plus a larger molecule.

In a condensation reaction, two molecules combine to form water plus a larger molecule.

For example, diethyl ether is a common but very flammable solvent. It is formed by the combination of two molecules of ethyl alcohol:

$$CH_3CH_2OH + HOCH_2CH_3 \rightarrow CH_3CH_2OCH_2CH_3 + H_2O$$

Two alcohols → Ether Water

Condensation reactions like this one are common both in nature and in the laboratory and factory. In Section 15.6 we'll see how nature uses condensation reactions to form molecules like proteins and sugars.

Example 15.7 Drawing Condensation Products

The industrial solvent MTBE is produced by a condensation reaction of two alcohols to form an ether. Given the precursors shown below, draw the structure of this product.

H_3C—C(CH_3)(CH_3)—OH + HO—CH_3 —Condensation→ + H_2O

t-Butyl alcohol, Methyl alcohol, MTBE, Water

In a condensation reaction, two molecules combine with the elimination of water. In this reaction we can remove the OH from one alcohol and the H (on the oxygen atom) from the other. The carbon atom that was attached to the OH forms a new bond to the oxygen atom in the other molecule. This reaction gives the products shown here.

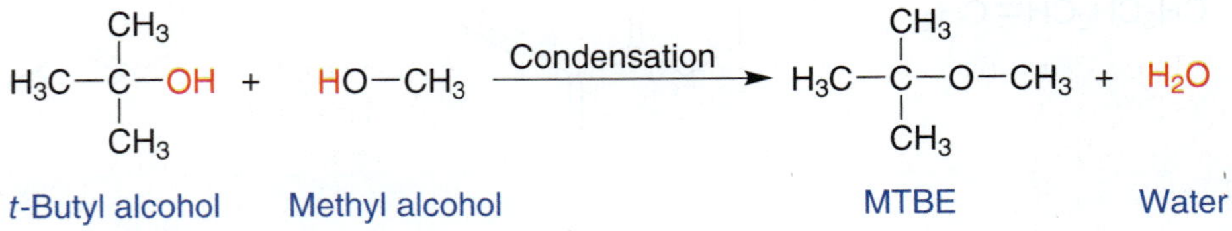

H_3C—C(CH_3)(CH_3)—OH + HO—CH_3 —Condensation→ H_3C—C(CH_3)(CH_3)—O—CH_3 + H_2O

t-Butyl alcohol, Methyl alcohol, MTBE, Water

Carbonyl Groups

After the alcohol group, perhaps the next most common functional group is the **carbonyl** group. A carbonyl is a carbon–oxygen double bond. The behavior of a carbonyl group depends largely on the other two atoms that are bonded to the carbon atom. For this reason, chemists use more specific functional group names that also include the atoms surrounding the carbonyl.

In an **aldehyde**, the carbonyl carbon bonds to a carbon atom on one side and a hydrogen atom on the other:

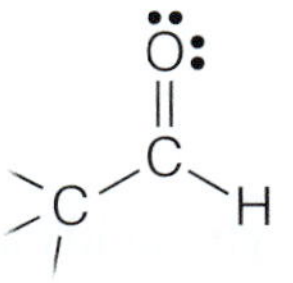

Small aldehydes are often quite fragrant, and they are welcome additions in the kitchen. For example, the main component of both cinnamon and vanilla contain aldehydes, in addition to other functional groups (**Figure 15.26**).

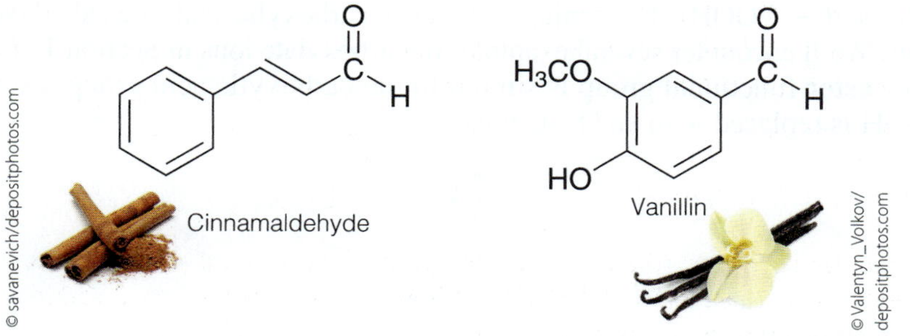

© savanevich/depositphotos.com
Cinnamaldehyde
Vanillin
© Valentyn_Volkov/depositphotos.com

Figure 15.26 The active ingredients of both cinnamon and vanilla contain aldehyde functional groups. What other functional groups are present in these molecules?

In a **ketone**, the carbonyl carbon is bonded to carbon atoms on both sides:

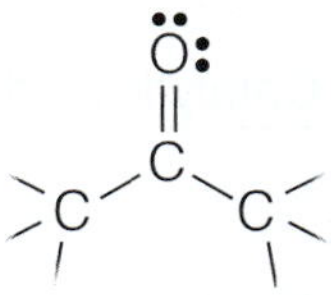

Ketones are more stable than aldehydes. The simplest ketone is *acetone*, CH_3COCH_3. This compound is a common solvent. Ketones are found across a wide variety of substances. For example, carvone is the "minty" component of spearmint oil (**Figure 15.27**).

In a **carboxylic acid**, a carbonyl is coupled with an alcohol (OH):

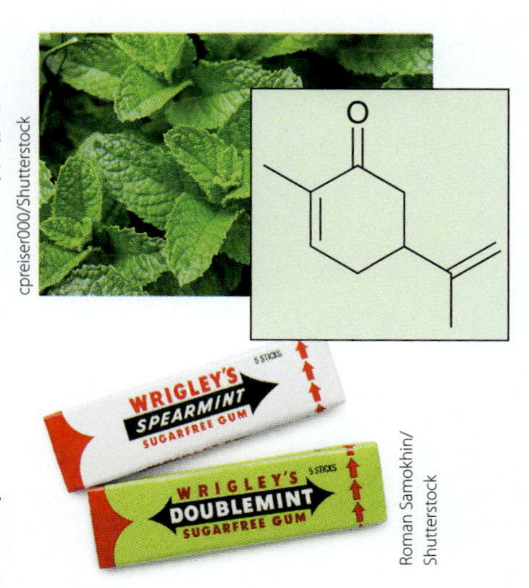

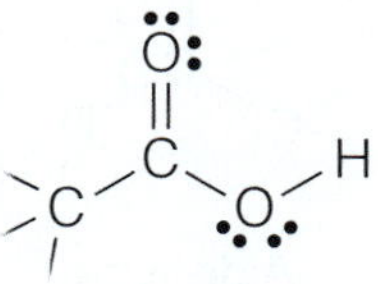

In condensed structures, we commonly represent this structure as —COOH or as —CO_2H. Carboxylic acids are present throughout plant and animal systems. For example, "fatty acids" are carboxylic acids that contain long carbon tails (**Figure 15.28**). These acids are abundant in animal fat. Vegetable oils are also composed of fatty acids, and fatty acids are a critical component of cell structure.

cpreiser000/Shutterstock
WRIGLEY'S SPEARMINT SUGARFREE GUM
WRIGLEY'S DOUBLEMINT SUGARFREE GUM
Roman Samokhin/Shutterstock

Figure 15.27 Carvone is the flavorful component in spearmint oil.

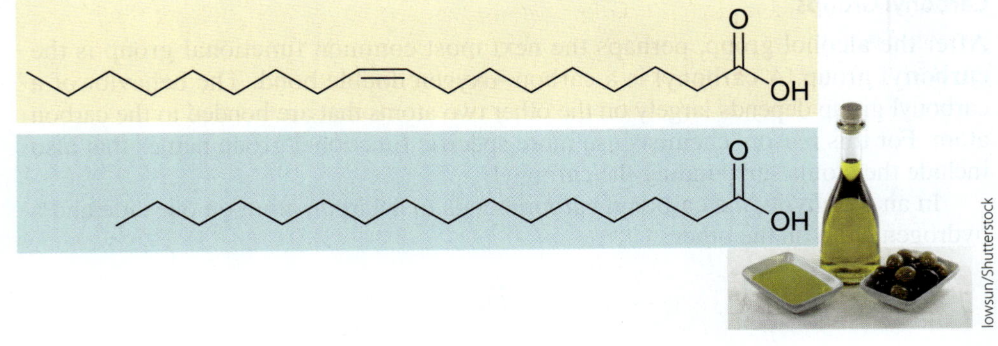

Figure 15.28 Cooking oils contain a mixture of fatty acids.

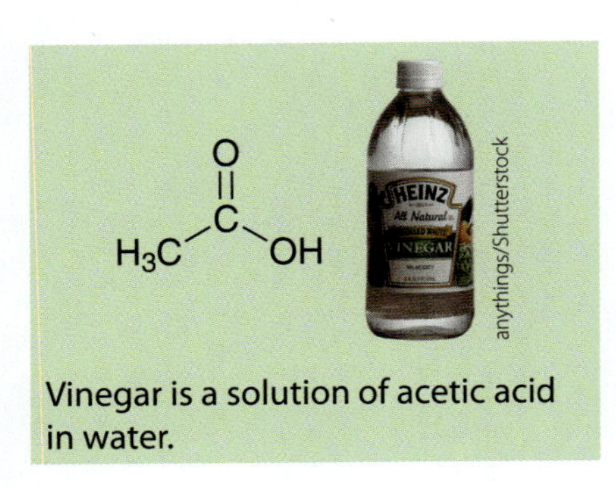

Vinegar is a solution of acetic acid in water.

As their name implies, carboxylic acids are acidic. For example, acetic acid ionizes in solution to produce H^+ and its conjugate base, the acetate ion:

$$CH_3CO_2H\ (aq) \rightleftharpoons CH_3CO_2^-\ (aq) + H^+\ (aq)$$

At the pH levels in biological systems, a large percentage of carboxylic acids are deprotonated (that is, the equilibrium lies toward the conjugate base, $—COO^-$, rather than the acid, —COOH). The conjugate bases of carboxylic acids are called *carboxylate ions*. We'll encounter several examples of carboxylate ions in Section 15.6.

The **ester** functional group is similar to the carboxylic acid group, except that the O—H is replaced with an O—C bond:

Esters commonly form through a condensation reaction involving a carboxylic acid with an alcohol. For example, ethyl acetate is a widely used solvent that is often present in fingernail polish remover. This ester may be prepared by the condensation reaction of acetic acid with ethyl alcohol:

$$H_3C{-}C({=}O){-}OH + HOCH_2CH_3 \xrightarrow{\text{Catalytic acid}} H_3C{-}C({=}O){-}OCH_2CH_3 + H_2O$$

A Summary of Oxygen-Containing Groups

So far in this section, we've looked at six main oxygen-containing groups: alcohols, ethers, aldehydes, ketones, carboxylic acids, and esters. These six structures are summarized in **Figure 15.29**. The oxygen-containing functional groups are all

Alcohol | Aldehyde | Carboxylic acid

Ether | Ketone | Ester

Figure 15.29 A summary of the oxygen-containing functional groups.

around us. Many complex molecules contain multiple oxygen-containing functional groups. I encourage you to practice identifying these functional groups in the examples that follow and in the end-of-chapter practice questions.

Example 15.8 Identifying Functional Groups

When wood burns, it produces a compound called syringol. Food manufacturers use this pleasant-smelling substance to produce "smoky" flavors. What functional groups are present in this syringol molecule?

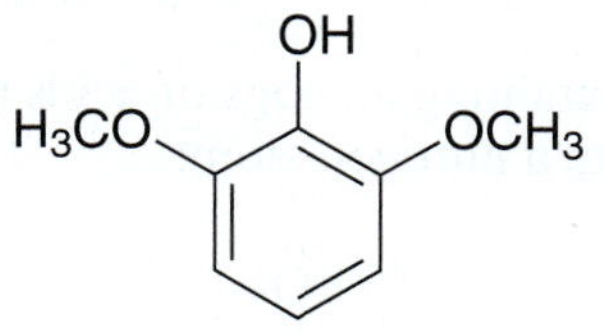

OH
H_3CO OCH_3

This molecule has three major functional groups: An aromatic ring (benzene), an alcohol group (the OH at the top), and two ether groups (the oxygen atoms on the left and right side, which are bonded to both the CH_3 and to a carbon atom on the benzene ring).

TRY IT

6. Curry is a delicious spice blend that is used in dishes across Southeast Asia. One of the key compounds in curry is called *curcumin*. A molecule of curcumin is shown below. What functional groups can you identify in this molecule?

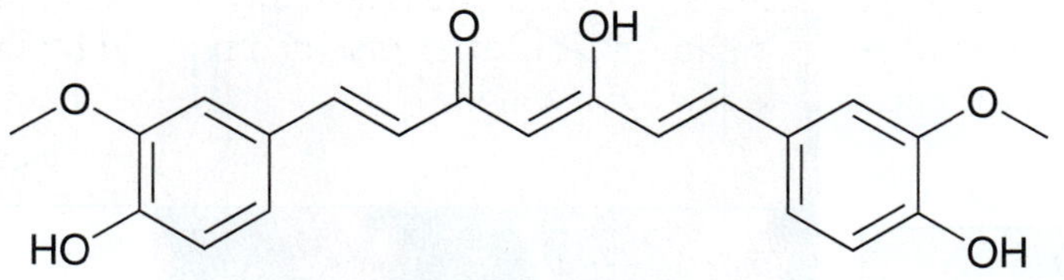

O OH O O HO OH

Curcumin

7. Butyl acetate is a sweet-smelling compound that occurs naturally in many fruits. This compound can be manufactured by the condensation of a carboxylic acid (acetic acid) with an alcohol (butyl alcohol). Given the reactant structures below, draw the structures of the two products, butyl acetate and water.

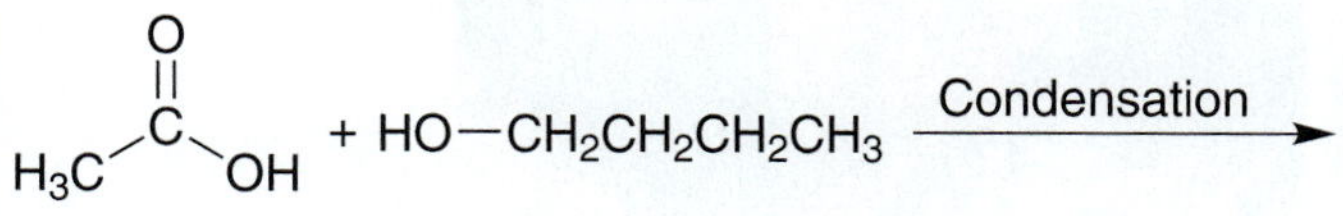

$H_3C{-}C(=O){-}OH$ + $HO{-}CH_2CH_2CH_2CH_3$ $\xrightarrow{\text{Condensation}}$

Acetic acid Butyl alcohol Butyl acetate Water

Check It
Watch Explanation

Nitrogen-Containing Functional Groups

Amines contain a nitrogen atom with three single bonds. In an amine, the nitrogen atom is bonded to one, two, or three carbon atoms. Amines are very common in biological molecules. For example, amino acids, as their name implies, contain both an amine group and a carboxylic acid group. (We'll look at amino acids in detail in Section 15.6.) Many other biologically important compounds also contain

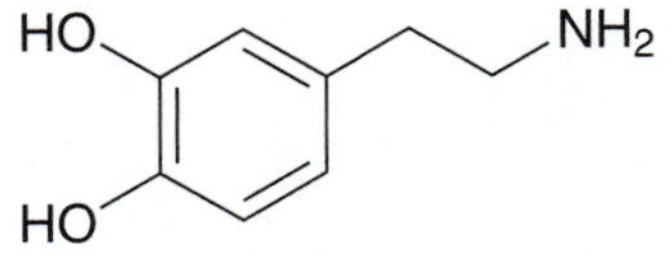

Dopamine

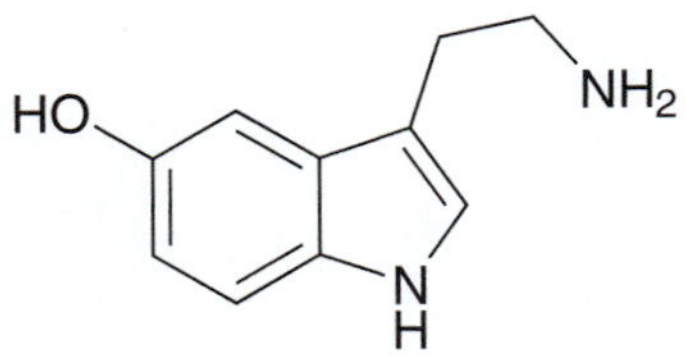

Serotonin

Figure 15.30 Dopamine and serotonin regulate brain function. Both contain amine functional groups.

amines: For example, the molecules dopamine and serotonin carry nerve signals in the brain. These important molecules are critical for motor function, mood modulation, perceptions of hunger and pain, and so on. Both of these molecules contain an amine functional group (**Figure 15.30**).

Amines are slightly basic compounds. Recall that ammonia reacts with water to produce the ammonium ion and hydroxide ion:

$$NH_3\,(aq) + H_2O\,(l) \rightleftharpoons NH_4^+\,(aq) + OH^-\,(aq)$$

Amines like methylamine (CH_3NH_2) react very similarly in water:

$$CH_3NH_2\,(aq) + H_2O\,(l) \rightleftharpoons CH_3NH_3^+\,(aq) + OH^-\,(aq)$$

Amides are nitrogen-containing analogs of acids and esters. These compounds contain a carbonyl adjacent to a nitrogen atom:

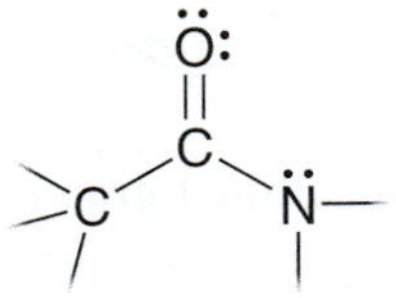

Amides are less reactive than carboxylic acids or esters. Amides form the linkages between building blocks in protein molecules (we'll discuss proteins in Section 15.6). Amides are also found in many smaller molecules, such as the insect repellant DEET and the molecule capsaicin, the substance that gives chili peppers their "hot" flavor (**Figure 15.31**). Amide bonds also make up the key link in materials like nylon.

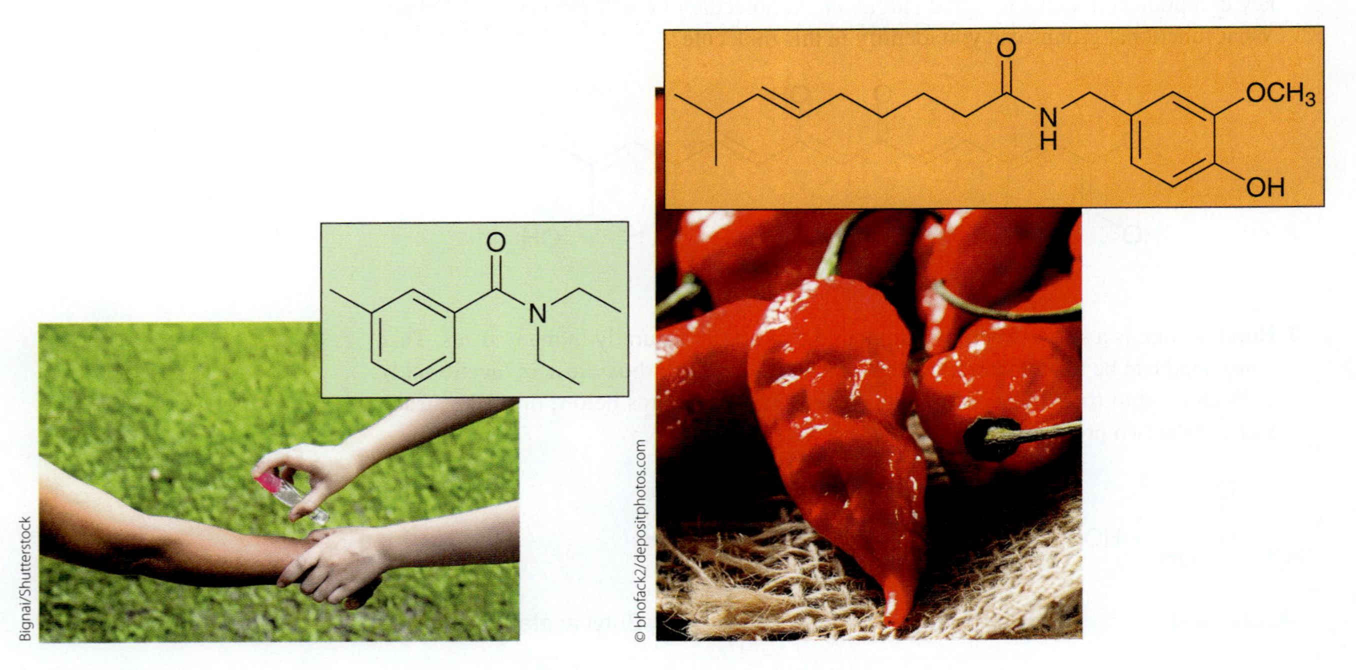

Figure 15.31 The insect repellant DEET and the hot flavoring capsaicin both contain amide functional groups.

Finally, the **nitrile** functional group is composed of a carbon–nitrogen triple bond. The simplest nitrile, HCN, is a highly toxic gas. However, as with other functional groups, the biological effects depend on the structure of the complete molecule. Citalopram (**Figure 15.32**) is a substance used to treat depression, anxiety, and obsessive-compulsive disorder. It contains both an amine and a nitrile functional group.

Citalopram

Figure 15.32 The antidepressant drug citalopram has several functional groups, including an ether, an amine, and a nitrile.

Example 15.9 Identifying Functional Groups

Dentists commonly use lidocaine to numb a patient's mouth before treatment. The structure of lidocaine is shown below. Identify and name the two nitrogen-containing functional groups present in lidocaine.

The nitrogen atom on the right-hand side of this molecule is surrounded by three carbon atoms. This is an amine group. The nitrogen atom on the left-hand side of the molecule is adjacent to a carbonyl—this is an amide group.

TRY IT

8. Tranilast is an anti-allergenic drug used in parts of Asia. This compound contains several of the functional groups we've discussed. Can you identify five different functional groups in this molecule?

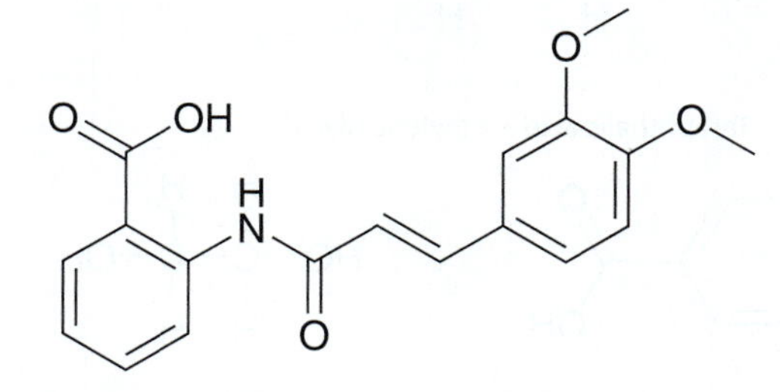

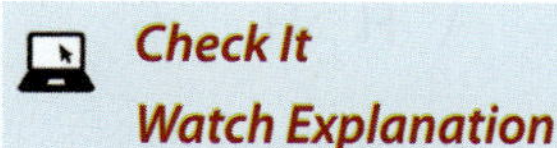

15.5 Polymers and Plastics

Polymers are molecules containing simple repeating units linked together in long chains of covalent bonds. *Biopolymers* are naturally occurring polymers. Many vitally important natural compounds have polymer structures. Biopolymers give wood its rigid structure. Proteins and even DNA are biopolymers. We will look at the structures of these compounds in more detail in the next section.

Plastics are synthetic polymers. From garbage bags to car parts to water bottles, plastics are a seemingly indispensable part of life. Plastics are produced from simple small molecules called **monomers** that assemble to form long chains (**Table 15.3**). Chemists normally represent polymers by drawing the structure of the repeating unit, surrounded by parentheses and the subscript *n* to indicate that this unit occurs over and over. For example, polypropylene is a common plastic used in toys, furniture, kitchen storage containers, and a host of other applications (**Figure 15.33**). It is composed of repeating three-carbon units. We represent the structure of polypropylene this way:

Polypropylene

TABLE 15.3 Common Monomers and Polymers

Polymer	Monomer(s)	Uses
Polyethylene	Ethylene	Grocery bags, packaging, etc.
Polypropylene	Propylene	Toys, furniture, fabrics, bottles, etc.
Poly(vinyl chloride)	Vinyl chloride	PVC piping (plumbing)
Polyethylene terephthalate (a polyester)	Terephthalic acid + ethylene glycol	Synthetic fabrics, resins, etc.
Polycarbonate	Bisphenol A + phosgene	Electronic components, DVDs, automotive headlights, bulletproof glass, etc.

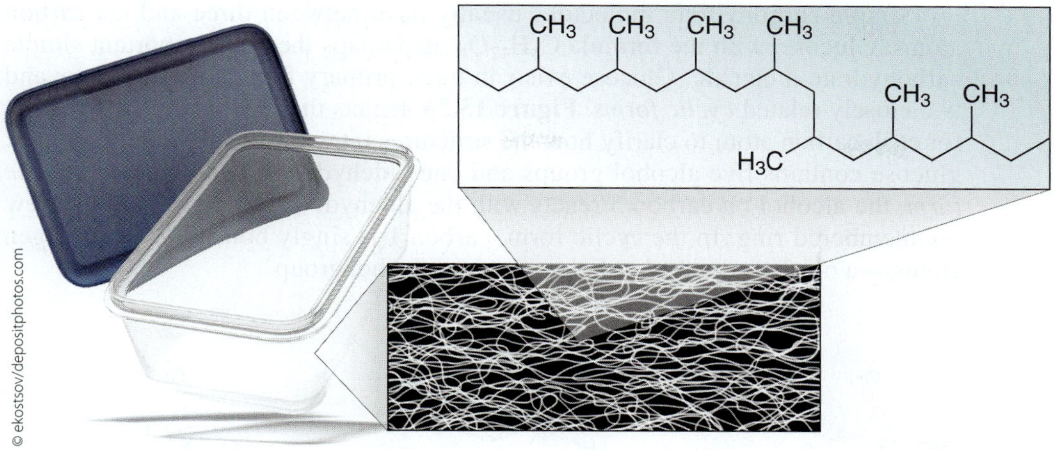

Figure 15.33 Plastics like polypropylene are composed of long molecular chains containing simple repeating units.

Polyester is a polymer made by combining a molecule with two carboxylic acids (terephthalic acid) with another molecule containing two alcohols (ethylene glycol) in a condensation reaction (**Figure 15.34**). This produces a repeating structure connected by ester groups. Water is a side product in this reaction.

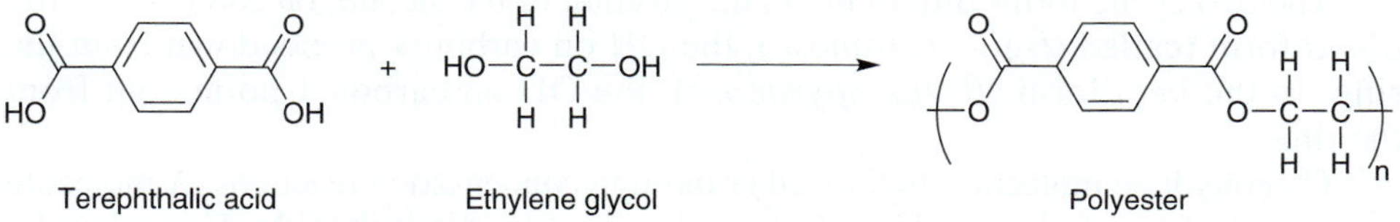

Terephthalic acid + Ethylene glycol → Polyester

Figure 15.34 Polyester is a polymer that forms through a condensation reaction.

Explore Figure 15.34

15.6 Biomolecules—An Introduction

So far, we've seen how atoms combine to form molecules. We've also looked at how different arrangements of atoms within a molecule, called functional groups, contribute to the overall structure and behavior of the substance. In this section, we'll briefly introduce the structures and properties of some of the most important biological molecules. These molecules, which form in living organisms, are essential to their structure and function.

Carbohydrates

Carbohydrates are compounds composed of carbon, hydrogen, and oxygen. These molecules play many important roles in plant and animal life. Most importantly, they store energy: Recall from Section 15.1 that plants store energy through the process of photosynthesis. This energy is stored in the form of carbohydrate molecules.

The term *carbohydrate* means "carbon and water." Carbohydrates have the general formula $C_m(H_2O)_n$, where the subscripts m and n denote the number of carbon atoms and H_2O units, respectively (**Figure 15.35**).

Figure 15.35 Table sugar is a compound called *sucrose*. This carbohydrate has the molecular formula $C_{12}H_{22}O_{11}$, also written as $C_{12}(H_2O)_{11}$.

Carbohydrates are composed of carbon, hydrogen, and oxygen. *Hydrocarbons* are composed of carbon and hydrogen only. ■

Simple carbohydrate molecules usually have between three and six carbon atoms. Glucose, with the formula $C_6H_{12}O_6$, is perhaps the most important simple carbohydrate molecule. Glucose exists in three primary forms: a *linear form* and two closely related *cyclic forms*. **Figure 15.36** depicts these forms, using numbers for each carbon atom to clarify how the structures interconvert. In the linear form, glucose contains five alcohol groups and one aldehyde. To convert to the *cyclic form*, the alcohol on carbon 5 reacts with the aldehyde (carbon 1) to form a new six-membered ring. In the cyclic form, carbon 1 is singly bonded to two oxygen atoms—a newly formed ether linkage and an alcohol group.

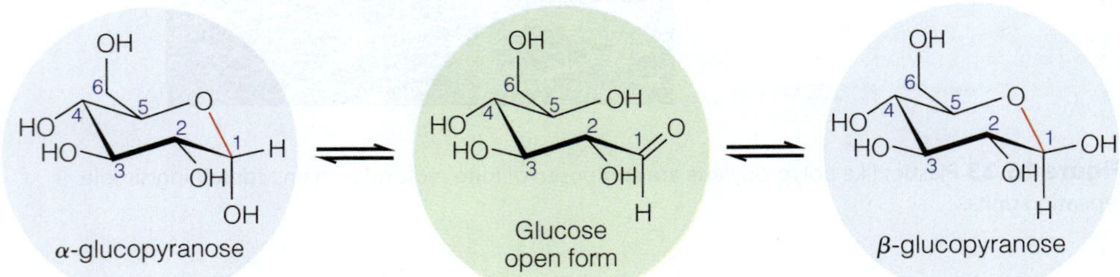

α-glucopyranose

Glucose open form

β-glucopyranose

Figure 15.36 Glucose converts between an open form (shown in green) and two closed, or cyclic, forms. The arrangement of atoms on carbon 1 is the only difference between the two closed forms. In these skeletal structures, the darker lines indicate the front of the ring—the part of the molecule that is closest to you. The bond that forms during the cyclization process is shown in red.

Explore Figure 15.36

The alpha and beta forms of glucopyranose are closely related compounds called *geometric isomers*. These compounds have the same bonding sequence, but they are locked into different spatial arrangements that give them slightly different properties.

The two cyclic forms differ only in the position of the alcohol on carbon 1: In the *alpha* form (called *α-glucopyranose*), the OH on carbon 1 points down from the ring. In the *beta* form (*β-glucopyranose*), the OH on carbon 1 points out from the ring.

Carbohydrate molecules link together through condensation reactions. A molecule composed of two linked carbohydrate molecules is a **disaccharide**. For example, **Figure 15.37** shows the formation of maltose (commonly called *malt sugar*) from two glucopyranose molecules. In this condensation, two alcohols combine to produce an ether linkage, plus water. The ether link between two simple sugars is called a *glycosidic bond*.

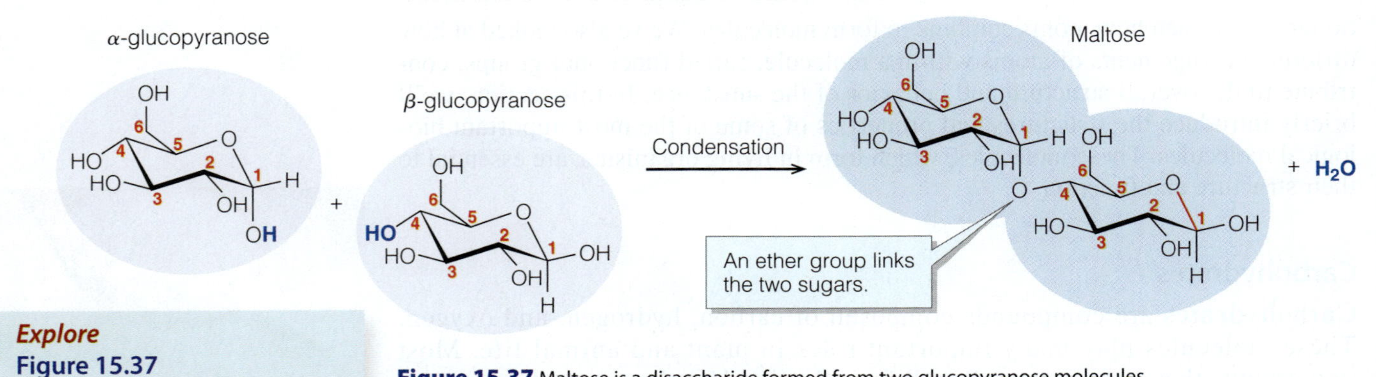

Explore Figure 15.37

Figure 15.37 Maltose is a disaccharide formed from two glucopyranose molecules.

Glycosidic bonds often form between carbon 1 of the first molecule and carbon 4 of the second molecule; this is called a 1,4-linkage. In the same way that the —OH group can be in the alpha or beta positions, the glycosidic bond also forms in the alpha or beta positions. These are referred to as *alpha-linkages* and *beta-linkages*. In the maltose example, the alpha alcohol on carbon 1 connects with the alcohol on carbon 4 of the second molecule. This is called an $\alpha(1 \rightarrow 4)$ linkage (read as "alpha 1-4").

Polysaccharides are naturally occurring polymers composed of many simple carbohydrates that are connected by glycosidic bonds. Two important polysaccharides, *cellulose* and *starch*, are built entirely from glucose molecules (**Figure 15.38**). In cellulose, the glucose molecules connect using only $\beta(1 \rightarrow 4)$ bonds. Plants create cellulose and use it in rigid cell walls. Cellulose also gives wood its rigid quality. In starch, the glucose molecules connect by $\alpha(1 \rightarrow 4)$ bonds. This material is less rigid than cellulose. Starches are found in many foods, such as potatoes, wheat, corn, and rice.

Polysaccharides are naturally occurring polymers of carbohydrates.

Our bodies contain enzymes that slowly break starch down into glucose molecules, then convert the glucose into energy through cellular respiration. However, humans cannot break down the $\beta(1 \rightarrow 4)$ bonds in cellulose. Because of this, wood, grass, and most leaves do not serve as nutrients.

Figure 15.38 Our bodies can break down the $\alpha(1 \rightarrow 4)$ bonds in starch, but not the $\beta(1 \rightarrow 4)$ bonds in cellulose. Because of this, we eat the potato but not the cutting board.

The carbohydrates in food that our bodies can't break down are called *fiber* or *roughage*. While we don't gain nutritional value from these compounds, they do keep the digestive tract working smoothly.

Amino Acids and Proteins

Proteins are large molecules that are vital to life. Some proteins function as *enzymes*: They catalyze very specific reactions. Other proteins transport molecules. Proteins are essential to the replication of our genetic code. And every move we make—every muscle contraction—arises from a change in the shape of a protein.

Proteins are biopolymers composed of building blocks called **amino acids**. Unlike most polymers, which contain simple repeating units, each protein contains a specific sequence of amino acid building blocks. This sequence gives the protein its unique properties.

Amino Acids

There are 20 fundamental amino acids (**Table 15.4**). Each amino acid contains a backbone consisting of an amine, a central carbon, and a carboxylic acid. The central carbon of each amino acid connects to a unique side chain (shown in blue).

TABLE 15.4 The 20 Amino Acids

Glycine (Gly)	Alanine (Ala)	Valine (Val)	Leucine (Leu)	Isoleucine (Ile)
Proline (Pro)	Phenylalanine (Phe)	Tyrosine (Tyr)	Serine (Ser)	Threonine (Thr)
Aspartic Acid (Asp)	Glutamic Acid (Glu)	Cysteine (Cys)	Methionine (Met)	Arginine (Arg)
Asparagine (Asn)	Glutamine (Gln)	Lysine (Lys)	Histidine (His)	Tryptophan (Trp)

Amino acids are sometimes represented in their neutral form, as shown on the left-hand side of the equilibrium below. However, because the amine is basic and the carboxylic acid is acidic, amino acids undergo an internal acid-base reaction to produce two charged sites. This equation shows the equilibrium for the simplest amino acid, glycine:

$$H_2N\text{–}CH_2\text{–}C(=O)\text{–}OH \rightleftharpoons H_3\overset{\oplus}{N}\text{–}CH_2\text{–}C(=O)\text{–}O^{\ominus}$$

Neutral form — Ionic form

In water (and in cells), amino acids exist almost completely in the ionic form. Because of the two charged sites on the ionic form, amino acids dissolve easily in water.

The English alphabet contains 26 letters that we use to create words, sentences, and even books. The amino acids are the "letters" from which complex proteins are made.

Peptides and Peptide Bonds

Amino acids link together through condensation reactions. In these reactions, the carbonyl of one amino acid connects with the amine of another to form an amide. The bond formed between amino acids is called a **peptide bond**, and groups of amino acids connected in this way are called *peptides*. For example, glycine and alanine can bond through this condensation reaction:

$$\underset{\text{Gly}}{H_3\overset{\oplus}{N}\text{–}CH(H)\text{–}C(=O)\text{–}O^{\ominus}} + \underset{\text{Ala}}{H_3\overset{\oplus}{N}\text{–}CH(CH_3)\text{–}C(=O)\text{–}O^{\ominus}} \longrightarrow \underset{\text{Gly-Ala}}{H_3\overset{\oplus}{N}\text{–}CH(H)\text{–}C(=O)\text{–}N(H)\text{–}CH(CH_3)\text{–}C(=O)\text{–}O^{\ominus}} + H_2O$$

The molecule formed from two amino acids is a *dipeptide*. Similarly, a chain of three amino acids is a *tripeptide*. Proteins are *polypeptides*—they are built from many amino acids. In the dipeptide structure above, notice that there is still an amine group and a carboxylic acid group on either end of the molecule. The amine end is called the *N-terminus*. The carboxylic acid end is called the *C-terminus* (**Figure 15.39**).

Amino acids connect through peptide bonds. ■

$$H_3\overset{\oplus}{N}\text{–}CH(H)\text{–}C(=O)\text{–}N(H)\text{–}CH(CH_3)\text{–}C(=O)\text{–}O^{\ominus}$$

N-terminus; The peptide bond connects the two building blocks.; C-terminus

Gly-Ala

Figure 15.39 This two-amino-acid structure contains an N-terminus, a C-terminus, and a peptide bond connecting the two amino acids.

Longer peptide sequences have a huge number of atoms present. For these molecules, using Lewis structures and even skeletal structures is cumbersome. For simplicity, chemists and biologists often describe peptides using three-letter designations to represent the amino acid sequence. In this convention, the structure is always written with the N-terminus on the left-hand side. For example, consider a tripeptide containing lysine (on the N-terminus), connected to glutamine, connected to tryptophan (the C-terminus). We can represent this complex structure using the designation Lys-Gln-Trp. Example 15.10 further illustrates this type of representation.

Example 15.10 Drawing Peptide Chains

Draw the tripeptide Cys-Ala-Thr.

Using Table 15.4, we draw the structures of the three amino acid building blocks, orienting the N-terminus of each molecule to the left:

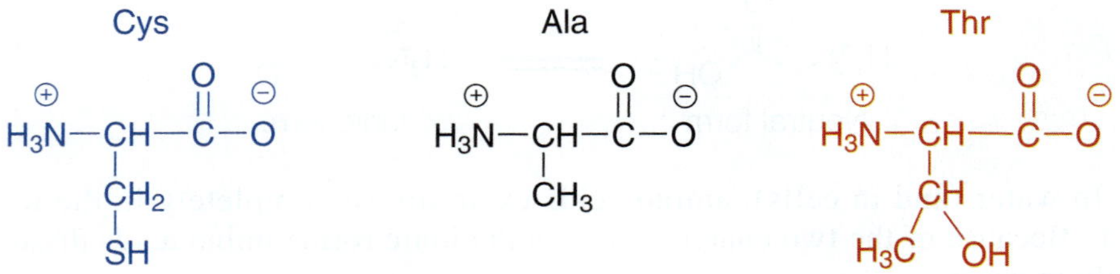

Cys

$$H_3\overset{\oplus}{N}-CH(CH_2-SH)-C(=O)-O^{\ominus}$$

Ala

$$H_3\overset{\oplus}{N}-CH(CH_3)-C(=O)-O^{\ominus}$$

Thr

$$H_3\overset{\oplus}{N}-CH(CH(H_3C)(OH))-C(=O)-O^{\ominus}$$

Next we connect the three structures. N is connected to the C=O of the structure to the left:

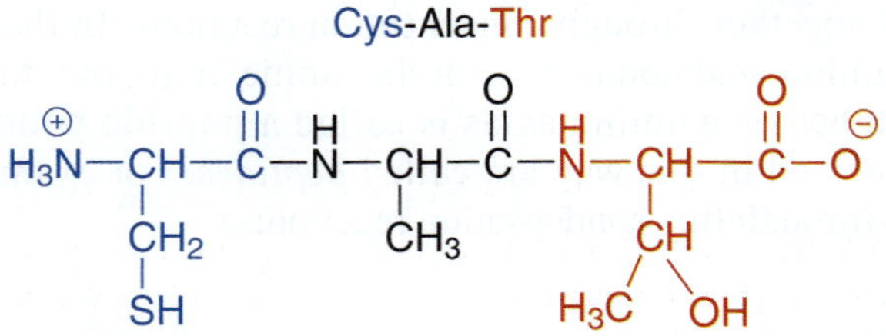

Cys-Ala-Thr

$$H_3\overset{\oplus}{N}-CH(CH_2-SH)-C(=O)-N(H)-CH(CH_3)-C(=O)-N(H)-CH(CH(H_3C)(OH))-C(=O)-O^{\ominus}$$

Again, notice that the first amino acid named in the sequence contains the N-terminus, and the last one named contains the C-terminus.

Check It

Watch Explanation

TRY IT

9. Draw the structure of the tripeptide Thr-Ala-Cys. How does this compare with the tripeptide Cys-Ala-Thr that we drew in Example 15.10?

10. Use the three-letter abbreviations to represent the tripeptide sequence shown here:

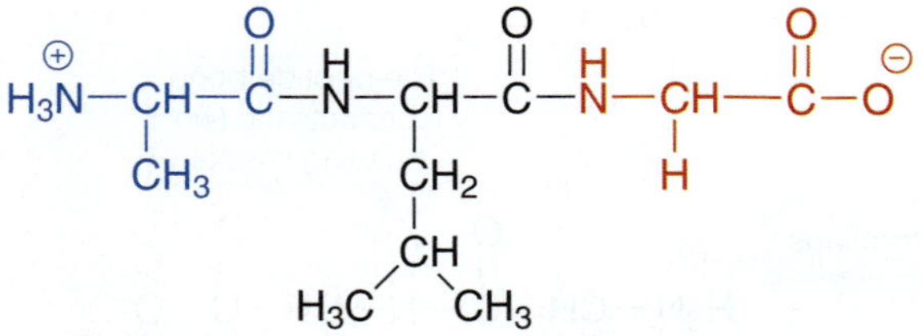

$$H_3\overset{\oplus}{N}-CH(CH_3)-C(=O)-N(H)-CH(CH_2-CH(H_3C)(CH_3))-C(=O)-N(H)-CH(H)-C(=O)-O^{\ominus}$$

Proteins are made of hundreds or even thousands of amino acids linked together through peptide bonds. These chains contain many O—H and N—H bonds that participate in strong hydrogen bonds between different parts of the chains, causing the proteins to fold in certain predictable patterns. The exact sequence of amino acids determines the way a protein chain will fold and therefore the shape of the protein. In turn, the shape of the protein determines how it can function.

For example, hemoglobin (**Figure 15.40**) is the protein that transports oxygen from the lungs and through the bloodstream to every cell in the body. A hemoglobin protein is composed of four different polypeptide chains. Each chain contains either 141 or 146 amino acid pieces. These chains bind with structures called *heme units* that contain Fe^{2+} ions. The Fe^{2+} ions bind to oxygen, transporting it throughout the body.

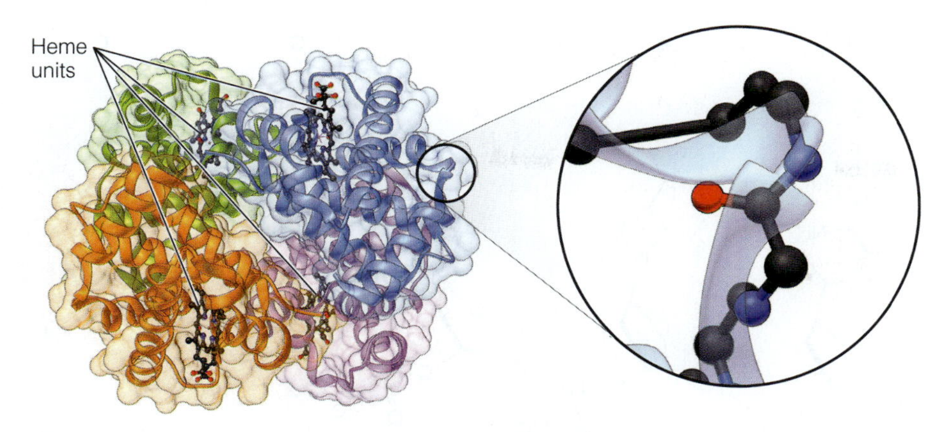

Figure 15.40 This image shows the structure of hemoglobin, the protein that transports oxygen throughout the body. The complex structure is composed of four peptide chains and four heme units. The colored ribbons in this structure represent the shapes of the polypeptide chains, as shown in the magnification at right. The space occupied by the protein is represented by a semitransparent surface.

DNA

We all begin life as a single cell. Embedded in that cell is a blueprint—a code for every feature and function of our bodies. This code, called our *genome* or simply our *genes*, is passed down from our parents. As we grow, this blueprint is duplicated. Every cell of our bodies—every strand of hair or scrape of skin—contains this blueprint in full.

Plants and animals store their genetic information in a set of polymeric molecules called **deoxyribonucleic acids (DNA)**. These molecules are massive: The human genetic code contains 6 billion different bits of coded information that are stored in just 46 molecules, called *chromosomes*.

The DNA coding in our chromosomes is built on four small molecules, called *bases*. The four bases are *adenine, thymine*, *guanine*, and *cytosine* (**Figure 15.41**). Each base is symbolized by the first letter of its name: A, T, G, and C. These four bases encode all the information needed to produce a living organism.

In a DNA molecule, these four bases bond to a simple sugar called *deoxyribose* and a phosphate linker. The base, the sugar, and the phosphate form a *nucleotide*—a single monomer in the DNA polymer (**Figure 15.42**). The nucleotides link together

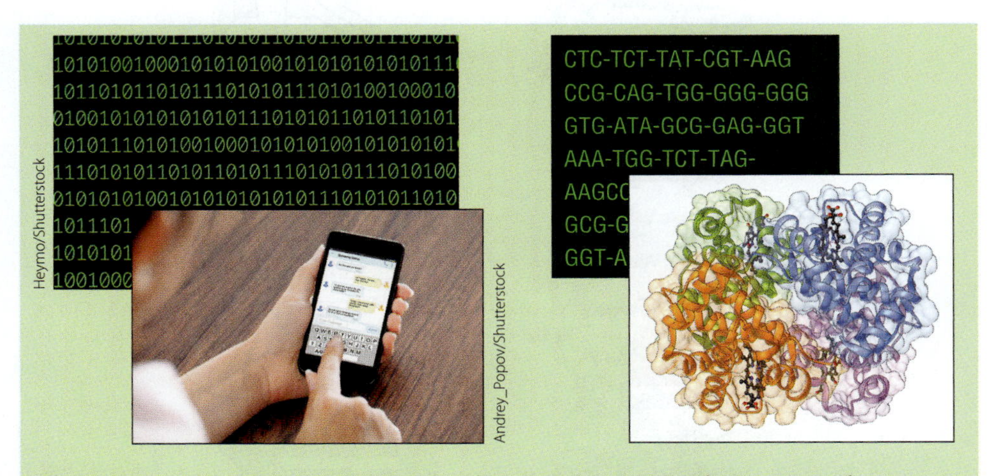

When you send a text message, you use letters to create words. But your phone uses a series of ones and zeros—a binary code—to store those letters. Much like letters are the building blocks for words, amino acids are the building blocks for proteins. And much like your phone stores letters in binary code, your DNA code stores the amino acid sequence for every protein of your body.

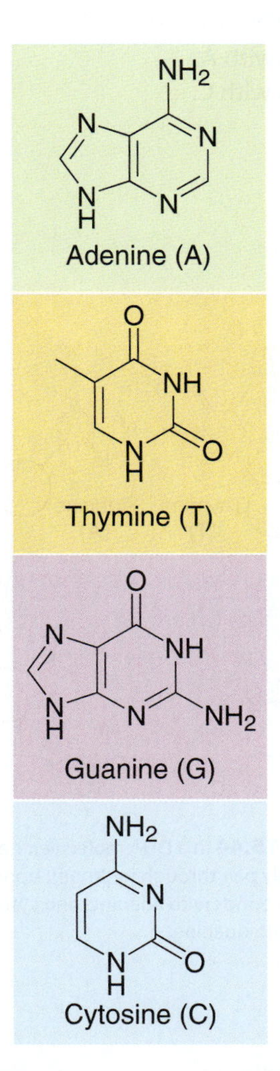

Figure 15.41 These four structures store the genetic code in DNA.

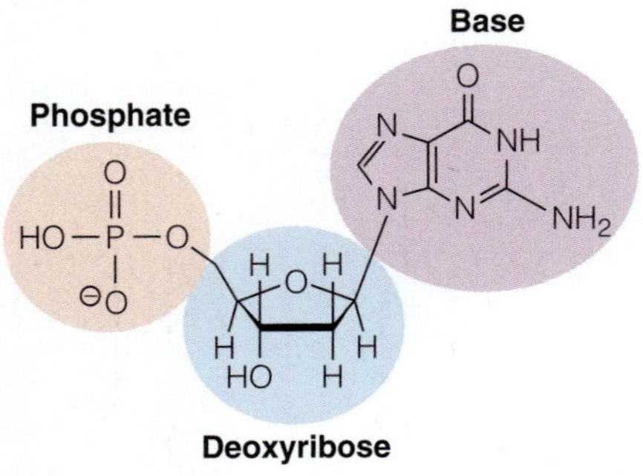

Base

Phosphate

Deoxyribose

Figure 15.42 A single nucleotide contains a base that is connected to a backbone composed of deoxyribose and a phosphate linker.

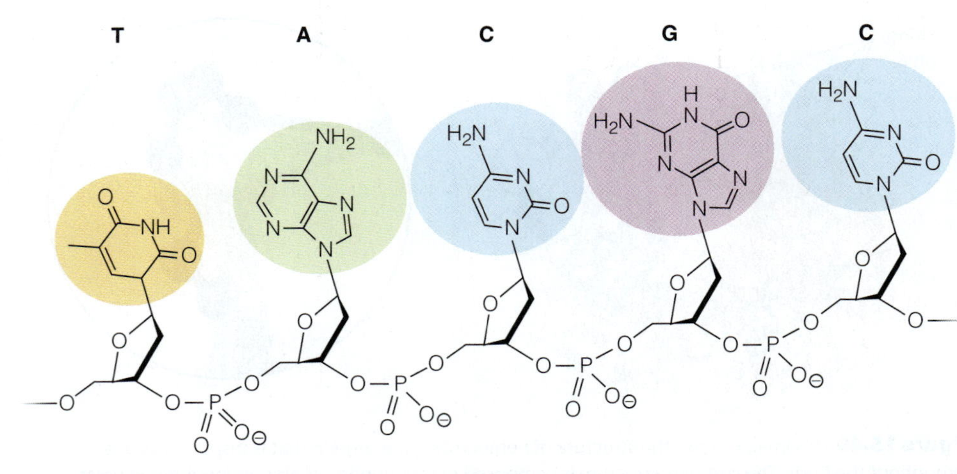

Figure 15.43 The phosphate linker and the deoxyribose provide the structural backbone for the DNA strand. The sequence of bases provides the "code" from which our bodies are built.

through condensation reactions to form a sugar–phosphate backbone (**Figure 15.43**). A single strand of DNA is composed of millions of these nucleotides, each representing a piece of the genetic code.

DNA bases pair together:

T pairs with A.

G pairs with C.

This elegant method for storing information is amazing, but it gets even better. Our bodies make copies of the DNA blueprint. Partial copies travel within the cells, and when cells divide, they produce a complete copy of the master blueprint. Here's how it works: Each base couples with one of the other bases by forming tight hydrogen bonds. Thymine bonds with adenine, and guanine bonds with cytosine (**Figure 15.44**). A complete DNA molecule contains two strands; each base in one strand is specifically paired to the matching base in the other strand. Because of these hydrogen bonds, DNA forms a unique shape called a *double helix*: The two backbones twist around each other with base pairs in between (**Figure 15.45**).

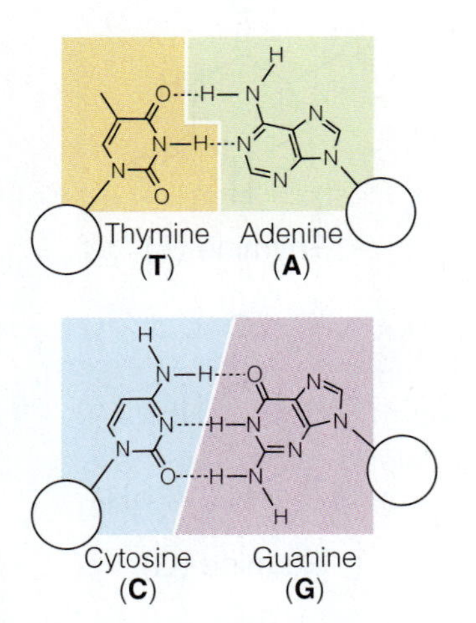

Figure 15.44 In a DNA molecule, the bases selectively pair through hydrogen bonds: Thymine bonds with adenine, and cytosine bonds with guanine.

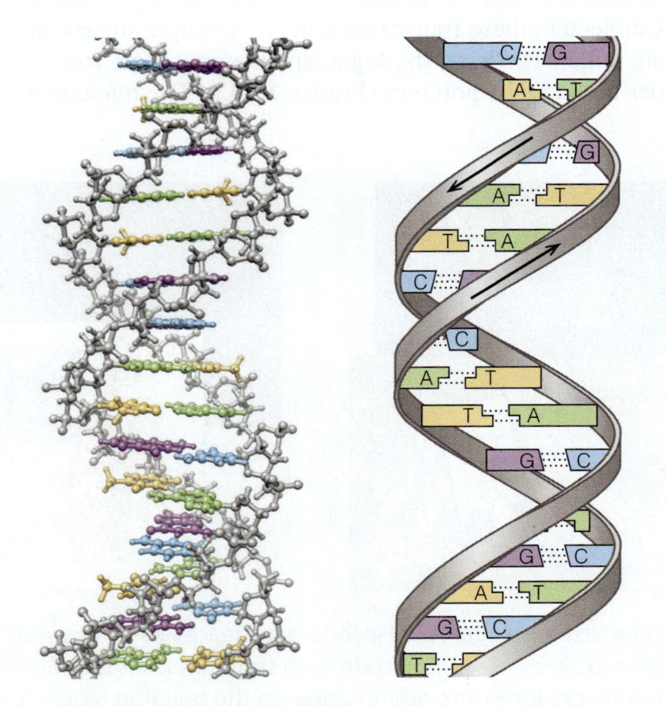

Figure 15.45 A DNA molecule contains two matching strands, wound together in a double helix. The image at left shows the detailed molecular structure, and the structure at right is a simplified representation.

Explore Figure 15.45

Because the two strands are complementary pairs, and each strand encodes the complete set of genetic instructions, the double-helix structure allows the cells to duplicate the genetic code. **Figure 15.46** illustrates how this works: When a cell is ready to divide, an enzyme (that is, a protein that serves a specific function) begins to "unzip" the double-helix structure. Other enzymes then build two new backbones with bases that match the two unzipped strands. When the unzipping is complete, two identical double-helix structures have been produced.

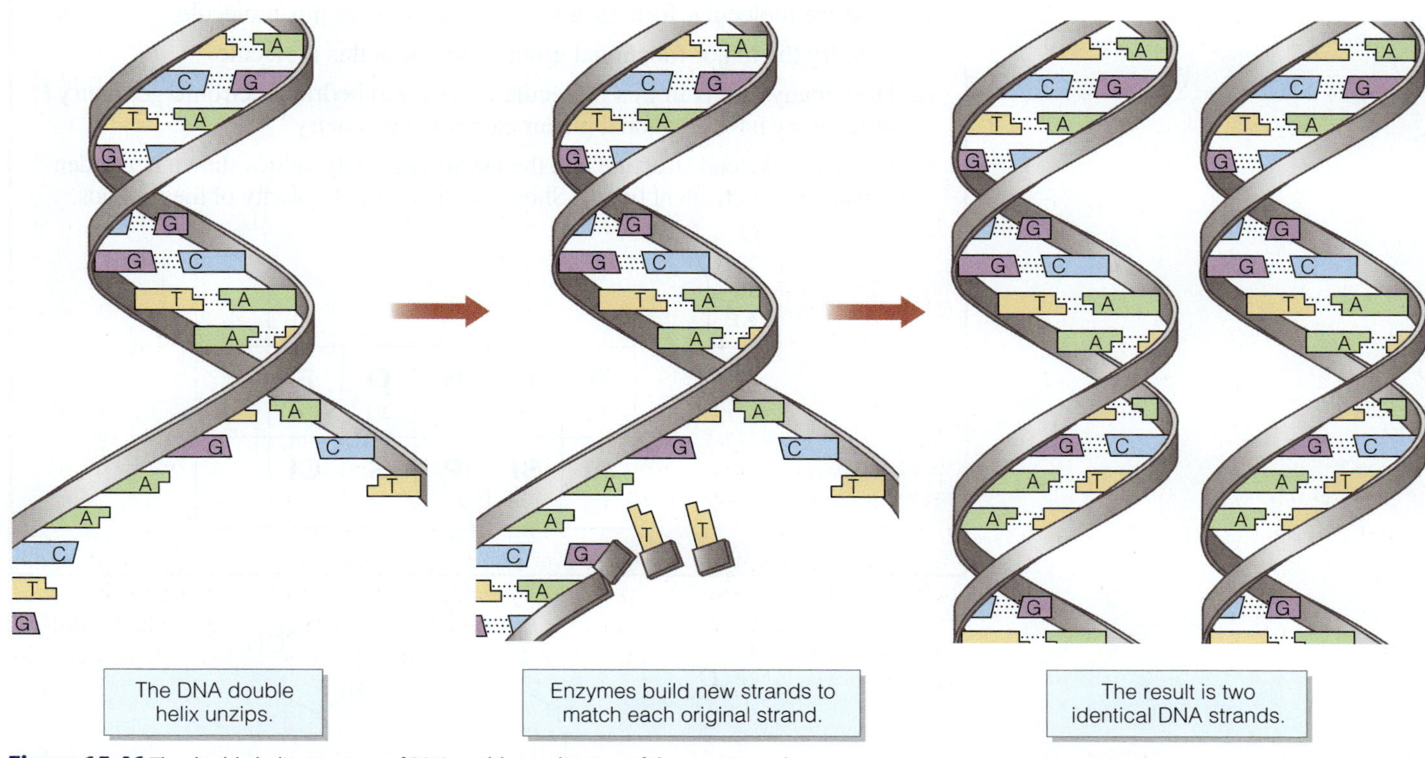

The DNA double helix unzips.
Enzymes build new strands to match each original strand.
The result is two identical DNA strands.

Figure 15.46 The double helix structure of DNA enables replication of the genetic code.

Explore **Figure 15.46**

In order for the cell to use the information stored on its blueprint, the DNA code is copied onto a related molecule called *ribonucleic acid* (RNA). RNA relays segments of information from the DNA to other parts of the cell, where it is used to assemble amino acids into proteins.

In 1984, the U.S. government launched the human genome project, a massive research effort involving laboratories across the world, with the goal of mapping every base pair in a human DNA sample. In 2003 the project was declared complete, and in 2007 scientists published the entire 6-billion-nucleotide sequence of a single person. Today, advanced genome sequencing techniques have become much faster and cheaper, and scientists routinely analyze gene sequences for living creatures past and present.

Capstone Video

Capstone Question

Hydroxychloroquine is a common drug used to treat malaria, lupus, arthritis, and other conditions. In 2020, it was one of the first treatments tested during the COVID-19 pandemic. Using the skeletal structure of hydroxychloroquine shown in **Figure 15.47**, answer each of the following:

a. Draw a complete Lewis structure for this molecule, showing all hydrogens and lone pairs.

b. Find the molecular formula and the molar mass of this molecule.

c. Identify the major functional groups present in this molecule.

d. How many atoms in this molecule have a tetrahedral electronic geometry? How many have a trigonal planar electronic geometry?

e. Using the skeletal structure and the electronegativity values shown here, identify any polar covalent bonds. Show the direction of polarity of these bonds.

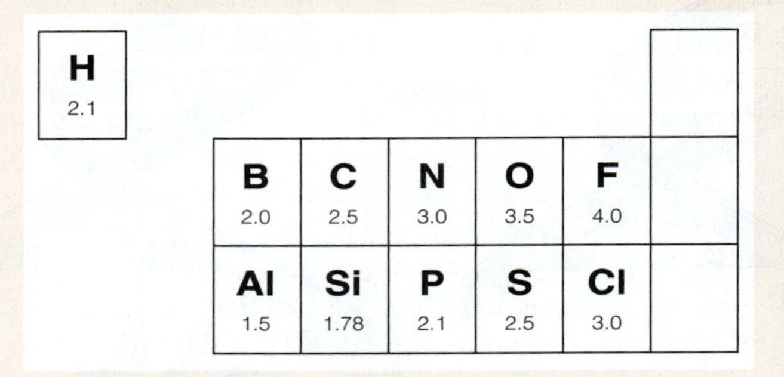

H
2.1
B
2.0
C
2.5
N
3.0
O
3.5
F
4.0
Al
1.5
Si
1.78
P
2.1
S
2.5
Cl
3.0

Figure 15.47 This is the skeletal structure of the anti-malarial drug hydroxychloroquine.

SUMMARY

Carbon compounds form the basis for life. Organic chemistry is the chemistry of carbon-based compounds. Organic chemistry is founded in the chemistry of living organisms, but it also includes carbon compounds from other sources.

Carbon is present in our atmosphere (as carbon dioxide), in the oceans (as bicarbonate ions), in plants and animals, and even in fossil fuels buried deep underground. The carbon cycle is a model that describes how carbon migrates between these different regions through processes such as photosynthesis, cell respiration, and exchange of carbon dioxide between the atmosphere and the sea.

Organic chemistry begins with covalent bonds. The number of covalent bonds a nonmetal atom forms depends on the number of electrons needed to complete its valence electron level. Hydrogen typically forms one bond, oxygen forms two, nitrogen three, and carbon four. All plant and animal life is composed primarily of these four atoms. Because of its ability to form four covalent bonds, carbon forms the backbone for most biological molecules.

We often represent simple molecules using Lewis structures. A complete Lewis structure depicts all the valence electrons, both bonded and unbonded. As molecules increase in complexity, Lewis structures quickly get cumbersome. We therefore use simplified structures to convey bonding information in ways that are less cluttered. In simplified Lewis structures, chemists show the bonded electrons as dashes but do not include nonbonded electrons. In condensed structures, atoms are listed in the sequence of their bonding, and the bonds are inferred from the arrangement.

In skeletal structures, lines represent carbon chains. Unless labeled otherwise, each vertex or end point in a skeletal structure represents a carbon atom. Hydrogen atoms bound to carbon are not drawn but are inferred from the structures. Hydrogen atoms bonded to other atoms (such as oxygen and nitrogen) are shown explicitly. Scientists often use these different drawing styles interchangeably, drawing structures with both skeletal and condensed elements.

Alkanes and cycloalkanes are composed only of carbon and hydrogen, and they contain all single bonds. In addition to C—H single bonds, most organic molecules contain groupings of atoms or bonds that behave in distinct patterns. These patterns of atoms and bonds are called *functional groups*. Molecules that are composed only of carbon and hydrogen are called *hydrocarbons*.

There are several hydrocarbon functional groups: Carbon–carbon double bonds are called *alkenes*. Carbon–carbon triple bonds are called *alkynes*. Aromatic compounds contain rings with alternating single and double bonds. Aromatic compounds are less reactive than simple alkenes. Benzene is the most common aromatic compound.

There are six major functional groups containing C, H, and O: alcohols, ethers, aldehydes, ketones, carboxylic acids, and esters. Each of these arrangements produces distinctive properties, and these functional groups are common in natural and synthetic molecules. The nitrogen-containing functional groups include amines, amides, and nitriles.

Alcohols, carboxylic acids, and amines commonly undergo condensation reactions. In these reactions, two smaller molecules combine to produce water plus a larger molecule.

Polymers are long-chain molecules composed of simple repeating units. Polymers that form in living systems are called *biopolymers*. Synthetic polymers are commonly called *plastics*.

In this chapter we briefly examined some of the major classes of biomolecules. Carbohydrates are compounds composed of carbon, hydrogen, and oxygen. Plants create carbohydrates through the process of photosynthesis. Simple carbohydrates often link together through glycosidic bonds. Cellulose and starch are two examples of polysaccharides—biopolymers made from simple carbohydrate monomers.

Proteins are large compounds that perform a wide variety of biological functions. Proteins are composed of 20 distinct building blocks, called *amino acids*. The sequence of amino acids helps determine the shape of proteins, which in turn determines how the protein functions.

The genetic information of every living creature is stored in its DNA. DNA molecules are massive, and they are composed of units of genetic code called *nucleotides*. Each nucleotide contains a base (that carries the specific genetic information), a sugar, and a phosphate linker. DNA uses only four bases: adenine, thymine, guanine, and cytosine. These bases form specific hydrogen-bonded pairs: Adenine binds with thymine, and guanine binds with cytosine. The DNA molecule contains a double-helix structure, in which the sugar and phosphate backbone circles around each base pair. The unique double-helix structure allows DNA to be duplicated, and it allows related molecules called *RNA* to transport fragments of the genetic code throughout the cells. Cells use the information stored in DNA to construct proteins.

This chapter has been a brief survey. Carbon and the other nonmetals combine in unique ways that enable an infinite combination of atoms and structures. These structures range from very small to astoundingly large molecules. The study of organic and biological chemistry is far too rich and vast for one person to understand. Yet this sea of knowledge revolves around a few simple rules: two bonds to oxygen, three bonds to nitrogen, and four bonds to carbon.

How Catalysts Work

Kristine Foley/Getty Images

Figure 15.48 In a wedding ceremony, the minister joins the bride and groom together.

At the beginning of this chapter, we introduced Robert Grubbs, a Nobel Prize–winning chemist who developed ruthenium catalysts that produce carbon–carbon bonds. His work led to cleaner and safer ways to manufacture medicines, plastics, lubricants, and many other valuable products. But how does the catalyst actually work?

Catalysts are like ministers who preside over weddings. They are not getting married, but they facilitate the bond between the bride and groom. Symbolically, the minister may even join the couple's hands together (**Figure 15.48**). Similarly, catalysts help new chemical bonds to form, but the catalysts are not changed in the overall process.

For example, chemists use ruthenium catalysts to prepare cyclic structures containing carbon–carbon double bonds, as shown in **Figure 15.49a**. Notice that the catalyst is not present in the starting material or in the products, but it is written over the arrow.

Figure 15.49b shows the step-by-step process through which the ruthenium catalyst works. In the first step, the catalyst (structure 1) reacts with a carbon–carbon double bond to form a four-atom ring (structure 2). This ring then breaks to form a new carbon–ruthenium double bond (structure 3). This cycle repeats, as ruthenium forms a new four-atom ring (structure 4) and then releases its bond to the larger structure, producing a new double bond. After the new bond forms, the catalyst returns to its original state. The bond has been formed; the ring is produced. You may kiss the bride.

Figure 15.49 (a) This reaction uses a ruthenium catalyst. (b) This reaction takes place in the step-by-step process shown.

Explore Figure 15.49

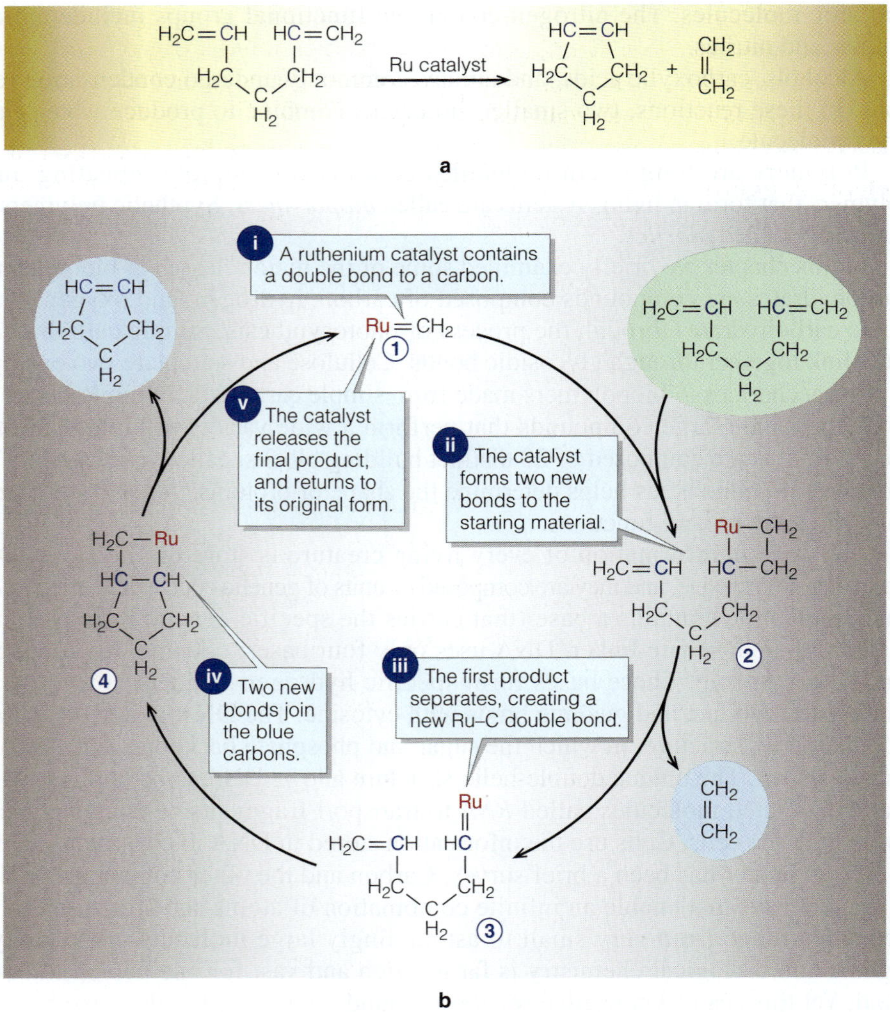

$H_2C{=}CH \quad HC{=}CH_2$ / $H_2C \quad CH_2$ / C / H_2 —Ru catalyst→ $HC{=}CH$ / $H_2C \quad CH_2$ / C / H_2 + $CH_2{=}CH_2$

a

i A ruthenium catalyst contains a double bond to a carbon.

$Ru{=}CH_2$ (1)

ii The catalyst forms two new bonds to the starting material.

(2)

iii The first product separates, creating a new Ru–C double bond.

(3)

iv Two new bonds join the blue carbons.

(4)

v The catalyst releases the final product, and returns to its original form.

b

Key Terms

5.1 Organic Chemistry and the Carbon Cycle

organic chemistry The chemistry of carbon-containing molecules.

carbon cycle A description of how Earth's carbon moves between rock and sediment, water and atmosphere, and plants and animals.

photosynthesis The sequence of chemical reactions by which green plants harvest the energy of the Sun.

cellular respiration The sequence of chemical reactions by which animals release energy stored in the chemical bonds of substances they consume.

5.2 Covalent Bonding with Carbon and Other Nonmetals

isomers Compounds having the same molecular formula but different bonding sequences.

5.3 Drawing Covalent Structures

condensed structure A way of representing chemical bonds that does not show most covalent bonds, but lists atoms in order of their connectivity.

skeletal structure A simplified representation for chemical structures in which the end of each line segment denotes a carbon, and C—H bonds are inferred rather than drawn explicitly.

5.4 Major Functional Groups

functional group A small group of atoms within a molecule that behaves in a characteristic manner.

hydrocarbon A compound that contains only carbon and hydrogen.

alkane A hydrocarbon composed entirely of single bonds.

cycloalkane An alkane that forms a cyclic structure (commonly called a *ring*).

alkene A functional group consisting of a carbon–carbon double bond; this term also refers to a simple molecule containing this functional group.

alkyne A functional group consisting of a carbon–carbon triple bond; this term also refers to a simple molecule containing this functional group.

benzene A very stable compound having the formula C_6H_6, in which the six carbon atoms form a ring with alternating single and double bonds; benzene is one of the simplest examples of an aromatic ring.

aromatic ring A ring structure that contains alternating single and double bonds and is generally less reactive than simple alkenes.

alcohol A functional group consisting of an oxygen atom that is singly bonded to a carbon atom and a hydrogen atom, giving the bonding sequence C—O—H.

ether A functional group composed of an oxygen atom singly bonded to two carbon atoms, giving the bonding sequence C—O—C.

condensation reaction A reaction in which two smaller molecules combine to produce water plus a larger molecule.

carbonyl A functional group consisting of a carbon–oxygen double bond.

aldehyde A functional group consisting of a carbonyl connected to a hydrogen atom.

ketone A functional group containing a carbonyl connected to two carbon atoms.

carboxylic acid A functional group containing a carbonyl bonded to an alcohol, commonly represented by the condensed formula —COOH.

ester A functional group containing a carbonyl bonded to an oxygen atom that is bonded to another carbon atom.

amine A functional group consisting of a nitrogen atom with three single bonds, usually to hydrogen or carbon atoms.

amide A functional group consisting of a carbonyl group bonded to a nitrogen atom.

nitrile A functional group consisting of a carbon–nitrogen triple bond.

5.5 Polymers and Plastics

polymer A molecule containing simple repeating units that are linked together in long covalent chains.

plastic A synthetic polymer.

monomer A small molecule that can connect with other molecules to form a polymer.

5.6 Biomolecules—An Introduction

carbohydrate A naturally occurring molecule composed of carbon, hydrogen, and oxygen and having the general formula $C_m(H_2O)_n$.

disaccharide A carbohydrate composed of two simpler carbohydrates that are linked together through a condensation reaction.

polysaccharide A carbohydrate composed of many simpler carbohydrates that are linked together.

protein A biopolymer composed of building blocks called amino acids; this type of molecule has many functions in living creatures.

amino acid Small molecules having both amine and carboxylic acid functional groups; plant and animal cells use 20 fundamental amino acids to create proteins.

peptide bond The carbon–nitrogen bond that connects two amino acids together; peptide bonds are formed by the condensation of a carboxylic acid and an amine to form an amide.

deoxyribonucleic acids (DNA) Massive molecules containing the genetic code of living creatures.

Additional Problems

15.1 Organic Chemistry and the Carbon Cycle

11. What are the two chemical products from cellular respiration?

12. What are the two chemical products from photosynthesis?

13. Is photosynthesis an endothermic or exothermic process? What about cellular respiration?

14. Write two chemical reactions to show how carbon dioxide reacts in water to produce the bicarbonate ion.

15. Describe a pathway in the carbon cycle by which carbon dioxide in the atmosphere becomes part of a coal deposit deep underground.

16. Describe a pathway in the carbon cycle by which the carbon atoms in a squirrel eventually become part of the carbon atoms in a tree.

17. Describe a pathway in the carbon cycle by which carbon dioxide from the atmosphere becomes trapped in the sediment on the ocean floor.

18. Describe two different pathways in the carbon cycle through which carbon atoms in a tree become part of the carbon dioxide in the atmosphere.

15.2 Covalent Bonding with Carbon and Other Nonmetals

19. Indicate the number of covalent bonds typically formed by each of these nonmetal atoms:

a. C b. N c. O d. F

20. Indicate the number of covalent bonds typically formed by each of these nonmetal atoms:

a. Si b. P c. S d. Cl

21. Draw proper Lewis structures for these compounds:

a. CH_4 b. CCl_4 c. C_2H_5Cl

22. Draw proper Lewis structures for these compounds:

a. CF_4 b. CH_2Cl_2 c. C_3H_8

23. Draw proper Lewis structures for the following compounds. Use double and triple bonds as needed.

a. C_2H_4 b. C_2H_4O c. C_2H_5N d. CHN

24. Draw proper Lewis structures for the following compounds. Use double and triple bonds as needed.

a. C_2H_2 b. C_3H_4 c. CH_2O d. HNO

25. Three of the following four structures have identical bonding sequences. Which of these compounds is not like the others—that is, which one is an isomer of the other three?

```
a.   H  H  F          b.  H    H
     |  |  |               \   |
  H—C——C——C—H           H—C——C—F
     |  |  |               /   |
     H  H  H              H    C
                             / | \
                            H  H  H

c.   H    H           d.   H    H
     |   /                 |   /
  H—C——C—H              H—C——C—H
     |   \                 |   \
     C    H                C    H
   / | \                 / | \
  H  H  F               F  H  H
```

26. Three of the following four structures have identical bonding sequences. Which of these compounds is not like the others—that is, which one is an isomer of the other three?

```
a.   H  H  Cl         b.  Cl    H
     |  |  |                \   |
  H—C——C——C—H            H—C——C—H
     |  |  |                /   |
     Cl H  H               H    C
                              / | \
                             H  H  Cl

c.   H    Cl          d.   Cl   H
     |   /                 |   /
  H—C——C—H              H—C——C—H
     |   \                 |   \
     C    H                C    Cl
   / | \                 / | \
  H  H  Cl              H  H  H
```

27. Two isomeric structures have the molecular formula C_2H_6O. Draw these two isomers.

28. Draw the four possible isomers having the structure C_4H_9Br.

29. Draw the three possible isomers having the molecular formula C_3H_8O.

30. Draw the four possible isomers having the structure C_3H_9N.

15.3 Drawing Covalent Structures

31. Show these condensed structures as Lewis structures:

a. $CH_3CH_2OCH_3$

b. $CH_3CH_2CH(CH_3)_2$

c. $CH_3(CH_2)_4CHCl_2$

32. Show these condensed structures as Lewis structures:

a. $H_2NCH_2CH_2OH$

b. $CH_3CH(OH)CH_3$

c. $CH_3(CH_2)_2CHBrCH_3$

33. Redraw these structures as condensed structures:

a.

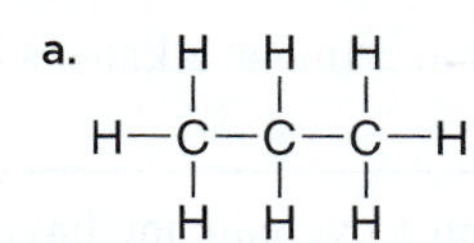

b.

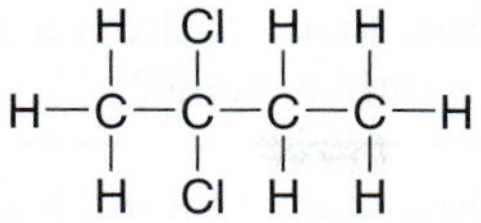

34. Redraw these structures as condensed structures:

a.

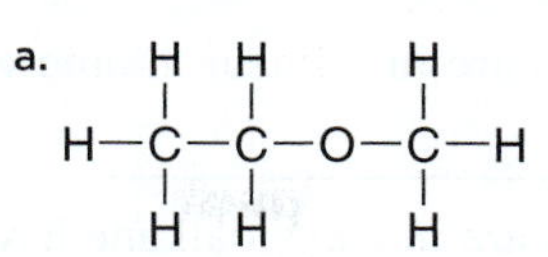

b.

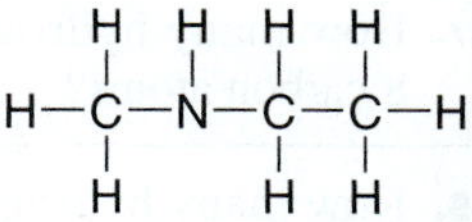

35. Write the molecular formula for each of these skeletal structures:

a.  b. 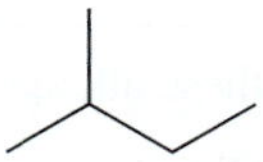 c.

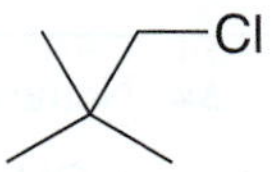

36. Write the molecular formula for each of these skeletal structures:

a. 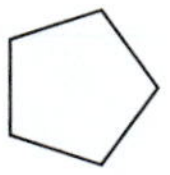 b.  c.

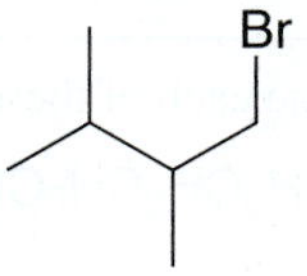

37. Write the molecular formula for each of these skeletal structures:

a. 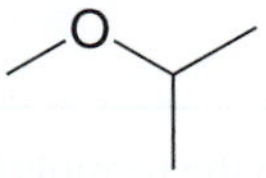 b. 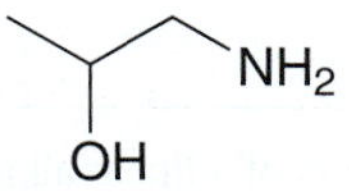 c.

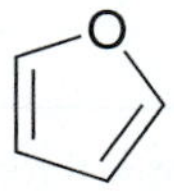

38. Write the molecular formula for each of these skeletal structures:

a. 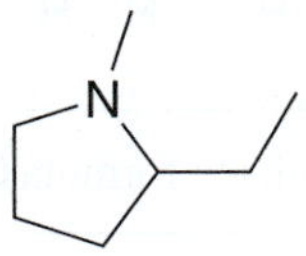 b.  c.

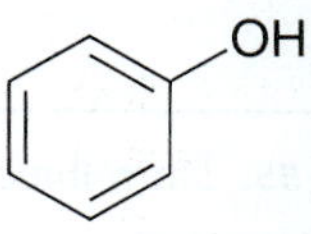

39. Draw these Lewis structures as skeletal structures:

a. 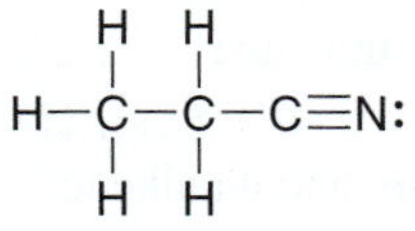 b. 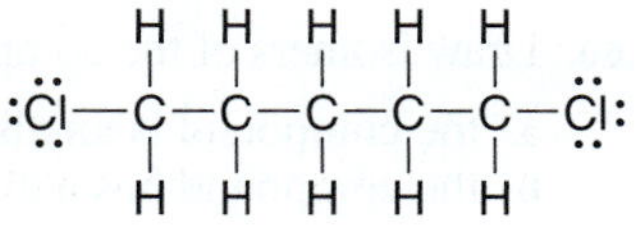 c.

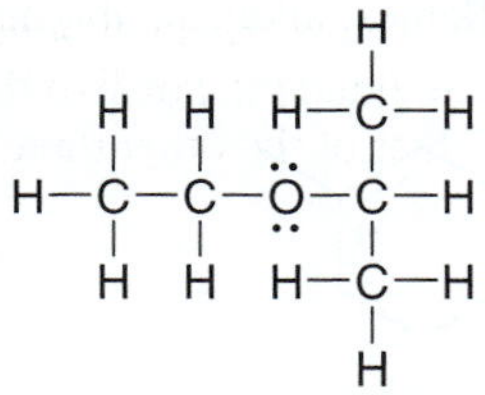

40. Draw these Lewis structures as skeletal structures:

a. 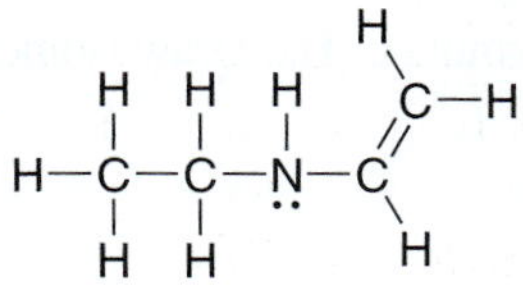 b. 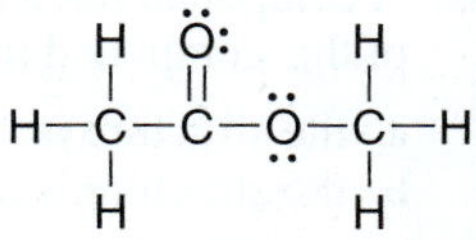 c.

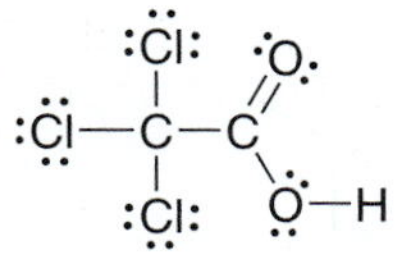

41. Draw these skeletal structures as Lewis structures:

a.  b. 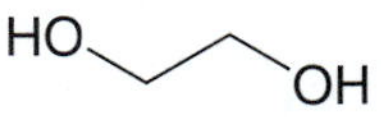 c.

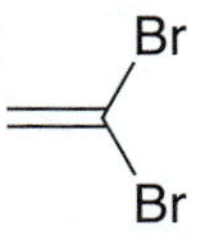

42. Show these skeletal structures as Lewis structures:

a. 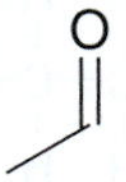 b. 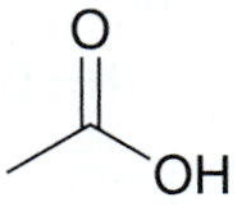 c.

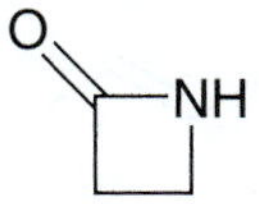

43. Draw these condensed structures as Lewis structures and as skeletal structures:

a. $CH_3CH_2NHCH_3$
b. CH_2CHCH_2OH
c. CH_3CH_2CN

44. Draw these condensed structures as Lewis structures and as skeletal structures:

a. CF_3CHCH_2
b. $(CH_3CH_2)_2NH$
c. $CHC(CH_2)_3Cl$

45. The skeletal structure below belongs to a class of compounds called penicillins, which were the first widely used antibiotics. What is the molecular formula of this compound?

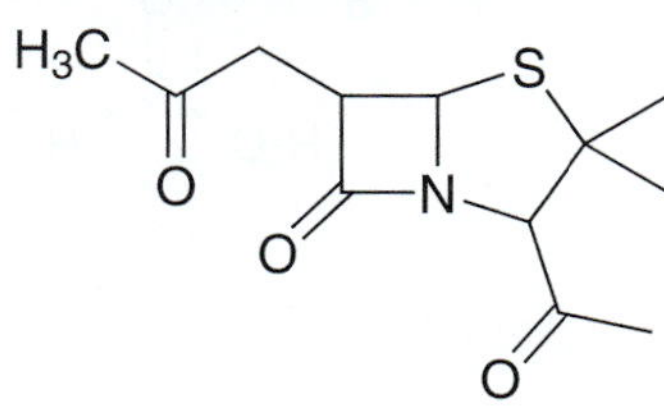

46. Nicotine is the highly addictive substance found in tobacco. The skeletal structure of nicotine is shown below. What is the molecular formula for this compound?

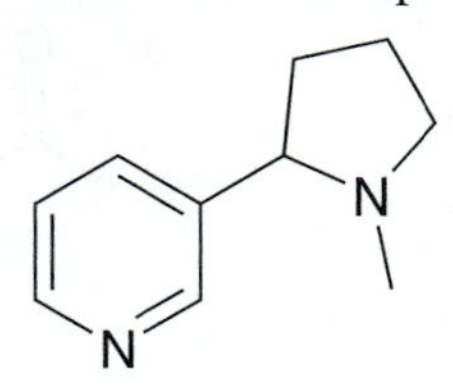

15.4 Major Functional Groups

47. How many hydrogen atoms are in a linear alkane with 8 carbon atoms?

48. How many hydrogen atoms are in a linear alkane with 12 carbon atoms?

49. How many hydrogen atoms are in a cycloalkane having 4 carbon atoms?

50. How many hydrogen atoms are in a cycloalkane having 8 carbon atoms?

51. What are the molecular formulas for pentane and for cyclopentane? Draw a skeletal structure for each of these molecules.

52. What are the molecular formulas for octane and for cyclooctane? Draw a skeletal structure for each of these molecules.

53. Name each of these alkanes:

a. $CH_3CH_2CH_2CH_3$

b.

c.

54. Name each of these alkanes or cycloalkanes:

a. $CH_3(CH_2)_5CH_3$

b.

c.

55. Draw three isomers of a linear alkane with the formula C_6H_{14}.

56. Draw three isomers of a linear alkane with the formula C_7H_{16}.

57. Draw three isomers of a cycloalkane with the formula C_7H_{14}. (*Hint*: Try different ring sizes.)

58. Draw three isomeric cycloalkanes having the formula C_8H_{16}. (*Hint*: Try different ring sizes.)

59. A compound has a molecular formula C_3H_6. Draw isomers of this compound in which

a. the structure is a cycloalkane.

b. the structure is an alkene.

60. Draw isomers of the compound C_5H_8 in which

a. the compound is an alkyne.

b. the compound has a cyclic structure and an alkene.

61. Identify these hydrocarbons as alkanes, alkenes, alkynes, or aromatic compounds:

a.

b.

c.

d.

62. Identify these hydrocarbons as alkanes, alkenes, alkynes, or aromatic compounds:

a.

b.

c.

d.

63. Identify the oxygen-containing functional groups in these small molecules:

a.

b.

c.

d.

64. Identify the oxygen-containing functional groups in these small molecules:

a. $CH_3CH_2CO_2H$

b.

c.

d. H_3C–C(=O)–H

65. Draw a structure having the formula C_4H_8O that contains:

a. an alcohol group
b. an aldehyde group
c. a ketone group
d. an ether group

66. Draw a structure having the formula $C_3H_6O_2$ that contains:

a. a carboxylic acid group
b. an ester group
c. both a ketone and an alcohol

67. The structure below is that of ascorbic acid, better known as vitamin C. Identify the major functional groups present in this molecule.

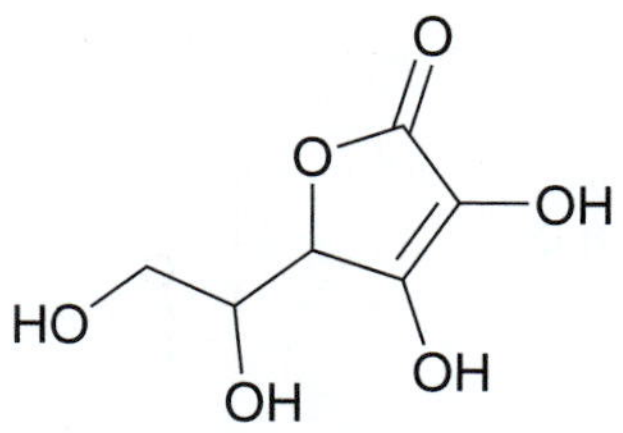

68. "Agent Orange" was an herbicide used in the Vietnam War to remove foliage that shielded enemy positions. The compound shown below made up 50% of the Agent Orange mixture. Identify the major functional groups present in this molecule.

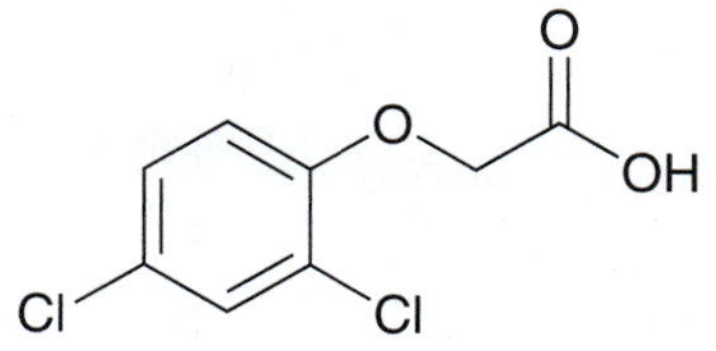

69. Identify the nitrogen-containing functional groups in these small molecules:

a.

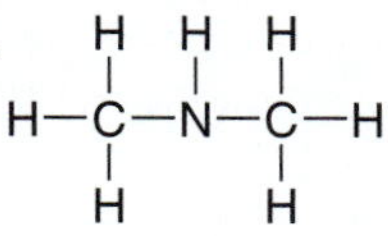

b.

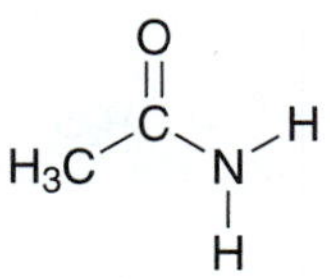

c.

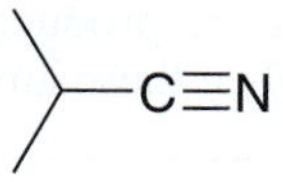

70. Identify the nitrogen-containing functional groups in these small molecules:

a.

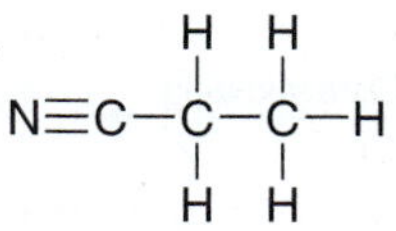

b.

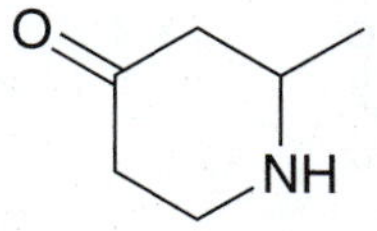

c.

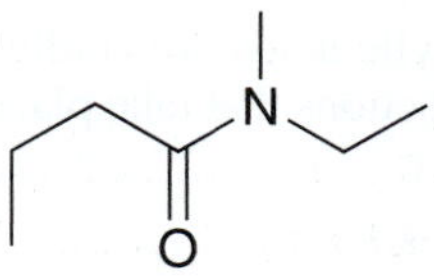

71. Albuterol is commonly used in breathing treatments to treat conditions such as bronchitis and asthma. Identify the major functional groups present in this molecule.

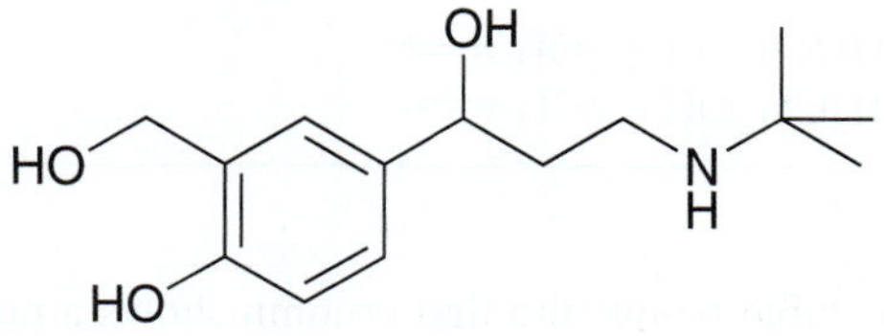

72. Strychnine is an infamous poison. This naturally occurring compound was extracted from certain shrubs and used for many centuries to kill animal pests. History records many cases of murder or attempted murder by strychnine poisoning; some have even postulated that this poison caused the death of Alexander the Great. The elaborate structure of this compound is shown below. What functional groups are present in this poison?

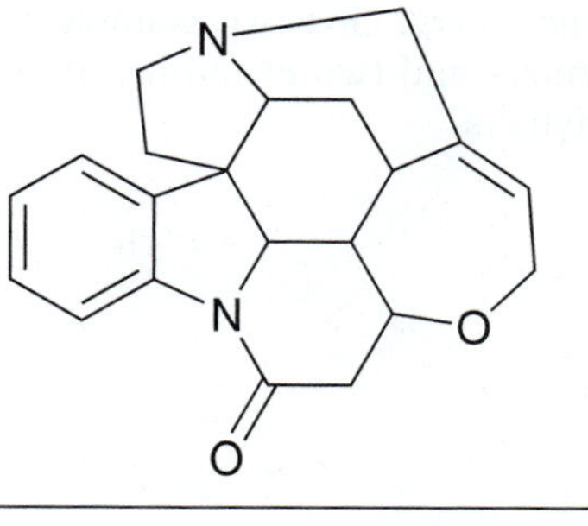

73. Draw the structure of the ether produced in this condensation reaction:

$CH_3CH_2CH_2OH$ + CH_3OH $\xrightarrow{\text{Catalytic acid}}$

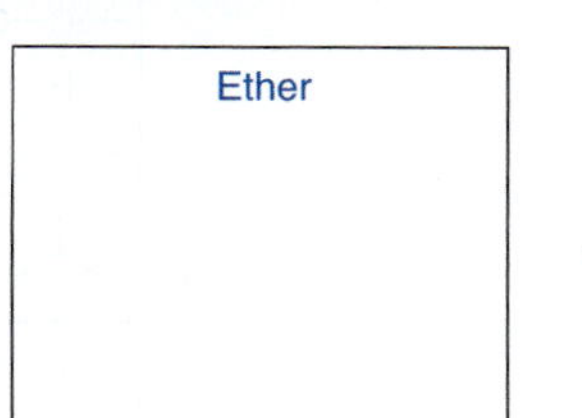

+ H_2O

74. Draw the structure of the ester produced in this condensation reaction:

$$CH_3CH_2\overset{O}{\overset{\|}{C}}OH + CH_3OH \xrightarrow{\text{Catalytic acid}} \boxed{\text{Ester}} + H_2O$$

75. Draw the structure of the amide produced in this condensation reaction:

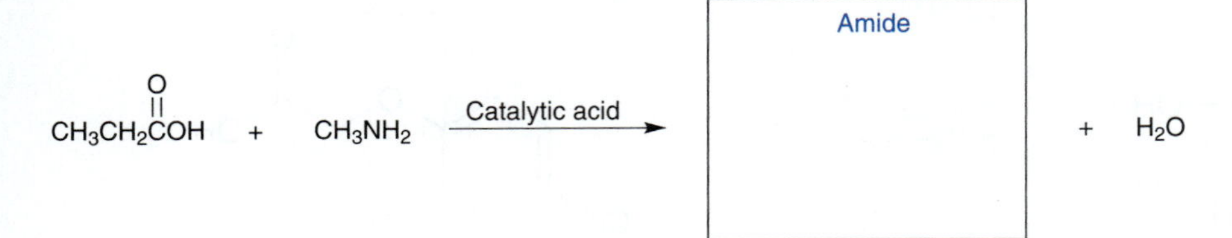

76. Draw the structure of the amide produced in this condensation reaction:

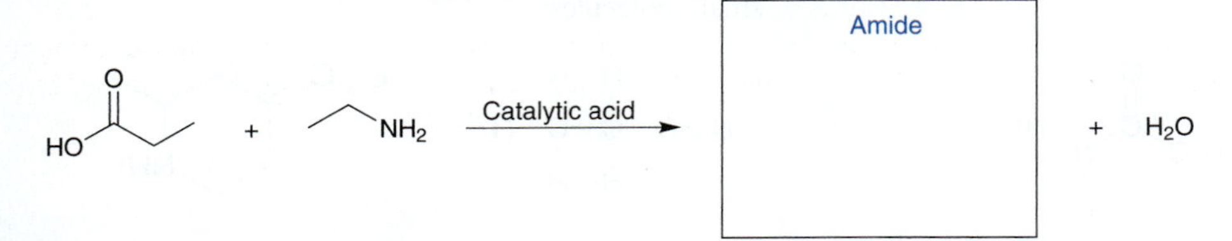

77. As their name implies, carboxylic acids are acidic. Show the products of the acid-base reactions that take place when these acids are mixed with water:

a. $CH_3CO_2H + H_2O \rightleftharpoons$
b. $CH_3CH_2CO_2H + H_2O \rightleftharpoons$

78. Amines are basic compounds. Show the products of the acid-base reaction that takes place when these amines are mixed with water:

a. $CH_3NH_2 + H_2O \rightleftharpoons$
b. $(CH_3)_3N + H_2O \rightleftharpoons$

79. Show the products from the acid-base reactions below. Remember that carboxylic acids are acidic and amines are basic.

a. $HCO_2H + NH_3 \rightleftharpoons$
b. $CH_3CO_2H + CH_3NH_2 \rightleftharpoons$

80. Show the products from the acid-base reactions below. In these reactions, the alcohols behave as acids, while the amines behave as bases.

a. $CH_3NH_2 + CH_3OH \rightleftharpoons$
b. $CH_3CH_2OH + NH_3 \rightleftharpoons$

15.5 Polymers and Plastics

81. What are polymers? Give two examples of naturally occurring polymers and two examples of synthetic (human-made) polymers.

82. In the table below, the first column shows a portion of a polymer structure. In the second column, draw a representation that shows the repeating unit of the polymer in parentheses. The first one is done as an example.

Partial Polymer Structure	Representation
H CH3 H CH3 H CH3 H CH3 H CH3 —C—C—C—C—C—C—C—C—C—C— H Cl H Cl H Cl H Cl H Cl	H CH3 (—C—C—)$_n$ H Cl
H H H H H H H H H H —C—C—C—C—C—C—C—C—C—C— H Br H Br H Br H Br H Br	
H H H H H H H H —C—C—O—C—C—O—C—C—O—C—C—O— H CH3 H CH3 H CH3 H CH3	

15.6 Biomolecules—An Introduction

83. What three elements are present in a carbohydrate?

84. What is the major function of carbohydrates in living structures?

85. What is the difference between alpha-glucopyranose and beta-glucopyranose?

86. What is the major difference between cellulose and starch?

87. Sucrose, or table sugar, is a disaccharide formed by the combination of two simple sugars, glucose and fructose, through a condensation reaction. The two alcohols that condense are highlighted in the structure below. Draw the structure of the sucrose molecule that results from this condensation.

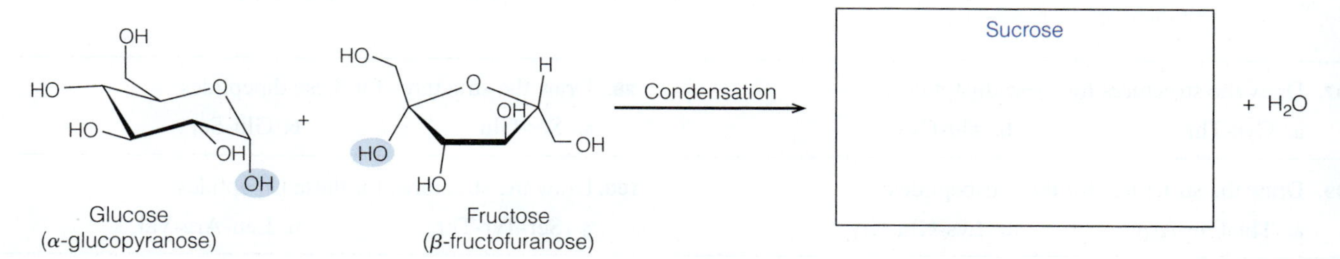

88. Lactose, or milk sugar, is a disaccharide formed by the combination of two simple sugars, galactose and glucose, through a condensation reaction. The two alcohols that condense are highlighted in the structure below. Draw the structure of the lactose molecule that results from this condensation.

OH OH HO O OH OH + HO OH O HO OH OH — Condensation → Lactose $+ H_2O$

Galactose
(β-glucopyranose)

Glucose
(β-glucopyranose)

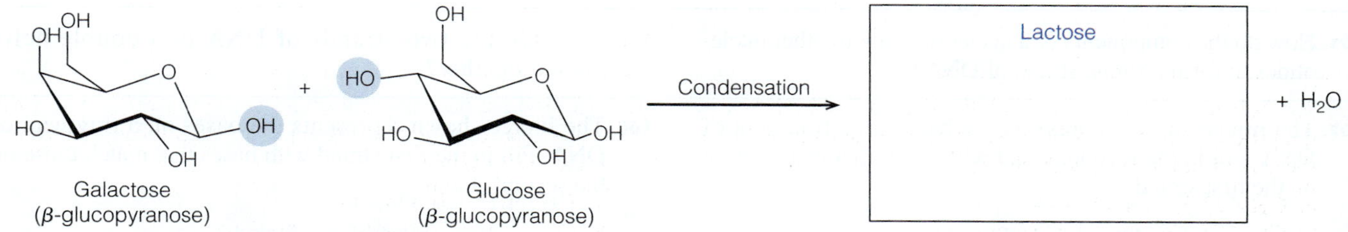

89. What types of roles do proteins play in living organisms?

90. What two functional groups are present in every amino acid?

91. How many of the amino acids in Table 15.4 have a benzene ring in their side chain?

92. How many of the amino acids in Table 15.4 have a second amine group in their side chain?

93. A tripeptide contains serine (the N-terminus), connected to proline, connected to arginine (the C-terminus). Give the representation of this structure using the three-letter amino acid abbreviations. Refer to Table 15.4 as needed.

94. A tripeptide contains lysine (the N-terminus), connected to glutamic acid, connected to isoleucine (the C-terminus). Give the representation of this structure using the three-letter amino acid abbreviations. Refer to Table 15.4 as needed.

95. Use the three-letter amino acid abbreviations to represent the tripeptide sequences shown here:

a. $H_3\overset{\oplus}{N}$—CH(—CH_2—CH_2—S—CH_3)—C(=O)—N(H)—CH(—H)—C(=O)—N(H)—CH(—CH(H_3C)(CH_2CH_3))—C(=O)—$O^{\ominus}$

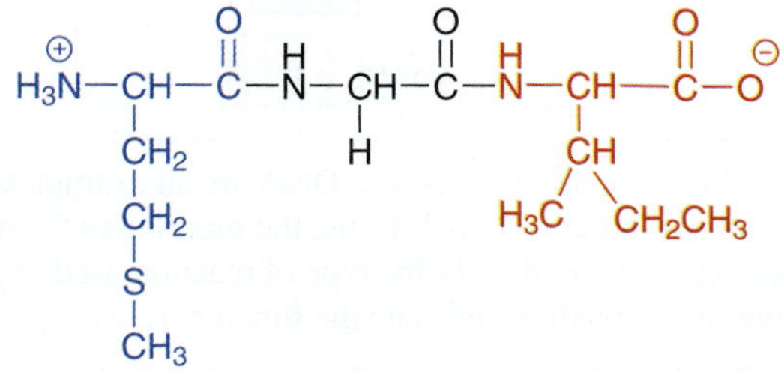

b. $H_3\overset{\oplus}{N}$—CH(—CH_2—SH)—C(=O)—N(H)—CH(—CH_2—C₆H₅)—C(=O)—N(H)—CH(—CH_2—OH)—C(=O)—$O^{\ominus}$

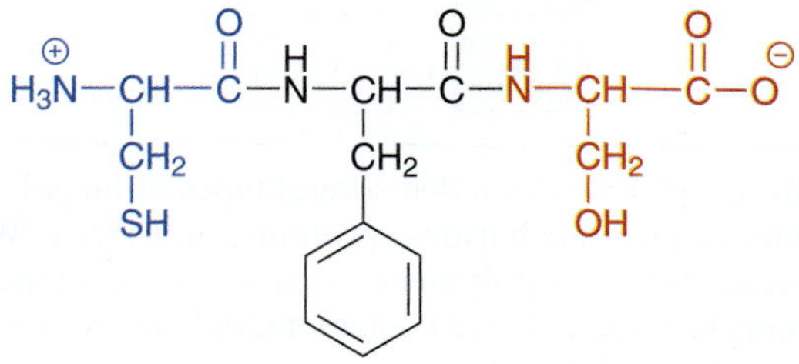

96. Use the three-letter amino acid abbreviations to represent the tripeptide sequences shown here:

a. $H_3\overset{\oplus}{N}-CH(CH_2CH_2C(=O)OH)-C(=O)-NH-CH(CH_2C(=O)OH)-C(=O)-NH-CH(CH_2CH_2C(=O)NH_2)-C(=O)-O^{\ominus}$

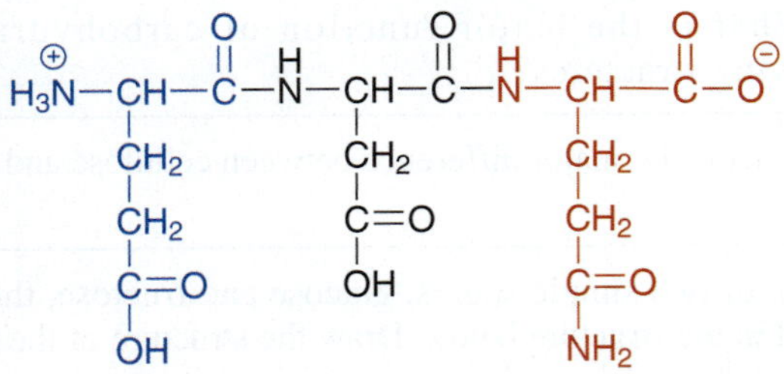

b. $H_3\overset{\oplus}{N}-CH(CH_2CH_2CH_2CH_2NH_2)-C(=O)-NH-CH(CH(OH)CH_3)-C(=O)-NH-CH(CH_2CH(CH_3)CH_3)-C(=O)-O^{\ominus}$

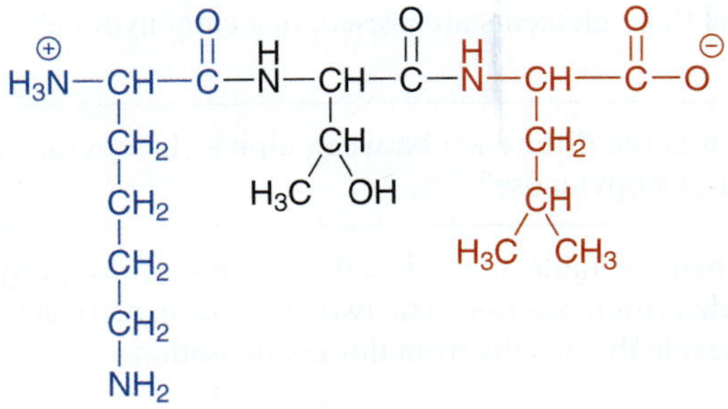

97. Draw the structures for these dipeptides:

a. Cys-Thr b. Thr-Cys

98. Draw the structures for these dipeptides:

a. Ser-Glu b. Glu-Ser

99. Draw the structures for these tripeptides:

a. Thr-Phe-Asp b. Lys-Gln-Gly

100. Draw the structures for these tripeptides:

a. Ser-Tyr-Thr b. Leu-Arg-Val

101. What types of interactions cause peptides to fold into different shapes?

102. Broadly, how does the amino acid sequence determine the properties of a protein?

103. Why is DNA important within an organism?

104. What are the three components of a nucleotide?

105. How do the components of a nucleotide link to other nucleotides to form a single strand of DNA?

106. How are the two strands of DNA in a double helix bound together?

107. The image shown represents the bases on a fragment of DNA. Fill in the second strand with bases that match those on the first strand.

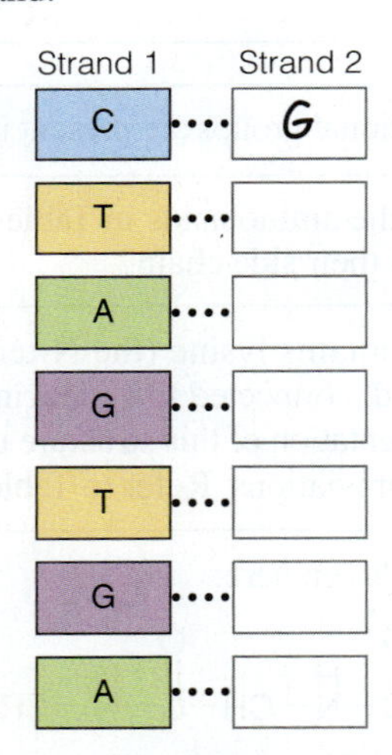

108. The image shown represents the bases on a fragment of DNA. Fill in the first strand with bases that match those on the second strand.

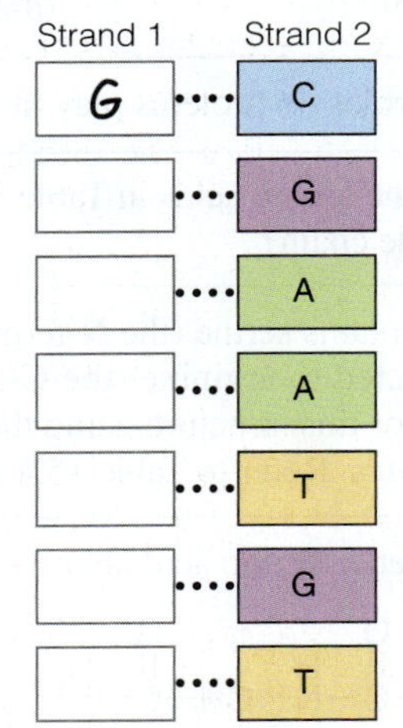

109. In this chapter we discussed several types of biopolymers, including polysaccharides, proteins, and DNA. Which of these classes of polymers contain only one repeating monomer? Which classes contain multiple monomers with information coded in the monomer sequence?

110. Cellulose, starch, proteins, and DNA are all examples of biopolymers. In the table below, list the monomers from which each one is built, identify the type of reaction used to join the monomers together, and state the function of each polymer.

Biopolymer	Monomer	Reaction That Joins Monomers	Function
Starch		Condensation	
Cellulose			
Protein			
DNA	Nucleotides		

Nuclear Chemistry

CHAPTER SIXTEEN

Figure 16.1 (a) On March 11, 2011, an earthquake struck northern Japan. The earthquake triggered a tsunami that slammed the coastline. (*continued*)

Fukushima

On March 11, 2011, an earthquake struck the Tōhoku region of northern Japan. Centered about 70 kilometers off Japan's east coast, the quake produced a tsunami — an immense wave over 40 meters high. Less than an hour after the earthquake, the tsunami slammed into the coastal areas, destroying nearly everything in its path and killing over 15,000 people (**Figure 16.1**).

Along the coast, nuclear power plants produced energy for the region. The plants were engineered to survive an earthquake, but they were not adequately prepared for the tsunami. As the waters rose around one of the power facilities — the Fukushima Daiichi plant — a new disaster began to unfold.

The heart of a nuclear power plant is the *reactor core*. The atomic changes that take place in the core generate huge amounts of heat energy, but they also produce very toxic by-products. Nuclear plants continuously pump cooling water through the core. Both the core and cooling water are sealed inside a central reactor building.

When the tsunami hit Fukushima Daiichi, rapidly rising water knocked out power to the cooling pumps as well as multiple backup systems. Inside the reactor building, temperatures and pressures began to rise. Engineers battled to restore cooling systems, while authorities evacuated the surrounding cities. The temperatures continued to rise. Steam and hydrogen gas began to fill the reactor building. The ultrahot nuclear fuel melted through the containment vessel, carving a path deep into the building's foundation. Then the hydrogen in the reactor exploded, ripping the top off the concrete building and spewing toxins into the air.

To cool the toxic fuel, emergency workers flooded the structures with water. The fuel was so hot that water boiled out of the buildings for weeks following the accident, releasing steam and toxic matter into the atmosphere. Cracks in the foundation released contaminated water into the sea. It was nine months before the fuel finally cooled, and the release of toxins subsided.

Figure 16.1 (continued) (b) The epicenter of the earthquake was about 70 km from the coast of Japan. (c) The Fukushima plant before the disaster: The four square buildings each housed a reactor core. (d) After the accident, emergency workers wore special protective gear. (e) A volunteer stands beside a radiation meter near the Fukushima site, four years after the disaster. The meter alerts residents to health risks resulting from the accident. (a) Handout/Reuters/Newscom; (c) AFLO/Newscom/AFLO photos/Fukushima Japan; (d) Sankei/Getty Images; (e) Christopher Jue/EPA/Shutterstock

The Fukushima accident stirred fears around the world. In Tokyo, reports of water contamination sparked a run on bottled water. Within days of the accident, the United States Food and Drug Administration blocked the import of some foods from Japan. Germany shut down half of its nuclear plants and drastically changed its long-term energy policies.

Seven years after the accident, contamination around the Fukushima site is still a problem, and many people have not returned to their homes. Yet despite global panic over the Fukushima meltdown, there were no immediate deaths from radiation exposure. In fact, very few people even got sick.

The events at Fukushima raise many important questions: What are the real risks of nuclear energy? What are the benefits? What level of exposure to these materials is acceptable? What levels are harmful?

In this chapter, we'll survey the key concepts of nuclear chemistry. We'll see how fundamental changes to atomic structure occur, and why structural changes are accompanied by such enormous changes in energy. From the dramatic and massive reactions of the stars to small and gradual changes on Earth, we'll explore many of the naturally occurring nuclear reactions that affect our lives. We'll also explore applications of nuclear chemistry, from commercial power generation to medicine to archaeology, and some of the public health and safety issues that arise when humans come in contact with the powerful forces stored inside the nucleus of the atom.

Intended Learning Outcomes

After completing this chapter and working the practice problems, you should be able to:

16.1 Nuclear Changes

- Write balanced nuclear equations to describe nuclear changes.

16.2 Radioactivity

- Define the different types of radioactive decay, and predict the products formed by these decays.
- Describe the rates of radioactive decay using half-lives.

16.3 Working with Radiation

- Describe how radioactive decay and human exposure are measured.
- Qualitatively describe the use of radioactive nuclides in medicine, geology, and archaeology.

16.4 Energy Changes in Nuclear Reactions

- Describe the interconversion of matter and energy in nuclear changes.
- Describe how mass defect and binding energy relate to the stability of a nucleus.

16.5 Nuclear Power: Fission and Fusion

- Describe the processes of nuclear fission and fusion.
- Broadly describe the risks, benefits, and technical challenges associated with the use of fission and fusion for power generation.

16.1 Nuclear Changes

We first looked at atomic structure in Chapter 3. Recall that atoms are composed of a very dense nucleus that contains protons and neutrons, surrounded by a cloud of negatively charged electrons (**Figure 16.2**).

Most everyday chemistry involves the electrons: Covalent and ionic bonds, acid-base behavior, and oxidation-reduction reactions all involve changes in electron structure. Atoms may gain, lose, or share electrons, but the atomic nuclei remain unchanged.

Nuclear chemistry describes a different type of change—those involving the nucleus of an atom. Nuclear changes involve massive amounts of energy. The heat produced by the Sun and the stars, the very hot temperatures of Earth's core, and the devastating power of the atomic bomb arise from changes in the atomic nucleus. Although most of the chemistry that occurs around us is nonnuclear in nature, nuclear changes are constantly occurring, and they are essential to life on Earth.

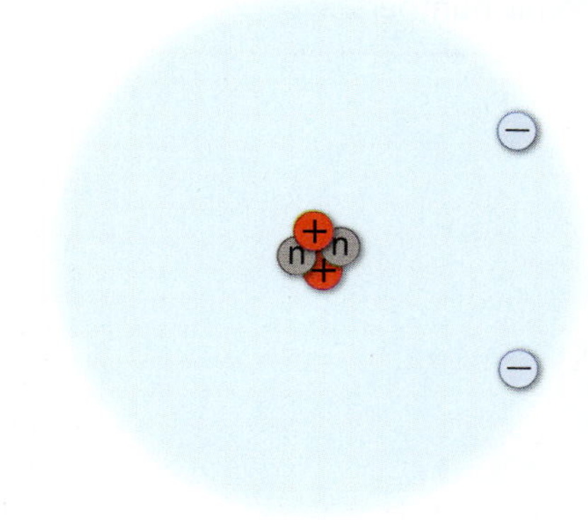

Figure 16.2 A nucleus is composed of protons and neutrons. The nucleus is surrounded by the electron cloud.

The Nucleus — A Review

Atoms are defined by the number of protons and neutrons in their nucleus. Collectively, protons and neutrons are called **nucleons**. Recall that the number of protons is the *atomic number*: This number gives the atom its unique properties. The *mass number* is the number of nucleons (the sum of protons and neutrons) in a particular nucleus. When writing atomic symbols, we show the mass number on the upper left-hand side of the atomic symbol. The atomic number is shown on the lower left-hand side (**Figure 16.3**).

A nucleon is a proton or a neutron.

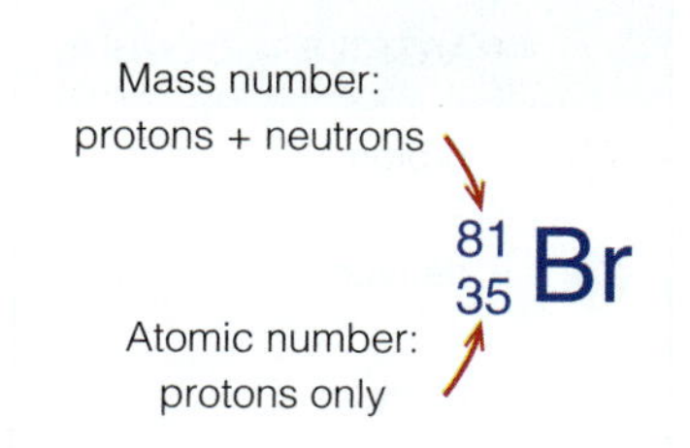

Figure 16.3 The mass number is written at the upper left of the symbol. The atomic number is written at the lower left.

An atom or nucleus containing a particular number of protons and neutrons is referred to as a **nuclide**. Nuclides that have the same atomic numbers are called **isotopes**. That is, isotopes are the same type of atom, but with different mass numbers. For example, lithium has two nuclides: ^{6}Li and ^{7}Li. These two nuclides are isotopes. The symbol ^{6}Li is read as "lithium-6," and it is commonly written this way as well. We will use both styles to represent nuclides in this chapter.

Example 16.1 Identifying Isotopes

Which of the following nuclides is an isotope of $^{11}_{5}B$?

a. *a nuclide with mass number 11 and atomic number 6*

b. *a nuclide with mass number 10 and atomic number 5*

c. *a nuclide with mass number 12 and atomic number 6*

Isotopes have the same atomic number but different mass numbers. Therefore answer (b) is correct.

Nuclides and Isotopes

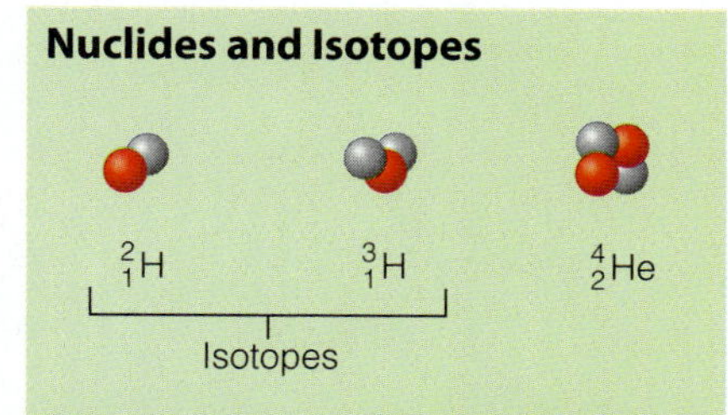

Three unique nuclides are shown here. The two with the same atomic number are isotopes.

Nuclear Reactions

In a **nuclear reaction**, the structure of an atomic nucleus changes. There are many types of nuclear reactions: Sometimes nuclei spontaneously decompose, releasing energy and smaller particles. Other reactions produce larger nuclei: For example, in the upper atmosphere, high-energy particles from the Sun collide with oxygen and nitrogen, transforming these atoms into different elements. In some laboratories, scientists emulate this process by firing high-energy particles into target elements to produce new nuclides for use in medicine. The heat from the Sun, the heat within Earth's core, and the energy produced from nuclear power plants all arise from nuclear changes. We'll look at these different changes in the sections that follow.

We often describe nuclear reactions using **nuclear equations**. In a balanced nuclear equation, the total mass numbers and the total atomic numbers are the same

Nuclear equations usually show both the mass numbers and the atomic numbers. ■

on both sides of the equation. For example, bismuth-212 slowly decomposes to form two new nuclides: thallium-208 and helium-4. We can write a balanced nuclear equation for this process as follows:

$$^{212}_{83}\text{Bi} \rightarrow {}^{208}_{81}\text{Tl} + {}^{4}_{2}\text{He}$$

Notice in this nuclear equation that the total atomic number (that is, the total number of protons) is the same on the left- and right-hand sides of the equation. Likewise, the total mass number is also the same on both sides of the equation (**Figure 16.4**).

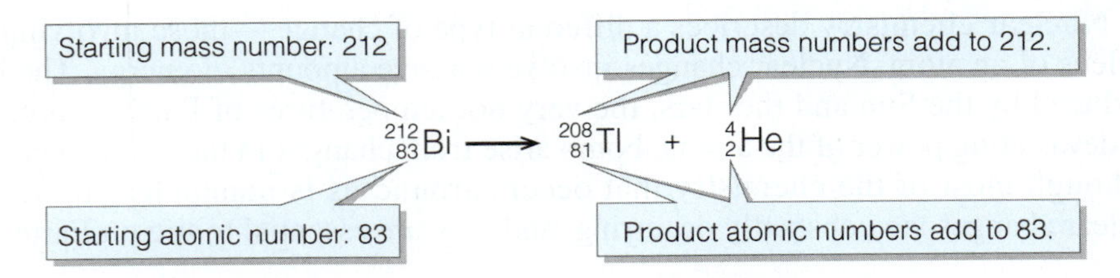

Figure 16.4 In a balanced nuclear equation, the sums of the mass numbers and the atomic numbers are the same on both sides of the equation.

In many nuclear equations, a nucleus gains or loses a fundamental atomic particle. In these cases, we use the symbol $^{1}_{1}\text{p}$ to represent a proton, $^{1}_{0}\text{n}$ to represent a neutron, and $^{0}_{-1}\text{e}$ to represent an electron (**Table 16.1**). For example, some medical tests use the nuclide indium-111 (^{111}In). Scientists produce this isotope in a laboratory using a device called a *cyclotron* (**Figure 16.5**). Using magnetic fields, this device accelerates a proton to very high speeds and then smashes it into a cadmium-111 nucleus. The result of this collision is a nucleus of indium-111 and a free neutron. We can depict the nuclear change using this equation:

$$^{1}_{1}\text{p} + {}^{111}_{48}\text{Cd} \rightarrow {}^{111}_{49}\text{In} + {}^{1}_{0}\text{n}$$

When writing nuclear equations, we can sometimes predict one product from the other products. Example 16.2 illustrates how this is done.

TABLE 16.1 Nuclear Symbols for Subatomic Particles

PARTICLE	SYMBOL
proton	$^{1}_{1}\text{p}$
neutron	$^{1}_{0}\text{n}$
electron	$^{0}_{-1}\text{e}$

Alain DENANTES/Gamma-Rapho via Getty Images

Figure 16.5 This cyclotron uses magnetic fields to accelerate charged particles. The collision of these high-energy particles results in a nuclear reaction, producing new isotopes for medicinal applications.

Example 16.2 Completing a Nuclear Equation

When molybdenum-100 is bombarded with a proton, it produces a new nuclide and emits two neutrons, as shown in the equation below. Determine the atomic number, mass number, and atomic symbol for the unidentified product.

$$^{1}_{1}\text{p} + ^{100}_{42}\text{Mo} \rightarrow ^{1}_{0}\text{n} + ^{1}_{0}\text{n} + \_\_\_\_\_$$

The total mass numbers on the left-hand side of the equation sum to 101. Because the right-hand side of the equation has two neutrons (each with a mass number of 1), the unknown mass number must be 99. Similarly, the atomic numbers on the left-hand side sum to 43. Since the neutrons have an atomic number of zero, the unknown atomic number must be 43. Atomic number 43 corresponds to the element technetium (Tc). Based on this, we can write a complete balanced equation.

$$^{1}_{1}\text{p} + ^{100}_{42}\text{Mo} \rightarrow ^{1}_{0}\text{n} + ^{1}_{0}\text{n} + ^{99}_{43}\text{Tc}$$

TRY IT

1. In Earth's upper atmosphere, a high-energy neutron collides with a nitrogen-14 nucleus, forming a carbon-14 nucleus and releasing one proton. Describe this process using a balanced nuclear equation.
2. When a proton collides with an ^{127}I nucleus in a cyclotron, it forms a new nucleus, ^{123}Xe, along with several neutrons. How many neutrons are released in this process? Write a balanced nuclear equation to describe this process.

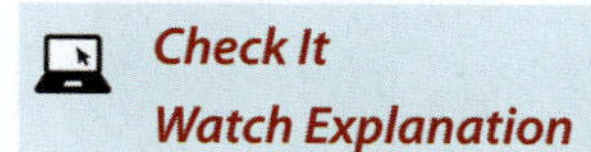

16.2 Radioactivity

Some of the most common nuclear reactions are **radioactive decays**. In these reactions, nuclei spontaneously transition to a more stable state. In some decays, nuclei eject particles from the nucleus. In other decays, the nucleus shifts slightly, releasing energy. The spontaneous release of particles and/or energy from the nucleus is called **radioactivity**. The energy released by these reactions is called *radiation*.

For example, the element radium slowly decomposes over thousands of years to form a new element, radon. Like bursting popcorn kernels, the radium atoms individually "pop," transforming into radon atoms (**Figure 16.6**). Each time an atom decays, the substance releases a small amount of energy. Because so many atoms are present, a sample of radium releases this energy over thousands of years.

The discovery of radioactivity was closely tied to the discovery of atomic structure introduced in Chapter 3. In 1897, Antoine Becquerel discovered that when he placed uranium near certain substances, such as zinc sulfide, the substances glowed. In the years that followed, Marie Curie and her husband Pierre (**Figure 16.7**) worked extensively to isolate and characterize radioactive substances. Through these experiments, they discovered the elements radium and polonium. Marie Curie was able to classify three types of radioactive decays, which she termed *alpha*, *beta*, and *gamma* radiation.

Ted Kinsman/Science Source

When radiation strikes zinc sulfide, it causes the compound to glow. After the discovery of radium in the early 1900s, trace amounts of this material were combined with zinc sulfide to create watch faces that glowed in the dark. Today, watch lights use much safer technologies.

Ted Kinsman/Science Source

Figure 16.6 This mineral contains the radioactive element radium. Radium releases particles and energy as it slowly decomposes to form a new element.

ullstein bild/Getty Images

Figure 16.7 Marie and Pierre Curie were jointly awarded the 1903 Nobel Prize in Physics. After Pierre's tragic death in 1906, Marie continued their work; in 1911 she received the Nobel Prize in chemistry.

Types of Radioactive Decay

Alpha Decay

In an **alpha (α) decay**, a nucleus ejects an *alpha particle* composed of two protons and two neutrons. An alpha particle is a helium nucleus. When a nucleus emits an alpha particle, its mass number decreases by four, and its atomic number decreases by two.

For example, when a $^{238}_{92}\text{U}$ atom undergoes an alpha decay (**Figure 16.8**), its mass number decreases by four, forming a new nuclide with a mass of 234. Its atomic number decreases by two, changing the atom from uranium (atomic number 92) to thorium (atomic number 90).

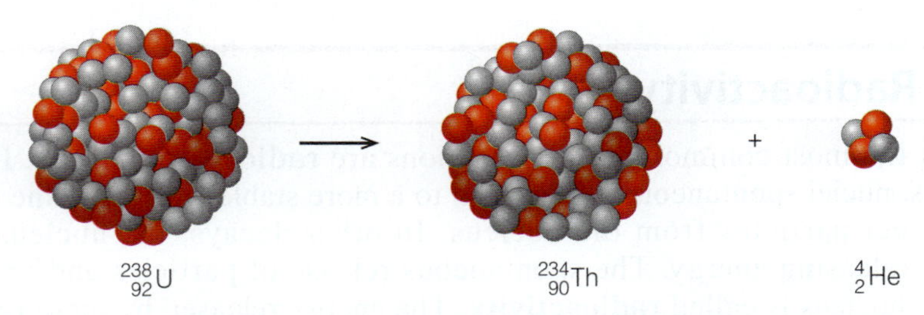

Figure 16.8 In an alpha decay, a larger nucleus ejects two protons and two neutrons (a helium nucleus) to form a smaller, more stable nucleus.

An alpha decay ejects $^{4}_{2}\text{He}$ from a nucleus. ■

We can represent this change using a nuclear equation:

$$^{238}_{92}\text{U} \rightarrow {}^{234}_{90}\text{Th} + {}^{4}_{2}\text{He}$$

Notice in this nuclear equation that the total mass numbers and the total atomic numbers are the same on both sides of the equation.

Example 16.3 Nuclear Equations and Alpha Decays

Write a nuclear equation showing the alpha decay of a ^{210}Po nucleus.

Polonium is atomic number 84. After undergoing an alpha decay, its mass decreases by four, and its atomic number decreases by two. The resulting nucleus has a mass number of 206 and an atomic number of 82. From the periodic table, we see that atomic number 82 is lead. Therefore, we can write the complete equation as follows.

$$^{210}_{84}\text{Po} \rightarrow {}^{206}_{82}\text{Pb} + {}^{4}_{2}\text{He}$$

Beta Decay

In a **beta (β) decay**, a neutron decays into two particles: a proton and an electron (**Figure 16.9**). The proton remains in the nucleus, while the electron is ejected. In a nuclear equation, we represent this as follows:

$$^{1}_{0}\text{n} \rightarrow {}^{1}_{1}\text{p} + {}^{0}_{-1}\text{e}$$

Neutron
Proton
Electron

Figure 16.9 In a beta decay, a neutron transforms into a proton and an electron.

Notice in this equation that our total mass numbers did not change (the tiny electron has a mass number of zero), and our total atomic number (that is, our total charge) also remained constant. The neutral particle split into two particles, one with a +1 charge and the other with a −1 charge.

When a nucleus undergoes a beta decay, the electron exits the nucleus, but the proton remains. The mass number of the nucleus does not change, but the atomic number increases by one. For example, radium-228 undergoes a beta decay to form actinium-228:

A beta decay increases the atomic number by one.

$$^{228}_{88}\text{Ra} \rightarrow {}^{228}_{89}\text{Ac} + {}^{0}_{-1}\text{e}$$

Example 16.4 illustrates this idea further.

Example 16.4 Nuclear Equations and Beta Decays

Write a nuclear equation showing the beta decay of thallium-209.

Thallium is atomic number 81. A beta decay increases the atomic number by one, but it doesn't change the mass number. The result is a nucleus with atomic number 82 (lead) and mass number 209. We can write the complete equation as follows.

$$^{209}_{81}\text{Tl} \rightarrow {}^{209}_{82}\text{Pb} + {}^{0}_{-1}\text{e}$$

Gamma Decay

In a **gamma (γ) decay**, a nucleus releases energy in the form of electromagnetic waves called *gamma rays*. Gamma decays do not involve the release of particles, so they do not change the atomic number or the mass number. A gamma decay may be thought of as the nuclear particles adjusting their arrangement to form a slightly more stable arrangement.

A gamma decay does not change the atomic number or mass number.

Radioactive Decay Series

Radioactive nuclei often decay multiple times before reaching a stable isotope. These decays occur through common sequences called a **decay series** or a *decay chain*. For example, radon is a naturally occurring gas that sometimes accumulates in low levels in basements or other low-lying areas. Radon-218 is a radioactive nuclide. Over time, the Rn nucleus undergoes a series of decays, forming several intermediates before eventually reaching a stable nucleus, lead-206. We can represent this series of decays as follows:

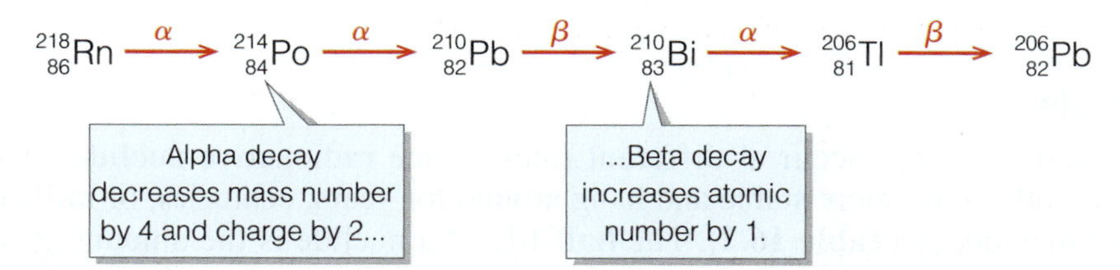
$^{218}_{86}\text{Rn} \xrightarrow{\alpha} {}^{214}_{84}\text{Po} \xrightarrow{\alpha} {}^{210}_{82}\text{Pb} \xrightarrow{\beta} {}^{210}_{83}\text{Bi} \xrightarrow{\alpha} {}^{206}_{81}\text{Tl} \xrightarrow{\beta} {}^{206}_{82}\text{Pb}$
Alpha decay decreases mass number by 4 and charge by 2...
...Beta decay increases atomic number by 1.

Three common decay series, known as the *thorium*, *uranium*, and *actinium series* (**Figure 16.10**), account for most of the abundant radioactive nuclides on Earth. The remaining naturally occurring radioactive nuclides result from *cosmic radiation*—high-energy particles from the Sun and other stars that react with elements on Earth to produce new radioactive nuclides.

In the 1980s, homes began to be tested for radon. These tests show elevated levels of radon in many basements and poorly ventilated areas.

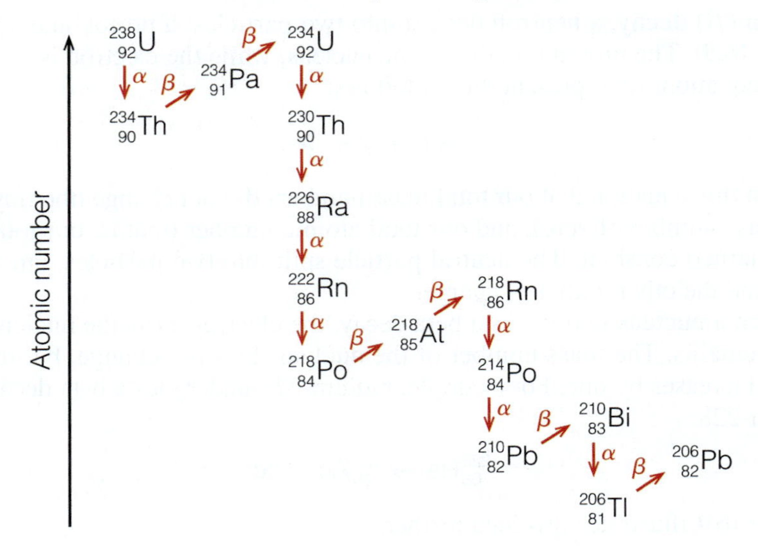

Figure 16.10 The uranium decay series begins with uranium-238 and ends with lead-206. This graphic shows one of several possible pathways.

Example 16.5 Predicting the Products from a Decay Series

Plutonium-241 is a radioactive substance produced in nuclear power plants. Over time it decays to form radium-225 through the sequence β, α, α, β, α, α. What intermediates are formed in this sequence of decays?

Each alpha decay decreases the mass number by four and the atomic number by two. In a beta decay the mass number is unchanged, but the atomic number increases by one. Based on this, we expect the mass numbers and atomic numbers to change as follows:

$$^{241}_{94}\text{Pu} \xrightarrow{\beta} {}^{241}_{95} \xrightarrow{\alpha} {}^{237}_{93} \xrightarrow{\alpha} {}^{233}_{91} \xrightarrow{\beta} {}^{233}_{92} \xrightarrow{\alpha} {}^{229}_{90} \xrightarrow{\alpha} {}^{225}_{88}$$

Using the periodic table, we can identify the atomic symbols that accompany each atomic number. This gives us a complete picture of the decay series, as shown.

$$^{241}_{94}\text{Pu} \xrightarrow{\beta} {}^{241}_{95}\text{Am} \xrightarrow{\alpha} {}^{237}_{93}\text{Np} \xrightarrow{\alpha} {}^{233}_{91}\text{Pa} \xrightarrow{\beta} {}^{233}_{92}\text{U} \xrightarrow{\alpha} {}^{229}_{90}\text{Th} \xrightarrow{\alpha} {}^{225}_{88}\text{Ra}$$

TRY IT

3. Write balanced equations showing these processes:

a. alpha decay of ^{40}Ca **b.** beta decay of ^{67}Cu

4. Radium-224 decays through the pathway α, α, α, β, α, β. What intermediates are formed in this decay series? What is the final stable nucleus formed?

Half-Life

Radioactive decays occur at different rates. Some radioactive nuclides decay quickly; others are more stable and hang around for years, centuries, or millennia before they decay (**Table 16.2**). The **half-life** of a nuclide is the amount of time required for one-half of a sample of a radioactive substance to decay into something else (**Figure 16.11**). For example, cesium-137 has a half-life of about 30 years. If you were given a 1.000-gram sample of cesium-137 today, in 30 years 0.500 grams

TABLE 16.2 Half-Lives of Common Radioactive Nuclides

Nuclide	Half-Life	Occurrence/Use
^{3}H	12.3 years	Energy (from the Sun)
^{10}Be	1.38 million years	Archaeology/geology
^{14}C	5,730 years	Archaeology
^{60}Co	5.3 years	Radiation therapy
^{99m}Tc*	6.0 hours	Medical imaging
^{125}I	59 days	Radiation therapy
^{131}Cs	9.7 days	Radiation therapy
^{222}Rn	3.8 days	Produced naturally from ^{238}U
^{238}U	4.5 billion years	Nuclear power
^{239}Pu	24,000 years	Nuclear power

*The "m" in the superscript means that this nuclide undergoes a gamma decay to form a more stable form of the same nuclide, ^{99}Tc.

would still be Cs-137—the rest would have decayed to form other substances. In 60 years (two half-lives), you would have 0.250 grams of Cs-137. In 90 years, the amount would be 0.125 g.

Consider another example: Iodine-131 is a radioactive nuclide that was released into the atmosphere during the Fukushima accident. However, iodine-131 has a short half-life of about 8 days. This means that 8 days after the initial release, one-half of the iodine-131 had decayed into something else. Mathematically, we can describe half-lives using the equation

$$N_t = N_0\left(\frac{1}{2}\right)^{\frac{t}{t_{1/2}}}$$

where N_t is the amount of a nuclide that remains, N_0 is the amount of nuclide originally present, t is the amount of time that has passed, and $t_{1/2}$ is the half-life of the nuclide. In this equation we can use any unit of measure for the amount, but the time passed and the half-life must use the same units. Example 16.6 illustrates this concept.

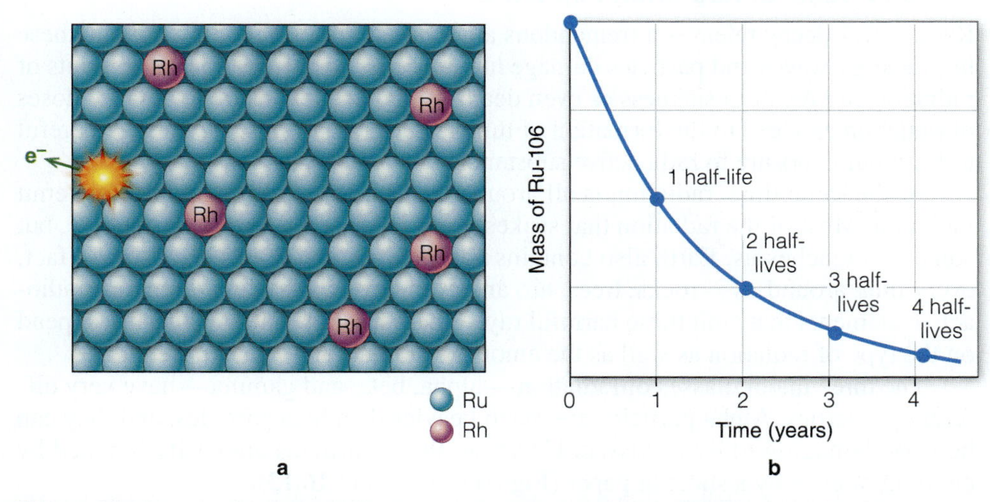

Figure 16.11 As a nuclide decays, it forms a new substance. (a) Ruthenium (Ru) becomes rhodium (Rh) through a beta decay. (b) The half-life is the amount of time it takes for one-half of a substance to decay to another substance. The half-life of ruthenium-106 is just over one year.

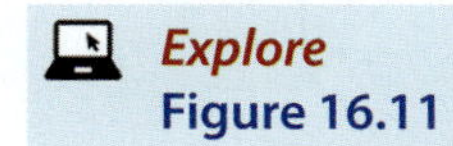

Example 16.6 Using Half-Lives

Samarium-153 EDTMP is the common name for a compound used to treat bone cancer. The compound contains samarium-153, which has a half-life of two days. If your laboratory received a sample containing 16.0 μg of ^{153}Sm, how many μg would be present after 10 days?

Each half-life decreases the amount of samarium present by one-half. After 10 days, five half-lives would have passed, so the amount of ^{153}Sm would divide in half five times:

$$16.0\ \mu g \xrightarrow{1} 8.0\ \mu g \xrightarrow{2} 4.0\ \mu g \xrightarrow{3} 2.0\ \mu g \xrightarrow{4} 1.0\ \mu g \xrightarrow{5} 0.5\ \mu g$$

Based on this, 0.5 μg of ^{153}Sm would be present after 10 days. We also could have used the half-life equation above to solve this problem. In this case, $t = 10$ days, $t_{1/2} = 2$ days, and $N_0 = 16.0$.

$$N_t = N_0\left(\frac{1}{2}\right)^{\frac{1}{t_{1/2}}} = (16.0\ \mu g)\left(\frac{1}{2}\right)^{\frac{10\ \text{days}}{2\ \text{days}}} = (16.0\ \mu g)\left(\frac{1}{2}\right)^{5} = 0.5\ \mu g$$

Because the half-life of this drug is so short, health professionals typically administer it within about 48 hours of its production.

TRY IT

5. Cobalt-58 is a radioactive isotope with a half-life of 71 days. If your laboratory received a sample containing 20.0 μg of ^{58}Co, how much ^{58}Co would be left after 7 months (213 days)?

6. Carbon-14 has a half-life of 5,730 years. If you have a 20-ng sample of ^{14}C today, how much will remain in 8,000 years? Estimate your answer using the definition of half-lives; then use the equation above to obtain a more precise answer.

16.3 Working with Radiation

Health Effects of Radiation Exposure

Radioactive decay releases a tremendous amount of energy. Like tiny bullets, these high-energy waves and particles damage human cells. Exposure to large amounts of radiation can result in sickness or even death. Long-term exposure to smaller doses of radiation can lead to the formation of tumors. Because of this, we must be careful to limit our exposure to radioactive substances.

At the same time, radiation is all around us. The Sun and stars constantly emit radiation. Most of the radiation that strikes Earth is absorbed in the atmosphere, but some still reaches us. Earth also contains naturally radioactive substances. In fact, everything around us—rocks, trees, air, and even our own bodies—contains radioactive elements that emit these harmful rays. The risks of radiation exposure depend on the type of radiation as well as the amount of radiation we're exposed to.

The three major classes of radiation—alpha, beta, and gamma—have very different properties. Alpha particles are much heavier than beta particles, and they can be very damaging to living tissue. However, these particles are easily blocked by clothing or even by a sheet of paper (**Figures 16.12** and **16.13**).

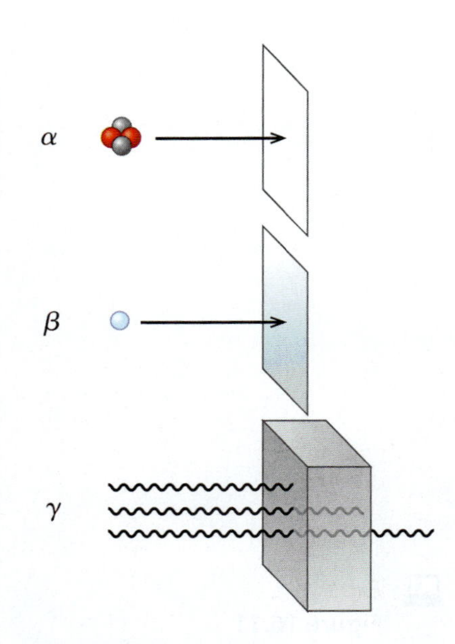

Figure 16.12 Alpha particles are the most damaging radiation, but they are also the easiest to block.

Beta radiation consists of fast-moving electrons that can go through clothing and into living tissue. A thin sheet of metal or other hard substance can block most beta radiation. Gamma radiation is the hardest to block. Heavy lead shields are often used to block gamma radiation. However, gamma rays—which have no mass—are also less damaging than alpha or beta particles.

Perhaps the most important factor in radiation safety is the need to understand *how much* radiation exposure we experience. Our bodies can easily absorb a small amount of radiation. In fact, radioactive substances are very useful—for power, for medicine, and even for archaeology. The important things are to monitor the amount of radiation exposure and to understand the health effects of different exposure levels.

AP Images/Toru Hanai

Figure 16.13 Workers in highly radioactive sites, like the Fukushima responders shown here, wear protective gear. This gear shields the workers from nearly all alpha particles and most beta particles

Measuring Radiation

There are many devices that measure radiation. One of the most common is the **Geiger counter**. A Geiger counter holds a small amount of a gas (such as helium) inside a metal tube containing positive and negative electrodes (**Figure 16.14**). When alpha, beta, or gamma radiation enters the tube, it strikes a gas atom, knocking off an outer electron of the atom. This ionized atom then migrates toward the negative electrode, creating a slight electric current that is amplified and measured as a "count." The counter measures the number of particles that strike the electrode in a given time period. Geiger counters often produce a clicking sound each time they register a count—this sound gives the user some idea of how much radiation is present.

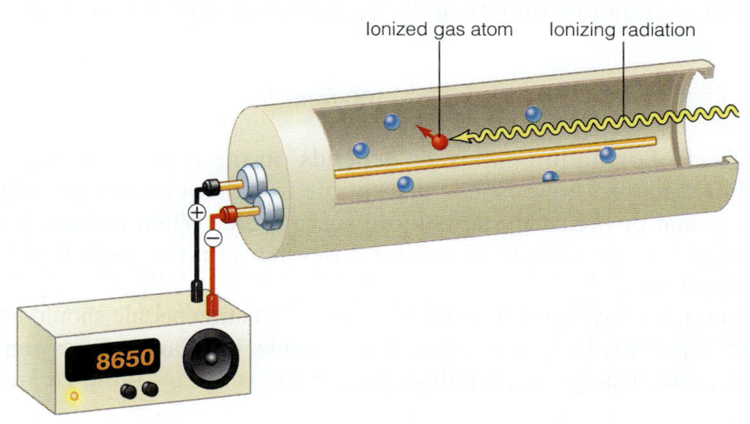

Figure 16.14 In a Geiger counter, incoming radiation ionizes atoms in a gas tube. The ions travel toward an electrode, where they produce a slight electrical current.

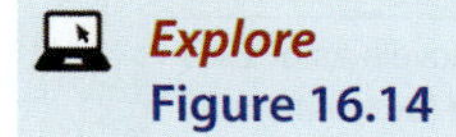

Explore
Figure 16.14

A similar device is a **scintillation counter**. This type of device uses a material that produces a tiny flash of light when radiation strikes it. A detector counts the number of flashes and creates a count of total radiation exposure.

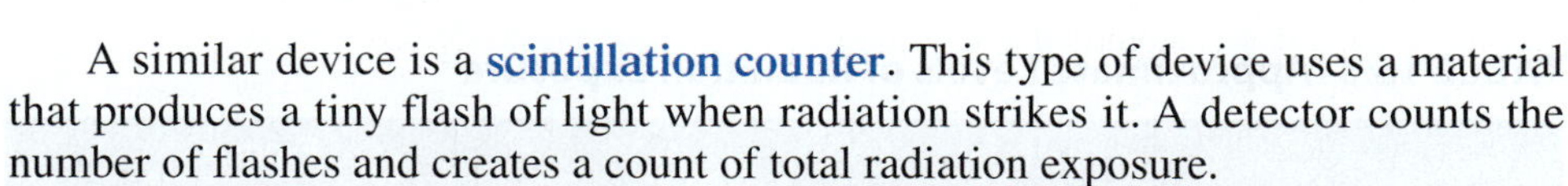

A more advanced instrument for measuring radiation is a **semiconductor counter**. This type of detector not only measures the number of decays but also determines which nuclides produced the decay. Different nuclides produce unique energy signatures when they decay. A semiconductor counter is able to measure the different energies of radiation absorbed and therefore identify which nuclides are present (**Figure 16.15**).

A **dosimeter** is a radiation detector that measures human exposure to radiation. People working in areas with high radiation risks often wear a dosimeter on their work uniform to monitor (and limit) their exposure to harmful radiation (**Figure 16.16**).

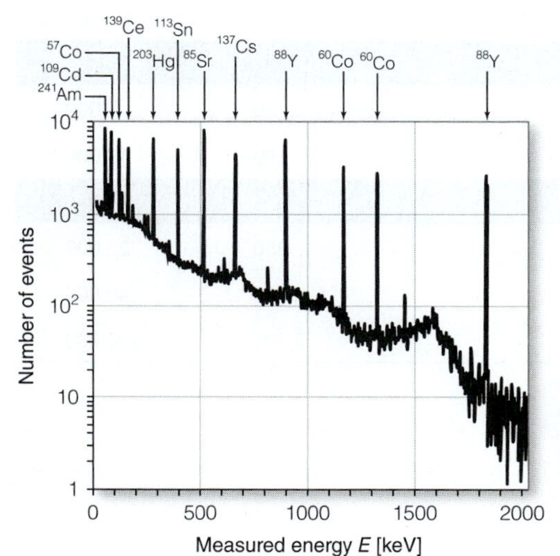

Figure 16.15 A semiconductor counter measures the energies of the radiation absorbed, and it can distinguish between different nuclide decays. Data from Claus Grupen, *Introduction to Radiation Protection: Practical Knowledge for Handling Radioactive Sources* (Berlin: Springer-Verlag, 2010).

Matthew Lloyd/Bloomberg via Getty Images

Figure 16.16 A worker wears a personal dosimeter that tracks his total radiation exposure.

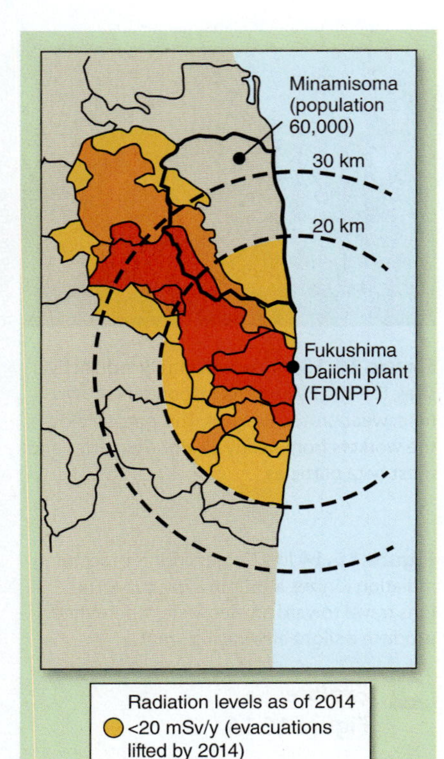

Three years after the disaster, Japanese researchers published a major study of radiation levels around the Fukushima site. Although radiation in many places has decreased to safe levels (< 20 mSv/yr), other areas near the reactor will remain uninhabitable for years.

The units sievert and rem measure human exposure to radiation. ■

Common Exposure Levels

Scientists use several units to measure radioactivity and radiation exposure. The simplest unit, the **becquerel (Bq)**, is the number of decays that occur each second. For example, a sample of radioactive material might undergo 300 decays each second; we express this rate as 300 Bq.

While becquerels are useful in describing how quickly a substance decays, this unit does not differentiate between alpha, beta, and gamma decays. Because each type of decay produces a different energy, the becquerel gives limited information on how these decays affect humans.

When describing human exposure to radiation, scientists commonly use a different unit, the **sievert (Sv)**. One sievert is one joule of energy per kilogram of mass:

$$1 \text{ sievert (Sv)} = 1 \text{ J/kg}$$

Because a sievert measures the actual amount of energy absorbed, it is a better indicator of radiation exposure than the becquerel. A similar unit to the sievert is the *rem*:

$$1 \text{ Sv} = 100 \text{ rem}$$

The effect of radiation exposure depends on the type of biological tissue exposed. For example, bone is less susceptible to radiation damage, while softer tissue like gonad or breast tissue is more susceptible. When assessing radiation exposure, health professionals take these differences into account to determine a total effective dose.

Many countries have guidelines for how much radiation people should be exposed to each year, especially in the workplace. For example, U.S. law requires employers to limit the total effective dose to 50 millisieverts (mSv) per year (**Table 16.3**).

TABLE 16.3 Approximate Levels of Radiation Exposure*

Event	Exposure mSv	Exposure mrem
Background radiation (cosmic rays, radiation in food, radon in the air, etc.; per year)	3	300
Dental X-ray	0.005	0.5
Plane flight (2 hours)	0.01	1
Mammogram	0.4	40
Radiation exposure near the Fukushima reactor site (2 km away, 1 month)**	10.0	1,000
Annual Limits		
U.S. maximum annual dose	50	5,000
Clear link to increase cancer risk (per year)	100	10,000
Short-Term Limits		
Short-term dose limit for emergency workers	250	25,000
Potential for radiation sickness (short-term levels)	400	40,000
Potential for death (short-term levels)	2,000	200,000

*Amounts will vary based on diet, location, and other factors.

**Based on measurements obtained in December 2015.

Uses of Radioactive Nuclides

Although it is important to limit exposure to radioactivity, radioactive nuclides also have several practical applications in fields ranging from medicine to geology and archaeology. In this section, we'll survey some applications of these nuclides.

Uses in Medicine

Nuclear medicine deals with the use of radioactive nuclides for medicinal purposes. The two primary applications for nuclear medicine are *imaging* and *radiation therapy*.

In radiation imaging, a patient is administered a dose of a compound containing a radioactive nuclide. As this material travels through the body, it emits gamma radiation. This radiation is detected using a *scintillation camera*, an instrument that produces an image based on the intensity of gamma radiation. For example, doctors sometimes use radiation imaging to locate internal infections. This involves "tagging" a patient's white blood cells with a radioactive isotope such as indium-111. The white blood cells congregate around infected sites. As they do, they provide a gamma ray "beacon" that pinpoints the location of the infection (**Figure 16.17**).

Radiation therapy is a common treatment for cancer. This technique uses the destructive properties of radiation to destroy cancer cells. In one variation of this technique, doctors implant "seeds" containing radioactive nuclei (such as Cs-131) near the tumor site (**Figure 16.18**). Radiation therapy is a messy technique: Although it may destroy cancer cells, it damages nearby healthy cells as well, and many patients battle radiation sickness throughout the treatment process.

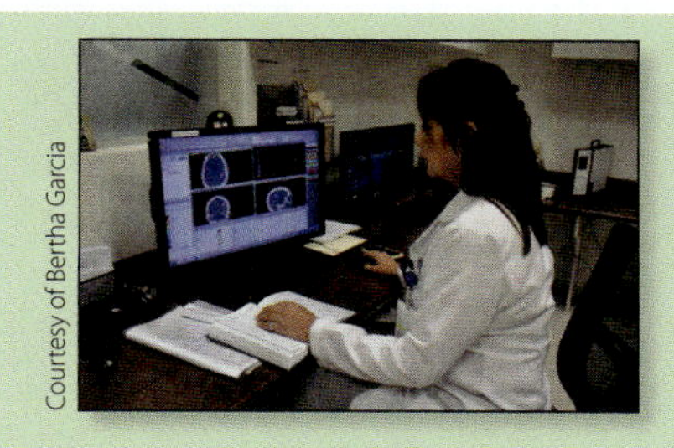

Health physicists study the health effects of radiation. Industries from medicine to mining involve exposure to radioactive substances, and professionals trained to address these safety concerns are often in high demand.

Figure 16.17 The dark spots in the rib cage are produced by gamma radiation emitted from white blood cells that have been tagged with radioactive indium. An image like this allows doctors to locate an internal infection.

Source: This research was originally published in JNMT. Love C and Palestro CJ. Radionuclide Imaging of Infection. *J Nucl Med Technol.* 2004; 32(2):47–57. Figure 3.

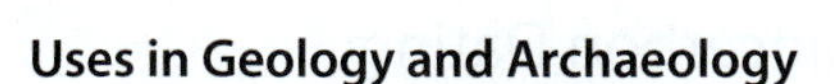

Figure 16.18 These "seeds" containing Cs-131 are implanted in a patient near the site of a tumor.

Uses in Geology and Archaeology

Because radioactive nuclei decompose at predictable rates, they are useful as a "clock" for dating archaeological artifacts and understanding Earth's history. Many different nuclei are used for this type of application, but we'll look at just a few examples here.

Carbon-14 Dating of Plant and Animal Remains

When high-energy radiation from the Sun enters the upper atmosphere, it produces several nuclear reactions. In one such reaction, a neutron collides with nitrogen-14 to produce carbon-14 and an additional proton. In Earth's atmosphere, a tiny fraction of the total carbon (about 1.5 atoms per trillion) is carbon-14.

$$^{1}_{0}n + ^{14}_{7}N \rightarrow ^{14}_{6}C + ^{1}_{1}p$$

Unlike the main isotopes of carbon (^{12}C and ^{13}C), carbon-14 is radioactive, with a half-life of about 5,730 years. As plants grow, they take in carbon dioxide from the atmosphere, including carbon-14. After a plant dies, the carbon-14 begins to decay. Based on the amount of carbon-14 left in a sample, scientists can determine the age of substances that were derived from plants. This includes animals (which directly or indirectly consume plant matter), paper, leather, or even food residues found in ancient pots (**Figure 16.19**).

Botond Horvath/Shutterstock

© The Israel Museum, Jerusalem, by ArdonBar-Hama/ Bridgeman Images

Figure 16.19 In 1948, a young shepherd discovered a collection of scrolls in the caves near Qumran, along the Dead Sea in Israel. The scrolls, including this text of the Book of Isaiah, were analyzed by carbon dating and found to be written approximately 2,200 years ago.

By dating fragments from a known time period (for example, a document from the Roman era), scientists have confirmed the validity of this process. However, carbon dating is considered reliable only for events occurring in the past 50,000 years.

Dating of Exposed Rock Surfaces Using Be-10 and Al-26

Quartz is a type of rock with the empirical formula SiO_2. When high-energy radiation strikes quartz, it produces small amounts of the radioactive nuclides beryllium-10 and aluminum-26. If quartz is exposed on Earth's surface, it is bombarded by solar radiation, and these nuclides gradually accumulate. If the quartz is buried, solar radiation is blocked, and the isotopes begin to decompose. Geologists use the amounts of Be-10 and Al-26 present in a quartz sample to determine how long a rock surface has been exposed or how long it has been buried. This technique enables geologists to track the movement of glaciers over long periods of time (**Figure 16.20**).

Fletcher & Baylis/Science Source

Figure 16.20 Tiny amounts of Be-10 and Al-26 trapped in the quartz structures within certain rocks help geologists determine how long the rock surfaces have been exposed.

Example 16.7 Using Half-Lives in Radiocarbon Dating

An archaeologist excavating an ancient city finds a sample of grain in a pot. Carbon-14 analysis shows that the grain has a ^{14}C concentration that is one-fourth of the amount in modern grain. How old is the grain in the pot?

Each half-life reduces the concentration of a radioisotope by one-half. Since only one-fourth of the amount of ^{14}C is present, two half-lives have passed:

$$\frac{1}{2} \times \frac{1}{2} = \frac{1}{4}$$

Because the half-life of ^{14}C is 5,730 years, the sample will be approximately 11,460 years old.

TRY IT

7. Iodine-131 decays by beta emission and has a half-life of about 8 days. Iodine-129 also decays by beta emission, but it has a half-life of about 16 million years. Which of these isotopes presents the greater radiation hazard?

8. The concentration of ^{14}C is usually measured in the number of decays observed. Living plants have a ^{14}C concentration of about 240 Bq/kg C. If a tree dies today, what concentration of ^{14}C would still be present after 3 half-lives?

16.4 Energy Changes in Nuclear Reactions

In the nucleus, positively charged protons and uncharged neutrons pack tightly together. Although the nucleus holds nearly all the mass of the atom, it is only a tiny fraction of the atom's volume (**Figure 16.21**).

Think about this for a moment: You know that positively charged particles repel each other, but in the nucleus, these particles pack tightly together. For example, a single atom of lead has 82 protons crammed into its tiny nuclear space. How is this possible?

The force of repulsion between charged particles is overcome by another force, called the **nuclear force**, that holds the nucleus together. Like other fundamental forces, such as gravity or the attraction and repulsion between charged particles, the nature of the nuclear force is not well understood. But just as we can measure energy changes caused by gravity or the attraction between charges, we can measure energy changes that occur in the nucleus. However, understanding these energy changes requires us to rethink some basic assumptions about the nature of matter.

Jim West/Alamy

Figure 16.21 Compared to the atom's volume, its nucleus is about the size of an insect inside a football stadium.

Mass Defect, Binding Energy, and Einstein's Famous Equation

Earlier in this book, we introduced the law of conservation of mass. This law states that in chemical changes, the total mass does not change—the mass before a chemical reaction is the same as the mass afterward. In a chemical reaction, the number and type of atoms do not change—only the arrangement of electrons changes.

The law of conservation of mass does not apply to nuclear changes. For example, a helium nucleus contains two protons and two neutrons. The masses of protons and neutrons have been precisely measured and are listed in **Table 16.4**. Based on the mass of the protons and neutrons, we would expect the mass of the helium nucleus to be 4.031880 atomic mass units (u):

Conservation of mass applies to chemical changes but not to nuclear changes.

$$\begin{aligned}\text{Expected mass of He nucleus} &= 2(\text{proton mass}) + 2(\text{neutron mass}) \\ &= 2(1.007276\text{ u}) + 2(1.008664\text{ u}) \\ &= 4.031880\text{ u}\end{aligned}$$

However, the mass of a helium nucleus has also been precisely measured: It is 4.001503 u. The actual mass is slightly less than the sum of the masses of the

TABLE 16.4 Subatomic Particles

Particle	Mass (u)	Charge	Location
Proton	1.007276	+1	Nucleus
Neutron	1.008664	none	Nucleus
Electron	0.000549	−1	Electron cloud

protons and neutrons! The difference between the masses of the individual particles and the mass of the complete nucleus is called the **mass defect**:

Expected mass:	4.031880 u
– Actual mass:	4.001503 u
Mass defect:	0.030377 u

Jack Orren Turner, Photographer/Library of Congress, Prints and Photographs; Division/ LC-USZ62-60242

Figure 16.22 Einstein's discoveries shook the way people think about matter and energy. His equation is vital to understanding the energy changes in nuclear reactions.

So what does the mass defect actually mean? In 1905 Albert Einstein (**Figure 16.22**) answered this question when he introduced the concept of *mass-energy equivalence*. This groundbreaking idea states that mass can be converted into energy, and energy can be converted into mass. The two are related by his famous equation:

$$E = mc^2$$

This equation states that energy (E) is equal to mass (m) times the speed of light (c) squared. When two protons and two neutrons combine to form a helium nucleus, a small fraction of the mass (the mass defect) is actually converted into energy. This energy, called the **binding energy**, is released when the protons and neutrons form a new nucleus. Stated differently, the binding energy is the energy *required* to break a nucleus into its individual particles.

For nuclear equations, the law of conservation of mass is replaced by a more complete law, called the **law of conservation of matter and energy**, which states that *in any change, the total matter and energy in the universe remain constant.*

In this brief survey, we will not work examples using Einstein's equation. However, it is important to recognize that nuclear reactions always involve a conversion between matter and energy. As we'll see in the sections that follow, these energy changes are vital to understanding and harnessing nuclear reactions.

In exothermic nuclear reactions, some mass is converted into energy. ■

Nuclide Stability

Every nuclide has a unique mass defect, and therefore a unique binding energy. Some nuclides have higher potential energy, while others are more stable. Nuclear scientists compare the stability of different nuclei using the *binding energy per nucleon*. The greater the binding energy per nucleon, the more stable the nucleus is. The most stable nucleus is ^{56}Fe. Nuclei that are near this size tend to be stable, but very large or very small nuclei are less stable (**Figure 16.23**). As we'll see in coming sections, these differences in stability create the potential for high-energy nuclear changes.

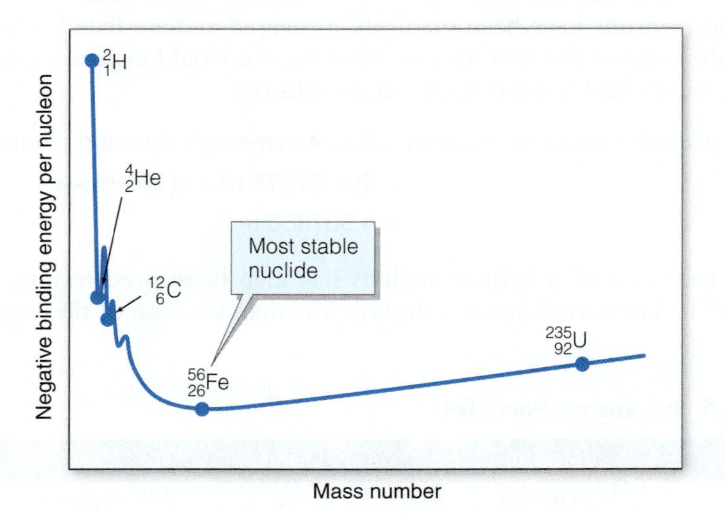

Figure 16.23 The most stable compounds have the highest binding energy per nucleon. In this graph, the negative binding energy is plotted so that the most stable isotope is at the bottom of the energy curve.

The stability of nuclides also depends on the ratio of protons to neutrons. Smaller atoms often have a 1:1 ratio of protons and neutrons in their nuclei. For example, most carbon atoms contain 6 protons and 6 neutrons. Larger atoms have more neutrons than protons, although the ratio is only slightly higher than 1:1 (**Table 16.5**). Notice in **Figure 16.24** that only a narrow band of neutron–proton ratios form stable nuclei.

TABLE 16.5 Proton–Neutron Ratios

Atom	Protons	Neutrons
Hydrogen	1	0
Helium	2	2
Carbon	6	6
Potassium	19	20
Iron	26	30
Gold	79	118
Lead	82	126
Uranium	92	146

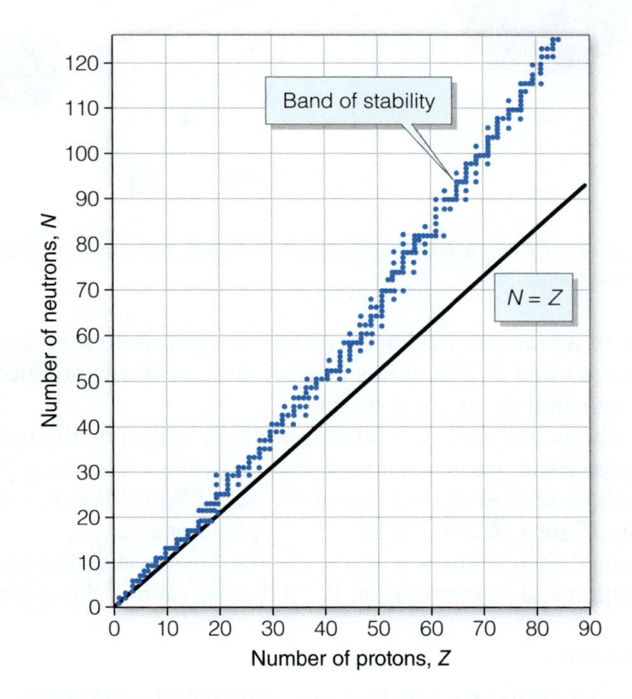

Figure 16.24 Most stable nuclei have slightly more neutrons than protons.

TRY IT

9. The alpha decay of polonium-210 (shown below) releases a tremendous amount of energy. Based on the law of conservation of matter and energy, which has the greater mass—the reactants or the products?

$$^{210}_{84}\text{Po} \rightarrow {}^{206}_{82}\text{Pb} + {}^{4}_{2}\text{He} + \text{energy}$$

Check It
Watch Explanation

16.5 Nuclear Power: Fission and Fusion

Fission

Nuclear power plants such as the Fukushima reactor depend on a process called **fission** to produce power. In a fission reaction, a large nucleus shatters into several smaller nuclei, releasing huge amounts of energy. Only a few nuclides are susceptible to fission reactions; the most common and important of these is uranium-235. When a high-energy neutron strikes a U-235 nucleus, the uranium shatters into smaller fragments (**Figure 16.25**).

Fission reactions create many different products. The following nuclear equations show two of the possible fragmentations:

$$^{1}_{0}\text{n} + {}^{235}_{92}\text{U} \rightarrow {}^{138}_{54}\text{Xe} + {}^{95}_{38}\text{Sr} + 3\,{}^{1}_{0}\text{n}$$
$$\rightarrow {}^{144}_{56}\text{Ba} + {}^{90}_{36}\text{Kr} + 2\,{}^{1}_{0}\text{n}$$

Explore
Figure 16.25

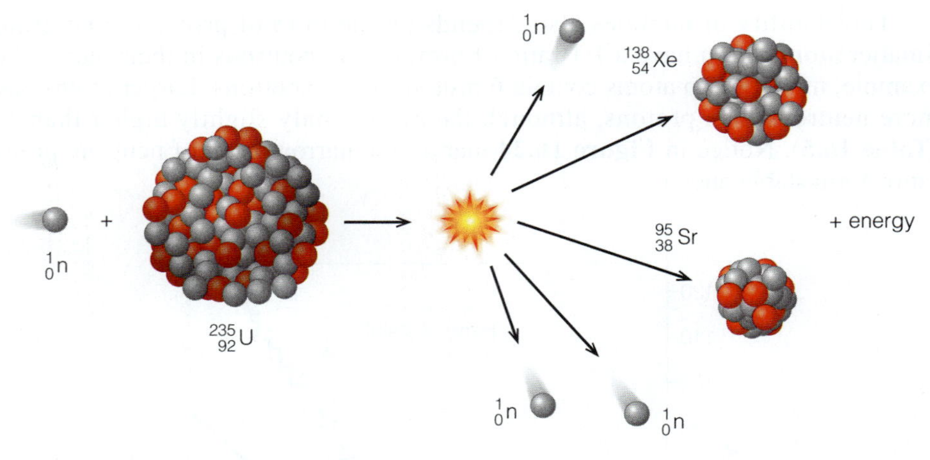

Figure 16.25 In a fission reaction, a heavy nucleus like ^{235}U shatters into smaller fragments, releasing a tremendous amount of energy.

The fission of a uranium nucleus produces several more high-energy neutrons. If there are enough other U-235 atoms nearby, these neutrons produce a chain reaction, releasing tremendous amounts of energy.

Because each nuclear fission produces multiple neutrons, the number of atoms involved in a fission chain reaction can multiply rapidly. The devastating power of atomic bombs, such as those unleashed on Japan at the end of World War II, rely on this unrestrained reaction (**Figure 16.26**). For peaceful applications, nuclear scientists and engineers carefully guard against runaway reactions. Two keys to avoiding runaway reactions are fuel composition and reactor design. We'll discuss each of these approaches next.

Everett Historical/Shutterstock

Figure 16.26 At the close of World War II, a weaponized nuclear chain reaction decimated the Japanese city of Hiroshima.

Uranium Enrichment

Naturally occurring uranium consists of two primary isotopes. Only the lighter isotope, U-235, is *fissile* (that is, it undergoes fission reactions). The heavier isotope, U-238, does not undergo fission reactions. The non-fissile U-238 is the major isotope: 99.3% of Earth's uranium is U-238. Only 0.7% of naturally occurring uranium is the fissile U-235 (**Figure 16.27**).

Peaceful nuclear power requires only about 5% enriched ^{235}U. Nuclear weapons require highly purified ^{235}U.

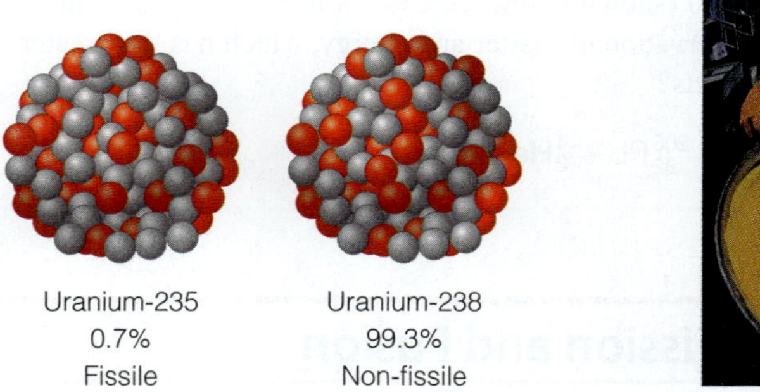

Figure 16.27 Uranium-235 undergoes nuclear fission, but it is only 0.7% of naturally occurring uranium.

Emory Kristof/National Geographic Image Collection

Figure 16.28 Purified U_3O_8 is called *yellowcake* because of its bright color.

To sustain a nuclear chain reaction, scientists take uranium through an *enrichment* process that increases the percentage of U-235 in a sample. Here's how it works: Uranium is mined as uranium oxide (U_3O_8). Purified U_3O_8—sometimes called *yellowcake* uranium because of its color and texture (**Figure 16.28**)—is converted to pure uranium metal and then to uranium hexafluoride, UF_6:

$$U_3O_8 \rightarrow 3\,U + 4\,O_2$$

$$U + 3\,F_2 \rightarrow UF_6$$

A liquid at room temperature, UF_6 has a low boiling point (57 °C) and is easily converted into gas. The UF_6 gas is then passed through a gas centrifuge (**Figure 16.29**). The centrifuge spins the gas samples, pushing the heavier $^{238}UF_6$ to the edges, while the lighter $^{235}UF_6$ collects in the middle. Because the mass difference is so small, this approach produces only a slight enrichment of UF_6. However, by repeating the process thousands of times, scientists eventually can obtain highly purified $^{235}UF_6$. This material can then be converted back into elemental uranium.

Nuclear reactors do not use highly purified uranium-235. The fuel rods used in nuclear reactors contain about 5% ^{235}U and 95% ^{238}U. The excess uranium-238 serves two purposes: First, it absorbs some of the neutrons, preventing a runaway reaction. Second, after absorbing a neutron, this nuclide undergoes two beta decays to form plutonium:

$$^{1}_{0}n + ^{238}_{92}U \rightarrow ^{239}_{92}U$$

$$^{239}_{92}U \xrightarrow{\beta} ^{239}_{93}Np + ^{0}_{-1}e$$

$$^{239}_{93}Np \xrightarrow{\beta} ^{239}_{94}Pu + ^{0}_{-1}e$$

Plutonium also undergoes nuclear fission. So even though U-238 does not undergo fission directly, it does slowly produce plutonium fuel.

In contrast to peaceful applications, weaponization of nuclear power requires highly purified U-235 or Pu-239. In recent years two nations, North Korea and Iran, have aggressively developed nuclear weapons programs. Despite international opposition, these nations installed large groups of centrifuges to carry out the enrichment process. North Korea became a nuclear power around 2009. It's likely that the spread of nuclear weapons will be a global concern for years to come.

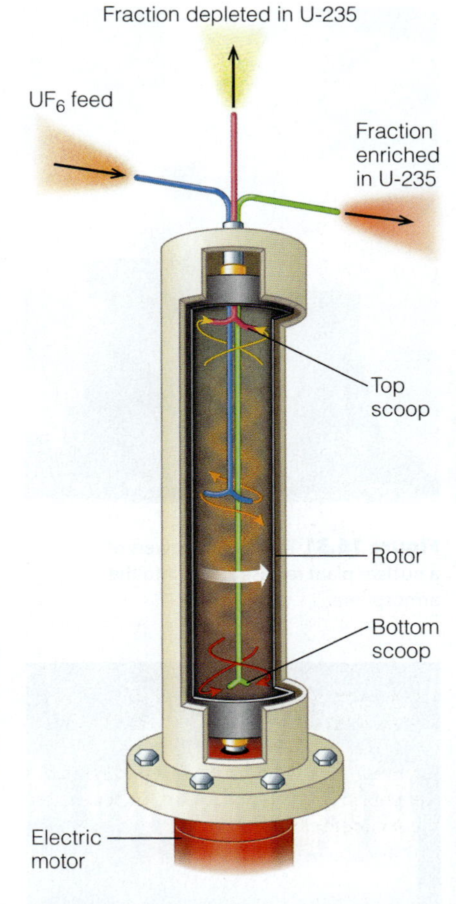

Figure 16.29 A gas centrifuge separates UF_6 based on molecular mass. When the centrifuge spins, the heavier $^{238}UF_6$ presses to the outside, while the lighter $^{235}UF_6$ concentrates in the middle.

Fission Reactor Design

A nuclear fission reactor produces heat that can be converted into electrical power. Because of the hazards of nuclear fission, reactors are designed to closely regulate the chain reaction and isolate the radioactive by-products. Nuclear reactors have three distinct stages, or loops: the *reactor loop*, the *power loop*, and the *cooling loop* (**Figure 16.30**).

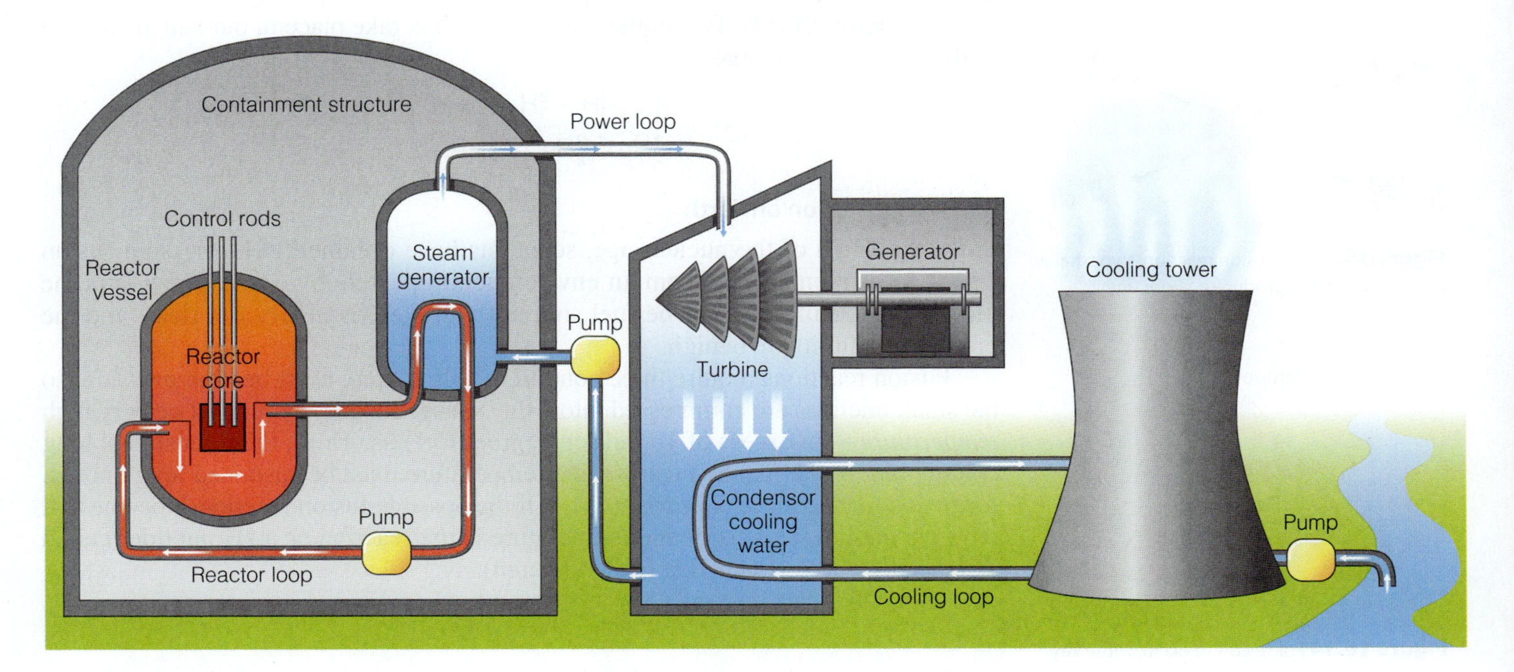

Figure 16.30 A schematic of a nuclear reactor. Water from the reactor core heats water in the power loop, converting it to steam, which turns a turbine to produce electricity. The cooling loop condenses the water in the power loop so it can be recycled. Excess heat energy is released through large cooling towers.

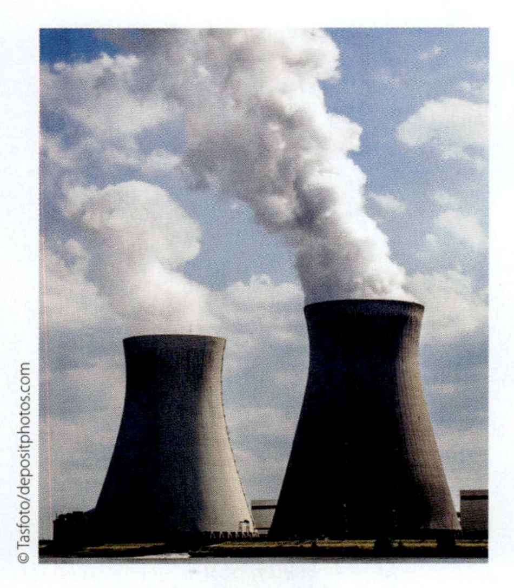

Figure 16.31 The cooling towers of a nuclear plant release steam into the atmosphere.

The reactor loop contains the reactor core, where fission takes place. The reactor core is immersed in water; pumps circulate this water to carry heat energy away from the core. A series of graphite *control rods* may be raised or lowered into the core. The control rods are able to absorb neutrons produced in the fission reaction. Inserting the control rods slows down the reaction; withdrawing the rods speeds it up.

Water in the reactor loop is contaminated with radioactive nuclides produced in the fission process. Because of this, the reactor loop is completely sealed off from the outside environment and housed in a reinforced *containment structure*. Inside the containment structure, the reactor loop comes in contact with water in the power loop. The reactor loop heats the water in the power loop, converting it to steam. The steam turns turbines, producing electric current.

After passing through the turbines, the steam is cooled and condensed by water in the third loop—the cooling loop. This water is drawn from a nearby source, such as an ocean or large river. The excess heat absorbed by the cooling loop is released through massive cooling towers (**Figure 16.31**). Although the towers appear to emit smoke, they release only water vapor into the air.

Notice in Figure 16.30 that no water from inside the containment structure is released into the environment. The use of three water loops prevents radioactive material from escaping into the environment.

Waste from Nuclear Fission

Figure 16.32 Spent nuclear fuel is stored deep underground.

Generating power through fission reactions has both benefits and drawbacks. On one hand, nuclear fission plants normally do not release toxic products into the environment. On the other hand, the by-products from nuclear fission reactions are highly radioactive and remain so for thousands of years. There is no way to get rid of these materials, so they must be stored in huge underground caverns (**Figure 16.32**). And, as the Fukushima disaster illustrates, there is always the risk of an accident with terrible environmental consequences.

Fusion

Energy produced by the Sun and other stars arises from a nuclear reaction called **fusion**, in which two smaller nuclei (usually hydrogen or helium) combine to form larger nuclei. Like fission reactions, fusion reactions produce huge amounts of heat energy (**Figure 16.33**). Two of the major events that take place in our sun are shown in these nuclear reactions:

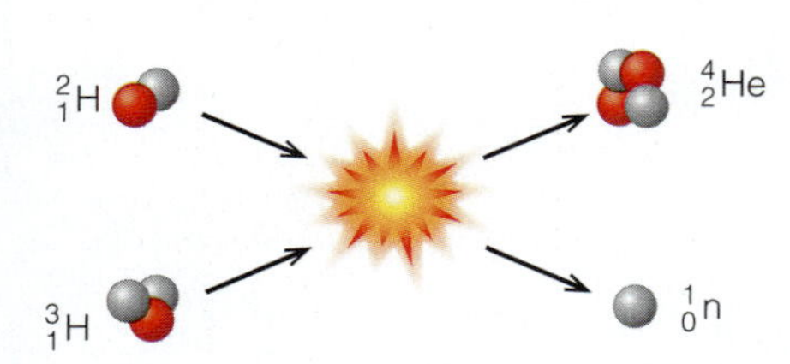

Figure 16.33 In a fusion reaction, two smaller nuclei combine to produce a larger one.

$$ {}^{1}_{1}H + {}^{2}_{1}H \rightarrow {}^{3}_{2}He $$
$$ {}^{3}_{2}He + {}^{3}_{2}He \rightarrow {}^{4}_{2}He + 2\,{}^{1}_{1}H $$

Replicating Fusion on Earth

Since the dawn of the nuclear age, scientists have dreamed of harnessing fusion power for use on Earth. From an environmental perspective, fusion power is the ultimate in clean energy: The fuel source (hydrogen) is nearly limitless, and the products (helium) are benign.

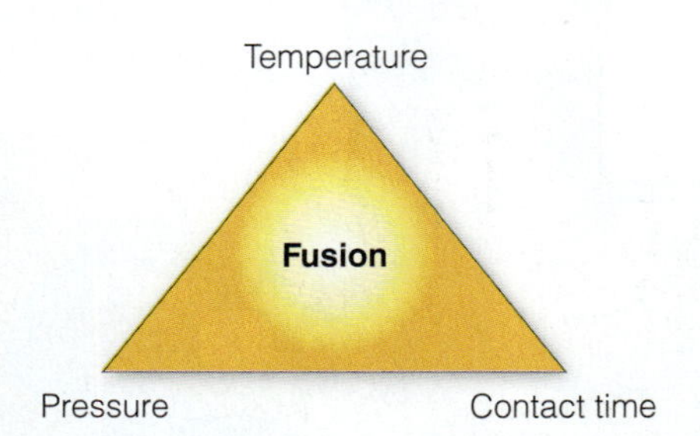

Figure 16.34 For fusion to occur, three key parameters must be met.

Fusion reactions require three conditions: First, there must be high *pressure*, so the small nuclides are compressed into a tiny space. Second, there must be very high *temperature*—in excess of 10 million degrees Celsius. Third, there must be a long *contact time*—that is, the pressure and temperature must be sustained long enough for fusion to take place (**Figure 16.34**). Of the possible fusion reactions, the one that requires the lowest density and temperature is the reaction of a ${}^{2}_{1}H$ nuclide (called *deuterium*) with a ${}^{3}_{1}H$ nuclide (called *tritium*):

$$ {}^{2}_{1}H + {}^{3}_{1}H \rightarrow {}^{4}_{2}H + {}^{1}_{0}n $$

In the stars, strong gravitational forces hold atoms tightly together, and fusion occurs easily. On Earth, obtaining the extremely high temperature and pressure requirements has proven very difficult—at least for peaceful purposes.

By the 1950s and 1960s, the United States and the Soviet Union had mastered the use of fusion for weapons purposes. *Thermonuclear weapons*, sometimes called *hydrogen bombs*, use a devastating two-stage detonation: The first stage is a fission reaction that compresses hydrogen fuel into a tiny area. The resulting heat, pressure, and contact time initiate a second detonation—a fusion reaction. **Figure 16.35** shows the mushroom cloud from a hydrogen bomb over 500 times more powerful than the fission bomb that was dropped on Hiroshima.

In 1961, the Soviet Union detonated the most powerful hydrogen bomb ever tested, releasing an energy equivalent to 50 million tons of TNT.

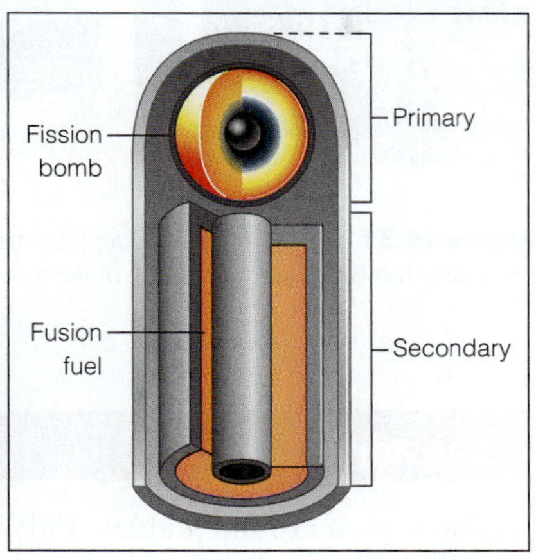

Figure 16.35 This mushroom cloud was caused by a hydrogen bomb test in 1952 that was over 500 times more powerful than the bomb dropped on Hiroshima. This type of weapon initiates a primary fission reaction that produces enough energy to generate a secondary fusion reaction.

Are We There Yet?

Despite advances in weapons technology, generations of scientists have struggled to harness fusion power for peaceful purposes. The core challenge is *plasma containment*. **Plasma** is a gas-like state that is produced at very high temperatures, and it is often considered the fourth state of matter. In the plasma state, electrons have such high energies that they do not remain bound to any one atom. The result is a gaseous "soup" of positive and negative charges.

Plasma is far too hot to contain in a solid structure. Instead it must be trapped inside a very strong electromagnetic field, sometimes called a *magnetic bottle*. To date the most effective magnetic bottle design is the *tokamak* design, developed by the Soviet Union in the 1950s. In this design, powerful magnetic fields create a donut-shaped pathway for plasma to circulate, as shown in **Figure 16.36**.

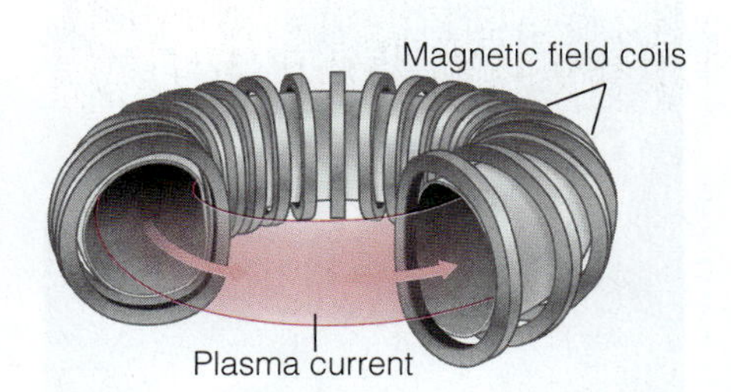

Figure 16.36 The tokamak design traps plasma inside a powerful magnetic coil. This image is a simulation of plasma movement within these coils.

Over the past 70 years, scientists have made steady improvements in the tokamak design. The plasma temperature, density, and contact time have steadily increased. Nuclear scientists have observed fusion reactions within tokamak reactors, but they have not yet reached the *break-even point*, where the energy output from a tokamak reactor exceeds the energy required to power the plasma containment vessel.

But that may be about to change. In 2013, scientists began construction of the massive *International Thermonuclear Experimental Reactor* (ITER) in southern France. Nuclear scientists expect ITER, which is scheduled to begin operating in 2025, to exceed the break-even point by a factor of 10—producing 500 megawatts of power for every 50 megawatts of power consumed (**Figure 16.37**). In the coming years, it will be interesting to see if this international effort is, in fact, able to meet this goal.

TRY IT

10. At first it might seem surprising that both breaking nuclei (fission) and forming nuclei (fusion) could release so much energy. Based on Figure 16.23, can you explain how both fission and fusion processes are energetically favorable? When does each of them occur?

Check It
Watch Explanation

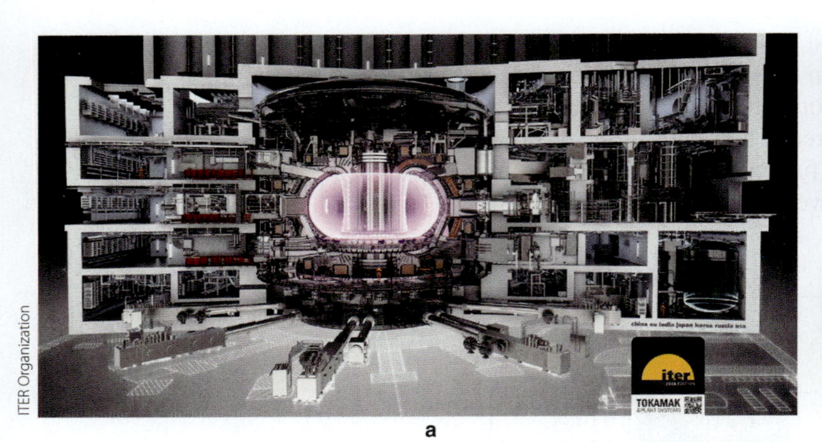

a

b

Figure 16.37 (a) The ITER reactor depicted here is currently under construction in southern France. (b) This photograph shows the construction of the massive reactor building.

Capstone Video

Capstone Question

The region around Naples, Italy, contains a network of underground tombs, called *catacombs*, that date to the time of the Roman Empire. Adorned with fantastic artwork (**Figure 16.38**), the catacombs are popular tourist destinations. However, the catacombs contain fairly high levels of radon, a naturally occurring radioactive gas that is produced in rock and accumulates in poorly ventilated underground areas.

In 2013, a team of researchers measured the radon concentrations in the San Gennaro Superiore catacombs. They reported the average concentration of radon emissions (C_{Rn}) to be 463 Bq/m^3. To determine the effective dose of radiation, they used a relationship like this:

$$\text{Effective Dose} = F \times C_{Rn} \times t$$

where F is a dose conversion factor equal to 8.0×10^{-9} Sv $\cdot$ Bq$^{-1} \cdot$ m$^3 \cdot$ hr^{-1}, and t is the amount of time a person is exposed to the radiation.

a. Using this relationship, calculate the effective dose in milliSieverts for a tourist visiting the tombs for one hour, and a tour guide who spends 317 hours/year in the catacombs.

b. The Italian government limits workplace radiation exposure to 3 mSv/year. How does the exposure for a tourist and a tour guide compare to the safety limits? How many hours per year would a person have to spend in the catacombs before reaching the exposure limit?

c. The radioactive nuclide ^{222}Rn forms when ^{226}Ra undergoes a radioactive decay. What type of decay is this? Write a balanced equation for this reaction.

d. Radon-222 has a half-life of 3.8 days. If you collected a sample of radon gas today, what percentage of the radon would remain after 19 days?

e. Radon is typically more hazardous than other elements in the uranium decay chain because it is a gas and can be inhaled into the lungs, where radioactive decay causes more damage. What is the electron configuration of neutral ^{222}Rn? Why is this substance a gas at room temperature, when all the other elements in the uranium series are solids?

Figure 16.38 The San Gennaro Superiore catacombs near Naples, Italy, are Roman-era tombs adorned with early Christian artwork. While they are a popular tourist site, guides must be aware of elevated radon levels within the tombs.

SUMMARY

Nuclear changes are all around us. Although most of the events we observe each day do not involve nuclear changes, these reactions are a vital part of our existence. Nuclear changes occur in a variety of settings, from fusion in the Sun to reactions in Earth's upper atmosphere to synthetic changes produced in laboratories and power plants.

We use nuclear equations to represent changes that take place in atomic nuclei. Because the identity of the atom is determined by the number of protons in a nucleus, many nuclear changes convert atoms from one element to another.

Radioactive nuclei decompose over time, releasing particles and energy to produce a more stable species. There are three major types of radioactive decay: alpha, beta, and gamma. An alpha decay is the emission of two protons and two neutrons—a helium nucleus. A beta decay is the emission of an electron from the nucleus—a process that converts a neutron into a proton. A gamma decay is a release of energy as the nucleus shifts to a more stable configuration. The high-energy particles and energy released by radioactive decay are called *radiation.*

Some radioactive nuclei decompose more quickly than others. The half-life is a measure of the rate of nuclear decay; it is the time required for one-half of a radioactive sample to decay to other substances.

While humans are constantly exposed to low levels of radiation, exposure to higher levels of radiation can be harmful or fatal. Many nations have guidelines for acceptable exposure levels.

Although radioactive substances carry some risks, they also are useful for many applications. Radioactive nuclei are used for medical imaging and treatment of cancer. Radioactive substances are also used in geology and archaeology to identify the ages of different artifacts. Carbon-14 is used to date materials derived from plant life. Be-10 and Al-26 are used to determine how long rock samples (particularly quartz) have been exposed to the atmosphere.

Nuclear changes involve a huge amount of energy. When protons and neutrons combine to form a new nucleus, a portion of the mass is converted into energy. The relationship between mass and energy is governed by Einstein's famous equation $E = mc^2$. The difference between the mass of a nucleus and the masses of the individual particles is the mass defect, and the energy released in a nuclear change is the binding energy.

Nuclear fission is the process by which larger nuclei, such as uranium and plutonium, are split into smaller nuclei, releasing huge amounts of heat energy. Nuclear power plants use the fission process. Nuclear fusion is the process by which smaller nuclei, such as hydrogen and helium, bind together to form larger ones. Nuclear fusion powers the Sun and the other stars.

Harnessing the nuclear fusion reaction for energy production has been a major goal of nuclear scientists for many years. A massive fusion reactor currently under construction in Europe may finally pass the break-even point, where the power produced by the fusion reaction equals the power required by the magnetic containment system, and begin to move toward producing commercial power.

Powering the Future

At the beginning of this chapter, we discussed the 2011 earthquake and tsunami that struck Japan, claiming over 15,000 lives and causing the biggest nuclear disaster in nearly 30 years. The nuclear meltdown contaminated the land and waters around the power plant. Much of the area around the Fukushima plant is now uninhabitable.

The Fukushima disaster provoked fears across the world. In the weeks that followed, policy experts debated the role of nuclear power in the world's future. Even today, that future is unclear.

What is clear is that the world needs power — lots of it. Over the past 80 years, world consumption of energy has steadily increased. As major countries like China and India continue to develop, the global appetite for power will only grow. But where will this power come from?

The traditional answer has been fossil fuels. Coal, oil, and natural gas are the workhorses of global energy (**Figure 16.39**). Recent discoveries of oil and natural gas ensure that we'll have enough fossil fuel energy for many years to come. But burning fossil fuels produces carbon dioxide, and many people worry about the effects of increased CO_2 levels in the atmosphere.

Nuclear fission solves this problem, but creates another one. Fission produces only a small amount of waste, but the waste produced is highly radioactive. This waste is not released into the atmosphere, but it must be stored somewhere — typically in secure underground locations. There is no way to get rid of nuclear waste. And in the event of an accident like Fukushima, the environmental effects can be devastating.

Many people tout renewable energy sources like solar, wind, hydroelectric, and geothermal energies. But these sources are expensive, geographically limited, and currently supply only a tiny fraction of the world's energy needs.

Perhaps nuclear fusion will provide a long-term solution to humanity's energy needs, but this resource is years away. Until then, the world will continue to wrestle with energy issues. It's an intricate dance of science, politics, and economics. There is no perfect answer.

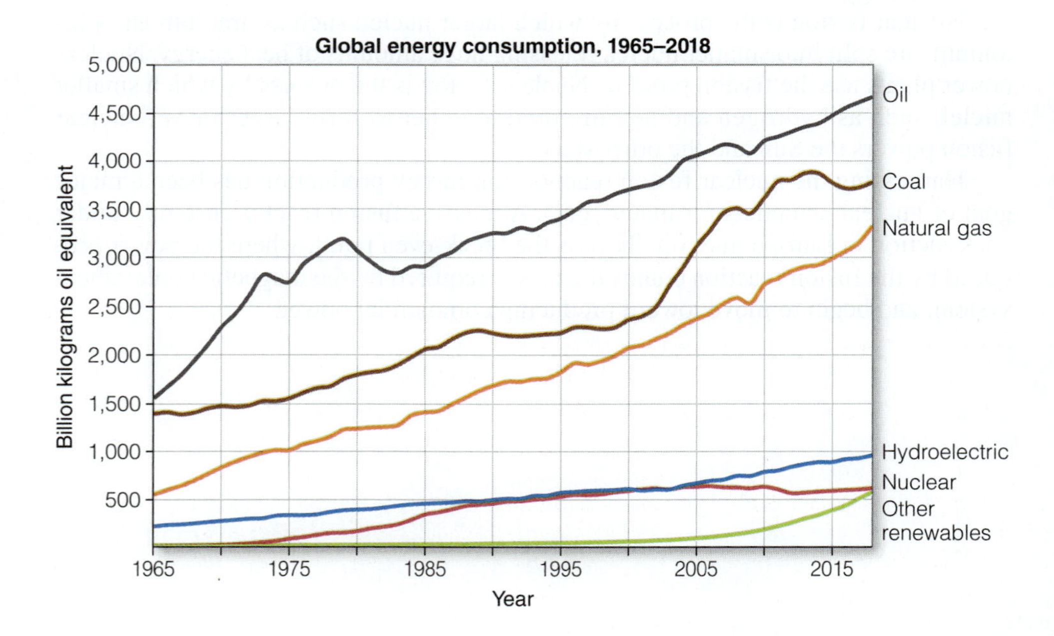

Figure 16.39 Global energy consumption continues to increase. Fossil fuels are the primary sources of energy, although hydroelectric and other renewables are having an increased impact in the past few years. Notice that nuclear energy usage has decreased since the Fukushima disaster.
Data from *BP Statistical Review of World Energy, June 2019*; retrieved April 7, 2020, from https://www.bp.com.

Key Terms

16.1 Nuclear Changes

nuclear chemistry The study of the properties and changes of atomic nuclei.

nucleon A particle that resides in the nucleus; either a proton or a neutron.

nuclide An atom or nucleus containing a particular number of protons and neutrons.

isotopes Nuclides having the same atomic numbers but different mass numbers.

nuclear reaction A reaction in which the structure of an atomic nucleus changes.

nuclear equations Equations that describe nuclear changes; in a balanced nuclear equation, the total mass numbers and total atomic numbers are the same on both sides of the equation.

16.2 Radioactivity

radioactive decay A nuclear change in which the nucleus spontaneously transitions to a more stable state; decays involve the release of energy and sometimes mass.

radioactivity The spontaneous release of particles and/or energy from the nucleus.

alpha (α) decay A radioactive decay in which the nucleus ejects an alpha particle consisting of two protons and two neutrons.

beta (β) decay A radioactive decay in which the nucleus ejects an electron (a beta particle); this change transforms a neutron into a proton.

gamma (γ) decay A radioactive decay in which the nucleus releases energy but no particles.

decay series A naturally occurring sequence of radioactive decays; also called a *decay chain*.

half-life The amount of time required for one-half of a radioactive substance to decay.

16.3 Working with Radiation

Geiger counter A device that measures radiation, including alpha- and beta-emissions and gamma rays.

scintillation counter A device used for measuring radiation; a scintillation counter measures the total number of radiative particles and waves striking the detector.

semiconductor counter A device that measures radiation; some semiconductor counters can distinguish between the decays of different elements based on the energy of the radiation.

dosimeter A radiation detector that measures human exposure to radiation.

becquerel (Bq) The number of radioactive decays that occur per second for a particular substance.

sievert (Sv) A measure of radiation useful for measuring human exposure; 1 Sv = 1 joule of energy per kilogram of mass.

nuclear medicine The use of radioactive nuclei for medicinal purposes.

16.4 Energy Changes in Nuclear Reactions

nuclear force The force that holds the nucleus together.

mass defect The difference between the masses of the individual particles and the mass of the complete nucleus.

binding energy The energy released when protons and neutrons combine to form a nucleus, or the energy required to break a nucleus into its component particles.

law of conservation of matter and energy In any change, the total matter and energy in the universe remain constant.

16.5 Nuclear Power: Fission and Fusion

fission A nuclear change in which a large nucleus shatters into several smaller nuclei, releasing large amounts of energy.

fusion A nuclear change in which two smaller nuclei (usually hydrogen or helium) combine to form larger nuclei.

plasma A gas-like state produced at very high temperatures and often considered the fourth state of matter; in this state, electrons do not remain bound to a single atom.

Additional Problems

16.1 Nuclear Changes

11. Which subatomic particle determines the elemental identity of an atom?

12. What is the difference between the atomic number and the mass number?

13. What is the difference between nucleons and nuclides?

14. What is the difference between nuclides and isotopes?

15. Identify the element that corresponds to each of these nuclides:

a. a nuclide with atomic number 15
b. a nuclide with atomic number 37
c. a nuclide with 6 protons and 7 neutrons
d. a nuclide with 9 protons and 10 neutrons

16. Identify the element that corresponds to each of these nuclides:

a. a nuclide with atomic number 18
b. a nuclide with atomic number 4
c. a nuclide with 7 protons and 8 neutrons
d. a nuclide with 21 protons and 24 neutrons

17. Determine the number of protons and neutrons in each of these nuclides:

a. ${}^{40}_{20}\mathrm{Ca}$ b. ${}^{41}_{20}\mathrm{Ca}$ c. ${}^{23}\mathrm{Na}$ d. ${}^{52}\mathrm{Cr}$

18. Determine the number of protons and neutrons in each of these nuclides:

a. ${}^{56}_{26}\mathrm{Fe}$ b. ${}^{54}_{26}\mathrm{Fe}$ c. ${}^{98}\mathrm{Tc}$ d. ${}^{159}\mathrm{Tb}$

19. Write atomic symbols, including atomic number and mass number, for these nuclides:

a. a nuclide with 15 protons and 17 neutrons
b. a nuclide with 4 protons and 4 neutrons
c. a nuclide with 9 protons and a mass number of 19
d. a nuclide with an atomic number 13 and mass number 28

20. Write atomic symbols, including atomic number and mass number, for these nuclides:

a. a nuclide with 31 protons and 38 neutrons
b. a nuclide with 26 protons and 30 neutrons
c. a nuclide with 46 protons and a mass number of 106
d. a nuclide with atomic number 16 and mass number 32

21. Write atomic symbols, including atomic number and mass number, for these nuclides:

a. helium-4
b. carbon-14
c. uranium-235

22. Write atomic symbols, including atomic number and mass number, for these nuclides:

a. magnesium-24
b. silicon-29
c. ytterbium-172

23. Indicate whether or not each of these nuclides is an isotope of carbon-12:

a. a nuclide with atomic number 5 and mass number 12
b. a nuclide in which 6 of the 13 nucleons are positively charged
c. a nuclide with 8 protons and 9 neutrons
d. a nuclide with 6 protons and 8 neutrons

24. Indicate whether or not each of these nuclides is an isotope of ${}^{16}_{8}O$:

a. a nuclide with atomic number 7 and mass number 16
b. a nuclide with atomic number 8 and mass number 17
c. a nuclide containing 18 nucleons, including 8 protons
d. a nuclide with 8 protons and 8 neutrons

25. Write a nuclear equation to describe this process: When samarium-152 is struck by a neutron, it forms samarium-153.

26. Write a nuclear equation to describe this process: A thorium-232 nucleus decomposes to form radium-228 and helium-4.

27. Fill in the missing nuclide to complete each nuclear equation:

a. ${}^{1}_{0}n + {}^{98}_{42}Mo \rightarrow$ ______
b. ${}^{222}_{86}Rn \rightarrow$ ______ $+ {}^{218}_{84}Po$
c. ${}^{235}_{92}U + {}^{1}_{0}n \rightarrow$ ______ $+ {}^{97}_{37}Rb + 2\,{}^{1}_{0}n$

28. Find the nuclide necessary to complete each nuclear equation:

a. ${}^{1}_{0}n + {}^{130}_{52}Te \rightarrow$ ______
b. ${}^{240}_{94}Pu \rightarrow {}^{4}_{2}He +$ ______
c. ${}^{235}_{92}U + {}^{1}_{0}n \rightarrow {}^{99}_{42}Mo +$ ______ $+ 2\,{}^{1}_{0}n$

29. Find the nuclide necessary to complete each nuclear equation:

a. ${}^{226}_{88}Ra \rightarrow {}^{4}_{2}He +$ ______
b. ${}^{1}_{1}p + {}^{111}_{48}Cd \rightarrow {}^{111}_{49}In +$ ______
c. ${}^{235}_{92}U + {}^{1}_{0}n \rightarrow {}^{156}_{62}Sm +$ ______ $+ 3\,{}^{1}_{0}n$

30. Find the nuclide necessary to complete each nuclear equation:

a. ${}^{214}_{84}Po \rightarrow$ ______ $+ {}^{210}_{82}Pb$
b. ${}^{1}_{1}p + {}^{100}_{42}Mo \rightarrow$ ______ $+ 2\,{}^{1}_{0}n$
c. ${}^{239}_{94}Pu + {}^{1}_{0}n \rightarrow$ ______ $+ {}^{140}_{54}Xe + 2\,{}^{1}_{0}n$

31. Palladium-103 is commonly used in treating prostate cancer. This isotope is produced by striking ${}^{102}Pd$ with a fast-moving neutron. Write a balanced nuclear equation to describe this process.

32. When a proton collides with ${}^{111}_{48}Cd$ in a particle accelerator, it forms a new nucleus, ${}^{111}_{49}In$, and emits a neutron. Write a balanced nuclear equation to describe this change.

33. When an alpha particle collides with ${}^{121}_{51}Sb$ in a particle accelerator, it forms a new nucleus, ${}^{123}I$, plus several neutrons. How many neutrons are released in this process? Write a balanced nuclear equation to describe this process.

34. When an alpha particle collides with ${}^{109}_{47}Ag$ in a particle accelerator, it forms a new nucleus, ${}^{111}In$, plus several neutrons. How many neutrons are released in this process? Write a balanced nuclear equation to describe this process.

35. A fast-moving neutron strikes a ${}^{235}U$ nucleus. The nucleus shatters, producing ${}^{155}Sm$, ${}^{78}Zn$, and three neutrons. Describe this change using a balanced nuclear equation.

36. Among the reactions that take place in the Sun is the combination of tritium (${}^{3}H$) and deuterium (${}^{2}H$) to produce ${}^{4}He$ plus a neutron. Describe this change using a balanced nuclear equation.

16.2 Radioactivity

37. What are the differences between alpha, beta, and gamma radiation?

38. Identify each of these events as alpha, beta, or gamma decays:

a. A cobalt-60 nucleus emits energy, but no mass.
b. Plutonium-239 ejects a helium nucleus to form uranium-235.
c. A nitrogen-16 nucleus ejects an electron to form oxygen-16.

39. Identify these reactions as alpha, beta, or gamma decays:

a. ${}^{201}_{79}\text{Au} \rightarrow {}^{201}_{80}\text{Hg} + {}^{0}_{-1}\text{e}$

b. ${}^{99m}_{42}\text{Mo} \rightarrow {}^{99}_{42}\text{Mo} + \text{energy}$

c. ${}^{233}_{92}\text{U} \rightarrow {}^{4}_{2}\text{He} + {}^{229}_{90}\text{Th}$

40. Identify these reactions as alpha, beta, or gamma decays:

a. ${}^{227}_{89}\text{Ac} \rightarrow {}^{223}_{87}\text{Fr} + {}^{4}_{2}\text{He}$

b. ${}^{227}_{89}\text{Ac} \rightarrow {}^{227}_{90}\text{Th} + {}^{0}_{-1}\text{e}$

c. ${}^{222}_{86}\text{Rn} \rightarrow {}^{222}_{86}\text{Rn} + \text{energy}$

41. Show the products formed by these alpha decays:

a. ${}^{226}_{88}\text{Ra} \xrightarrow{\alpha}$ b. ${}^{229}_{90}\text{Th} \xrightarrow{\alpha}$ c. ${}^{210}_{84}\text{Po} \xrightarrow{\alpha}$

42. Show the products formed by these alpha decays:

a. ${}^{238}_{92}\text{U} \xrightarrow{\alpha}$ b. ${}^{241}_{95}\text{Am} \xrightarrow{\alpha}$ c. ${}^{223}_{87}\text{Fr} \xrightarrow{\alpha}$

43. Show the products formed by these beta decays:

a. ${}^{8}_{3}\text{Li} \xrightarrow{\beta}$ b. ${}^{25}_{11}\text{Na} \xrightarrow{\beta}$ c. ${}^{223}_{87}\text{Fr} \xrightarrow{\beta}$

44. Show the products formed by these beta decays:

a. ${}^{28}_{13}\text{Al} \xrightarrow{\beta}$ b. ${}^{207}_{81}\text{Tl} \xrightarrow{\beta}$ c. ${}^{231}_{90}\text{Th} \xrightarrow{\beta}$

45. Write balanced equations showing these processes:

a. alpha decay of ${}^{231}_{90}\text{Th}$

b. beta decay of ${}^{228}_{89}\text{Ac}$

c. gamma decay of ${}^{99m}_{43}\text{Tc}$

46. Write balanced equations showing these processes:

a. alpha decay of ${}^{232}_{90}\text{Th}$

b. beta decay of ${}^{120}_{48}\text{Cd}$

c. gamma decay of ${}^{124m}_{54}\text{Xe}$

47. Astatine-219 sometimes undergoes this series of decays: α, β, β. What elements are formed in this decay series?

48. Polonium-218 is a radioactive nuclide that sometimes decays through this pathway: β, β, α, α, β, β. What nuclides form in this decay series?

49. Thorium-227 is a radioactive isotope that sometimes decays through this pathway: $\alpha, \alpha, \alpha, \beta, \alpha, \beta$. What nuclides form in this decay series? What stable nucleus results from this decay?

50. Lead-210 is a radioactive isotope that sometimes decays through this pathway: $\beta, \alpha, \beta, \beta, \beta, \alpha$. What nuclides form in this decay series? What stable nucleus results from this decay?

16.3 Working with Radiation

51. Why is alpha radiation more damaging to tissue than gamma radiation? Why can we shield ourselves more easily from alpha radiation than from gamma radiation?

52. List two factors that determine how damaging radiation is to human tissue.

53. List three devices commonly used to measure radiation.

54. How are a Geiger counter and a scintillation counter similar? How are they different?

55. What advantage does a semiconductor counter have over a traditional Geiger counter?

56. Although becquerels (Bq) and sieverts (Sv) are both used to measure radiation, they are important in very different circumstances. What is the difference between these units? When is each of them important?

57. Convert these radiation exposure measurements to rem:

a. exposure for a full-body CT scan (0.01275 Sv)

b. exposure from living in a brick house for one year (0.0007 Sv)

c. radiation from outer space for a person living near sea level on the Atlantic Coast of the United States (0.41 mSv)

58. Convert these radiation exposure measurements to millisieverts (mSv):

a. exposure from a chest X-ray (10 mrem)

b. living within 50 miles of a coal-fired power plant (3×10^{-5} rem)

c. exposure from smoking a pack of cigarettes each day (36 mrem)

59. Some people worry about the cancer risk associated with a dental X-ray. The X-ray provides an exposure of 0.005 mSv while the threshold exposure for an increased risk of cancer is 100 mSv. Ignoring other sources of radiation, how many dental X-rays would be required before the radiation exposure reached the threshold for a cancer risk?

60. Health officials encourage women over age 50 to get a mammogram (an X-ray test for breast cancer) every two years. While X-rays may detect tumors, they also can cause them. The radiation exposure from a mammogram is 40 mrem, and the lowest annual exposure that has been clearly linked to an increased risk of cancer is 100 mSv. Ignoring other sources of radiation, how many mammograms would be required in a year before the radiation exposure reached the threshold for a cancer risk?

61. According to U.S. guidelines, companies cannot expose employees to more than 50 mSv per year while they are working. For people flying in an airplane, radiation exposure is 0.005 mSv/hour. At this rate, what is the annual radiation exposure for a flight attendant who spends 1,200 hours in the air?

62. Fukushima City in Japan is a city of approximately 300,000 people, located about 60 km northwest of the site of the 2011 nuclear meltdown. In April 2020, radiation measurements in this city showed radiation levels of 0.09 μSv/hr. On the same day, measurements about 10 kilometers from the meltdown site were 5.75 μSv/hr. An exposure level of 100 mSv/year has been linked to an increased risk of cancer. Based on the measured levels, is a resident of Fukushima City at increased risk for cancer? How does the annual exposure level in Fukushima City compare to the exposure levels closer to the site?

63. In nuclear medicine, radioactive nuclei are used primarily for imaging and for radiation therapy. Give an example of a nucleus used for each procedure, and briefly describe the role it plays.

64. Most radioisotopes used in medicinal imaging are gamma emitters rather than alpha or beta emitters. Why do you think this is so?

65. How is carbon-14 formed in Earth's atmosphere? How are scientists able to use this process to determine the age of plant matter?

66. Although carbon-14 is taken up by plants, scientists also use it to determine the age of animal remains, such as bone or leather. Why are they able to do this?

67. Scientists measure the number of carbon-14 decays to determine the percentage of carbon-14 in a sample. The carbon-14 in living plant matter decays at a rate of about 240 Bq (that is, 240 decays/second). What will the rate of decay be after one half-life? After two half-lives?

68. A sample of leather is analyzed by carbon-14 dating. The concentration of ^{14}C in this sample is found to be about 1/16th of what a modern leather sample would contain. About how old is the sample? The half-life of ^{14}C is 5,370 years.

69. How are beryllium-10 and aluminum-26 formed in quartz rock (SiO_2)? What information can geologists mine from the concentration of these isotopes in quartz?

70. Recently, a team of geologists measured the levels of beryllium-10 in quartz rock from Mt. Darling, Antarctica (**Figure 16.40**). They found that the concentrations of Be-10 decreased as they sampled farther down the mountain. What does this finding suggest about the ice on Mt. Darling?

Figure 16.40 An aerial view of Mt. Darling, Antarctica.

16.4 Energy Changes in Nuclear Reactions

71. What is the mass defect of a nucleus?

72. Chemical reactions (those involving the gain, loss, or sharing of electrons) follow the law of conservation of mass. Nuclear reactions (those involving changes in the nuclear structure) do not. Why is this so?

73. What is the relationship between the mass defect and the energy released in a nuclear reaction?

74. What is the binding energy of a nucleus?

75. The most stable nucleus (in terms of binding energy per nucleon) is ^{56}Fe. Based on this, what type of nuclear reactions would you anticipate from nuclei that are much lighter, and those that are much heavier, than this nucleus?

76. In general, how does the neutron–proton ratio change from small nuclides to larger nuclides?

77. The nuclear reaction below absorbs energy. Based on the law of conservation of matter and energy, which has more mass, the reactants or the products?

$$\text{Energy} + {}^{2}_{1}\text{H} + {}^{14}_{7}\text{N} \rightarrow {}^{13}_{7}\text{N} + {}^{3}_{1}\text{H}$$

78. The alpha decay below releases a tremendous amount of energy. Based on the law of conservation of matter and energy, which has more mass, the reactants or the products?

$${}^{240}_{94}\text{Pu} \rightarrow {}^{4}_{2}\text{He} + {}^{236}_{92}\text{U} + \text{energy}$$

16.5 Nuclear Power: Fission and Fusion

79. In a nuclear fission reaction, a heavy nucleus splits into several lighter nuclei. Complete these nuclear equations to show the missing products from the fission of a uranium-235 nucleus:

a. ${}^{1}_{0}\text{n} + {}^{235}_{92}\text{U} \rightarrow {}^{95}_{38}\text{Sr} + \_\_\_\_\_\_ + 2\,{}^{1}_{0}\text{n}$

b. ${}^{1}_{0}\text{n} + {}^{235}_{92}\text{U} \rightarrow \_\_\_\_\_\_ + {}^{138}_{54}\text{Xe} + 3\,{}^{1}_{0}\text{n}$

c. ${}^{1}_{0}\text{n} + {}^{235}_{92}\text{U} \rightarrow {}^{93}_{37}\text{Rb} + \_\_\_\_\_\_ + 2\,{}^{1}_{0}\text{n}$

80. In a nuclear fission reaction, a heavy nucleus splits into several lighter nuclei. Complete these nuclear equations to show the missing products from the fission of a plutonium-239 nucleus:

a. ${}^{1}_{0}\text{n} + {}^{239}_{94}\text{Pu} \rightarrow {}^{100}_{40}\text{Zr}\ \_\_\_\_\_\_ + 3\,{}^{1}_{0}\text{n}$

b. ${}^{1}_{0}\text{n} + {}^{239}_{94}\text{Pu} \rightarrow \_\_\_\_\_\_ + {}^{134}_{52}\text{Te} + 3\,{}^{1}_{0}\text{n}$

c. ${}^{1}_{0}\text{n} + {}^{239}_{94}\text{Pu} \rightarrow {}^{134}_{53}\text{I} + \_\_\_\_\_\_ + 3\,{}^{1}_{0}\text{n}$

81. What is a nuclear chain reaction? What subatomic particles are responsible for propagating a chain reaction?

82. Describe the isotope distribution of naturally occurring uranium. In nuclear reactions, how do these isotopes differ?

83. Approximately 95% of the uranium in a nuclear fuel pellet is the non-fissile isotope ${}^{238}\text{U}$. Why don't nuclear reactors use highly purified ${}^{235}\text{U}$ for fuel? What purposes does the ${}^{238}\text{U}$ serve?

84. Describe the differences in the composition of uranium fuel used for nuclear weapons and uranium fuel used for peaceful nuclear power.

85. Write two balanced chemical reactions showing the conversion of uranium oxide (U_3O_8) to uranium hexafluoride.

86. What is a gas centrifuge? How do these machines separate the isotopes of uranium?

87. In a nuclear reactor, uranium-238 nuclei absorb neutrons to become U-239. This unstable isotope undergoes two beta decays to produce plutonium-239. Write two balanced nuclear equations showing the products formed in this decay sequence.

88. In fission reactors, what feature controls the chain reaction to prevent it from becoming too hot and melting through the containment vessel?

89. Describe the steps required to convert nuclear energy to electrical energy in a power generator.

90. Describe the role of the reactor loop, the power loop, and the cooling loop in the function of a nuclear reactor.

91. Nuclear reactions produce highly radioactive by-products. How does the reactor design prevent the cooling water from being contaminated with radioactive by-products?

92. What is the difference between nuclear fission and nuclear fusion?

93. What three conditions must be met for a fusion reaction to occur?

94. Fill in the missing reactants or products to complete these fusion reactions:

a. $\_\_\_\_\_\_ + {}^{2}_{1}\text{H} \rightarrow {}^{3}_{2}\text{H}$

b. ${}^{3}_{2}\text{H} + {}^{3}_{2}\text{H} \rightarrow \_\_\_\_\_\_ + 2\,{}^{1}_{1}\text{H}$

c. ${}^{2}_{1}\text{H} + {}^{3}_{1}\text{H} \rightarrow {}^{4}_{2}\text{He} + \_\_\_\_\_\_$

95. One of the most common fusion reactions involves tritium (${}^{3}_{1}\text{H}$). Nuclear scientists produce tritium by bombarding lithium-6 nuclei with neutrons. When a neutron strikes lithium-6, the nucleus splits to produce helium-4 and tritium. Write a balanced nuclear equation for this reaction.

96. What major technical challenge has hindered the development of nuclear fusion?

97. What is a plasma? How does a plasma differ from a gas?

98. What is a tokamak? What potential does the tokamak hold for controlled nuclear fusion?

99. What is the break-even point in nuclear fusion?

100. The nuclear fission reaction below releases a tremendous amount of energy. Based on this, which has the greater mass—the starting materials, or the products?

$${}^{1}_{0}\text{n} + {}^{235}_{92}\text{U} \rightarrow {}^{144}_{56}\text{Ba} + {}^{90}_{36}\text{Kr} + 2\,{}^{1}_{0}\text{n} + \text{energy}$$

USEFUL TABLES AND FIGURES

Common Unit Conversions

Length	Volume
1 m = 3.281 ft	1 L = 1 dm^3
1 km = 0.621 mi	1 mL = 1 cm^3
1 cm = 0.394 in	1 L = 0.264 gal
Mass or Weight	
1 kg = 2.205 lb	
Energy	**Pressure**
1 J = 1 $kg{\cdot}m^2/s^2$	1 atm = 760 mm Hg (torr)
1 cal = 4.184 joules	1 atm = 14.70 psi
1,000 cal = 1 kcal = 1 Calorie	1 atm = 101.3 kPa
1 BTU = 1,055 J	1 atm = 1.013 bar

Common Metric Prefixes

Prefix	Symbol	Meaning	
tera-	T	10^{12}	1,000,000,000,000
giga-	G	10^{9}	1,000,000,000
mega-	M	10^{6}	1,000,000
kilo-	k	10^{3}	1,000
deci-	d	10^{-1}	$\frac{1}{10}$
centi-	c	10^{-2}	$\frac{1}{100}$
milli-	m	10^{-3}	$\frac{1}{1,000}$
micro-	μ	10^{-6}	$\frac{1}{1,000,000}$
nano-	n	10^{-9}	$\frac{1}{1,000,000,000}$
pico-	p	10^{-12}	$\frac{1}{1,000,000,000,000}$

Mathematical Relationships and Constants

Density	
$d = \frac{m}{V}$	
Temperature	
$°F = \frac{9}{5}°C + 32$ $°C = \frac{5}{9}(°F - 32)$	$K = °C + 273.15$
Energy, frequency, and wavelength	
$C = \lambda v$	$c = 3.00 \times 10^{8}$ m/s
$E = hv$	$h = 6.63 \times 10^{-34}$ J·s
Avogadro's number	
6.02×10^{-23} particles = 1 mole	
Energy and temperature	
$\Delta E = q + w$	$q = C\Delta T$
$q = ms\Delta T$	$s_{water} = 4.184$ J/g·°C
Gas laws	
$\frac{P_1V_1}{T_1} = \frac{P_2V_2}{T_2}$	
$PV = nRT$	$R = 0.0821$ L·atm/mol·K
Concentration	
$\text{ppm} = \frac{\text{mass of solute}}{\text{mass of solution}} \times 10^{6}$	$\text{mass\%} = \frac{\text{mass of solute}}{\text{mass of solution}} \times 100\%$
$\text{ppb} = \frac{\text{mass of solute}}{\text{mass of solution}} \times 10^{9}$	$\text{volume\%} = \frac{\text{volume of solute}}{\text{volume of solution}} \times 100\%$
$\text{Molarity} = \frac{\text{moles of solute}}{\text{liters of solution}}$	$\text{mass/volume\%} = \frac{\text{mass of solute}}{\text{volume of solution}} \times 100\%$
Acids and bases	
$[H^{+}][OH^{-}] = 1 \times 10^{-14}$	$pH = -\log[H^{+}]$

The Activity Series

Reducing strength (increasing upward)

Element	Half-Reaction
Lithium	$Li \rightarrow Li^{+} + e^{-}$
Potassium	$K \rightarrow K^{+} + e^{-}$
Barium	$Ba \rightarrow Ba^{2+} + 2\ e^{-}$
Calcium	$Ca \rightarrow Ca^{2+} + 2\ e^{-}$
Magnesium	$Mg \rightarrow Mg^{2+} + 2\ e^{-}$
Aluminum	$Al \rightarrow Al^{3+} + 3\ e^{-}$
Zinc	$Zn \rightarrow Zn^{2+} + 2\ e^{-}$
Chromium	$Cr \rightarrow Cr^{3+} + 3\ e^{-}$
Iron	$Fe \rightarrow Fe^{2+} + 2\ e^{-}$
Cobalt	$Co \rightarrow Co^{2+} + 2\ e^{-}$
Nickel	$Ni \rightarrow Ni^{2+} + 2\ e^{-}$
Tin	$Sn \rightarrow Sn^{2+} + 2\ e^{-}$
Lead	$Pb \rightarrow Pb^{2+} + 2\ e^{-}$
Hydrogen	$H_2 \rightarrow 2\ H^{+} + 2\ e^{-}$
Copper	$Cu \rightarrow Cu^{2+} + 2\ e^{-}$
Silver	$Ag \rightarrow Ag^{2+} + e^{-}$
Platinum	$Pt \rightarrow Pt^{2+} + 2\ e^{-}$
Gold	$Au \rightarrow Au^{+} + e^{-}$

Elemental Color Codes

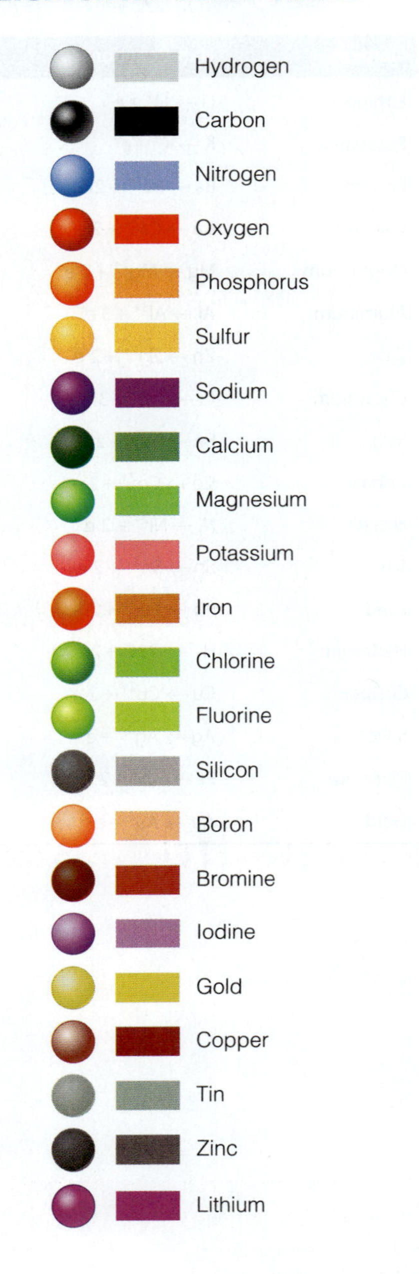

Solubility Rules

		Ammonium NH_4^+	Sodium Na^+	Potassium K^+	Silver Ag^+	Magnesium Mg^{2+}	Calcium Ca^{2+}	Barium Ba^{2+}	Iron(II) Fe^{2+}	Zinc Zn^{2+}	Copper(II) Cu^{2+}	Lead(II) Pb^{2+}	Iron(III) Fe^{3+}	Aluminum Al^{3+}
Nitrate	NO_3^-	Soluble	Soluble	Soluble	Soluble	Soluble	Soluble	Soluble	Soluble	Soluble	Soluble	Soluble	Soluble	Soluble
Chlorate	ClO_3^-	Soluble	Soluble	Soluble	Soluble	Soluble	Soluble	Soluble	Soluble	Soluble	Soluble	Soluble	Soluble	Soluble
Perchlorate	ClO_4^-	Soluble	Soluble	Soluble	Soluble	Soluble	Soluble	Soluble	Soluble	Soluble	Soluble	Soluble	Soluble	Soluble
Acetate	$C_2H_3O_2^-$	Soluble	Soluble	Soluble	Soluble	Soluble	Soluble	Soluble	Soluble	Soluble	Soluble	Soluble	Soluble	Soluble
Chloride	Cl^-	Soluble	Soluble	Soluble	Insoluble	Soluble	Soluble	Soluble	Soluble	Soluble	Soluble	Insoluble	Soluble	Soluble
Bromide	Br^-	Soluble	Soluble	Soluble	Insoluble	Soluble	Soluble	Soluble	Soluble	Soluble	Soluble	Insoluble	Soluble	Soluble
Iodide	I^-	Soluble	Soluble	Soluble	Insoluble	Soluble	Soluble	Soluble	Soluble	Soluble	Soluble	Insoluble	Soluble	Soluble
Sulfate	SO_4^{2-}	Soluble	Soluble	Soluble	Insoluble	Soluble	Insoluble	Insoluble	Soluble	Soluble	Soluble	Insoluble	Soluble	Soluble
Hydroxide	OH^-	Soluble	Soluble	Soluble	Insoluble	Insoluble	Insoluble	Soluble	Insoluble	Insoluble	Insoluble	Insoluble	Insoluble	Insoluble
Sulfite	SO_3^{2-}	Soluble	Soluble	Soluble	Insoluble	Insoluble	Insoluble	Insoluble	Insoluble	Insoluble	Insoluble	Insoluble	Insoluble	Insoluble
Carbonate	CO_3^{2-}	Soluble	Soluble	Soluble	Insoluble	Insoluble	Insoluble	Insoluble	Insoluble	Insoluble	Insoluble	Insoluble	Insoluble	Insoluble
Phosphate	PO_4^{3-}	Soluble	Soluble	Soluble	Insoluble	Insoluble	Insoluble	Insoluble	Insoluble	Insoluble	Insoluble	Insoluble	Insoluble	Insoluble

Soluble
Insoluble

Pauling's Table of Electronegativities

H 2.1																	
Li 1.0	Be 1.5											B 2.0	C 2.5	N 3.0	O 3.5	F 4.0	
Na 0.9	Mg 1.2											Al 1.5	Si 1.8	P 2.1	S 2.5	Cl 3.0	
K 0.8	Ca 1.0	Sc 1.3	Ti 1.5	V 1.6	Cr 1.6	Mn 1.5	Fe 1.8	Co 1.8	Ni 1.8	Cu 1.9	Zn 1.7	Ga 1.6	Ge 1.8	As 2.0	Se 2.4	Br 2.8	
R 0.8	Sr 1.0	Y 1.2	Zr 1.4	Nb 1.6	Mo 1.8	Tc 1.9	Ru 2.2	Rh 2.2	Pd 2.2	Ag 1.9	Cd 1.7	In 1.7	Sn 1.8	Sb 1.9	Te 2.2	I 2.5	
Cs 0.7	Ba 0.9	La 1.1	Hf 1.3	Ta 1.5	W 1.7	Re 1.9	Os 2.2	Ir 2.2	Pt 2.2	Au 2.4	Hg 1.9	Tl 1.8	Pb 1.8	Bi 1.9	Po 2.0	At 2.2	
Fr 0.7	Ra 0.9																

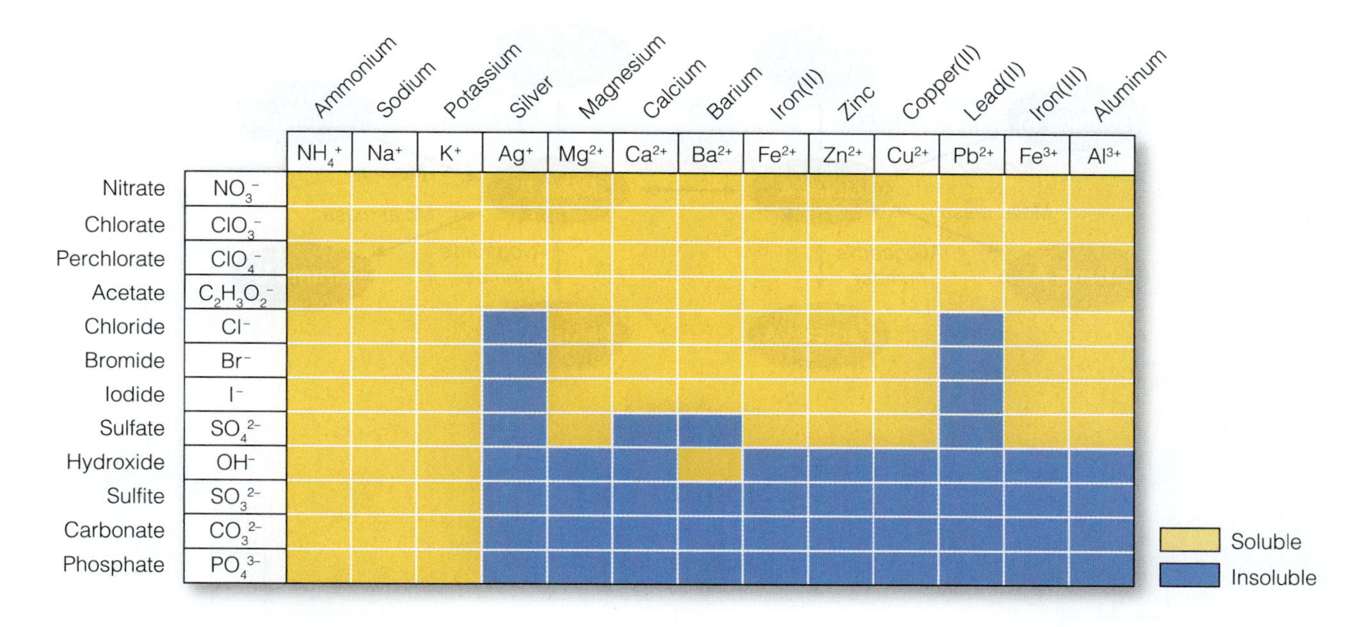

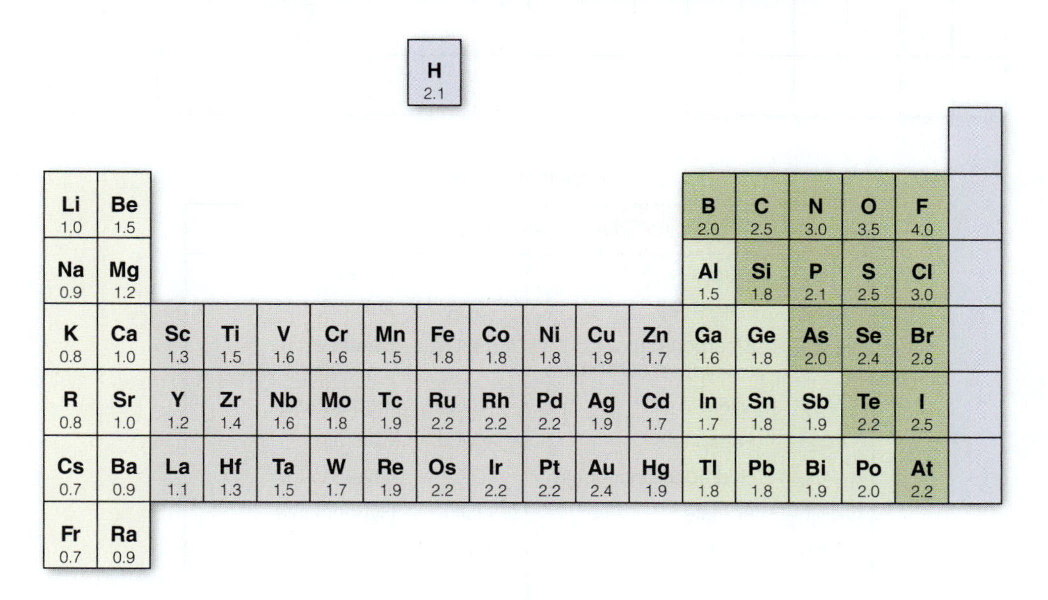

The Complete Mole Map

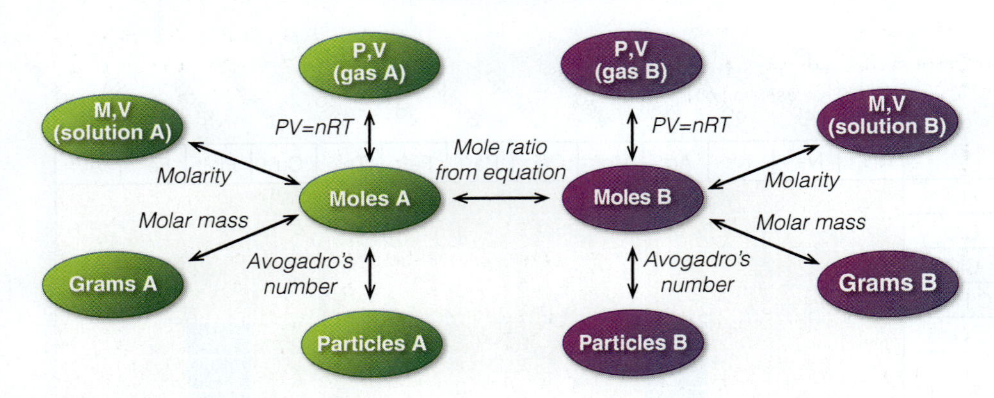

Common Ions

Monatomic atoms

H^+																	
Li^+	Be^{2+}													N^{3-}	O^{2-}	F^-	
Na^+	Mg^{2+}											Al^{3+}		P^{3-}	S^{2-}	Cl^-	
K^+	Ca^{2+}				Cr^{2+} Cr^{3+}	Mn^{2+} Mn^{3+}	Fe^{2+} Fe^{3+}	Co^{2+} Co^{3+}		Cu^+ Cu^{2+}	Zn^{2+}					Br^-	
Rb^+	Sr^{2+}									Ag^+			Sn^{2+} Sn^{4+}			I^-	
													Pb^{2+} Pb^{4+}				

Polyatomic atoms

NH_4^+	Ammonium		
NO_3^-	Nitrate	SO_4^{2-}	Sulfate
CO_3^{2-}	Carbonate	SO_3^{2-}	Sulfite
HCO_3^-	Bicarbonate (Hydrogen carbonate)	HSO_4^-	Bisulfate (Hydrogen sulfate)
NO_2^-	Nitrite	ClO_4^-	Perchlorate
PO_4^{3-}	Phosphate	ClO_3^-	Chlorate
HPO_4^{2-}	Hydrogen phosphate	ClO_2^-	Chlorite
$C_2H_3O_2^-$	Acetate	ClO^-	Hypochlorite
OH^-	Hydroxide	CrO_4^{2-}	Chromate
CN^-	Cyanide	$Cr_2O_7^{2-}$	Dichromate
O_2^{2-}	Peroxide	MnO_4^-	Permanganate

ANSWERS TO ODD-NUMBERED PROBLEMS

Chapter 1: Foundations

11. Matter is anything that has mass and takes up volume.
- **a.** Air is made of matter.
- **b.** A football is made of matter.
- **c.** Sunlight is not made of matter.
- **d.** Water is made of matter.

13. Composition refers to the simple components that make up the material. Structure refers to both the composition and arrangement of those simpler substances.

15. Mixtures in which the components are evenly blended throughout are homogeneous mixtures. In contrast, heterogeneous mixtures contain regions with significantly different composition.
- **a.** Earth's atmosphere is an example of a homogeneous mixture.
- **b.** Fluorine gas is an example of an element.
- **c.** Carbon monoxide is an example of a compound.

17. **a.** element
- **b.** heterogeneous mixture
- **c.** homogeneous mixture

19. Particles in a solid vibrate within their spaces in the ordered framework. As a solid is heated to a liquid, the atoms vibrate faster. The more heat we add, the faster the atoms move, until eventually they begin to break out of the rigid framework and travel freely past each other. The substance is now in the liquid state. The particles in a liquid move randomly but remain close to each other. If we continue heating the liquid, the atoms move faster and faster, until they begin to break out of the liquid phase and enter the gas phase. Particles in the gas phase move about freely, interacting very little with each other.

21. **a.** solid
- **b.** gas
- **c.** solid

23. **a.** physical
- **b.** physical
- **c.** chemical
- **d.** chemical
- **e.** physical

25. **a.** physical
- **b.** chemical
- **c.** physical

27. physical change

29. physical change

31. When a substance is heated, the particles within the substance vibrate or move faster and faster.

33. **a.** The rock on the top of the hill has more potential energy due to its position. The rock at the bottom of the hill is more stable because it has no more potential energy to release.
- **b.** Wood has more potential energy because it has more stored chemical energy in it than the ashes have. As such, the ashes are more stable because they have no more potential energy to release.
- **c.** A tightly wound spring has more potential energy because it has stored energy. The relaxed spring is more stable because it has no potential energy to release.

35. **a.** a mixture
- **b.** a compound
- **c.** chemical change
- **d.** The product, in container (b), has less potential energy and is more stable.

37. Because it absorbs energy from the sun, the growth is endothermic.

39. A hypothesis is a tentative explanation that has not been tested, whereas a theory is an idea supported by experimental evidence.

41. This is an example of a theory since it is supported by evidence and explains the world around us.

43. To test the hypothesis, you could assess the students several times during the 9–11 a.m. time frame and at other time frames during the day. Then, you would review the results and compare how the students did in each time frame. Did the students do better during the 9–11 a.m. slot or not? If they did, then your hypothesis was proven correct. To increase the reliability of your results, you should test the students several times during each time frame and make sure the assessment is of similar difficulty each time.

Chapter 2: Measurement

23. **a.** 1.5×10^6 km
- **b.** 3.98×10^{-4} g
- **c.** 1.2×10^9 J
- **d.** 2.019×10^{-2} s

25. **a.** 0.0005192 kg
- **b.** 1,230 J
- **c.** 420,000,000 °C
- **d.** 602,000,000,000,000,000,000,000 atoms

27. **a.** 5.21×10^{10} min
b. 8.3×10^{8} g
c. 4.35×10^{-7} m

29. **a.** 5.3×10^{-8} **b.** 1.2×10^{2} **c.** 1.2×10^{9}
d. 1.94×10^{-5}

31. **a.** ampere **b.** Kelvin **c.** millisecond
d. milligram **e.** kiloampere

33. **a.** 1,000,000 g
b. 1,000 calories
c. 1,000,000 microseconds
d. 1,000,000,000 A

35. A number rounded farthest to the right from the decimal is the more precise measurement. Therefore, a number rounded to the nearest nanometer would be more precise than a number rounded to the nearest gigameter.

37. Since we don't know the true value of the compound's mass, we cannot say with confidence which balance is more accurate. However, the second balance (the one that measured the compound as 11.1925 grams) is more precise, since it measured the mass to the ten-thousandths place as compared to the first balance (hundredths place).

39. Since the ruler is marked in increments of 1 cm, our estimation needs to go only as low as the tenths place. Since the test tube is slightly longer than 5 cm, we can safely estimate a measurement of 5.1 cm.

41. **a.** 4 significant digits
b. 4 significant digits
c. 5 significant digits
d. 2 significant digits

43. 20.31 meters has a total of 4 significant digits.

45. **a.** 5.200×10^{6} ng
b. 3.920×10^{-4} nL
c. $\$1.8000 \times 10^{6}$

47. Measured numbers have a degree of uncertainty, but exact numbers do not. For example, if you had 4 quarters in your pocket, you counted an exact number. There was nothing to measure.

49. **a.** exact
b. measured
c. measured
d. exact

51. **a.** 6.12×10^{3} ft^2
b. 31.43 cm
c. 7.0 g/cm^3

53. Each person should receive $250.00 (a sum of money reported in 5 significant digits).

55. 50.4 cm

57. 103.5 g

59. **a.** 2.13×10^{7} ng
b. 17.3974 km
c. 310 μL

61. **a.** 2.321×10^{-5} L
b. 5×10^{1} kg
c. 5.40×10^{2} nm

63. **a.** 81 μL
b. 0.112507 kg
c. 54.0 nm

65. 5.4×10^{-1} MΩ

67. **a.** 38.6 cm
b. 1.976×10^{5} J
c. 1.09×10^{3} kg
d. 5.71 km

69. 13 mm

71. 2.41×10^{2} meters/second

73. 14.6 km/L

75. 0.17 g/lb

77. 0.7 g/m^3

79. 1×10^{6} m^2

81. 27 ft^3

83. 1×10^{6} mL

85. 9.9 kW

87. 8.90×10^{3} kg/m^3

89. 0.643 g/mL

91. 1.0 g/mL

93. **a.** The solid will sink in water since its density is 1.05 g/cm^3.
b. The solid will float in water since its density is 0.96 g/cm^3.

95. **a.** 0.259 cm^3
b. 2.3×10^{2} cm^3

97. Iron: 1.6×10^{2} g
Aluminum: 54 g
Lead: 2.3×10^{2} g

99. Absolute zero is the lowest possible temperature at which particles have zero kinetic energy. Absolute zero is 0 K. Absolute zero is –273.15 °C. Absolute zero is –460 °F.

101. 2.5 °F is the larger temperature change, since $1.8 \times 1.3 = 2.34$ °F.

103. 22 °C; 295 K

105. Carpet is cheaper at $0.644/ft^2.
For an area of 600 ft^2, carpet would cost $386 and hardwood would cost $840. Hardwood will cost more.

107. **a.** 1.1×10^{3} mussels/m^2
b. low: 6.5×10^{11} mussels; high: 7.8×10^{11} mussels

Chapter 3: Atoms

15. The atom is the fundamental unit of matter.

17. **a.** Democritus
b. Antoine Lavoisier
c. John Dalton

19. Because the bottle is sealed (nothing can escape), the mass of the substances before the reaction is the same as the mass of the substances at the end of the reaction. This follows the law of conservation of mass, which states that matter is neither created nor destroyed in a chemical reaction.

21. According to the law of conservation of mass, matter cannot be created nor destroyed, so the mass of the products will be equal to the mass of the reactants. A total of 300 g of sodium chloride and water will be produced from the 200 g of sodium hydroxide and 100 g hydrogen chloride.

23. 40.3 g magnesium oxide

25. **a.** possible
b. not possible
c. not possible
d. possible

27. Reaction (a) does not follow the law of conservation of mass. The reactants contain four orange spheres and eight green spheres, whereas the products contain three orange spheres and nine green spheres. In the other two reactions, the number of orange and green spheres is the same for the reactants and products.

29. **a.** cesium
b. titanium
c. potassium
d. boron
e. xenon

31. **a.** tungsten
b. boron
c. uranium
d. sulfur
e. carbon

33. **a.** F **b.** Ge **c.** Sn **d.** Mg

35. **a.** I **b.** Cu **c.** Ag **d.** H

37. Elements within the same column (group) on the periodic table have similar properties.

39. **a.** metalloid
b. main-group metal
c. transition metal
d. main-group metal
e. inner-transition metal
f. nonmetal
g. main-group metal
h. transition metal

41. The blocks of elements on the left and right sides of the periodic table are the main-group elements.

43. **a.** metal
b. nonmetal
c. metal
d. metal

45. These are Group 1A elements known as the alkali metals.

47. **a.** halogen
b. noble gas
c. alkali metal
d. alkaline earth metal

49. Beryllium, because it is located within the same group (column) as magnesium

51. Argon and sulfur, because these two elements are not found in the same group as chlorine

53. **a.** V **b.** Cl **c.** Ne

55. positive, negative, and neutral

57. **1.** b **2.** c **3.** d **4.** a

59. In 1909, Ernest Rutherford and his students were studying the behavior of positively charged particles called *alpha particles*. They were interested in the pattern these heavy particles made as they passed through a thin gold film. Rutherford expected the alpha particles to pass through the film and hit behind it. However, a small number of the particles reflected off the film and back toward the alpha particle source.

61. The electron cloud occupies a larger volume than the nucleus, but the nucleus has more mass.

63. Dalton's idea that atoms could not be broken down into smaller pieces was proven incorrect with the discovery of subatomic particles.

65. The proton and the neutron each have a mass of about one atomic mass unit. Both of these particles are located in the nucleus. The proton has a +1 charge, and the electron has a −1 charge.

67. **a.** Ti = 22 **b.** S = 16
c. Te = 52 **d.** Li = 3

69. **a.** Al **b.** Rn **c.** Es

71. The atomic number is equal to the number of protons. The mass number is equal to the number of protons plus the number of neutrons.
a. atomic number = 14; mass number = 30
b. atomic number = 27; mass number = 59
c. atomic number = 20; mass number = 46

73. **a.** $^{39}_{19}K$
b. $^{40}_{18}Ar$
c. $^{257}_{100}Fm$

75. 28

77.

Atom	Symbol	Protons	Neutrons	Atomic Number	Mass Number
Hydrogen	H	1	0	1	1
Sulfur	S	16	16	16	32
Tellurium	Te	52	76	52	128
Helium	He	2	2	2	4
Zirconium	Zr	40	51	40	91

79. Total value = \$1.48
Weighted average value of coins = \$0.07

81. $^{69}_{31}$Ga and $^{71}_{31}$Ga

83. Both isotopes of silver have 47 protons. However, ^{107}Ag has 60 neutrons and ^{109}Ag has 62 neutrons.

85. Average atomic mass Br = 80.0 u

87. Average atomic mass Pb = 207.2174 u

89. The mass number of an isotope is the sum of the protons and neutrons. The mass that is usually displayed on periodic tables is the average atomic mass. This is a weighted average of the masses of the different isotopes of an element.

91. The plum pudding model described small, negatively charged electrons spread throughout a positive atomic substance. The Bohr model correctly described the dense, positively charged nucleus with orbiting electrons. The Bohr model described the electrons in fixed orbits around the nucleus, like planets orbiting the sun.

93. **a.** 1
b. 2
c. 6
d. 34

95. **a.** Li number of protons = 3; number of electrons = 3
b. Ge number of protons = 32; number of electrons = 32
c. Bi number of protons = 83; number of electrons = 83
d. Ba number of protons = 56; number of electrons = 56

97. Beryllium ion = +2 charge

99. **a.** −3 charge
b. +2 charge
c. 0 (neutral) charge

101. A neutral silver atom has 47 protons and will therefore have 47 electrons. A + 1 charged silver ion will have one less electron than its neutral counterpart; 46 electrons.

103. Titanium +2 has 22 protons and therefore 20 electrons.
Titanium +4 has 22 protons and therefore 18 electrons.

105. Gold +1 has 79 protons and therefore 78 electrons.
Mercury +2 has 80 protons and therefore 78 electrons.

107. F −1 has 9 protons and therefore 10 electrons.
Cl −1 has 17 protons and therefore 18 electrons.
Br −1 has 35 protons and therefore 36 electrons.

Chapter 4: Light and Electronic Structure

13. **a.** Ultraviolet is higher in energy than infrared.
b. Green light is higher in energy than red light.
c. Microwave is higher in energy than radio.

15. **a.** The low-frequency wave has less energy.
b. Infrared light is lower in energy than yellow light.

17. **a.** Red light has a lower frequency than green light.
b. The 400 nm light does *not* have a lower frequency than green light.
c. Light with a wavelength of 1.0×10^{-6} m has a lower frequency than green light.

19. **a.** radio wave
b. microwave
c. infrared light

21. **a.** 400 m
b. 1.5×10^{7} m
c. 1,200 m

23. Green.

25. 2.76×10^{-19} J

27. 9.47×10^{-15} J

29. $\nu = 7.04 \times 10^{13}\ s^{-1}$
$\lambda = 4.26 \times 10^{-6}$ m
This radiation falls in the infrared range.

31. Rather than producing a continuous spectrum of colors, gas lamps produce colors of only certain energies, called *spectral lines*. For example, the light from a hydrogen lamp contains only four lines of color: red, light blue, deep blue, and violet. These distinctive patterns are called *line spectra*. Scientists often use line spectra as "fingerprints" to identify elements. For example, when scientists analyze light from the Sun, they find lines that correspond to the spectral lines of hydrogen and helium. From this, they know that the Sun and other stars are composed largely of these two elements.

33. When an electron absorbs light (i.e., energy), it jumps to a higher energy level, producing an excited state electron.

35. Electrons absorb energy that causes them to jump to high energy levels. As these electrons return to the ground state, they release energy in the form of light.

37. **a.** absorbed
b. emitted
c. absorbed
d. emitted

39.

Transition	Wavelength	Region
5 → 4	4,057 nm	infrared
4 → 2	487 nm	visible
2 → 1	122 nm	ultraviolet

41. In quantum mechanics, we never talk about the exact position of an electron. According to Heisenberg, this is impossible for us to know. Instead, we talk about the *most probable locations* of the electrons or the *energies* that the electrons possess. These probable locations are orbitals.

43. 2

45. **a.** Level 1: The *s* orbital is present.
b. Level 2: The *s* and *p* sublevel orbitals are present.
c. Level 3: The *s*, *p*, and *d* sublevel orbitals are present.
d. Level 4: The *s*, *p*, *d*, and *f* sublevel orbitals are present.

47. A total of 6 electrons can reside in a *p* sublevel.

49. An orbital is the region where the electrons are most likely to be found. Each sublevel contains a different number of these orbitals.

51. Two electrons can occupy the lowest energy level since it's an *s* orbital. The second lowest energy level holds the *s* and *p* orbitals, which can hold up to 8 electrons.

53. **a.** This is possible since level 1 is the *s* orbital.
b. This is possible since level 4 includes the *s* orbital.
c. This is not possible since the *p* orbital is not part of level 1.

55.

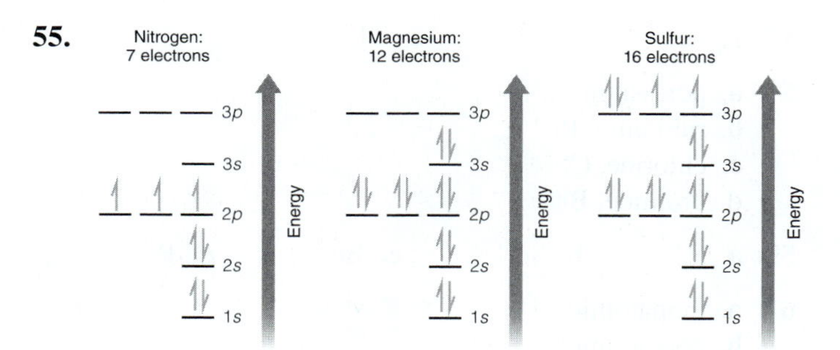

57. **a.** $1s^1$
b. $1s^2 2s^2$
c. $1s^2 2s^2 2p^2$
d. $1s^2 2s^2 2p^6$

59. The valence level is the highest-occupied energy level. The *s* and *p* sublevels are involved with the atom's valence.

61. The octet rule states that an atom is stabilized by having its highest-occupied (valence) energy level filled. The noble gas family illustrates the octet rule.

63. **a.** $[He]2s^1$
b. $[Ne]3s^2$
c. $[Ne]3s^2 3p^5$

65. F: $[He]2s^2 2p^5$
Cl: $[Ne]3s^2 3p^5$
Br: $[Ar]4s^2 3d^{10} 4p^5$
I: $[Kr]5s^2 4d^{10} 5p^5$

67. The electronic configuration for Mg is $1s^2 2s^2 2p^6 3s^2$. It has 2 valence electrons ($3s^2$) and 10 inner-shell electrons ($1s^2 2s^2 2p^6$).

69. **a.** The inner-shell electrons are $1s^2$ and the valence electrons are $2s^2 2p^3$.
b. The inner-shell electrons are $1s^2 2s^2 2p^6 3s^2 3p^6$ and the valence electrons are $4s^1$.
c. The inner-shell electrons are [Ar] and the valence electrons are $4s^2 3d^{10} 4p^1$.
d. The inner-shell electrons are [Xe] and the valence electrons are $6s^2 4f^{14} 3d^4$.

71. For anions, you add the same electrons to the outer sublevels as the charge. For cations, you remove the same number from the valence electrons.
a. O^{2-} $1s^2 2s^2 2p^6$
b. K^+ $1s^2 2s^2 2p^6 3s^2 3p^6$
c. Br^- $1s^2 2s^2 2p^6 3s^2 3p^6 4s^2 3d^{10} 4p^6$
d. N^{3-} $1s^2 2s^2 2p^6 3s^2 3p^6$

73. **a.** O would need to gain 2 electrons and be O^{2-}.
b. F would need to gain 1 electron and be F^-.
c. Na would need to lose 1 electron and be Na^+.
d. Mg would need to lose 2 electrons and be Mg^{2+}.

75. Li^+, Be^{2+}, and H^-

77. Fluorine has an electronic configuration of $1s^2 2s^2 2p^5$ and needs only to gain one electron to have the noble gas–like configuration of neon ($1s^2 2s^2 2p^6$). Sodium has an electron configuration of $1s^2 2s^2 2p^6 3s^1$ and needs only to lose an electron to have neon's electronic configuration. F gaining one electron and Na losing one electron gives a filled $2s^2$ and $2p^6$ sublevel set (i.e., obeys the octet rule).

79. K^+ and S^{2-} will both have the stability of a filled valence level having an electronic configuration of Ar ($1s^2 2s^2 2p^6 3s^2 3p^6$).

81. Their outer electrons are found in an *s* orbital.

83. Si, P, and Ar all have valence electrons in energy level 3 (all row 3 elements).

85. The inner electron configuration of elements in row 2 is $1s^2$.

87. Two, from their *s* valence orbital

89. **a.** 2 **b.** 3
c. 7 **d.** 8

91. **a.** Mg: $3s^2$
b. N: $2s^2 2p^3$
c. P: $3s^2 3p^3$
d. As: $4s^2 4p^3$
e. Cl: $3s^2 3p^5$
f. Ca: $4s^2$

93. **a.** K **b.** N **c.** Pd

95. **a.** Following the 4*s* is the 3*d* subshell.
b. Following the 4*d* is the 5*p* subshell.
c. Following the 3*p* is the 4*s* subshell.

97. **a.** $4p^2$ **b.** $5p^4$
c. $4d^5$ **d.** $2p^1$

99. **a.** group 2A, the alkaline earth metals
b. group 8A, the noble gases
c. group 2A, the alkaline earth metals

101. The valence electron configuration of group 6A (the chalcogens) is s^2p^4.

103. Group 12 is the last column of the *d* block. The electronic configuration is d^{10}.

Chapter 5: Chemical Bonds and Compounds

15. Li has 1 valence electron.
C has 4 valence electrons.
Si has 4 valence electrons.
Kr has 8 valence electrons.
Se has 6 valence electrons.

17. Na• •N̈• H• •Äs• •S̈b•

19. Mg: $1s^2 2s^2 2p^6 3s^2$; the $3s^2$ are the valence electrons.
N: $1s^2 2s^2 2p^3$; the $2s^2 2p^3$ are the valence electrons.
P: $1s^2 2s^2 2p^6 3s^2 3p^3$; the $3s^2 3p^3$ are the valence electrons.
I: $1s^2 2s^2 2p^6 3s^2 3p^6 4s^2 3d^{10} 4p^6 5s^2 4d^{10} 5p^5$; the $5s^2 5p^5$ are the valence electrons.

21. **a.** It does not. A single electron occupies the 3*s* orbital.
b. It does. It's isoelectronic with Ne.
c. It does not. It needs one electron to complete an octet.
d. It does. It's isoelectronic with Ne.

23. the alkali metals

25. It is isoelectronic with Ar: $1s^2 2s^2 2p^6 3s^2 3p^6$.

27. **a.** $1s^2 2s^1$
b. $1s^2$
c. $1s^2 2s^2 2p^6 3s^1$
d. $1s^2 2s^2 2p^6$

29. **a.** +2 **b.** +2 **c.** +1 **d.** +1

31. +1 and + 2

33. **a.** sodium ion
b. magnesium ion
c. chromium(II) ion
d. chromium(III) ion

35. **a.** iron(II) ion
b. iron(III) ion
c. rubidium ion
d. barium ion

37. **a.** Sr^{2+}
b. Zn^{2+}
c. Cu^{2+}
d. Mn^{3+}

39. Group 7A, the halogens, forms only –1 ions. Group 6A, the chalcogens, forms only –2 ions.

41. One more electron would make it $[He]2s^2 2p^6$.

43. **a.** $1s^2 2s^2 2p^6 3s^2 3p^5$
b. $1s^2 2s^2 2p^6 3s^2 3p^6$
c. $1s^2 2s^2 2p^6 3s^2 3p^6 4s^2 3d^{10} 4p^5$
d. $1s^2 2s^2 2p^6 3s^2 3p^6 4s^2 3d^{10} 4p^6$

45. **a.** lose 1 electron to fulfill its octet
b. gain 2 electrons to fulfill its octet
c. lose 2 electrons to fulfill its octet
d. gain 1 electron to fulfill its octet

47. **a.** lose 2 electrons to fulfill its octet
b. gain 1 electron to fulfill its octet
c. gain 2 electrons to fulfill its octet
d. lose 2 electrons to fulfill its octet

49. **a.** F^-
b. I^-
c. O^{2-}
d. Se^{2-}

51. **a.** fluoride ion
b. sulfide ion
c. oxide ion
d. iodide ion

53. **a.** +2 **b.** –2 **c.** –1

55. **a.** –1 **b.** +1 **c.** –1

57. **a.** potassium, K^+
b. rubidium, Rb^+
c. chloride, Cl^-
d. bromide, Br^-

59. **a.** F^- **b.** Sr^{2+} **c.** Be^{2+} **d.** P^{3-}

61. **a.** monatomic
b. polyatomic
c. polyatomic
d. monatomic

63. The commonly used suffixes for oxyanions are *–ite* and *–ate*.

65. **a.** NH_4^+ **b.** CO_3^{2-} **c.** OH^- **d.** $C_3H_3O_2^-$

67. **a.** ClO_3^- **b.** SO_3^{2-} **c.** ClO^- **d.** MnO_4^-

69. **a.** Sn^{4+} **b.** Cu^{2+} **c.** F^- **d.** SO_4^{2-}

71. **a.** Zn^{2+} **b.** CrO_4^{2-} **c.** SO_3^{2-} **d.** P^{3-}

73. **a.** $LiCl$ **b.** $CaBr_2$ **c.** CaO **d.** Fe_3P_2

75. **a.** $AlCl_3$ **b.** FeS **c.** $CaSO_4$ **d.** Al_2O_3

77. **a.** $Cr(C_2H_3O_2)_3$
b. $Zn(ClO_3)_2$
c. $AgNO_3$
d. $PbCO_3$

79. **a.** $Cr(ClO)_3$
b. $KMnO_4$
c. $NaCN$
d. $Pb(ClO_4)_2$

81. **a.** +2 since the anion is −1 each
b. +4 since the anion is −1 each
c. +2 since the anion is −1 each
d. +2 since the anion is −2

83. **a.** sodium bromide
b. potassium oxide
c. iron(III) bromide
d. copper(II) sulfide

85. **a.** iron(II) carbonate
b. aluminum nitrite
c. barium nitrate
d. ammonium sulfate

87. Since they do not gain or lose electrons to fulfill their valence, the nonmetals will share electrons with other nonmetals to fulfill their octet.

89. Two electrons are shared per covalent bond. In drawing structures we typically represent these two electrons as a line.

91. There are five water molecules in this figure. Each water molecule has two covalent bonds (O—H twice per molecule).

93. The molecular formula for acetic is $C_2H_4O_2$. The empirical formula is CH_2O.

95. The molecular formula for freon 112 is $C_2Cl_4F_2$. The empirical formula is CCl_2F.

97. Prefixes are necessary when naming binary covalent compounds because they often combine in more than one ratio.

99. **a.** sulfur dichloride
b. nitrogen trifluoride
c. dinitrogen tetroxide
d. tetraphosphorus decoxide

101. **a.** $AsBr_3$ **b.** N_2O_5 **c.** S_2O_2

103. Binary ionic compounds are composed of a metal and nonmetal, whereas binary covalent compounds are comprised of two nonmetals.

105. **a.** ionic bonds
b. covalent bonds
c. ionic bonds

107. **a.** ionic lattice
b. discrete molecule
c. discrete molecule
d. ionic lattice

109. **a.** ionic compound (metal + nonmetal); sodium bromide
b. covalent compound (only nonmetals); phosphorus tribromide
c. ionic compound (metal + nonmetal); magnesium bromide
d. covalent compound (only nonmetals); sulfur dibromide

111. **a.** covalent compound (only nonmetals); silicon tetrachloride
b. ionic compound (metal + nonmetal); aluminum chloride
c. covalent compound (only nonmetals); boron tribromide
d. ionic compound (metal + polyatomic ion); sodium sulfite

113. **a.** $MnCl_3$ **b.** PCl_3 **c.** SO_2 **d.** TiO_2

115. *Electrolyte* is the term used for a compound that, when dissolved in water, will conduct electricity. Ionic compounds are likely to be electrolytes.

117. When sodium sulfate dissociates in water, it produces Na^+ ions and SO_4^{2-} ions.

119. **a.** KCl is likely to dissociate in water because it's an ionic compound. Ionic compounds tend to dissociate when dissolved in water.
b. $CaBr_2$ is likely to dissociate in water because it's an ionic compound. Ionic compounds tend to dissociate when dissolved in water.
c. CO_2 will not dissociate into ions when dissolved in water since it's not an ionic compound (i.e., it's a covalent compound).
d. C_2H_6O will not dissociate into ions when dissolved in water since it's not an ionic compound (i.e., it's a covalent compound).

121. Acids are covalent compounds that produce H^+ ions in aqueous solution. Most acids contain a covalent bond between hydrogen and a species that can form a stable anion. When dissolved in water, this bond breaks to produce a hydrogen cation and a corresponding anion. Their ability to dissociate makes them differ from most covalent compounds.

123. **a.** hydrochloric acid
b. hydrobromic acid
c. hydroiodic acid

125. **a.** nitric acid
b. nitrous acid
c. perchloric acid
d. chlorous acid

127. formic acid

129. **a.** ionic; sodium nitrite
b. covalent; dinitrogen tetroxide
c. acid; nitrous acid
d. ionic; potassium nitrite

Chapter 6: Chemical Reactions

19. $H_2CO_3 \rightarrow H_2O + CO_2$

21. $2\ CH_4S + H_2O_2 \rightarrow C_2H_6S_2 + 2\ H_2O$

23. **a.** $H_2 + Br_2 \rightarrow 2\ HBr$
b. $2\ Na + F_2 \rightarrow 2\ NaF$

25. 30 units of A will produce 15 units of C.

27. a. 30 molecules of H_2 will yield 60 molecules of HCl.
b. 6 molecules of Cl_2 are needed to make 12 molecules of HCl.

29. a. 15 molecules of Cl_2 are needed to react with 10 Al atoms.
b. 10 Al atoms will produce 10 units of $AlCl_3$.

31. a. $2\ PCl_3 + 3\ F_2 \rightarrow 2\ PF_3 + 3\ Cl_2$
b. $2\ SO_2 + O_2 \rightarrow 2\ SO_3$
c. $2\ B + 3\ F_2 \rightarrow 2\ BF_3$

33. a. $PCl_3 + 3\ H_2O \rightarrow H_3PO_3 + 3\ HCl$
b. $3\ O_2 + 2\ H_2S \rightarrow 2\ SO_2 + 2\ H_2O$
c. $4\ HCl + MnO_2 \rightarrow MnCl_2 + 2\ H_2O + Cl_2$

35. a. $Hg(NO_3)_2 + 2\ KCl \rightarrow 2\ KNO_3 + HgCl_2$
b. $MgBr_2 + 2\ NaOH \rightarrow 2\ NaBr + Mg(OH)_2$
c. $2\ AgC_2H_3O_2 + BaCl_2 \rightarrow 2\ AgCl + Ba(C_2H_3O_2)_2$

37. a. $N_2 + 3\ H_2 \rightarrow 2\ NH_3$
b. $2\ C_2H_6 + 7\ O_2 \rightarrow 4\ CO_2 + 6\ H_2O$
c. $6\ HCl + 2\ Al \rightarrow 2\ AlCl_3 + 3\ H_2$

39. a. $4\ Na + O_2 \rightarrow 2\ Na_2O$
b. $Pb(NO_3)_2 + 2\ KCl \rightarrow PbCl_2 + 2\ KNO_3$
c. $2\ C_2H_2 + 5\ O_2 \rightarrow 4\ CO_2 + 2\ H_2O$

41. a. $Zn\ (s) + CuCl_2\ (aq) \rightarrow Cu\ (s) + ZnCl_2\ (aq)$
b. $C_3H_8\ (g) + 5\ O_2\ (g) \rightarrow 3\ CO_2\ (g) + 4\ H_2O\ (g)$
c. $Fe\ (s) + Cl_2\ (g) \rightarrow FeCl_2\ (s)$

43. a. aqueous hydrogen chloride (usually called *hydrochloric acid*) and aqueous sodium carbonate
b. solid iron and aqueous nitric acid
c. aqueous zinc chloride and solid lead(II) perchlorate

45. a. decomposition
b. synthesis
c. double displacement

47. a. synthesis
b. single displacement
c. double displacement

49. Ionic compounds are formed when metals and nonmetals react.

51. a. Mg is oxidized; O is reduced.
b. Fe is oxidized; S is reduced.
c. Ca is oxidized; Br is reduced.

53. a. $Be + Cl_2 \rightarrow BeCl_2$
b. $2\ K + Cl_2 \rightarrow 2\ KCl$
c. $Co\ (s) + Cl_2\ (g) \rightarrow CoCl_2$
d. $2\ Cu\ (s) + O_2\ (g) \rightarrow 2\ CuO$

55. a. $2\ Mg + O_2 \rightarrow 2\ MgO$
b. $C_2H_4 + 3\ O_2 \rightarrow 2\ H_2O + 2\ CO_2$
c. $2\ C_4H_{10}\ (g) + 13\ O_2\ (g) \rightarrow 10\ H_2O + 8\ CO_2$

57. Oxygen (O_2) is always involved in combustion reactions.

59. Hydrocarbons are compounds made of only hydrogen and carbon atoms. Combustion of hydrocarbons produces water vapor and carbon dioxide gas.

61. a. $2\ C_2H_2 + 5\ O_2 \rightarrow 2\ H_2O + 4\ CO_2$
b. $C_4H_8 + 6\ O_2 \rightarrow 4\ H_2O + 4\ CO_2$
c. $2\ C_9H_{18} + 27\ O_2 \rightarrow 18\ H_2O + 18\ CO_2$

63. a. $KCl\ (s) \rightarrow K^+\ (aq) + Cl^-\ (aq)$
b. $Li_2SO_4\ (s) \rightarrow 2\ Li^+\ (aq) + SO_4^{2-}\ (aq)$
c. $(NH_4)_2CO_3\ (s) \rightarrow 2\ NH_4^+\ (aq) + CO_3^{2-}\ (aq)$

65. a. $NaHCO_3\ (s) \rightarrow Na^+\ (aq) + HCO_3^-\ (aq)$
b. $AlCl_3\ (s) \rightarrow Al^{3+}\ (aq) + 3\ Cl^-\ (aq)$
c. $Cr(ClO_2)_3\ (s) \rightarrow Cr^{3+}\ (aq) + 3\ ClO_2^-\ (aq)$

67. a. Compounds containing the ammonium ion are always soluble.
b. Compounds containing ions with a charge of 2 or 3 tend to be insoluble.
c. Compounds containing the nitrate ion are always soluble.

69. a. This compound is soluble because it contains the ammonium and chlorate ions.
b. Compounds containing ions with a charge of 2 or 3 tend to be insoluble.
c. This compound is soluble because it contains the perchlorate ion.

71. a. soluble
b. soluble
c. insoluble
d. soluble
e. soluble
f. soluble

73. Generally a salt with a +1 cation and a −2 anion is more likely to be soluble than a salt with a +2 cation and a −2 anion due to the larger charge on the +2 cation.

75. a. $Ag^+\ (aq) + NO_3^-\ (aq) + K^+\ (aq) + Cl^-\ (aq) \rightarrow K^+\ (aq) + Cl^-\ (aq) + AgCl\ (s)$
b. $Ba^{2+}\ (aq) + 2\ ClO_4^-\ (aq) + 2\ K^+\ (aq) + SO_4^{2-}\ (aq) \rightarrow BaSO_4\ (s) + 2\ K^+\ (aq) + 2\ ClO_4^-\ (aq)$

77. The spectator ions are NO_3^- and Cs^+.

79. The spectator ions are $Na^+\ (aq)$ and $NO_3^-\ (aq)$. The net ionic equation is

$$Br^-\ (aq) + Ag^+\ (aq) \rightarrow AgBr\ (s)$$

81. a. $Fe^{2+}\ (aq) + 2\ Cl^-\ (aq) + 2\ K^+\ (aq) + 2\ OH^-\ (aq) \rightarrow Fe(OH)_2\ (s) + 2\ K^+\ (aq) + 2\ Cl^-\ (aq)$
b. $Cl^-\ (aq)$ and $K^+\ (aq)$
c. $Fe^{2+}\ (aq) + 2\ OH^-\ (aq) \rightarrow Fe(OH)_2\ (s)$
d. The driving force is the formation of $Fe(OH)_2\ (s)$.

83. a. $2\ KCl\ (aq) + Pb(NO_3)_2\ (aq) \rightarrow PbCl_2\ (s) + 2\ KNO_3\ (aq)$
b. $3\ KOH\ (aq) + FeCl_3\ (aq) \rightarrow Fe(OH)_3\ (s) + 3\ KCl\ (aq)$
c. $3\ BaCl_2\ (aq) + 2\ K_3PO_4\ (aq) \rightarrow Ba_3(PO_4)_2\ (s) + 6\ KCl\ (aq)$

85. **a.** $2\ Fe(C_2H_3O_2)_3\ (aq) + 3\ BaS\ (aq) \rightarrow Fe_2S_3\ (s) + 3\ Ba(C_2H_3O_2)_2\ (aq)$
b. $2\ K_3PO_4\ (aq) + 3\ CuSO_4\ (aq) \rightarrow 3\ K_2SO_4\ (aq) + Cu_3(PO_4)_2\ (s)$
c. no reaction
d. $FeSO_4\ (aq) + Ba(OH)_2\ (aq) \rightarrow BaSO_4\ (s) + Fe(OH)_2\ (s)$
(Note that the reaction in part d forms two insoluble products.)

87. **a.** $LiI\ (aq) + AgNO_3\ (aq) \rightarrow LiNO_3\ (aq) + AgI\ (s)$
$Li^+\ (aq) + I^-\ (aq) + Ag^+\ (aq) + NO_3^-\ (aq) \rightarrow Li^+\ (aq) + AgI\ (s) + NO_3^-\ (aq)$
$I^-\ (aq) + Ag^+\ (aq) \rightarrow AgI\ (s)$
b. $Pb(C_2H_3O_2)_2\ (aq) + ZnSO_4\ (aq) \rightarrow Zn(C_2H_3O_2)_2\ (aq) + PbSO_4\ (s)$
$Pb^{2+}\ (aq) + 2\ C_2H_3O_2^-\ (aq) + Zn^{2+}\ (aq) + SO_4^{2-}\ (aq) \rightarrow PbSO_4\ (s) + Zn^{2+}\ (aq) + 2\ C_2H_3O_2^-\ (aq)$
$Pb^{2+}\ (aq) + SO_4^{2-}\ (aq) \rightarrow PbSO_4\ (s)$
c. $Pb(NO_3)_2\ (aq) + CaI_2\ (aq) \rightarrow PbI_2\ (s) + Ca(NO_3)_2\ (aq)$
$Pb^{2+}\ (aq) + 2\ NO_3^-\ (aq) + Ca^{2+}\ (aq) + 2\ I^-\ (aq) \rightarrow PbI_2\ (s) + 2\ NO_3^-\ (aq) + Ca^{2+}\ (aq)$
$Pb^{2+}\ (aq) + 2\ I^+\ (aq) \rightarrow PbI_2\ (s)$

89. Acids are covalent compounds that ionize to produce H^+ ions and a stable anion when dissolved in water. Bases are compounds that produce hydroxide (OH^-) ions in aqueous solution.

91. **a.** $HCl\ (aq) \rightarrow H^+\ (aq) + Cl^-\ (aq)$
b. $HNO_3\ (aq) \rightarrow H^+\ (aq) + NO_3^-\ (aq)$

93. **a.** $H_2SO_4 + 2\ NaOH \rightarrow Na_2SO_4 + 2\ H_2O$
b. $3\ HNO_3 + Al(OH)_3 \rightarrow Al(NO_3)_3 + 3\ H_2O$
c. $3\ H_2SO_4 + 2\ Cr(OH)_3 \rightarrow Cr_2(SO_4)_3 + 6\ H_2O$

95. **a.** $HCl\ (aq) + KOH\ (aq) \rightarrow KCl\ (aq) + H_2O\ (l)$
b. $H_2SO_4\ (aq) + 2\ LiOH\ (aq) \rightarrow Li_2SO_4\ (aq) + 2\ H_2O\ (l)$
c. $Ba(OH)_2\ (aq) + 2\ HNO_2\ (aq) \rightarrow Ba(NO_2)_2\ (aq) + 2\ H_2O\ (l)$

97. $NaOH\ (aq) + HI\ (aq) \rightarrow NaI\ (aq) + H_2O\ (l)$
$Na^+\ (aq) + OH^-\ (aq) + H^+\ (aq) + I^-\ (aq) \rightarrow Na^+\ (aq) + I^-\ (aq) + H_2O\ (l)$

99. $Pb^{2+}\ (aq) + 2\ I^-\ (aq) \rightarrow PbI_2\ (s)$

101. **a.** synthesis, oxidation-reduction
b. single displacement, oxidation-reduction
c. double displacement, precipitation
d. single displacement, oxidation-reduction
e. double displacement, acid-base neutralization
f. combustion, oxidation-reduction

103. **a.** $AgC_2H_3O_2\ (aq) + NaCl\ (aq) \rightarrow AgCl\ (s) + NaC_2H_3O_2\ (aq)$
b. $2\ Na\ (s) + Cl_2\ (g) \rightarrow 2\ NaCl\ (s)$
c. $2\ HCl\ (aq) + Ba(OH)_2\ (aq) \rightarrow 2\ H_2O\ (l) + BaCl_2\ (aq)$
d. $C_{10}H_{20}\ (l) + 15\ O_2\ (g) \rightarrow 10\ CO_2\ (g) + 10\ H_2O\ (g)$

Chapter 7: Mass Stoichiometry

17. **a.** Formula mass for $BaSO_4$: 233.39 u
b. Formula mass for K_2S: 110.26 u
c. Formula mass for $C_6H_{12}O_6$: 180.18 u
d. Formula mass for $FeCl_3$: 162.20 u

19. **a.** Percent by mass for $C_6H_{12}O_6$: 53.28%
b. Percent by mass for C_2H_6O: 34.72%
c. Percent by mass for $C_{16}H_{19}N_3O_5S$: 21.89%

21. % carbon = 59.99%
% hydrogen = 4.49%
% oxygen = 35.52%

23. % silver = 63.50%; mass is 63.50 g Ag.

25. Scientists determine formula mass using a technique called mass spectrometry.

27. Nitroglycerin and trinitrotoluene have a formula mass of 227 u, and both could be the unknown that was analyzed.

29. The sample is morphine because its elemental analysis (% C and % H) matches the percent composition of morphine (71.55% C and 6.72% H).

31. **a.** 1 mole C_3H_8 = 44.14 g
b. 1 mole $CaCl_2$ = 110.98 g
c. 1 mole C_2H_6O = 46.08 g
d. 1 mole $C_{12}H_{22}O_{11}$ = 342.34 g

33. **a.** Report in grams/mole because this is a measured quantity.
b. Because it is a single molecule, it's more appropriate to use atomic mass units (u).

35. **a.** 1.13×10^3 g KCl
b. 38.4 g $MgSO_4$
c. 833 g Ne

37. **a.** 295 g Au
b. 1.65×10^3 g BF_3
c. 0.524 g He

39. **a.** 7.507 moles Mg
b. 0.413 moles Cl_2
c. 1.468 moles Pb

41. **a.** 0.226 moles Fe
b. 0.864 moles LiF
c. 1.2×10^{-5} moles Br_2

43. **a.** 1.02 moles Na; 6.14×10^{23} atoms Na
b. 0.588 moles P; 3.54×10^{23} atoms P
c. 16.01 moles C; 9.64×10^{24} atoms C

45. **a.** 0.627 moles H_2O; 3.77×10^{23} molecules H_2O
b. 0.563 moles PCl_3; 3.39×10^{23} molecules PCl_3
c. 2.94×10^3 moles CO_2; 1.77×10^{27} molecules CO_2

47. **a.** 1.2×10^{24} molecules HCl
b. 7.2×10^{22} atoms Au
c. 1.291×10^{24} atoms Zn

49. 0.241 moles Cu; 1.45×10^{23} atoms Cu

51. 1.235×10^{26} molecules H_2O; 3,698 g H_2O

53. 0.05302 moles Zn; 3.467 g Zn

55. 0.2279 moles H_2; 0.4604 g H_2

57. 3.988×10^{24} molecules $C_{12}H_{22}O_{11}$

59. 20 moles O_2 consumed
20 moles PbO_2 produced

61. 6.0 moles HCl consumed
3.0 moles $ZnCl_2$ produced
3.0 moles H_2 produced

63. **a.** 1 mole H_2O consumed
b. 0.5 mole H_2 produced
c. 28.0 moles K required
d. 3,014.2 moles KOH produced

65. **a.** 0.217 moles Zn consumed
b. 0.434 moles HCl consumed
c. 0.217 moles $ZnCl_2$ consumed
d. 29.6 g $ZnCl_2$ produced

67. **a.** 72.6 g CuBr consumed
b. 32.2 g Cu produced
c. 46.6 g $MgBr_2$ produced

69. 91.1 g $FeBr_2$ produced

71. 212.8 g NH_3 required

73. 52.6 g CaO produced

75. 84.04 g CaO produced; 65.96 g CO_2 produced

77. 1.4×10^2 g $Fe(OH)_2$ produced

79. **a.** 35.9 g Cl_2 consumed
b. 3.05×10^{23} molecules Cl_2 consumed
c. 65.9 g $SnCl_4$ produced

81. Limiting reagent is SO_2. Excess reagent is O_2.

83. **a.** Compound A is the limiting reagent.
b. 2.7 moles of C are expected to form.

85. Because a lake will have a large abundance of water, CaO is likely to be the limiting reagent.

87. The limiting reagent is Fe, and a total of 0.10 mole FeS can form in this reaction.

89. The limiting reagent is K, and a total of 0.29 g KOH can form in this reaction.

91. **a.** The limiting reagent is Na, producing 10.0 moles of NaI.
b. 5.0 moles I_2 used
c. 1.0 moles I_2 excess

93. **a.** The limiting reagent is $SOCl_2$.
b. The amount of excess reagent used in the reaction is 0.420 moles.
c. The amount of excess reagent left over is 0.69 moles H_2O.

95.

	Si +	O_2 →	SiO_2
Starting Moles	1.0 mol	4.0 mol	0 mol
Change	−1.0 mol	−1.0 mol	+1.0 mol
Ending Moles	0 mol	3.0 mol	1.0 mol
Ending Grams	0.0 g	96 g	60 g

97. **a.** The limiting reagent is KOH.
b. The mass of water produced in the reaction is 48.2 g.
c. The amount of excess reagent left over is 9.0 g H_3PO_4.

99. Sometimes part of the material adheres to the container walls. Other times, product is lost during the purification process. Sometimes a competing reaction forms unwanted side products in a reaction. All of these can lower the percent yield.

101. 92.30%

103. **a.** The limiting reagent is A.
b. The theoretical yield of D is 1.5 moles D.
c. 80%

105. **a.** Theoretical yield = 28.5 g C_6H_5Br
b. Percent yield = 57.2%

107. **a.** The limiting reagent is Na_3PO_4.
b. The theoretical yield of $Mg_3(PO_4)_2$ is 2.75 g.
c. Percent yield = 77.8%

109. The law of conservation of mass only appears not to be followed. All gases have mass. When the wood is burned, water vapor and carbon dioxide are produced in addition to the ashes. The remaining 463 g (500 g − 37 g ashes = 463 g) is lost to the environment in the form of water vapor and carbon dioxide, common products from combustion reactions. The law of conservation of mass is still obeyed, and if we were to capture these gases and measure their masses, we would be able to verify this law.

111. **a.** 3.219×10^5 g C_2H_6O; 408.0 L C_2H_6O
b. 6.156×10^5 g $C_4H_8O_2$; 615.6 kg $C_4H_8O_2$; 682.5 L $C_4H_8O_2$

Chapter 8: Energy

13. Energy is the ability to do work. Changes in energy take place in two forms: *heat* and *work*. Work is the transfer of energy from one form to another, whereas heat causes the particles within the substance to vibrate or move faster.

15. **a.** Energy is released when the anvil falls off the cliff.
b. The chemical energy stored in the spaghetti is released when digested.
c. As the string releases its tension, the energy is released. If the string has an arrow on it, it will transfer this energy to the arrow.
d. The chemical energy stored in dynamite is released when the dynamite detonates.

17. a. 4.61×10^7 J
b. 1.085×10^6 J
c. 1.02×10^6 BTU

19. 2.46×10^6 BTU; 2.60×10^6 kJ

21. 4.39×10^5 J

23. The two ways in which energy transfers take place are through heat and work.

25. This is an exothermic reaction because heat is released with the products. Some of the potential energy of the reactants was converted to heat energy in the reaction. Therefore, the products have less potential energy than the reactants.

27. a. endothermic
b. exothermic
c. exothermic
d. endothermic

29. a. $H_2O\ (l) \rightarrow H_2O\ (s) + \text{heat}$
b. $2\ Mg\ (s) + O_2\ (g) \rightarrow 2\ MgO\ (s) + \text{heat}$
c. $Ca(OH)_2\ (s) + \text{heat} \rightarrow CaO\ (s) + H_2O\ (l)$

31. a. exothermic
b. exothermic
c. endothermic

33. If the calcium sulfate is the system, then water is the surroundings. Exothermic.

35. a. exothermic
b. endothermic

37. a. $\Delta E = q + w$, so 20 kJ + 30 kJ = 50 kJ
b. $\Delta E = q + w$, so 30 kJ + 45 kJ = 75 kJ

39. $\Delta E = q + w$, so $q = \Delta E - w$ = 47.0 kJ – 15.0 kJ = 32.0 kJ lost as heat

41. The potential energy was released as heat, which warmed the surroundings. It was also lost as work, which caused the ground to shake. The total energy of the universe has been left unchanged.

43.

	System: Ice Bottle	Surroundings: Water in Cooler
Change	+182,000 J	−182,000 J
Result	Ice melts + temperature increases	Water cools, temperature decreases
Heat Absorbed or Released?	Absorbed	Released

45. *Heat* refers to the *total* kinetic energy transferred from one substance or object to another. Temperature is a measure of the *average* kinetic energy of the particles in a substance.

47. The specific heat is the amount of heat required to raise the temperature of one gram of material by one degree Celsius. Heat capacity is the amount of heat required to raise the temperature of an object regardless of its mass.

49. Because iron has the greater specific heat, it releases more heat energy.

51. 2.37 J/(g·°C)

53. 7.7 °C

55. 46.5 cal

57. 16 kJ/°C

59. 122 kJ

61. 1260 °C

63. 0.05 °C

65. A coffee cup is made of disposable foam, like Styrofoam™. Because disposable foam is a very good insulator, nearly all of the heat involved in the change is absorbed or released by the water. By measuring the temperature change for the water, we can determine the amount of heat that is gained or lost by the system. In bomb calorimetry, the test substance is placed in a heavy steel container. Outside this container is a second container, which is filled with water and equipped with a thermometer. A pair of ignition wires detonates the sample. The gases produced cannot expand, so all of the energy is released as heat into the bomb calorimeter. A more precise measurement is obtained with a bomb calorimeter than with a coffee cup.

67. The water absorbed 1,300 calories. This same amount of energy was released by the metal.

69. $q_{water} = 3{,}500$ J
$q_{metal} = -3{,}500$ J

71. a. $q_{water} = 1{,}700$ J
$q_{metal} = -1{,}700$ J
b. $s_m = 0.22 \dfrac{\text{cal}}{\text{g·°C}}$
c. The unknown metal has a specific heat of 0.220 cal/(g·°C).
d. The unknown metal is most likely aluminum.

73. $q_{solid} = -11.9$ kJ
Heat of solution = −57.5 kJ/mol

75. 64.85 °C

77. 4.81 kcal in a 1.0-g sample

79. The value of the fuel blend (in kJ/g) = 350.8 kJ/g

81. a. 5.550×10^3 kJ released
b. 488 kJ released

83. 629 g propane

85. 17.51 kJ/g

87. a. −3,809.4 kJ
b. −5,079.2 kJ
c. −634.90 kJ

89. a. −99.62 kJ
b. −67.36 kJ
c. −23.22 kJ

91. -5.370×10^4 kJ

93. 157 g NaOH

95. −154 kJ

97. 6.34×10^4 g Al

99. 19.49 °C

Chapter 9: Covalent Bonding and Molecules

15. The octet rule states that atoms are stabilized by completely filled *s* and *p* sublevels. This corresponds to 8 electrons in the valence shell. Nonmetals can fulfill the octet rule by sharing valence electrons.

17. There are a total of 6 bonded electrons in this molecule (2 electrons per bond) and a total of 6 nonbonded electrons.

19. These atoms fulfill the octet rule.

Atom	Covalent Bonds	Lone Pairs	Valence Electrons
H	1	0	2
O	2	2	8
O	2	2	8
H	1	0	2

21. a. complete octet (central atom has 8 electrons around it)
b. incomplete octet (central atom has fewer than 8 electrons around it)
c. expanded octet (central atom has more than 8 electrons around it)

23. boron

25. a. 8
b. 8
c. 26
d. 14

27. a. 40
b. 8
c. 32
d. 24

29. a. H—H b. :Cl—Cl: c. H—Cl:

31. a. :N≡N: b. H—N=O c. :Cl—S—Cl:

33. a. :Br—Cl: b. F—Si(F)(F)—F c. S=C=S

35. Carbon: 0
Oxygen (double bonded): 0
Oxygen (single bonded): −1 each

37. a. −1 b. 0 c. +1

39. a. —O: 6 − 1 − 6 = −1
b. =O 6 − 2 − 4 = 0
c. ≡O: 6 − 3 − 2 = +1

41. a. H_3C—C(=O)—O: 6 − 1 − 6 = −1
b. H_3C—NH_3 5 − 4 − 0 = +1
c. O=N—O: 6 − 1 − 6 = −1

43. a. 5 + 4(1) − 1 = 8
b. 6 + 3(6) + 2 = 26
c. 3 + 4(1) + 1 = 8
d. 6 + 1 + 1 = 8

45. a. H—N⊕(H)(H)—H b. ⊖:O—N=O c. ⊖:O—N⊕(—O:⊖)=O

47. a. S=O b. O=S=O c. :F—P(F)(F)(F)—F:

49. a. ⊖:O—S(=O)(=O)—O:⊖ b. ⊖:O—S(—O:⊖)=O c. O=Cl(=O)—O:⊖

51. H—C(—O—H)=N—H

53. Resonance structures are a set of Lewis structures that show how electrons are distributed around a molecule or ion. Resonance structures are used when a single Lewis structure cannot sufficiently show the distribution of the electrons in the molecule or ion. Chemists use a double-headed arrow to indicate resonance structures.

55.

4 − 4 − 0 = 0 4 − 4 − 0 = 0

:S—C≡N: ⟷ S=C=N

6 − 1 − 6 = −1 5 − 3 − 2 = 0 6 − 2 − 4 = 0 5 − 2 − 4 = −1

57. Because there are four oxygen atoms per Lewis structure and the charge is distributed evenly across all four oxygen atoms, each O atom contributes ¼ negative charge.

59.

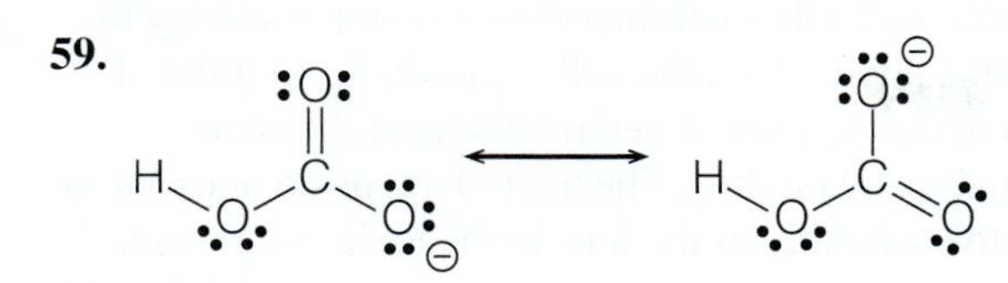

61. **a.** tetrahedral, 109.5°
b. linear, 180°
c. trigonal planar, 120°

63. **a.** charge sets = 2; electronic geometry = linear
b. charge sets = 3; electronic geometry = trigonal planar
c. charge sets = 4; electronic geometry = tetrahedral

65. **a.** charge sets = 3; electronic geometry = trigonal planar
b. charge sets = 2; electronic geometry = linear
c. charge sets = 4; electronic geometry = tetrahedral

67. **a.** electronic geometry = trigonal planar; molecular geometry = bent
b. electronic geometry = tetrahedral; molecular geometry = bent
c. electronic geometry = tetrahedral molecular geometry = tetrahedral

69. **a.** electronic geometry = tetrahedral; molecular geometry = trigonal pyramidal
b. electronic geometry = linear; molecular geometry = linear
c. electronic geometry = tetrahedral; molecular geometry = trigonal pyramidal

71. **a.** BCl_3 — Trigonal planar

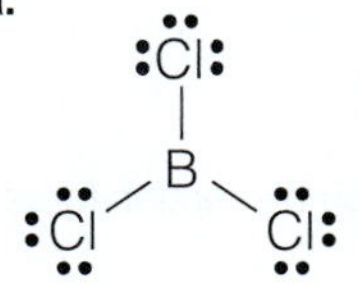

b. PCl_3 — Trigonal pyramidal

c. O=C=O — Linear

73. **a.** $SiCl_4$ — Tetrahedral

b. O=S=O with =O — Trigonal planar

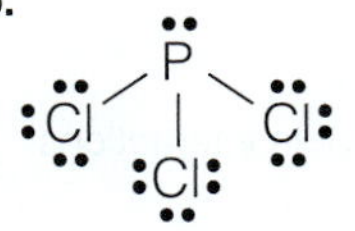

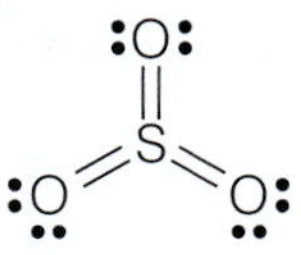

75. **a.**

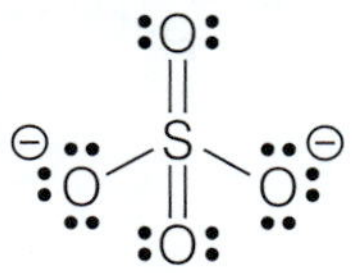

Electronic geometry = Tetrahedral
Molecular geometry = Tetrahedral

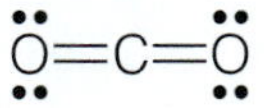

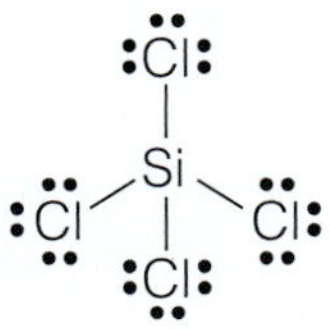

b.

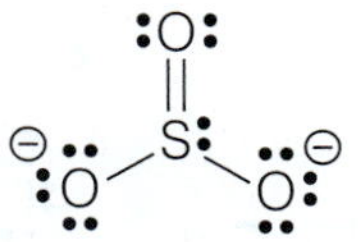

Electronic geometry = Tetrahedral
Molecular geometry = Trigonal pyramidal

77. **a.**

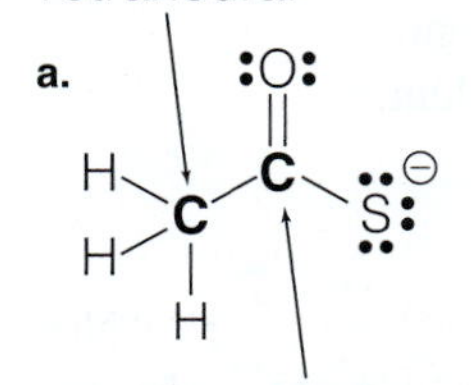

b. H—C≡C—C(H)(H)—O—H

79. Both C=O carbons have a trigonal planar electronic geometry. The highlighted N atom has a tetrahedral electronic geometry.

81. **a.** Cl **b.** C **c.** O **d.** Ca

83. Ordered from most to least electronegative:
a. O, Si, Ti, Ca
b. O, Br, I, Cs
c. F, N, In, K

85. **a.** covalent
b. ionic
c. polar covalent
d. polar covalent

87. **a.** C—O
b. Ca—Cl
c. W—S

89. **a.** δ^- N—H δ^+ **b.** δ^- O—H δ^+ **c.** δ^+ P—Cl δ^-

91. **a.** Sn—Cl (arrow toward Cl) **b.** S—Ge (arrow toward S) **c.** I—Cl (arrow toward Cl)

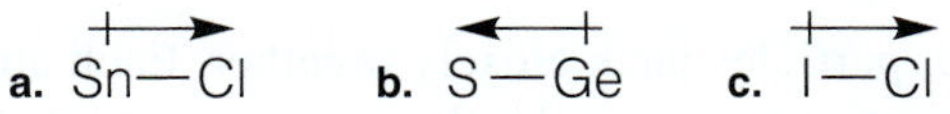

93. **a.** $H_2C{=}O$: O δ^-, C δ^+ **b.** N≡C—H: N δ^-, C δ^+ **c.** H—P(H)—Cl: P δ^+, Cl δ^-

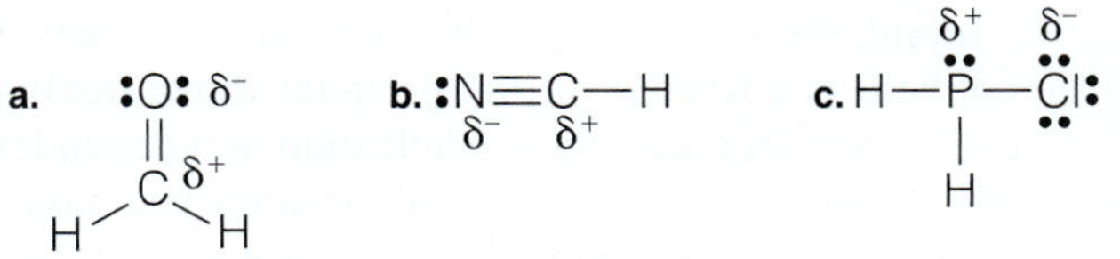

95. a. H—S—C≡N: b.

97. a. b. c.

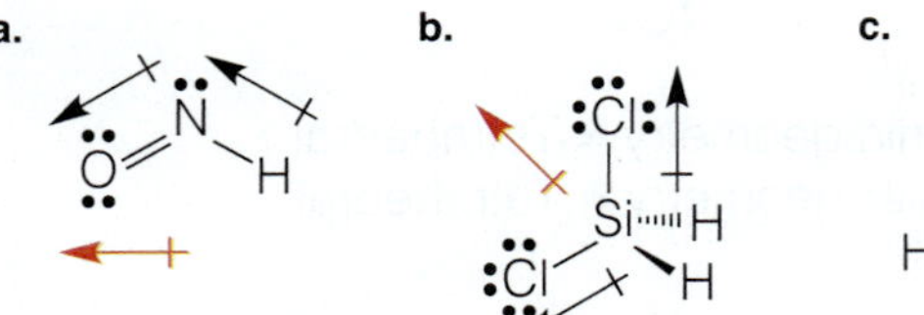

99. a. b. c.

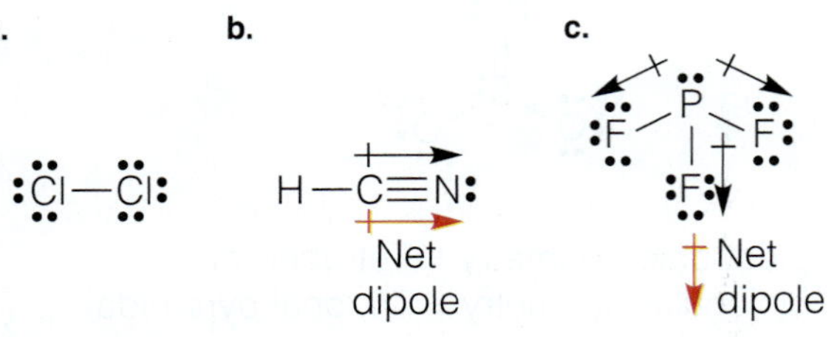
Net
dipole
Net
dipole

a. This bond is not polar and has no net dipole.
b. The C—N bond is polar covalent.
c. All P—F bonds are polar covalent.

101.
a. b. c.

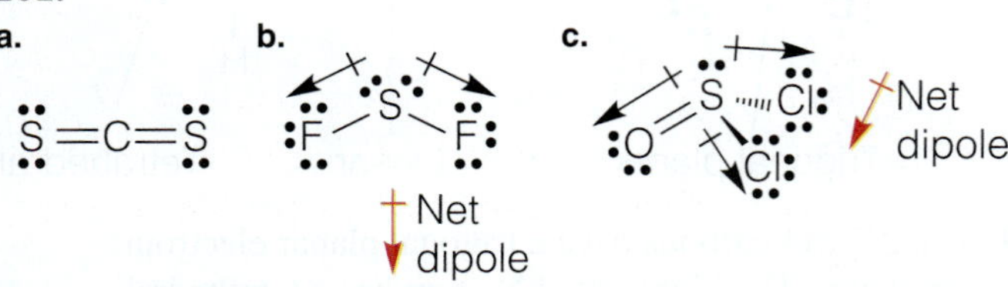
Net
dipole
Net
dipole

a. The C—S bonds are not polar. This molecule has no net dipole.
b. The S—F bonds are polar covalent.
c. The S—Cl bonds and the S=O bond are all polar covalent.

103.

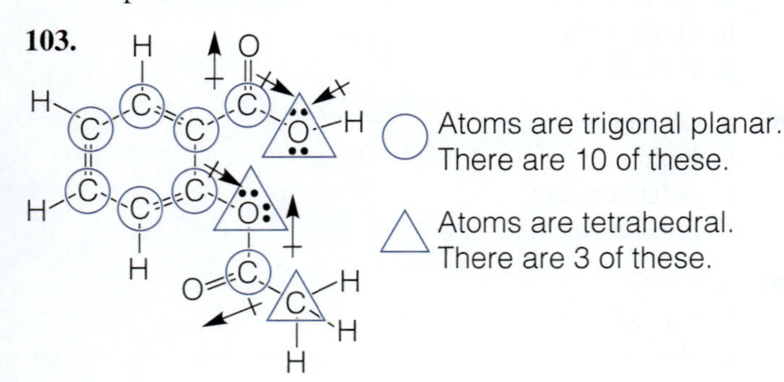
Atoms are trigonal planar.
There are 10 of these.
Atoms are tetrahedral.
There are 3 of these.

Chapter 10: Solids, Liquids, and Gases

21. In a solid, particles pack closely together. Each atom is held in a fixed position by the atoms around it. In a liquid, the particles are close together, but they are not held in a fixed position; particles move freely past each other. In a gas, the particles move independently of each other—they are spaced far apart and have little or no interaction with the other gas particles as they move.

23. In an ionic compound, alternating cations and anions are packed tightly together in rigid frameworks called lattices. The interactions between the oppositely charged ions are strong, so it takes a tremendous amount of energy to disrupt the lattice. Metallic solids usually form ordered lattices of tightly packed neutral atoms of the same elemental metal or alloy. The neutral atoms do not bind as tightly to each other as the ions in the ionic compound.

25. Intramolecular forces are the bonds within a molecule. Covalent bonds are the intramolecular forces present in a molecular solid. Intermolecular forces are the weaker forces between different molecules. Dipole–dipole interactions, hydrogen bonds, and dispersion forces are all intermolecular forces in a molecular solid.

27. a. ionic b. molecular
c. metallic d. ionic

29. a. ionic
b. covalent
c. covalent
d. covalent
N_2 will have the lowest boiling point; it is a covalent compound with no dipole. LiF will have the highest boiling point; it is an ionic compound.

31. London dispersion is the strongest type of intermolecular force in this molecule.

33.

The strongest type of intermolecular interaction in formaldehyde is dipole–dipole.

35.

a. no net dipole; London dispersion forces

b. net dipole; dipole-dipole interactions

c. net dipole; hydrogen bonding

37.

39. Polymers are compounds that contain long chains of covalently bonded atoms. Polymers are composed of small, repeating units that are bonded together, whereas a molecular solid is composed of molecules held together by intermolecular forces.

41. In a gas, the particles are spaced far apart. They move freely, interacting very little with the particles around them. Gas particles travel in a straight line until they bounce off another particle or off the walls of the container.

43. Pressure inside a container results from the force the gas particles exert on the inside of the container when they collide with the walls of the container. Pressure from outside a container also results from gas molecules colliding with the outside walls of the container.

45. If the pressure in your mouth is less than the outside pressure, the outside pressure will force the liquid up the straw and into your mouth. If the pressure in the mouth is greater than the outside pressure, the mouth pressure will force the liquid down through the straw.

47. 765.3 mm Hg; lower than standard atmospheric pressure

49. 692 torr; lower than standard atmospheric pressure

51. 1 atmosphere = 1.013×10^3 mbars = 760 torr = 29.9 inches Hg

53. **a.** 930 mbars
b. 2.4×10^3 torr
c. 20.6 psi

55. 95.0 psi

57. As the volume decreases, pressure increases. If the temperature decreases, pressure decreases.

59. 0.84 L

61. 1.79 L

63. 1.65 L

65. 137 K

67. Absolute zero is the lowest possible temperature, corresponding to 0 K or −273.15 °C; at this temperature, all motion stops, and the particles in a substance have zero kinetic energy.

69. 1,050 mb

71. 427 K

73. 0.40 bars

75. 2.36 L

77. 44.8 L

79. 2.2×10^2 moles

81. 22 atm

83. 3.7×10^2 K

85. 23.0 atm

87. 3.00 L

89. 2.09 atm

91. 1.57×10^{23} molecules CO_2

93. **a.** 0.094 moles
b. 20.2 g/mol
c. This formula mass corresponds to neon gas.

95. the pressure caused by one gas in a mixture

97. The partial pressure of methane = 18 psi.
The partial pressure of oxygen = 27 psi.

99. 9.34 atm

101. 4 torr

103. Nitrogen gas (N_2) would move faster.

105. Diffusion is the spread of particles through random motion; lighter gases diffuse more quickly than heavier gases. In contrast, effusion is the process of a gas escaping from a container; lighter gases effuse more quickly than heavier gases.

107. 2.5 moles H_2; 56 L H_2

109. **a.** 60.0 moles H_2O
b. 45.0 moles CO_2
c. 1.01×10^3 L CO_2
d. More moles of gas are produced (7 moles total versus 6 moles).

111. **a.** 0.086 mole CH_4
b. 0.17 mole O_2
c. 0.17 mole H_2O
d. 3.8 g CO_2

113. 0.18 atm

115. 1.10×10^3 L CO_2

117. A mixture containing 96% air and 4% propane (answer a) would completely react with almost no excess of either gas.

Chapter 11: Solutions

19. Potassium nitrate is the solute because it is dissolved into the solvent (water). The small amount of potassium nitrate dissolved in 2.0 liters of water, in comparison to what can dissolve in one liter (300 g), tells us this solution is dilute.

21. 28.6% $CaCl_2$

23. 0.80 L cleaner; 3.2 L water

25. 24 mL alcohol

27. **a.** 5.20% g/mL
b. 200% g/mL
c. 18.9% g/mL

29. 200 g NaCl

31. **a.** 5.0 ppm
b. 0.0050 ppm
c. 5.0×10^{-6} ppm

33. 10 mg oxygen

35. 0.03 ppm

37. 4.0 ppm; 0.00040% (g/mL)

39. 0.005 ppm; 5 ppb

41. 5×10^{-6} ppm; 5×10^{-3} ppb

43. **a.** 0.132 M KI
b. 0.056 M CO_2
c. 1.31 M $Mg(C_2H_3O_2)_2$

45. **a.** 10 moles NaOH
b. 1.2 moles KCl
c. 0.0003 moles Na_2CO_3

47. 14 L

49. 1.8×10^2 g NaCl
Measure out the desired amount of NaCl and add it to a 3.0-L volumetric flask. Partially fill the flask with water and mix until the NaCl dissolves completely. Dilute the solution, slowly adding water until the correct volume is reached.

51. 0.38 M HCl

53. Add 2.0 mL of the stock solution to a 100-mL volumetric flask and dilute with water to the correct volume.

55. **a.** electrolyte
b. electrolyte
c. not an electrolyte
d. not an electrolyte

57.

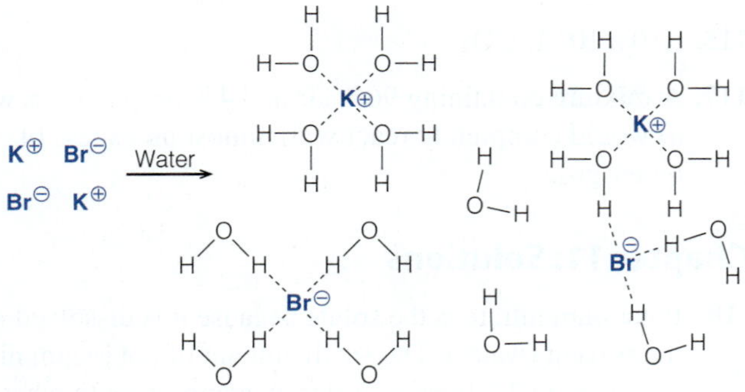

K⊕ Br⊖ / Br⊖ K⊕ —Water→ (K⊕ and Br⊖ ions each surrounded by H_2O molecules)

59. **a.** $Ba(OH)_2\ (s) \rightarrow Ba^{2+}\ (aq) + 2\ OH^-\ (aq)$
b. $(NH_4)_2SO_4\ (s) \rightarrow 2NH_4^+\ (aq) + SO_4^{2-}\ (aq)$

61. **a.** $MgBr_2\ (s) \rightarrow Mg^{2+}\ (aq) + 2\ Br^-\ (aq)$
b. $AlCl_3\ (s) \rightarrow Al^{3+}\ (aq) + 3\ Cl^-\ (aq)$
c. $Ca(NO_3)_2\ (s) \rightarrow Ca^{2+}\ (aq) + 2\ NO_3^-\ (aq)$

63. **a.** 1 M **b.** 4 M **c.** 9 M

65. **a.** 2.0 M
b. 0.7 M
c. 2.6×10^{-4} M

67. 1.84 moles Cl^-

69. 2.89 moles Cl^-; 3.6 M Cl^-

71. 0.50 M Mg^{2+}; 3.1 M K^+; 2.1 M SO_4^{2-}

73. A colligative property is one that depends on the number of particles dissolved in the solution, but not on the type of particles dissolved.

75. $MgCl_2$ has a total ion concentration of 3.0 moles/L; KBr has a total ion concentration of 2.0 moles/L. Therefore, 1.0 M $MgCl_2$ has the lower freezing point.

77. $NaNO_3$ has a total ion concentration of 4.0 moles/L; $AlCl_3$ has a total ion concentration of 6.0 mole/L. Because the $AlCl_3$ solution has the higher total ion concentration, it has the lower freezing point, the higher boiling point, and the higher osmotic pressure.

79. A 3.0-M NaCl (*aq*) solution will provide the greater number of ions (mole/L); therefore, it will have the higher osmotic pressure. Water will flow from the 1.0-M *aq*. Na_2SO_4solution into the 3.0-M NaCl (*aq*) solution in order to dilute it.

81. 0.154 M
a. hypertonic
b. hypertonic
c. hypertonic
d. isotonic

83. The driving force for a precipitation reaction is the formation of the precipitate.

85. Complete ionic equation:
$Ca^{2+}\ (aq) + 2\ Br^-\ (aq) + Pb^{2+}\ (aq) + 2\ NO_3^-\ (aq) \rightarrow PbBr_2\ (s) + Ca^{2+}\ (aq) + 2\ NO_3^-\ (aq)$
Net ionic equation:
$2\ Br^-\ (aq) + Pb^{2+}\ (aq) \rightarrow PbBr_2\ (s)$

87. **a.** metal-displacement reaction
b. neutralization reaction
c. precipitation reaction

89. **a.** $2\ FeCl_3\ (aq) + 3\ ZnSO_4\ (aq) \rightarrow 3\ ZnCl_2\ (aq) + Fe_2(SO_4)_3\ (s)$
b. $HBr\ (aq) + LiOH\ (aq) \rightarrow H_2O\ (l) + LiBr\ (aq)$

91. $2\ AgClO_3\ (aq) + MgCl_2\ (aq) \rightarrow 2\ AgCl\ (s) + Mg(ClO_3)_2\ (aq)$

93. 9.53 moles NO_3^-

95. 0.400 moles Al

97. 0.33 moles H_2

99. 0.011 M NaOH

101. 38.1 g Cu

103. 1.9×10^{-5} moles Br^-; 3.8×10^{-5} M Br^-

105. 0.00805 M Ca^{2+}

107. 0.10 L KOH

109. a. Calcium chloride is an electrolyte as it dissociate in solution to increase the ability of water to conduct electricity.

b.

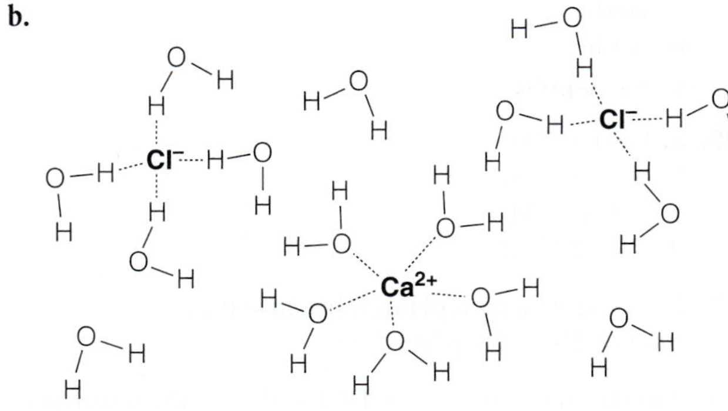

c. 40.0 mg/L

d. 1.0×10^6 g $CaCO_3$

Chapter 12: Acids and Bases

23. a. $HBr\ (aq) \rightarrow H^+\ (aq) + Br^-\ (aq)$

b. $HBr\ (aq) + H_2O\ (l) \rightarrow H_3O^+\ (aq) + Br^-\ (aq)$

c. $H_2SO_4\ (aq) \rightarrow H^+\ (aq) + HSO_4^-\ (aq)$

d. $H_2CO_3\ (aq) + H_2O\ (l) \rightarrow H_3O^+\ (aq) + HCO_3^-\ (aq)$

25. a. $NaOH\ (s) \rightarrow Na^+\ (aq) + OH^-\ (aq)$

b. $Ca(OH)_2\ (s) \rightarrow Ca^{2+}\ (aq) + 2\ OH^-\ (aq)$

27. a. $HCl\ (aq) + H_2O\ (l) \rightarrow H_3O^+\ (aq) + Cl^-\ (aq)$

b. $HNO_3\ (aq) + H_2O\ (l) \rightarrow H_3O^+\ (aq) + NO_3^-\ (aq)$

c. $HClO_4\ (aq) + H_2O\ (l) \rightarrow H_3O^+\ (aq) + ClO_4^-\ (aq)$

29. a. $NaOH\ (s) \rightarrow Na^+\ (aq) + OH^-\ (aq)$

b. $Mg(OH)_2\ (s) \rightarrow Mg^{2+}\ (aq) + 2\ OH^-\ (aq)$

c. $CsOH\ (s) \rightarrow Cs^+\ (aq) + OH^-\ (aq)$

31. a. Diprotic because it has two available H^+

b. Monoprotic because it has one available H^+

c. Monoprotic because it has one available H^+

d. Diprotic because it has two available H^+

33. In the Arrhenius definition, acids produce H^+ ions in water, which may also be represented by the hydronium ion, H_3O^+. Bases produce hydroxide (OH^-) ions in water. A Brønsted-Lowry acid is any compound that gives up an H^+ ion (called a *proton*) in a chemical reaction. Conversely, a Brønsted-Lowry base is any compound that accepts an H^+ ion.

35.

H—Br —Water→ [H^+ surrounded by water molecules] [Br^- surrounded by water molecules]

The acid in solution ionizes to produce an H^+ cation and a corresponding anion. In the aqueous solution, the H^+ ion is stabilized by multiple water molecules. This can be shown as the H^+ ion attached to a water molecule to form H_3O^+. Both mean the same thing.

37. a. HF is the acid and KOH is the base.

b. HBr is the acid and H_2O is the base.

c. H_2SO_4 is the acid and H_3N is the base.

39. Strong acids completely ionize in water. Most acids only partially ionize in water—these are weak acids.

41. a. HCl

b. HBr

c. H_2SO_4

d. H_3PO_4 (weak acid)

43. a. weak acid **b.** weak acid

c. weak acid **d.** strong acid

45. The ionization of weak acids is an example of an equilibrium reaction. This type of reaction occurs in both the forward and backward directions. The acid can dissociate into its ions, but the dissociated ions can also re-form the acid.

47. a. NH_4^+ (H—N⊕—H with H above and below) **b.** NH_2^- (⊖:N—H with H below)

49. a. acid = $HClO_2$; base = H_2O; conjugate acid = H_3O^+; conjugate base = ClO_2^-

b. acid = H_2CO_3; base = H_2O; conjugate acid = H_3O^+; conjugate base = HCO_3^-

c. acid = $CH_3NH_3^+$; base = H_2O; conjugate acid = H_3O^+; conjugate base = CH_3NH_2

51. a. acid = HCl; base = CH_3OH; conjugate acid = $CH_3OH_2^+$; conjugate base = Cl^-

b. acid = CH_3OH; base = PBr_3; conjugate acid = $HPBr_3^+$; conjugate base = CH_3O^-

c. $H_2S\ (g) + HF\ (aq) \rightleftharpoons H_3S^+\ (aq) + F^-\ (aq)$

Base Acid conjugate acid conjugate base

53. $HF\ (aq) + H_2O\ (l) \rightleftharpoons H_3O^+\ (aq) + F^-\ (aq)$

$F^-\ (aq) + H_2O\ (l) \rightleftharpoons HF\ (aq) + OH^-\ (aq)$

55. $H_2CO_3\ (aq) + H_2O\ (l) \rightleftharpoons H_3O^+\ (aq) + HCO_3^-\ (aq)$

$HNO\ (aq) + H_2O\ (l) \rightleftharpoons H_3O^+\ (aq) + NO_2^-\ (aq)$

$NO_2^-\ (aq) + H_2O\ (l) \rightleftharpoons HNO_2\ (aq) + OH^-\ (aq)$

57. $H_3PO_4\ (aq) + H_2O\ (l) \rightleftharpoons H_3O^+\ (aq) + H_2PO_4^-\ (aq)$

$H_2PO_4^-\ (aq) + H_2O\ (l) \rightleftharpoons H_3O^+\ (aq) + HPO_4^{2-}\ (aq)$

$HPO_4^{2-}\ (aq) + H_2O\ (l) \rightleftharpoons H_3O^+\ (aq) + PO_4^{3-}\ (aq)$

59. a. $HClO\ (aq) + H_2O\ (l) \rightleftharpoons H_3O^+\ (aq) + ClO^-\ (aq)$

b. $ClO^-\ (aq) + H_2O\ (l) \rightleftharpoons HClO\ (aq) + OH^-\ (aq)$

61. Conjugate acid: CH_6N^+; conjugate base: Cl^-

63. H_3O^+ (H—O⊕—H with H below) ⊖:O—H

Conjugate acid Conjugate base

53. **a.** $K = \frac{[C]}{[A][B]^2}$

b. $K = \frac{[C][D]^2}{[A]^3[B]}$

55. **a.** $K = \frac{[H^+][NO_2^-]}{[HNO_2]}$

b. $K = \frac{[[Cr(OH)_4]^-]}{[Cr^{3+}][OH^-]^4}$

57. **a.** The reaction favors the reactants because the K value is less than 1.

b. The reaction favors the products because the K value is greater than 1.

c. The reaction favors the products because the K value is greater than 1.

59. $K = 1.8 \times 10^{-4}$ (equilibrium lies to the left)

61. **a.** $K = \frac{[H_3O^+][HO_2^-]}{[H_2O_2]}$

b. $K = \frac{[H^+][B(OH)_4^-]}{[B(OH)_3]}$

63. $K = \frac{[H_3O^+][N_3^-]}{[HN_3]} = \frac{[0.0043][0.0043]}{[0.994]} = 1.9 \times 10^{-5}$

65. $[HF] = 1.5 \times 10^{-7}$ M

67. $[H_3O^+] = 6.4 \times 10^{-4}$ M

69. **a.** $K = [Pb^{2+}][I^-]^2$

b. $K = \frac{1}{[Fe^{3+}][OH^-]^3}$

71. The solubility product is the equilibrium constant for a solution of an ionic compound. A small solubility product indicates the ionic compound is not very soluble in the solution.

73. $K_{sp} = [Cd^{2+}][CO_3{}^{2-}] = 1.0 \times 10^{-12}$

75. $[Ag^+] = 1.33 \times 10^{-5}$ M

77. $[Pb^{2+}] = 1.4 \times 10^{-3}$ M and $[I^-] = 2.7 \times 10^{-3}$ M

79. **a.** $K = \frac{P_{NO_2}^2}{P_{N_2O_2}}$

b. $K = \frac{P_{Cl_2}^2 P_{H_2O}^2}{P_{HCL}^4 P_{O_2}}$

81. 16 atm

83. Le Chatelier's principle states that when a change in concentration, temperature, pressure, or other key factor takes place, the equilibrium shifts to minimize that change, and a new equilibrium is established.

85. **a.** Adding more A will shift the reaction equilibrium to the right.

b. Removing B will shift the reaction equilibrium to the left.

c. Adding C will shift the reaction equilibrium to the left.

d. Removing C will shift the reaction equilibrium to the right.

87. **a.** Adding Br_2 will shift the reaction equilibrium to the right.

b. Removing Cl_2 will shift the reaction equilibrium to the left.

c. Removing BrCl will shift the reaction equilibrium to the right.

d. Adding a catalyst will speed up the reaction rate, but it will not affect the equilibrium.

89. More CH_3I will be produced if methanol is the solvent because it will be present in a large amount and drive the reaction to the right. If water is the solvent, it will drive the reaction to the left.

91. Adding OH^- shifts the equilibrium to the left to accommodate the increased amount of OH^-.

93. This is an endothermic reaction because heat is on the reactant side. Increasing the heat in an endothermic reaction will shift the equilibrium to the right. Removing heat will shift the equilibrium to the left.

95. This is an exothermic reaction because of the negative enthalpy value. Therefore, heating the mixture will shift the equilibrium to the left.

97. The reactants contain two moles of gas, and the products contain one mole of gas. Increasing the pressure will cause the equilibrium to shift toward the side with fewer moles of gas. Therefore, the reaction will shift to the right.

99. **a.** Equilibrium shifts to the right.

b. Equilibrium shifts to the left.

c. Equilibrium shifts to the right.

101. $K = \frac{[H^+][ClO^-]}{[HClO]}$

$[H^+] = [ClO^-] = 1.7 \times 10^{-4}$ M

pH = 3.76

Chapter 14: Oxidation-Reduction Reactions

13. **a.** oxidized

b. reduced

c. reduced

d. oxidized

15. Scientists use oxidation numbers to keep track of electron changes in redox reactions.

17. **a.** +2

b. +1

c. +4

d. +1

19. **a.** Cu = +1, Cl = −1
b. Cu = +2, Cl = −1
c. Si = +4, O = −2
d. H = +1, O = −2

21. **a.** Be = +2, Br = −1
b. B = +3, Br = −1
c. N = −3, H = +1
d. Li = +1, H = −1

23. **a.** +2
b. −1
c. O = −2, Br = +1
d. H = +1, N = −3

25. **a.** +3
b. O = −2, C = +4
c. O = −2, I = +5
d. F = −1, B = +3

27. **a.** oxidized = Mg; reduced = O
b. oxidized = Fe; reduced = S
c. oxidized = Ca; reduced = Br

29. **a.** oxidized = H_2; reduced = O
b. oxidized = C; reduced = O
c. oxidized = H_2; reduced = S

31. **a.** oxidizing agent = O_2
b. oxidizing agent = O_2
c. oxidizing agent = F_2

33. **a.** reducing agent = Ca
b. reducing agent = Be
c. reducing agent = S

35. Ionic compounds are formed from the reaction of metals and nonmetals.

37. **a.** $Fe\ (s) + Cl_2\ (g) \rightarrow FeCl_2\ (s)$
oxidized = Fe; reduced = Cl_2
b. $O_2\ (g) + 2\ Hg\ (l) \rightarrow 2\ HgO\ (s)$
oxidized = Hg; reduced = O_2

39. **a.** $Be\ (s) + Cl_2\ (g) \rightarrow BeCl_2\ (s)$
b. $2\ K\ (s) + Cl_2\ (g) \rightarrow 2\ KCl\ (s)$
c. $Co\ (s) + Cl_2\ (g) \rightarrow CoCl_2\ (s)$
d. $2\ Cu\ (s) + O_2\ (g) \rightarrow 2\ CuO\ (s)$

41. oxygen

43. **a.** $2\ Mg\ (s) + O_2\ (g) \rightarrow 2\ MgO\ (s)$
b. $C_2H_4\ (g) + 3\ O_2\ (g) \rightarrow 2\ CO_2\ (g) + 2\ H_2O\ (g)$
c. $2\ C_4H_{10}\ (g) + 13\ O_2\ (g) \rightarrow 8\ CO_2\ (g) + 10\ H_2O\ (g)$

45. **a.** $Co\ (s) + SnCl_2\ (aq) \rightarrow CoCl_2\ (aq) + Sn\ (s)$
b. $AuNO_3\ (aq) + Ag\ (s) \rightarrow Au\ (s) + AgNO_3\ (aq)$

47. The color fades due to an oxidation-reduction reaction between nickel(II) chloride and solid aluminum:

$3\ NiCl_2\ (aq) + 2\ Al\ (s) \rightarrow 2\ AlCl_3\ (aq) + 3\ Ni\ (s)$

As the reaction proceeds, the green nickel(II) chloride is replaced by $AlCl_3$ and the aluminum gets a coating of solid nickel on it.

49. **a.** Complete ionic equation:
$Mg\ (s) + Cu^{2+}\ (aq) + 2\ NO_3^-\ (aq) \rightarrow Mg^{2+}\ (aq) + 2\ NO_3^-\ (aq) + Cu\ (s)$
Net ionic equation:
$Mg\ (s) + Cu^{2-}\ (aq) \rightarrow Mg^{2+}\ (aq) + Cu\ (s)$
b. Complete ionic equation:
$Al\ (s) + Cr^{3+}\ (aq) + 3\ Cl^-\ (aq) \rightarrow Al^{3+}\ (aq) + 3\ Cl^-\ (aq) + Cr\ (s)$
Net ionic equation:
$Al\ (s) + Cr^{3+}\ (aq) \rightarrow Al^{3+}\ (aq) + Cr\ (s)$
c. Complete ionic equation:
$2\ Cr\ (s) + 3\ Pt^{2+}\ (aq) + 6\ Br^-\ (aq) \rightarrow 2\ Cr^{3+}\ (aq) + 6\ Br^-\ (aq) + 3\ Pt\ (s)$
Net ionic equation:
$2\ Cr\ (s) + 3\ Pt^{2+}\ (aq) \rightarrow 2\ Cr^{3+}\ (aq) + 3\ Pt\ (s)$

51. **a.** $Mg\ (s) + Pb(NO_3)_2\ (aq) \rightarrow Mg(NO_3)_2\ (aq) + Pb\ (s)$
b. $2\ Al\ (s) + 3\ PtBr_2\ (aq) \rightarrow 3\ Pt\ (s) + 2\ AlBr_3\ (aq)$

53. Aluminum and calcium can reduce Cu^{2+} to elemental copper.

55. Aluminum can reduce Fe^{2+} and Pb^{2+}.

57. **a.** This reaction will occur spontaneously.
b. This reaction will occur spontaneously.
c. This reaction will not occur.
d. This reaction will not occur.

59. **a.** $Sn^{2+}\ (aq) + Ni\ (s) \rightarrow Ni^{2+}\ (aq) + Sn\ (s)$
b. No reaction occurs.
c. $Co\ (s) + Pb(C_2H_3O_2)_2\ (aq) \rightarrow Pb\ (s) + Co(C_2H_3O_2)_2\ (aq)$

61. **a.** oxidized by reaction with water
b. oxidized by reaction with water
c. oxidized by reaction with water
d. not oxidized by reaction with water

63. **a.** $2\ Na\ (s) + 2\ H_2O\ (l) \rightarrow 2\ NaOH\ (aq) + H_2\ (g)$
b. $2\ K\ (s) + 2\ H_2O\ (l) \rightarrow 2\ KOH\ (aq) + H_2\ (g)$
c. $Ca\ (s) + 2\ H_2O\ (l) \rightarrow Ca(OH)_2\ (aq) + H_2\ (g)$

65. Lithium, aluminum, and nickel are oxidized by reaction with acid.

67. **a.** $Mg\ (s) + 2\ HCl\ (aq) \rightarrow MgCl_2\ (aq) + H_2\ (g)$
b. $Zn\ (s) + 2\ HCl\ (aq) \rightarrow ZnCl_2\ (aq) + H_2\ (g)$
c. $Fe\ (s) + 2\ HCl\ (aq) \rightarrow FeCl_2\ (aq) + H_2\ (g)$

69. **a.** $Ca\ (s) + Fe^{2+}\ (aq) \rightarrow Fe\ (s) + Ca^{2+}\ (aq)$
b. $Ca\ (s) + 2\ H^+\ (aq) \rightarrow Ca^{2+}\ (aq) + H_2\ (g)$
c. No reaction will occur.
d. $Zn\ (s) + Fe^{2+}\ (aq) \rightarrow Fe\ (s) + Zn^{2+}\ (aq)$

71. **a.** Oxidation: $Mg\ (s) \rightarrow Mg^{2+}\ (aq) + 2e^-$
Reduction: $Co^{2+}\ (aq) + 2e^- \rightarrow Co\ (s)$
b. Oxidation: $Cu\ (s) \rightarrow Cu^{2+}\ (aq) + 2e^-$
Reduction: $2\ Au^+\ (aq) + 2e^- \rightarrow 2\ Au\ (s)$

73. Oxidation: $Fe\ (s) \rightarrow Fe^{2+}\ (aq) + 2e^-$
Reduction: $Co^{2+}\ (aq) + 2e^- \rightarrow Co\ (s)$

75. **a.** Oxidation: $Ni\ (s) \rightarrow Ni^{2+}\ (aq) + 2e^-$
Reduction: $2\ Ag^+\ (aq) + 2e^- \rightarrow 2\ Ag\ (s)$
b. Oxidation: $Zn\ (s) \rightarrow Zn^{2+}\ (aq) + 2e^-$
Reduction: $Co^{2+}\ (aq) + 2e^- \rightarrow Co\ (s)$

77. The electrode where oxidation occurs is the anode. The electrode where reduction occurs is the cathode. The current in an electrochemical cell arises from a redox reaction at the two electrodes. The salt bridge is essential to the function of the cell. A salt bridge contains an ionic compound like sodium nitrate. As the electrons move from the anode to the cathode, the anode cell builds up a positive charge while the cathode builds up a negative charge. The salt bridge provides spectator ions that can balance these charges.

79. Anode: $Zn\ (s) \rightarrow Zn^{2+}\ (aq) + 2e^-$
Cathode: $Fe^{2+}\ (aq) + 2e^- \rightarrow Fe\ (s)$

81. For this reaction to occur, two half-reactions must take place:
Anode: $Zn\ (s) \rightarrow Zn^{2+}\ (aq) + 2e^-$
Cathode: $Cu^{2+}\ (aq) + 2e^- \rightarrow Cu\ (s)$
If the Zn^{2+} solution were replaced with pure water, the reaction would still take place.
However, if the Cu^{2+} solution were replaced with pure water, the reaction would not take place, since Cu^{2+} is a reactant.

83. $Al\ (s) + 3\ Ag^+\ (aq) \rightarrow Al^{3+}\ (aq) + 3\ Ag\ (s)$

85. $2\ Li\ (s) + 2\ H^+\ (aq) \rightarrow 2\ Li^+\ (aq) + H_2\ (s)$

87. $3\ Ca\ (s) + 2\ Cr^{3+}\ (aq) \rightarrow 2\ Cr\ (s) + 3\ Ca^{2+}\ (aq)$

89. $3\ Ba\ (s) + 2\ Fe^{3+}\ (aq) \rightarrow 2\ Fe\ (s) + 3\ Ba^{2+}\ (aq)$

91. Oxidation: $Zn\ (s) \rightarrow Zn^{2+}\ (aq) + 2e^2$
Reduction: $2\ H^+\ (aq) + 2e^- \rightarrow H_2\ (g)$
$Zn\ (s) + 2\ H^+\ (aq) \rightarrow H_2\ (g) + Zn^{2+}\ (aq)$

93. Oxidation: $Cr\ (s) \rightarrow Cr^{3+}\ (aq) + 3e^-$
Reduction: $Pb^{2+}\ (aq) + 2e^- \rightarrow Pb\ (s)$
$2\ Cr\ (s) + 3\ Pb^{2+}\ (aq) + 6\ NO_3^-\ (aq)$
$\rightarrow 3\ Pb\ (s) + 2\ Cr^{3+}\ (aq) + 6\ NO_3^-\ (aq)$

95. Electroplating is a technique that uses electrical potential energy (such as from a battery) to produce a thin layer of a metal such as gold, silver, copper, or chromium on the outside surface of another material.

In an electrochemical cell, a redox reaction creates an electric current. In electroplating, the reverse occurs: an electric current drives a redox reaction to take place.

97. The gold in solution (Au^+) is reduced to Au metal on the ring. Therefore, the ring is the cathode. The zinc counter electrode is the anode.

99. A fuel cell is a device that converts the energy of a combustion reaction directly into electrical energy by separating the oxidation and reduction half-reactions. The most common fuel cells involve the reaction of hydrogen with oxygen. A hydrogen fuel cell consists of two electrodes separated by a proton exchange membrane that allows protons, but not electrons, to flow through it.

101. **a.** synthesis
b. single displacement
c. combustion

103. Reactants:
H (in $C_6H_{12}O_6$) oxidation number = +1
O (in $C_6H_{12}O_6$) oxidation number = −2
C (in $C_6H_{12}O_6$) oxidation number = 0
O (in O_2) oxidation number = 0

Products:
O (in CO_2) oxidation number = −2
C (in CO_2) oxidation number = +4
H (in H_2O) oxidation number = +1
O (in H_2O) oxidation number = −2

In this reaction, carbon is oxidized from an oxidation number of 0 to an oxidation number of +4. Also, oxygen (in O_2) is reduced from an oxidation state of 0 to an oxidation state of −2.

105. 54.0 g $Fe(NO_3)_2$

Chapter 15: Organic Chemistry and Biomolecules

11. Cellular respiration produces carbon dioxide and water.

13. Photosynthesis is an endothermic process, and cellular respiration is an exothermic process.

15. Carbon dioxide is absorbed by plants to make glucose and oxygen. The plants eventually die and begin to decay. After a long period of decay, the carbon-containing compounds of the dead plants are converted into coal.

17. The ocean exchanges carbon dioxide with the atmosphere. In water, carbon dioxide reacts to form carbonic acid and then bicarbonate ions. Many creatures produce hard calcium carbonate shells from the bicarbonate ions dissolved in the water. Over time, these shells accumulate on the seafloor, forming sediments that are rich in carbon.

19. **a.** 4 **b.** 3 **c.** 2 **d.** 1

21. **a.**
```
   H
   |
H—C—H
   |
   H
```
b.
```
     ..
    :Cl:
 ..  |   ..
:Cl—C—Cl:
 ..  |   ..
    :Cl:
     ..
```
c.
```
   H  H
   |  |   ..
H—C—C—Cl:
   |  |   ..
   H  H
```

23. a. 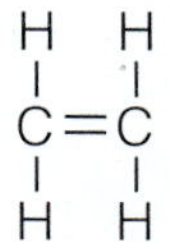 b. 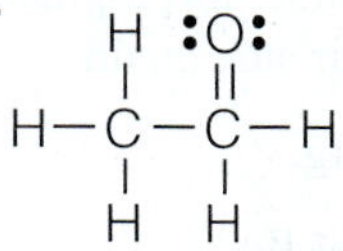 c. 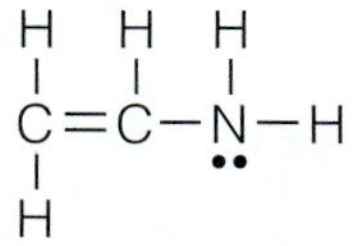 d. 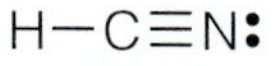 $H{-}C{\equiv}N{:}$

25. The compound in (b) is an isomer. The F atom is attached to a different carbon.

27. 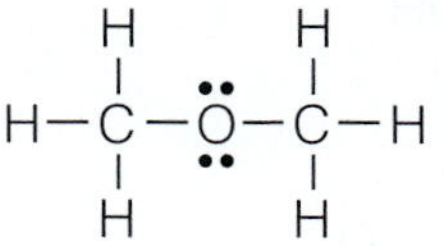 and

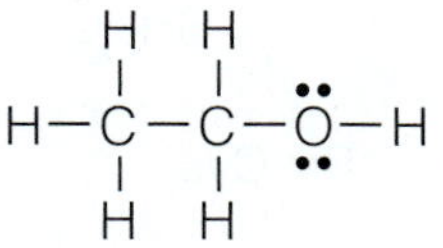

29.

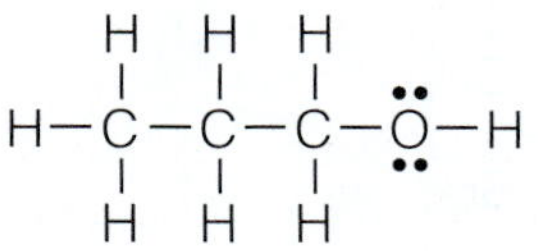

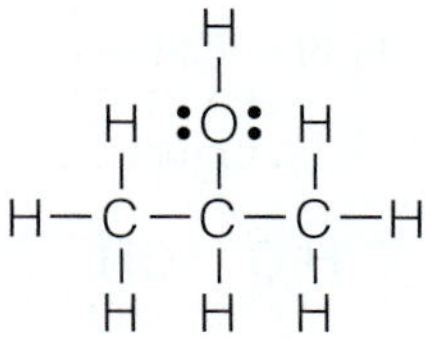

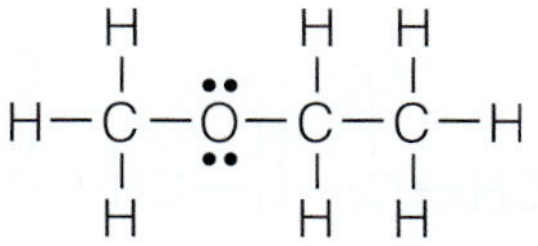

31. a. 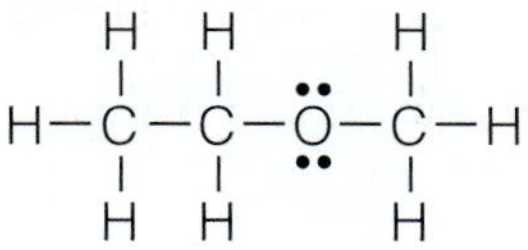 b. 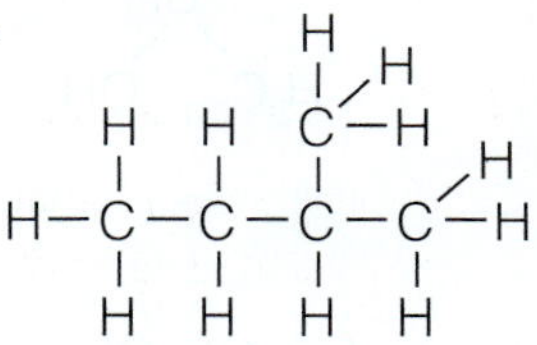 c.

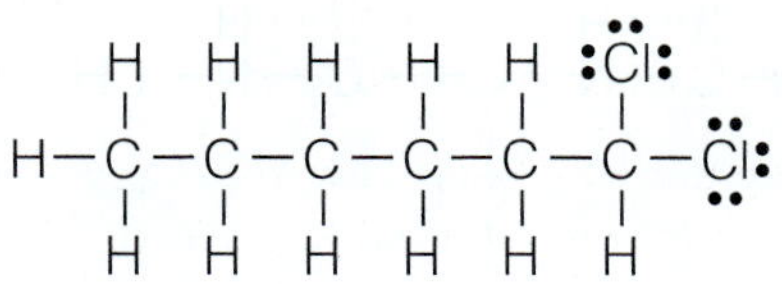

33. a. $CH_3CH_2CH_3$
b. $CH_3CCl_2CH_2CH_3$

35. a. C_5H_{12}
b. C_5H_{12}
c. $C_5H_{11}Cl$

37. a. $C_4H_{10}O$
b. C_3H_9NO
c. C_4H_4O

39. a. 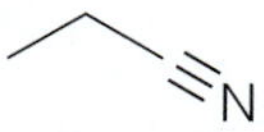 b.  c.

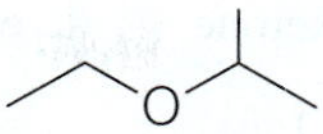

41. a. 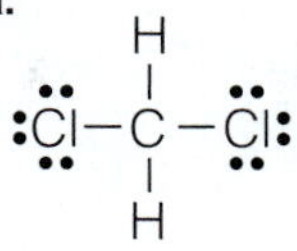 b. 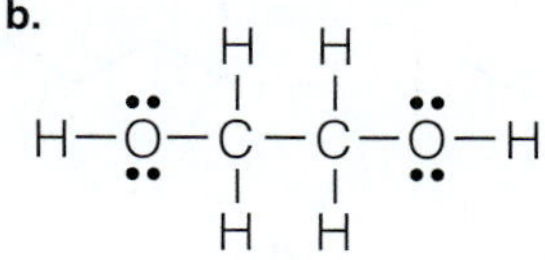 c.

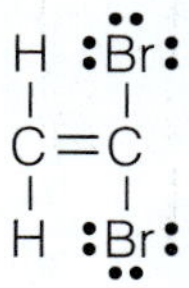

43. a. 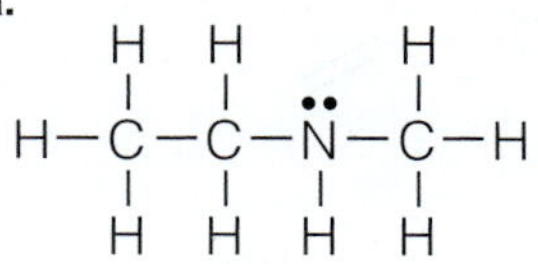 and b. and c. 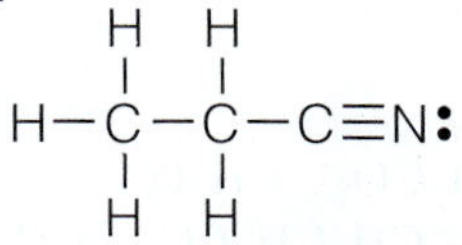 and

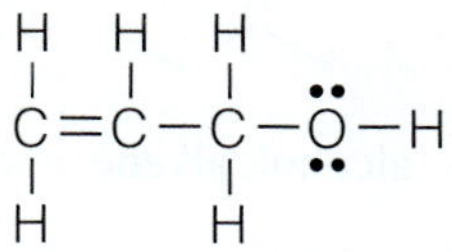

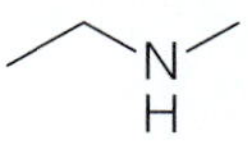

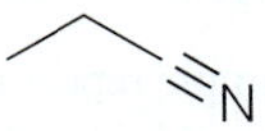

45. $C_{11}H_{15}O_4SN$

47. 18

49. 8

51. C_5H_{12} C_5H_{10}

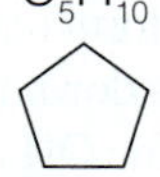

Pentane Cyclopentane

53. a. butane b. hexane c. heptane

55.

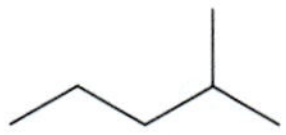

57.

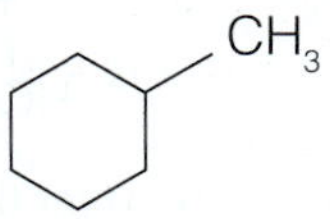

CH_3

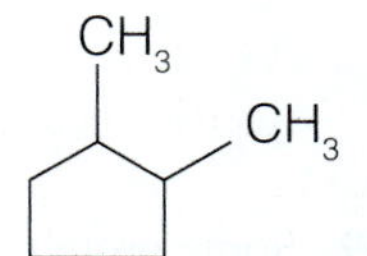

59. a.  b.

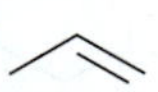

61. a. alkane b. aromatic c. alkene d. alkyne

63. a. alcohol b. ether c. ketone d. ester

65. a. 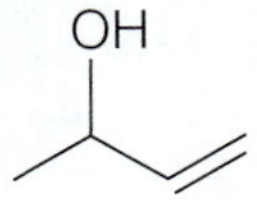 b. 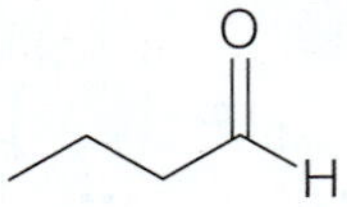 c. 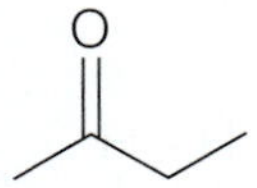 d.

67. alcohol, alkene, and ester

69. a. amine b. amide c. nitrile

71. alcohol, aromatic, amine

73. $CH_3CH_2CH_2OCH_3$

75.

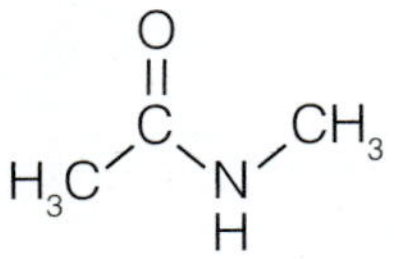

77. a. $CH_3CO_2H + H_2O \rightleftharpoons CH_3CO_2^- + H_3O^+$
b. $CH_3CH_2CO_2H + H_2O \rightleftharpoons CH_3CH_2CO_2^- + H_3O^+$

79. a. $HCO_2H + NH_3 \rightleftharpoons NH_4 + HCO_2^-$
b. $CH_3CO_2H + CH_3NH_2 \rightleftharpoons CH_3CO_2^+ + CH_3NH_3^+$

81. Polymers are molecules containing simple repeating units linked together in long chains of covalent bonds. Naturally occurring polymers include proteins and DNA. Two examples of synthetic polymers are polypropylene and polyester.

83. carbon, hydrogen, and oxygen

85. The two cyclic forms differ only in the position of the alcohol on carbon 1: In the *alpha* form (called *α-glucopyranose*), the OH on carbon 1 points down from the ring. In the *beta* form (*β-glucopyranose*), the OH on carbon 1 points out from the ring.

87.

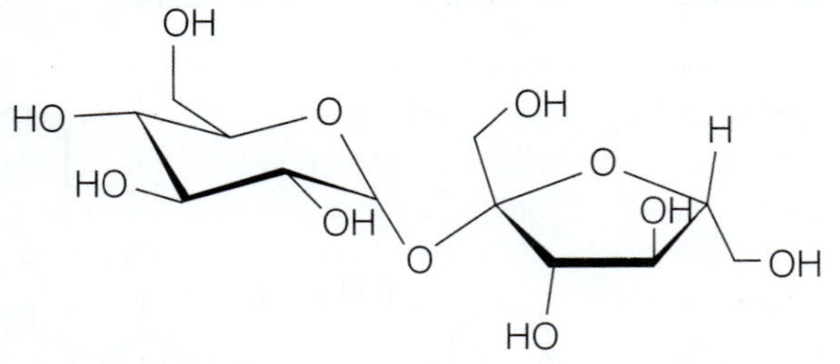

89. Some proteins function as *enzymes* that catalyze very specific reactions. Other proteins transport molecules. Proteins are essential to the replication of our genetic code. And every move we make—every muscle contraction—arises from a change in the shape of a protein.

91. Phenylalanine, tryptophan, and tyrosine have the benzene ring in their side chain.

93. Ser-Pro-Arg

95. a. Met-Gly-Ile
b. Cys-Phe-Ser

97. a. Cys-Thr

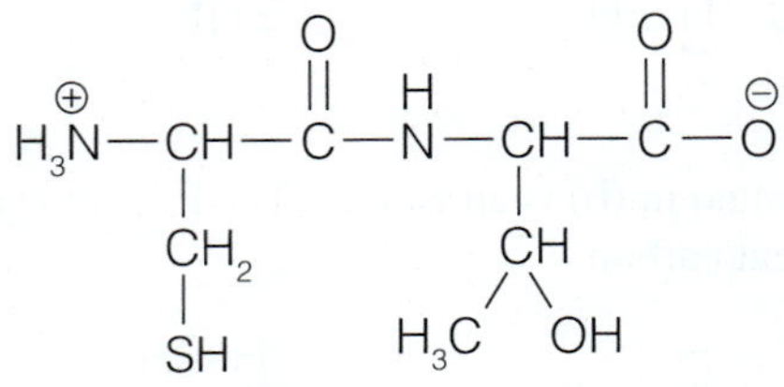

b. Thr-Cys

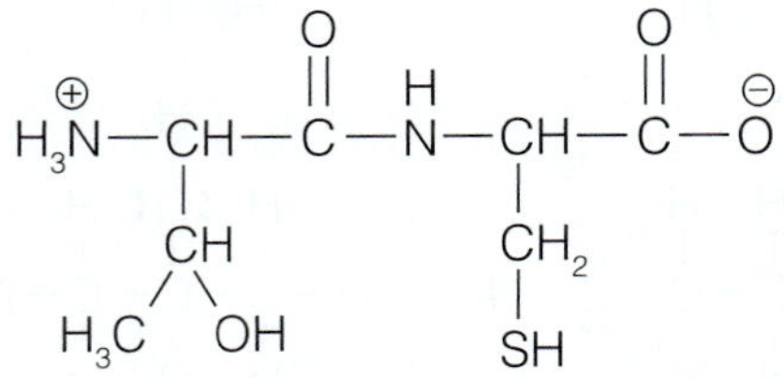

99. a. Thr-Phe-Asp

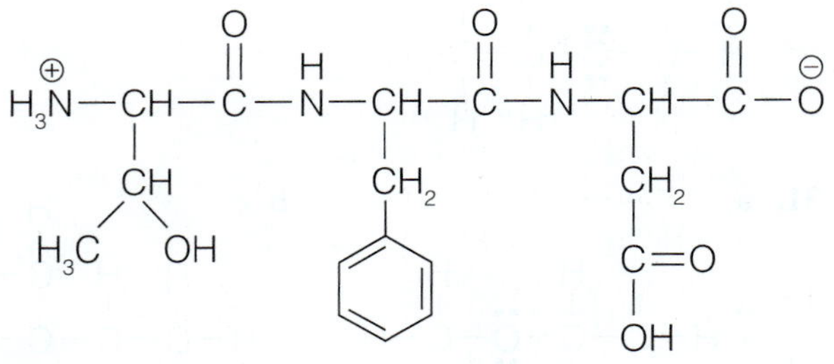

b. Lys-Gln-Gly

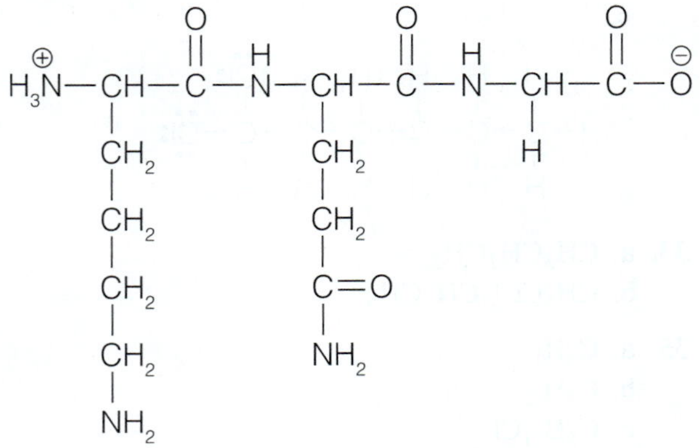

101. Hydrogen bonding interactions affect the shape of proteins.

103. DNA stores all of our genetic information needed for replication and reproduction.

105. The nucleotides link together through condensation reactions to form a sugar–phosphate backbone.

107.

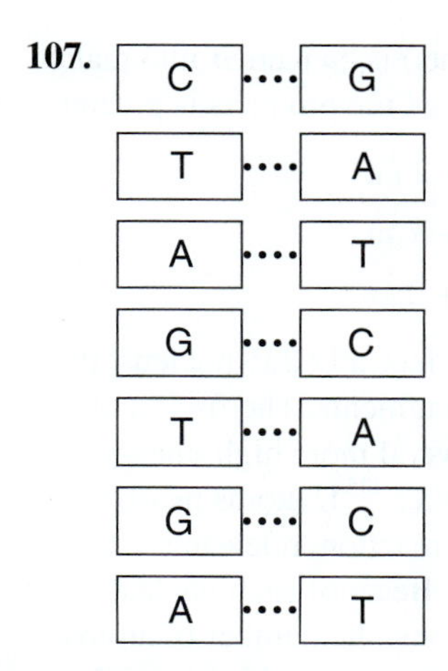

109. Polysaccharides contain one repeating monomer. Proteins and DNA contain multiple monomers with coded information in the monomer sequence.

Chapter 16: Nuclear Chemistry

11. The number of protons determines the elemental identity of an atom.

13. Collectively, protons and neutrons are called nucleons. A nuclide is an atom or nucleus containing a particular number of protons and neutrons.

15. **a.** P **b.** Rb **c.** C **d.** F

17. **a.** protons = 20 and neutrons = 20
b. protons = 20 and neutrons = 21
c. protons = 11 and neutrons = 12
d. protons = 24 and neutrons = 28

19. **a.** $^{32}_{15}P$ **b.** $^{8}_{4}Be$ **c.** $^{19}_{9}F$ **d.** $^{28}_{13}Al$

21. **a.** $^{4}_{2}He$ **b.** $^{14}_{6}C$ **c.** $^{235}_{92}F$

23. **a.** This is not an isotope of carbon-12.
b. This is an isotope of carbon-12 (it has the same number of protons, 6).
c. This is not an isotope of carbon-12.
d. This is an isotope of carbon-12 (it has the same number of protons, 6).

25. $^{1}_{0}n + ^{152}_{62}Sm \rightarrow ^{153}_{62}Sm$

27. **a.** $^{1}_{0}n + ^{98}_{42}Mo \rightarrow ^{99}_{42}Mo$
b. $^{222}_{86}Rn \rightarrow ^{4}_{2}He + ^{218}_{84}Po$
c. $^{235}_{92}U + ^{1}_{0}n \rightarrow ^{137}_{55}Cs + ^{97}_{37}Rb + 2\,^{1}_{0}n$

29. **a.** $^{226}_{88}Ra \rightarrow ^{4}_{2}He + ^{222}_{88}Rn$
b. $^{1}_{1}p + ^{111}_{48}Cd \rightarrow ^{111}_{49}In + ^{1}_{0}n$
c. $^{235}_{92}U + ^{1}_{0}n \rightarrow ^{156}_{62}Sm + ^{77}_{30}Zn + 3\,^{1}_{0}n$

31. $^{102}_{46}Pd + ^{1}_{0}n \rightarrow ^{103}_{46}Pd$

33. $^{4}_{2}He + ^{121}_{51}Sb \rightarrow ^{123}_{53}I + 2\,^{1}_{0}n$

35. $^{235}_{92}U + ^{1}_{0}n \rightarrow ^{155}_{62}Sm + ^{78}_{30}Zn + 3\,^{1}_{0}n$

37. In an alpha (α) decay, a nucleus ejects an alpha particle composed of two protons and two neutrons. An alpha particle is a helium nucleus. In a beta (β) decay, a neutron decays into two particles: a proton and an electron. The proton remains in the nucleus, while the electron is ejected. In a gamma (γ) decay, a nucleus releases energy in the form of electromagnetic waves called *gamma rays*. Gamma decays do not involve the release of particles, so they do not change the atomic number or the mass number.

39. **a.** beta (β) decay
b. gamma (γ) decay
c. alpha (α) decay

41. **a.** $^{226}_{88}Ra \xrightarrow{\alpha} ^{4}_{2}He + ^{222}_{86}Rn$
b. $^{228}_{90}Th \xrightarrow{\alpha} ^{4}_{2}He + ^{224}_{88}Ra$
c. $^{210}_{84}Po \xrightarrow{\alpha} ^{4}_{2}He + ^{206}_{82}Pb$

43. **a.** $^{8}_{3}Li \xrightarrow{\beta} ^{8}_{4}Be + ^{0}_{-1}e$
b. $^{25}_{11}Na \xrightarrow{\beta} ^{25}_{12}Mg + ^{0}_{-1}e$
c. $^{223}_{87}Fr \xrightarrow{\beta} ^{223}_{88}Ra + ^{0}_{-1}e$

45. **a.** $^{231}_{90}Th \xrightarrow{\alpha} ^{4}_{2}He + ^{227}_{88}Ra$
b. $^{228}_{89}Ac \xrightarrow{\beta} ^{228}_{90}Th + ^{0}_{-1}e$
c. $^{99m}_{43}Tc \xrightarrow{\gamma} ^{99}_{43}Tc + \text{energy}$

47. Astatine-219 first undergoes alpha decay to produce $^{215}_{83}Bi$.
$^{219}_{85}At \xrightarrow{\alpha} ^{4}_{2}He + ^{215}_{83}Bi$
Next, $^{215}_{83}Bi$ undergoes beta decay to $^{215}_{84}Po$.
$^{215}_{83}Bi \xrightarrow{\beta} ^{215}_{84}Po + ^{0}_{-1}e$
Finally, the second beta decay produces $^{215}_{85}At$.
$^{215}_{84}Po \xrightarrow{\beta} ^{215}_{85}At + ^{0}_{-1}e$

49. The first three alpha decays of thorium-227 produce $^{223}_{88}Ra$, $^{219}_{86}Rn$, and finally $^{215}_{84}Po$.
$^{227}_{90}Th \xrightarrow{\alpha} ^{4}_{2}He + ^{223}_{88}Ra$
$^{223}_{88}Ra \xrightarrow{\alpha} ^{4}_{2}He + ^{219}_{86}Rn$
$^{219}_{86}Rn \xrightarrow{\alpha} ^{4}_{2}He + ^{215}_{84}Po$
The fourth decay, a beta decay, produces $^{215}_{85}At$.
$^{215}_{84}Po \xrightarrow{\beta} ^{215}_{85}At + ^{0}_{-1}e$
The final alpha decay yields $^{211}_{83}Bi$.
$^{215}_{85}At \xrightarrow{\alpha} ^{4}_{2}He + ^{211}_{83}Bi$
The final beta decay produces $^{211}_{84}Po$.
$^{211}_{83}Bi \xrightarrow{\beta} ^{211}_{84}Po + ^{0}_{-1}e$

51. Because alpha particles are much heavier than gamma particles, they can be very damaging to our body tissue. However, these particles are easily blocked by clothing or even by a sheet of paper. Gamma radiation is the hardest to block. Heavy lead shields are often used to block gamma radiation. Even so, gamma rays are less damaging than alpha particles.

53. Geiger counter, scintillation counter, semiconductor counter, dosimeter

55. A semiconductor counter is able to measure the different energies of radiation absorbed, and therefore it can identify which nuclides are present. A Geiger counter cannot do this.

57. **a.** 1.275 rem **b.** 0.07 rem **c.** 0.041 rem

59. 20,000 X-rays

61. 6 mSv/yr

63. In radiation imaging, a patient is administered a dose of a compound containing a radioactive nuclide. As this material travels through the body, it emits gamma radiation. This radiation is detected using a scintillation camera, an instrument that produces an image based on the intensity of gamma radiation. Indium-111 has been used in radiation imaging to detect internal infections. The patient's white blood cells are tagged with indium-111. The white blood cells congregate around the infected sites. The gamma rays that are emitted help pinpoint the location of the infection.

Radiation therapy is a common treatment for cancer. This technique uses the destructive properties of radiation to destroy cancer cells. In one variation of this technique, doctors implant "seeds" containing radioactive nuclei (such as Cs-131) near the tumor site.

65. When high-energy radiation from the sun enters the upper atmosphere, it produces several nuclear reactions. In one such reaction, a neutron collides with nitrogen-14 to produce carbon-14 and an additional proton:

$$^{1}_{0}\text{n} + ^{14}_{7}\text{N} \rightarrow ^{14}_{6}\text{C} + ^{1}_{1}\text{p}$$

Carbon-14 is radioactive and has a half-life of about 5,730 years. As plants grow, they take in carbon dioxide from the atmosphere, including carbon-14. After a plant dies, the carbon-14 begins to decay. Based on the amount of carbon-14 left in a sample, scientists are able to determine the age of substances that were derived from plants.

67. After one half-life, the rate of decay will be one-half the original rate of decay (120 Bq). After two half-lives, the rate of decay will be one-quarter of the original rate of decay (60 Bq).

69. When high-energy radiation strikes quartz, it produces small amounts of the radioactive nuclides beryllium-10 and aluminum-26. If quartz is exposed on Earth's surface, it is bombarded by solar radiation, and these nuclides gradually accumulate. If the quartz is buried, solar radiation is blocked and the isotopes begin to decompose. Geologists use the amounts of Be-10 and Al-26 present in a quartz sample to determine how long a rock surface has been exposed, or how long it has been buried.

71. The mass defect is the difference between the masses of the individual particles and the mass of the complete nucleus. The difference lies in a small fraction of the mass that is converted into energy.

73. When the individual particles combine to form a nucleus, a small fraction of the mass (the mass defect) is converted into energy. This energy is released when the protons and neutrons form a new nucleus.

75. Lighter nuclei will combine to form heavier, more stable nuclei, and heavier nuclei will break apart to form lighter, more stable nuclei.

77. In this reaction, energy (reactants) is converted into matter (products). Therefore, the mass of the products is greater.

79. **a.** $^{1}_{0}\text{n} + ^{235}_{92}\text{U} \rightarrow ^{95}_{38}\text{Sr} + ^{139}_{54}\text{Xe} + 2\ ^{1}_{0}\text{n}$

b. $^{1}_{0}\text{n} + ^{235}_{92}\text{U} \rightarrow ^{94}_{38}\text{Sr} + ^{139}_{54}\text{Xe} + 3\ ^{1}_{0}\text{n}$

c. $^{1}_{0}\text{n} + ^{235}_{92}\text{U} \rightarrow ^{93}_{37}\text{Rb} + ^{141}_{55}\text{Cs} + 2\ ^{1}_{0}\text{n}$

81. When a high-energy neutron strikes a U-235 nucleus, the uranium shatters into smaller fragments. The fission of a uranium nucleus produces several more high-energy neutrons. If there are enough other ^{235}U atoms nearby, these neutrons produce a chain reaction, releasing tremendous amounts of energy. Because each nuclear fission produces multiple neutrons, the number of atoms involved in a fission chain reaction can multiply rapidly.

83. The use of highly purified uranium-235 in a nuclear reactor would cause a runaway reaction. The excess uranium-238 serves two purposes: First, it absorbs some of the neutrons released in the fission of uranium-235, preventing a runaway reaction. Second, after absorbing a neutron, U-238 undergoes two beta decays to form plutonium, which also undergoes nuclear fission.

85. $U_3O_8 \rightarrow 3\ U + 4\ O_2$

$U + 3\ F_2 \rightarrow UF_6$

87. $^{239}_{92}\text{U} \xrightarrow{\beta} ^{239}_{93}\text{Np} + ^{0}_{-1}\text{e}$

$^{239}_{93}\text{Np} \xrightarrow{\beta} ^{239}_{94}\text{Pu} + ^{0}_{-1}\text{e}$

89. Fission takes place in the reactor core, which generates a tremendous amount of heat. The reactor core is immersed in water; pumps circulate this water through the reactor loop to carry heat energy away from the core. The reactor loop comes in contact with water in the power loop. The reactor loop heats the water in the power loop, converting it to steam. The steam turns turbines, producing electric current.

91. The reactor loop, which contains the water contaminated with the radioactive by-products, is completely sealed off from the outside environment and housed in a reinforced containment structure.

93. First, there must be high pressure, so the small nuclides are compressed into a tiny space. Second, there must be very high temperature—more than 10 million degrees Celsius. Third, there must be a long contact time—that is, the pressure and temperature must be sustained long enough for fusion to take place.

95. $^{1}_{0}\text{n} + ^{6}_{3}\text{Li} \rightarrow ^{4}_{2}\text{He} + ^{3}_{1}\text{H}$

97. Plasma is a gas-like state that is produced at very high temperatures, and it is often considered the fourth state of matter. In the plasma state, electrons have such high energies that they do not remain bound to any one atom. The result is a gaseous "soup" of positive and negative charges.

99. The break-even point is the level at which the energy output from a tokamak reactor exceeds the energy required to power the plasma containment vessel.

GLOSSARY

A

absolute zero The lowest possible temperature, corresponding to 0 K or −273.15 °C; at this temperature, the particles in a substance have zero kinetic energy.

accuracy A measure of how reliable measurements are—that is, how closely they reflect the true value.

acid A compound that produces H^+ or H_3O^+ ions in water (Arrhenius definition); a compound that donates H^+ ions (Brønsted-Lowry definition).

acid rain The effect observed when nonmetal oxides combine with moisture in the atmosphere to produce acidic rainfall.

activation energy The energy barrier for a reaction; this energy determines how quickly a reaction occurs.

activity series A table that lists metals by their reducing powers; used to predict whether metal displacement reactions will occur.

actual yield The amount that a chemist actually recovers from an experiment.

alcohol A functional group consisting of an oxygen that is singly bonded to a carbon and a hydrogen, giving the bonding sequence C—O—H.

aldehyde A functional group consisting of a carbonyl connected to a hydrogen atom.

alkali metals Metal elements in column 1 (or 1A) of the periodic table; these metals are very reactive.

alkaline earth metals Metal elements in column 2 (or 2A) of the periodic table; these metals are very reactive.

alkane A hydrocarbon composed entirely of single bonds.

alkene A functional group consisting of a carbon–carbon double bond; this term also refers to a simple molecule containing this functional group.

alkyne A functional group consisting of a carbon–carbon triple bond; this term also refers to a simple molecule containing this functional group.

alpha (α) decay A radioactive decay in which the nucleus ejects an alpha particle consisting of two protons and two neutrons.

amide A functional group consisting of a carbonyl group bonded to a nitrogen atom.

amine A functional group consisting of a nitrogen atom with three single bonds, usually to hydrogen or carbon atoms.

amino acid Small molecules having both amine and carboxylic acid functional groups; plant and animal cells use 20 fundamental amino acids to create proteins.

anion A negatively charged ion.

aromatic ring A ring structure that contains alternating single and double bonds and is generally less reactive than simple alkenes.

Arrhenius definition Describes an acid as a compound that produces H^+ or H_3O^+ ions in water, and a base as a compound that produces OH^- ions in water.

atmosphere (atm) A unit of gas pressure; 1 atm = 760 mm Hg.

atomic mass unit (u or amu) A unit of mass equal to 1.66×10^{-27} kg.

atomic number The number of protons in an atom; also the number of electrons in a neutral atom.

atomic theory A theory describing matter in terms of fundamental units called atoms.

atoms The fundamental units of matter.

Avogadro's law If pressure and temperature are constant, the volume of a gas is proportional to the number of moles of gas present.

Avogadro's number The number of particles in a mole; 6.02×10^{23}.

B

balanced equation A chemical equation in which the number and type of atoms is the same for the reactants and the products.

barometer A device used to measure atmospheric pressure.

base A compound that produces OH^- ions in water (Arrhenius definition); compounds that accept H^+ ions (Brønsted-Lowry definition).

battery A device that produces electric current through a redox reaction in which the two half-reactions occur at separate sites; also called an *electrochemical cell*.

becquerel (Bq) The number of radioactive decays that occur per second for a particular substance.

benzene A very stable compound having the formula C_6H_6, in which the six carbon atoms form a ring with alternating single and double bonds; benzene is one of the simplest examples of an aromatic ring.

beta (β) decay A radioactive decay in which the nucleus ejects an electron (a beta particle); this change transforms a neutron into a proton.

binding energy The energy released when protons and neutrons combine to form a nucleus, or the energy required to break a nucleus into its component particles.

Bohr model An early model of atomic structure that treated the atom like a tiny solar system, with the nucleus at the center, and the electrons orbiting the nucleus.

boiling point elevation A colligative property of water; the presence of solute in an aqueous solution raises the boiling point above that of pure water.

bomb calorimetry A technique for measuring heat changes using a sealed container; commonly used to measure high-energy reactions.

Boyle's law The pressure and volume of an ideal gas are inversely related; the product of PV is constant at constant temperature.

Brønsted-Lowry definition Describes an acid as a compound that donates H^+ ions, and a base as a compound that accepts H^+ ions.

buffer A solution containing a mixture of acidic and basic components; it resists changes in pH.

C

calorimetry An experimental technique used to measure heat changes.

carbohydrate A naturally occurring molecule composed of carbon, hydrogen, and oxygen and having the general formula $C_m(H_2O)_n$.

carbon cycle A description of how Earth's carbon moves between rock and sediment, water and atmosphere, and plants and animals.

carbonyl A functional group consisting of a carbon–oxygen double bond.

carboxylic acid A functional group containing a carbonyl bonded to an alcohol, commonly represented by the condensed formula —COOH.

catalyst A species that is not part of a balanced equation but causes a reaction to go more quickly.

cation A positively charged ion.

cellular respiration The sequence of chemical reactions by which animals release energy stored in the chemical bonds of substances they consume.

Celsius scale (°C) A temperature scale commonly used throughout the world. On the Celsius scale, water freezes at 0 °C and boils at 100 °C. Sometimes called the *centigrade scale*.

charge A characteristic property of subatomic particles that affects how particles interact with each other.

Charles's law The volume of a gas is directly proportional to its temperature; the relationship between V and T is constant at constant pressure.

chemical changes Changes that produce new substances; also called *chemical reactions*.

chemical equation A symbolic representation of a chemical change. Such equations consist of reactants and products separated by an arrow.

chemical formula A representation of the type and amount of each element present in a compound.

chemical properties Properties of a substance that cannot be measured without changing the identity of a substance.

chemical reactions Changes that produce new substances; also called *chemical changes*.

chemistry The study of matter and its changes.

coefficient In a chemical formula, the numbers written before each reagent or product to indicate the ratios in which components of the reaction are consumed or produced.

coffee cup calorimetry A technique for measuring heat changes that uses an insulated container (such as a Styrofoam™ coffee cup) to measure heat changes.

colligative properties Properties that depend on the number of particles dissolved in the solution, but not on the type of particles dissolved.

combined gas law A combination of Boyle's law and Charles's law; it states that for an ideal gas, the quantity PV/T is constant; usually expressed by the equation $P_1V_1/T_1 = P_2V_2/T_2$, where the subscripts 1 and 2 denote two different conditions.

combustion A reaction in which oxygen gas combines with elements or compounds to produce oxide compounds.

complete ionic equation An equation that shows all ions present in a solution.

composition The components that make up a material.

compounds Pure substances composed of more than one element in a fixed ratio.

concentration The amount of solute present in solution.

condensation A transition from the gas phase to the liquid phase.

condensation reaction A reaction in which two smaller molecules combine to produce water plus a larger molecule.

condensed structure A way of representing chemical bonds that does not show most covalent bonds, but lists atoms in order of their connectivity.

conjugate acid The acid on the right-hand side of a chemical equation in an acid-base equilibrium; the acid formed when a base reacts with H^+.

conjugate base The base on the right-hand side of a chemical equation in an acid-base equilibrium; the base formed when an acid releases an H^+.

conversion factors Fractions that are used to convert from one unit to another. A conversion factor contains equivalent amounts of different units in the numerator and the denominator.

covalent bond A bond in which two electrons are shared between atoms; covalent bonds typically form between nonmetals.

covalent compounds Compounds formed by covalent bonds; these compounds form discrete groups of atoms called molecules.

covalent networks Long two- or three-dimensional sequences of covalent bonds, resulting in very large single molecules.

cycloalkane An alkane that forms a cyclic structure (commonly called a *ring*).

D

decay series A naturally occurring sequence of radioactive decays; also called a *decay chain*.

decomposition A reaction in which a single reactant forms two or more products.

density A physical property of a substance, defined as the mass per unit volume.

deoxyribonucleic acids (DNA) Massive molecules containing the genetic code of living creatures.

diffusion The spread of particles through random motion; lighter gases diffuse more quickly than heavier gases.

dipole–dipole interaction An intermolecular force between two molecules containing net dipoles.

disaccharide A carbohydrate composed of two simpler carbohydrates that are linked together through a condensation reaction.

dissociation The process by which ions are pulled apart from a solid lattice when an ionic compound dissolves in water.

dosimeter A radiation detector that measures human exposure to radiation.

double displacement A reaction in which two compounds swap cation-anion pairs to form two new compounds.

***d* sublevel** A sublevel that contains five orbitals and can hold up to 10 electrons; the *d* sublevel is present in energy levels 3 and higher.

E

effusion The process of a gas escaping from a container; lighter gases effuse more quickly than heavier gases.

electrochemistry The study of chemical processes that involve the movement of electrons.

electrode In electrochemical cells and related devices, a site where an oxidation or reduction half-reaction takes place.

electrolyte solution An aqueous solution containing dissociated ions; this type of solution conducts electricity more effectively than pure water.

electromagnetic radiation A form of energy produced when charged particles move or vibrate relative to each other; electromagnetic radiation exists as waves.

electromagnetic spectrum All forms of electromagnetic energy, ranging from low-energy waves (TV and radio) to visible light to high-energy waves such as gamma rays.

electron A negatively charged subatomic particle; the electrons occupy the space around the nucleus.

electron cloud The space around the nucleus; the electron cloud accounts for nearly the entire volume of the atom.

electron configuration The number of electrons in each energy level and sublevel.

electronegativity A measure of how strongly atoms pull bonded electrons.

electronic geometry A description of the arrangement of electrons around a central atom.

electroplating A technique that uses electrical potential energy (such as from a battery) to produce a thin layer of a metal such as gold, silver, copper, or chromium on the outside surface of another material.

element A substance made of only one type of atom.

elemental analysis A technique used to determine the percent composition of a substance.

empirical formula A chemical formula that gives the smallest whole-number ratio of atoms in a compound.

endothermic change A physical or chemical change that absorbs energy from the surroundings.

energy The ability to do work.

equilibrium The state in which forward and reverse reactions take place at the same rate, so the concentrations of reactants and products do not change.

equilibrium constant (*K*) The ratio of products to reactants when a reaction is at equilibrium.

equilibrium expression An equation that describes the balance between reactants and products in an equilibrium.

equilibrium reaction A reaction that occurs in both the forward and backward directions.

ester A functional group containing a carbonyl bonded to an oxygen that is bonded to another carbon.

ether A functional group composed of an oxygen singly bonded to two carbon atoms, giving the bonding sequence C—O—C.

exact numbers Numbers for which there is no uncertainty. Counted integers and defined relationships (such as metric prefixes) are exact numbers.

excess reagent In a chemical reaction, a reagent that is present in larger stoichiometric quantities than the other reagents; an excess reagent is not completely consumed.

exothermic change A physical or chemical change that releases heat energy to the surroundings.

expanded octet A bonding arrangement in which an atom has 10 or 12 valence electrons; this is possible only with elements in rows 3–7 of the periodic table.

F

Fahrenheit scale (°F) A temperature scale commonly used in the United States. On the Fahrenheit scale, water freezes at 32 °F and boils at 212 °F.

fission A nuclear change in which a large nucleus shatters into several smaller nuclei, releasing large amounts of energy.

formal charge A method of identifying charged sites on a molecule or ion. The formal charge of an atom is the number of electrons in the valence of that atom in its neutral, unbonded state, minus the number of covalent bonds, minus the number of unshared electrons.

formula mass The mass of a molecule or formula unit.

formula unit In ionic compounds, the smallest number of ions necessary to form a compound; the combination of atoms described by an empirical formula.

freezing A transition from the liquid phase to the solid phase.

freezing point depression A colligative property of water; the presence of solute in an aqueous solution lowers the freezing point below that of pure water.

frequency (ν) The number of waves that pass through a point in one second; typically measured in hertz.

***f* sublevel** A sublevel that contains seven orbitals and can hold up to 14 electrons; the *f* sublevel is present in energy levels 4 and higher.

fuel cell A device that converts the energy of a combustion reaction directly into electrical energy by separating the oxidation and reduction half-reactions.

fuel value The amount of heat energy that can be released by a combustion reaction of a certain substance.

functional group A small group of atoms within a molecule that behaves in a characteristic manner.

fusion A nuclear change in which two smaller nuclei (usually hydrogen or helium) combine to form larger nuclei.

osmotic pressure The tendency of water to move toward regions of greater concentrations; an imbalance in concentration inside and outside of a living cell can cause the cell to swell or shrink.

oxidation The loss of electrons.

oxidation number A bookkeeping tool for tracking electron changes in oxidation-reduction reactions.

oxidation-reduction reaction A chemical change in which one species loses electrons (oxidation) while another gains electrons (reduction); also called a *redox reaction*.

oxyacid A covalent compound that dissociates in aqueous solution to form H^+ and an oxyanion.

oxyanion A negatively charged polyatomic ion containing oxygen.

P

partial pressure The pressure caused by one gas in a mixture.

parts per billion (ppb) A measure of concentration used for very dilute solutions; 1 ppm = 1 g of solute per 1,000,000,000 grams of solution, or 1 μg of solute per liter of solution.

parts per million (ppm) A measure of concentration used for dilute solutions; 1 ppm = 1 g of solute per 1,000,000 grams of solution, or 1 mg of solute per liter of solution.

peptide bond The carbon–nitrogen bond that connects two amino acids together; peptide bonds are formed by the condensation of a carboxylic acid and an amine to form an amide.

percent by mass A measure of concentration; the mass of solute divided by the total mass of the solution, expressed as a percentage.

percent by volume A measure of concentration; the volume of solute divided by the total volume of the solution, expressed as a percentage.

percent composition The percentage (by mass) of each element in a compound.

percent yield A measure of the efficiency of a reaction; the actual yield divided by the theoretical yield, expressed as a percentage.

period A horizontal row on the periodic table; a period encompasses a range of behavior from metallic to nonmetallic.

periodic table of the elements A chart that organizes all the known elements based on their masses and properties.

phenolphthalein A pH indicator that is bright pink in base, but colorless in acid.

pH indicator A compound that changes color depending on the pH; the color changes indicate the approximate pH of the solution.

photon A small increment or packet of electromagnetic energy (often visible light).

photosynthesis The sequence of chemical reactions by which green plants harvest the energy of the Sun.

pH paper Paper containing a blend of indicators that can be used to estimate pH based on color.

pH scale A scale that indicates the relative concentration of acid or base in an aqueous solution; pH is defined as the negative log of the hydronium concentration.

physical changes Changes that do not alter the identity of the substance.

physical properties The properties of a substance that can be measured without changing the identity of the substance.

plasma A gas-like state produced at very high temperatures and often considered the fourth state of matter; in this state, electrons do not remain bound to a single atom.

plastic A synthetic polymer.

polar covalent bond A covalent bond in which the atoms do not share electrons evenly; in this type of bond, one atom has a slight positive charge, while the other has a slight negative charge.

polyatomic ion A group of covalently bonded atoms with an overall charge.

polymer A molecule containing simple repeating units that are linked together in long covalent chains.

polyprotic acids Acids that can release more than one H^+ into aqueous solution.

polysaccharide A carbohydrate composed of many simpler carbohydrates that are linked together.

potential energy Energy that is stored.

precipitate A solid product formed from the combination of two solutions.

precipitation reaction A type of chemical change in which two aqueous solutions combine to produce an insoluble product.

precision A measure of how finely a measurement is made, or how close a group of measurements are to each other. Precision is often denoted by significant digits.

pressure The force that an object exerts divided by the area over which it is applied; for gases, pressure describes the force that gases exert on their surroundings.

principal quantum number An integer number that identifies the energy level an electron occupies.

product The compounds produced in a chemical change, shown on the right-hand side of a chemical equation.

protein A biopolymer composed of building blocks called amino acids; this type of molecule has many functions in living creatures.

proton A positively charged subatomic particle that resides in the nucleus of the atom.

***p* sublevel** A sublevel that contains three orbitals and can hold up to six electrons; the *p* sublevel is present in energy levels 2 and higher.

pure substances Substances composed of only one element or only one compound.

Q

quantum model The modern description of electronic behavior that treats electrons as particles and as waves.

R

radioactive decay A nuclear change in which the nucleus spontaneously transitions to a more stable state; decays involve the release of energy and sometimes mass.

radioactivity The spontaneous release of particles and/or energy from the nucleus.

reactant The starting material in a chemical change, shown on the left-hand side of a chemical equation.

reaction energy diagram A depiction of the energetic changes that accompany a chemical reaction.

reaction enthalpy (ΔH_{rxn}) The amount of heat energy that is absorbed or released in a chemical reaction at constant pressure.

reaction rate The speed at which a chemical reaction takes place.

reduction The gain of electrons.

resonance structures A set of Lewis structures that show how electrons are distributed around a molecule or ion. Resonance structures are used when a single Lewis structure cannot adequately depict the structure.

S

scientific law A statement that describes observations that are true in widely varying circumstances. Scientific laws often describe mathematical relationships. However, they do not explain why something occurs; they only observe that it occurs.

scientific method A cyclical process of making observations, formulating new ideas, and then testing those ideas through experiments.

scientific notation A way to show very large and very small numbers in a concise format. Scientific notation expresses numbers as the product of two values, called the *coefficient* and the *multiplier*.

scintillation counter A device used for measuring radiation; a scintillation counter measures the total number of radiative particles and waves striking the detector.

self-ionization A process that occurs in water when a water molecule fragments to produce H^+ and OH^- ions.

semiconductor counter A device that measures radiation; some semiconductor counters can distinguish between the decays of different elements based on the energy of the radiation.

sievert (Sv) A measure of radiation useful for measuring human exposure; 1 Sv = 1 joule of energy per kilogram of mass.

significant digits The digits contained in a measured value. The number of significant digits indicate how precisely a measurement is made. Also called *significant figures*.

single displacement A reaction in which one element replaces another element in a compound.

skeletal structure A simplified representation for chemical structures in which the end of each line segment denotes a carbon, and C—H bonds are inferred rather than drawn explicitly.

solid A state of matter having a definite shape and a definite volume. The particles in a solid are held in fixed positions.

solubility product (K_{sp}) The equilibrium constant for the solution of a slightly soluble ionic compound.

soluble Having the ability to be dissolved in a liquid.

solute A substance that is dissolved in a solution.

solution A homogeneous mixture; for example, a solid mixed in a liquid.

solvent The major component of a solution.

specific heat The amount of heat required to raise the temperature of one gram of a substance by one degree Celsius; sometimes called *specific heat capacity*.

spectator ion An ion that is present in a solution but not directly involved in a chemical change.

s sublevel A sublevel that contains one orbital and can hold up to two electrons; the *s* sublevel is present in every energy level.

states of matter The classification of matter as a solid, liquid, or gas (also called the *phases of matter*).

stoichiometry problem A problem that relates the amount of one reagent or product to another in a chemical reaction, using a balanced equation.

strong acid An acid that completely ionizes in water.

structure The arrangement of simple units within a substance. In chemistry, structure refers to both the composition and arrangement of simple units within a substance.

subatomic particles The particles from which atoms are composed. The three major subatomic particles are protons, neutrons, and electrons.

sublevel A set of electron orbitals that occurs in an electron energy level; the four main sublevels are *s*, *p*, *d*, and *f*.

subscript In a chemical formula, subscript numbers show the number of each atom or ion present.

surroundings In thermodynamics, everything that exists around the system being studied.

synthesis A reaction in which two reactants join together to form a single product; also called a *combination reaction*.

system In thermodynamics, the part of the universe being studied.

T

temperature A measure of the average kinetic energy of the molecules in a substance.

tetrahedral A geometry in which four atoms or electron sets are separated by 109.5° angles.

theoretical yield The amount of product that can form in a chemical reaction, based on the balanced equation and the amount of starting materials present.

theory An idea that has been tested and refined; also a way of thinking about a particular topic.

thermodynamics The scientific field that deals with energy and temperature changes.

titration An analytical technique that can precisely measure the concentration of an acid or base by measuring the volume required for a neutralization to occur.

transition metals The metals in columns 3–12 of the periodic table; these metals are harder and less reactive than those in columns 1–2.

transition state The highest-energy arrangement of atoms that occurs during a chemical reaction.

trigonal planar A geometry in which three atoms or electron sets are separated by 120° angles.

U

uncertainty principle The idea that it is impossible to know the exact velocity and location of a particle; this principle becomes important when studying electrons.

units of measurement Quantities with accepted values.

V

valence level The highest-occupied electron energy level in an atom.

valence shell electron pair repulsion (VSEPR) model A way of predicting the geometry of molecules based on the number of electron sets around a central atom.

vaporization A transition from the liquid phase to the gas phase.

visible spectrum The narrow range of electromagnetic energy that we perceive as light.

W

wavelength (λ) The distance from a point on one wave to the same point on the next wave.

weak acid An acid that only partially ionizes in water.

work The transfer of energy from one form to another.

INDEX

A

B

C

D

E

F

I

N

O

P

Q

R

X

Y

Z